Bioactive Peptides from Food

Food Analysis & Properties

Series Editor
Leo M. L. Nollet
University College Ghent, Belgium

This CRC series **Food Analysis and Properties** is designed to provide a state-of-art coverage on topics to the understanding of physical, chemical and functional properties of foods: including (1) recent analysis techniques of a choice of food components; (2) developments and evolutions in analysis techniques related to food; (3) recent trends in analysis techniques of specific food components and/or a group of related food components.

Flow Injection Analysis of Food Additives
Edited by Claudia Ruiz-Capillas and Leo M. L. Nollet

Marine Microorganisms: Extraction and Analysis of Bioactive Compounds
Edited by Leo M. L. Nollet

Multiresidue Methods for the Analysis of Pesticide Residues in Food
Edited by Horacio Heinzen, Leo M.L. Nollet, and Amadeo R. Fernandez-Alba

Spectroscopic Methods in Food Analysis
Edited by Adriana S. Franca and Leo M.L. Nollet

Phenolic Compounds in Food: Characterization and Analysis
Edited by Leo M.L. Nollet, Janet Alejandra Gutierrez-Uribe

Testing and Analysis of GMO-containing Foods and Feed
Edited by Salah E. O. Mahgoub, Leo M. L. Nollet

Fingerprinting Techniques in Food Authenticity and Traceability
Edited by K.S. Siddiqi and Leo M.L. Nollet

Hyperspectral Imaging Analysis and Applications for Food Quality
Edited by Nrusingha Charan Basantia, Leo M.L. Nollet, Mohammed Kamruzzaman

Ambient Mass Spectroscopy Techniques in Food and the Environment
Edited by Leo M.L. Nollet and Basil K. Munjanja

Food Aroma Evolution: During Food Processing, Cooking and Aging
Edited by Matteo Bordiga, Leo M. L. Nollet

Mass Spectrometry Imaging in Food Analysis
Edited by Leo M. L. Nollet

Proteomics for Food Authentication
Edited by Leo M. L. Nollet, Otles, Semih

Analysis of Nanoplastics and Microplastics in Food
Edited by Leo M. L. Nollet and Khwaja Salahuddin Siddiqi

Chiral Organic Pollutants: Monitoring and Characterization in Food and the Environment
Edited by Edmond Sanganyado, Basil Munjanja, and Leo M. L. Nollet

Sequencing Technologies in Microbial Food Safety and Quality
Edited by Devarajan Thangadurai, Leo M.L. Nollet, Saher Islam, and Jeyabalan Sangeetha

Nanoemulsions in Food Technology: Development, Characterization, and Applications
Edited by Javed Ahmad and Leo M.L. Nollet

Mass Spectrometry in Food Analysis
Edited by Leo M.L. Nollet and Robert Winkler

Bioactive Peptides from Food: Sources, Analysis, and Functions
Edited by Leo M.L. Nollet and Semih Ötleş

Nutriomics: Well-being through Nutrition
Edited by Devarajan Thangadurai, Saher Islam, Leo M.L. Nollet, and Juliana Bunmi Adetunji

For more information, please visit the Series Page: https://www.crcpress.com/Food-Analysis--Properties/book-series/CRCFOODANPRO

Bioactive Peptides from Food
Sources, Analysis, and Functions

Edited by
Leo M.L. Nollet and Semih Ötleş

CRC Press
Taylor & Francis Group
Boca Raton London New York

CRC Press is an imprint of the
Taylor & Francis Group, an **informa** business

First edition published 2022
by CRC Press
6000 Broken Sound Parkway NW, Suite 300, Boca Raton, FL 33487-2742

and by CRC Press
2 Park Square, Milton Park, Abingdon, Oxon, OX14 4RN
CRC Press is an imprint of Taylor & Francis Group, LLC

© 2022 selection and editorial matter, Leo M.L. Nollet and Semih Ötleş individual chapters, the contributors

Reasonable efforts have been made to publish reliable data and information, but the author and publisher cannot assume responsibility for the validity of all materials or the consequences of their use. The authors and publishers have attempted to trace the copyright holders of all material reproduced in this publication and apologize to copyright holders if permission to publish in this form has not been obtained. If any copyright material has not been acknowledged please write and let us know so we may rectify in any future reprint.

Except as permitted under U.S. Copyright Law, no part of this book may be reprinted, reproduced, transmitted, or utilized in any form by any electronic, mechanical, or other means, now known or hereafter invented, including photocopying, microfilming, and recording, or in any information storage or retrieval system, without written permission from the publishers.

For permission to photocopy or use material electronically from this work, access www.copyright.com or contact the Copyright Clearance Center, Inc. (CCC), 222 Rosewood Drive, Danvers, MA 01923, 978-750-8400. For works that are not available on CCC please contact mpkbookspermissions@tandf.co.uk

Trademark notice: Product or corporate names may be trademarks or registered trademarks and are used only for identification and explanation without intent to infringe.

Library of Congress Cataloging-in-Publication Data

Names: Nollet, Leo M. L., 1948- editor. | Ötles, Semih, editor.
Title: Bioactive peptides from food : sources, analysis, and functions / edited by Leo M.L. Nollet, Semih Otles.
Description: First edition. | Boca Raton : Taylor & Francis, 2022. | Series: Food analysis & properties | Includes bibliographical references and index.
Identifiers: LCCN 2021046280 (print) | LCCN 2021046281 (ebook) | ISBN 9780367608538 (hardback) | ISBN 9780367617783 (paperback) | ISBN 9781003106524 (ebook)
Subjects: LCSH: Peptides--Physiological effect. | Bioactive compounds.
Classification: LCC QP552.P4 B536 2022 (print) | LCC QP552.P4 (ebook) | DDC 572/.65--dc23/eng/20211123
LC record available at https://lccn.loc.gov/2021046280
LC ebook record available at https://lccn.loc.gov/2021046281

ISBN: 978-0-367-60853-8 (hbk)
ISBN: 978-0-367-61778-3 (pbk)
ISBN: 978-1-003-10652-4 (ebk)

DOI: 10.1201/9781003106524

Typeset in Sabon
by Deanta Global Publishing Services, Chennai, India

Contents

Series Preface		ix
Preface		xi
Editors		xiii
List of Contributors		xv

SECTION 1   BIOACTIVE PEPTIDES

Chapter 1 Bioactive Peptides: An Overview 3
Alessandro Colletti and Arrigo F.G. Cicero

SECTION 2   SOURCES OF BIOACTIVE PEPTIDES

Chapter 2 Bioactive Peptides from Algae 31
Rahel Suchintita Das, Brijesh K. Tiwari, and Marco Garcia-Vaquero

Chapter 3 Meat 55
Sajad Ahmad Mir, Zahida Naseem, Danish Rizwan, and Farooq Ahmad Masoodi

Chapter 4 Dairy Products 75
Fernanda Gobbi Amorim, Larissa Zambom Côco, Francielle Almeida Cordeiro, Bianca Prandi Campagnaro, and Rafaela Aires

Chapter 5 Seafood 97
Jesús Aarón Salazar Leyva, Rosa Stephanie Navarro Peraza, Israel Benítez García, Emmanuel Martínez Montaño, and Idalia Osuna Ruíz

Chapter 6 Cereals 129
Hitomi Kumagai

Chapter 7 Soybean 141
Toshihiro Nakamori

Chapter 8	Bioactive Peptides in Pulses *Anne Pihlanto, Elena Maestri, Markus Nurmi, Nelson Marmiroli, and Minna Kahala*	153
Chapter 9	Other Nonconventional Sources *Semih Ötleş and Vasfiye Hazal Özyurt*	181

SECTION 3 FROM PROTEINS TO PEPTIDES

Chapter 10	Enzymatic Hydrolysis of Proteins *Mohammad Zarei, Belal J. Muhialdin, Kambiz Hassanzadeh, Chay Shyan Yea, and Raman Ahmadi*	189
Chapter 11	Protein Hydrolysates in Animal Nutrition: Industrial Production, Bioactive Peptides, and Functional Significance *Yongqing Hou, Zhenlong Wu, Zhaolai Dai, Genhu Wang, and Guoyao Wu*	209
Chapter 12	Fermentation Process: The Factory of Bioactive Peptides *Belal J. Muhialdin, Mohammad Zarei, Kambiz Hassanzadeh, Chay Shyan Yea, and Raman Ahmadi*	233
Chapter 13	Subcritical Water Extraction and Microwave-Assisted Extraction *Leo M.L. Nollet*	259
Chapter 14	Fractionation and Purification of Bioactive Peptides *Chay Shyan Yea, Raman Ahmadi, Mohammad Zarei, and Belal J. Muhialdin*	267

SECTION 4 ANALYSIS OF BIOACTIVE PEPTIDES

Chapter 15	Liquid Chromatography-Mass Spectrometry (LC-MS) Analysis of Bioactive Peptides *Evelien Wynendaele, Kevin Van der Borght, Nathan Debunne, and Bart De Spiegeleer*	301
Chapter 16	Bioinformatic Analysis *Semih Ötleş, Bahar Bakar, and Burcu Kaplan Türköz*	321

SECTION 5 CHEMICAL SYNTHESIS OF PEPTIDES

Chapter 17	Chemical Synthesis of Peptides *Javed Ahamad, Raja Kumar Parabathina, and Javed Ahmad*	349

SECTION 6   FUNCTIONS OF BIOACTIVE PEPTIDES

Chapter 18   Antihypertensive Activity — 365
Ritam Bandopadhyay, Pragya Shakti Mishra, and Awanish Mishra

Chapter 19   Bioactive Peptides in Neurodegenerative Diseases — 391
Kambiz Hassanzadeh, Marco Feligioni, Mohammad Zarei, Belal J. Muhialdin, Rita Maccarone, Massimo Corbo, and Lucia Buccarello

Chapter 20   Antimicrobial Activity of Bioactive Peptides and Their Applications in Food Safety: A Review — 415
Mohammad Hossein Maleki, Hooman Jalilvand Nezhad, Nima Keshavarz Bahadori, Milad Daneshniya, and Zahra Latifi

Chapter 21   Opioid Activity — 427
Sureal Ahmad Sheikh, Pragya Shakti Mishra, Ritam Bandopadhyay, and Awanish Mishra

Chapter 22   Immunomodulating Activity — 441
Lourdes Santiago-López, A. Alejandra López-Pérez, Lilia M. Beltrán-Barrientos, Adrián Hernández-Mendoza, Belinda Vallejo-Cordoba, and Aarón F. González-Córdova

Chapter 23   Bioactive Peptides: Cytomodulatory Activity — 461
Carlotta Giromini and Mariagrazia Cavalleri

Chapter 24   Other Biological Functions — 485
Leo M.L. Nollet

SECTION 7   REGULATORY STATUS OF BIOACTIVE PEPTIDES

Chapter 25   Regulatory Status of Bioactive Peptides — 509
Faraat Ali and Javed Ahmad

Index — 521

Series Preface

There will always be a need to analyze food compounds and their properties. Current trends in analyzing methods include automation, increasing the speed of analyses, and miniaturization. Over the years, the unit of detection has evolved from micrograms to pictograms.

A classical pathway of analysis is sampling, sample preparation, cleanup, derivatization, separation, and detection. At every step, researchers are working and developing new methodologies. A large number of papers are published every year on all facets of analysis. So, there is a need for books that gather information on one kind of analysis technique or on the analysis methods for a specific group of food components.

The scope of the CRC Series on Food Analysis & Properties aims to present a range of books edited by distinguished scientists and researchers who have significant experience in scientific pursuits and critical analysis. This series is designed to provide state-of-the-art coverage on topics such as:

1. Recent analysis techniques on a range of food components.
2. Developments and evolution in analysis techniques related to food.
3. Recent trends in analysis techniques for specific food components and/or a group of related food components.
4. The understanding of physical, chemical, and functional properties of foods.

The book *Bioactive Peptides from Food* is volume number 19 of this series.

I am happy to be a series editor of such books for the following reasons:

- I am able to pass on my experience in editing high-quality books related to food.
- I get to know colleagues from all over the world more personally.
- I continue to learn about interesting developments in food analysis.

Much work is involved in the preparation of a book. I have been assisted and supported by a number of people, all of whom I would like to thank. I would especially like to thank the team at CRC Press/Taylor & Francis, with a special word of thanks to Steve Zollo, senior editor.

Many, many thanks to all the editors and authors of this volume and future volumes. I very much appreciate all their effort, time, and willingness to do a great job.

I dedicate this series to:

- My wife, for her patience with me (and all the time I spend on my computer).
- All patients suffering from prostate cancer; knowing what this means, I am hoping they will have some relief.

Preface

In the market of functional foods, the products targeting health-promoting and mental well-being have prompted the food industry to increase research and development of these new foods and food products.

The growth rate of the functional food market is estimated at more then 6% for the period 2021–2027. Recent scientific studies suggest that food proteins not only serve as nutrients, but can also modulate the body's physiological functions. Bioactive peptides in food proteins have attracted a lot of scientific interest due to their purposive biofunctional attributes. Bioactive peptides are organic substances formed by amino acids joined by covalent bonds known as amide or peptide bonds, and are released mainly by enzymatic hydrolysis. Bioactive peptides have different human health effects depending on the sequence of amino acids which they contain in their structures.

The book is divided into seven sections:

Section 1: Bioactive Peptides, Section 2: Sources of Bioactive Peptides, Section 3: From Proteins to Peptides, Section 4: Analysis of Bioactive Peptides, Section 5: Chemical Synthesis of Peptides, Section 6: Functions of Bioactive Peptides, Section 7: Regulatory Status of Bioactive Peptides.

The book considers fundamental concepts, sources (from algae, meat, dairy products, seafood to cereals, soybean, and pulses), hydrolysis (enzymatic, chemical, fermentation), fractionation, purification, analysis (chemical, bioinformatics), chemical synthesis, functions (antihypertensive activity, antioxidant activity, antimicrobial activity, opioid activity, immunomodulating activity, cytomodulatory activity), and regulatory status of nutraceutical bioactive peptides.

The book addresses food scientists, technologists, chemists, nutrition researchers, producers, and processors working in the whole food science and technology field as well as those who are interested in the development of innovative functional products.

We would like to acknowledge and thank all chapter authors for their fruitful collaboration in bringing together different topics of bioactive peptides in one integral and valuable reference.

We would especially like to thank the team at CRC Press/Taylor & Francis, with a special word of thanks to Steve Zollo, senior editor.

Leo M.L. Nollet
Semih Ötleş

Editors

Leo M. L. Nollet earned an MS (1973) and PhD (1978) in biology from the Katholieke Universiteit Leuven, Belgium. He is an editor and associate editor of numerous books. He edited for M. Dekker, New York – now CRC Press of Taylor & Francis Publishing Group – the first, second, and third editions of Food Analysis by HPLC and Handbook of Food Analysis. The last edition is a two-volume book. Dr. Nollet also edited the Handbook of Water Analysis (first, second, and third editions) and Chromatographic Analysis of the Environment, third and fourth editions (CRC Press). With F. Toldrá, he coedited two books published in 2006, 2007, and 2017: Advanced Technologies for Meat Processing (CRC Press) and Advances in Food Diagnostics (Blackwell Publishing – now Wiley). With M. Poschl, he coedited the book Radionuclide Concentrations in Foods and the Environment, also published in 2006 (CRC Press). Dr. Nollet has also coedited with Y. H. Hui and other colleagues on several books: Handbook of Food Product Manufacturing (Wiley, 2007), Handbook of Food Science, Technology, and Engineering (CRC Press, 2005), Food Biochemistry and Food Processing (first and second editions; Blackwell Publishing – now Wiley – 2006 and 2012), and the Handbook of Fruits and Vegetable Flavors (Wiley, 2010). In addition, he edited the Handbook of Meat, Poultry, and Seafood Quality, first and second editions (Blackwell Publishing – now Wiley – 2007 and 2012). From 2008 to 2011, he published five volumes on animal product-related books with F. Toldrá: Handbook of Muscle Foods Analysis, Handbook of Processed Meats and Poultry Analysis, Handbook of Seafood and Seafood Products Analysis, Handbook of Dairy Foods Analysis (2nd edition in 2021), and Handbook of Analysis of Edible Animal By-Products. Also, in 2011, with F. Toldrá, he coedited two volumes for CRC Press: Safety Analysis of Foods of Animal Origin and Sensory Analysis of Foods of Animal Origin. In 2012, they published the Handbook of Analysis of Active Compounds in Functional Foods. In a coedition with Hamir Rathore, Handbook of Pesticides: Methods of Pesticides Residues Analysis was marketed in 2009; Pesticides: Evaluation of Environmental Pollution in 2012; Biopesticides Handbook in 2015; and Green Pesticides Handbook: Essential Oils for Pest Control in 2017. Other finished book projects include Food Allergens: Analysis, Instrumentation, and Methods (with A. van Hengel; CRC Press, 2011) and Analysis of Endocrine Compounds in Food (Wiley-Blackwell, 2011). Dr. Nollet's recent projects include Proteomics in Foods with F. Toldrá (Springer, 2013) and Transformation Products of Emerging Contaminants in the Environment: Analysis, Processes, Occurrence, Effects, and Risks with D. Lambropoulou (Wiley, 2014). In the series Food Analysis & Properties, he edited (with C. Ruiz-Capillas) Flow Injection Analysis of Food Additives (CRC Press, 2015) and Marine Microorganisms: Extraction and Analysis of Bioactive Compounds (CRC Press, 2016). With A.S. Franca, he coedited Spectroscopic Methods in Food Analysis (CRC Press, 2017), and with Horacio Heinzen and Amadeo R. Fernandez-Alba he coedited Multiresidue Methods for the Analysis of Pesticide Residues in Food (CRC Press, 2017). Further volumes in the series Food Analysis & Properties are Phenolic Compounds in Food: Characterization and Analysis (with Janet Alejandra Gutierrez-Uribe, 2018), Testing and Analysis of GMO-containing Foods and Feed (with Salah E. O. Mahgoub, 2018), Fingerprinting Techniques in Food Authentication and Traceability (with K.S. Siddiqi, 2018), Hyperspectral Imaging Analysis and Applications for Food Quality (with N.C. Basantia, Leo M.L. Nollet, Mohammed Kamruzzaman, 2018), Ambient Mass Spectroscopy Techniques in Food and the Environment (with Basil K. Munjanja, 2019), Food Aroma Evolution: During Food Processing, Cooking, and Aging (with M. Bordiga, 2019), Mass Spectrometry Imaging in Food Analysis (2020), Proteomics in Food Authentication (with S. Ötleş, 2020), Analysis of Nanoplastics and Microplastics in Food (with K.S. Siddiqi, 2020), Chiral Organic Pollutants, Monitoring and Characterization in Food and the Environment (with Edmond Sanganyado and Basil K. Munjanja, 2020), Sequencing Technologies in Microbial Food Safety and Quality (with Devarajan Thangardurai, Saher Islam, Jeyabalan Sangeetha, 2021), and Nanoemulsions in Food Technology: Development, Characterization, and Applications (with Javed Ahmad, 2021).

A native of Izmir, Turkey, Professor **Semih Ötleş** obtained his B.Sc. degree from the Department of Food Engineering (Ege University) in 1980. During his assistantship at Ege University, in 1985, he received an M.S. in Food Chemistry, and in 1989, after completing his thesis research on the instrumental analysis and chemistry of vitamins in foods he received a Ph.D. in Food Chemistry from Ege University. In 1991–92, he completed postdoctoral training on meat proteins including OECD – Postdoctoral Fellowship, in the Research Center Melle at Gent University, Belgium. Afterword, he joined the Department of Food Engineering at Ege University as a scientist of Food Chemistry, being promoted to Associate Professor in 1993 and to Professor in 2000. He was Vice Dean at Engineering Faculty (2003–2009), Head of the Department of Nutrition and Dietetics (2008–2011), and Vice Rector in Ege University (2012–2016). The research activities of Professor Ötleş have been focused on instrumental methods of food analysis. Ötleş began a series of projects on the separation and instrumental analysis techniques, first for analysis of vitamins in foods, then protein chemistry, carbohydrates, carotenoids, proteomics, and, most recently, bioactive peptides. Other activies span the fields of instrumental food analysis like UPLC, GC, GC/MS, LC/MS/MS analysis, hyphenated techniques, soy chemistry, aromatics, medical and functional foods and nutraceutical chemistry. Included are multi-residue analysis of various foods, n-3 fatty acids in fish oils, and medical and functional foods.

Contributors

Salazar Leyva Jesús Aarón
Polytechnic University of Sinaloa
Sinaloa, Mexico

Javed Ahamad
Department of Pharmacognosy
Faculty of Pharmacy
Tishk International University
Kurdistan Region, Iraq

Javed Ahmad
Department of Pharmaceutics
College of Pharmacy
Najran University
Najran, Kingdom of Saudi Arabia

Raman Ahmadi
Department of Food Science and
 Technology
Faculty of Agriculture
University of Tabriz
Tabriz, Iran
and
Drug Applied Research Center
Tabriz University of Medical Sciences
Tabriz, Iran

Rafaela Aires
Laboratory of Translational Physiology
Health Sciences Center
Federal University of Espirito Santo
Vitoria, Brazil

Faraat Ali
Department of Inspection and Enforcement
Laboratory Services
Botswana Medicines Regulatory Authority
 (BoMRA)
Gaborone, Botswana

Fernanda Gobbi Amorim
Faculty of Sciences
Department of Chemistry
 (Sciences)
Molecular Systems (MolSys)
Liège, Belgium

Nima Keshavarz Bahadori
Young Researchers and Elite
 Club
Qazvin Branch
Islamic Azad University
Qazvin, Iran
and
Department of Food Science and
 Technology
Faculty of Industrial and Mechanical
 Engineering
Qazvin Branch
Islamic Azad University
Qazvin, Iran

Bahar Bakar
Department of Food Engineering
Graduate School of Natural and Applied
 Sciences
Ege University
Izmir, Turkey

Ritam Bandopadhyay
Department of Pharmacology
School of Pharmaceutical Sciences
Lovely Professional University
Phagwara, India

Lilia M. Beltrán-Barrientos
Laboratorio de Química y Biotecnología de Productos Lácteos
Centro de Investigación en Alimentación y Desarrollo A.C. (CIAD)
Sonora, México

Kevin Van der Borght
DruQuaR (Drug Quality & Registration)
Faculty Pharmaceutical Sciences
Ghent University
Ghent, Belgium

Lucia Buccarello
Laboratory of Neuronal Cell Signalling
EBRI Rita Levi-Montalcini Foundation
Rome, Italy

Bianca Prandi Campagnaro
Laboratory of Translational Physiology and Pharmacology
Pharmaceutical Sciences Graduate Program
Vila Velha University
Vila Velha, Brazil

Mariagrazia Cavalleri
Department of Health, Animal Science and Food Safety
University of Milan,
Milan, Italy

Arrigo F.G. Cicero
Italian Nutraceutical Society (SINut)
and
Medical and Surgical Sciences Department
University of Bologna
Bologna, Italy

Larissa Zambom Côco
Laboratory of Translational Physiology and Pharmacology
Pharmaceutical Sciences Graduate Program
Vila Velha University
Vila Velha, Brazil

Alessandro Colletti
Department of Science and Drug Technology
University of Turin
Turin, Italy
and
Italian Nutraceutical Society (SINut)

Massimo Corbo
Department of Neurorehabilitation Sciences
Casa di Cura del Policlinico
Milan, Italy

Francielle Almeida Cordeiro
School of Pharmaceutical Sciences of Ribeirão Preto
Department of Biomolecular Sciences Chemistry
University of São Paulo
Ribeirão Preto, Brazil

Zhaolai Dai
College of Animal Science and Technology
China Agricultural University
Beijing, China

Milad Daneshniya
Department of Food Science and Technology
Faculty of Industrial and Mechanical Engineering
Qazvin Branch
Islamic Azad University
Qazvin, Iran

Rahel Suchintita Das
Section of Food and Nutrition
School of Agriculture and Food Science
University College Dublin
Dublin, Ireland
and
TEAGASC, Food Research Centre
Dublin, Ireland

Martínez Montaño Emmanuel
Polytechnic University of Sinaloa
Sinaloa, Mexico

Marco Feligioni
Laboratory of Neuronal Cell
  Signalling
EBRI Rita Levi-Montalcini
  Foundation
Rome, Italy
and
Fondazione Pisana per la Scienza
Pisa, Italy
and
Department of Neurorehabilitation
  Sciences
Casa di Cura del Policlinico
Milan, Italy

García Israel Benitez Garcia
Polytechnic University of Sinaloa
Sinaloa, Mexico

Marco Garcia-Vaquero
Section of Food and Nutrition
School of Agriculture and Food Science
University College Dublin
Dublin, Ireland

Carlotta Giromini
Department of Health, Animal Science and
  Food Safety
University of Milan,
Milan, Italy

Aarón F. González-Córdova
Laboratorio de Química y Biotecnología de
  Productos Lácteos
Centro de Investigación en Alimentación y
  Desarrollo A.C. (CIAD)
Sonora, México

Kambiz Hassanzadeh
Laboratory of Neuronal Cell Signalling
EBRI Rita Levi-Montalcini Foundation
Rome, Italy
and
Department of Biotechnology and Applied
  Clinical Sciences
University of L'Aquila
L'Aquila, Italy

Adrián Hernández-Mendoza
Laboratorio de Química y Biotecnología de
  Productos Lácteos
Centro de Investigación en Alimentación y
  Desarrollo A.C. (CIAD)
Sonora, México

Yongqing Hou
Hubei Key Laboratory of Animal Nutrition
  and Feed Science
Hubei Collaborative Innovation Center for
  Animal Nutrition and Feed Safety
Wuhan Polytechnic University
Wuhan, China

Osuna Ruíz Idalia
Polytechnic University of Sinaloa
Sinaloa, Mexico

Minna Kahala
Natural Resources Institute
  Finland
Jokioinen, Finland

Hitomi Kumagai
Department of Chemistry and
  Life Science
College of Bioresource Sciences
Nihon University
Fujisawa-shi, Japan

Zahra Latifi
Young Researchers and Elite Club
Sari Branch
Islamic Azad University
Mazandaran, Iran
and
Quality Engineering Research Center and Scientific Association of Food Science and Technology
Islamic Azad University
Noor Branch
Mazandaran, Iran

A. Alejandra López-Pérez
Instituto Tecnológico de Sonora
Sonora, México

Rita Maccarone
Department of Biotechnology and Applied Clinical Sciences
University of L'Aquila
L'Aquila, Italy

Elena Maestri
SITEIA.PARMA
Interdepartmental Centre for Food Safety
Technologies and Innovation for Agri-Food and Department of Chemistry, Life Sciences and Environmental Sustainability
University of Parma
Parma, Italy

Mohammad Hossein Maleki
Young Researchers and Elite Club
Qazvin Branch
Islamic Azad University
Qazvin, Iran
and
Department of Food Science and Technology
Faculty of Industrial and Mechanical Engineering
Qazvin Branch
Islamic Azad University
Qazvin, Iran

Nelson Marmiroli
SITEIA.PARMA
Interdepartmental Centre for Food Safety
Technologies and Innovation for Agri-Food and Department of Chemistry, Life Sciences and Environmental Sustainability
University of Parma
Parma, Italy

Farooq Ahmad Masoodi
Department of Food Science and Technology
University of Kashmir
Hazratbal Srinagar, India

Sajad Ahmad Mir
Department of Food Science and Technology
University of Kashmir
Hazratbal Srinagar, India

Awanish Mishra
Department of Pharmacology and Toxicology
National Institute of Pharmaceutical Education and Research (NIPER)
Guwahati, India

Pragya Shakti Mishra
Department of Nuclear Medicine
Sanjay Gandhi Postgraduate Institute of Medical Sciences
Lucknow, India

Belal J. Muhialdin
Department of Food Science
Faculty of Food Science and Technology
University Putra Malaysia
Selangor, Malaysia

Toshihiro Nakamori
Global R&D Strategy Group
Fuji Oil Holdings Inc.
Osaka, Japan

Zahida Naseem
Department of Food Science and
 Technology
University of Kashmir
Hazratbal Srinagar, India

Hooman Jalilvand Nezhad
Young Researchers and Elite Club
Qazvin Branch
Islamic Azad University
Qazvin, Iran
and
Department of Food Science and Technology
Faculty of Industrial and Mechanical
 Engineering
Qazvin Branch
Islamic Azad University
Qazvin, Iran

Leo M.L. Nollet
University College Ghent
Ghent, Belgium

Markus Nurmi
Natural Resources Institute Finland
Jokioinen, Finland

Semih Ötleş
Department of Food Engineering
Faculty of Engineering
Ege University
Izmir, Turkey

Vasfiye Hazal Özyurt
Department of Gastronomy and Culinary
 Arts
Mugla Sıtkı Kocman University
Mugla, Turkey

Raja Kumar Parabathina
Department of Biomedical Sciences
Faculty of Public Health & Medical
 Sciences
Mettu University
Ethiopia

Anne Pihlanto
Natural Resources Institute Finland
Jokioinen, Finland

Danish Rizwam
Department of Food Science and
 Technology
University of Kashmir
Hazratbal Srinagar, India

Lourdes Santiago-López
Laboratorio de Química y Biotecnología de
 Productos Lácteos
Centro de Investigación en Alimentación y
 Desarrollo A.C. (CIAD)
Sonora, México

Sureal Ahmad Sheikh
School of Pharmaceutical Sciences
Lovely Professional University
Punjab, India

Bart De Spiegeleer
DruQuaR (Drug Quality & Registration)
Faculty Pharmaceutical Sciences
Ghent University
Ghent, Belgium

Navarro Peraza Rosa Stephanie
Polytechnic University of Sinaloa
Sinaloa, Mexico

Brijesh K. Tiwari
TEAGASC, Food Research Centre
Dublin, Ireland

Burcu Kaplan Türköz
Department of Food Engineering
Faculty of Engineering
Ege University
Izmir, Turkey

Belinda Vallejo-Cordoba
Laboratorio de Química y Biotecnología de
 Productos Lácteos
Centro de Investigación en Alimentación y
 Desarrollo A.C. (CIAD)
Sonora, México

Genhu Wang
Research and Development Division
Shanghai Gentech Industries Group
Shanghai, China

Guoyao Wu
Hubei Key Laboratory of Animal
  Nutrition and Feed Science
Hubei Collaborative Innovation Center
  for Animal Nutrition and Feed Safety
Wuhan Polytechnic University
Wuhan, China
and
College of Animal Science and
  Technology
China Agricultural University
Beijing, China
and
Department of Animal Science
Texas A&M University
College Station, Texas

Zhenlong Wu
College of Animal Science and
  Technology
China Agricultural University
Beijing, China

Evelien Wynendaele
DruQuaR (Drug Quality &
  Registration)
Faculty Pharmaceutical Sciences
Ghent University
Ghent, Belgium

Chay Shyan Yea
Department of Food Science
Faculty of Food Science and
  Technology
University Putra Malaysia
Selangor, Malaysia

Mohammad Zarei
Virginia Seafood Agricultural Research
  and Extension Center
Virginia Tech
Hampton, USA

SECTION 1

Bioactive Peptides

CHAPTER 1

Bioactive Peptides
An Overview

Alessandro Colletti and Arrigo F.G. Cicero

CONTENTS

Introduction	3
Sources	5
Animal Sources	5
Vegetal Sources	6
New Sources	6
From Production to Commercialization	7
Stability and Bioavailability	8
Applications	9
Anti-Inflammatory Activity	9
Antihypertensive Activity	14
Lipid-Lowering Activity	15
Anticancer Activity	16
Immunomodulatory Activity	17
Other Biological Activities	17
Discussion	18
Conclusions	19
References	19

INTRODUCTION

Bioactive peptides (BP) are a heterogeneous class of molecules contained in a wide range of plants and animals (Karaś et al., 2019). The first bioactive peptide was identified in 1950 by Olaf Mellander, who isolated BP from casein (which improves bone calcification in rachitic children) (Mellander, 1950).

BP can be defined as protein fragments that include between 2 and 20 amino acids able to modulate physiological functions in humans (Aluko, 2012). In general, the protein in BP is inactive in the precursor molecule, but becomes active after release to the active site (Karaś, 2019). The precursor protein undergoes enzymatic or chemical hydrolysis in the gastrointestinal tract through microbiota fermentation processes and the BP can thus be absorbed through specific peptide transporters (Mazorra-Manzano et al., 2017). This is the most important difference that allows the classification of BP into exogenous and endogenous molecules, obtained via gastrointestinal digestion and artificially, respectively (Udenigwe and Aluko, 2012).

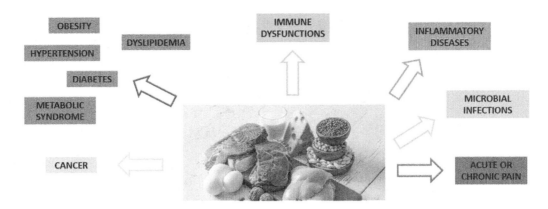

FIGURE 1.1 Bioactive peptides and clinical applications.

Even though BP have been commonly used as additives to give body to broths, soups, and sauces for many years, recent research shows that BP have important health benefits. Today the estimated peptide-based product market is around $40 billion per year (Lemes et al., 2016), demonstrating the strong demand from health-conscious consumers. BP are routinely used in different fields, including pharmacology, nutraceuticals, cosmetology, and human and pet food. For the most part, they are classified as functional foods, but some countries have different rules (Ozuna et al., 2015). However, one of the most important applications of BP regards the nutraceutical sector. It is well-known that BP regulate many important body functions (Figure 1.1), exercising antimicrobial, antioxidant, anti-inflammatory, anticoagulant, anticancer, lipid-lowering, antihypertensive, and antihyperglycemic effects through various pathways of action (depending on their structure and amino acid composition) that are, in many ways, still unknown (Jakubczyk et al., 2020).

Recently, new extraction and isolation techniques have been developed that use both vegetable (e.g., bromelain, ficin, papain) and animal (e.g., trypsin, chymotrypsin, pepsin, milk-peptides) matrices, to obtain BP with specific activities and improve the bioactivity as well as the bioavailability (Arulrajah et al., 2020). In effect, the bioactivity and the bioavailability of isolated peptides majorly depend on the degree of hydrolysis during the isolation process; this is the reason BP require both *in vitro* and *in vivo* studies before their commercialization (Girgih et al., 2014). Currently, at least 1500 BP have been reported in the BIOPEP database, but only a few of them have been studied for clinical applications (Singh et al., 2014).

An important source of BP are by-products, defined as "materials generated by a production chain" particularly interesting to reduce food waste and make suitable use of resources (zero-waste approach) (Colletti et al., 2020). Several by-products of the various phases of food production have been studied to find ways to limit food production's environmental and economic impact, and researchers have experimented with new processes for the recovery of BP components (Calcio Gaudino et al., 2020). For example, it can be estimated that about 170,000 tons of soybean curd residue, also known as "okara", are produced from 1 million tons of soymilk (protein content 3.5%) (Wang et al., 1996). okara contains around 15–30% of proteins, among which the low molecular weight fraction (less than 1 kDa) consists of digestible bioactive peptides, has the potential to inhibit the angiotensin-converting enzyme (ACE) and shows great antioxidant activity (Jimenez-Escrig et al., 2010).

The aim of this chapter is to describe the main animal and plant sources of BP, a summary of the production methods, the bioactivity and bioavailability of BP, and their limitations in clinical practice. Finally, the nutraceutical applications as well as the future perspectives will also be reviewed.

SOURCES

BP are mostly present inside bioactive proteins of different origins. To date, bovine milk, cheese, dairy products, meat, eggs, and fish, such as tuna, sardine, salmon and herring, are the greatest animal sources of BP, while wheat, maize, soy, rice, mushrooms, pumpkin, sorghum, and amaranth represent the greatest vegetal sources of BP derived from foods (Cicero et al., 2016).

BP from both animal and vegetal sources can be released during gastrointestinal digestion by trypsin or other microbial enzymes; and then, in addition to supplying the required raw materials for protein biosynthesis (nutritional value) and presenting a source of energy, they exhibit distinct biological activities (Sánchez and Vázquez, 2017).

ANIMAL SOURCES

Milk and dairy products are some of the most important sources of BP, which are released during gastrointestinal digestion or food processing, presumably confirming the importance of breastfeeding in the first months of life (Moller et al., 2008). Bovine and maternal milk contains different peptides with immunomodulatory (such as lactoferrin and immunoglobulins), neurotrophic (opioid peptides derived mostly from the hydrolysis of casein), cytotoxic, anti-carcinogenic, antibacterial, and anti-thrombotic activities described in the next chapters (Sánchez and Vázquez 2017). Both human and bovine colostrum are also rich sources of growth factors and BP, which appear to play a significant role in postnatal development (Chai et al., 2020). Other important, but less known, sources of BP are buffalo, donkey, camel, goat, mare, sheep, and yak milk (El-Salam and El-Shibiny 2013). Endogenous peptides derived from donkey milk (EWFTFLKEAGQGAKDMWR, GQGAKDMWR, REWFTFLK, and MPFLKSPIVPF) have been investigated in cardiovascular disease prevention (Zenezini Chiozzi et al., 2016). The enzymatic proteolysis of whey proteins gives rise to several antihypertensive peptides, including α-lactorphin (Tyr-Gly-Leu-Phe) and β-lactorphin (Tyr-Leu-Leu-Phe), Tyr-Pro, Lys-Val-Leu-Pro-Val-Pro-Gln, α-lactalbumin, and β-lactoglobulin (Sipola et al., 2002). The use of lactic acid bacteria represents a new strategy to obtain BP from the fermentation processes of milk proteins. As an example, *L.helveticus* LBK-16H fermented milk contains the BP Val-Pro-Pro and Ile-Pro-Pro which are well-known for their antihypertensive activities, acting as ACE inhibitory molecules (Cicero et al., 2016).

Eggs are also a good source of many BP used in medicine and the food industry (Sun et al., 2016). The peptide Arg-Val-Pro-Ser-Leu from egg white protein exhibits antihypertensive properties (Yu Z. et al., 2011). From boiled eggs, more than 60 peptides have been identified and some of them demonstrate anti-inflammatory and antioxidant effects through *in vitro* studies (Remanan and Wu 2014).

BP derived from meat products have the potential to be studied as functional foods and nutraceuticals. Both meat and fish BP have been shown to exhibit antimicrobial, antiproliferative and cardiovascular-protection activities *in vitro* and *in vivo* even if a limited number of nutraceuticals containing meat-derived BP are commercially available (Cicero et al., 2016). Arg-Pro-Arg peptide from pork meat showed the greatest antihypertensive activity *in vivo* in addition to Lys-Ala-Pro-Val-Ala and Pro-Thr-Pro-Val-Pro (Escudero et al., 2012).

VEGETAL SOURCES

Vegetal sources are also rich in proteins and BP. Food processing as well as the gastrointestinal hydrolysis of soybean seeds and soymilk generate many BP with established lipid-lowering, antihypertensive, anticancer, anti-inflammatory, and antimicrobial activity (Capriotti et al., 2015). Soy-fermented foods, natto and tempeh, were digested starting from glycinin (a soy protein) with a variety of endoproteases to generate oligopeptides with ACE inhibitory and anti-thrombotic activities (Gibbs et al., 2004).

Other rich sources of BP are cereal grains, including wheat, barley, rice, rye, oat, millet, sorghum, and corn, which have all been studied in regard to cardiovascular disease prevention (ACE inhibitory peptides, dipeptidyl peptidase inhibitor, peptides with antithrombotic, antioxidant, and hypotensive activities) (Malaguti et al., 2014). Even cocoa, roasted malt, coffee-fermented beer, and aged sake contain antioxidant BP, although the clinical studies are limited (Yamamoto et al., 2016).

Vegetable waste represents an important source of by-products obtained by the various phases of food production. Researchers have studied by-products to find ways to limit environmental and economic impacts (circular economy) caused by food processing, experimenting with new processes for the recovery of BP components (Mirabella et al., 2014). The process of olive oil extraction generates solid and liquid waste, which contain proteins and BP hydrolysates (prepared by treatment of different proteases) and have numerous potential health benefits (Esteve et al., 2015). Another example is okara, derived from soybean curd residue, characterized by a protein component of low molecular weight fraction (less than 1 kDa) of digestible peptides able to inhibit the ACE and exhibits great antioxidant activity (Jimenez-Escrig et al., 2010).

NEW SOURCES

New sources of peptides are gaining in popularity as demonstrated by the attention and the studies on edible insects (Nongonierma et al., 2017). In this context, Nongonierma et al. described for the first time the generation of BP from edible insects of the orders Orthoptera, Coleoptera, Blattodea, Lepidoptera, Isoptera, and Hymenoptera, characterized by a large number of beneficial activities, including antimicrobial, antidiabetic, antioxidant, anti-inflammatory, and ACE inhibitory properties (Vercruysse et al., 2005). For example, the peptides KHV, ASL, and GNPWM derived from *B. mori* has been shown to inhibit ACE, reducing both systolic and diastolic blood pressure (Tao et al., 2017). Other insects, such as cricket (*Gryllodes sigillatus*), mealworm (*Tenebrio molitor*), and desert locust (*Schistocerca gregaria*) have been shown to modulate the activity of other enzymes such as DPP-IV, α-glucosidase, and lipase, although clinical trials are still lacking (Zielińska et al., 2020).

BP derived from by-products of seafood proteins exhibit various bioactivities including antioxidant (e.g. GSGGL, GPGGFI, and FIGP peptides from *N. septentrionalis* skin), neuroprotective (e.g., seafood by-product–derived collagen peptides), antidiabetic (e.g.,

PYSFK, GFGPEL, and VGGRP peptides from grass carp skin), ACE inhibitory (e.g., GASSGMPG and LAYA from *G. macrocephalus* skin gelatin via pepsin hydrolysis), DPP-IV inhibitory (e.g., peptides obtained from Atlantic salmon), immunomodulatory (e.g., BP from skipjack tuna bones), antibacterial (e.g., BP from crustaceans), antiproliferative, and anticancer (e.g., hydrolysates prepared from rainbow trout) properties (Jakubczyk et al., 2020).

Finally, different plant seeds are inexpensive sources of antioxidant (e.g., wild hazelnut peptides), antibacterial (e.g., SMRKPPG identified from peony seed), anticancer (e.g., the protein hydrolysate purified from amaranth seeds) antidiabetic (e.g., BP from wild hazelnut) and antihypertensive (e.g., flaxseed-derived peptides) peptides (Jakubczyk et al., 2020).

FROM PRODUCTION TO COMMERCIALIZATION

BP can be obtained using several methods, which generally include the extraction and isolation of proteins from a food source, followed by the isolation of a specific protein isolate that undergoes *in vitro*, *in vivo* or *in silico* enzymatic hydrolysis. Different enzymes can be used to isolate short-chain peptides. Once the protein hydrolysate is obtained, the separation and purification of peptides is necessary in addition to the determination of bioactivity. Subsequently, the peptide fraction of interest is analyzed to determine the sequence of amino acids and, thus, the peptide synthesis. BP can be synthetized, and studies on both bioavailability and bioactivity represent the main step before proceeding with clinical trials (Figure 1.2) (Sun et al., 2020).

The enzymatic hydrolysis, microbial fermentation, and chemical synthesis are the three main methods of producing BP. Chemical hydrolysis consists of the formation of peptide hydrolysates using an acid or a base at high temperatures and resulting in the cleavage of the peptides. However, although this technique is a robust method of hydrolysis, it is considered unspecific, poorly reproducible, and leads to amino acid denaturation (Zumwalt et al., 1987). Enzymatic digestion is a process that produces peptides from soy, corn, potato, peanut, milk, whey, egg, and meat proteins with target functionalities including antioxidant, anti-inflammatory, antihypertensive, antidiabetic, antimicrobial,

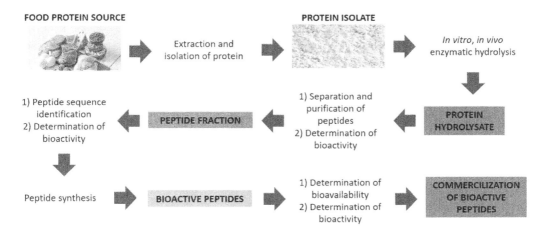

FIGURE 1.2 Scheme of BP isolation, preparation, and commercialization.

and anticancer activities. The efficacy of protein hydrolysates depends on several factors, such as the protein substrate pretreatment, type of proteases used, and the hydrolysis conditions applied (Zarei et al., 2012). Microbial fermentation has been demonstrated to produce antioxidant peptides using purified enzymes or fungal proteases, such as those found in natto and tempeh. BP produced by microbial fermentation can differ in type, amount, and activity depending on the cultures used (Hayes et al., 2007). Recently, the fermentation of milk proteins with specific microbial strains (e.g., *Lactobacillus bulgaricus*, *Lactobacillus helveticus* MB2-1, and *Lactobacillus plantarum* B1-6 and 70810) have been used to produce antihypertensive and lipid-lowering peptides (Rui et al., 2015).

According to the BIOPEP database, BP with antihypertensive, antioxidant, and antidiabetic properties account for the majority, although a set of other interesting activities, such as antimicrobial, antiproliferative, and lipid-lowering, have been described as well (Bechaux et al., 2019). The main limiting factor in the generation of BP is not the level of the raw material but the technological processes and difficulty in efficiently generating a specific BP sequence without altering their functionality and bioaccessibility (Bechaux et al., 2019). In this context, new separation and purification techniques (well-described in the next chapters) have been developed to overcome these problems (Piovesana et al., 2018).

STABILITY AND BIOAVAILABILITY

Once a peptide is ingested, it faces different physicochemical environments, which can negatively influence its stability and bioactivity. The stability and bioavailability of BP are strongly variable, depending on the degree of hydrolysis during the isolation process (amino acid composition) and the gastrointestinal environment, peptide size, and the hydrophobicity (Toldrá et al., 2020). In addition, the food matrix could interact negatively or positively with the chemical structure of BP, modifying both the stability and bioavailability (Sun et al., 2020). BP can undergo a chemical hydrolysis starting from the stomach where either acidic conditions or gastric enzymes interact and hydrolyze proteins and peptides. In addition, even the presence of pancreatic enzymes, the proteases from the microbiota, and the drastic change in pH (from ~2 of the stomach to 7 of the large intestine) could influence the hydrolysis of BP into the intestine (Langguth et al., 1997). Finally, BP can be degraded also at the brush border, cytosol of enterocytes, and even in the lysosomes and other cell organelles (Renukuntla et al., 2013).

In this context, *in vitro* studies using simulated gastrointestinal digestion systems are needed to investigate the stability and bioaccesibility of many peptides from food proteins (Li et al., 2020). However, observations coming from *in vitro* studies show better results than the *in vivo* data due to the latter's low bioavailability, which is strongly influenced by the absorption of and the susceptibility to physiological enzymes, which break down the peptides into inactive fragments (Fitzgerald et al., 2004). For example, the peptides MAP1 and MAP2 derived from milk proteins showed *in vitro* ACE inhibitory activity. However, only MAP1 has an acceptable stability and can reach the desired cellular sites of action, acting as an antihypertensive peptide (Boelsma et al., 2010).

In fact, orally administered BP can encounter over 40 different types of proteases during their passage to the small intestine and over 60 lysosomal peptidases, increasing the risk of degradation and reduction of their bioactivity (Woodley, 1994). Cysteine, basic amino acids, as well as aromatic amino acids are particularly sensitive to the endogenous oxidation and nitrosation processes (Bechaux et al., 2018).

Bioactive peptides are transported into the blood through several options, including the intestinal peptide transporters, which help in carrying the smaller peptides across the intestine, and the paracellular transport (passive transport) used for hydrophobic oligopeptides, which can be easily absorbed into blood circulation (solutes with a molecular weight higher than 3500 Daltons cannot be transported via this route) (Darewicz et al., 2011). Another type of intestinal transport is transcytosis by which particles are taken up by cells (Hebden et al., 1999).

Hayakawa and Lee have shown that in some areas of the jejunum and ileum, the aminopeptidase activity is low compared with other parts of the gastrointestinal tract, constituting a potential targeting site for BP delivery (Hayakawa and Lee, 1992). In this context, novel strategies (Table 1.1) have been developed to improve oral bioavailability of BP, using enzyme inhibitors and/or absorption enhancers (which protect BP from enzymatic degradation and improve their hydrophilicity or surface charge) as well as site-specific release forms to avoid or limit the intestinal enzymatic hydrolysis (Philippart et al., 2016).

APPLICATIONS

The most important application for BP is in the nutraceutical field (Cicero et al., 2017). Nutraceuticals are molecules of either animal or vegetal origin, which exert a positive impact on human health if incorporated into food or pharmaceutical formulations (sachets, capsules, tablets, or others) (Lee et al., 2013). In particular, food-based BP can be classified into different categories including traditional or functional foods (e.g., yogurt, milk, soy, or cheese, which are naturally rich in substances such as BP that can be useful for maintaining health), novel foods (new foods such as BP that don't have a significant history of widespread consumption and which would require safety studies), and fortified foods (foods with added substances such as BP to enhance their nutritional or physiological effects) (Chakrabarti et al., 2018). Recently however, nutraceuticals and functional BP-based foods have attracted much attention for their capacity to promote health and mitigate diseases, such as type II diabetes, cancer, obesity, diarrhea, thrombosis, dental carries, immunodeficiency, cardiovascular diseases, and other risk factors, including hypertension and hypercholesterolemia (Lee et al., 2013).

BP are a heterogeneous class of peptides, which can be used also in food manufacturing, because in addition to providing valuable nutritional and therapeutic properties, they retard oxidative degradation of lipids, thereby improving the quality and nutritional value of food (Shori and Baba, 2014).

Last but not least, BP are proposed also by the cosmetic industries as cosmetic ingredients, for their demonstrated capacity to modulate cell proliferation, cell migration, inflammation, angiogenesis, melanogenesis, and protein synthesis and regulation. In this regard, both *in vitro* and *in vivo* studies have shown the possible applications of BP in dermatology, including wound healing, acne, pigmentation problems of the skin, and dermatitis (Ledwoń et al., 2020).

ANTI-INFLAMMATORY ACTIVITY

Bioactive peptides obtained from animal and vegetable sources have also shown antiinflammatory activities. However, the mechanisms of action are mostly unknown, and

TABLE 1.1 Strategies of Delivery Systems for Bioactive Peptides

Delivery System	Mechanism of Targeting	Advantages	Drawbacks	Ref.
Absorption enhancers	Use of bile salts to form reverse micelles, disrupt membrane, and open up tight junctions. It consists of enzyme inhibition and mucolytic activity. Use of chelators (EDTA, citric acid, salicylates) that interfere with calcium ions and disrupt intracellular junctions. It decreases trans epithelial electrical resistance. Use of surfactants. Perturbation of intercellular lipids, lipid order, orientation, and fluidity. Inhibition of efflux mechanisms. Use of fatty acids and derivates to increase fluidity of phospholipid membranes, contraction of actin myofilaments, and opening of tight junctions Use of cationic polymers (chitosan). Combine effect of mucoadhesion and opening of tight junctions via ionic interactions with the cell membrane. Use of anionic polymer (polyacrilic acid). Combine effect of enzyme inhibition and opening of tight junctions through removal of extracellular calcium ions. Use of acylcarnitines for membrane disruption and opening of tight junctions with a calcium independent mechanism	Increased membrane permeation	Can transport undesirable molecules present in gastrointestinal tract	Brayden and Mrsny 2011, Renukuntla et al. 2013
Cell penetrating peptides	Proteins were enabled to be delivered into cells or tissues by hybridizing with target molecules	Enhance bioavailability and targeting of proteins	Toxic effect	Morishita et al. 2002
Emulsion	Solid in oil in water emulsion Oil in water emulsion Enteric coated oil in water emulsion	Protect drug from acid and luminal proteases in the GIT and enhance permeation through intestinal mucosa	Physiochemical instability in long-term storage and requirement for storage at low temperatures	Park et al. 2011, Toorisaka et al. 2003

(*Continued*)

TABLE 1.1 (Continued)

Delivery System	Mechanism of Targeting	Advantages	Drawbacks	Ref.
Enzyme inhibitors	Sodium glycocholate, aprotinin, amastatin, bestatin, boroleucin, puromycin, polymer inhibitor conjugates	Resist to enzymatic hydrolysis of the stomach and intestine	Inducing several side effects and high toxicity in case of prolonged use. May affect the absorption of other molecules, alter the metabolic pattern of gastrointestinal tract	Harish et al. 2010
Hydrogel	Three-dimensional mesh-like networks containing hydrophilic polymers that imbibe large amounts of water and form a gel-like matrix as a result of physical or chemical crosslinking of individual polymer chains	Biodegradable and biocompatible, can be used as potential carriers of peptides	May contain pathogen, may alter bioactivities of peptides	Renukuntla et al. 2013
Liposomes	Colloidal vehicles composed of a lipid bilayer containing hydrophilic and hydrophobic substances with unique structure that provide attractive properties	Improve efficiency of peptides, decrease toxicity, passive targeting	Physiochemically unstable, captured by the reticuloendothelial system	Saez et al. 2007
	Double liposomes, fusogenic liposomes, crosslinked liposomes	Improve physical stability and increase membrane permeation	Low stability of liposomes	Park et al. 2011
Membrane transporter and receptor targeting	Covalent conjugation of carrier molecules to peptides. Transport carriers enable the recognition by a specific endogenous membrane transporter or a receptor (vitamin B12, transferrin, invasins, viral hemaggulitinin, toxin, and lectin)	Enhance the intercellular delivery system to target cells, enhance oral absorption, enhance intestinal absorption and improve bioavailability	Limited to transporting of small drugs	Bai et al. 2005, Morishita et al. 2002

(*Continued*)

TABLE 1.1 (Continued)

Delivery System	Mechanism of Targeting	Advantages	Drawbacks	Ref.
Microencapsulation	Coating of solid, liquid or gaseous materials with a film of polymer or fats to generate free-flowing micrometric particles. Double emulsion/solvent evaporation procedure. Diffusion through the matrix and polymer degradation leading to particle erosion	Improve product stability, low fluctuation of peptide concentration	Difficult to scale up, incomplete release of peptides, peptide instability when encapsulated	Saez et al. 2007
	Eudragit S100 microspheres particulate carrier system pH-sensitive microspheres	Prevent proteolytic degradation in stomach and upper portion of small intestine. Restrict release of drug to favorable area of gastrointestinal tract	Concerns of protein stability during processing, release, and storage	Park et al. 2011
Microneedles	Ingestible capsule containing radially protruding microneedles that could be used as a platform for oral delivery of peptides	Acceptable bioavailability, safely pass through the GIT, no intestinal obstruction	Low retention time, low patient compliance	Traverso et al. 2015
Mucoadesive polymers	Hydrogel micro particles, lectin-conjugated alginate microparticles, thiolated polymers, gastrointestinal mucoadhesive patch system, polymers that can bind to biological substrates	Increase membrane permeation, site-specific delivery, improve bioavailability of peptides	Limited by the natural mucus turnover in the intestine	Şenel and Hincal 2001
	Mucoadhesive polymer–inhibitor conjugates	Resists enzymatic degradation, site-specific drug delivery	Limited by high cost of enzyme inhibitors	

(Continued)

TABLE 1.1 (Continued)

Delivery System	Mechanism of Targeting	Advantages	Drawbacks	Ref.
N-Methylation	Addition of methyl group in the N extremities of peptides	Improve stability of peptides and their resistance to enzymes of the GIT. Enhance solubility	Limited by transporters that modulate their passive diffusion, nonspecific	Gao et al. 2001, Clement et al. 2002
Nanoparticles	Chitosan nanoparticles Polystyrene nanoparticles, colloidal particle carriers	Prevent enzymatic degradation and increase intestinal epithelial absorption	Low loading efficiency of hydrophilic drugs, difficulty of precise size control, and avoidance of particle aggregation	Toorisaka et al. 2003
	Ligands attached to nanoparticles for surface modification with functional groups	High degree of engineering precision, control the size of the nanoparticles	Not available	Yun et al. 2013
PEGylation	Addition of polyethylglycol groups	Increases residence time, greater thermal and enzymatic stability	Decrease biological activity and heterogeneity	Segura-Campos et al. 2011
Prodrug approach	Phenyl propionic acid	Prodrug permeability improved 1608-fold than parent drug	Lack of methodology, structural complexity, stability problem of protein	Hsieh et al. 2009

Adapted from Bechaux et al., 2019.

only a few of these peptides have been investigated through both *in vitro* and *in vivo* studies (Marcone et al., 2017). Guha and colleagues suggest the possible anti-inflammatory action of BP, reducing the transcription of kinase factors, such as NF-kB and MAPK or other cytosolic compounds. It is not well established if BP act directly on the cell membrane or indirectly with different receptors, but the most realistic hypothesis is that BP operate in both ways depending on the type of peptide (Malinowski et al., 2014). Other suggested mechanisms of action include the inhibition of the pro-inflammatory JNK-MAPK pathway, reducing the formation of atherosclerotic plaque (obtained with IPP and VPP peptides) (Aihara et al., 2009) and the modulation of the expression of intestinal chemokines and cytokines (obtained with beans, milk, and soy peptides) (Majumder et al., 2016). Several BP act through different pathways as in the case of lunasin that have shown to inhibit IL-6, IL-1β PGE2 production, the expression of COX-2 and inducible NOS, as well as the activation of the NF-kB Akt-mediated route (Cicero et al., 2016). Even the polypeptide DMPIQAFLLYQEPVLGPVR, derived from β-casein, and a tripeptide derived from ovotransferrin (present in the albumen of eggs) inhibit the NF-kB pathway, reducing the transcription of vascular cell adhesion molecule 1 (VCAM1) and intracellular adhesion molecule 1 (ICAM-1) (Majumder et al., 2016; Malinowski et al., 2014). The γ-glutamyl cysteine peptide extracted from beans have shown to inhibit the phosphorylation of JNK and IkB, while valineproline-tyrosine (VPY) reduced the secretion of IL-8 and TNF-α (Majumder et al., 2016). Another interesting peptide fraction which exhibits anti-inflammatory properties has been extracted from milk fermented with *Lactobacillus plantarum* strains, resulting in an equivalent of sodium diclofenac in preclinical studies (Aguilar-Toalá et al., 2017).

BP derived from milk has shown to reduce postprandial inflammation in obese people expressed as plasma monocyte chemoattractant protein-1 (MCP-1) and chemokine ligand 5 (CCL5) (Holmer-Jensen et al., 2011).

Although preliminary data suggest an interesting anti-inflammatory activity of BP related to the modulation of transcription factors and the inhibition of the expression of pro-inflammatory cytokines and chemokines, data on humans are still lacking, and long-term randomized control trials (RCTs) are urgently needed to investigate and consolidate these results.

ANTIHYPERTENSIVE ACTIVITY

Hypertension is one of the most important cardiovascular risk factors, and data indicate that the lifetime risk of developing hypertension is a staggering 90%. In addition, the estimated global burden of hypertension will increase to 1.56 billion afflicted individuals by 2025 (McInnes, 2005). Hypertension accounts about 7.6 million premature deaths and 92 million DAILYs (disability-adjusted life-years: 1 DAILY = 1 lost year of healthy life) (Lawes et al., 2008). In this context, the European guidelines for hypertension include the nutraceutical approach for both prehypertensive subjects with borderline values of blood pressure and hypertensive patients in combination with conventional treatments (Borghi and Cicero, 2017; Cicero and Colletti, 2015).

Numerous bioactive peptides from different sources (milk, fish, plants, meat) have demonstrated to possess antihypertensive properties, acting through several pathways such as the inhibition of the renin-angiotensin system, increase in the activity of certain vasodilating agents (nitric oxide), and the reduction of the activity of the sympathetic system (Aluko, 2015). The most common target of BP-lowering molecules is the renin-angiotensin system, with the specific inhibition of renin or ACE, respectively responsible for the conversion of angiotensinogen into angiotensin I, and the angiotensin I into angiotensin II (Cicero et al., 2013). One of the richest sources of antihypertensive peptides is milk, which is particularly rich in tripeptides (e.g., valine-proline-proline (VPP) and isoleucine-proline-proline (IPP)) and polypeptides (e.g., FFVAPFPEVFGK and YLGYLEQLLR) (Cicero et al., 2013). Numerous RCTs have investigated the effects of BP on blood pressure. For example, in a meta-analysis of 18 RCTs, the lactotripeptides (LTP) IPP and VPP (dosages from 5 to 100 mg/day) have shown to reduce systolic blood pressure by −3.73 mmHg (95% CI: −6.70, −1.76) and diastolic blood pressure by −1.97 mmHg (95% CI: −3.85, −0.64) (Cicero et al., 2011). The effects were more evident in Asian people, suggesting a possible genetic/population-dependent effect. LTP demonstrated to positively improve the arterial stiffness measured as pulse wave velocity in mildly hypertensive subjects (Cicero et al., 2016).

The enzymatic or pepsin hydrolysis of whey proteins can generate several BP with antihypertensive properties. The decapeptide DRVYIHPFHL, octapeptide DRVYIHPF, and heptapeptide RVYIHPF have shown to inhibit the renin-angiotensin-aldosterone (RAS) system (Yadav et al., 2015). Even the casein proteins and BP isolated from the whey of cow's milk demonstrated significant blood pressure-lowering properties both in prehypertensive and hypertensive subjects (Nongonierma and FitzGerald, 2015). The enzymatic hydrolysis of whey proteins also produces the ACE-inhibitor peptides, α-lactalbumin and β-lactoglobulin, and lactorphins, which lower the blood pressure by normalizing the endothelial function (Dong et al., 2013).

Other antihypertensive tripeptides (LKP, IKP, LRP) are extracted from fish (e.g., bonito, tuna, sardine). These BP have shown to increase the endothelial NO levels and aorta vasodilatation in rats (Cheung et al., 2015). Even the peptides MVGSAPGVL and LGPLGHQ (from *Okamejei kenojei*) and AHIII (from *Styela clava*) exhibited antihypertensive activity in preclinical studies.

Finally, different BP extracted from numerous plant species (e.g., soy, barley, oak, pea) seem to reduce blood pressure through various mechanisms of action even if it's not always possible to discriminate between the effect of plant proteins and other components of a vegetable matrix which could contribute to the antihypertensive effects (Nirupama et al., 2015).

In conclusion, BP derived from both vegetal and animal sources have demonstrated to mildly reduce blood pressure in humans. However, long-term RCTs in normotensive and prehypertensive patients are needed for further confirmation.

LIPID-LOWERING ACTIVITY

Elevated plasma concentrations of total cholesterol (TC) and LDL cholesterol (LDL-C) and, under certain conditions, low concentrations of HDL cholesterol (HDL-C) are among the main modifiable risk factors for cardiovascular diseases (Ford et al., 2010). An examination of data from 18,053 individuals aged >20 years who participated in national health and nutrition surveys in the United States from 1999 to 2006 showed that the unadjusted prevalence of hypercholesterolemia varies from 53.2 to 56.1%. Indeed, a recent report from the American Heart Association (AHA) has confirmed that in the United States only 75.7% of children and 46.6% of adults have targeted TC levels (TC<170 mg/dL for children and <200 mg/dL for adults, in subjects not treated pharmacologically) (Mozaffarian et al., 2016). These percentages are comparable with western countries (Baigent et al., 2005; Zdrojewski et al., 2016).

BP from soy, lupine, and milk proteins are known to possess all the properties of the lipid-lowering nutraceuticals class. A meta-analysis of 35 RCTs investigated the effects of soy protein (B-conglycinin globulin) on cholesterolemia. The results showed a reduction in LDL-C of 3% (−4.83 mg/L; 95% CI: −7.34, −2.31), in TC of 2% (−5.33 mg/L; 95% CI: −8.35, −2.30), and in triacylglycerol of 4% (−4.92 mg/L; 95% CI: −7.79, −2.04). The period of treatment was from 4 weeks to 1 year, and the lipid-lowering effects were greater in moderately hypercholesterolemic patients (−7.47 mg/L; 95% CI: −11.79, −3.16) compared with healthy people (−2.96 mg/L; 95%CI: −5.28, −0.65) (Tokede et al., 2015). Even the lunasin peptide extracted from soy has shown lipid-lowering activities even in animal models (Lule et al., 2015). BP improve the lipid profile probably acting as hydroxymethylglutaril-CoA (HMG-CoA) reductase inhibitors, upregulators of LDL receptors, regulators of both the sterol regulatory element-binding protein 2 (SREBP2) pathway and the fecal excretion of bile salts (Lammi et al., 2014). The hydrolysate extract of *Mucuna pruriens* and BP from cowpea demonstrated to reduce LDL-C and TC, interacting with the micelle formation and the absorption of exogenous cholesterol (Marques et al., 2015; Herrera Chalé et al., 2016).

In summary, several lipid-lowering peptides have shown to improve the lipid profile in mildly dyslipidemic patients. Nevertheless, the evaluations of both the pharmacodynamic and pharmacokinetic profiles and the long-term effective dosages in humans are still necessary.

ANTICANCER ACTIVITY

One of the most important areas of research for bioactive peptides concerns their role as anticancer agents. BP from plants, milk, egg, and marine organisms have demonstrated to possess a cytotoxic activity in numerous cancer cell lines, with the advantages of low toxicity, high tissue penetration for their small size, cell diffusion, and permeability (Otvos, 2008). BP can act via different mechanisms of action, including the inhibition of cell migration, tumor angiogenesis and gene transcription/cell proliferation, and altering the tubulin structures of cells (Tyagi et al., 2015).

One of the most important anticancer peptides is lunasin, testing against breast, skin, colon, prostate cancers, and leukemia and lymphoma cell lines (Hernández-Ledesma et al., 2013). Lunasin acts by suppressing the transformation of cells induced by chemical carcinogens (in mouse fibroblast, and human breast Michigan Cancer Foundation-7 (MCF-7) cells), inducing cell cycle arrest in G2/M phase and apoptosis through the activation of caspase-3 (in L1210 leukemic and human colon adenocarcinoma cells) and inhibiting the metastasis of human colon cancer cells (Dia de Mejia, 2011). *In vivo* studies, especially on mice models, showed the reduction of lymphoma volume and liver metastasis of colon cancer (Chang et al., 2014).

The peptide Glutammate-Glycine-Arginine-Proline-Arginine from rice, at the dose of 600–700 µg/mL, have caused 84% inhibition of the growth of colon cancer cells (Caco-2 and human colorectal adenocarcinoma cell line, HCT-116), 80% inhibition of the growth of breast cancer cells (MCF-7, MDA-MB-231), and 84% inhibition of the growth of liver cancer cells (HepG-2) (Kannan et al., 2010).

Other rich sources of anticancer BP are legume seeds, which contain different protease inhibitors, such as the Bowman-Birk inhibitor that demonstrated (in vitro) a preventive effect against prostate, breast, and colon cancers (Park et al., 2005). The Bowman-Birk inhibitor in clinical trials was approved by the Food and Drug Administration (FDA), especially in people with oral leukoplakia or benign prostatic hyperplasia (Malkowicz et al., 2001). Plant-derived lectins (from tepary bean and mistletoe) have demonstrated cytotoxic effects on the cervical carcinoma cell line, C33-A, and human colon carcinoma cell line, Sw480 (Valadez-Vega et al., 2011).

Among the BP of animal origin, lactoferrin from milk demonstrates in studies *in vitro* and *in vivo* to inhibit the growth of breast cancer (MDA-MB-231) and nasopharyngeal carcinoma cells by, respectively, arresting the cell cycle at the G1/S transition and suppressing Akt signaling (Deng et al., 2013). It acts primarily by inducing the process of apoptosis, modulating the gene expression, and reducing the angiogenesis (Varadhachary et al., 2004). Hydrolysates from egg proteins, lysozyme, and ovomucin are able to inhibit the proliferation of different tumor cells, improving the effectiveness of chemotherapy (colorectal cancer, B16 melanoma) (Azuma et al., 2000; Sava, 1989). Even many marine peptides (e.g., from tuna, sponges, squid) induce antiproliferative activity against different cancer types although these results require confirmation *in vivo* (Cicero et al., 2017).

In summary, BP can be considered an interesting category of molecules, which include also anticancer peptides with demonstrated cytotoxic and anti-tumoral activity *in vitro* and in animal models. However, human RCTs are still lacking, and the clinical setting must be thoroughly investigated before considering any kind of prescription.

IMMUNOMODULATORY ACTIVITY

BP can also modulate immune responses. Studies *in vitro* and *in vivo* show that αS1-casein and β-casein were able to stimulate phagocytes, the production of IgG in lymphocytes, and the proliferation of T-lymphocytes (Hata et al., 1998). Even the peptides from fish have shown immunomodulatory effects, especially enhancing the macrophage and natural killer activity as well as lymphocyte proliferation (e.g., BP from chum Salmon or from Atlantic cod) (Yang et al., 2009). Another possible mechanism of immune regulation is the modulation of gut-associated immunity modulating the phagocytic activity and the IgA-secreting cells in the small intestine lamina propria (Duarte et al., 2006). Studies in mice have shown the ability of oyster hydrolysates to enhance the proliferation of lymphocytes, the activity of natural killer cells, and the phagocytic rate of macrophages (Wang et al., 2010). Similar results were observed with tryptic hydrolysates of soybean and rice proteins (Kitts and Weiler, 2003).

In summary, BP exhibit immunomodulatory effects acting on both specific and nonspecific immunity. RCTs are needed to investigate the efficacy and the safety profile of these molecules.

OTHER BIOLOGICAL ACTIVITIES

BP from different sources such as wheat gliadin, soy proteins, egg yolk proteins, porcine myofibrillar proteins, and pea and aquatic by-product proteins are well-known to possess antioxidant properties and to protect against oxidative stress, which is typical of the major chronic diseases (Wang et al., 2007; Malaguti et al., 2014).

BP antioxidant mechanisms of action include the metal ion chelation against enzymatic and nonenzymatic peroxidation of lipids and essential fatty acids as well as the reduction of reactive species of oxygen acting like free radical scavengers. In this regard, carnosine and anserine, which represent two abundant peptides in meats, have shown to reduce and prevent stress-related diseases (Hipkiss and Brownson, 2000). The antioxidant properties of lunasin have been confirmed *in vitro*, to protect cell antioxidant defenses of human Caco-2 cells treated with hydrogen peroxide and tert-butylhydroperoxide and to scavenge both peroxyl and superoxide radicals (Garcia-Nebot et al., 2014).

Even BP isolated from oyster, shrimp, squid, and blue mussel have shown antioxidant properties although data on humans are still lacking (Harada et al., 2010).

Other peptides such as α-lactorphin and β-lactorphin (opioid peptides) exert analgesic activity, acting as agonists of opiate receptors. Studies *in vitro* investigated the effects of BP from α-lactalbumin and β-lactoglobulin. These peptides have shown analgesic activity at micromolar concentrations (Pihlanto-Leppälä, 2000).

Other analgesic peptides are those found in rice albumin, gluten, and vegetal sources, such as spinach (Teschemacher, 2003).

Finally, some BP-like lactoferrin (fragment 17-41) or peptides derived from bovine meat (GFHI, DFHING, FHG, GLSDGEWQ) are under investigation with studies both *in vitro* and *in vivo* for their antimicrobial effects. BP have shown a large spectrum of action against viruses, bacteria, protozoa, and fungi. As an example, the fragment 17-41 of lactoferrin, also known as lactoferricin, has a bactericidal activity altering the membrane permeability of bacteria through the interaction of the lipid A part of bacterial lipopolysaccharides (Orsi 2004). Preliminary data have shown that BP are in general well-tolerated, do not induce pathogen resistance, and their activity is against both Gram-positive and Gram-negative bacteria (Jang et al., 2008).

DISCUSSION

The interest in BP is growing year by year, confirming the current progress in clinical field and the attention of food and nutraceutical companies (Chakrabarti et al., 2018). BP include peptides composed of amino acids, which are well-known to regulate different aspects of cellular function and communications (Craik et al., 2013).

Despite several studies conducted on human populations having demonstrated the effectiveness of BP, some limitations should be considered, including the small sample size on clinical trials, the short period of treatments, and the great heterogeneity of the obtained results (Cicero et al., 2017). Currently, the best evidence in humans regards the supplementation of BP in cardiovascular disease prevention and, in particular, as lipid-lowering or antihypertensive agents (Cicero et al., 2013). In these contexts, BP demonstrate to be a valid nutraceutical option if combined with other molecules, confirming their efficacy, tolerability, and safety profile (Cicero et al., 2016). The antioxidant, anti-inflammatory, analgesic activities of BP have been investigated in different studies although long-term clinical trials are still lacking and their mechanisms of action need to be clarified (Craik et al., 2013). Interesting preliminary data include the potential actions of BP in a variety of cancer cell lines but, once again, a limited number of phase I studies have been carried out (Cicero et al., 2017). Finally, the immunomodulatory actions of different peptides on both specific and nonspecific immunity are under investigation with preclinical studies (Franck et al., 2002).

The evidence on BP presents other limitations beyond the paucity of clinical trials. First, pharmacokinetic data are often lacking but required to determine the dosage and the frequency of administration as well as to understand the different inter- and intra-individual variability (Yoshikawa, 2015). The inter-variability (age, sex, diseases, concomitant therapy, etc.) of BP effects is one of the most important aspects that should be considered: as an example, lactotripeptides (IPP and VPP) supplemented in European or Asian subjects demonstrated differences on blood pressure reduction, probably due to a genetic/population-dependent effect on Asian people (Cicero et al., 2011).

Pharmacokinetic studies are important to the study of biopharmaceutical aspects of BP and, consequently, to develop the best formulative strategies (e.g., lipid microparticle systems, micelle, emulsion, microencapsulation, etc.) to improve the oral bioavailability and the biological effects (Cao et al., 2019). However, these studies are often limited because BP have a short life (less than 2 hours) as well as a low plasma concentration (pmol/mL), limiting the measurement of oral bioavailability (Iwai et al., 2005).

Oral bioavailability of BP is, generally, extremely low and the chemical structure of peptides must be considered: in this context, *in vitro* studies using simulated gastrointestinal digestion systems are needed to investigate the stability and bioaccesibility of many BP (Amigo and Hernández-Ledesma, 2020). In fact, processes such as digestion can transform BP into inactive peptides, or reduce bioactivity, or prevent peptides that cannot reach the bloodstream in sufficient quantities from exerting their effects. Recently, the development of nonconventional dosage forms has improved BP bioavailability, reducing both microbiota and chemical degradation, which represent the main limitation of oral peptide supplementation (Sayd et al., 2018).

A similar discourse should be made for the pharmacodynamics of bioactive peptides: in fact, the putative mechanisms of action are often unclear, and individual peptides appear to act through several pathways and thus, to possess several pleiotropic activities. In addition, it is difficult to attribute the action of a single peptide when the entire complex of protein hydrolysates is studied (Rutherfurd-Markwick et al., 2012). A possible solution includes the *in vitro* activity guided fractionation, characterized by the combination of the analytical separation of protein digested fraction and the *in vivo* evaluation of the activity of specific BP (Sato et al., 2013).

Finally, another important aspect which could be investigated is the different transporter used for intestinal absorption: it is known that the specific duodenal transporter which absorbs the major part of BP can be saturated, depending on the dosages, and any mechanism of competition between two or more peptides for the same transporter can reduce bioavailability (Sato et al., 2013).

CONCLUSIONS

BP studies are encouraging and demonstrate their potential role as nutraceuticals and food supplements. BP have shown to prevent and/or treat (in addition to conventional therapies) several diseases and/or risk factors. However, human pharmacokinetic and bioactivity studies are critical to a better understanding of these compounds. Long-term RCTs are also needed to test the efficacy as well as their potential immunogenicity. Finally, the isolation, extraction, and production techniques of BP must be standardized and applicable from a laboratory scale to industrial scale; in this context, an economic analysis of the potential exploitation of BP must be considered.

REFERENCES

Aguilar-Toalá, J.E.; Santiago Lopez, L.; Peres, C.M.; Peres, C.; Garcia, H.S.; Vallejo-Cordoba, B. et al. 2017 Assessment of multifunctional activity of bioactive peptides derived from fermented milk by specific Lactobacillus plantarum strains. *J Dairy Sci.* 100(1):65–75.

Aihara, K.; Ishii, H.; Yoshida, M. 2009 Casein-derived tripeptide, Val-Pro-Pro (VPP), modulates monocyte adhesion to vascular endothelium. *J Atheroscler Thromb.* 16(5):594–603.

Aluko, R. 2012 *Functional Foods and Nutraceuticals*, Springer, pp. 37–61.

Aluko, R.E. (2015). Antihypertensive peptides from food proteins. *Annu. Rev Food Sci Technol.* 6:235–262.

Amigo, L.; Hernández-Ledesma, B. 2020 Current Evidence on the Bioavailability of Food Bioactive Peptides. *Molecules.* 25(19):4479.

Arulrajah, B.; Muhialdin, B.J.; Zarei, M.; Hasan, H.; Saari, N. 2020 Lacto-fermented Kenaf (Hibiscus cannabinus L.) seed protein as a source of bioactive peptides and their applications as natural preservatives. *Food Control.* 110:106969.

Azuma, N.; Suda, H.; Iwasaki, H.; Yamagata, N.; Saeki, T.; Kanamoto, R. et al. 2000 Antitumorigenic effects of several food proteins in a rat model with colon cancer and their reverse correlation with plasma bile acid concentration. *J Nutr Sci Vitaminol.* 46(2):91–96.

Bai, Y.; Ann, D.K.; Shen, W.C. 2005 Recombinant granulocyte colony-stimulating factor-transferrin fusion protein as an oral myelopoietic agent. *Proc. Natl. Acad. Sci. U. S. A.* 102(20):7292–7296.

Baigent, C.; Keech, A.; Kearney, P.M.; Blackwell, L.; Buck, G.; Pollicino, C. et al. 2005 Cholesterol Treatment Trialists' (CTT) Collaborators. Efficacy and safety of cholesterol-lowering treatment: Prospective metaanalysis of data from 90,056 participants in 14 randomised trials of statins. *Lancet.* 366(9493)1267–1278.

Bechaux, J.; de La Pomélie, D.; Théron, L.; Santé Lhoutellier, V.; Gatellier, P. 2018 Iron-catalysed chemistry in the gastrointestinal tract: Mechanisms, kinetics and consequences. A review. *Food Chem.* 268:27–39.

Bechaux, J.; Gatellier, P.; Le Page, J.F.; Drillet, Y.; Sante-Lhoutellier, V. 2019 A comprehensive review of bioactive peptides obtained from animal byproducts and their applications. *Food Funct.* 10(10):6244–6266.

Boelsma, E.; Kloek, J. 2010 IPP-rich milk protein hydrolysate lowers blood pressure in subjects with stage 1 hypertension, a randomized controlled trial. *Nutr J.* 9:52.

Borghi, C.; Cicero, A.F. 2017. Nutraceuticals with clinically detectable blood pressure lowering effect: A review of available randomized clinical trials and their meta-analyses. *Br J Clin Pharmacol.* 83(1):163–171.

Brayden, D.J.; Mrsny, R.J. 2011 Oral peptide delivery: Prioritizing the leading technologies. *Ther. Deliv.* 2(12):1567–1573.

Calcio Gaudino, E.; Colletti, A.; Grillo, G.; Tabasso, S.; Cravotto, G. 2020 Emerging processing technologies for the recovery of valuable bioactive compounds from potato peels. *Foods.* 9(11):1598.

Cao, S.J.; Xu, S.; Wang, H.M.; Ling, Y.; Dong, J.; Xia, R.D. et al. 2019 Nanoparticles: Oral delivery for protein and peptide drugs. *AAPS Pharm Sci Tech.* 20(5):190.

Capriotti, A.L.; Caruso, G.; Cavaliere, C.; Samperi, R.; Ventura, S.; Chiozzi, R.Z. et al. 2015 Identification of potential bioactive peptides generated by simulated gastrointestinal digestion of soybean seeds and soy milk proteins. *J Food Comp Anal.* 44:205–213.

Chai, K.F.; Voo, A.Y.H.; Chen, W.N. 2020 Bioactive peptides from food fermentation: A comprehensive review of their sources, bioactivities, applications, and future development. *Compr Rev Food Sci Food Saf.* 19(6):3825–3885.

Chakrabarti, S.; Guha, S.; Majumder, K. 2018 Food-derived bioactive peptides in human health: Challenges and opportunities. *Nutrients.* 10(11):1738.

Chang, H.C.; Lewis, D.; Tung, C.Y.; Han, L.; Henriquez, S.M.P.; Voiles, L. et al. 2014 Soypeptide lunasin in cytokine immunotherapy for lymphoma. *Cancer Immunol Immunother.* 63(3):283–295.

Cheung, R.C.; Ng, T.B.; Wong, J.H. 2015 Marine peptides: Bioactivities and applications. *Mar Drugs.* 13(7):4006–4043.

Cicero, A.F.; Colletti, A.; Rosticci, M.; Cagnati, M.; Urso, R.; Giovannini, M. et al. 2016 Effect of Lactotripeptides (Isoleucine-Proline-Proline/Valine-Proline-Proline) on blood pressure and arterial stiffness changes in subjects with suboptimal blood pressure control and metabolic syndrome: A double-blind, randomized, crossover clinical trial. *Metab Syndr Relat Disord.* 14(3):161–166.

Cicero, A.F.G.; Aubin, F.; Azais-Braesco, V.; Borghi, C. 2013 Do the lactotripeptides isoleucine–proline–proline and valine–proline–proline reduce systolic blood pressure in European subjects? A meta-analysis of randomized controlled trials. *Am J Hypertens.* 26(3):442–449.

Cicero, A.F.G.; Fogacci, F.; Colletti, A. 2017. Potential role of bioactive peptides in prevention and treatment of chronic diseases: A narrative review. *Br J Pharmacol.* 174(11):1378–1394.

Cicero, A.F.G.; Gerocarni, B.; Laghi, L.; Borghi, C. 2011 Blood pressure lowering effect of lactotripeptides assumed as functional foods: A meta-analysis of current available clinical trials. *J Hum Hypertens.* 25(7):425–436.

Cicero, A.F.G.; Colletti, A. 2015 Nutraceuticals and blood pressure control: Results from clinical trials and meta-analyses. *High Blood Press Cardiovasc Prev.* 22(3):203–213.

Clement, S.; Still, J.G.; Kosutic, G.; McAllister, R.G. 2002 Oral insulin product hexyl-insulin monoconjugate 2 (HIM2) in type 1 diabetes mellitus: The glucose stabilization effects of HIM2. *Diabetes Technol. Ther.* 4(4):459–466.

Colletti, A.; Attrovio, A.; Boffa, L.; Mantegna, S.; Cravotto, G. 2020 Valorisation of by-products from Soybean (Glycine max (L.) Merr.) processing. *Molecules.* 25(9):2129.

Craik, D.J.; Fairlie, D.P.; Liras, S.; Price, D. 2013 The future of peptidebased drugs. *Chem Biol Drug Des.* 81(1):136–147.

Darewicz, M.; Dziuba, B.; Minkiewicz, P.; Dziuba, J. 2011 The preventive potential of milk and colostrum proteins and protein fragments. *Food Rev Int.* 27(4):357–388.

Deng, M.; Zhang, W.; Tang, H.; Ye, Q.; Liao, Q.; Zhou, Y. et al. 2013 Lactotransferrin acts as a tumor suppressor in nasopharyngeal carcinoma by repressing AKT through multiple mechanisms. *Oncogene.* 32(36):4273–4283.

Dia, V.P.; de Mejia, E.G. 2011 Lunasin potentiates the effect of oxaliplatin preventing outgrowth of colon cancer metastasis, binds to α5β1 integrin and suppresses FAK/ERK/NF-κB signaling. *Cancer Lett.* 313(2):167–180.

Dong, J.Y.; Szeto, I.M.; Makinen, K.; Gao, Q.; Wang, J.; Qin, L.Q. et al. 2013 Effect of probiotic fermented milk on blood pressure: A meta-analysis of randomised controlled trials. *Br J Nutr.* 110(7):1188–1194.

Duarte, J.; Vinderola, G.; Ritz, B.; Perdigon, G.; Matar, C. 2006 Immunomodulating capacity of commercial fish protein hydrolysate for diet supplementation. *Immunobiology.* 211(5):341–350.

El-Salam, M.H.A.; El-Shibiny, S. 2013 Bioactive peptides of buffalo, camel, goat, sheep, mare, and yak milks and milk products. *Food Rev Int.* 29(1):1–23.

Escudero, E.; Toldra, F.; Sentandreu, M.A.; Nishimura, H.; Arihara, K. 2012 Antihypertensive activity of peptides identified in the in vitro gastrointestinal digest of pork meat. *Meat Science.* 91(3):382–384.

Esteve, C.; Marina, M L.; García, M.C. 2015 Novel strategy for the revalorization of olive (Olea europaea) residues based on the extraction of bioactive peptides. *Food Chemistry.* 167:272–280.

Fitzgerald, R.J.; Murray, B.A.; Walsh, D.J. 2004 Hypotensive peptides from milk proteins. *J Nutr.* 134(4):980S–988S.

Ford, E.S.; Li, C.; Pearson, W.S.; Zhao, G.; Mokdad, A.H. 2010 Trends in hypercholesterolemia, treatment and control among United States adults. *Int J Cardiol.* 140(2):226–235.

Franck, P.; Moneret Vautrin, D.A.; Dousset, B.; Kanny, G.; Nabet, P.; Guénard-Bilbaut, L. et al. 2002 The allergenicity of soybean-based products is modified by food technologies. *Int Arch Allergy Immunol.* 128(3):212–219.

Gao, J.; Sudoh, M.; Aubé, J.; Borchardt, R.T. 2001 Transport characteristics of peptides and peptidomimetics: I. N-methylated peptides as substrates for the oligopeptide transporter and P-glycoprotein in the intestinal mucosa. *J Pept Res.* 57(5):361–373.

Garcia-Nebot, M.J.; Recio, I.; Hernandez-Ledesma, B. 2014 Antioxidant activity and protective effects of peptide lunasin against oxidative stress in intestinal Caco-2 cells. *Food Chem Toxicol.* 65:155–161.

Gibbs, B.F.; Zougman, A.; Masse, R.; Mulligan, C. 2004 Production and characterization of bioactive peptides from soy hydrolysate and soy-fermented food. *Food Research International.* 37(2):123–131.

Girgih, A.T.; He, R.; Malomo, S.; Offengenden, M.; Wu, J.; Aluko R.E. 2014 Structural and functional characterization of hemp seed (Cannabis sativa L.) protein-derived antioxidant and antihypertensive peptides. *J Funct Foods.* 6:384–394.

Guha, S.; Majumder, K. 2019 Structural-features of food-derived bioactive peptides with anti-inflammatory activity: A brief review. *J Food Biochem.* 43:e12531.

Harada, K.; Maeda, T.; Hasegawa, Y.; Tokunaga, T.; Tamura, Y.; Koizumi, T. 2010 Antioxidant activity of fish sauces including puffer (Lagocephalus wheeleri) fish sauce measured by the oxygen radical absorbance capacity method. *Mol Med Rep.* 3(4):663–668.

Harish, I.; Anand, K.; Manish, V. 2010 Oral insulin – A review of current status. *Diabetes, Obes. Metab.* 12(3):179–185.

Hata, I.; Higashiyama, S.; Otani, H. 1998 Identification of a phosphopeptide in bovine alpha s1-casein digest as a factor influencing proliferation and immunoglobulin production in lymphocyte cultures. *J Dairy Res.* 65(4):569–578.

Hayakawa, E.; Lee, V.H.L. 1992 Aminopeptidase activity in the jejunal and heal peyer's patches of the albino rabbit. *Pharm. Res.* 9(4):535–540.

Hayes, M.; Ross, R.P.; Fitzgerald, G.F.; Stanton, C. 2007 Putting microbes to work: Dairy fermentation, cell factories and bioactive peptides. Part I: Overview. *Biotechnol J.* 2(4):426–434.

Hebden, J.M.; Wilson, C.G.; Spiller, R.C.; Gilchrist, P.J.; Blackshaw, E.; Frier, M.E. et al. 1999 Regional differences in quinine absorption from the undisturbed human colon assessed using a timed release delivery system. *Pharm. Res.* 16(7):1087–1092.

Hernández-Ledesma, B.; Hsieh, C.C.; de Lumen, B.O. 2013 Chemopreventive properties of peptide lunasin: A review. *Protein Pept Lett.* 20(4):424–432.

Herrera Chalé, F.; Ruiz Ruiz, J.C.; Betancur Ancona, D.; Acevedo Fernández, J.J.; Segura Campos, M.R. 2016 The hypolipidemic effect and antithrombotic activity of Mucuna pruriens protein hydrolysates. *Food Funct.* 7(1):434–444.

Hipkiss, A.R.; Brownson, C. 2000 A possible new role for the anti-ageing peptide carnosine. *Cell Mol Life Sci.* 7(5):747–753.

Holmer-Jensen, J.; Karhu, T.; Mortensen, L.S.; Pedersen, S.B.; Herzig, K.H.; Hermansen, K. 2011 Differential effects of dietary protein sources on postprandial low-grade inflammation after a single high fat meal in obese non-diabetic subjects. *Nutr J.* 10:115.

Hsieh, P-W.; Hung, C-F.; Fang, J-Y. 2009 Current prodrug design for drug discovery. *Curr Pharm Des*. 15(19):2236–2250.

Iwai, K.; Hasegawa, T.; Taguchi, Y.; Morimatsu, F.; Sato, K.; Nakamura, Y. et al. 2005 Identification of food-derived collagen peptides in human blood after oral ingestion of gelatin hydrolysates. *J Agric Food Chem*. 53(16):6531–6536.

Jakubczyk, A.; Karaś, M.; Rybczyńska-Tkaczyk, K.; Zielińska, E.; Zieliński, D. 2020 Current trends of bioactive peptides-new sources and therapeutic effect. *Foods*. 9(7):846.

Jang, A.; Jo, C.; Kang, K.S.; Lee, M. 2008 Antimicrobial and human cancer cell cytotoxic effect of synthetic angiotensin-converting enzyme (ACE) inhibitory peptides. *Food Chem*. 107(1):327–336.

Jimenez-Escrig, A.; Alaiz, M.; Vioque, J.; Ruperez, P. 2010 Health-promoting activities of ultra-filtered okara protein hydrolysates released by in vitro gastrointestinal digestion: Identification of active peptide from soybean lipoxygenase. *Eur Food Res Technol*. 230(4):655–663.

Kannan, A.; Hettiarachchy, N.S.; Lay, J.O.; Liyanage, R. 2010 Human cancer cell proliferation inhibition by a pentapeptide isolated and characterized from ricebran. *Peptides*. 31(9):1629–1634.

Karaś, M. 2019 Influence of physiological and chemical factors on the absorption of bioactive peptides. *Int J Food Sci Technol*. 54(5):1486–1496.

Karaś, M.; Jakubczyk, A.; Szymanowska, U.; Krystyna, J.; Lewicki, S.; Złotek, U. 2019 Different temperature treatments of millet grains affect the biological activity of protein hydrolysates and peptide fractions. *Nutrients*. 11(3):550.

Kitts, D.D.; Weiler, K. 2003 Bioactive proteins and peptides from food sources. Applications of bioprocesses used in isolation and recovery. *Curr Pharm Des*. 9(16):1309–1323.

Lammi, C.; Zanoni, C.; Scigliuolo, G.M.; D'Amato, A.; Arnoldi, A. 2014 Lupin peptides lower low-density lipoprotein (LDL) cholesterol through an up-regulation of the LDL receptor/sterol regulatory element binding protein 2 (SREBP2) pathway at HepG2 cell line. *J Agric Food Chem*. 62(29):7151–7159.

Langguth, P.; Bohner, V.; Heizmann, J.; Merkle, H.P.; Wolffram, S.; Amidon, G.L. et al. 1997 The challenge of proteolytic enzymes in intestinal peptide delivery. *J Controlled Release*. 46(1–2):39–57.

Lawes, C.M.; Vanders, H.S.; Rodgers, A. 2008 Global burden of blood pressure related disease, 2001. *Lancet*. 371(9623):1513–1518.

Ledwoń, P.; Errante, F.; Papini, A.M.; Rovero, P.; Latajka, R. 2020 Peptides as active ingredients: A challenge for cosmeceutical industry. *Chem Biodivers*. 18(2):e2000833.

Lee, J.K.; Li-Chan, E.C.Y.; Jeon, J.K.; Byun, H.G. 2013 Development of functional materials from seafood by-products by membrane separation technology. In: S.K. Kim (ed.) *Seafood Processing By-Products*. Springer, 35–62.

Lemes, A.C.; Sala, L.; Ores, J.D.C.; Braga, A.R.C.; Egea, M.B.; Fernandes, K.F. 2016 A review of the latest advances in encrypted bioactive peptides from protein-rich waste. *Int J Mol Sci*. 17(6):950.

Li, T.; Shi, C.; Zhou, C.; Sun, X.; Ang, Y.; Dong, X. et al. 2020 Purification and characterization of novel antioxidant peptides from duck breast protein hydrolysates. *LWT Food Sci Technol*. 125(2):109215.

Lule, V.K.; Garg, S.; Pophaly, S.D.; Tomar, S.K. 2015 Potential health benefits of lunasin: A multifaceted soy-derived bioactive peptide. *J Food Sci*. 80(3):R485–R494.

Majumder, K.; Chakrabarti, S.; Morton, J.S.; Panahi, S.; Kaufman, S.; Davidge, S.T. et al. 2013 Egg-derived tri-peptide IRW exerts antihypertensive effects in spontaneously hypertensive rats. *PLoS One*. 8(11):e82829.

Majumder, K.; Mine, Y.; Wu, J. 2016 The potential of food proteinderived anti-inflammatory peptides against various chronic inflammatory diseases. *J Sci Food Agric.* 96(7):2303–2311.

Malaguti, M.; Dinelli, G.; Leoncini, E.; Bregola, V.; Bosi, S.; Cicero, A.F.G. et al. 2014.Bioactive peptides in cereals and legumes: Agronomical, biochemical and clinical aspects. *Int J Mol Sci.* 15(11):21120–21135.

Malinowski, J.; Klempt, M.; Clawin-Rädecker, I.; Lorenzen, P.C.; Meisel, H. 2014 Identification of a NFκB inhibitory peptide from tryptic β-casein hydrolysate. *Food Chem.* 165:129–133.

Malkowicz, S.B.; McKenna, W.G.; Vaughn, D.J.; Wan, X.S.; Propert, K.J.; Rockwell, K. et al. 2001 Effects of bowman-birk inhibitorconcentrate (BBIC) in patients with benign prostatic hyperplasia. *Prostate.* 48(1):16–28.

Marcone, S.; Belton, O.; Fitzgerald, D.J. 2017. Milk-derived bioactive peptides and their health promoting effects: A potential role in atherosclerosis. *Br J Clin Pharmacol.* 83(1):152–162.

Marques, M.R.; Soares Freitas, R.A.; Corrêa Carlos, A.C.; Siguemoto, É.S.; Fontanari, G.G.; Arêas, J.A. 2015 Peptides from cowpea present antioxidant activity, inhibit cholesterol synthesis and its solubilisation into micelles. *Food Chem.* 168:288–293.

Mazorra-Manzano M.A.; Ramírez-Suarez J.C.; Yada R.Y. 2017 Plant proteases for bioactive peptides release: A review. *Crit Rev Food Sci Nutr.* 58(13):2147–2163.

McInnes, G.T. 2005 Lowering blood pressure for cardiovascular risk reduction. *J Hypertens Suppl.* 23(1):S3–8.

Mellander, O. 1950 The physiological importance of the casein phosphopeptide calcium salts. II. Peroral calcium dosage of infants. Some aspects of the pathogenesis of rickets. *Acta Soc Bo Pol.* 55(5–6):247–257.

Mirabella, N.; Castellani, V.; Sala, S. 2014 Current options for the valorization of food manufacturing waste: A review. *J Cleaner Prod.* 65:28–41.

Moller, N.P.; Scholz-Ahrens, K.E.; Roos, N.; Schrezenmeir, J. 2008 Bioactive peptides and proteins from foods: Indication for health effects. *European Journal of Nutrition.* 47(4):171–182.

Morishita, M.; Lowman, A.M; Takayama, K.; Nagai, T.; Peppas, N.A. 2002 Elucidation of the mechanism of incorporation of insulin in controlled release systems based on complexation polymers, *J Control Rel.* 81(1–2):25–32.

Nirupama, G.; Mohammad, B.; Hossain, D.K.R.; Nigel, P.B. 2015 A review of extraction and analysis of bioactives in oat and barley and scope for use of novel food processing technologies. *Molecules.* 20(6):10884–10909.

Nongonierma, A.B.; FitzGerald, R.J. 2015 Bioactive properties of milk proteins in humans: A review. *Peptides.* 73:20–34.

Nongonierma, A.B.; FitzGerald, R.J. 2017 Unlocking the biological potential of proteins from edible insects through enzymatic hydrolysis: A review. *Innov Food Sci Emerg Technol.* 43:239–252.

Orsi, N. 2004 The antimicrobial activity of lactoferrin: Current status and perspectives. *Biometals.* 17(3):189–96.

Otvos, L. Jr. 2008 Peptide-based drug design: Here and now. *Methods Mol Biol.* 494:1–8.

Ozuna, C.; Paniagua-Martínez, I.; Castaño-Tostado, E.; Ozimek, L.; Amaya-Llano, S.L. 2015 Innovative applications of high-intensity ultrasound in the development of functional food ingredients: Production of protein hydrolysates and bioactive peptides. *Food Res Int.* 77(4):685–696.

Park, J.H.; Jeong, H.J.; de Lumen, B.O. 2005 Contents and bioactivities of lunasin, bowman-birk inhibitor, and isoflavones in soybean seed. *J Agric Food Chem.* 53(20):7686–7690.

Park, K.; Kwon, I.C.; Park, K. 2011 Oral protein delivery: Current status and future prospect. *React Funct Polym.* 71(3):280–287.

Philippart, M.; Schmidt, J.; Bittner, B. 2016 Oral delivery of therapeutic proteins and peptides: An overview of current technologies and recommendations for bridging from approved intravenous or subcutaneous administration to novel oral regimens. *Drug Res.* 66(3):113–120.

Pihlanto-Leppälä, A. 2000 Bioactive peptides derived from bovine whey proteins: Opioid and ace-inhibitory peptides. *Trends Food Sci Techn.* 11(9–10):347–356.

Piovesana, S.; Capriotti, A.L.; Cavaliere, C.; La Barbera, G.; Montone, C.M.; Zenezini Chiozzi, R. et al. 2018. Recent trends and analytical challenges in plant bioactive peptide separation, identification and validation. *Anal Bioanal Chem.* 410(15):3425–3444.

Remanan, M.K.; Wu, J. 2014 Antioxidant activity in cooked and simulated digested eggs. *Food & Function.* 5(7):1464–1474.

Renukuntla, J.; Vadlapudi, A.D.; Patel, A.; Boddu, S.H.; Mitra, A.K. 2013 Approaches for enhancing oral bioavailability of peptides and proteins. *Int J Pharm.* 447(1–2):75–93.

Rui, X.; Wen, D.; Li, W.; Chen, X.; Jiang, M.; Dong, M. 2015 Enrichment of ACE inhibitory peptides in navy bean (Phaseolus vulgaris) using lactic acid bacteria. *Food & Function.* 6(2):622–629.

Rutherfurd-Markwick, K.J. 2012 Food proteins as a source of bioactive peptides with diverse functions. *Br J Nutr.* 108(Suppl. 2):S149–S157.

Saez, V.; Ramón, J.; Peniche, C. 2007 Microspheres as delivery systems for the controlled release of peptides and proteins. *Biotecnol Aplic.* 24:108–116.

Sánchez, A.; Vázquez, A. 2017 Bioactive peptides: A review. *Food Quality and Safety.* 1(1):29–46.

Sato, K.; Egashira, Y.; Ono, S.; Mochizuki, S.; Shimmura, Y.; Suzuki, Y. et al. 2013 Identification of a hepatoprotective peptide in wheat gluten hydrolysate against D-galactosamine-induced acute hepatitis in rats. *J Agric Food Chem.* 61(26):6304–6310.

Sava, G. 1989 Reduction of B16 melanoma metastases by oral administration of eggwhitelysozyme. *Cancer Chemother Pharmacol.* 25(3):221–222.

Sayd, T.; Dufour, C.; Chambon, C.; Buffière, C.; Remond, D.; Santé-Lhoutellier, V. 2018 Combined in vivo and in silico approaches for predicting the release of bioactive peptides from meat digestion. *Food Chem.* 249:111–118.

Segura-Campos, M.; Chel-Guerrero, L.; Betancur-Ancona, D.; Hernandez-Escalante, V.M. 2011 Bioavailability of bioactive peptides. *Food Rev Int.* 27(3):213–226.

Şenel, S.; Hıncal, A.A. 2001 Drug permeation enhancement via buccal route: Possibilities and limitations. *J Controlled Release.* 72(1–3):133–144.

Shori, A.B.; Baba, A.S. 2014 Comparative antioxidant activity, proteolysis and in vitro α-amylase and α-glucosidase inhibition of Allium sativum yogurts made from cow and camel milk. *Journal of Saudi Chemical Society.* 18(5):456–463.

Singh, B.P.; Vij, S.; Hati, S. 2014 Functional significance of bioactive peptides derived from soybean. *Peptides.* 54:171–179.

Sipola, M.; Finckenberg, P.; Vapaatalo, H.; Pihlanto-Leppälä, A.; Korhonen, H.; Korpela, R. et al. 2002 Alpha-lactorphin and beta-lactorphin improve arterial function in spontaneously hypertensive rats. *Life Sciences.* 71(11):1245–1253.

Sun, X.; Acquah, C.; Aluko, R.E.; Udenigwe, C.C. 2020 Considering food matrix and gastrointestinal effects in enhancing bioactive peptide absorption and bioavailability. *J Funct Foods.* 64:103680.

Sun, X.; Chakrabarti, S.; Fang, J.; Yin, Y.; Wu, J. 2016 Low-molecularweight fractions of alcalase hydrolyzed egg ovomucin extract exert antiinflammatory activity in human dermal fibroblasts through the inhibition of tumor necrosis factor-mediated nuclear factor κB pathway. *Nutrition Research.* 36(7):648–657.

Tao, M.; Wang, C.; Liao, D.; Liu, H.; Zhao, Z.; Zhao, Z. 2017 Purification, modification and inhibition mechanism of angiotensin I-converting enzyme inhibitory peptide from silkworm pupa (Bombyx mori) protein hydrolysate. *Process Biochem.* 54:172–179.

Teschemacher, H. 2003 Opioid receptor ligands derived from food proteins. *Curr Pharm Des.* 9(16):1331–44.

Tokede, O.A.; Onabanjo, T.A.; Yansane, A.; Gaziano, J.M.; Djoussé, L. 2015 Soya products and serum lipids: A meta-analysis of randomised controlled trials. *Br J Nutr.* 114(6):831–843.

Toldrá, F.; Gallego, M.; Reig, M.; Aristoy, M. 2020 Bioactive peptides generated in the processing of dry-cured ham. *Food Chem.* 321:126689.

Toorisaka, E.; Hiroshige, O.; Arimori, K.; Kamiya, N.; Goto, M. 2003 Hypoglycemic effect of surfactant-coated insulin solubilized in a novel solid-in-oil-in-water (S/O/W) emulsion. *Int J Pharm.* 252(1–2):271–274.

Traverso, G.; Schoellhammer, C.M.; Schroeder, A.; Maa, R.; Lauwers, G.Y.; Polat, B.E. et al. 2015 Microneedles for drug delivery via the gastrointestinal tract. *J Pharm Sci.* 104(2):362–367.

Tyagi, A.; Tuknait, A.; Anand, P.; Gupta, S.; Sharma, M.; Mathur, D. et al. 2015 CancerPPD: A database of anticancer peptides and proteins. *Nucl Acids Res.* 43:D837–D843.

Udenigwe C.C.; Aluko R.E. 2012 Food protein-derived bioactive peptides: Production, processing, and potential health benefits. *J Food Sci.*77(1):R11–R24.

Valadez-Vega, C.; Alvarez-Manilla, G.; Riverón-Negrete, L.; GarcíaCarrancá, A.; Morales-González, J.A.; Zuñiga-Pérez, C. et al. 2011 Detection of cytotoxic activity of lectin on human colon adenocarcinoma (Sw480) and epithelial cervical carcinoma (C33-A). *Molecules.* 16(3):2107–2118.

Varadhachary, A.; Wolf, J.S.; Petrak, K.; O'Malley, B.W. Jr; Spadaro, M.; Curcio, C. et al. 2004 Oral lactoferrin inhibits growth of established tumors and potentiates conventional chemotherapy. *Int J Cancer.* 111(3):398–403.

Vercruysse, L.; Smagghe, G.; Herregods, G.; Van Camp, J. 2005 ACE inhibitory activity in enzymatic hydrolysates of insect protein. *J Agric Food Chem.* 53(13):5207–5211.

Wang, H.J.; Murphy, P.A. 1996 Mass balance study of isoflavones during soybean processing. *J Agric Food Chem.* 44(8):2377–2383.

Wang, J.; Zhao, M.; Zhao, Q.; Jiang, Y. 2007 Antioxidant properties of papain hydrolysates of wheat gluten in different oxidation systems. *Food Chem.* 101(4):1658–1663.

Wang, Y.K.; He, H.L.; Wang, G.F.; Wu, H.; Zhou, B.C.; Chen, X.L. et al. 2010 Oyster (Crassostrea gigas) hydrolysates produced on a plant scale have antitumor activity and immunostimulating effects in BALB/c mice. *Mar Drugs.* 8(2):255–268.

Woodley, J.F. 1994 Enzymatic barriers for GI peptide and protein delivery. *Crit Rev Ther Drug Carrier Syst.* 11(2–3):61–95.

Writing Group Members; Mozaffarian, D.; Benjamin, E.J.; Go, A.S.; Arnett, D.K; Blaha, M.J. et al. 2016 Heart disease and stroke statistics—2016 update: A report from the American Heart Association. *Circulation*. 133(4):e38–e360.

Yadav, J.S.; Yan, S.; Pilli, S.; Kumar, L.; Tyagi, R.D.; Surampalli, R.Y. 2015 Cheese whey: A potential resource to transform into bioprotein, functional/nutritional proteins and bioactive peptides. *Biotechnol Adv*. 33(6 Pt 1):756–774.

Yamamoto, K.; Hayashi, M.; Murakami Y.; Araki, Y.; Otsuka, Y.; Kashiwagi, T. et al. 2016 Development of LC-MS/MS analysis of cyclic dipeptides and its application to tea extract. *Biosci Biotechnol Biochem*. 80(1):172–177.

Yang, R.; Zhang, Z.; Pei, X.; Han, X.; Wang, J.; Wang, L. et al. 2009 Immunomodulatory effects of marine oligopeptide preparation from chum salmon (Oncorhynchus keta) in mice. *Food Chem*. 113(2):464–470.

Yoshikawa, M. 2015 Bioactive peptides derived from natural proteins with respect to diversity of their receptors and physiological effects. *Peptides*. 72:208–225.

Yu, Z.;Yongguang, Y.; Zhao, W.; Wang, F.; Yu, Y.; Liu, B. et al. 2011 Characterization of ACE-inhibitory peptide associated with antioxidant and anticoagulation properties. *J Food Sci*. 76(8):C1149–C1155.

Yun, Y.; Cho Y.W.; Park, K. 2013 Nanoparticles for oral delivery: Targeted nanoparticles with peptidic ligands for oral protein delivery. *Adv Drug Delivery Rev*. 65(6):822–832.

Zarei, M.; Ebrahimpour, A.; Abdul Hamid, A.; Anwar, F.; Saari, N. 2012 Production of defatted palm kernel cake protein hydrolysate as a valuable source of natural antioxidants. *Int J Mol Sci*. 13(7):8097–8111.

Zdrojewski, T.; Solnica, B.; Cybulska, B.;Bandosz, P.; Rutkowski, M.; Stokwiszewski, J et al. Prevalence of lipid abnormalities in Poland. The NATPOL 2011 survey. *Kardiol Pol*. 74(3):213–223.

Zenezini Chiozzi, R.; Capriotti, A.L.; Cavaliere, C.; La Barbera, G.; Piovesana, S.; Samperi, R. et al. 2016 Purification and identification of endogenous antioxidant and ACE-inhibitory peptides from donkey milk by multidimensional liquid chromatography and nanoHPLC-high resolution mass spectrometry. *Anal Bioanal Chem*. 408(20):5657–5666.

Zielińska, E.; Karaś, M.; Baraniak, B.; Jakubczyk, A. 2020 Evaluation of ACE, α-glucosidase, and lipase inhibitory activities of peptides obtained by in vitro digestion of selected species of edible insects. *Eur Food Res Technol*. 246:1361–1369.

Zumwalt, R.W.; Absheer, J.S.; Kaiser, F.E.; Gehrke, C.W. 1987 Acid hydrolysis of proteins for chromatographic analysis of amino acids. *J Assoc Off Anal Chem*. 70(1):147–151.

SECTION 2

Sources of Bioactive Peptides

CHAPTER 2

Bioactive Peptides from Algae

Rahel Suchintita Das, Brijesh K. Tiwari, and Marco Garcia-Vaquero

CONTENTS

Introduction	31
Algal Proteins	32
Macroalgal Proteins	32
Microalgal Proteins	34
Methods for Generation of Bioactive Peptides	35
Protein Extraction	35
Protein Hydrolysis	35
Peptide Purification	37
Peptide Identification	38
Validation Activity of Peptides	38
Hydrolysates and Peptides from Algae	39
Antioxidant Peptides	39
Antihypertensive Peptides	40
Immunomodulatory Peptides	40
Antidiabetic Peptides	46
Anticancer Peptides	46
Other Biological Activities	47
Challenges and Future Applications of Bioactive Peptides	47
Acknowledgments	48
References	48

INTRODUCTION

Marine ecosystems are one of the richest yet, comparatively, least-explored reservoirs of natural bioactive molecules. The hostile environments stimulate marine organisms, including macroalgae and microalgae, to develop complex and distinctive metabolic pathways through which they produce bioactive compounds to aid in their nutrition and as self-defense and survival mechanisms (Sánchez & Vázquez, 2017). The popular perception of algae has evolved such that it is now considered an alternative and sustainable source of nutritious food components, including protein, which is the ideal starting material for the generation of bioactive peptides. As algae are exposed to more extreme conditions, compared to terrestrial crops grown inland, their proteins also have significantly different amino acid compositions and peptide sequences compared to terrestrial plants (Wang et al., 2017b); thus, they offer excellent opportunities for the discovery of new bioactive peptides and the novel health benefits associated with those molecules.

DOI: 10.1201/9781003106524-4

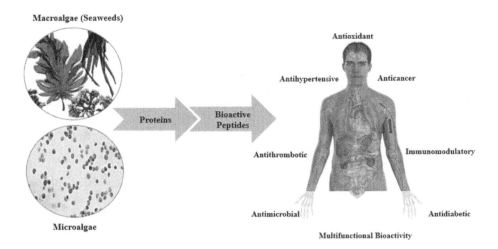

FIGURE 2.1 Physiological bio-functionalities or bioactivities of peptides generated from algal proteins as reviewed in the scientific literature (Murray and FitzGerald, 2007).

Bioactive peptides (or cryptides) are sequences of between 3 and 20 amino acid residues that are inert within their parent protein and display their biological activities once they are released following hydrolytic processes (Garcia-Vaquero et al., 2019). These compounds can be considered crucial sources of nitrogen and amino acids when added to the human diet, while also displaying multiple physiological functions including immunomodulatory, antibacterial, antithrombotic, and antihypertensive activities (see Figure 2.1).

Short- and medium- size peptides are better absorbed and can modulate specific metabolic pathways by binding or inhibiting targeted receptors, ultimately leading to a positive impact on human health (Hou et al., 2017). The toxic or adverse effects of bioactive peptides are normally considered to be minimal or nonexistent and so, since certain peptides can pass through cell membranes, they are useful as carriers for targeted drug delivery. Thus, these compounds offer an excellent opportunity to prevent pathological conditions through efficient dietary management and establish their use as a safe and side effect-free alternative to conventional drugs. Therefore, the discovery of bioactive peptides and the exploration of their mechanism of action will aid in developing innovative functional foods and nutraceuticals.

This chapter highlights the prospects of macroalgae and microalgae as viable protein sources and substrates for the generation of bioactive peptides, followed by methods of production, purification, and validation of these algal compounds. Moreover, the physiological bio-functionalities exhibited by these peptides generated from macroalgae and microalgae, as well as the current challenges and future prospects in this field of research are also highlighted.

ALGAL PROTEINS

Macroalgal Proteins

Macroalgae can be divided into 3 main taxonomic groups on basis of their pigmentation: Chlorophyta (green algae), Rhodophyta (red algae) and Phaeophyta (brown algae).

TABLE 2.1 Crude Protein Values of Selected Seaweed Species

Species	Crude Protein	Mean Value
Rhodophyta		
Gelidium microdon	14.61	15.18
	15.75	
Osmundea pinnatifida	20.32	20.64
	20.97	
Porphyra sp.	25.64	25.80
	25.97	
Pterocladiella capillacea	20.56	20.52
	20.48	
Sphaerococcus coronopifolius	19.60	19.56
	19.51	
Phaeophyta		
Cystoseira abies-marina	6.94	6.81
	6.69	
Fucus spiralis	10.56	10.77
	10.97	
Chlorophyta		
Ulva compressa	27.52	26.62

Table modified from Patarra et al. (2011) and reproduced with permission from Springer.

In general, red seaweeds contain high levels of protein with levels up to 47% dry weight (DW), green seaweeds contain moderate amounts ranging from 9 to 26% DW, while brown macroalgae contain lower protein contents (3–15% DW). As an example, the crude protein of selected seaweed species is summarized in Table 2.1. Macroalgae have a highly variable protein composition, in terms of both total protein content and amino acid profiles, depending on the seaweed species and on environmental factors such as the place of collection, season, light intensity, temperature, and nutrient concentration in the sea water (Garcia-Vaquero and Hayes, 2016).

Moreover, the chemical analysis of the biomass also contributes to some variation in the protein contents reported in the scientific literature. Angell et al. (2016) reported that studies used either direct extraction procedures (42% of all studies) or applied an indirect nitrogen-to-protein conversion factor of 6.25 (52% of all studies) when estimating the protein contents from seaweeds. Meta-analysis of the true protein content, defined as the sum of the proteomic amino acids, revealed that direct extraction procedures underestimated protein content by 33%, while the most commonly used indirect nitrogen-to-protein conversion factor of 6.25 overestimated protein content by 43%. An overall median nitrogen-to-protein conversion factor of 4.97 and an overall mean nitrogen-to-protein conversion factor of 4.76 was established by evaluating the variation in these factors for 103 species across 44 studies that span three phyla, multiple geographic regions, and a range of nitrogen contents. The overall median value of 5 was suggested as the most accurate universal seaweed nitrogen-to-protein (SNP) conversion factor by Angell et al. (2016).

In terms of amino acid composition, in general, most seaweeds contain all the essential amino acids and are a rich source of aspartic and glutamic acids. Overall, brown seaweeds contain high levels of the aforementioned acidic amino acids, ranging from 18 to 44% in *Fucus* sp., *Sargassum* sp., *Laminaria digitata*, and *Ascophyllum nodosum* (Fleurence et al., 2018). Biris-Dorhoi et al. (2020) determined that 1 g of algae meal from *Enteromorpha intestinalis*, *Palmaria palmata*, and *Vertebrata lanosa* contained equal to or higher amounts of all of the essential amino acids when compared to the same amounts rice, corn, and wheat. However, threonine, lysine, tryptophan, sulphur-containing amino acids (cysteine and methionine), and histidine are the least frequent amino acids detected in macroalgal proteins, although their contents are still higher in seaweeds compared to terrestrial plants (Galland-Irmouli et al., 1999).

Microalgal Proteins

Microalgae are a heterogeneous group encompassing Bacilliariophyta (Diatoms), Dinophyta (Dinoflagellates), Rhaphidophyta, Haptophyta, Chlorophyta, Prasinophyta, Prymnesiophyta, Cryptophyta, and Chrysophyta (golden, green, and yellow-brown flagellates), and photoautotrophic prokaryotics, such as cyanobacteria (El Gamal, 2010; Rasmussen and Morrissey, 2007). Currently, commercially cultivated microalgae primarily focus on the cultivation of the *Chlorella, Spirulina, Dunaliella, Nannochloris, Nitzschia, Crypthecodinium, Schizochytrium, Tetraselmis,* and *Skeletonema* species (Lee et al., 2009).

Microalgae proteins are present in various parts of the cell, including cytoplasm, organelles, plastids, cell walls, and nucleus and have diverse chemical forms, such as enzymes and pigment-, lipid-, or carbohydrate-conjugates. The main microalgal proteins include native proteins (particularly lectins and phycobiliproteins), endogenous peptides (linear, cyclic, depsipeptides (peptides with one or more amide bonds replaced with an ester bond)), free amino acids, and mycosporine-like amino acids (Dominguez, 2013). The biochemical composition of different microalgal species is summarized in Table 2.2.

As seen in Table 2.2, the high protein content of various microalgal species makes them excellent candidates as alternative sources of protein. Stack et al. (2020) reported that most protein content of microalgae in the literature is determined through total nitrogen values using a conversion factor of 6.25. Specific conversion factors of 4.58 and 5.95 have also been reported for microalgae, although no universal microalgal conversion

TABLE 2.2 General Composition of Selected Microalgae Sp. Expressed as % on DW

Microalgae sp.	Protein	Carbohydrates	Lipids
Chlamydomonas reinhardtii	48	17	21
Chlorella vulgaris	51–58	12–17	14–22
Dunaliella salina	57	32	6
Euglena gracilis	39–61	14–18	14–20
Porphyridium cruentum	28–39	40–57	9–14
Scenedesmus obliquus	50–56	10–17	12–14
Spirogyra sp.	6–20	33–64	11–21
Arthrospira maxima	60–71	13–16	6–7
Spirulina platensis	46–63	8–14	4–9
Synechococcus sp.	63	15	11

Table modified from Becker (2007) and reproduced with permission from Elsevier.

factor is suitable for all microalgal species, culture conditions, or even growth stages during their cultivation cycle (López et al., 2010; Lourenço et al., 2004).

The amino acid profiles of *C. vulgaris* and *D. bardawil* are marginally deficient for isoleucine and tryptophan, respectively; while *S. obliquus*, *A. maxima*, and *A. platensis* are deficient for two or three essential amino acids (see Table 2.3). Since microalgae display different amino acid profiles under different cultured conditions, optimization of microalgae cultivation conditions may facilitate the production of proteins with the desired amino acid profile.

METHODS FOR GENERATION OF BIOACTIVE PEPTIDES

Methodologies for the production of bioactive peptides involve four major steps: (1) identifying suitable protein sources; (2) peptide release by enzymatic hydrolysis; (3) purification/isolation of peptides; and (4) identification/validation of biological activities of peptides.

Protein Extraction

The process of generation of bioactive peptides from algae generally starts with the extraction, fractionation, isolation, and concentration of proteins (Garcia-Vaquero et al., 2019; Cermeño et al., 2020). Both seaweed and microalgae proteins have been extracted using conventional methods using multiple solutions including distilled water, buffers, acid or alkaline solutions, and lysis/surfactant-containing solutions (Harrysson et al., 2018; Mittal et al., 2019; Zhao et al., 2019) followed by several rounds of centrifugation and recovery of protein using ultrafiltration, precipitation and/or chromatographic techniques (Hayes et al., 2018). Methods such as mechanical grinding (Barbarino and Lourenço, 2005), osmotic shock (Joubert and Fleurence, 2008), ultrasound treatment (Keris-Sen et al., 2014), pulsed electric fields (Prabhu et al., 2019), microwave-assisted extraction (Passos et al., 2015), high hydrostatic pressure (O'Connor et al., 2020), and enzymes (Hardouin et al., 2014) help in breaking down the algal cell wall, increasing the availability of proteins.

Protein Hydrolysis

Bioactive peptides, which are inactive when encrypted in parent protein sequences, can be released by (i) enzymatic hydrolysis via gastrointestinal protease action, (ii) microbial enzymatic hydrolysis during fermentation, (iii) *in vitro* enzymatic hydrolysis using exogenous enzymes, (iv) chemical/physical hydrolytic processes.

Among these approaches, enzymatic hydrolysis methods are preferred, especially in the nutraceutical industries. The main advantages of these enzymatic methods are: they do not require harsh chemicals and physical treatments, so there are no undesirable residual chemicals in the final peptides (Clemente, 2000; Hannu Korhonen & Pihlanto, 2006); there is a large range of proteolytic enzymes from different sources (animal, plant, and microbial) currently available, and the functionality and the nutritive value are retained in the final product (Gao et al., 2006; Cermeño et al., 2020). However, the use of enzymes requires pH adjustment (using acid or alkalis), adding inorganic salts which are difficult to

TABLE 2.3 Amino Acid Profile of Different Microalgae Compared to Conventional Protein Sources and the WHO/FAO (1973) Reference Levels in g per 100 g Protein

Source	Ile	Leu	Val	Lys	Phe	Tyr	Met	Cys	Try	Thr	Ala	Arg	Asp	Glu	Gly	His	Pro	Ser
WHO/FAO	4.0	7.0	5.0	5.5	6.0		3.5		1.0	5.0								
Egg	6.6	8.8	7.2	5.3	5.8	4.2	3.2	2.3	1.7	5.0	–	6.2	11.0	12.6	4.2	2.4	4.2	6.9
Soybean	5.3	7.7	5.3	6.4	5.0	3.7	1.3	1.9	1.4	4.0	5.0	7.4	1.3	19.0	4.5	2.6	5.3	5.8
Chlorella vulgaris	3.8	8.8	5.5	8.4	5.0	3.4	2.2	1.4	2.1	4.8	7.9	6.4	9.0	11.6	5.8	2.0	4.8	4.1
Dunaliella bardawil	4.2	11.0	5.8	7.0	5.8	3.7	2.3	1.2	0.7	5.4	7.3	7.3	10.4	12.7	5.5	1.8	3.3	4.6
Scenedesmus obliquus	3.6	7.3	6.0	5.6	4.8	3.2	1.5	0.6	0.3	5.1	9.0	7.1	8.4	10.7	7.1	2.1	3.9	3.8
Arthrospira maxima	6.0	8.0	6.5	4.6	4.9	3.9	1.4	0.4	1.4	4.6	6.8	6.5	8.6	12.6	4.8	1.8	3.9	4.2
Spirulina platensis	6.7	9.8	7.1	4.8	5.3	5.3	2.5	0.9	0.3	6.2	9.5	7.3	11.8	10.3	5.7	2.2	4.2	5.1
Aphanizomenon sp.	2.9	5.2	3.2	3.5	2.5	–	0.7	0.2	0.7	3.3	4.7	3.8	4.7	7.8	2.9	0.9	2.9	2.9

Table modified from Becker (2007) and reproduced with permission from Elsevier.

eliminate from the final product (Wang et al., 2017a); and the resulting hydrolysates may have altered bioactive properties depending on multiple processing conditions, including specificity of the enzymes used and degree of hydrolysis of the protein (Zou et al., 2016). *In vitro* hydrolysis using exogenous enzymes to produce bioactive peptides is preferred compared to microbial fermentation due to the short reaction time, ease of scalability, and predictability of the processes. There are usually five main independent variables to control when performing an enzymatic hydrolysis: enzyme concentration, pH, extraction temperature, extraction time, and water/material ratio (Wang et al., 2017b).

A protein extract generated from the macroalga *Ulva lactuca* was hydrolysed by the addition of 1% Papain® (stirring at 300 rpm, pH 6 and temperature 60 °C for 24 h period). The hydrolysate exhibited ACE and renin inhibitory activities of 82.37 and 3.70%, respectively (Garcia-Vaquero et al., 2019). *Nannochloropsis oculata* pretreated with cellulase was subjected to digestion with commercial proteolytic enzymes such as Pepsin®, Trypsin®, α-Chymotrypsin®, Papain, Alcalase®, and Neutrase® at an enzyme/substrate ratio of 1/100 (w/w) for 24 h. The Pepsin hydrolysate exhibited the highest ACE inhibitory activity (Samarakoon et al., 2013).

Peptide Purification

To determine the sequence of a peptide and to perform individual activity assays, an effective purification and concentration of the target molecules is essential. Purifying peptides and obtaining them in adequate quantities is a complex and expensive process. Further steps to enrich and concentrate peptides include several common fractionation and purification techniques, particularly membrane-separation techniques, such as ultrafiltration and nanofiltration within specific molecular-weight ranges (Korhonen and Pihlanto, 2007).

In general, *in vitro* bioactivity assay-directed purification is the approach adopted to fractionate and purify peptides with specific biofunctional activities. If specific peptides are to be purified, specific chromatographic techniques should be used in combination with membrane processing, such as ion-exchange, affinity, gel-permeation platforms (Kim, 2013; Pouliot et al., 2006), and reverse-phase liquid chromatography (Chabeaud et al., 2009). The choice of the processing technique chiefly depends on the peptide structural feature of interest, but consideration should be given to feasibility for industry scale-up of the processing as well (Ejike et al., 2017). The degree to which a peptide needs to be purified depends on the potency of the bioactive component. On an industrial level, the cost of purifying specific bioactive peptides must be considered against the value of the purified or semipurified product. As a result, high-cost semi- and preparative- scale chromatography may only be utilized if highly purified peptides are required for commercialization (Pouliot et al., 2006). Each purification technology has its own advantages and disadvantages, which the researcher should consider clearly before the purification of peptides (Wang et al., 2017b).

A variety of membrane separation approaches have been utilized to enrich and concentrate peptides. During membrane filtration, the sample is pumped across a semipermeable membrane with a specific molecular weight cutoff (MWCO), mostly under pressure, generating permeate and retentate streams. This separation allows non-hydrolyzed proteins and/or large molecular mass molecules to be efficiently removed from the sample and only peptides or smaller molecular mass components pass in to the permeate stream (Cermeño et al., 2020).

Membrane processing is commonly used as a preliminary enrichment step prior to using specific chromatographic fractionation techniques, separating peptides on the basis of: (1) molecule size, such as size exclusion chromatography (SEC) or gel permeation or gel filtration chromatography (GP or GFP); (2) hydrophobicity of the molecules, such as reversed-phase chromatography (RP); and (3) charge of the peptides, i.e. ion exchange chromatography (IEX). Crude or peptide-enriched protein hydrolysates are loaded onto chromatographic columns and diverse fractions corresponding with the eluted peaks are pooled and confirmed for the desired bioactivity which is followed by the selection of the most potent fractions and further processed.

Some authors working in the purification of algal peptides used only one purification strategy. For example, Mao et al. (2017) used Sephadex G-15 gel filtration column to purify peptides with antiproliferation activities resulting from hydrolysates of red macroalga *Pyropia haitanensis*. Fitzgerald et al. (2012) used RP chromoatographic techniques to purify renin inhibitory peptides from *Palmaria palmata* hydrolysates. However, other authors used several chromatographic procedures in tandem to effectively purify algal peptides prior to peptide identification. Suetsuna and Chen (2001) purified ACE inhibitory peptides generated from *Spirulina platensis* by IEX and GFP. Further purification of the peptides was done by RP chromatographic techniques, concluding in the identification of multiple peptides.

Peptide Identification

As reviewed in detail by Cermeño et al. (2020), currently the most commonly used peptide identification technique is mass spectrophotometry (MS). Liquid chromatography (LC) coupled with an electrospray ionization (ESI) source is frequently used for peptide identification, owing to its comparatively low cost. Specific systems allow the sequencing of peptides using tandem MS/MS and LC–ESI–MS/MS and can be coupled with different mass analyzers including quadrupole (Q), time of flight (TOF), ion trap, Q-TOF, TOF-TOF, and Q-ion trap. Q-TOF was used by Admassu et al. (2018) to identify bioactive peptides from the seaweed *Porphyra* spp. and by Chen et al. (2019) for sequencing bioactive peptides from microalgae, *Isochrysis zhanjiangensis*.

Peptide sequencing can be achieved by the use of certain software systems that process the data from the MS detector into MS or MS/MS spectra, such as PEAKSTM and MascotTM. Some of these softwares facilitate the peptide analysis by database-driven analysis and de novo sequencing (Cermeño et al., 2020). Harnedy et al. (2015) identified the DPP-IV inhibitory peptides, ILAP, LLAP, and MAGVDHI, from the macroalgae *P. palmata* using the Mascot search engine coupled with the SwissProt Eukaryota database along with the de novo sequencing tool, PEAKS Studio 6.0 software. Automatic Edman degradation, which is nowadays replaced by LC-MS, was used to elucidate the sequences of the ACE inhibitory peptides, ALLAGDPSVLEDR and VVGGTGPVDEWGIAGAR, from the red algae *B. fuscopurpurea* alone, along with MALDI-TOF (Suetsuna et al., 2004) and with fast atom bombardment (FAB)-MS techniques (Suetsuna and Nakano, 2000) on peptides from *P. palmata* and *U. pinnatifida*, respectively.

Validation Activity of Peptides

It is essential to perform *in vitro* as well as *in vivo* studies to validate the bioactive potential of the peptides and substantiate the health claims of these compounds (see Figure 2.2).

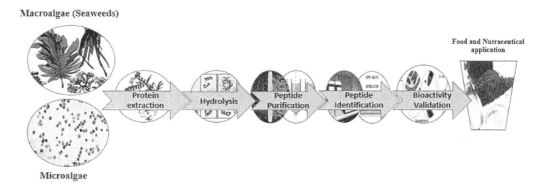

FIGURE 2.2 Flow diagram of algal bioactive peptide generation and applications.

It should also be noted that peptides that have been selected for one specific activity could possibly demonstrate positive results for other bioactivities and therefore these peptides can be referred to as multifunctional peptides (Lammi et al., 2019). A valid research approach should evaluate the stability of the peptide toward luminal and brush border peptidases and assess its absorption, distribution, metabolism, and excretion properties. *In vitro* studies exploring potential biological activities of peptides should use physiologically relevant concentrations and time (kinetics) as mandatory criteria (Foltz et al., 2010). Effect of food processing and factors associated with it, such as temperature and pressure on the stability and potency of the peptides, should also be taken in to account when using bioactive peptides as functional foods (Maestri et al., 2019).

HYDROLYSATES AND PEPTIDES FROM ALGAE

Macroalgal and microalgal bioactive peptides identified in the scientific literature display a wide array of biological functions and numerous health benefits, including antioxidant, antihypertensive, immunomodulatory, antidiabetic, anticancer, and other biological activities. A summary of the main macroalgal and microalgal bioactive peptides described in the recent scientific literature, as well as their associated biological activities and key information related to their generation purification and identification are summarized in Table 2.4.

Antioxidant Peptides

In humans, the accumulation of free radicals and reactive oxygen species (ROS), formed either from normal cell metabolism *in situ* or from exposure to sources of harmful radiation or pollution, can lead to oxidative stress and, thus, associated health disorders, such as cancer, diabetes mellitus, aging, hypertension, and neurodegenerative and inflammatory diseases (Lafarga et al., 2020).

Numerous studies have been conducted on marine algae to explore their antioxidant potential. Heo et al. (2005) reported that hydrolysates of seven brown macroalgae (*Ecklonia cava*, *Ishige okamurae*, *Sargassum fullvelum*, *Sargassum horneri*, *Sargassum coreanum*, *Sargassum thunbergii*, and *Scytosiphon lomentaria*) using five proteolytic enzymes (Protamex®, Kojizyme®, Alcalase®, Flavourzyme®, and Neutrase®) displayed free-radical-scavenging activity against 1,1-diphenyl-2-picrylhydrazyl (DPPH), superoxide anion radical (O_2), hydroxyl radical (HOÆ), and hydrogen peroxide (H_2O_2). Xia et al. (2019) obtained DPPH-scavenging peptides from *Dunaliella salina* protein hydrolysates generated by ultrasound extraction, *in vitro* gastrointestinal digestion, and membrane ultrafiltration. The authors reported that the fraction of 500–1000 Daltons (Da) had the maximum DPPH-scavenging activity and revealed four novel peptides (ILTKAAIEGK, IIYFQGK, NDPSTVK, and TVRVRPPQR).

Another powerful antioxidant bioactive peptide (VECYGPNRPQF) was identified from *C. vulgaris* (Sheih et al., 2009). This peptide displayed antioxidant activity against the HOÆ, O_2, DPPH, and 2,2′-azinobis (ABTS) radicals. In a recent study conducted by Pereira et al. (2019) on *Spirulina* sp. LEB 18, the antioxidant activity of protein hydrolysates obtained from the microalgae was shown to be retained following heat treatment at 18, 63, and 100 °C. Furthermore, the antioxidant capacity of these protein hydrolysates to inhibit DPPH was increased by approximately 25% upon exposure to acidified media (pH 4 and pH 6), and these compounds were also bioavailable post-digestion.

Antihypertensive Peptides

Hypertension is recognized as a major noncommunicable disease and a risk factor for cardiovascular disease-related mortality. One of the key therapeutic approaches in the management of hypertension is inhibition of renin, ACE-I, and other enzymes involved in the renin-angiotensin-aldosterone system (Mercier et al., 2014). Pharmaceutical companies have marketed various ACE inhibitors Captopril®, Enalapril®, Alcacepril®, and Lisinopril® for the management of hypertension. However, these drugs have undesirable side effects including skin rashes, cough, and taste disturbance (Atkinson and Robertson, 1979). Thus, natural inhibitors, such as bioactive peptides, could offer an alternative to manage hypertension on a daily basis by their incorporation in the diet, while avoiding some of the adverse side effects of these pharmacological treatments.

Garcia-Vaquero et al. (2019) reported that a Papain hydrolysate generated from *Ulva* sp. and their purified fractions had ACE-I inhibitory activities ranging from 96.91 to 98.06%. A total of 48 novel peptides were identified from these four fractions by LC-MS/MS (Garcia-Vaquero et al., 2019). Fitzgerald et al. (2014) reported that the renin inhibitory peptide, IRLIIVLMPILMA, was identified from a Papain hydrolysate of dulse (*Palmaria palmata*). This peptide had an *in vitro* renin IC_{50} of 3.34 mM. It also showed *in vivo* antihypertensive activities after oral administration in spontaneously hypertensive rats at a dosage of 50 mg/kg body weight. From *P. palmata*, Furuta et al. (2016) isolated nine ACE inhibitory peptides (YRD, AGGEY, VYRT, VDHY, IKGHY, LKNPG, LDY, LRY, and FEQDWAS), and once synthesized, the peptide LRY had a remarkably high ACE inhibitory activity, with an IC_{50} of 0.044 μmol.

Immunomodulatory Peptides

Immunomodulation comprises of all the routes to modify/regulate the immune response of a living system for therapeutic objectives, including the activation of the immune system to diminish inflammatory reactions and fight diseases, such as microbial infections and cancer, through immunotherapy (Riccio and Lauritano, 2020). The occurrence and severity of inflammatory diseases and allergies are soaring and multiple compounds, such

TABLE 2.4 Summary of Macroalgal and Microalgal Bioactive Peptides and Associated Biological Activities, as well as Key Information Related to Their Generation Purification and Identification

	Algae sp.	Protein Extraction	Protein Hydrolysis	Peptide Purification	Peptide Identification	Peptide Sequences	Bioactivity of Peptide	Reference
Macroalgae	*Ulva lactuca*	Ultrasound, ammonium sulphate, dialysis 3.5 kDa	Papain (pH 6, 60 °C, E:S 1%, 24 h, 300rpm stirring)	10, 3 and 1 kDa UFH membrane filtration, RP-HPLC (C18)	Nano LC ESI QTOF MS/MS	Several	ACE and renin inhibition	Garcia-Vaquero et al. (2019)
	Pyropia haitanensis	Ultrasonication (450 W for 25 min, every 6 s with 9 s interval)	Pepsin and Papain (Pepsin- pH 1-2.5, 33-45 °C, E/S:1–5% (w/w), 2-10 h. pH 5.5-7.5, 50-70 °C, E/S 1-5% (w/w), 2-10 h)	0, 5, and 3 kDa molecular weight cutoff ultrafiltration. Sephadex G-15 gel filtration chromatography	MS and MS/MS MALDI-TOF-MS	QTDDNHSNVLWAGFSR	Anti-proliferation activity	Mao et al. (2017)
	Pyropia sp.	KCl-HCl buffer (0.02 M, pH 2.0) at 4% (w/v), 4°C)	Pepsin (E:S 1:40 (w/w) 45 °C, 5 h)	(i) SEC using a Sephadex G-100 column (1.4 × 200 mm) in 0.05 M TrisBuffer at 0.5 mL/min. (ii) IEC using a DEAE Sephadex column (1.4 × 50 mm) in 0.05 M Tris buffer at 0.5 mL/min (iii) RP-HPLC using a Sephadex G-25 column (1.4 × 200 mm) in 0.05 M Tris buffer at 0.5 mL/min.	C-18 RP column of UFLC	NMEKGSSSVVSSRM(+15.99)KQ	Anticoagulant	Indumathi and Mehta (2016)

(*Continued*)

TABLE 2.4 (Continued)

Algae sp.	Protein Extraction	Protein Hydrolysis	Peptide Purification	Peptide Identification	Peptide Sequences	Bioactivity of Peptide	Reference
Saccharina longicruris	Sodium phosphate and ammonium sulphate	Trypsin (pH 7, 30 °C, E:S 5%, 24 h)	10 kDa centrifugal filter (prior to and following hydrolysis), IEF fractionation	Nano LC ES MS/MS	QVHPDTGISK, LPDAALNR, EAESSLTGGNGCAK, ISAILPSR, IGNGGELPR, MALSSLPR, ILVLQSNQIR, TITLDVEPSDTIDGVK, ISGLIYEETR	Antibacterial	Beaulieu, Bondu, Doiron, Rioux, and Turgeon (2015)
Palmaria palmata	Ultrasonication and aqueous extraction	Papain (60 °C, pH 6, 24 h)	RP-HPLC	Nano UPLC ESI QTOF MS/MS	IRLIIVLMPILMA	Renin inhibitory, antihypertensive in sontaneous hypertensive rats	Fitzgerald et al. (2012), Fitzgerald, Aluko, Hossain, Rai, and Hayes (2014)
P. palmata	Aqueous extraction	Thermolysin, Pepsin, Trypsin and Chymotrypsin	Sequential membrane filtration	RP-HPLC	YRD, AGGEY, VYRT, VDHY, IKGHY, LKNPG, LDY, LRY, FEQDWAS	ACE inhibitory	Furuta et al. (2016)
P. palmata	Aqueous extraction	Alcalase, Flavourzyme, corolase (pH 7, 50 °C, E:S 1%, 4 h)	-	SPE (C18) and SP RP-HPLC (C18)	-	Antidiabetic	Harnedy and FitzGerald (2013)

(*Continued*)

TABLE 2.4 (Continued)

	Algae sp.	Protein Extraction	Protein Hydrolysis	Peptide Purification	Peptide Identification	Peptide Sequences	Bioactivity of Peptide	Reference
	P. palmata	Alkaline and aqueous extracts isoelectric precipitation	Corolase PP (pH 7.0, 50 °C, E:S 1%, 4 h)	SPE (C18), SP-RP-HPLC (C18)	UPLC ESI QTOF MS/MS	SDITRPGGQM	Antioxidant	Harnedy, O'Keeffe, and FitzGerald (2017)
	Porphyra columbina	Hot and cold water extraction	55 °C; pH 4.3; E/S ratio; 50 g kg/l and 55 °C; pH, 7; E/S ratio, 20 g kg/l for fungal protease concentrate and Flavourzyme, respectively	–	–	–	Immunomodulatory	Cian et al. (2012)
	Porphyra sp.	Enzymatic and iso-electric precipitation	Pepsin (pH 2, 37 °C for 4 h)	Ultrafiltration (UF), ephadex gel chromatography, RP-HPLC.	ESI-Q-TOF- MS	GGSK, ELS	Antidiabetic	Admassu et al. (2018)
Microalgae	*Chlorella ellipsoidea*	–	Pepsin, Trypsin, and α-Chymotrypsin (E:S (1:100, w/w), 12 h)	Sephadex G-25 gel filtration column	RP-HPLC Q-TOF ESI MS	LNGDVW	Antioxidant	Ko, Kim, and Jeon (2012)
	Chlorella pyrenoidosa	Aquous extraction along with low-temperature high-pressure continuous flow cell breakage	Papain, Trypsin, and Alcalase	Ultrafiltered (10, 5 and 3 kDa)	Ion exchange chromatography, Gel filtration chromatography	CPAP	Antitumor	Z. Wang and Zhang (2016).
	Chlorella vulgaris	–	Pepsin (E/S ratio of 2% (w/w), 50 °C, 15 h	Gel filtration chromatography	Q-TOF MS/MS	VECYGPNRPQF	Antioxidant	Sheih et al. (2009)

(*Continued*)

TABLE 2.4 (Continued)

Algae sp.	Protein Extraction	Protein Hydrolysis	Peptide Purification	Peptide Identification	Peptide Sequences	Bioactivity of Peptide	Reference
Tetradesmus obliquus	10% (w/v) trichloroacetic acid in acetone precipitation	Alcalase (E:S 1:10, pH 8 at 60 °C for 4 h)	Solid phase extraction (SPE) on C18 cartridges, two-dimensional chromatography	RP nanoHPLC-MS/MS ion trap-Orbitrap mass spectrometer	WPRGYFL, GPDRPKFLGPF, WYGPDRPKFL, SDWDRF	Antioxidant and ACE-inhibitory activity	Montone et al., 2018
Spirulina platensis	Ultrasonication and repeated freezing-and-thawing methods and 50% saturated NH_4SO_4 precipitation.	Alkaline protease and Papain	Sephadex G-25 chromatography, RP-HPLC, Superdex 75 10/300 GL chromatography	LC-MS/MS	KLVDASHRLATGDVAVRA	Antimicrobial	Sun et al. (2016)
S. platensis	-	Protease K (700 U/g) was then added to hydrolyze the *S. platensis* for 3 h at 55 °C	3 kDa ultrafiltration, gel filtration chromatography, RP-HPLC	Q-TOF ESI, HPLC with MALDI-TOF MS	PNN	Antioxidant	Yu et al. (2016)
S. platensis	Freeze-thawing and ultrasonication	Trypsin, Alcalase, Papain, and Pepsin	Ultrafiltrated (10, 5 and 3 kDa)	Gel filtration chromatography	-	Antitumor	B. Zhang and Zhang (2013)
Spirulina sp.	-	Serine endopeptidase from *Bacillus licheniformis* (pH 9.5, 60 °C)	-	-	-	Antioxidant	Pereira et al. (2019)
Spirulina maxima	Refreezing and thawing, sonication	Trypsin and Chymotrypsin (E/S of 1/100 for each enzyme) for 4 h, 37 °C, Pepsin (E/S of 1/100, pH 2, 4 h, 37 °C)	Ultrafiltration (UF)	Anion-exchange chromatography, Gel-permeation chromatography, RP-HPLC, Q-TOF ESI, MS/MS	LDAVNR, MMLDF	Anti-inflammatory and anti-allergic	Vo et al. (2013)

(Continued)

TABLE 2.4 (Continued)

Algae sp.	Protein Extraction	Protein Hydrolysis	Peptide Purification	Peptide Identification	Peptide Sequences	Bioactivity of Peptide	Reference
Nannochloropsis oculata	-	Alcalase, Neutrase, Flavourzyme, Trypsin, and Protamex (E:S- 1/100 (w/w), pH 6-8, 40-50 °C, 8 h)	Preparative HPLC, RP-HPLC on C18 column	Q-TOF ESI, MS/MS	-	Anti-osteoporotic	Nguyen et al. (2013).
Chlamydomonas sp.	-	Microbial hydrolysis with *Candida utiliz* and *Bacillus subtillis* for 36 h at 37 °C	Anion exchange column	Q-TOF LC/MS/MS	PQPKVLDS	Anti- *H. pylori* activity	Himaya et al. (2013)
Dunaliella salina	NaOH solubilisation and ultrasonication	Pepsin and Trypsin	Dialysis, ultrafiltration	RPLC with an C18, column coupled online with a Q Exactive HF mass spectrometer	IITKAAIEGK, IIYFQGK, NDPSTVK, TVRVRPPQR	Antioxidant	Xia et al. (2019)

as corticosteroids, antihistamines and nonsteroidal anti-inflammatory drugs, are being explored to prevent and treat these pathologies (Ben-Aharon et al., 2018). However, these drugs have major side-effects linked with their long-term use (Vo et al., 2013). Hence, it is imperative to explore natural and safe alternatives to these drugs.

Cian et al. (2012) reported the immunomodulating effect of enzymatic hydrolysates from phycobiliproteins from *Porphyra columbina* on rat splenocytes. These hydrolysates were rich in low molecular weight peptides containing Asp, Ala and Glu (Cian et al., 2012). In the case of microalgae, anti-inflammatory and anti-allergic peptides have been isolated from *Spirulina maxima* using gastrointestinal endopeptidases (Trypsin, α-Chymotrypsin, and Pepsin). The purification of the hydrolysate resulted in the identification of 2 peptides (LDAVNR and MMLDF) that were effective against histamine production from mast cells, suppressed cytokine generation from endothelial cells and inhibited reactive oxygen species production from mast cells and endothelial cells (Vo et al., 2013).

Antidiabetic Peptides

Diabetes mellitus is a group of metabolic diseases characterized by hyperglycemia resulting from defects in insulin secretion, insulin action, or both, leading to abnormally high blood glucose levels. Chronic hyperglycemia has been linked to long-term dysfunction, and failure of various organs, particularly the eyes, kidneys, nerves, heart, and blood vessels. Several pathogenic processes are involved in the development of diabetes. These range from autoimmune destruction of the ß-cells of the pancreas with consequent insulin deficiency- type I (insulin-dependent) to abnormalities that result in resistance to insulin action type II (non-insulin-dependent) (Mellitus, 2005).

Harnedy and FitzGerald (2013) have identified potent antidiabetic peptides with dipeptidyl peptidase (DPP) IV inhibitory activity from hydrolysates of *Palmaria palmata* using the enzymes Alcalase®, Flavourzyme®, and Corolase®. Inhibition of excessive rise in blood sugar due to uncontrolled carbohydrate breakdown by enzymes such as α-amylase and α-glucosidase are promising therapies in diabetes mellitus management (Telagari and Hullatti, 2015). α-amylase inhibitory peptides GGSK and ELS were identified from *Porphyra* species and had IC_{50} values of 2.58 ± 0.08 mM and 2.62 ± 0.05 mM, respectively (Admassu et al., 2018).

Anticancer Peptides

Various peptides against several cancer types/cancer cell lines have also been identified in recent scientific literature, mainly from microalgae. CPAP peptide, isolated from *Chlorella pyrenoidosa*, induced apoptosis, cell membrane shrinkage, and necrotic death in human liver cancer HepG2 cells. Micro- and nano-encapsulation of this peptide retained its functionality even after simulated gastrointestinal enzymatic treatment, and, thus, these strategies can be used for gradual controlled release of peptides in the intestine (Wang and Zhang, 2016). An antitumor polypeptide Y2 fraction was obtained from *Spirulina platensis* and had cytotoxicity toward MCF-7 and HepG2 cells. Nano-encapsulation of this hydrolysate also retained the antitumor activity of the peptides (Zhang and Zhang, 2013). Infections of *Helicobacter pylori* can lead to chronic inflammation and increases the risk of developing duodenal and gastric ulcer disease and gastric cancer (Wroblewski et al.,

2010). A peptide isolated from *Chlamydomonas* sp. hydrolysates, H-P-6 (PQPKVLDS), effectively suppressed *H. pylori*–dependent hyper-proliferation and migration of gastric epithelial cells (Himaya et al., 2013).

Other Biological Activities

Recent scientific literature has also focused on the discovery of peptides with other biological properties including antimicrobials, anti-osteoporosis and anti-thrombosis peptides.

Antimicrobial peptides are small molecules and have a broad spectrum of antimicrobial activity. Sun et al. (2016) identified the peptide KLVDASHRLATGDVAVRA from *S. platensis* protein hydrolysates. This peptide had significant antibacterial effects against Gram-negative (*Escherichia coli*) and Gram-positive (*Staphylococcus aureus*) bacteria, with minimum inhibitory concentration values of 8 and 16 mg/mL, respectively (Sun et al., 2016).

There have also been advances in the discovery of anti-osteoporosis peptides. An osteoblast activating peptide was isolated from biodiesel by-products of *Nannochloropsis oculata* (Nguyen et al., 2013). This peptide promoted osteoblast differentiation by increasing expression of several osteoblast phenotype markers such as alkaline phosphatase, osteocalcin, collagen type I, bone morphogenetic proteins (BMP-2 and BMP2/4) and bone mineralization in both human osteoblastic cell (MG-63) and murine mesenchymal stem cell (D1) (Nguyen et al., 2013).

Thrombosis is a common pathology underlying ischemic heart disease, ischemic stroke, and venous thromboembolism (Raskob et al., 2014). Blood coagulation is a complex chain process involving a sequence of stimulus responses along with coagulation factors and enzymes (Ngo et al., 2012). A potent and novel anticoagulant peptide was identified from *Porphyra yezoensis* (Indumathi and Mehta, 2016). The purified peptide showed a dose-dependent prolongation of activated partial thromboplastin time when performing *in vitro* clotting time tests using human plasma. The authors suggested that the peptide interacts with the clotting factors involved in the fundamental pathway of blood coagulation (Indumathi and Mehta, 2016).

CHALLENGES AND FUTURE APPLICATIONS OF BIOACTIVE PEPTIDES

Obtaining algae with consistent characteristics can be a challenge to obtain reproducible and consistent results and biological properties from peptides when isolated or generated from natural sources. However, aquaculture could partially solve these issues by curtailing the effects of variation in seasonality, temperature, light, and nutrients on the original algal biomass (Pimentel et al., 2019). The aforementioned processes currently in place to generate, identify, and purify peptides are time-consuming and could be quite expensive for large-scale production, although recent alternatives such as *in silico* testing and predictions are promising tools when approaching the discovery of new bioactive peptides (Garcia-Vaquero et al., 2019).

Another challenge in the field of bioactive peptides is the limited number of studies from the *in vitro* testing stage that continue to animal and later human trials, necessary for the confirmation of the effects of these peptides. Further research is required to assess their bioavailability *in vivo*, as well as their stability and final available concentrations (Li et al., 2019). Methods, such as micro/nano-encapsulation of compounds, could be explored as a viable approach to overcome these difficulties.

Moreover, from a regulatory aspect, robust practical evidence is required to allow the safe and explicit use of bioactive peptides and sustain their related health claims. Thus, improving the cost-efficiency and standardization of the production of these peptides will be key aspects for the future commercialization of these compounds. The wide diversity of the algal biomass and the endless bioactive potential of these natural resources offer huge industrial and potential exploitation opportunities in the field of nutraceuticals. Furthermore, the recent technological developments in processing technologies as well as the wide range of possible applications to alleviate multiple ailments through multifunctional peptides unravel huge potential for future research to explore algae in the field of bioactive peptides.

ACKNOWLEDGMENTS

Rahel Suchintita Das is in receipt of a PhD grant from the UCD Ad Astra Studentship (R20909). The researchers also acknowledge funding from the transnational ERA-NET Co-fund JPI HDHL and the Department of Agriculture Food and the Marine (DAFM) as part of the project AMBROSIA.

REFERENCES

Admassu, H., Gasmalla, M. A., Yang, R., & Zhao, W. (2018). Identification of bioactive peptides with α-amylase inhibitory potential from enzymatic protein hydrolysates of red seaweed (Porphyra spp). *Journal of Agricultural and Food Chemistry*, 66(19), 4872–4882.

Angell, A. R., Mata, L., de Nys, R., & Paul, N. A. (2016). The protein content of seaweeds: A universal nitrogen-to-protein conversion factor of five. *Journal of Applied Phycology*, 28(1), 511–524.

Atkinson, A., & Robertson, J. (1979). Captopril in the treatment of clinical hypertension and cardiac failure. *The Lancet*, 314(8147), 836–839.

Barbarino, E., & Lourenço, S. O. (2005). An evaluation of methods for extraction and quantification of protein from marine macro-and microalgae. *Journal of Applied Phycology*, 17(5), 447–460.

Beaulieu, L., Bondu, S., Doiron, K., Rioux, L.-E., & Turgeon, S. L. (2015). Characterization of antibacterial activity from protein hydrolysates of the macroalga Saccharina longicruris and identification of peptides implied in bioactivity. *Journal of Functional Foods*, 17, 685–697.

Becker, E. W. (2007). Micro-algae as a source of protein. *Biotechnology Advances*, 25(2), 207–210.

Ben-Aharon, O., Magnezi, R., Leshno, M., & Goldstein, D. A. (2018). Association of immunotherapy with durable survival as defined by value frameworks for cancer care. *JAMA oncology*, 4(3), 326–332.

Biris-Dorhoi, E.-S., Michiu, D., Pop, C. R., Rotar, A. M., Tofana, M., Pop, O. L., & Farcas, A. C. (2020). Macroalgae—A sustainable source of chemical compounds with biological activities. *Nutrients*, 12(10), 3085.

Cermeño, M., Kleekayai, T., Amigo-Benavent, M., Harnedy-Rothwell, P., & FitzGerald, R. J. (2020). Current knowledge on the extraction, purification, identification, and validation of bioactive peptides from seaweed. *Electrophoresis*, 41(20), 1694–1717.

Chabeaud, A., Vandanjon, L., Bourseau, P., Jaouen, P., & Guérard, F. (2009). Fractionation by ultrafiltration of a saithe protein hydrolysate (Pollachius virens): Effect of material and molecular weight cutoff on the membrane performances. *Journal of Food Engineering*, 91(3), 408–414.

Chen, M.-F., Zhang, Y. Y., Di He, M., Li, C. Y., Zhou, C. X., Hong, P. Z., & Qian, Z.-J. (2019). Antioxidant peptide purified from enzymatic hydrolysates of Isochrysis Zhanjiangensis and its protective effect against ethanol induced oxidative stress of HepG2 cells. *Biotechnology and Bioprocess Engineering*, 24(2), 308–317.

Cian, R. E., Martínez-Augustin, O., & Drago, S. R. (2012). Bioactive properties of peptides obtained by enzymatic hydrolysis from protein byproducts of Porphyra columbina. *Food Research International*, 49(1), 364–372.

Clemente, A. (2000). Enzymatic protein hydrolysates in human nutrition. *Trends in Food Science & Technology*, 11(7), 254–262.

Dominguez, H. (2013). *Functional ingredients from algae for foods and nutraceuticals*: Elsevier.

Ejike, C. E., Collins, S. A., Balasuriya, N., Swanson, A. K., Mason, B., & Udenigwe, C. C. (2017). Prospects of microalgae proteins in producing peptide-based functional foods for promoting cardiovascular health. *Trends in Food Science & Technology*, 59, 30–36.

El Gamal, A. A. (2010). Biological importance of marine algae. *Saudi Pharmaceutical Journal*, 18(1), 1–25.

Fitzgerald, C., Aluko, R. E., Hossain, M., Rai, D. K., & Hayes, M. (2014). Potential of a renin inhibitory peptide from the red seaweed Palmaria palmata as a functional food ingredient following confirmation and characterization of a hypotensive effect in spontaneously hypertensive rats. *Journal of Agricultural and Food Chemistry*, 62(33), 8352–8356.

Fitzgerald, C. n., Mora-Soler, L., Gallagher, E., O'Connor, P., Prieto, J., Soler-Vila, A., & Hayes, M. (2012). Isolation and characterization of bioactive pro-peptides with in vitro renin inhibitory activities from the macroalga Palmaria palmata. *Journal of Agricultural and Food Chemistry*, 60(30), 7421–7427.

Fleurence, J., Morançais, M., & Dumay, J. (2018). Seaweed proteins. In *Proteins in food processing* (pp. 245–262): Elsevier.

Foltz, M., van der Pijl, P. C., & Duchateau, G. S. (2010). Current in vitro testing of bioactive peptides is not valuable. *The Journal of Nutrition*, 140(1), 117–118.

Furuta, T., Miyabe, Y., Yasui, H., Kinoshita, Y., & Kishimura, H. (2016). Angiotensin I converting enzyme inhibitory peptides derived from phycobiliproteins of dulse Palmaria palmata. *Marine Drugs*, 14(2), 32.

Galland-Irmouli, A.-V., Fleurence, J., Lamghari, R., Luçon, M., Rouxel, C., & Barbaroux, O., Guéant, J.-L. (1999). Nutritional value of proteins from edible seaweed Palmaria palmata (Dulse). *The Journal of Nutritional Biochemistry*, 10(6), 353–359.

Gao, M.-T., Hirata, M., Toorisaka, E., & Hano, T. (2006). Acid-hydrolysis of fish wastes for lactic acid fermentation. *Bioresource Technology*, 97(18), 2414–2420.

Garcia-Vaquero, M., Mora, L., & Hayes, M. (2019). In vitro and in silico approaches to generating and identifying angiotensin-converting enzyme I inhibitory peptides from green macroalga Ulva lactuca. *Marine Drugs*, 17(4), 204.

Garcia, J., Palacios, V., & Roldán, A. (2016). Nutritional potential of four seaweed species collected in the barbate estuary (Gulf of Cadiz, Spain). *J. Nutr Food Sci*, 6(2). http://dx.doi.org/10.4172/2155-9600.1000505

Hardouin, K., Burlot, A.-S., Umami, A., Tanniou, A., Stiger-Pouvreau, V., Widowati, I., & Bourgougnon, N. (2014). Biochemical and antiviral activities of enzymatic hydrolysates from different invasive French seaweeds. *Journal of Applied Phycology*, 26(2), 1029–1042.

Harnedy, P. A., & FitzGerald, R. J. (2013). In vitro assessment of the cardioprotective, antidiabetic and antioxidant potential of Palmaria palmata protein hydrolysates. *Journal of Applied Phycology*, 25(6), 1793–1803.

Harnedy, P. A., O'Keeffe, M. B., & FitzGerald, R. J. (2017). Fractionation and identification of antioxidant peptides from an enzymatically hydrolysed Palmaria palmata protein isolate. *Food Research International*, 100, 416–422.

Harnedy, P. A., O'Keeffe, M. B., & FitzGerald, R. J. (2015). Purification and identification of dipeptidyl peptidase (DPP) IV inhibitory peptides from the macroalga Palmaria palmata. *Food Chemistry*, 172, 400–406.

Harrysson, H., Hayes, M., Eimer, F., Carlsson, N.-G., Toth, G. B., & Undeland, I. (2018). Production of protein extracts from Swedish red, green, and brown seaweeds, Porphyra umbilicalis Kützing, Ulva lactuca Linnaeus, and Saccharina latissima (Linnaeus) JV Lamouroux using three different methods. *Journal of Applied Phycology*, 30(6), 3565–3580.

Hayes, M., Bastiaens, L., Gouveia, L., Gkelis, S., Skomedal, H., Skjanes, K., & Dodd, J. (2018). *Novel proteins for food, pharmaceuticals and agriculture*. Chichester, UK: John Wiley & Sons, Ltd.

Heo, S.-J., Park, E.-J., Lee, K.-W., & Jeon, Y.-J. (2005). Antioxidant activities of enzymatic extracts from brown seaweeds. *Bioresource Technology*, 96(14), 1613–1623.

Himaya, S., Dewapriya, P., & Kim, S.-K. (2013). EGFR tyrosine kinase inhibitory peptide attenuates Helicobacter pylori-mediated hyper-proliferation in AGS enteric epithelial cells. *Toxicology and Applied Pharmacology*, 269(3), 205–214.

Hou, Y., Wu, Z., Dai, Z., Wang, G., & Wu, G. (2017). Protein hydrolysates in animal nutrition: Industrial production, bioactive peptides, and functional significance. *Journal of Animal Science and Biotechnology*, 8(1), 1–13.

Indumathi, P., & Mehta, A. (2016). A novel anticoagulant peptide from the Nori hydrolysate. *Journal of Functional Foods*, 20, 606–617.

Joubert, Y., & Fleurence, J. (2008). Simultaneous extraction of proteins and DNA by an enzymatic treatment of the cell wall of Palmaria palmata (Rhodophyta). *Journal of Applied Phycology*, 20(1), 55–61.

Keris-Sen, U. D., Sen, U., Soydemir, G., & Gurol, M. D. (2014). An investigation of ultrasound effect on microalgal cell integrity and lipid extraction efficiency. *Bioresource Technology*, 152, 407–413.

Kim, S.-K. (2013). *Marine proteins and peptides: Biological activities and applications*: John Wiley & Sons.

Ko, S.-C., Kim, D., & Jeon, Y.-J. (2012). Protective effect of a novel antioxidative peptide purified from a marine Chlorella ellipsoidea protein against free radical-induced oxidative stress. *Food and chemical toxicology*, 50(7), 2294–2302.

Korhonen, H., & Pihlanto, A. (2006). Bioactive peptides: Production and functionality. *International Dairy Journal*, 16(9), 945–960.

Korhonen, H., & Pihlanto, A. (2007). Technological options for the production of health-promoting proteins and peptides derived from milk and colostrum. *Current Pharmaceutical Design*, 13(8), 829–843.

Lafarga, T., Acién-Fernández, F. G., & Garcia-Vaquero, M. (2020). Bioactive peptides and carbohydrates from seaweed for food applications: Natural occurrence, isolation, purification, and identification. *Algal Research*, 48, 101909.

Lammi, C., Aiello, G., Boschin, G., & Arnoldi, A. (2019). Multifunctional peptides for the prevention of cardiovascular disease: A new concept in the area of bioactive food-derived peptides. *Journal of Functional Foods*, 55, 135–145.

Lee, S.-H., Chang, D.-U., Lee, B.-J., & Jeon, Y.-J. (2009). Antioxidant activity of solubilized Tetraselmis suecica and Chlorella ellipsoidea by enzymatic digests. *Preventive Nutrition and Food Science*, 14(1), 21–28.

Li, Y., Lammi, C., Boschin, G., Arnoldi, A., & Aiello, G. (2019). Recent advances in microalgae peptides: Cardiovascular health benefits and analysis. *Journal of Agricultural and Food Chemistry*, 67(43), 11825–11838.

López, C. V. G., García, M. D. C. C., Fernández, F. G. A., Bustos, C. S., Chisti, Y., & Sevilla, J. M. F. (2010). Protein measurements of microalgal and cyanobacterial biomass. *Bioresource Technology*, 101(19), 7587–7591.

Lourenço, S. O., Barbarino, E., Lavín, P. L., Lanfer Marquez, U. M., & Aidar, E. (2004). Distribution of intracellular nitrogen in marine microalgae: Calculation of new nitrogen-to-protein conversion factors. *European Journal of Phycology*, 39(1), 17–32.

Maestri, E., Pavlicevic, M., Montorsi, M., & Marmiroli, N. (2019). Meta-Analysis for correlating structure of bioactive peptides in foods of animal origin with regard to effect and stability. *Comprehensive Reviews in Food Science and Food Safety*, 18(1), 3–30.

Mao, X., Bai, L., Fan, X., & Zhang, X. (2017). Anti-proliferation peptides from protein hydrolysates of Pyropia haitanensis. *Journal of Applied Phycology*, 29(3), 1623–1633.

Mellitus, D. (2005). Diagnosis and classification of diabetes mellitus. *Diabetes care*, 28(S37), S5–S10.

Mercier, K., Smith, H., & Biederman, J. (2014). Renin-angiotensin-aldosterone system inhibition: Overview of the therapeutic use of angiotensin-converting enzyme inhibitors, angiotensin receptor blockers, mineralocorticoid receptor antagonists, and direct renin inhibitors. *Primary Care: Clinics in Office Practice*, 41(4), 765–778.

Mittal, R., Sharma, R., & Raghavarao, K. (2019). Aqueous two-phase extraction of R-Phycoerythrin from marine macro-algae, Gelidium pusillum. *Bioresource Technology*, 280, 277–286.

Murray, B., & FitzGerald, R. (2007). Angiotensin converting enzyme inhibitory peptides derived from food proteins: Biochemistry, bioactivity and production. *Current Pharmaceutical Design*, 13(8), 773–791.

Ngo, D.-H., Vo, T.-S., Ngo, D.-N., Wijesekara, I., & Kim, S.-K. (2012). Biological activities and potential health benefits of bioactive peptides derived from marine organisms. *International journal of Biological Macromolecules*, 51(4), 378–383.

Nguyen, M. H. T., Qian, Z.-J., Nguyen, V.-T., Choi, I.-W., Heo, S.-J., Oh, C. H., ... Jung, W.-K. (2013). Tetrameric peptide purified from hydrolysates of biodiesel byproducts of Nannochloropsis oculata induces osteoblastic differentiation through MAPK and Smad pathway on MG-63 and D1 cells. *Process Biochemistry*, 48(9), 1387–1394.

O'Connor, J., Meaney, S., Williams, G. A., & Hayes, M. (2020). Extraction of protein from four different seaweeds using three different physical pretreatment strategies. *Molecules*, 25(8), 2005.

Passos, F., Carretero, J., & Ferrer, I. (2015). Comparing pretreatment methods for improving microalgae anaerobic digestion: Thermal, hydrothermal, microwave and ultrasound. *Chemical Engineering Journal, 279*, 667–672.

Patarra, R. F., Paiva, L., Neto, A. I., Lima, E., & Baptista, J. (2011). Nutritional value of selected macroalgae. *Journal of Applied Phycology, 23*(2), 205–208.

Pereira, A. M., Lisboa, C. R., Santos, T. D., & Costa, J. A. V. (2019). Bioactive stability of microalgal protein hydrolysates under food processing and storage conditions. *Journal of Food Science and Technology, 56*(10), 4543–4551.

Pimentel, F. B., Alves, R. C., Harnedy, P. A., FitzGerald, R. J., & Oliveira, M. B. P. (2019). Macroalgal-derived protein hydrolysates and bioactive peptides: Enzymatic release and potential health enhancing properties. *Trends in Food Science & Technology, 93*, 106–124.

Pouliot, Y., Gauthier, S., & Groleau, P. (2006). Membrane-based fractionation and purification strategies for bioactive peptides. *Nutraceutical Proteins and Peptides in Health and Disease*, 639–658.

Prabhu, M. S., Levkov, K., Livney, Y. D., Israel, A., & Golberg, A. (2019). High-voltage pulsed electric field preprocessing enhances extraction of starch, proteins, and ash from marine macroalgae ulva ohnoi. *ACS Sustainable Chemistry & Engineering, 7*(20), 17453–17463.

Raskob, G. E., Angchaisuksiri, P., Blanco, A. N., Buller, H., Gallus, A., Hunt, B. J., McCumber, M. (2014). Thrombosis: A major contributor to global disease burden. *Arteriosclerosis, Thrombosis, and Vascular Biology, 34*(11), 2363–2371.

Rasmussen, R. S., & Morrissey, M. T. (2007). Marine biotechnology for production of food ingredients. *Advances in Food and Nutrition Research, 52*, 237–292.

Riccio, G., & Lauritano, C. (2020). Microalgae with immunomodulatory activities. *Marine Drugs, 18*(1), 2.

Samarakoon, K. W., Kwon, O.-N., Ko, J.-Y., Lee, J.-H., Kang, M.-C., Kim, D., & Jeon, Y.-J. (2013). Purification and identification of novel angiotensin-I converting enzyme (ACE) inhibitory peptides from cultured marine microalgae (Nannochloropsis oculata) protein hydrolysate. *Journal of Applied Phycology, 25*(5), 1595–1606.

Sánchez, A., & Vázquez, A. (2017). Bioactive peptides: A review. *Food Quality and Safety, 1*(1), 29–46.

Sheih, I.-C., Wu, T.-K., & Fang, T. J. (2009). Antioxidant properties of a new antioxidative peptide from algae protein waste hydrolysate in different oxidation systems. *Bioresource Technology, 100*(13), 3419–3425.

Stack, J., Le Gouic, A. V., & FitzGerald, R. J. (2020). Bioactive proteins and peptides from microalgae. *Encyclopedia of Marine Biotechnology, 3*, 1443–1474.

Suetsuna, K., & Chen, J.-R. (2001). Identification of antihypertensive peptides from peptic digest of two microalgae, Chlorella vulgaris and Spirulina platensis. *Marine Biotechnology, 3*(4), 305–309.

Suetsuna, K., Maekawa, K., & Chen, J.-R. (2004). Antihypertensive effects of Undaria pinnatifida (wakame) peptide on blood pressure in spontaneously hypertensive rats. *The Journal of Nutritional Biochemistry, 15*(5), 267–272.

Suetsuna, K., & Nakano, T. (2000). Identification of an antihypertensive peptide from peptic digest of wakame (Undaria pinnatifida). *The Journal of Nutritional Biochemistry, 11*(9), 450–454.

Sun, Y., Chang, R., Li, Q., & Li, B. (2016). Isolation and characterization of an antibacterial peptide from protein hydrolysates of Spirulina platensis. *European Food Research and Technology, 242*(5), 685–692.

Telagari, M., & Hullatti, K. (2015). In-vitro α-amylase and α-glucosidase inhibitory activity of Adiantum caudatum Linn. and Celosia argentea Linn. extracts and fractions. *Indian Journal of Pharmacology*, 47(4), 425.

Vo, T.-S., Ryu, B., & Kim, S.-K. (2013). Purification of novel anti-inflammatory peptides from enzymatic hydrolysate of the edible microalgal Spirulina maxima. *Journal of Functional Foods*, 5(3), 1336–1346.

Wang, X., Yu, H., Xing, R., Chen, X., Liu, S., & Li, P. (2017a). Optimization of the extraction and stability of antioxidative peptides from mackerel (Pneumatophorus japonicus) protein. *BioMed Research International*, 2017. https://doi.org/10.1155/2017/6837285

Wang, X., Yu, H., Xing, R., & Li, P. (2017b). Characterization, preparation, and purification of marine bioactive peptides. *BioMed Research International*, 2017. https://doi.org/10.1155/2017/9746720

Wang, Z., & Zhang, X. (2016). Inhibitory effects of small molecular peptides from Spirulina (Arthrospira) platensis on cancer cell growth. *Food & Function*, 7(2), 781–788.

Wroblewski, L. E., Peek, R. M., & Wilson, K. T. (2010). Helicobacter pylori and gastric cancer: Factors that modulate disease risk. *Clinical Microbiology Reviews*, 23(4), 713–739.

Xia, E., Zhai, L., Huang, Z., Liang, H., Yang, H., Song, G., & Tang, H. (2019). Optimization and identification of antioxidant peptide from underutilized Dunaliella salina protein: Extraction, In vitro gastrointestinal digestion, and fractionation. *BioMed Research International*, 2019. https://doi.org/10.1155/2019/6424651

Yu, J., Hu, Y., Xue, M., Dun, Y., Li, S., Peng, N., & Zhao, S. (2016). Purification and identification of antioxidant peptides from enzymatic hydrolysate of Spirulina platensis. *Journal of Microbiology and Biotechnology*, 26(7), 1216–1223.

Zhang, B., & Zhang, X. (2013). Separation and nanoencapsulation of antitumor polypeptide from Spirulina platensis. *Biotechnology Progress*, 29(5), 1230–1238.

Zhao, M., Sun, L., Fu, X., & Chen, M. (2019). Phycoerythrin-phycocyanin aggregates and phycoerythrin aggregates from phycobilisomes of the marine red alga Polysiphonia urceolata. *International Journal of Biological Macromolecules*, 126, 685–696.

Zou, T.-B., He, T.-P., Li, H.-B., Tang, H.-W., & Xia, E.-Q. (2016). The structure-activity relationship of the antioxidant peptides from natural proteins. *Molecules*, 21(1), 72.

CHAPTER 3

Meat

Sajad Ahmad Mir, Zahida Naseem, Danish Rizwan, and Farooq Ahmad Masoodi

CONTENTS

Introduction	55
History of Bioactive Peptide Discovery	56
Sources of Bioactive Peptides	57
Bioactive Peptides from Meat and Meat Products	57
Trimmings and Cuttings	58
Bones	58
Blood	59
Production of Bioactive Peptides	60
Types of Bioactive Peptides	61
Antihypertensive Peptides	61
Antihypertensive Peptides from Myosin Sources	62
Antihypertensive Peptides from Troponin Sources	62
Antithrombotic Peptides	63
Antioxidant Peptides	63
Antimicrobial Peptides	64
Antidiabetic Peptides	64
Conclusion	65
References	65

INTRODUCTION

The general consensus is that diet represents a crucial factor in terms of human health status. Food proteins have long been recognized for their nutritional and functional properties. The nutritional properties of proteins are associated with their amino acid content in conjunction with the physiological utilization of specific amino acids upon digestion and absorption (Korhonen and Pihlanto, 2006; Friedman, 1996). On the other hand, the functional properties of proteins relate to their contribution to the physiochemical and sensory properties of foods (Vercruysse et al., 2005). Proteins of animal origin have been recognized for their nutritional properties as an essential source of amino acids upon digestion, but both digestion and industrial processing may liberate peptides, which have biological functions, from the parent protein (Albenzio et al., 2017). The value of proteins as an essential source of amino acids is well documented, but recently it has been recognized that dietary proteins exert many other functionalities *in vivo* by means of biologically active peptides. Inactive within the sequence of the parent protein, such peptides

DOI: 10.1201/9781003106524-5

can be released by digestive enzymes during gastrointestinal transit or by fermentation or ripening during food processing (Korhonen, 2009; Chakrabarti et al., 2014; Dziuba and Dziuba, 2014) and play an important role in human health by affecting the digestive, endocrine, cardiovascular, immune, and nervous systems (Bhat et al., 2015).

Bioactive peptides are specific protein fragments which, above and beyond their nutritional capabilities, have a positive impact on the body's function or condition which may ultimately influence health.

Bioactive peptides have been defined as "food derived components (genuine or generated) that, in addition to their nutritional value exert a physiological effect in the body" (Vermeirssen et al., 2007). These bioactive peptides are usually 2–20 amino acid residues in length, although, some have been reported to be >20 amino acid residues such as the soy-derived peptide, lunasin, composed of 43 amino acid residues (Dia and de Mejia, 2013). Bioactive peptides are inactive or latent in the parent protein but are released in an active form after proteolytic digestion (Hayes et al., 2007a).

Until now, diverse bioactive substances had been studied, but an increasing interest is directed toward bioactive peptides of animal origin, particularly milk- and egg-derived because, presently, products like bovine milk, egg, cheese, and dairy products are extremely important sources of bioactive peptides derived from food (Urista et al., 2011; Dave et al., 2014). The two criteria which act as a basis for using food proteins as bioactive peptide sources are:

(i) Aim to use underutilized food proteins or food industry by-products rich in proteins for value addition.
(ii) Exploit a particular protein containing a specific peptide sequence or amino acid residue for its pharmaceutical potential (Udenigwe and Aluko, 2011).

There has been an increase in the knowledge regarding bioactive peptides particularly during the last two decades, and various peptides exhibiting diverse activities, such as antithrombotic, antihypertensive, antioxidative, antibacterial, and cholesterol-lowering properties, have been reported (Mils et al., 2011; Stuknyteet al., 2011; Suarez-Jimenez et al., 2012; Zambrowicz et al., 2013; Bah et al., 2013; Singh et al., 2014; Dave et al., 2014; Chakrabarti et al., 2014; Dziuba and Dziuba, 2014; Li and Yu, 2014). The activity of peptides is based on the immanent amino acid composition and sequence. Peptides become active only when they are released from the parent protein molecule where they are encrypted. Bioactive peptides usually contain 2–20 amino acid residues, although, some are reported to have more than 20 amino acid residues. Once liberated as independent entities, bioactive peptides act as potential metabolism modulators and regulatory compounds with hormone-like activities (Korhonen and Pihlanto, 2003a).

Meat and fish have been known as valuable sources of protein for populations around the globe; furthermore, meat and fish proteins come across as potential, novel sources of bioactive peptides. To date, bioactive peptides displaying antihypertensive, antioxidant, antimicrobial, and antiproliferative effects have been isolated from hydrolysates of meat and fish proteins (Kim et al., 2009; Matsui et al., 2006; Cinq-Mars et al., 2008; Jang et al., 2008).

HISTORY OF BIOACTIVE PEPTIDE DISCOVERY

The first food-derived bioactive peptide was identified in 1950 when Mellander reported that phosphorylated casein-derived peptides enhanced vitamin D-independent bone calcification in rachitic infants (Mellander, 1950). However, interest in this field has increased considerably in the last 20 years, with research mainly focused on the identification of bioactive peptides from dairy proteins (Rutherfurd-Markwick and Moughan, 2005).

SOURCES OF BIOACTIVE PEPTIDES

There are different sources of bioactive peptides e.g., algae, dairy products, seafood, cereals, soy, mushrooms, sorghum, and meat.

Bioactive Peptides from Meat and Meat Products

Meat is an excellent source of well-balanced essential amino acids, particularly those containing sulphur, as it contains an appreciable number of proteins with high biological value. Meat is also an exceptional source of valuable nutrients, such as minerals and vitamins (Biesalski, 2005; Chan, 2004; Mulvihill, 2004). Apart from the basic nutrients, meat is also a source of bioactive compounds, such as conjugated linoleic acid, which have gathered much attention recently (Arihara and Ohata, 2008). Cattle, sheep, goats, pigs, and poultry are the most common sources of meat for human consumption (Hoffman and Cawthorn, 2013). Meat is mainly composed of water, protein, and lipids, but also contains carbohydrates, vitamins, and other bioactive components in minimal concentrations (Pereira and Vicente, 2013).

Meat is one of the most valuable livestock products and has been defined by the European Commission (EC) as edible parts removed from the carcass of domestic animals including bovine, porcine, ovine and caprine animals, poultry, and wild game (Regulation EC 835/2004). Proteins are the most important component of meat and they can be broadly classified as water soluble, insoluble, or soluble in varying concentrations of salt. The meat industry involves the slaughter of animals and the generation of end- and by-products. What constitutes by-products depends on several factors including traditions, culture, and religion, but they generally include skin, bones, meat trimmings, blood, fatty tissues, horns, feet, hooves, or internal organs (Di Bernardini, et al., 2011; Toldrá et al., 2012). These meat by-products are rich in lipids, carbohydrates, and proteins, hence can be utilized in a wide range of applications, particularly as raw materials to develop high value-added ingredients for the functional foods market (Lafarga and Hayes, 2014). For example, collagen is used as a functional ingredient, as it has a positive influence on the delivery and bioactivity of bone morphogenic protein-2 and ectopic bone formation, enhancing bone healing (Bhakta et al., 2013).

Due to the presence of high-quality proteins, meat represents the most investigated source for the isolation of novel bioactive peptides. Different mechanisms generate bioactive peptides from meat and meat by-products. During meat post-mortem aging, the proteolytic activity due to endogenous enzymes (calpains and cathepsins) is a key process that affects the destructuration of proteins and, consequently, the production and release of a large number of peptides and free amino acids (Sentandreu et al., 2002; Toldrá et al., 2012). Changes of temperature and pH can affect the content of bioactive peptides during meat storage owing to the alteration in the activity of endogenous enzymes and the destruction of pH or heat-sensitive amino acids (Korhonen et al., 1998; Leygonie et al., 2012). Bioactive peptides are produced naturally in the gastrointestinal tract of mammals during the digestion process, wherein the dietary meat proteins are metabolized (Bauchart et al., 2007; Adje et al., 2011a and b). During gastrointestinal proteolysis, ingested meat proteins are acted upon by stomach-secreted digestive enzymes, such as pepsin, trypsin, chymotrypsin, elastase, and carboxypeptidase, secreted into the small intestine, and resulting in the consequent generation of biologically active peptides (Pihlanto and Korhonen, 2003). Several bioactive peptides have been obtained through enzymatic hydrolysis from meat collagen or slaughtered by-products (trimmings, organs, hemoglobin), as reported in many research investigations (Lafarga and Hayes, 2014; Vercruysse et al., 2005). Freezing and cooking can also impact the isolation and availability of bioactive peptides from meat. Freezing can lead to denaturation of meat proteins, such as myosin, due to different chemical and physical stresses, including ice formation, pH variations, and cold temperature (Christensen et al., 2013), which in turn increases formation of bioactive peptides. Cooking has a positive effect on the production of peptides and their biological activities (Fu et al., 2017; Leygonie et al., 2012) due to alterations in the native state of protein molecules because of the denaturation and rupture of intramolecular forces of proteins caused by heat (Yu et al., 2017). It has been reported that the content of the three bioactive peptides, namely carnosine, anserine, and glutathione, decreased and were eventually lost during ripening and cooking (Bauchart et al., 2006).

Trimmings and Cuttings

Trimmings refer to portions of meat remaining after the preparation of primary cuts from the carcass and include fat, gristles, and meat. They can also include mechanically recovered meat. These trimmings may be chopped or disintegrated and blended properly to obtain a uniform protein content ranging between 14 to 17% and 7.5 to 12% for pork and veal, respectively (Field, 1988). The meat industries make sure to utilize these trimmings and cuttings by converting them into subsidiary quality meat products such as hot dogs. However, these by-products are as good a source for bioactive peptides as the primal cuts and can be used for generation of these biologically active peptides. The meat product industry is also an important producer of trimmings that could be exploited as an excellent source of ACE inhibitory and antioxidant peptides as indicated by various scientific studies carried out on dry-cured ham during the last decade. A lot of trimmings are generated in this type of industry when deboning and slicing of hams is done for its distribution in small weight packages. A research study carried out on peptide fractions extracted from Spanish dry-cured ham indicated their antihypertensive and antioxidant potential (Escudero et al., 2012). The fractions were tested *in vitro* and *in vivo* for their antihypertensive activity by evaluating the changes in spontaneously hypertensive rats. It was noted that one of the analyzed peptide fractions isolated from Spanish dry-cured ham lowered systolic blood pressure (SBP) by 38.38 mmHg.

Bones

Bones are one of the most important sources of collagen and gelatin, about 10–13% in content. Collagen and gelatin have been described as proteins containing biologically active peptides, with promising health benefits for humans (Alemán et al., 2013). However, very few studies have been mentioned in the literature that describes the purification and identification of bioactive peptides from these by-products. The antioxidant peptides QYDQGV, YEDCTDCGN, and AADNANELFPPN have been identified from an aqueous extract of water buffalo horn, commonly used in Chinese medicine. Results indicated these bioactive peptides were capable of reducing the DPPH radical and protecting the cerebral microvascular endothelial cells of rats against hydrogen peroxide (H_2O_2)-induced injury (Liu et al., 2010). And extensive research of marine by-products, where the backbones have been hydrolyzed using different enzymes to produce bioactive peptides that promote human health and prevent chronic disease (Šližytė et al., 2009; Ravallec et al., 2001; Kim et al., 2000) have been undertaken. For example, the antioxidant peptide, VKAGFAWTANQQLS, was obtained from tuna backbone and hydrolyzed using various proteases such as alcalase, neutrase, papain, pepsin, and trypsin (Je et al., 2007).

Collagen is the most abundant protein in vertebrates, as it is the main fibrous protein constituent in bones, cartilage, and skin (Gómez-Guillén et al., 2011). The nutritional value of collagen is very low, but it is an important protein and an equally important source of bioactive peptides in the food industry due to its abundance of non-essential amino acids such as glycine and proline. Collagen is widely used in pharmaceutical companies as it has been proven that orally administered collagen peptides have beneficial effects on bone metabolism. Administration of collagen hydrolysates obtained from chicken legs have been noted to improve bone mineral density in rats, there by exerting a positive effect on osteoporosis by increasing the organic substance content of bone (Watanabe-Kamiyana et al., 2010). It has also been seen that ingestion of chicken bone collagen hydrolysates helps in preventing atherosclerosis through their lipid-lowering effects as well as inhibiting the expression of inflammatory cytokines (Zhang et al., 2010). Most of the research studies about collagen peptides are focused on their bioactive properties especially antioxidant and ACE inhibitory activity. Four peptides with antioxidant activity have been identified from hydrolyzed porcine skin collagen, hydrolyzed by use of various protease treatments. One of the antioxidative peptides, Gln-Gly-Ala-Arg, was synthesized and the antioxidant confirmed *in vitro* (Li et al., 2007). On the other hand, four peptides showing noteworthy *in vitro* and *in vivo* ACE inhibitory potential against spontaneous hypertensive rats were reported from chicken skin collagen hydrolysates, hydrolyzed using an enzyme derived from *Aspergillus* species (Saiga et al., 2008).

Blood

Blood is a body fluid that constitutes a rich protein by-product, representing up to 4% of animal weight (Anderson, 1988), with hemoglobin accounting for more than half of the proteins present (Hsieh and Ofori, 2011). It is composed of red blood cells (erythrocytes), white blood cells (leukocytes), and platelets suspended in blood plasma, which contains proteins such as fibrinogen, globulins, and albumins (Bah et al., 2013). Albumin is the main protein in plasma and also a prime element in the regulation of fluid distribution, colloidal osmotic pressure, and the transport of small metabolites in blood (Rondeau and Bourdon, 2011). Blood is mostly obtained from bovine and porcine sources, and the studies related to its value as a generator of bioactive peptides are mainly focused on two fractions – cellular fractions, especially hemoglobin cells, and the plasma fraction. Hemoglobin and plasma hydrolysates mainly exhibit antihypertensive, antioxidant, antimicrobial, and opioid activity (Chang et al., 2007). Peptides GFPTTKTYFPHF and VVYPWT, corresponding to the 34–46 fragment of the α-chain and the 34–39 fragment of the β-chain of porcine hemoglobin, obtained by hydrolysis using pepsin enzyme, exhibited ACE inhibitory activity, showing IC_{50} values of 4.92 and 6.02 µM, respectively (Yu et al., 2006). The antimicrobial activity of peptides derived from the hemoglobin chain is, by far, the most studied property.

Peptides obtained from the hydrolysis of bovine α-chain hemoglobin, using enzyme pepsin, showed antibacterial activity against *Kocuria luteus*, *Listeria innocua*, *Escherichia coli*, and *Staphylococcus aureus* and also showed ACE inhibitory activity in an ICμ range from 42.55 to 1,095 μM (Adje et al., 2011a; Adje et al., 2011b).

PRODUCTION OF BIOACTIVE PEPTIDES

There are a number of methods by which peptides with biological activity can be produced from precursor proteins. The most common ones are (1) enzymatic hydrolysis with digestive enzymes, (2) by means of the microbial activity of fermented foods, (3) through the action of enzymes derived from proteolytic microorganisms, and (4) by use of chemicals (Korhonen and Pihlanto, 2003). The existing literature presents enzymatic hydrolysis of whole protein as the most commonly used and explored technique. It is an alluring protein extraction method selected by food and pharmaceutical industries to produce bioactive peptides owing to the subtle processing conditions employed. Moreover, only a minimal number of by-products are generated in this process (Ryan et al., 2011; Borrajo et al., 2019). A number of bioactive peptides have been isolated from meat using digestive enzymes such as pepsin, trypsin and chymotrypsin. Various proteases from microbial origin (Alcalase®, Flavourzyme®, Neutrase®, collagenase, or proteinase K), animal origin (pepsin, trypsin, chymotrypsin, elastase, and carboxypeptidase), and plant origin (ficin, bromelain, and papain) have also been employed to generate bioactive peptides from meat sources (Jang et al., 2008). Chemical hydrolysis using acid/alkali is widely used for cleaving the peptide bonds of proteins. The commonly employed chemicals for hydrolysing proteins include hydrochloric acid, sulfonic acid, sodium hydroxide, and potassium hydroxide (Borrajo et al., 2019; Wang et al., 2017). Acids have been used for hydrolysing fish proteins (Wisuthiphaet et al., 2016). Several researchers have described the isolation of bioactive peptides via bacterial fermentation of milk proteins, where *Lactobacilli* have been commonly used as a fermentation microorganism (Hayes et al., 2007b; Kudoh et al., 2001; Pihlanto et al., 2010). Apart from *Lactobacillus*, few other microorganisms, such as *Chryseobacterium* and *Monascus purpureus*, have been evaluated for their proteolytic activity (Fontoura et al., 2014; Yu et al., 2017). However, the microbial fermentation of meat proteins has been less successful, presumably due to the poor proteolytic activity of the *Lactobacilli* used in meat fermentations (Arihara and Ohata, 2006; Hammes et al., 2003). High hydrostatic pressure processing (HHP), ultrasounds, pulsed electric fields, and other novel, green processing methods have been evaluated to enhance the degree of hydrolysis during production of bioactive peptides. High pressure processing, for example, has been recently employed alone and in combination with other processes, such as enzymatic hydrolysis, to enhance proteolysis. The high pressure-assisted enzymatic hydrolysis involves exposing of cleavage sites by HHP to accelerate enzymatic action. (Bamdad et al., 2017; Al-Ruwaih et al., 2019). Ultrasound is another novel, eco-friendly processing technique which is based on the principal of acoustic wave generation leading to acoustic cavitation. The cavitation process results in the generation of strong forces at the microscopic level. This technique has been employed, in combination with enzymatic hydrolysis, to produce bioactive peptides, as it alone is not enough to cleave the peptide bonds of proteins. It has also been used to overcome the shortcomings (long hydrolysis time and low protein conversion rate) of the traditional enzymatic hydrolysis process (Ulug et al., 2021; Qu et al., 2012). Recombinant DNA technology is one of the methods which has gained attention and popularity recently. It uses cloning and gene expression to generate recombinant peptides. This technique of peptide synthesis has the potential to function as a cost-effective alternative for industrial scale production (Chauhan and Kanwar, 2020).

The purity of bioactive peptides for use in therapeutics and clinical research should be 95% or above. This purity greatly depends on the separation technique employed. Generally, after synthesis, bioactive peptides are subjected to washing and centrifugation to get rid of all the unwanted residues, followed by filtration and freeze-drying. Moreover, some advanced techniques, such as reverse phase high performance liquid chromatography, electrophoresis, and various chromatographic procedures, are used for purifying the bioactive peptides (Sewald and Jakubke, 2002; Dagan et al., 2002; Chauhan and Kanwar, 2020). On an industrial level, membrane technology processes, including reverse osmosis, ultra-filtration, nano-filtration, and microfiltration are employed (Conde et al., 2013).

TYPES OF BIOACTIVE PEPTIDES

Antihypertensive Peptides

Bioactive peptides have been derived from a wide range of foods. The most extensively researched of these are the peptides with antihypertensive potential, particularly those capable of inhibiting the activity of Angiotensin I-converting enzyme (ACE) (Vercruysse et al., 2005, Pihlanto et al., 2008). Angiotensin I-converting enzyme (ACE) inhibitory peptides were first isolated from snake venom (Ferreira et al., 1970), and, since then, numerous synthetic ACE inhibitors have been produced, with captopril being the most common. However, captopril and other synthetic ACE inhibitors are known to induce various side-effects, including coughing, taste disturbances, and skin rashes (Qian et al., 2007, Vermeirssen et al., 2002). The fact that hypertension affects one-third of the western worlds' population and is a known risk factor for stroke and cardiovascular disease, along with the side effects associated with synthetically produced ACE inhibitors has contributed to the ongoing search for food-derived antihypertensive peptides and their exploitation in functional foods and nutraceuticals (Shalaby et al., 2006; Lopez-Fandino et al., 2006).

ACE, or kininase II, is a dipeptidyl carboxy peptidase (EC 3.4.15.1) found in various tissues in the body and is integral to the moderation of blood pressure and normal heart function (Shalaby et al., 2006). In the rennin-angiotensin system, ACE catalyses the conversion of the inactive form of angiotensin I (Ang I) to the potent vasoconstrictor angiotensin II (Ang II). Additionally, ACE is involved in the deactivation of the hypotensive peptide, bradykinin (Ondetti et al., 1977). This is as a result of ACE cleaving the C-terminal dipeptide from Ang I (His-Leu) and bradykinin. Ang II is well-documented as a potent vasoconstrictor, which acts directly on vascular smooth muscle cells. Angiotensin II is also responsible for the expansion of vascular volume via sodium and fluid retention (Brown and Vaughan, 1998; Folkow et al., 1961; Biron et al., 1961; Padfield and Morton, 1977). Bradykinin is responsible for uterine and ileal smooth muscle contraction, enhanced vascular permeability, activation of peripheral and C fibers, and increased mucous secretion (Brown and Vaughan, 1998; Proud et al., 1988).

The generation of force (i.e., contraction) by skeletal muscle is the responsibility of the muscle proteins, actin, and myosin. Both of these proteins are associated with two distinct types of muscle filaments. Myosin proteins are associated with the thick filaments of skeletal muscle while actin is associated with the thin filaments of skeletal muscle. Also present in the thin filaments are the proteins troponin and tropomyosin (Vercruysse et al., 2005; Lawrie and Ledward, 2006). In striated muscle, the myosin thick filaments extend along the thin filaments of actin and are known as sarcomeres. Sarcomeres are further longitudinally repeated to form myofibrils. Contraction of striated muscle is due to the interaction of the thick and thin filaments. Nerve impulses release Ca^{2+} from the sarcoplasmic reticulum, which pass through the myofibrils resulting in the sliding of myosin and actin filaments over one another. This causes the sarcomere to shorten, resulting in muscle contraction (Spudich and Watt, 1971). Interestingly, ACE inhibitory peptides have been identified in the hydrolysates of actin, myosin, and troponin (Katayama et al., 2004; Muguruma et al., 2009; Katayama et al., 2007).

Antihypertensive Peptides from Myosin Sources

Arihara and colleagues (2001) identified two ACE inhibitory pentapeptides from the thermolysin digestion of porcine myosin. These were named mayopentapeptides A and B. The amino acid sequence of mayopentapeptide A was found to be MNPPK, which corresponded to positions 79–83 on the myosin heavy chain, while mayopentapeptide B, with the amino acid sequence ITTNP, corresponded to positions 306–310 on the myosin heavy chain. The antihypertensive activities of mayopentapeptide A and mayopentapeptide B were investigated in spontaneously hypertensive rats (SHRs). Administration of myopentapeptide A and mayopentapeptide B at a concentration of 1 mg per kilogram of animal weight resulted in a maximum decrease in SBP of 23.4 and 3.0 mmHg and 21.0 and 3.1 mmHg after six h, respectively. After 24 h of administration, the SBP of both test groups was found to be significantly lower than that of the control group, indicating that both mayopentapeptide A and mayopentapeptide B are potent antihypertensive peptides *in vivo* (Nakashima et al., 2002).

Another ACE inhibitory octapeptide, VKKVLGNP, was discovered corresponding to positions 47–54 on the myosin light chain (Katayama et al., 2007) with IC_{50} value of 28.5 µM. The administration of this purified peptide to SHRs at a rate of 10 mg per kilogram of weight resulted in the lowering of SBP. It was noted that the SBP of hypertensive rats continued to decrease up to 3 h post-administration and returned back to the pre-administration value after a time period of 9 h.

Porcine skeletal muscle was also found to be a notable source of antihypertensive peptides when crude myosin B present in this muscle was hydrolyzed with pepsin (Muguruma et al., 2009). Indeed, a novel peptide M6 with the amino acid sequence KRVITY was isolated from myosin B hydrolysate. This peptide corresponded to positions 191–196 on the myosin heavy chain and exhibited antihypertensive properties. It was noted that oral administration of M6 peptide resulted in immediate lowering of SBP of SHRs with a maximum decrease of 23 mmHg 6 h after administration, followed by test animals returning to the same SBP value as that of the control after 9 h. This reduction in SBP indicated the potency of M6 peptide as hypo tensor *in vivo*. Furthermore, the M6 peptide was found to retain its ACE inhibitory activity after heat treating the myosin B (98 °C for 10 min) prior to hydrolysis by pepsin. Retention of ACE inhibitory activity after such thermal treatment correlates to the retention of biological activity when subjected to the cooking process.

Antihypertensive Peptides from Troponin Sources

ACE inhibitory peptide with amino acid sequence, RMLGQTPTK, extracted from porcine skeletal muscle, has been found to be released by Troponin C, a regulatory protein. Another ACE inhibitory peptide with the amino acid sequence, KRQKYDI, has been extracted from crude troponin of porcine origin by the action of pepsin enzyme (Katayama et al., 2008).

Antithrombotic Peptides

Venous thromboembolism (VTE) is a common and underestimated condition and one of the main causes of death and disability in high-income countries (Franchini and Mannucci, 2009). VTE is related either to trauma, long immobilization (for example long-haul flights), or abnormalities in blood coagulation and is also common after certain types of operations, such as hip fracture surgery or total knee replacement (Franchini and Mannucci, 2009; Paraskevas, Nicolaides and Mikhailidis, 2013). Blood coagulation is a complex process with several coagulation factors and enzymes involved and with significant molecular similarities to the milk-clotting process defined by the interaction of κ-casein with chymosin (Jolles, 1975). There are only a handful of reported antithrombotic peptides identified to date, with most of them generated as a result of enzymatic hydrolysis of κ-casein. The antithrombotic activity of the peptide, YQEPVLGPVRGPFPIIV, derived from bovine casein by the action of *Lactobacillus casei* has been observed by Rojas-Ronquillo et al. (2012). When tested *in vitro*, this peptide presented antithrombotic activity and also inhibited ACE-I. The peptide maintained its bioactivity after enzymatic treatment with pepsin and trypsin. The effects of a papain-hydrolyzed pork meat diet on the plasma and liver cholesterol levels *in vivo* in rats has been studied by Morimatsu et al. (1996). This study showed that plasma and liver cholesterol concentrations were significantly lower in rats fed the papain-hydrolyzed pork meat as compared to those fed with untreated meat.

Antioxidant Peptides

Oxidation in the body and in food stuffs has a very important role to play and has been widely recognized. Oxidative metabolism is essential for the survival of the cell; however, it is linked to a side-effect, that is, the production of free radicals and other reactive oxygen species that are capable of causing detrimental changes in the cells. When an excess of free radicals is formed, they can overwhelm protective enzymes, like superoxide dismutase, catalase, and peroxidase. This causes destructive and lethal cellular effects, like apoptosis, by oxidizing cellular proteins, membrane lipids, DNA, and enzymes, thus shutting down the cellular process (Sharma et al., 2011). A free radical is a chemical entity capable of independent existence with one or more unpaired orbital electrons, generated during normal cell metabolism or because of exposure to external factors including cigarette smoking, air pollutants, X-rays, or industrial chemicals (Pham-Huy, et al, 2008). The most common reactive oxygen species (ROS) include free radicals like hydroxyl radical (OH·) and superoxide anion radical (O_2^-), and non-free radicals, such as hydrogen peroxide (H_2O_2) and oxygen singlet ($^1O_2$) (Di Bernardini et al., 2011). When ROS are produced in excessive quantity and not readily eliminated from the body, a phenomenon called oxidative stress is generated, where the presence of these ROS oxidise lipids, proteins, and nucleic acids result in cellular damage and apoptosis. Oxidative stress plays a key role in the development of chronic and degenerative illnesses such as cancer, rheumatoid arthritis, osteoporosis, or cardiovascular and neurodegenerative diseases (Pham-Huy et al., 2008; Tierney et al., 2010).

The human body has an endogenous antioxidant system which includes enzymes such as catalase or superoxide dismutase, non-enzymatic compounds like vitamin C, and a number of antioxidant peptides such as anserine and carnosine to counteract oxidative stress (Xiong, 2010). Antioxidants are either produced by the human body or incorporated through diet. Peptides generated from the digestion of various proteins are reported to have antioxidant properties and can be incorporated into food products to provide them with their antioxidant benefits.

Antioxidant peptides can also be naturally present in meat or produced during processing. The antioxidant activity of hydrolysates derived from protease-treated porcine myofibrillar proteins was reported by Saiga et al. (2003). Five peptides with amino acid sequences, IEAEGE, DSGVT, VPSIDDQEELM, DAQEKLE and EELDNALN, were evaluated for their antioxidant potential, and it was noted that peptide with sequence DAQEKLE showed the highest antioxidant activity (Saiga et al., 2003). Proteins from meat by-products, especially blood proteins, have been widely used for the generation of bioactive peptides. Li et al. (2007) hydrolyzed porcine skin collagen with different protease treatments to generate four antioxidant peptides with bioactivities. Three antioxidant peptides from aqueous extract of water buffalo horn were generated by Liu et al. (2010) and were identified as having the amino acid sequences: QYDQGV, YEDCTDCGN, and AADNANELFPPN. These novel antioxidant peptides protected rat cerebral cells against H_2O_2 induced injury (Liu et al., 2010).

Antimicrobial Peptides

Antimicrobial peptides (AMPs) have been identified in a range of foods, found in bacteria, fungi, animals, and plants. They act as defence peptides in hosts, forming an important component of the innate immunity. Proteolytic digestion of foods also gives rise to peptides with antimicrobial potential (Pellegrini, 2003; Lei et al., 2019). Bovine cruor (coagulated blood), one of the by-products generated from slaughterhouses, acts as a good source of peptides with antimicrobial properties. α-137-141 peptide demonstrated inhibitory effect against microorganisms, such as yeast, molds, and bacteria (particularly coliforms) (Przybylski et al., 2016). Peptides derived from hydrolysis of sarcoplasmic proteins from beef show antimicrobial effect. A peptide with amino acid sequence, GLSDGEWQ, demonstrated antimicrobial effect against bacterial strains namely, *Bacillus cereus, E. coli*, and *Listeria monocytogenes* (Jang et al., 2008). Tripeptide (KYR) isolated from α-chain of hemoglobin exhibited inhibitive action against gram-positive as well as gram-negative bacteria, particularly *E. coli, Salmonella enteritidis, Listeria innocua*, and *S. aureus* (Catiau et al., 2011).

Antidiabetic Peptides

Diabetes mellitus, an endocrine metabolic disorder, is quite prevalent globally and has attracted a great deal of research attention over the years (Gondi et al., 2015). Different methods apart from routine medication have been suggested for efficient treatment and management of diabetes, all of them focusing on controlling the blood sugar levels (Ramadhan et al., 2017). Bioactive peptides of food origin with dipeptidyl peptidase-4 inhibitor (DPP-4i) activity show significant activity as antidiabetic peptides, as these peptides are capable of regulating the sugar level in the blood. The antidiabetic peptides inhibit the dipeptidyl peptidase-4 (DPP-4) enzyme leading to the release of certain hormones termed *incretins*, which are essential for regulating the glucose level in the blood stream (Bechaux et al., 2019). The antidiabetic activity of fish muscle proteins of raw goby fish was evaluated by Nasri et al., (2015). The muscle proteins were enzymatically hydrolyzed and the hydrolysates were found to be effective in controlling diabetes in rats by reducing the glucose levels in the bloodstream. ER, a dipeptide, obtained from Bovine serum albumin by enzymatic hydrolysis using papain, showed DPP-4 inhibitory activity (Lafarga et al., 2016).

CONCLUSION

A tremendous number of bioactive peptides are entrapped within the chemical structure of proteins in the food matrix. These peptides are released by hydrolysis of food proteins by various processes including enzymatic hydrolysis, fermentation, chemical hydrolysis, and other novel processing techniques. The bioactive peptides derived from meat and its by-products demonstrate significant levels of biological activity, such as antithrombotic, antioxidative, antihypertensive, etc., that are favorable for human health. The peptides of food origin, especially that of meat and meat by-products, present an excellent research challenge to optimally exploit these peptides for designing and developing health-promoting functional foods.

REFERENCES

Adje, E. Y., Balti, R., Kouach, M., Dhulster, P., Guillochon, D., Nedjar-Arroume, N. (2011a). Obtaining antimicrobial peptides by controlled peptic hydrolysis of bovine hemoglobin. *International Journal of Biological Macromolecules*, 49(2), 143–153.

Adje, E., Balti, R., Kouach, M., Guillochon, D., Nedjar-Arroume, N. (2011b). α 67–106 of bovine hemoglobin: A new family of antimicrobial and angiotensin I-converting enzyme inhibitory peptides. *European Food Research and Technology*, 232(4), 637–646.

Albenzio, M., Santillo, A., Caroprese, M., Malva, A. D., Marino, R. (2017). Bioactive peptides in animal food products. *MDPI Foods*, 6(5), 35.

Alemán, A., Gómez-Guillén, M. C., Montero, P. (2013). Identification of ACE inhibitory peptides from squid skin collagen after *in vitro* gastrointestinal digestion. *Food Research International*, 54(1), 790–795.

Al-Ruwaih, N., Ahmed, J., Mulla, M. F., Arfat, Y. A. (2019). High-pressure assisted enzymatic proteolysis of kidney beans protein isolates and characterization of hydrolysates by functional, structural, rheological and antioxidant properties. *Lebensmittel-Wissenschaft und--Technologie- Food Science and Technology*, 100, 231–236.

Anderson, B. A. (1988). Composition and nutritional value of edible meat by-products. In: *Edible Meat By-Products* (A. M. Pearson, T. R. Dutson, Eds.), pp. 15–45. London, UK: Elsevier Applied Science.

Arihara, K., Nakashima, Y., Mukai, T., Ishikawa, S., Itoh, M. (2001). Peptide inhibitors for angiotensin I-converting enzyme from enzymatic hydrolysates of porcine skeletal muscle proteins. *Meat Science*, 57(3), 319–324.

Arihara, K., Ohata, M. (2006). Functional properties of bioactive peptides derived from meat proteins. In: *Advanced Technologies for Meat Processing* (F. Toldra, Ed.), pp. 245–274. New York: Springer.

Arihara, K., Ohata, M. (2008). Bioactive compounds in meat. In: *Meat Biotechnology* (F. Toldrá, Ed.), pp. 231-249. New York: Springer.

Bah, C.S., Bekhit, A.E.-D.A., Carne, A., McConnell, M.A. (2013). Slaughterhouse blood: An emerging source of bioactive compounds. *Comprehensive Reviews in Food Science and Food Safety*, 12, 314–331. https://doi.org/10.1111/1541-4337.12013

Bamdad, F., Bark, S., Kwon, C. H., Suh, J. W., Sunwoo, H. (2017). Anti-inflammatory and antioxidant properties of peptides released from β-lactoglobulin by high hydrostatic pressure-assisted enzymatic hydrolysis. *Molecules*, 22(6), 949.

Bauchart, C., Morzel, M., Chambon, C., Mirand, P. P., Reynès, C., Buffère, C., Rémond, D. (2007). Peptides reproducibly released by *in vivo* digestion of beef meat and trout flesh in pigs. *British Journal of Nutrition*, 98(6), 1187–1195.

Bauchart, C., Remond, D., Chambon, C., Mirand, P. P., Savary-Auzeloux, I., Reynes, C., Morzel, M. (2006). Small peptides (<5 kDa) found in ready-to-eat beef meat. *Meat Science*, 74(4), 658–666.

Bechaux, J., Gatellier, P., Le Page, J. F., Drillet, Y., Sante-Lhoutellier, V. (2019). A comprehensive review of bioactive peptides obtained from animal by products and their applications. *Food and Function*, 10(10), 6244–6266.

Bhakta, G., Lim, Z. X. H., Rai, B., Lin, T., Hui, J. H., Prestwich, G. D., Van Wijnen, A. J., Nurcombe, V., Cool, S. M. (2013). The influence of collagen and hyaluronan matrices on the delivery and bioactivity of bone morphogenetic protein-2 and ectopic bone formation. *Acta Biomaterialia*, 9(11), 9098–9106.

Bhat, Z. F., Kumar, S., Bhat, H. F. (2015). Bioactive peptides of animal origin: A review. *Journal of Food Science and Technology*, 52(9), 5377–5392.

Biesalski, H. K. (2005). Meat as a component of a healthy diet-are there any risks or benefits if meat is avoided in the diet. *Meat Science*, 70(3), 509–524.

Biron, P., Koiw, E., Nowaczynski, W., Brouillet, J., Genest, J. (1961). The effects of intravenous infusions of valine-5 angiotensin II and other pressor agents on urinary electrolytes and corticosteroids, including aldosterone. *Journal of Clinical Investigation*, 40(2), 338–347.

Borrajo, P., Pateiro, M., Barba, F. J., Mora, L., Franco, D., Toldrá, F., Lorenzo, J. M. (2019). Antioxidant and antimicrobial activity of peptides extracted from meat by-products: A review. *Food Analytical Methods*, 12(11), 2401–2415.

Brown, N. J., Vaughan, D. E. (1998). Angiotensin-converting enzyme inhibitors. *Circulation*, 97(14), 1411–1420.

Catiau, L., Traisnel, J., Delval-Dubois, V., Chihib, N. E., Guillochon, D., Nedjar-Arroume, N. (2011). Minimal antimicrobial peptidic sequence from hemoglobin alpha-chain: KYR. *Peptides*, 32(4), 633–638.

Chakrabarti, S., Jahandideh, F., Wu, J. (2014). Food-derived bioactive peptides on inflammation and oxidative stress. *BioMed Research International*. doi:10.1155/2014/608979. https://doi.org/10.1155/2014/608979

Chan, W. (2004). Macronutrients in meat. In: *Encyclopedia of Meat Sciences* (W. K. Jensen, C. Devine, M. Dikeman, Eds.), pp. 614–618. Oxford: Elsevier.

Chang, C. Y., Wu, K. C., Chiang, S. H. (2007). Antioxidant properties and protein compositions of porcine hemoglobin hydrolysates. *Food Chemistry*, 100(4), 15371543.

Chauhan, V., Kanwar, S. S. (2020). Chapter 4. Bioactive peptides: Synthesis, functions and biotechnological applications. In: *Biotechnological Production of Bioactive Compounds* (Madan L. Verma, Anuj K. Chandel, Eds.), pp. 107–137. Amsterdam: Elsevier. ISBN 9780444643230.

Christensen, L., Ertbjerg, P., Løje, H., Risbo, J., Vanden Berg, F. W., Christensen, M. (2013). Relationship between meat toughness and properties of connective tissue from cows and young bulls heat treated at low temperatures for prolonged times. *Meat Science*, 93(4), 787–795.

Cinq-Mars, C. D., Hu, C., Kitts, D. D., Li-Chan, E. C. Y. (2008). Investigations into inhibitor type and mode, simulated gastrointestinal digestion, and cell transport of the angiotensin I-converting enzyme-inhibitory peptides in pacific hake (*Merluccius productus*) fillet hydrolysate. *Journal of Agricultural and Food Chemistry*, 56(2), 410–419.

Conde, E., Reinoso, B. D., González-Muñoz, M. J., Moure, A., Domínguez, H., Parajó, J. C. (2013). Recovery and concentration of antioxidants from industrial effluents and from processing streams of underutilized vegetal biomass. *Food and Public Health*, 3, 69–91.

Dagan, A., Efron, L., Gaidukov, L., Mor, A., Ginsburg, H. (2002). In-vitro antiplasmodium effects of dermaseptin S4 derivatives. *Antimicrobial Agents and Chemotherapy*, 46(4), 1059–1066.

Dave, L. A., Montoya, C. A., Rutherfurd, S. M., Moughan, P. J. (2014). Gastrointestinal endogenous proteins as a source of bioactive peptides - An In-silico study. *PLOS ONE*, 9(6), e98922.

Di Bernardini, R., Harnedy, P., Bolton, D., Kerry, J., O'Neill, E., Mullen, A. M., Hayes, M. (2011). Antioxidant and antimicrobial peptidic hydrolysates from muscle protein sources and by-products. *Food Chemistry*, 124(4), 1296–1307.

Dia, V. P., de Mejia, E. G. (2013). Mode of administration affected the capability of soybean-derived peptide lunasin to prevent metastasis of human colon cancer cells in a mouse model. *The FASEB Journal*, 27(1), 863.13-863.13. https://doi.org/10.1096/fasebj.27.1_supplement.863.13

Dziuba, B., Dziuba, M. (2014). Milk proteins-derived bioactive peptides in dairy products: Molecular, biological and methodological aspects. *Acta Scientiarum Polonorum. Technologia Alimentaria*, 13(1), 5–25.

Escudero, E., Toldrá, F., Sentandreu, M. A., Nishimura, H., Arihara, K. (2012). Antihypertensive activity of peptides derived from the in-vitro gastrointestinal digestion of pork meat. *Meat Science*, 91, 306–311.

Ferreira, S. H., Bartelt, D. C., Greene, L. J. (1970). Isolation of bradykinin-potentiating peptides from *Bothrops jararaca* venom. *Biochemistry*, 9(13), 2583–2593.

Field, R. A. (1988). Mechanically separated meat, poultry and fish. In: *Edible Meat by Products* (A. M. Pearson, T. R. Dutson, Eds.), pp. 83–126. London, UK: Elsevier Applied Science.

Folkow, B., Johansson, B., Mellander, S. (1961). The comparative effects of angiotensin and noradrenaline on consecutive vascular sections. *Acta Physiologica Scandinavica*, 53, 99–104.

Fontoura, R., Daroit, D. J., Correa, A. P., Meira, S. M., Mosquera, M., Brandelli, A. (2014). Production of feather hydrolysates with antioxidant, angiotensin-I converting enzyme-and dipeptidyl peptidase-IV inhibitory activities. *New Biotechnology*, 31(5), 506–513.

Franchini, M., Mannucci, P. M. (2009). A new era for anticoagulants. *European Journal of Internal Medicine*, 20(6), 562–568.

Friedman, M. (1996). Nutritional value of proteins from different food sources: A review. *Journal of Agricultural and Food Chemistry*, 44(1), 6–29.

Fu, Y., Jette, F. Y., Therkildsen, M. (2017). Bioactive peptides in beef: Endogenous generation through postmortem aging. *Meat Science*, 123, 134–142.

Gómez-Guillén, M. C., Giménez, B., López-Caballero, M. E., Montero, M. P. (2011). Functional and bioactive properties of collagen and gelatin from alternative sources: A review. *Food Hydrocolloids*, 25(8), 1813–1827.

Gondi, M., Basha, S. A., Bhaskar, J. J., Salimath, P. V., Rao, U. J. (2015). Antidiabetic effect of dietary mango (*Mangifera indica* L.) peel in streptozotocin-induced diabetic rats. *Journal of the Science of Food and Agriculture*, 95(5), 991–999.

Hammes, W. P., Haller, D., Ganzle, M. G. (2003). Fermented meat. In *Handbook of Fermented Functional Foods* (E. R. Farnworth, Ed.), pp. 251–269. New York: CRC.

Hayes, M., Ross, R. P., Fitzgerald, G. F., Stanton, C. (2007a). Putting microbes to work: Dairy fermentation, cell factories and bioactive peptides. Part I: Overview. *Journal of Biotechnology*, 2(4), 426–434.

Hayes, M., Stanton, C., Slattery, H., O'Sullivan, O., Hill, C., Fitzgerald, G. F., Ross, R. P. (2007b). Casein fermentate of *Lactobacillus animalis* DPC6134 contains a range of novel pro-peptide angiotensin-converting enzyme inhibitors. *Applied and Environmental Microbiology*, 73(14), 4658–4667.

Hoffman, L. C., Cawthorn, D. (2013). Exotic protein sources to meet all needs. *Meat Science*, 95(4), 764–771.

Hsieh, Y. H., Ofori, J. A. (2011). Food-grade proteins from animal by-products. Their usage and detection methods. In: *Handbook of Analysis of Edible Animal By-Products* (L. M. L. Nollet, F. Toldrá, Eds.), pp. 3–11. Boca Raton, FL: CRC Press.

Jang, A., Jo, C., Kang, K. S., Lee, M. (2008). Antimicrobial and human cancer cell cytotoxic effect of synthetic angiotensin-converting enzyme (ACE) inhibitory peptides. *Food Chemistry*, 107(1), 327–336.

Je, J. Y., Qian, Z. J., Byun, H. G., Kim, S. K. (2007). Purification and characterization of an antioxidant peptide obtained from tuna backbone protein by enzymatic hydrolysis. *Process Biochemistry*, 42(5), 840–846.

Jolles, P. (1975). Structural aspects of milk-clotting process – Comparative features with blood-clotting process. *Molecular and Cellular Biochemistry*, 7(2), 73–85.

Katayama, K., Anggraeni, H. E., Mori, T., Ahhmed, A. A., Kawahara, S., Sugiyama, M., Nakayama, T., Maruyama, M., Mugurumat, M. (2008). Porcine skeletal muscle troponin is a good source of peptides with angiotensin-I converting enzyme inhibitory activity and antihypertensive effects in spontaneously hypertensive rats. *Journal of Agricultural and Food Chemistry*, 56(2), 355–360.

Katayama, K., Mori, T., Kawahara, S., Miake, K., Kodama, Y., Sugiyama, M., Kawamura, Y., Nakayama, T., Maruyama, M., Muguruma, M. (2007). Angiotensin-I converting enzyme inhibitory peptide derived from porcine skeletal muscle myosin and its antihypertensive activity in spontaneously hypertensive rats. *Journal of Food Science*, 72(9), S702–S706.

Katayama, K., Tomatsu, M., Kawahara, S., Yamauchi, K., Fuchu, H., Kodama, Y., Kawamura, Y., Muguruma, M. (2004). Inhibitory profile of nonapeptide derived from porcine troponin C against angiotensin I-converting enzyme. *Journal of Agricultural and Food Chemistry*, 52(4), 771–775.

Kim, E. K., Lee, S. J., Jeon, B. T., Moon, S. H., Kim, B., Park, T. K., Han, J. S., Park, P. J. (2009). Purification and characterisation of antioxidative peptides from enzymatic hydrolysates of venison protein. *Food Chemistry*, 114(4), 1365–1370.

Kim, S. K., Choi, Y. R., Park, P. J., Choi, J. H., Moon, S. H. (2000). Screening of biofunctional peptides from cod processing wastes. *Journal of Korean Society of Agricultural Chemistry and Biotechnology*, 33, 198–204.

Korhonen, H. (2009). Milk-derived bioactive peptides: from science to applications. *Journal of Functional Foods*, 1(2), 177–187.

Korhonen, H., Pihlanto, A. (2003). Food-derived bioactive peptides opportunities for designing future foods. *Current Pharmaceutical Design*, 9(16), 1297–1308.

Korhonen, H., Pihlanto, A. (2006). Bioactive peptides: Production and functionality. *International Dairy Journal*, 16(9), 945–960.

Korhonen, H., Pihlanto-Leppälä, A., Rantamäki, P., Tupasela, T. (1998). Impact of processing on bioactive proteins and peptides. *Trends in Food Science and Technology*, 9(8–9), 307–319.

Kudoh, Y., Matsuda, S., Igoshi, K., Oki, T. (2001). Antioxidative peptide from milk fermented with *Lactobacillus delbrueckii* subsp. bulgaricus IFO13953. *Journal of the Japanese Society for Food Science and Technology*, 48(1), 44–50.

Lafarga, T., Aluko, R. E., Rai, D. K., O'Connor, P., Hayes, M. (2016). Identification of bioactive peptides from a papain hydrolysate of bovine serum albumin and assessment of an antihypertensive effect in spontaneously hypertensive rats. *Food Research International*, 81, 91–99.

Lafarga, T., Hayes, M. (2014). Bioactive peptides from meat muscle and by-products: Generation, functionality and application as functional ingredients. *Meat Science*, 98(2), 227–239.

Lawrie, R. A., Ledward, D. A. (2006). Lawrie's meat science. In: *Lawrie's Meat Science*, 7th ed. (R. A. Lawrie, Ed.), pp. 41–73. Cambridge, UK: Woodhead Publishing.

Lei, J., Sun, L., Huang, S., Zhu, C., Li, P., He, J., Mackey, V., Coy, D. H., He, Q. (2019). The antimicrobial peptides and their potential clinical applications. *American Journal of Translational Research*, 11(7), 3919–3931.

Leygonie, C., Britz, T. J., Hoffman, L. C. (2012). Impact of freezing and thawing on the quality of meat: Review. *Meat Science*, 91(2), 93–98.

Li, B., Chen, F., Wang, X., Ji, B., Wu, Y. (2007). Isolation and identification of antioxidative peptides from porcine collagen hydrolysate by consecutive chromatography and electrospray ionization–mass spectrometry. *Food Chemistry*, 102(4), 1135–1143.

Li, Y., Yu, J. (2014). Research progress in structure-activity relationship of bioactive peptides. *Journal of Medicinal Food*. doi:10.1089/jmf.2014.0028.

Liu, R., Wang, M., Duan, J. A., Guo, J. M., Tang, Y. P. (2010). Purification and identification of three novel antioxidant peptides from *Cornu Bubali* (Water Buffalo Horn). *Peptides*, 31(5), 786–793.

Lopez-Fandino, R., Otte, J., Van Camp, J. (2006). Physiological, chemical and technological aspects of milk-protein-derived peptides with antihypertensive and ACE inhibitory activity. *International Dairy Journal*, 16(11), 1277–1293.

Matsui, T., Matsumoto, K., Mahmud, T. H. K., Arjumand, A. (2006). Antihypertensive peptides from natural resources. In: *Advances in Phytomedicine*, Volume 2, pp. 255–271. Oxford, UK: Elsevier.

Mellander, O. (1950). The physiological importance of the casein phosphopeptide calcium salts. II. Peroral calcium dosage of infants. *Acta Societatis Medicorum Upsaliensis*, 55(5–6), 247–255.

Mils, S., Ross, R. P., Hill, C., Fitzgerald, G. F., Stanton, C. (2011). Milk intelligence: Mining milk for bioactive substances associated with human health. *International Dairy Journal*, 21(6), 377–340.

Morimatsu, F., Ito, M., Budijanto, S., Watanabe, I., Furukawa, Y., Kimura, S. (1996). Plasma cholesterol-suppressing effect of papain-hydrolyzed pork meat in rats fed hypercholesterolemic diet. *Journal of Nutrition Science and Vitaminology*, 42(2), 145–153.

Muguruma, M., Ahhmed, A. M., Katayama, K., Kawahara, S., Maruyama, M., Nakamura, T. (2009). Identification of pro-drug type ACE inhibitory peptide sourced from porcine myosin B: Evaluation of its antihypertensive effects *in vivo*. *Food Chemistry*, 114(2), 516–522.

Mulvihill, B. (2004). Micronutrients in meat. In: *Encyclopedia of Meat Sciences* (W. K. Jensen, C. Devine, M. Dikeman, Eds.), pp. 618–623. Oxford: Elsevier.

Nakashima, Y., Arihara, K., Sasaki, A., Mio, H., Ishikawa, S., Itoh, M. (2002). Antihypertensive activities of peptides derived from porcine skeletal muscle myosin in spontaneously hypertensive rats. *Journal of Food Science*, 67(1), 434–437.

Nasri, R., Abdelhedi, O., Jemil, I., Daoued, I., Hamden, K., Kallel, C., Elfeki, A., Lamri-Senhadji, M., Boualga, A., Nasri, M., Karra-Châabouni, M. (2015). Ameliorating effects of goby fish protein hydrolysates on high-fat-high-fructose diet-induced hyperglycemia; oxidative stress and deterioration of kidney function in rats. *Chemico-Biological Interactions*, 242, 71–280.

Ondetti, M. A., Rubin, B., Cushman, D. W. (1977). Design of specific inhibitors of angiotensin-converting enzyme: New class of orally active antihypertensive agents. *Science*, 196(4288), 441–444.

Padfield, P. L., Morton, J. J. (1977). Effects of angiotensin II on arginine-vasopressin in physiological and pathological situations in man. *Journal of Endocrinology*, 74(2), 251–259.

Paraskevas, K. I., Nicolaides, A. N., Mikhailidis, D. P. (2013). Statins and venous thromboembolism: The jury is still out. *Angiology*, 64(7), 489–491.

Pellegrini, A. (2003). Antimicrobial peptides from food proteins. *Current Pharmaceutical Design*, 9(16), 1225–1238.

Pereira, P., Vicente, A. (2013). Meat nutritional composition and nutritive role in the human diet. *Meat Science*, 93(3), 586–592.

Pham-Huy, L. A., He, H., Pham-Huy, C. (2008). Free radicals, antioxidants in disease and health. *International Journal of Biomedical Sciences*, 4(2), 89–96.

Pihlanto, A., Akkanen, S., Korhonen, H. J. (2008). ACE inhibitory and antioxidant properties of potato (*Solanum tuberosum*). *Food Chemistry*, 109(1), 104–112.

Pihlanto, A., Korhonen, H. (2003). Bioactive peptides and proteins. *Advances in Food and Nutrition Research*, 47, 175–276.

Pihlanto, A., Virtanen, T., Korhonen, H. (2010). Angiotensin I converting enzyme (ACE) inhibitory activity and antihypertensive effect of fermented milk. *International Dairy Journal*, 20(1), 3–10.

Proud, D., Kaplan, A. P. (1988). Kinin formation: Mechanisms and role in inflammatory disorders. *Annual Review of Immunology*, 6, 49–83.

Przybylski, R., Firdaous, L., Châtaigné, G., Dhulster, P., Nedjar, N. (2016). Production of an antimicrobial peptide derived from slaughterhouse by-product and its potential application on meat as preservative. *Food Chemistry*, 211, 306–313.

Qian, Z. J., Jung, W. K., Lee, S. H., Byun, H. G., Kim, S. K. (2007). Antihypertensive effect of an angiotensin I-converting enzyme inhibitory peptide from bullfrog (*Rana catesbeiana* Shaw) muscle protein in spontaneously hypertensive rats. *Process Biochemistry*, 42(10), 1443–1448.

Qu, W., Ma, H., Jia, J., He, R., Luo, L., Pan, Z. (2012). Enzymolysis kinetics and activities of ACE inhibitory peptides from wheat germ protein prepared with SFP ultrasound-assisted processing. *Ultrasonics Sonochemistry*, 19(5), 1021–1026.

Ramadhan, A. H., Nawas, T., Zhang, X., Pembe, W. M., Xia, W., Xu, Y. (2017). Purification and identification of a novel antidiabetic peptide from Chinese giant salamander (*Andrias davidianus*) protein hydrolysate against α-amylase and α-glucosidase. *International Journal of Food Properties*, 20(sup3), S3360–S3372.

Ravallec, P. R., Charlot, C., Pires, C., Braga, V., Batista, I., Wormhoudt, A. V., Gal, Y. L., Fouchereau-Peron, M. (2001). The presence of bioactive peptides in hydrolysates prepared from processing waste of sardine (*Sardina pilchardus*). *Journal of the Science of Food and Agriculture*, 81(11), 1120–1125.

Regulation (EC) No 835/2004 of the European Parliament and of the Council of 29 April 2004. Laying down specific hygiene rules for on the hygiene of foodstuffs [2004] OJEU L139/55.

Rojas-Ronquillo, R., Cruz-Guerrero, A., Flores-Nájera, A., Rodríguez-Serrano, G., GómezRuiz, L., Reyes-Grajeda, J. P., Jiménez-Guzmán, J., García-Garibay, M. (2012). Antithrombotic and angiotensin-converting enzyme inhibitory properties of peptides released from bovine casein by *Lactobacillus casei* Shirota. *International Dairy Journal*, 26(2), 147–154.

Rondeau, P., Bourdon, E. (2011). The glycation of albumin: Structural and functional impacts. *Biochimie*, 93(4), 645–658.

Rutherfurd-Markwick, K. J., Moughan, P. J. (2005). Bioactive peptides derived from food. *Journal of AOAC International*, 88(3), 955–966.

Ryan, J. T., Ross, R. P., Bolton, D., Fitzgerald, G. F., Stanton, C. (2011). Bioactive peptides from muscle sources: Meat and fish. *Nutrients*, 3(9), 765–791.

Saiga, A., Iwai, K., Hayakawa, T., Takahata, Y., Kitamura, S., Nishimura, T., Morimatsu, F. (2008). Angiotensin I-converting enzyme-inhibitory peptides obtained from chicken collagen hydrolysate. *Journal of Agricultural and Food Chemistry*, 56(20), 9586–9591.

Saiga, A., Tanabe, S., Nishimura, T. (2003). Antioxidant activity of peptides obtained from porcine myofibrillar proteins by protease treatment. *Journal of Agriculture and Food Chemistry*, 51(12), 3661–3667.

Sentandreu, M. A., Coulis, G., Ouali, A. (2002). Role of muscle endopeptidases and their inhibitors in meat tenderness. *Trends in Food Science and Technology*, 13(12), 400–421.

Sewald, N., Jakubke, H. (2002). *Peptides: Chemistry and Biology*. Weinheim: Wiley-VCH Verlag GmbH and Co. KGaA, 543.

Shalaby, S. M., Zakora, M., Otte, J. (2006). Performance of two commonly used angiotensin-converting enzyme inhibition assays using FA-PGG and HHL as substrates. *Journal of Dairy Research*, 73(2), 178–186.

Sharma, S., Singh, R., Rana, S. (2011). Bioactive peptides: A review. *International Journal Bioautomation*, 15(4), 223–250.

Singh, B. P., Vij, S., Hati, S. (2014). Functional significance of bioactive peptides derived from soybean. *Peptides*, 54, 171–179.

Šližytė, R., Mozuraitytė, R., Martínez-Alvarez, O., Falch, E., Fouchereau-Peron, M., Rustad, T. (2009). Functional, bioactive and antioxidative properties of hydrolysates obtained from cod (*Gadus morhua*) backbones. *Process Biochemistry*, 44(6), 668–677.

Spudich, J. A., Watt, S. (1971). The regulation of rabbit skeletal muscle contraction. I. Biochemical studies of the interaction of the tropomyosin-troponin complex with actin and the proteolytic fragments of myosin. *Journal of Biological Chemistry*, 246(15), 4866–4871.

Stuknyte, M., De Noni, I., Gugliemetti, S., Minuzzo, M., Mora, D. (2011). Potential immunomodulatory activity of bovine casein hydrolysates produced after digestion with proteinase of lactic acid bacteria. *International Dairy Journal*, 21(10), 763–769.

Suarez-Jimenez, G. M., Burgos-Hernandez, A., Ezquerra-Brauer, J. M. (2012). Bioactive peptides and depsipeptides with anticancer potential: Sources from marine animals. *Marine Drugs*, 10(5), 963–986.

Tierney, M. S., Croft, A. K., Hayes, M. (2010). A review of antihypertensive and antioxidant activities in macroalgae. *Botanica Marina*, 53(5), 387–408.

Toldrá, F., Aristoy, M. C., Mora, L., Reig, M. (2012). Innovations in value-addition of edible meat by-products. *Meat Science*, 92(3), 290–296.

Udenigwe, C. C., Aluko, R. E. (2011). Food protein-derived bioactive peptides: Production, processing, and potential health benefits. *Journal of Food Science*, 77: R11–R24. doi:10.1111/j.1750-3841.2011.02455.

Ulug, S. K., Jahandideh, F., Wu, J. (2021). Novel technologies for the production of bioactive peptides. *Trends in Food Science and Technology*, 108, 27–39.

Urista, M. C., Fernández, Á. R., Rodriguez, F. R., Cuenca, A. A., Jurado, T. A. (2011). Review: Production and functionality of active peptides from milk. *Food Science and Technology International*, 17(4), 293–317.

Vercruysse, L., Van Camp, J., Smagghe, G. (2005). ACE inhibitory peptides derived from enzymatic hydrolysates of animal muscle protein: A review. *Journal of Agricultural and Food Chemistry*, 53(21), 8106–8115.

Vermeirssen, V., Camp, J. V., Verstraete, W. (2007). Bioavailability of angiotensin I converting enzyme inhibitory peptides. *British Journal of Nutrition*, 92, 357–366.

Vermeirssen, V., Van Camp, J., Verstraete, W. (2002). Optimisation and validation of an angiotensin-converting enzyme inhibition assay for the screening of bioactive peptides. *Journal of Biochemical and Biophysical Methods*, 51(1), 75–87.

Wang, X., Yu, H., Xing, R., Li, P. (2017). Characterization, preparation, and purification of marine bioactive peptides. *BioMed Research International*, 2017, 1–16.

Watanabe-Kamiyama, M. I., Shimizu, M., Kamiyama, S., Taguchi, Y., Sone, H., Morimatsu, F., Shirakawa, H., Furukawa, Y., Komai, M. (2010). Absorption and effectiveness of orally administered low molecular weight collagen hydrolysate in rats. *Journal of Agricultural and Food Chemistry*, 27(58(2)), 835–841.

Wisuthiphaet, N., Klinchan, S., Kongruang, S. (2016). Fish protein hydrolysate production by acid and enzymatic hydrolysis. King Mongkut's University of Technology. *North Bangkok International Journal of Applied Science and Technology*, 9, 261–270.

Xiong, Y. L. (2010). Antioxidant peptides. In: *Bioactive Proteins and Peptides as Functional Foods and Nutraceuticals* (Y. Mine, E. Li-Chan, B. Jiang, Eds.), pp. 29–42. Ames, IA: Wiley-Blackwell.

Yu, H. C., Hsu, J. L., Chang, C. I., Tan, F. J. (2017). Antioxidant properties of porcine liver proteins hydrolyzed using *Monascus purpureus*. *Food Science and Biotechnology*, 26(5), 1217–1225.

Yu, Y., Hu, J., Miyaguchi, Y., Bai, X., Du, Y., Lin, B. (2006). Isolation and characterization of angiotensin I-converting enzyme inhibitory peptides derived from porcine hemoglobin. *Peptides*, 27(11), 2950–2956.

Zambrowicz, A., Timmer, M., Eckert, E., Trziszka, T. (2013). Evaluation of the ACE inhibitory activity of egg-white proteins degraded with pepsin. *Polish Journal of Food and Nutrition Sciences*, 63(2), 103–108.

Zhang, Y., Kouguchi, T., Shimizu, K., Sato, M., Takahata, Y., Morimatsu, F. (2010). Chicken collagen hydrolysate reduces proinflammatory cytokine production in C57BL/6.KOR-ApoEshl mice. *Journal of Nutritional Science and Vitaminology*, 56(3), 208–210.

CHAPTER 4

Dairy Products

Fernanda Gobbi Amorim, Larissa Zambom Côco, Francielle Almeida Cordeiro, Bianca Prandi Campagnaro, and Rafaela Aires

CONTENTS

Introduction	75
Milk	76
Yogurt	79
Cheese	85
Kefir	86
References	87

INTRODUCTION

Since prehistoric times, milk has been considered a highly nutritious, but perishable, food. Over time, the emergence of different methods for preserving milk allowed for increased consumption of this healthy food (Doyle and Beuchat, 2007). Milk represents a secretion from different mammals and is a very flexible raw material for the processing of a wide range of dairy products. This processing can be done through changes of temperature (i.e., pasteurization), fermentation by microorganisms, or separation processes (i.e., coagulation) which may impact the taste and textures of dairy products. Therefore, each type of dairy food has its own properties. The production of dairy food also leads to differences in microflora composition and the generation of bioactive molecules which make these types of food very nutritious (Doyle and Beuchat, 2007; Nollet and Toldrá, 2009).

Due to the extensive nutritional value of dairy foods, the interest in studying their properties and applications has grown over the years in the scientific field. A search in the periodic database (PubMed) with the keyword "dairy" shows that more than 84,000 scientific papers have been published. The first study was published in 1886 and over just the last 10 years, the papers related to this subject represent almost 42,000. In addition, dairies have been explored in different clinical trials (according to www.clinicaltrials.gov) which comprises around 870 records between the interventional and observational clinical studies.

The increased insertion of dairy products in the human food routine and greater scientific interest can be explained by the rich source of bioactive molecules present in them. Bioactive molecules are any components present in food, animals, or plants that influence another organism (Grajek, Olejnik, and Sip, 2005; Fernandes, Coelho, and de las Mercedes Salas-Mellado, 2019). In this way, nature has been a source of bioactive molecules for millennia and extremely useful for the design and development of new drugs (Cragg and Newman, 2013). The main property which makes dairy products a subject of interest is the presence of potential bioactive peptides in them. Dairy-derived bioactive peptides are released as a result of the following processes: gastrointestinal digestion, food processing, or enzymatic and bacterial fermentation. These processes produce diverse beneficial effects such as antihypertensive, anti-inflammatory, immunomodulating, antimicrobial, antidiabetic, anticancer, antioxidant, and antithrombotic (Marcone, Belton, and Fitzgerald, 2017; Amorim et al., 2019).

Many peptides derived from foods are biologically active encrypted fragments within the primary sequence of mature proteins. Digestion of proteins *in vivo* or *in vitro* produces free amino acids and peptides that enter the bloodstream and exert their systemic effects. Bioactive peptides can be produced *in vivo*, for example, by gastrointestinal digestion, and *in vitro* by chemical processes such as heating, protein hydrolysis, or fermentation (Dhaval, Yadav, and Purwar, 2016). Proteolysis products can generate peptides or proteins with biological actions different from their precursor, which explains why they are called "cryptic peptides" or "crypteins". The latter term appeared in 2006, and the process of producing crypteins has been observed in endocrine signaling, the extracellular matrix, the complement cascade, as well as milk (Autelitano et al., 2006). Some varieties of proteases may be involved in the natural production of crypteins that appear to play a diverse role in modulating processes such as angiogenesis, immune function, and cell growth (Autelitano et al., 2006). The generation of these crypteins represents an important mechanism for increasing the functional diversity of proteins and provides new possibilities for bioactive molecules with potential therapeutic actions and/or biotechnological tools (Autelitano et al., 2006).

Therefore, this chapter will focus on the bioactive peptides found in the most important dairy products and the milk itself. It is expected that the updates provided herein concerning each dairy food will emphasize the importance of studies, as well as their consumption.

Milk

Milk is an essential food for the survival, development, and health of mammals and is composed of numerous components such as lactose, lipids, proteins, vitamins, minerals, oligosaccharides, intrinsic immunological factors, immunoglobulins, enzymes, growth factors, and bioactive compounds (Punia et al., 2020; Bhattacharya et al., 2019). Milk production and consumption have been increasing considerably, with bovine milk being the most consumed followed by other kinds of milk, such as buffalo, goat, sheep, and camel milk (Gerosa and Skoet, 2012; Khan et al., 2018).

Milk has gained prominence because it is a main source of peptides, which have important biological activities, such as antioxidants, cardiovascular, neural, immune, and gastrointestinal (Kulinich and Liu, 2016; Bhattacharya et al., 2019) (Figure 4.1).

The encrypted peptides are short sequences of amino acids present in proteins and are inactive until they are hydrolyzed by enzymes during food digestion, microbial fermentation, or through the processing of dairy products (Nagpal et al., 2011; Baum et al., 2013; Capriotti et al., 2016; Rai, Sanjukta, and Jeyaram, 2017). The main source of production of these bioactive peptides are the proteins present in milk (Punia et al., 2020). There are three main groups of milk proteins that participate in the formation of peptides:

1. Casein (CN): composed of αs-casein (αs1 and αs2), β-casein, and κ-casein;
2. Whey proteins: composed of β-lactoglobulin, α-lactalbumin, lactoferrin, immunoglobulins, serum albumin, glycomacropeptides, enzymes, and growth factor;

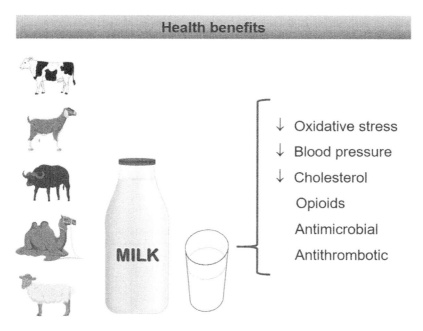

FIGURE 4.1 Health benefits of milk consumption (Created by smart.servier.com and pixabay.com).

3. Proteins of milk fat globule membrane (MFGM): mainly composed of mucin-1 and xanthine dehydrogenase/oxidase (Vargas-Bello-Pérez, Márquez-Hernández, and Hernández-Castellano, 2019).

The proteins in the milk of different mammals are practically the same, however, the quantities of the protein components may vary, as is observed in Figure 4.2.

The search for dairy peptides, including milk, and their benefits has been growing enormously in recent years (Park and Nam, 2015; Claeys et al., 2014). A search in a public repository such as PubMed with the keyword of "bioactive peptides from milk" retrieves a total of 844 articles published in the last 10 years (2010–2020). After a refined search for peptides according to their animal sources, the most studied mammals are bovine (207), followed by goats (37), buffalo (20), camels (19), and sheep (16).

Bioactive peptides have many biological activities, such as antioxidant, angiotensin-converting enzyme (ACE) inhibitory, opioids, hypocholesterolemic, antithrombotic, and antimicrobial. The main peptides studied are represented in Table 4.1 according to their main sources, sequences, and activities. Milk-derived peptides that exhibit antioxidant activity typically contain 5 to 11 amino acids, including hydrophobic amino acids, such as proline, histidine, and tryptophan (Sah et al., 2018). Casein-derived peptides seem to have an antioxidant activity by scavenging free radicals from the radical's superoxide anion and hydroxyl radical and through the inhibition of enzymatic and nonenzymatic lipid peroxidation (Rival et al., 2001; Suetsuna, Ukeda, and Ochi, 2000).

In 2018, Tonolo et al. described the *in vivo* antioxidant activity of four synthetic peptides derived from bovine β-casein and k-casein. All the milk bioactive peptides studied were able to preserve cell viability against induced oxidative stress in Caco-2 cells. These findings indicate that they might have a role in the reduction of oxidative stress

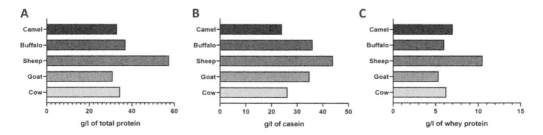

FIGURE 4.2 Distribution of total proteins (A), casein proteins (B), and whey proteins (C) presented according the ruminant sources (Vargas-Bello-Pérez, Márquez-Hernández, and Hernández-Castellano, 2019; Claeys et al., 2014).

TABLE 4.1 Origin and Amino Acid Sequences of Biologically Functional Peptides from Milk

Protein Sources	Peptide Sequences	Biological Activities	References
Casein	TPTPGL	Antioxidant	(Suetsuna, Ukeda, and Ochi, 2000)
Casein	VLGAMAPL	Antioxidant	(Rival et al., 2001)
Casein	AVPYPQR, KVLPVPEK, and ARHPHPHLSFM	Antioxidant	(Tonolo et al., 2018)
Casein	YQKFPQYLQY	Angiotensin-converting enzyme inhibitory	(Xue et al., 2018)
Casein	KDQDK, QVTSTEV, and TAQVTSTEV	Antithrombotic	(Qian et al., 1995)
Casein	VPP and IPP	Reduction of blood pressure	(Ishida et al., 2011)
Casein	FFVAPFEVFGK	Reduction of blood pressure	(Townsend et al., 2004)
Casein and whey protein	TPPF	Opioids	(Capriotti et al., 2016)
Whey protein	IIAGL	Hypo-cholesterolemic	(Nagaoka et al., 2001)
Whey protein	KRDS	Antimicrobial	(Ramesh, Kamini, and Puvanakrishnan, 2004)

by maintaining sulfhydryl groups that exert effects against the decrease of glutathione (Tonolo et al., 2018).

In addition to the antioxidant effects, peptides derived from bovine milk can contribute to the reduction of blood pressure by ACE inhibition. Ishida and Shibata (2011) performed a double-blind, placebo-controlled, and randomized clinical trial with 48 subjects in order to evaluate the effects of ingesting an excess of tablets containing casein hydrolysates, incorporating ACE-inhibitory peptides such as Val-Pro-Pro (VPP) and Ile-Pro-Pro (IPP), in subjects with blood pressure ranging from normal to mild hypertension. The authors found a significant decrease in the systolic blood pressure in the active group compared with the placebo group. These findings highlight a potential application of these bioactive peptides as an antihypertensive agent without side effects (Ishida et al. 2011).

Another biological effect that bioactive peptides from milk present is antithrombotic activity (Zambrowicz et al., 2013). This was observed in the casoplatelins, peptides derived from κ-Casein. Iwaniak and Minkiewicz (2007) described the effects of casopiastrin, a casoplatelin (fragment 106–116) released after the proteolytic action of trypsin. This peptide showed an antithrombotic activity by inhibition of the fibrinogen binding process whereas another peptide (fragment 103–111) of this protein inhibited platelet aggregation (Iwaniak and Minkiewicz, 2007).

The opioid activity attributed to some peptides derived from casein and whey proteins is determined through the presence of tyrosine residues in the N-terminal position. Some authors describe that the presence of this particular residue act as opioid ligands with agonistic or antagonistic activities (Chrzanowska, 1998; Zambrowicz et al., 2013; Capriotti et al., 2016). Normally, in addition to the presence of tyrosine in the N-terminus, these peptides have other aromatic amino acid residues (Phe or Tyr) in the 3rd or 4th position of their primary sequences (Park and Nam, 2015).

Peptides derived from lactoferrin, a whey protein, have been described with potential antimicrobial activity (Ramesh, Kamini, and Puvanakrishnan, 2004). The most studied peptides from this group are lactoferricin (Lfc) and lactoferrampin (Lfa), located in the N(1) domain of bovine lactoferrin. These peptides present cationic and hydrophobic characteristics which may be important to their interactions in the microorganisms' cell membranes (Chrzanowska, 1998; Ramesh, Kamini, and Puvanakrishnan, 2004). Biasibetti et al. (2020) showed that Lfc alone inhibited *Staphylococcus intermedius* and *Malassezia pachydermatis*, while Lfa acted against *Candida albicans*. The combination of activities of bovine Lfc and Lfa is shown to have a synergistic effect on antimicrobial activity *in vivo* which may be through the aggregation of peptides in the cytoplasmic membranes and the permeabilization of the bacterial membrane, which culminates in cell death (McCann et al., 2005; Ramesh, Kamini, and Puvanakrishnan, 2004; Biasibetti et al., 2020).

These bioactive peptides derived from ruminant milk have shown a broad range of benefits for their consumers, making them potential health-promoting substances. In addition, these peptides open new perspectives for their application in functional foods or may be helpful in the prototype design of new drugs in the pharmaceutical industry.

Yogurt

Yogurt is a popular dairy food full of vitamins, minerals, and proteins with probiotic properties which provide health benefits (Morelli, 2014). The earliest known evidence of yogurt consumption goes back to the eleventh century in the Middle East (Gurakan and Altay, 2010). Currently, the consumption of this fermented form of milk is high due to its nutritional and probiotic properties as well as the commercial options in flavor, viscosity, and percentage of fat and sugar, among others (Chandan, 2006). In the last 20 years, evidence from clinical and randomized controlled trials (Figure 4.3) show the health benefits of yogurt, such as cardio-metabolic protection (Ramchandran and Shah, 2011; Cormier et al., 2016).

Fermented dairy foods have been associated with multiple health benefits, particularly linked to probiotic strains and bioactives produced by the metabolic activity of milk components (Manzanarez-Quín et al., 2021). Probiotics are preparations containing live

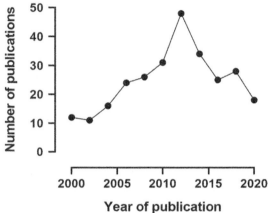

FIGURE 4.3 The time course of published clinical and randomized controlled trials related to the research field "yogurt" in the last two decades (PubMed Platform https://www.ncbi.nlm.nih.gov/pubmed).

microorganisms, or their components, that can confer health benefits to the host (Lopitz-Otsoa et al., 2006). Although probiotics are not limited to lactic acid bacteria of dairy origin, the term "dairy probiotics" refers to the majority of probiotics that can actively ferment milk or be viable in fermented milk in high numbers (Zoumpopoulou et al., 2018; Manzanarez-Quín et al., 2021).

Health effects have been associated with probiotic properties and bioactive peptides that are released during milk fermentation into the gastrointestinal tract (Fernandez et al., 2017). Table 4.2 summarizes the health benefits of probiotic yogurt evidenced in clinical and randomized controlled trials according to their type of disease in which they were applied, the probiotic composition, and the observed effects.

During yogurt production, several factors can influence the quality and quantity of final product constituents, such as the type of milk, nutrient supplementation, temperature, and starter culture (Gurakan and Altay, 2010). The first stage of production consists of the choice of milk. For instance, yogurts prepared with goat milk have a greater amount of fat than cow milk. Traditionally, however, artisanal manufacturing and the dairy industry use cow milk (Chandan, 2006).

Before the fermentation step, the milk is sterilized of impurities and pathogenic microorganisms by centrifugal clarification and thermalization processes (Sfakianakis and Tzia, 2014). Then, the standardization of milk components is performed, consisting of the manipulation of milk constituents in accordance with the amount of fat and nutrients desired in the final product (Sfakianakis and Tzia, 2014). The milk can be supplemented by derivatives such as milk and casein powder (and isolated whey), providing a concentrated source of proteins and bioactive compounds (Gurakan and Altay, 2010). Larger amounts of fat, proteins, carbohydrates, and minerals impact the increase of density, viscosity, and consistency of the final product. In addition, in the phase pasteurization, overheating of milk (80–95°C) favors the elimination of pathogenic microorganisms that are still present as well as the inactivation of enzymes and whey protein denaturation (Gurakan and Altay, 2010).

TABLE 4.2 Health Benefits of Probiotic Yogurt

Diseases	Probiotic Compositions	Effects	References
H. pylori infection	*Bifidobacterium lactis* *Lactobacillus acidophilus* *Lactobacillus paracasei*	Eradication rate of 100%	(Srinarong et al., 2014)
Diabetes type 2	*Lactobacillus acidophilus* *Bifidobacterium lactis*	Reduction of fasting glucose; Improved total antioxidant status	(Ejtahed et al., 2012)
Metabolic syndrome	*Lactobacillus acidophilus* *Bifidobacterium lactis*	Reduction of fasting glucose; Improved serum endothelial function markers	(Rezazadeh et al., 2019)
Antibiotic-associated diarrhea	*Lactobacillus rhamnosus* *Bifidobacterium lactis* *Lactobacillus acidophilus*	Reduction of diarrhea cases	(Fox et al., 2015)
Upper respiratory tract infections (URTI)	*Lactobacillus paracasei*	Reduction of URTI events	(Pu et al., 2017)
Nonalcoholic fatty liver disease	*Lactobacillus acidophilus* *Bifidobacterium lactis*	Improved hepatic enzymes and serum total cholesterol	(Nabavi et al., 2014)

One of the most crucial phases in the manufacture of yogurt is the incubation with a starter culture, the fermentation stage. To be considered "yogurt", two starter species of acid bacteria are required: *Streptococcus salivarius ssp. thermophilus* and *Lactobacillus delbrueckii ssp. bulgaricus* (Sfakianakis and Tzia, 2014). The viscous texture and flavor result from the biochemical action of acid bacteria consuming milk sugars (lactose) as an energy source and releasing lactic acid, leading to a reduction in pH and coagulation of milk proteins in a process called "denaturation" (Sfakianakis and Tzia, 2014). These starter bacteria strains do not support enzymatic digestion in the gastrointestinal tract, thus the fortification with others strains, such as *Lactobacillus acidophilus*, *Lactobacillus casei*, and *Bifidobacterium sp.*, is performed to guarantee the probiotic source (Homayouni et al., 2012). The supplementation can be performed before the fermentation process or after the manufacturing of the final product (Gurakan and Altay, 2010). The maintenance of live microorganisms after refrigeration is supported by acidified pH, which also contributes to the prevention of pathogenic bacteria proliferation (Sfakianakis and Tzia, 2014).

Qualitatively, the yogurt presents a similar components profile to milk. However, it exhibits quantitative differences due to the production process (Wang et al., 2013). In the standardization of milk components and pasteurization phases, the evaporation of water from milk promotes an increase in the concentration of nutrients (Baspinar and Güldaş, 2021). In addition, the proteolytic activity of bacterial culture contributes to the increase of yogurt peptide components (Adolfsson, Meydani, and Russell, 2004). The milk-derived constituents may be affected by the type of mammal (cow, buffalo, goat, etc.), animal feeding, storage conditions, climate, season, and other factors. These factors, in turn, also affect the final composition of yogurt (Adolfsson, Meydani, and Russell, 2004;

Chandan, 2006). Other parameters that influence the composition of yogurt include (1) starter culture, (2) supplementation of bacteria species, (3) temperature in the fermentation phases, (4) type of milk solids added before fermentation, and (5) duration of the fermentation process (Gurakan and Altay, 2010).

A common characteristic of milk-derived products is the formation of a food matrix; however, the physical structures differ between them. The bacterial fermentation of milk to yogurt forms a semi-solid matrix (gel structure), improving viscosity, osmolality, energy density, and nutritional value (Fernandez and Marette, 2017). Figure 4.4 summarizes the nutrients of the yogurt matrix. The gel structure favors a longer gastrointestinal transit time and forms a physical barrier against degradation of nutrients and bacteria, thereby increasing the bioavailability and absorption of nutrients and probiotic action. The yogurt matrix also elicits higher nutrient–nutrient and nutrient–bacteria interactions (Fernandez et al., 2017).

In addition to matrix nutrients, yogurt has a higher concentration of bioactive peptides and free amino acids than milk (Cavalheiro et al., 2020). The proteolysis in the yogurt can occur in several ways (Fernandez et al., 2017):

a) Proteolytic enzymes from lactic acid bacteria (LAB) in the fermentation;
b) Digestive enzymes (such as trypsin and proteases) in the gastrointestinal tract;
c) Proteolytic enzymes from microorganisms in the gastrointestinal tract.

In the fermentation process, the heat treatment and acidification pH by LAB result in denaturation and coagulation of milk proteins, especially the casein (Adolfsson, Meydani, and Russell, 2004). Regarding the formation of peptides by LAB, smaller structures from proteins (oligo, di, or tripeptides) are released by proteinases. These produced fragments are hydrolyzed by endopeptidases and aminopeptidases to free amino acids and peptides, which are easily absorbed, thus promoting biological activity and physiological effects (Fernandez et al., 2017).

Concerning the biological activities of yogurt, it is expected that this dairy product shares some similarities with those effects observed in milk consumption. The bioactive

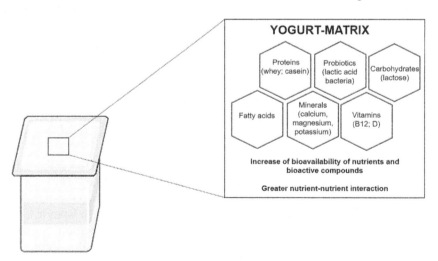

FIGURE 4.4 Components and functions of yogurt matrix (Created by smart.servier.com).

peptides from probiotic yogurt exhibit ACE-inhibitory, antioxidant, antibacterial, and immunomodulatory activities, among others shown in Table 4.3.

Most of the peptides identified by mass spectrometry analysis are antihypertensive and ACE inhibitors originating from casein (Kunda et al., 2012; Jin et al., 2016). In fact, studies in animals and humans have shown a decrease in blood pressure after yogurt consumption, which can be attributed in part to ACE inhibitors and antihypertensive peptides (Lana et al., 2018; Ramchandran and Shah, 2011). Yogurts produced from the fermentation of milk by the *Lactobacillus helveticus*, *Saccharomyces cerevisiae*, *Lactobacillus delbrueckii ssp. Bulgaricus*, and *Lactococcus lactis ssp. cremoris* strains have exhibited high production of ACE-inhibitory peptides (Takano, 1998; Gobbetti et al., 2000).

The *in vitro* ACE inhibition activity and antihypertensive effects of the peptides from yogurt have already been reported in spontaneously hypertensive rats (SHR) (Yamamoto, Maeno, and Takano, 1999b; Donkor et al., 2007; Jin et al., 2016). In SHR, Tyr-Pro (YP) from yogurt (*Lactobacillus helveticus*) reduced blood pressure by 30 mmHg (Yamamoto, Maeno, and Takano, 1999a). These reports highlight the potential use of probiotic yogurt as a coadjuvant therapy for hypertension.

Probiotic yogurt is also a rich source of peptides with antioxidant activity (Aloğlu and Öner, 2011; Kunda et al., 2012). The antioxidant effects of peptide fractions from probiotic yogurt have been confirmed by *in vitro* assays, such as DPPH, ABTS, and Fe^{2+} chelation (Farvin et al., 2010). These findings help to clarify the antioxidant effects associated with yogurt consumption. In the cytoprotective and genotoxicity/DNA fragmentation assays, yogurt showed protective effects against oxidative damage (Yamamoto et al., 2019). Yogurt consumption reduced oxidative stress in patients with type 2 diabetes mellitus and an experimental model of colitis (Ejtahed et al., 2012; Yoon et al., 2019).

Several strategies have been investigated to guarantee bacterial survival and high bioavailability of bioactive peptides after fermentation. The bioavailability of peptides can be affected by several factors, such as (1) the fermentation process, (2) gastrointestinal digestion, (3) proteolytic activity of bacterial strain, and (4) yogurt storage (Donkor et al., 2007; Papadimitriou et al., 2007; Jin et al., 2016).

The profile of bioactive peptides from yogurt changes during gastrointestinal digestion (Jin et al., 2016). Stimulated gastric and pancreatic digestion increase the number of peptides originated from yogurt with ACE and dipeptidyl peptidase IV (DPP-IV) inhibitory activities. The proportion of peptides released from α_{s1}-CN, α_{s2}-CN as a result of *in vitro* gastrointestinal digestion is greater than that from yogurt fermentation, while β-CN is the protein more cleaved in both processes. In contrast, peptides from κ-CN are more abundant in yogurt fermentation. Encrypted ACE inhibitor peptides not cleaved in the fermentation process, such as YFYPEL [α_{s1}-CN f(143–149)], FVAPFPEV [α_{s1}-CN f(24–31)] and VAPFPEVF [α_{s1}-CN f(25–32)], are resealed by digestive proteases, thereby increasing the beneficial health effects of yogurt (Jin et al., 2016).

Yogurt exhibits quantitative differences in bioactive peptide content due to some bacterial strains which exhibit greater proteolytic activity than others. For example, yogurt fermentation with *Lactobacillus delbrueckii ssp. bulgaricus* promotes higher peptide and free amino acid formation than *Streptococcus thermophilus* and *Lactobacillus paracasei ssp. Paracasei* (Papadimitriou et al., 2007).

The viability of bacterial strains and bioactivity profiles change during cold storage of yogurt. Generally, the survival and activity of bacterial strains increase in refrigerated storage; however, this is dependent on species and the association between them. For instance, the viability of *Lactobacillus acidophilus* is affected by the presence of

TABLE 4.3 Biological Activities of Peptides from Yogurt

Activities	Protein Fragments	Peptide Sequences	Analysis	References
ACE-inhibition	α_{s1}-CN f(24–32) β-CN f(194–209) β-CN f(85–93) α_{s2}-CN(f19–23) β-CN(f74–76) β-CN(f69–73)	FVAPFPEVF; QEPVLGPVRGPFPIIV; PPFLQPEVM; TYKEE; IPP; SLPQN	*In vitro* ACE assay *In vitro* ACE assay	(Jin et al., 2016) (Donkor et al., 2007)
Antihypertensive	α_{s1}-CN f(146–147) β-CN f(114–115) k-CN f(58–59)	TP	Blood pressure in SHR	(Yamamoto, Maeno, and Takano, 1999a)
Antioxidant	Water-soluble peptide extract Isolated peptide fractions	Not determined	DPPH and ABTS assays DPPH, Fe^{2+} chelation and TBARS assays	(Aloğlu and Öner, 2011) (Farvin et al., 2010)
DPP-IV inhibition	α_{s1}-CN f(24–32) β-CN f(85–93)	FVAPFPEVF; PPFLQPEVM	*in vitro* DPP-IV assay	(Jin et al., 2016)
Immunomodulator	Water-soluble peptide extract	Not determined	*In vitro* immunomodulatory activity	(Theodorou and Politis, 2016)
Antibacterial	Water-soluble peptide extract	Not determined	*In vitro* antibacterial assay	(Sah et al., 2016)
Opioid agonist	β-CN f(94–123)	GVSKVKEAMAPKHKEMPFP KYPVEPFTESQ	Mucin production in HT29-MTX cells	(Plaisancie et al., 2013)

DPP-IV: Dipeptidyl peptidase IV; ABTS: 2,2c-Azino-bis (3-ethylbenzothiazoline-6-sulfonic acid); DPPH: 2,2-diphenyl-1-picrylhydrazyl; HT29-MTX: human intestinal mucus-producing cells; SHR: spontaneously hypertensive rats; TBARS: thiobarbituric acid reactive substance

Lactobacillus delbrueckii ssp. *bulgaricus* (Dave and Shah, 1997). Poor survival of Bifidobacterium species is observed after cold storage (Dave and Shah, 1997; Donkor et al., 2007). In parallel to the bacterial viability, the released peptides increase during the cold storage of yogurts. The probiotic yogurts showed greater ACE-inhibitory activity during the initial stage of storage (first three weeks) but decreased afterward (Donkor et al., 2007).

In comparison with milk, there are few studies that have explored the characterization of bioactive peptides from yogurt. On the other hand, the interest in strategies to improve proteolysis via LAB as well as preserve viability of these bioactive compounds after the fermentation process is quite evident.

Cheese

Cheese is one of the most functional and consumed foods around the world. Its properties go beyond mineral salts, vitamins, short-chain fatty acids, proteins, and peptides (Raikos and Dassios, 2014; Martín-Del-Campo et al., 2019). However, the quantity and properties of these substances will depend on the type of cheese and its time of ripening. For example, white cheese has around 28% protein, while parmesan has 40% (López-Expósito, Amigo, and Recio, 2012).

The protein casein is hydrolysed by proteases in a range of peptides during cheese ripening. These proteases and peptidases are derived from microbial sources, milk, or started culture, and the peptides generated will interact directly with consumers' receptors or as a precursor of bioactive endogenous molecules (López-Expósito, Amigo, and Recio, 2012). After cleavage, these peptides will interact with endogenous receptors in the consumer organism (López-Expósito and Recio, 2008; Baptista and Gigante, 2021).

White-brined cheese is a famous cheese derived from South-Eastern European countries, the ripening of which occurs in brine for 2–3 months. This process causes intense proteolysis and, because of the high salt concentrations and mass transfer, the level of bioactive peptides is different when compared with Gouda or pasta filata cheeses. White-brined cheese peptides have ACE-inhibitory, antioxidant, antimicrobial, and antitumoral activities (Barać et al., 2017).

Water-soluble peptides from buffalo cheese demonstrated antimicrobial activity against *Enterococcus faecalis* ATCC 6057 and *Bacillus subtilis* ATCC 6633 microorganisms, as well as ACE-inhibitory and antioxidant activities. Moreover, two peptides were identified (GLPQGVLNENLL and LYQEPVLGPVRGPFPI) in *Bubalus bubalis* cheese, confirming that these dairy products can contribute to the health of its consumers (da Silva et al., 2019).

The starter culture is also important in cheese peptide composition. White-brined goat milk when incubated in a starter culture containing *Lactococcus lactis* ssp. *lactis* and *cremoris* combined with *Lactobacillus casei*, *L. plantarum*, and *L. bulgaricus*, showed a significant increase of antioxidant and ACE-inhibitory activities when compared with white-brined goat milk without this adjunct culture (Kocak et al., 2020). Baptista et al., (2020) evaluated the adjunct culture of *Lactobacillus helveticus* in Prato cheese production and observed an increase of β-CN(f193-206) and β-CN(b194-209) bioactive peptides and decrease of α_{S1}-CN(f1-9) peptide release, showing that starter and adjunct culture influence bioactive peptide quantity (Baptista et al., 2020). Egger et al. (2021) produced Raclette-type cheeses with raw milk and heat-treated, with 120 days, and starter culture with or without *Lactobacillus helveticus* strain mixture. The authors found a higher number of small peptides in digested raw milk when compared with pasteurized cheese, which can be beneficial to cheese consumers (Egger et al., 2021).

The number of ACE-inhibitory peptide fragments was also diverse in Manchego cheese when the milk was pasteurized and the commercial mixed-strain starter culture was modified or removed. The authors also identified a difference in peptide amount according to the ripening time after analysis of the mass spectrometry technique. The higher concentration of peptides was obtained in four months of ripening and raw milk (Gómez-Ruiz, Ramos, and Recio, 2004; 2007).

The mass spectrometry technique has been widely used in the identification of compounds derived from casein hydrolysis, such as α_{s-1}- and β-casein. VPP and IPP, which are β-casein antihypertensive peptides, were found in cheddar, gouda, gorgonzola, brie cheese, and others. AYFYPEL and RYLGY, two α_{s-1}-casein peptides, were found in cheddar, gorgonzola, brie cheese, and others (López-Expósito and Recio, 2008; Gómez-Ruiz, Ramos, and Recio, 2007; Stuknytė et al., 2015; Sieber et al., 2010).

CPPs (casein phosphopeptides) are peptides derived from casein phosphorylation derived from hydrolysis. These peptides seem to be related to the absorption of minerals. CPP concentration will depend on cheese type and ripening and, in recent years, many of them have been isolated and characterized. CPPs were found and isolated in emmental, cheddar, comté and parmigiano-reggiano cheeses (Sforza et al., 2003; Gagnaire et al., 2011).

Parmigiano-reggiano is a long-ripening Italian cheese made from raw cow milk and whey-started microorganisms. A considerable number of peptides were found in parmigiano-reggiano after ripening, highlighting isracidin, a peptide with antimicrobial, antihypertensive, and antioxidant activities (Martini, Conte, and Tagliazucchi, 2020). Iwaniak et al. (2021) also evaluated ACE- and DPP-IV-inhibitory activity with different β-casein concentrations and 1–60 days of ripening. The results showed an increase of these inhibitors after 60 days of ripening, confirming that the longer the ripening time, the greater the quantitative ACE-inhibition peptides in cheeses (Iwaniak and Minkiewicz, 2007).

In vivo studies with bioactive peptides present in cheeses have been shown to decrease gut inflammation and diseases, reduce blood pressure, and inhibit formation of kidney stones. Moreover, some producers have changed parameters and are designing cheese production for more functional cheeses. Among these modifications is the combination of lactic acid bacteria and yeasts in fermentation besides prebiotic use in production (Chourasia et al., 2021; Sprong, Schonewille, and van der Meer, 2010).

Although much is known about the nutritional properties of cheeses, it is still a vast field to be explored. Many types of cheese have not had their compounds isolated and detailed, which would be an important step toward improving production processes and increasing the concentration of peptides with therapeutic properties (Barać et al., 2017). In addition, it is important not only to isolate and characterize these peptides but also to go further in the *in vivo* studies to reveal new potential activities.

Kefir

Recently, the influence of the gut microbiome over the proper functioning of the human organism has been gaining prominence as an object of many studies. The nutritional supplementation based on functional food (nutraceuticals) has shown an impressive impact in preventing and treating gut dysbiosis. Among a great diversity of nutraceuticals, the prebiotics, probiotics, and synbiotics (a mixture of prebiotics and probiotics) have been the most studied (Vasquez et al., 2020).

One of the most popular options of probiotics is the synbiotic milk kefir, which is fermented milk with grains (prebiotic matrix) containing a mixture of probiotic microorganisms such as bacteria, lactobacilli, and yeasts, such as *Lactobacillus brevis*, *Lactobacillus helveticus*, *Lactobacillus kefir*, *Leuconostoc mesenteroides*, *Kluyveromyces lactis*, *Kluyveromyces marxianus*, and *Pichia fermentans*. The complex symbiotic relationship of these microorganisms in the grains may confer the major difference related to yogurts, which are mainly fermented with just species of thermophile bacteria (Chandan and Kilara, 2013). Also, kefir shows different times and temperatures (around 24–48 h and 20–30°C) of fermentation when compared to yogurts. Kefir has a fermentation time which can vary around 24–48 h with two consecutive temperatures, the first fermentation at 20–30°C and the second one at 4°C. On the other hand, for the yogurt production, the pasteurized milk is heated to around 85°C for 30 min and cooled to 40 °C with the addition of the bacteria followed by fermentation of 4 h (Chandan and Kilara, 2013; Ferreira et al., 2010; Amorim et al., 2019). Since kefir has differences concerning its manufacturing and the presence of a prebiotic matrix, this probiotic was categorized separately. The final product composition of kefir can vary according to the origin of the kefir, the substrate used in the fermentation process, and the culture maintenance methods used (Amorim et al., 2019; Vasquez et al., 2020).

Due to the bioactive peptides released during the fermentation of this synbiotic food, kefir has shown to be able to impact the metabolic system through the improvement of digestion and tolerance to lactose (Hertzler and Clancy, 2003), antihypertensive (Klippel et al., 2016; Amorim et al., 2019; Friques et al., 2015), and control of plasma glucose and cholesterol (Hadisaputro et al., 2012; Santanna et al., 2017). Also, kefir has shown to present antibacterial, antiviral (Hamida et al., 2021; Rodrigues et al., 2005), and healing effects (Huseini et al., 2012), probably due to its anti-inflammatory, anti-allergic, and antioxidant properties (Liu, Chen, and Lin, 2005; Ozcan et al., 2019; Lee et al., 2007). In addition to these several positive benefits to health, kefir also has potential antitumoral effects (Gao et al., 2013; Guzel-Seydim, Seydim, and Greene, 2013; Liu et al., 2002; Liu, Chen, and Lin, 2005).

The bioactive peptides from kefir responsible for those effects are still unexplored in the literature. However, Amorim et al. (2019) described 35 peptides in the kefir proteomic with a potential action as antihypertensives peptides. The putative activity as ACE-inhibitory peptides were proven by *in vitro* and *in vivo* analysis after an *in silico* prediction with the ACE enzyme by molecular docking (Amorim et al., 2019). The presence of other bioactive peptides was supported by the previous studies and the majority are released from mainly casein (Ebner et al., 2015; Quirós et al., 2005; Izquierdo-González et al., 2019; Ferreira et al., 2010). Although there are several studies exploring the potential effects of the raw kefir, there is a gap in the studies which performed characterization and exploration of the isolated molecules. The improvement of analytical techniques may enable the in-depth identification and the bioprospection of these bioactive peptides.

REFERENCES

Adolfsson, O., S. N. Meydani, and R. M. Russell. 2004. "Yogurt and gut function." *The American Journal of Clinical Nutrition* 80(2):245–256.

Aloğlu, H. Ş., and Z. Öner. 2011. "Determination of antioxidant activity of bioactive peptide fractions obtained from yogurt." *Journal of Dairy Science* 94(11):5305–5314.

Amorim, F. G., L. B. Coitinho, A. T. Dias, A. G. F. Friques, B. L. Monteiro, L. C. D. Rezende, T. M. C. Pereira, B. P. Campagnaro, E. De Pauw, E. C. Vasquez, and L. Quinton. 2019. "Identification of new bioactive peptides from Kefir milk through proteopeptidomics: Bioprospection of antihypertensive molecules." *Food Chemistry* 282:109–119. doi: 10.1016/j.foodchem.2019.01.010.

Autelitano, D. J., A. Rajic, A. I. Smith, M. C. Berndt, L. L. Ilag, and M. Vadas. 2006. "The cryptome: A subset of the proteome, comprising cryptic peptides with distinct bioactivities." *Drug Discovery Today* 11(7–8):306–314. doi: 10.1016/j.drudis.2006.02.003.

Baptista, D. P., and M. L. Gigante. 2021. "Bioactive peptides in ripened cheeses: Release during technological processes and resistance to the gastrointestinal tract." *Journal of the Science of Food and Agriculture.* doi: 10.1002/jsfa.11143.

Baptista, D. P., F. Negrão, M. N. Eberlin, and M. L. Gigante. 2020. "Peptide profile and angiotensin-converting enzyme inhibitory activity of Prato cheese with salt reduction and Lactobacillus helveticus as an adjunct culture." *Food Research International* 133:109190. doi: 10.1016/j.foodres.2020.109190.

Barać, M., M. Pešić, T. Vučić, M. Vasić, and M. Smiljanić. 2017. "White cheeses as a potential source of bioactive peptides." *Mljekarstvo/Dairy* 67 (1):3–16.

Baspinar, B., and M. Güldaş. 2021. "Traditional plain yogurt: A therapeutic food for metabolic syndrome?" *Critical Reviews in Food Science and Nutrition* 61:1–15. doi: 10.1080/10408398.2020.1799931

Baum, F., M. Fedorova, J. Ebner, R. Hoffmann, and M. Pischetsrieder. 2013. "Analysis of the endogenous peptide profile of milk: Identification of 248 mainly casein-derived peptides." *Journal of Proteome Research* 12(12):5447–5462. doi: 10.1021/pr4003273.

Bhattacharya, M., J. Salcedo, R. C. Robinson, B. M. Henrick, and D. Barile. 2019. "Peptidomic and glycomic profiling of commercial dairy products: Identification, quantification and potential bioactivities." *NPJ Science of Food* 3:4. doi: 10.1038/s41538-019-0037-9.

Biasibetti, E., S. Rapacioli, N. Bruni, and E. Martello. 2020. "Lactoferrin-derived peptides antimicrobial activity: An in vitro experiment." *Natural Product Research* 1–5. doi: 10.1080/14786419.2020.1821017.

Capriotti, A. L., C. Cavaliere, S. Piovesana, R. Samperi, and A. Laganà. 2016. "Recent trends in the analysis of bioactive peptides in milk and dairy products." *Analytical and Bioanalytical Chemistry* 408(11):2677–2685. doi: 10.1007/s00216-016-9303-8.

Cavalheiro, F. G., D. P. Baptista, B. D. Galli, F. Negrão, M. N. Eberlin, and M. L. Gigante. 2020. "High protein yogurt with addition of *Lactobacillus helveticus*: Peptide profile and angiotensin-converting enzyme ACE-inhibitory activity." *Food Chemistry* 333:127482.

Chandan, R. C. 2006. "History and consumption trends." In: *Manufacturing Yogurt and Fermented Milks* (R. C. Chandan, C. H. White, A. Kilara, Y. H. Hui, Eds.) pp. 3–30. Oxford: Blackwell Publishing.

Chandan, R. C., and A. Kilara. 2013. *Manufacturing Yogurt and Fermented Milks.* Wiley Online Library.

Chourasia, R., M. M. Abedin, L. Chiring Phukon, D. Sahoo, S. P. Singh, and A. K. Rai. 2021. "Biotechnological approaches for the production of designer cheese with improved functionality." *Comprehensive Reviews in Food Science and Food Safety* 20(1):960–979. doi: 10.1111/1541-4337.12680.

Chrzanowska, J. 1998. "Enzymatyczne modyfikacje Bialek mleka." *Zeszyty Naukowe Akademii Rolniczej we Wrocławiu. Technologia Żywności* 12:23–38.

Claeys, W. L., Claire Verraes, Sabine Cardoen, Jan De Block, André Huyghebaert, Katleen Raes, Koen Dewettinck, and Lieve Herman. 2014. "Consumption of raw or heated milk from different species: An evaluation of the nutritional and potential health benefits." *Food Control* 42:188–201.

Cormier, Hubert, Élisabeth Thifault, Véronique Garneau, Angelo Tremblay, Vicky Drapeau, Louis Pérusse, and Marie-Claude Vohl. 2016. "Association between yogurt consumption, dietary patterns, and cardio-metabolic risk factors." *European Journal of Nutrition* 55(2):577–587.

Cragg, G. M., and D. J. Newman. 2013. "Natural products: A continuing source of novel drug leads." *Biochimica et Biophysica Acta* 1830 (6):3670–3695. doi: 10.1016/j.bbagen.2013.02.008.

da Silva, Diego Dias, Meire dos Santos Falcão de Lima, Milena Fernandes da Silva, Girliane Regina da Silva, Júlia Furtado Campos, Wendell Wagner Campos Albuquerque, Maria Taciana Holanda Cavalcanti, and Ana Lúcia Figueiredo Porto. 2019. "Bioactive water-soluble peptides from fresh buffalo cheese may be used as product markers." *LWT* 108:97–105.

Dave, Rajiv I., and Nagendra P. Shah. 1997. "Viability of yoghurt and probiotic bacteria in yoghurts made from commercial starter cultures." *International Dairy Journal* 7(1):31–41.

Dhaval, Anusha, Neelam Yadav, and Shalini Purwar. 2016. "Potential applications of food derived bioactive peptides in management of health." *International Journal of Peptide Research and Therapeutics* 22(3):21.

Donkor, O. N., A. Henriksson, T. K. Singh, Todor Vasiljevic, and Nagendra P. Shah. 2007. "ACE-inhibitory activity of probiotic yoghurt." *International Dairy Journal* 17(11):1321–1331.

Doyle, Michael P., and Larry R. Beuchat. 2007. *Food Microbiology: Fundamentals and Frontiers*, 3rd ed. American Society of Microbiology.

Ebner, J., A. Asci Arslan, M. Fedorova, R. Hoffmann, A. Kucukcetin, and M. Pischetsrieder. 2015. "Peptide profiling of bovine kefir reveals 236 unique peptides released from caseins during its production by starter culture or kefir grains." *Journal of Proteomics* 117:41–57. doi: 10.1016/j.jprot.2015.01.005.

Egger, L., O. Ménard, L. Abbühl, D. Duerr, H. Stoffers, H. Berthoud, M. Meola, R. Badertscher, C. Blaser, D. Dupont, and R. Portmann. 2021. "Higher microbial diversity in raw than in pasteurized milk Raclette-type cheese enhances peptide and metabolite diversity after in vitro digestion." *Food Chemistry* 340:128154. doi: 10.1016/j.foodchem.2020.128154.

Ejtahed, Hanie S., Javad Mohtadi-Nia, Aziz Homayouni-Rad, Mitra Niafar, Mohammad Asghari-Jafarabadi, and Vahid Mofid. 2012. "Probiotic yogurt improves antioxidant status in type 2 diabetic patients." *Nutrition* 28(5):539–543.

Farvin, K. H. S., Caroline P. Baron, Nina Skall Nielsen, Jeanette Otte, and Charlotte Jacobsen. 2010. "Antioxidant activity of yoghurt peptides: Part 2–characterisation of peptide fractions." *Food Chemistry* 123(4):1090–1097.

Fernandes, Sibele Santos, Michele Silveira Coelho, and Myriam de las Mercedes Salas-Mellado. 2019. "Bioactive compounds as ingredients of functional foods: Polyphenols, carotenoids, peptides From animal and plant sources new." In: *Bioactive Compounds* (M. R. S. Campos, Ed.) pp. 129–142. Elsevier.

Fernandez, Melissa Anne, and André Marette. 2017. "Potential health benefits of combining yogurt and fruits based on their probiotic and prebiotic properties." *Advances in Nutrition* 8(1):155S–164S.

Fernandez, Melissa Anne, Shirin Panahi, Noémie Daniel, Angelo Tremblay, and André Marette. 2017. "Yogurt and cardiometabolic diseases: A critical review of potential mechanisms." *Advances in Nutrition* 8(6):812–829.

Ferreira, I. M., O. Pinho, D. Monteiro, S. Faria, S. Cruz, A. Perreira, A. C. Roque, and P. Tavares. 2010. "Short communication: Effect of kefir grains on proteolysis of major milk proteins." *Journal of Dairy Science* 93(1):27–31. Doi: 10.3168/jds.2009-2501.

Fox, Michael J., Kiran DK Ahuja, Iain K. Robertson, Madeleine J. Ball, and Rajaraman D. Eri. 2015. "Can probiotic yogurt prevent diarrhoea in children on antibiotics? A double-blind, randomized, placebo-controlled study." *BMJ Open* 5(1):e006474. doi: 10.1136/bmjopen-2014-006474

Friques, A. G., C. M. Arpini, I. C. Kalil, A. L. Gava, M. A. Leal, M. L. Porto, B. V. Nogueira, A. T. Dias, T. U. Andrade, T. M. Pereira, S. S. Meyrelles, B. P. Campagnaro, and E. C. Vasquez. 2015. "Chronic administration of the probiotic kefir improves the endothelial function in spontaneously hypertensive rats." *Journal of Translational Medicine* 13:390. doi: 10.1186/s12967-015-0759-7.

Gagnaire, V., S. Carpino, C. Pediliggieri, J. Jardin, S. Lortal, and G. Licitra. 2011. "Uncommonly thorough hydrolysis of peptides during ripening of Ragusano cheese revealed by tandem mass spectrometry." *Journal of Agriculture and Food Chemistry* 59(23):12443–12452. doi: 10.1021/jf2027268.

Gao, Jie, Fengying Gu, Hui Ruan, Qihe Chen, Jie He, and Guoqing He. 2013. "Induction of apoptosis of gastric cancer cells SGC7901 in vitro by a cell-free fraction of Tibetan kefir." *International Dairy Journal* 30(1):5. doi: 10.1016/j.idairyj.2012.11.011.

Gerosa, Stefano, and Jakob Skoet. 2012. "Milk availability: Trends in production and demand and medium-term outlook."*AgEcon*. https://ageconsearch.umn.edu/record/289000

Gobbetti, M., P. Ferranti, E. Smacchi, F. Goffredi, and F. Addeo. 2000. "Production of angiotensin-I-converting-enzyme-inhibitory peptides in fermented milks started by *Lactobacillus delbrueckii* subsp. bulgaricus SS1 and Lactococcus lactissubsp. cremoris FT4." *Applied and Environmental Microbiology* 66(9):3898–3904.

Grajek, W., A. Olejnik, and A. Sip. 2005. "Probiotics, prebiotics and antioxidants as functional foods." *Acta Biochimica Polonica* 52(3):665–671.

Gurakan, G. C., and N. Altay. 2010. "Yogurt microbiology and biochemistry." In: *Development and Manufacture of Yogurt and Other Functional Dairy Product*. F. Yildiz (Ed.), 98–116. Boca Raton, FL: CRC Press, Taylor & Francis Group.

Guzel-Seydim, Z. B., A. C. Seydim, and A. K. Greene. 2013. "Comparison of amino acid profiles of milk, yoghurt and Turkish kefir." *Milchwissenschaft* 58:3.

Gómez-Ruiz, J. A., M. Ramos, and I. Recio. 2004. "Identification and formation of angiotensin-converting enzyme-inhibitory peptides in Manchego cheese by high-performance liquid chromatography-tandem mass spectrometry." *Journal of Chromatography. Part A* 1054(1–2):269–277. doi: 10.1016/j.chroma.2004.05.022.

Gómez-Ruiz, J. A., M. Ramos, and I. Recio. 2007. "Identification of novel angiotensin-converting enzyme-inhibitory peptides from ovine milk proteins by CE-MS and chromatographic techniques." *Electrophoresis* 28(22):4202–4211. doi: 10.1002/elps.200700324.

Hadisaputro, S., R. R. Djokomoeljanto, Judiono, and M. H. Soesatyo. 2012. "The effects of oral plain kefir supplementation on proinflammatory cytokine properties of the hyperglycemia Wistar rats induced by streptozotocin." *Acta Medica Indonesiana* 44(2):100–104.

Hamida, Reham Samir, Ashwag Shami, Mohamed Abdelaal Ali, Zakiah Nasser Almohawes, Afrah E. Mohammed, and Mashael Mohammed Bin-Meferij. 2021. "Kefir: A protective dietary supplementation against viral infection." *Biomedicine and Pharmacotherapy* 133:110974.

Hertzler, S. R., and S. M. Clancy. 2003. "Kefir improves lactose digestion and tolerance in adults with lactose maldigestion." *Journal of the American Dietetic Association* 103(5):582–587. doi: 10.1053/jada.2003.50111.

Homayouni, Aziz, Maedeh Alizadeh, Hossein Alikhah, and Vahid Zijah. 2012. "Functional dairy probiotic food development: Trends, concepts, and products." In: *Immunology and Microbiology:"Probiotics"*. E. Rigobelo (Ed.), 197–212. Rijeka: InTech.

Huseini, H. F., G. Rahimzadeh, M. R. Fazeli, M. Mehrazma, and M. Salehi. 2012. "Evaluation of wound healing activities of kefir products." *Burns* 38(5):719–723. doi: 10.1016/j.burns.2011.12.005.

Ishida, Y., Y. Shibata, I. Fukuhara, Y. Yano, I. Takehara, and K. Kaneko. 2011. "Effect of an excess intake of casein hydrolysate containing Val-Pro-Pro and Ile-Pro-Pro in subjects with normal blood pressure, high-normal blood pressure, or mild hypertension." *Bioscience, Biotechnology and Biochemistry* 75(3):427–433. doi: 10.1271/bbb.100560.

Iwaniak, Anna, and Piotr Minkiewicz. 2007. "Proteins as the source of physiologically and functionally active peptides." *Acta Scientiarum Polonorum. Technologia Alimentaria* 6(3):5–15.

Iwaniak, A., D. Mogut, P. Minkiewicz, J. Żulewska, and M. Darewicz. 2021. "Gouda cheese with modified content of β-casein as a source of peptides with ACE- and DPP-IV-inhibiting bioactivity: A study based on in silico and in vitro protocol." *International Journal of Molecular Sciences* 22:2949. doi: 10.3390/ijms22062949.

Izquierdo-González, Juan J., Francisco Amil-Ruiz, Sabina Zazzu, Rosa Sánchez-Lucas, Carlos A. Fuentes-Almagro, and Manuel J. Rodríguez-Ortega. 2019. "Proteomic analysis of goat milk kefir: Profiling the fermentation-time dependent protein digestion and identification of potential peptides with biological activity." *Food Chemistry* 295:456–465.

Jin, Yan, Yu Yang, Yanxia Qi, Fangjun Wang, Jiaze Yan, and Hanfa Zou. 2016. "Peptide profiling and the bioactivity character of yogurt in the simulated gastrointestinal digestion." *Journal of Proteomics* 141:24–46.

Khan, M. U., M. Pirzadeh, C. Y. Förster, S. Shityakov, and M. A. Shariati. 2018. "Role of milk-derived antibacterial peptides in modern food biotechnology: Their synthesis, applications and future perspectives." *Biomolecules* 8(4). doi: 10.3390/biom8040110.

Klippel, B. F., L. B. Duemke, M. A. Leal, A. G. Friques, E. M. Dantas, R. F. Dalvi, A. L. Gava, T. M. Pereira, T. U. Andrade, S. S. Meyrelles, B. P. Campagnaro, and E. C. Vasquez. 2016. "Effects of kefir on the cardiac autonomic tones and baroreflex sensitivity in spontaneously hypertensive rats." *Frontiers in Physiology* 7:211. doi: 10.3389/fphys.2016.00211.

Kocak, Ali, Tuba Sanli, Elif Ayse Anli, and Ali Adnan Hayaloglu. 2020. "Role of using adjunct cultures in release of bioactive peptides in white-brined goat-milk cheese." *LWT* 123:109127.

Kulinich, A., and L. Liu. 2016. "Human milk oligosaccharides: The role in the fine-tuning of innate immune responses." *Carbohydrate Research* 432:62–70. doi: 10.1016/j.carres.2016.07.009.

Kunda, Pradeep B., Fernando Benavente, Sergio Catalá-Clariana, Estela Giménez, José Barbosa, and Victoria Sanz-Nebot. 2012. "Identification of bioactive peptides in a functional yogurt by micro liquid chromatography time-of-flight mass spectrometry assisted by retention time prediction." *Journal of Chromatography. Part A* 1229:121–128.

Lana, Alberto, Jose R. Banegas, Pilar Guallar-Castillón, Fernando Rodríguez-Artalejo, and Esther Lopez-Garcia. 2018. "Association of dairy consumption and 24-hour blood pressure in older adults with hypertension." *The American Journal of Medicine* 131(10):1238–1249.

Lee, Mee-Young, Kyung-Seop Ahn, Ok-Kyung Kwon, Mee-Jin Kim, Mi-Kyoung Kim, In-Young Lee, Sei-Ryang Oh, and Hyeong-Kyu Lee. 2007. "Anti-inflammatory and anti-allergic effects of kefir in a mouse asthma model." *Immunobiology* 212(8):647–654.

Liu, J. R., S. Y. Wang, Y. Y. Lin, and C. W. Lin. 2002. "Antitumor activity of milk kefir and soy milk kefir in tumor-bearing mice." *Nutrition and Cancer* 44(2):183–187. doi: 10.1207/s15327914nc4402_10.

Liu, Je-Ruei, Ming-Ju Chen, and Chin-Wen Lin. 2005. "Antimutagenic and antioxidant properties of milk– kefir and soymilk– kefir." *Journal of Agricultural and Food Chemistry* 53(7):2467–2474.

Lopitz-Otsoa, F., A. Rementeria, N. Elguezabal, and J. Garaizar. 2006. "Kefir: una comunidad simbiótica de bacterias y levaduras con propiedades saludables." *Revista Iberoamericana de Micología* 23(2):67–74. doi: 10.1016/S1130-1406(06)70016-X.

López-Expósito, I., and I. Recio. 2008. "Protective effect of milk peptides: Antibacterial and antitumor properties." *Advances in Experimental Medicine and Biology* 606:271–293. doi: 10.1007/978-0-387-74087-4_11.

López-Expósito, Iván, Lourdes Amigo, and Isidra Recio. 2012. "A mini-review on health and nutritional aspects of cheese with a focus on bioactive peptides." *Dairy Science and Technology* 92(5):419–438.

Manzanarez-Quín, C. G., L. M. Beltrán-Barrientos, A. Hernández-Mendoza, A. F. González-Córdova, and B. Vallejo-Cordoba. 2021. "Invited review: Potential anti-obesity effect of fermented dairy products." *Journal of Dairy Science* 104(4):3766–3778. doi: 10.3168/jds.2020-19256.

Marcone, Simone, Orina Belton, and Desmond J. Fitzgerald. 2017. "Milk-derived bioactive peptides and their health promoting effects: A potential role in atherosclerosis." *British Journal of Clinical Pharmacology* 83(1):152–162.

Martín-Del-Campo, S. T., P. C. Martínez-Basilio, J. C. Sepúlveda-Álvarez, S. E. Gutiérrez-Melchor, K. D. Galindo-Peña, A. K. Lara-Domínguez, and A. Cardador-Martínez. 2019. "Production of antioxidant and ACEI peptides from cheese whey discarded from Mexican white cheese production." *Antioxidants (Basel)* 8(6). doi: 10.3390/antiox8060158.

Martini, Serena, Angela Conte, and Davide Tagliazucchi. 2020. "Effect of ripening and in vitro digestion on the evolution and fate of bioactive peptides in Parmigiano-Reggiano cheese." *International Dairy Journal* 105:104668.

McCann, K. B., B. J. Shiell, W. P. Michalski, A. Lee, J. Wan, H. Roginski, and M. J. Coventry. 2005. "Isolation and characterisation of antibacterial peptides derived from the f (164–207) region of bovine αS2-casein." *International Dairy Journal* 15(2):133–143.

Morelli, Lorenzo. 2014. "Yogurt, living cultures, and gut health." *The American Journal of Clinical Nutrition* 99(5):1248S–1250S.

Nabavi, S., M. Rafraf, M. H. Somi, A. Homayouni-Rad, and M. Asghari-Jafarabadi. 2014. "Effects of probiotic yogurt consumption on metabolic factors in individuals with nonalcoholic fatty liver disease." *Journal of Dairy Science* 97(12):7386–7393.

Nagaoka, S., Y. Futamura, K. Miwa, T. Awano, K. Yamauchi, Y. Kanamaru, K. Tadashi, and T. Kuwata. 2001. "Identification of novel hypocholesterolemic peptides derived from bovine milk beta-lactoglobulin." *Biochemical and Biophysical Research Communications* 281(1):11–17. doi: 10.1006/bbrc.2001.4298.

Nagpal, R., P. Behare, R. Rana, A. Kumar, M. Kumar, S. Arora, F. Morotta, S. Jain, and H. Yadav. 2011. "Bioactive peptides derived from milk proteins and their health beneficial potentials: An update." *Food and Function* 2(1):18–27. doi: 10.1039/c0fo00016g.

Nollet, Leo M. L., and Fidel Toldrá. 2009. *Handbook of Dairy Foods Analysis*. Boca Raton, FL: CRC Press.

Ozcan, Tulay, Saliha Sahin, Arzu Akpinar-Bayizit, and Lutfiye Yilmaz-Ersan. 2019. "Assessment of antioxidant capacity by method comparison and amino acid ymbioticzation in buffalo milk kefir." *International Journal of Dairy Technology* 72(1):65–73.

Papadimitriou, Christos G., Anna Vafopoulou-Mastrojiannaki, Sofia Vieira Silva, Ana-Maria Gomes, Francisco Xavier Malcata, and Efstathios Alichanidis. 2007. "Identification of peptides in traditional and probiotic sheep milk yoghurt with angiotensin I-converting enzyme (ACE)-inhibitory activity." *Food Chemistry* 105(2):647–656. Doi: 10.1016/j.foodchem.2007.04.028.

Park, Y. W., and M. S. Nam. 2015. "Bioactive peptides in milk and dairy products: A review." *Korean Journal for Food Science of Animal Resources* 35(6):831–840. Doi: 10.5851/kosfa.2015.35.6.831.

Plaisancie, Pascale, Jean Claustre, Monique Estienne, Gwenaele Henry, Rachel Boutrou, Armelle Paquet, and Joelle Léonil. 2013. "A novel bioactive peptide from yoghurts modulates expression of the gel-forming MUC2 mucin as well as population of goblet cells and Paneth cells along the small intestine." *The Journal of Nutritional Biochemistry* 24(1):213–221.

Pu, Fangfang, Yue Guo, Ming Li, Hong Zhu, Shijie Wang, Xi Shen, Miao He, Chengyu Huang, and Fang He. 2017. "Yogurt supplemented with probiotics can protect the healthy elderly from respiratory infections: A randomized controlled open-label trial." *Clinical Interventions in Aging* 12:1223.

Punia, H., J. Tokas, A. Malik, S. Sangwan, S. Baloda, N. Singh, S. Singh, A. Bhuker, P. Singh, S. Yashveer, S. Agarwal, and V. S. Mor. 2020. "Identification and detection of bioactive peptides in milk and dairy products: Remarks about agro-foods." *Molecules* 25(15). Doi: 10.3390/molecules25153328.

Qian, Z. Y., P. Jollès, D. Migliore-Samour, F. Schoentgen, and A. M. Fiat. 1995. "Sheep kappa-casein peptides inhibit platelet aggregation." *Biochimica et Biophysica Acta* 1244(2–3):411–417. Doi: 10.1016/0304-4165(95)00047-f.

Quirós, A., B. Hernández-Ledesma, M. Ramos, L. Amigo, and I. Recio. 2005. "Angiotensin-converting enzyme inhibitory activity of peptides derived from caprine kefir." *Journal of Dairy Science* 88(10):3480–3487. Doi: 10.3168/jds.S0022-0302(05)73032-0.

Rai, A. K., S. Sanjukta, and K. Jeyaram. 2017. "Production of angiotensin I converting enzyme inhibitory (ACE-I) peptides during milk fermentation and their role in reducing hypertension." *Critical Reviews in Food Science and Nutrition* 57(13):2789–2800. Doi: 10.1080/10408398.2015.1068736.

Raikos, V., and T. Dassios. 2014. "Health-promoting properties of bioactive peptides derived from milk proteins in infant food: A review." *Dairy Science and Technology* 94:91–101. Doi: 10.1007/s13594-013-0152-3.

Ramchandran, Lata, and Nagendra P. Shah. 2011. "Yogurt can beneficially affect blood contributors of cardiovascular health status in hypertensive rats." *Journal of Food Science* 76(4):H131–H136.

Ramesh, C. V., N. R. Kamini, and R. Puvanakrishnan. 2004. "A novel synthetic peptide derivative from lactoferrin exhibiting antimicrobial activity." *Biotechnology and Applied Biochemistry* 40(3):271–275. Doi: 10.1042/ba20040008.

Rezazadeh, Leila, Bahram Pourghassem Gargari, Mohammad Asghari Jafarabadi, and Beitullah Alipour. 2019. "Effects of probiotic yogurt on glycemic indexes and endothelial dysfunction markers in patients with metabolic syndrome." *Nutrition* 62:162–168.

Rival, S. G., S. Fornaroli, C. G. Boeriu, and H. J. Wichers. 2001. "Caseins and casein hydrolysates. 1. Lipoxygenase inhibitory properties." *Journal of Agriculture and Food Chemistry* 49(1):287–294. Doi: 10.1021/jf000392t.

Rodrigues, K. L., L. R. Caputo, J. C. Carvalho, J. Evangelista, and J. M. Schneedorf. 2005. "Antimicrobial and healing activity of kefir and kefiran extract." *International Journal of Antimicrobial Agents* 25(5):404–408. Doi: 10.1016/j.ijantimicag.2004.09.020.

Sah, B. N. P., T. Vasiljevic, S. McKechnie, and O. N. Donkor. 2018. "Antioxidative and antibacterial peptides derived from bovine milk proteins." *Critical Reviews in Food Science and Nutrition* 58(5):726–740. Doi: 10.1080/10408398.2016.1217825.

Sah, Baidya Nath Prasad, Todor Vasiljevic, Sandra McKechnie, and O. N. Donkor. 2016. "Antibacterial and antiproliferative peptides in ymbiotic yogurt—Release and stability during refrigerated storage." *Journal of Dairy Science* 99(6):4233–4242.

Santanna, A. F., P. F. Filete, E. M. Lima, M. L. Porto, S. S. Meyrelles, E. C. Vasquez, D. C. Endringer, D. Lenz, D. S. P. Abdalla, T. M. C. Pereira, and T. U. Andrade. 2017. "Chronic administration of the soluble, nonbacterial fraction of kefir attenuates lipid deposition in LDLr-/- mice." *Nutrition* 35:100–105. Doi: 10.1016/j.nut.2016.11.001.

Sfakianakis, Panagiotis, and Constatnina Tzia. 2014. "Conventional and innovative processing of milk for yogurt manufacture; development of texture and flavor: A review." *Foods* 3(1):176–193.

Sforza, S., L. Ferroni, G. Galaverna, A. Dossena, and R. Marchelli. 2003. "Extraction, semi-quantification, and fast on-line identification of oligopeptides in Grana Padano cheese by HPLC-MS." *Journal of Agriculture and Food Chemistry* 51(8):2130–2135. Doi: 10.1021/jf025866y.

Sieber, Robert, Ueli Bütikofer, Charlotte Egger, Reto Portmann, Barbara Walther, and Daniel Wechsler. 2010. "ACE-inhibitory activity and ACE-inhibiting peptides in different cheese varieties." *Dairy Science and Technology* 90(1):47–73.

Sprong, R. C., A. J. Schonewille, and R. van der Meer. 2010. "Dietary cheese whey protein protects rats against mild dextran sulfate sodium-induced colitis: Role of mucin and microbiota." *Journal of Dairy Science* 93(4):1364–1371. Doi: 10.3168/jds.2009-2397.

Srinarong, Chanagune, Sith Siramolpiwat, Arti Wongcha-um, Varocha Mahachai, and Ratha-korn Vilaichone. 2014. "Improved eradication rate of standard triple therapy by adding bismuth and probiotic supplement for Helicobacter pylori treatment in Thailand." *Asian Pacific Journal of Cancer Prevention: APJCP* 15(22):9909–9913.

Stuknytė, M., S. Cattaneo, F. Masotti, and I. De Noni. 2015. "Occurrence and fate of ACE-inhibitor peptides in cheeses and in their digestates following in vitro static gastrointestinal digestion." *Food Chemistry* 168:27–33. Doi: 10.1016/j.foodchem.2014.07.045.

Suetsuna, K., H. Ukeda, and H. Ochi. 2000. "Isolation and characterization of free radical scavenging activities peptides derived from casein." *Journal of Nutritional Biochemistry* 11(3):128–131. Doi: 10.1016/s0955-2863(99)00083-2.

Takano, Toshiaki. 1998. "Milk derived peptides and hypertension reduction." *International Dairy Journal* 8(5–6):375–381.

Theodorou, Georgios, and Ioannis Politis. 2016. "Effects of peptides derived from traditional Greek yoghurt on expression of pro-and anti-inflammatory genes by ovine monocytes and neutrophils." *Food and Agricultural Immunology* 27(4):484–495.

Tonolo, F., M. Sandre, S. Ferro, A. Folda, V. Scalcon, G. Scutari, E. Feller, O. Marin, A. Bindoli, and M. P. Rigobello. 2018. "Milk-derived bioactive peptides protect against oxidative stress in a Caco-2 cell model." *Food and Function* 9(2):1245–1253. Doi: 10.1039/c7fo01646h.

Townsend, R. R., C. B. McFadden, V. Ford, and J. A. Cadée. 2004. "A randomized, double-blind, placebo-controlled trial of casein protein hydrolysate (C12 peptide) in human essential hypertension." *American Journal of Hypertension* 17(11 Pt 1):1056–1058. Doi: 10.1016/j.amjhyper.2004.06.018.

Vargas-Bello-Pérez, E., R. I. Márquez-Hernández, and L. E. Hernández-Castellano. 2019. "Bioactive peptides from milk: Animal determinants and their implications in human health." *Journal of Dairy Research* 86(2):136–144. Doi: 10.1017/s0022029919000384.

Vasquez, Elisardo C., Rafaela Aires, Alyne M. M. Ton, and Fernanda G. Amorim. 2020. "New insights on the beneficial effects of the probiotic kefir on vascular dysfunction in cardiovascular and neurodegenerative diseases." *Current Pharmaceutical Design* 26(30):3700–3710(11).

Wang, Huifen, Kara A. Livingston, Caroline S. Fox, James B. Meigs, and Paul F. Jacques. 2013. "Yogurt consumption is associated with better diet quality and metabolic profile in American men and women." *Nutrition Research* 33(1):18–26.

Xue, L., X. Wang, Z. Hu, Z. Wu, L. Wang, H. Wang, and M. Yang. 2018. "Identification and characterization of an angiotensin-converting enzyme inhibitory peptide derived from bovine casein." *Peptides* 99:161–168. doi: 10.1016/j.peptides.2017.09.021.

Yamamoto, N., M. Maeno, and T. Takano. 1999a. "Purification and characterization of an antihypertensive peptide from a yogurt-like product fermented by Lactobacillus helveticus CPN4." *Journal of Dairy Science* 82(7):1388–1393. doi: 10.3168/jds.S0022-0302(99)75364-6.

Yamamoto, Naoki, Momoka Shoji, Hiroki Hoshigami, Kohei Watanabe, Tappei Takatsuzu, Shin Yasuda, Keiji Igoshi, and Hideki Kinoshita. 2019. "Antioxidant capacity of soymilk yogurt and exopolysaccharides produced by lactic acid bacteria." *Bioscience of Microbiota. Bioscience of Microbiota, Food and Health* 40(4):97–104.

Yamamoto, Naoyuki, Masafumi Maeno, and Toshiaki Takano. 1999b. "Purification and characterization of an antihypertensive peptide from a yogurt-like product fermented by *Lactobacillus helveticus* CPN4." *Journal of Dairy Science* 82(7):1388–1393.

Yoon, Ji-Woo, Sung-Il Ahn, Jin-Woo Jhoo, and Gur-Yoo Kim. 2019. "Antioxidant activity of yogurt fermented at low temperature and its anti-inflammatory effect on DSS-induced colitis in mice." *Food Science of Animal Resources* 39(1):162.

Zambrowicz, A., M. Timmer, A. Polanowski, G. Lubec, and T. Trziszka. 2013. "Manufacturing of peptides exhibiting biological activity." *Amino Acids* 44(2):315–320. doi: 10.1007/s00726-012-1379-7.

Zoumpopoulou, G., A. Tzouvanou, E. Mavrogonatou et al. 2018. "Probiotic features of lactic acid bacteria isolated from a diverse pool of traditional greek dairy products regarding specific strain-host interactions." *Probiotics & Antimicrobial Proteins* 10:313–322. doi: 10.1007/s12602-017-9311-9.

CHAPTER 5

Seafood

Jesús Aarón Salazar Leyva, Rosa Stephanie Navarro Peraza, Israel Benítez García, Emmanuel Martínez Montaño, and Idalia Osuna Ruíz

CONTENTS

Seafood: a Plentiful Source of Bioactive Peptides	97
Main Seafood Sources of Bioactive Peptides	97
Naturally Active Peptides	98
Peptides That Are Inactive within a Parent Protein	98
Development of Bioactive Peptides from Seafood	99
Isolating Naturally Bioactive Peptides from Seafood	99
Producing Bioactive Peptides from Seafood	100
Autolysis	100
Exogenous Enzymatic Hydrolysis	101
Fermentative Processes	103
Purification of Seafood Bioactive Peptides	103
Bioactive Properties of Seafood Peptides	104
Nutraceutical Application of Bioactive Seafood Peptides	116
References	116

SEAFOOD: A PLENTIFUL SOURCE OF BIOACTIVE PEPTIDES

Around 70% of the Earth's surface is covered by oceans, which harbor a great diversity of living organisms, far exceeding the one in the terrestrial compartment (Haard, Simpson, and Sikorski, 1994; Suarez-Jimenez, Burgos-Hernandez, and Ezquerra-Brauer, 2012). Global seafood consumption (including finfish, crustaceans, mollusks, and other aquatic animals) increased at an average annual rate of 3.1% over the past 56 years, whereas other animal-protein food (meat, dairy, milk, etc.) only increased by 2.1% per year (FAO, 2020). The global rise in demand of seafood is related to population growth and the increasing awareness of the health benefits associated with seafood consumption (Chakrabarti, Guha, and Majumder, 2018; Jakubczyk et al., 2020). It is therefore important to explore the reserve of nutrients and active biomolecules offered by the sea life used as food.

Main Seafood Sources of Bioactive Peptides

Bioactive peptides occur in food either as natural compounds or as part of a parent polypeptide, where the peptide is inactive (Chakrabarti, Guha, and Majumder, 2018).

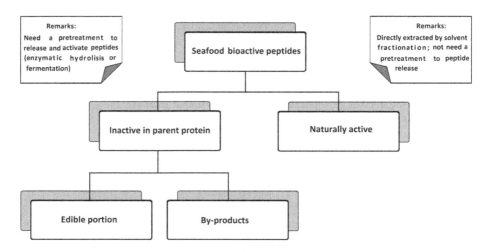

FIGURE 5.1 Classification of seafood bioactive peptides.

Biologically active peptides found in seafood can be classified into: 1) naturally active peptides and 2) inactive peptides within a parent protein (Fig. 5.1). The two types of seafood bioactive peptides shown in Fig. 5.1 are briefly discussed below.

Naturally Active Peptides

This type of bioactive peptides can be obtained directly from seafood biomass using solvent gradient extraction. Methanol, acetone, chloroform, and carbon tetrachloride are the solvents most commonly used. Once crude peptides are in solution, they can be screened for biological activity and then subjected to further purification (Kiran et al., 2014; Sable, Parajuli, and Jois, 2017). Several natural peptides have been extracted from various marine animal species not commonly regarded as seafood, including tunicates, mollusks, and sponges. In fact, many of these species are considered to be poisonous (Verdes et al., 2016). For instance, the peptide Didemnin B, an anticancer agent obtained from the tunicate, *Trididemnum solidum*, was the first marine natural product to reach the clinical trial stage. Due to several toxicological issues, Phase-2 trials of this peptide were ineffective (Shin et al., 1991; Benvenuto et al., 1992; Kucuk et al., 2000; Sable, Parajuli, and Jois, 2017).

Additional important natural seafood peptides are those exhibiting antimicrobial activity. The existence of these peptides is related to the hostile environments where aquatic organisms live, with frequent exposure to microorganisms (Sable, Parajuli, and Jois, 2017). The hemolymph of numerous marine invertebrates (oyster, crab, lobster, and shrimp) has been documented to contain antimicrobial peptides (Destoumieux-Garzón et al., 2016). In addition, body parts such as the skin and gills of some fish species are known to be rich in peptides with antimicrobial activity. For example, chrysophsin-1, an amphipathic peptide produced by gill cells of the red seabream, exhibits activity against a wide range of Gram-positive and Gram-negative bacteria (Mason et al., 2007).

Peptides That Are Inactive within a Parent Protein

Seafood proteins are rich sources of peptide fragments contained in their primary structure where their biological activity is either inactive or enclosed within the parent protein. Such peptides can be extracted by applying various pretreatments to seafood biomass (e.g., enzymatic hydrolysis and microbial fermentation) to release their bioactive properties (Jakubczyk et al., 2020). The edible portion of seafood (or by-products) produced during processing are major sources of these peptides.

Regarding the use of edible seafood portions to release bioactive peptides, seafood is an excellent source of high-quality, well-balanced, and easily digestible protein containing all essential amino acids. In fact, protein content is higher in seafood than in terrestrial animals (17.3% vs. 13.8% on average, respectively; fresh weight) (Mazorra-Manzano et al., 2018; Tacon and Metian, 2018). Recent studies have demonstrated the reliability of releasing peptides from edible parts of several seafood species, including yellowfin tuna (*Thunnus albacares*), oyster (*Crassostrea gigas*), or white shrimp (*Litopenaeus vannamei*), which show antimicrobial, antihypertensive, and antioxidant activity, respectively (Cerrato et al., 2020; Guo et al., 2020; Latorres et al., 2018).

According to FAO (2020), the continued expansion of various fishery-related activities has led to increased generation of by-products which, depending on the species, may account for up to 70% of the processed material. Seafood by-products are generally not used for direct human consumption and historically have been discarded as waste, used as ingredients for animal feed, or applied as silage or fertilizer. However, alternative uses of seafood by-products (e.g., fins, heads, viscera, trimmings, scales, skin, roe, etc.) have drawn attention in recent years. Several studies have shown that this protein-rich biomass is a source of enclosed bioactive peptides that can be released using various processing technologies (Al Khawli et al., 2019; Ucak et al., 2021). Recently, efforts have been made to apply the marine biorefinery concept – processing routes aimed at obtaining value-added compounds from seafood by-products and optimize their valorization by means of eco-efficiency and benchmark strategies (García-Santiago et al., 2020).

DEVELOPMENT OF BIOACTIVE PEPTIDES FROM SEAFOOD

Isolating Naturally Bioactive Peptides from Seafood

All living organisms contain native peptides not only as part of protein structures but as free peptides in bodily fluids (e.g., lymphatic fluids or blood plasma) and excretions (e.g., mucus). Marine organisms are a major source of non-proteinogenic peptides that include cyclic and linear peptide structures, as well as depsipeptides (peptides in which one or more amide groups are replaced by the respective ester group) (Rangel et al., 2017). Since these free peptides are not part of proteins, no hydrolytic processes are required to extract them. Instead, various purification steps are performed to obtain the peptides of interest (Sperstad et al., 2011; Cheung, Ng, and Wong, 2015). These include (1) extraction using solvents of different polarity or aqueous buffers in order to preserve the functionality of the molecules; (2) fractionation using salts, such as ammonium sulfate or acetone; (3) purification by preparative chromatography (molecular exclusion and ion exchange) and desalination methods; and (4) analytical purification using reversed-phase high performance liquid chromatography (RP-HPLC). Once the peptide fractions of interest have been isolated, they can be subjected to structural and functional characterization. Edman degradation or protease fragmentation and, eventually, mass spectrometry analysis can be used to identify the primary structure of peptides while circular dichroism and nuclear magnetic resonance (NMR) allow elucidating the secondary and tertiary structure, respectively (Lazcano-Pérez et al., 2012).

Examples of bioactive non-proteinogenic peptides extracted from marine organisms include those isolated from marine snails of the genus *Conus*. These snails produce a range of poisonous peptides containing 10–40 amino acids known as conotoxins or conopeptides, whose key feature is their highly conserved cysteine framework, which is also used to categorize them (Jin et al., 2019). These peptides are secreted and used by the organism as a defense mechanism and to immobilize and facilitate the digestion of prey. These peptides have been shown to possess various therapeutic properties (Robinson et al., 2017) including antibacterial (Lebbe et al., 2016), anticancer (Kumari et al., 2019), analgesic (Jagonia et al., 2019), etc.

Antimicrobial peptides (3159, 655, 636 Daltons (Da)) have been obtained from the hemolymph of mollusks such as the bivalve *Crassostrea gigas*. The purification process encompassed C-18 solid-phase extraction, size-exclusion chromatography, and RP-HPLC, followed by MS/MS and Edman degradation. However, the sequence of these peptides could not be identified after Edman degradation. A particular case was the 635 Da peptide (Cg-636) whose primary structure consisted mainly of XPPXXIV, but the amino acid X was identified as a nonstandard residue (Defer et al., 2013).

Naturally bioactive peptides from marine organisms have proved to be promising sources for synthesizing synthetic analogs. For example, Epinecidin-1 is an antimicrobial peptide identified in the fish *Epinephelus coioides*; its synthetic analog (Epi-1) possesses anticancer and immunomodulatory properties (Neshani et al., 2019). A similar case is pleurocidin, an α-helical cationic peptide with antimicrobial properties isolated from the mucus of the fish *Pleuronectes americanus*. Its analog Plc-2 (KHVGKAALTHYL) possesses elevated antimicrobial and antifungal properties (Souza et al., 2013).

Producing Bioactive Peptides from Seafood

Several strategies are used to produce bioactive peptides from edible parts or by-products of seafood, including chemical, enzymatic, or fermentative processes (Figure 5.2). Chemical hydrolysis using high concentrations of acids (HCl or H_2SO_4) or alkalis (often NaOH or KOH) at high temperature, is a low-cost, easy-to-perform process suitable to produce protein hydrolysates, fish protein concentrates, and gelatin, particularly at industrial scale (Ghaly et al., 2013). However, some reactions (e.g., racemization) that are difficult to control may damage or destroy some amino acids (e.g., Ser, Thr, Arg, and Cys with alkaline hydrolysis or Trp with acid hydrolysis) during production, thus reducing the quality of the hydrolysate (Fernandes, 2016; Zamora-Sillero et al., 2018). In addition, acid hydrolyzation might leave a high NaCl content in the product, which reduces its usability as food (Petrova et al., 2018).

Enzymatic hydrolysis (EH) of proteins is the method most commonly used for producing seafood protein hydrolysates (SPH) under controlled, optimized conditions (Navarro-Peraza et al., 2020). Enzymatic hydrolysis has several advantages over chemical hydrolysis, including the ability to modify and enhance the bioactive and techno-functional properties of seafood proteins (Villamil et al., 2017; Ishak and Sarbon, 2018). Moreover, as no side reactions occur during hydrolysis, the nutritional value of the original protein source can be preserved.

Autolysis

Autolysis is a process in which endogenous proteolytic enzymes, mainly acidic proteases (e.g., pepsins and cathepsins) act upon tissular proteins; organic acids (e.g., formic,

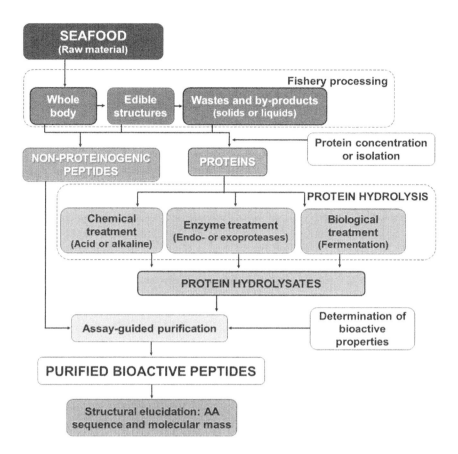

FIGURE 5.2 Flow chart of the process for obtaining bioactive peptides from seafood.

acetic, phosphoric, citric, and propionic acids) and antioxidants (e.g., ethoxyquin, butylated hydroxytoluene, sodium benzoate) are usually added to prevent bacterial spoilage and lipid rancidity (Olsen and Toppe, 2017). Autolysis is a cost-effective method for producing acidic silages containing bioactive peptides or other nutritional components, such as essential fatty acids and minerals. A major shortcoming of this process is the varying nutritional quality and functionality of the end product (Zamora-Sillero, Gharsallaoui, and Prentice, 2018) due to the difficulty of obtaining homogenous hydrolysates from the raw material as the content of endogenous enzymes depends on several factors, such as raw material freshness, seasonality, and physiological state of the capture, etc. (Van'T Land et al., 2017).

Autolytic technologies have been used to digest protein-rich structures from fishery by-products and discards and obtain protein hydrolysates with bioactive properties such as antioxidant (Ozyurt et al., 2018; Özyurt et al., 2018) and antimicrobial (Murthy, Rai, and Bhaskar, 2014) activity.

Exogenous Enzymatic Hydrolysis

Protein hydrolysis with commercial exogenous enzymes breaks down proteins into peptides and free amino acids. This process allows a better control of the resulting hydrolysate and its biochemical, techno-functional, and bioactive properties (critical quality attributes, CQA), some of which might not have been present in the original source (Li-Chan, 2015). The key variables involved in enzymatic hydrolysis production are known as critical process parameters (CPP). CPP comprise characteristics of the substrate to be hydrolyzed (e.g., protein concentration, amino acid profile), the particular enzyme (e.g., enzyme type, purity, catalytic properties, stability, optimum pH, temperature, etc.), and the process (enzyme/substrate ratio, temperature, pH, incubation time, and degree of hydrolysis [DH]) (Li-Chan, 2015).

DH is a crucial parameter, as it represents the number of cleaved peptide bonds as a proportion of the total number of peptide bonds in the native protein (Fernandes, 2016). As DH increases, the molecular weight of the product peptide decreases, hence increasing their solubility. The selection of the proteolytic enzyme is another key factor in this process since its specificity affects the size and composition of the peptides produced, which, in turn, influence the functional and bioactive properties of the resulting hydrolysates (Zamora-Sillero, Gharsallaoui, and Prentice, 2018). Many commercial proteases are currently available, which are classified based on their source (animal, plant, bacterial, or fungal), catalytic action (endo- or exoproteases), and active site (serine proteases, aspartic proteases, cysteine proteases, and metalloproteases) (dos Santos Aguilar and Sato, 2018).

Today, microorganisms are the main producers of proteases due to their easy manipulation and economic advantages, in addition to their biochemical diversity and possibility of genetic manipulation (Fernandes, 2016; dos Santos Aguilar and Sato, 2018). The microbial proteases most commonly used are neutrase (obtained from *Bacillus amyloliquefaciens*), alcalase (from *Bacillus licheniformis*), Flavourzyme (from *Aspergillus oryzae*), Protamex (from *Bacillus sp.*), proteinase A and K (from *Saccharomyces cerevisiae* and *Engyodontium album*, respectively), and thermolysin (from *Bacillus thermoproteolyticus*) (Fernandes, 2016). Several studies have evaluated the use of aspartic (pepsins) and serine (trypsin, chymotrypsin, collagenase, and elastase) proteases extracted and purified from viscera (stomach, pancreas, pyloric caeca, intestines) of teleost fish as enzyme sources to hydrolyze seafood protein (Rios-Herrera et al., 2019; Caruso et al., 2020).

The use of plant proteases to produce protein hydrolysates from seafood protein has recently drawn attention. Cysteine proteases obtained from papaya (papain), pineapple (bromelin), and fig (ficin) are those most commonly used in the food processing industry (Mazorra-Manzano, Ramírez-Suarez, and Yada, 2018). Cysteine proteases from fruits of Bromeliaceae plants were recently used to produce protein hydrolysates from fish by-products showing antioxidant properties (Romero-Garay et al., 2020). Other serine proteases (e.g., cucumisin), mainly obtained from Cucurbitaceae species, have been used to produce seafood protein hydrolysates. For example, Alavi et al. (2019) used cantaloupe crude enzyme extracts to obtain protein hydrolysates from kilka fish (*Clupeonella cultriventriscaspia*), which showed antioxidant and functional properties (protein solubility, oil and water holding, and emulsifying properties).

Enzymatic hydrolysis has been used to transform seafood proteins into value-added products by improving their functional, physicochemical, nutritional, and bioactive properties (Zamora-Sillero et al., 2018). Production of seafood protein hydrolysates (SPH) or bioactive peptides is one of the leading applications of enzymatic hydrolysis and has drawn the attention of food researchers worldwide. SPH are produced by the enzymatic hydrolysis of whole bodies, edible structures (mainly muscular tissue), and by-products from the seafood processing industry (Petrova, Tolstorebrov, and Eikevik, 2018). The latter includes bones (Yang, Zhao et al., 2019), heads (Zhang, Zhao, Zhao et al., 2019), dark muscle (Bui et al., 2020), frames (Idowu et al., 2019), exoskeletons (da Silva et al., 2017), skin (Tkaczewska et al., 2020), viscera (Villamil, Váquiro, and Solanilla, 2017), and roe (Rajabzadeh et al., 2018). Hydrolysis produces peptides of different size, depending on the parental structure used as substrate, enzyme source, reaction time, and degree of hydrolysis (Fernandes, 2016). Also, SPH have been recently obtained from fishery waste effluents (liquid waste from the fishery processing industry), but information on this topic is still scarce compared to the data available on solid by-products. Fishery effluents from washing, thawing, softening, and cooking operations, but also those derived from fishmeal production (stickwater), contain a considerable amount of soluble proteins (Navarro-Peraza et al., 2020) which, after recovery and concentration, can be used to produce protein hydrolysates, some of them with bioactive properties, such as antioxidant (Amado et al., 2013), antihypertensive (Amado et al., 2014), and antiproliferative (Hung et al., 2014) activity.

Fermentative Processes

Microbial fermentation using fungi or lactic acid bacteria (LAB) is another strategy for producing bioactive peptides from parent protein in seafood (mainly fish by-products). However, as seafood contains low amounts, carbohydrates must be added, such as fermentable sugars (e.g., molasses; Özyurt et al., 2018) or even monosaccharides (e.g., glucose; Murthy et al., 2014). Microorganisms used in fermentative process are characterized by their ability to synthetize and secrete *in situ* organic acids (mainly lactic acid) or hydrolases (chitinases, proteases, and lipases), which are used to produce protein hydrolysates and recover other valuable molecules such as lipids (Özyurt et al., 2018), chitin/chitosan and astaxanthin (Bruno et al., 2019), peptones (Vázquez et al., 2019), and other compounds.

LAB belonging to the genera *Bacillus*, *Lactobacillus*, *Pediococcus*, *Streptococcus*, and *Enterococcus*, which characteristically produce lactic acid as the main final product of fermentation (Marti-Quijal et al., 2020), have been widely used to produce protein hydrolysates. Fermented food products obtained from fish, crustaceans, or mollusks have been identified as rich sources of bioactive compounds with antioxidant (Djellouli et al., 2020), anticoagulant, ACE inhibitory (Kleekayai et al., 2015), antimicrobial (Murthy, Rai, and Bhaskar, 2014), antifungal (Song et al., 2017), anti-osteoporosis, or insulin-secretion stimulation (Chen, Chen et al., 2020) activity. For instance, antioxidant activity (total antioxidant activity and DPPH radical scavenging effect) was recorded in fermented fish silage produced from two discard fish species (*Equulites klunzingeri* and *Carassius gibelio*) using molasses and inoculation with *Lactobacillus brevis*, *Lactobacillus plantarum*, *Pediococcus acidilactici*, *Streptococcus* spp., and *Enterococcus gallinarum* (Ozyurt et al., 2018). Products with angiotensin-converting enzyme (ACE) inhibitory activity were obtained by fermenting meat of skipjack tuna (*Katsuwonus pelamis*) with salt (Wenno, Suprayitno, and Hardoko, 2016). Furthermore, Gao et al., (2016) developed bio-converted products from shrimp by-products and two fungi species (*Boletus edulis* and *Suillus bovinus*) using a liquid fermentation process. The bio-converted products showed angiotensin I-converting enzyme (ACE) inhibitory and antioxidant activities.

Purification of Seafood Bioactive Peptides

Target peptides produced by hydrolysis are recovered and purified at the industrial scale using membrane separation processes such as ultrafiltration, nanofiltration, or a combination of several methods (Kim and Venkatesan, 2014). Membrane separation technologies (MST) use hydrostatic pressure to force a liquid to pass through a semipermeable membrane. The fluid that passes through the membrane is known as "permeate" and the retained liquid is the "retentate". MST are performed using the cross-flow technique to prevent the retentate from accumulating on the membrane surface while the permeate passes through it (Lee et al., 2014). Membranes are commonly made of synthetic materials (e.g., nylon, polyethersulfone, polyvinylidene difluoride, polytetrafluoroethylene, mixed cellulose ester, cellulose acetate, polypropylene, or other materials) that allow concentrating and separating multicomponent solutions based on their molecular size and shape without inflicting thermal damage (Calabrò and Basile, 2011; Petrova, Tolstorebrov, and Eikevik, 2018). The membrane pore size usually ranges from 0.1 to 5000 nm, and membranes are classified based on the particle size separated: microfiltration membranes separate 50 nm or larger particles; ultrafiltration (UF) membranes separate 3 nm or larger particles; nanofiltration (NF) membranes separate 1 nm or larger particles; and reverse-osmosis membranes separate particles larger than 0.1 nm (Halim, Yusof, and Sarbon, 2016; Ishak and Sarbon, 2018).

The retention capacity of membranes is determined by the molecular weight cut-off (MWCO), which is mainly related to the membrane pore size. MST minimizes protein denaturation. UF and NF membranes have been widely used to fractionate hydrolysates from edible components and by-products (either solid or liquid) of seafood to obtain molecules (such as peptides) within a specific molecular weight range that might show enhanced bioactive properties, such as antioxidant, antihypertensive, and antiproliferative activity (Pezeshk et al., 2019). Peptide fractions with a molecular weight of 1–4 kDa have been identified as most promising for nutritional and pharmaceutical purposes (Abejón et al., 2018). MST have also been used in the treatment of effluents from seafood processing to recover value-added compounds, such as protein, and to produce fish protein concentrates and hydrolysates, etc. (Amado et al., 2013; Navarro-Peraza et al., 2020).

BIOACTIVE PROPERTIES OF SEAFOOD PEPTIDES

Several authors have identified – through *in vitro*, *in vivo*, or *in silico* studies – various biological properties in peptides from seafood or its by-products. Antibacterial, antioxidant, antimutagenic, antihypertensive, antidiabetic, antiproliferative, anti-inflammatory, and cytotoxic bioactivity has been identified in seafood peptides of low molecular weight (<5 kDa). Such peptides have been produced primarily by enzymatic hydrolysis, using either muscle or by-products of various aquatic species intended for human consumption (predominantly wild or farmed fish species vs. other seafood types, such as crustaceans and mollusks) as base biomass (Table 5.1). The wild fish most extensively studied in this regard include species of the family Scombridae such as mackerel and tuna; small Clupeidae species such as anchovy, sardine, and herring; various species of the cutlassfish family, Trichiuridae; and seafood species including clam, mussel, oyster, scallop, shrimp, and squid; farmed fish species include sea bream, tilapia, and carp (Table 5.1).

The species, source, and physiological condition of aquatic organisms are key drivers for their potential use as raw materials in the production of bioactive peptides. The pre-slaughter starvation of farmed seabream fish (*Sparus aurata*) has been reported to increase the formation of antihypertensive peptides produced by simulating the digestion of myosin in fillets (Lippe et al., 2021).

Additional beneficial bioactive properties that have been reported include osteogenic activity in proteins of blue mussel (*Mytilus edulis*) hydrolyzed with pepsin and trypsin

TABLE 5.1 Relevant Studies Related to the Production of Bioactive Seafood Peptides

Seafood	Production Method	Bioactivity Demonstrated	Study Type	Peptide Sequence or General Composition (Molecular Weight)	Reference
EDIBLE PORTION/ PRODUCT					
Barred mackerel (*Scomberomorus commerson*)	Alcalase, Actinidin	Antimicrobial Antioxidant Cytotoxicity	in vitro	Gly, Ala and Pro, mainly Fraction with peptides < 3 kDa	Mirzapour-Kouhdasht et al. (2021)
Mackerel (*Scomber japonicus*)	Protamex	Antioxidant	in vitro	IANLAATDIIF (1194.63 Da), ALSTWTLQLGSTSFSASPM (2097.923 Da), and LGTLLFIAIPI (1282.72 Da)	Bashir et al. (2020)
Yellowfin tuna (*Thunnus albacares*)	Simulated GI[a]	Antimicrobial	in vitro in silico	Fraction with peptides < 5 kDa, Probably KKLGELLK and KLGELLK, mixtures of anionic peptides mainly	Cerrato et al. (2020)
Tuna fish	Pepsin, trypsin, and chymotrypsin	Antihyperuricemic (xanthine oxidase inhibitior)	in vitro in silico	EEAK	Yu et al (2021)
Sardinelle (*Sardinella aurita*)	*Bacillus subtilis* A26 proteases	Antimicrobial Antioxidant Antihypertensive[b]	in vitro	Medium-chain hydrophobic peptides	Jemil et al. (2017)
Anchovy fish meal	Alkaline protease NS37071	Antioxidant Antimicrobial	in vitro	TPSAGK (559 Da), TPSNLGGK (772 Da), LE (260 Da) and LEE (389 Da)	Wang et al. (2018)
Thai fish sauces from Indina anchovy (*Stolephorus indicus*)	Industrial fermentation	Antioxidant Antihypertensive Cytoprotective	in vitro	Glu was the predominant free amino acid, in peptides Phe, Trp, and Tyr, mainly	Hamzeh et al. (2020)
Fermented fish sauce (Budu)	Industrial fermentation	Antioxidant	in vitro	LDDPVFIH and VAAGRTDAGVH	Najafian and Babji (2019)
Fermented fish (pekasam)	*Lactobacillus plantarum* IFRPD P15 strain	Antioxidant	in vitro	AIPPHPYP and IAEVFLITDPK	Najafian and Babji (2018)

(*Continued*)

TABLE 5.1 (Continued)

Seafood	Production Method	Bioactivity Demonstrated	Study Type	Peptide Sequence or General Composition (Molecular Weight)	Reference
Pacific herring (*Clupea pallasii*)	Trypsin	Antioxidant	*in vitro*	LHDELT (726.35 Da) and KEEKFE (808.40 Da)	Wang et al. (2019)
Herring	Enzyme mix	Anti-inflammatory	*in vitro*	IVPAS (485.2855 Da) and FDKPVSPLL (1014.5751 Da)	Durand et al. (2020)
Pacific thread herring (*Ophistonema libertate*)	Alcalase	Antioxidant	*in vitro*	Peptides <1350 Da and high concentrations of anionic and cationic amino acids	Sandoval-Gallardo et al. (2020)
Lizard fish (*Saurida elongata*)	Neutral protease	Antihypertensive	*in vitro*	RYRP (592 Da)	Sun et al. (2017)
Cutlassfish (*Trichiurus lepturus*)	Pepsin	Antihypertensive	*in vitro*	FSGGE (496.44 Da)	Kim et al. (2020)
Hairtail (*Trichiurus japonicus*)	Papain + alcalase	Antioxidant	*in vitro*	QNDER (660.3 Da), KS (233.0 Da), KA (217.1 Da), AKG (274.1 Da), TKA (318.0 Da), VK (245.1 Da), MK (277.0 Da), and IYG (351.0 Da)	Yang et al. (2019)
Leatherjacket Fish (*Meuchenia sp.*)	Papain, bromelain, and flavourzyme	Antihypertensive	*in vitro*	EPLYV, DPHI, AER, EQIDNLQ, and WDDME (Fractions <5000 Da)	Salampessy et al. (2017)
Spotless smoothhound (*Mustelus griseus*)	Trypsin	Antioxidant	*in vitro*	GAERP (528.61 Da), GEREANVM (904.91 Da), and AEVG (374.33 Da)	Tao et al. (2018)
Skate (*Raja porosa*)	Trypsin and alcalase	Antioxidant	*in vitro*	FIMGPY (726.90 Da), GPAGDY (578.58 Da) and IVAGPQ (583.69 Da)	Pan et al. (2016)
Boarfish (*Capros aper*)	Alcalase 2.4L and flavourzyme 500L	Antidiabetic[c] Insulinotropic	*in vitro*	IPVDM and LPVDM	Harnedy-Rothwell et al. (2020)
Croceine croaker (*Pseudosciaena crocea*)	Pepsin, Alcalase	Antioxidant	*in vitro*	YLMSR (651.77 Da), VLYEE (668.82 Da), and MILMR (662.92 Da)	Chi et al. (2015)

(*Continued*)

TABLE 5.1 (Continued)

Seafood	Production Method	Bioactivity Demonstrated	Study Type	Peptide Sequence or General Composition (Molecular Weight)	Reference
Crimson snapper (*Lutjanus erythropterus*) scales	Papain	Antioxidant	in vitro in vivo	Rich in hydrophobic amino acids and polar amino acids (concentrated at below 3000 Da)	Chen et al. (2020)
Black eelpout (*Lycodes diapterus*)	Pepsin	Antioxidant	in vitro	DLVKVEA (784 Da)	Lee and Byun (2019)
Fish collagen	Protamex protease	Antihypertensive	in vitro	GHVGAAGS (653.25 Da)	Yu et al. (2020)
Surimi from the olive flounder (*Paralichthys olivaceus*)	-	Antihypertensive	in vitro	IVDR (501.4 Da), WYK (495.5 Da), and VASVI (487.4 Da)	Oh et al. (2020)
Monkfish muscle (*Lophius litulon*)	Pepsin, trypsin, and In vit. GI methods	Antioxidant Cytoprotective function on HepG$_2$ Cells Damage by H$_2$O$_2$	in vitro	EDIVCW (763.82 Da), MEPVW (660.75 Da), and YWDAW (739.75 Da)	Hu et al. (2020)
Oyster (*Crassostrea gigas*)	Pepsin, and trypsin	Antihypertensive	in vitro	Peptides with a molecular weight < 2000 Da	Guo et al. (2020)
Oyster (*Crassostrea gigas*)	Sequential trypsin-papain	Promoting effect on male hormone production	in vitro in vivo	Mainly hydrophobic peptides (<3000 Da)	Zhang et al. (2021)
Portuguese Oyster (*Crassostrea angulata*)	Pepsin, bromelain, and papain	Antihypertensive	in vitro in silico	IDSLEGSVSR (1061.53 Da) LTQENFDLQHQVQELDAANAGLAK (2652.32 Da)	Gomez et al. (2019)
Pearl oyster (*Pinctada fucata martensii*)	Alkaline protease	Antihypertensive	in vitro in vivo	HLHT (507.117 Da) and GWA (334.202 Da)	Liu et al. (2019)
Scallops (*Chlamys farreri*)	Simulated GI: pepsin and pancreatin	Antioxidant	in vitro	VPSIDDQEELM (1251.46 Da), DAQEKLE (831.87 Da), and EELDNALN (916.93 Da)	Wu et al. (2018)

(*Continued*)

TABLE 5.1 (Continued)

Seafood	Production Method	Bioactivity Demonstrated	Study Type	Peptide Sequence or General Composition (Molecular Weight)	Reference
Blue mussel (*Mytilus edulis*)	Pepsin	Antioxidant Hepatoprotective	*in vitro*	PIIVYWK (1004.57 Da), TTANIEDRR (1074.54 Da), and FSVVPSPK (860.09 Da)	Park et al. (2016)
Blue mussel (*Mytilus edulis*)	Trypsin	Antithrombotic	*in vitro* *in silico*	ELEDSLDSER	Qiao et al. (2018)
Blue mussel (*Mytilus edulis*)	Trypsin	Anticoagulant	*in vitro* *in silico*	VQQELEDAEFRADSAEGSLQK (2331.0806 Da), RMEADIAAMQSDLDDALNGQR (2319.064 Da) and AAFLLGVNSNDLLK (1473.8192Da)	(Qiao, Tu, Chen et al. 2018)
Blue mussel (*Mytilus edulis*)	Pepsin	Antiadipogenic	*in vitro*	< 1000 Dal	Oh et al.(2020)
Blue mussel (*Mytilus edulis*)	Fermented with *Bacillus natto* strain	Antihypertensive Antiinflammatory	*in vitro* *in vivo*	VISDEDGVTH (1076.15 Da)	Chen et al (2018).
Ark shell (*Scapharca subcrenata*)	Pepsin	Antioxidant	*in vitro*	MCLDSCLL (897.5 Da) and HPLDSLCL (897.5 Da)	Jin et al. (2018)
Ark shell (*Scapharca subcrenata*)	Pepsin	Osteogenic	*in vitro* *in vivo*	AWLNH (640 Da) and PHDL (480.2 Da)	Oh et al. (2019)
Ark shell (*Scapharca subcrenata*)	Pepsin	Antiadipogenic	*in vitro*	Asp, Glu, mainly. Thr, Ser, Gly, Ala, Val, Leu, Lys, and Arg (Fraction with peptide <500 Da)	Hyung et al. (2017)
Blood cockle (*Tegillarca granosa*)	Alcalase, neutrase	Antioxidant	*in vitro*	EPLSD (59.55 Da), WLDPDG (701.69 Da), MDLFTE (754.81 Da), WPPD (513.50 Da), EPVV (442.48 Da), and CYIE (526.57 Da)	Yang et al. (2019)

(*Continued*)

TABLE 5.1 (Continued)

Seafood	Production Method	Bioactivity Demonstrated	Study Type	Peptide Sequence or General Composition (Molecular Weight)	Reference
Squid (*Todarodes pacificus*)	Papain, ficin, and prolyl endopeptidase	Antihypertensive	*in vitro* *in silico*	IIY and NPPK	Yu et al. (2019)
Kilka (*Clupeonella cultriventris caspi*)	Alcalase, neutrase, protamex, and pepsin	Antihypertensive Antimicrobial	*in vitro*	Hydrolysates with low molecular weight peptides (~ 10 D)	Qara and Najafi (2018)
White shrimp (*Litopenaeus vannamei*)	Alcalase 2.4 L and Protamex.	Antioxidant	*in vitro*	High levels of hydrophobic amino acids, Asp and Glu	Latorres et al. (2018)
Silver carp (*Hypophthalmichthys molitrix*)	Neutrase, alcalase, pepsin, trypsin, and flavourzyme	Antidiabetic Antihypertensive	*in vitro*	AALEQ TER, LLDLGVP, and KAVGEPPLF	Zhang et al. (2019)
Largemouth bass (*Micropterus salmoides*)	Pepsin, trypsin, and chymotrypsin	Antidiabetic	*in vitro*	VSM, ISW, VSW, ICY, ISD, and ISE	Wang et al. (2020)
Tilapia (*Oreochromis niloticus*)	Alcalase	Antihypertensive	*in vitro*	TL, TI, IK, LR, LD, IQ, DI, AILE, ALLE, ALIE, and AIIE	Yesmine et al. (2017)
Tilapia (*Oreochromis niloticus*)	*Virgi bacillus halodenitrificans* SK1-3-7 proteinases	Antihypertensive	*in vitro*	MILLLFR (905.83 Da)	Toopcham et al. (2015)
Large yellow croaker (*Pseudosciaena crocea*)	Neutral protease	Antioxidant	*in vitro*	SRCHV (601.3Da) and PEHW (568.3 Da)	Zhang et al. (2017)
Edible seahorse (*Hippocampus abdominalis*)	Alcalase	Cytoprotective	*in vitro*	HGSH (436.43 Da) and KGPSW (573.65 Da)	Oh et al. (2021)
BY-PRODUCTS			*in vitro*		
Dark meat of yellowfin tuna (*Thunnus albacares*)	Papain	Antioxidant	*in vitro*	Less than 3000 Da	Unnikrishnan et al. (2020)

(*Continued*)

TABLE 5.1 (Continued)

Seafood	Production Method	Bioactivity Demonstrated	Study Type	Peptide Sequence or General Composition (Molecular Weight)	Reference
Tuna dark muscle	Orientase and protease XXIII	Antioxidant	in vitro	LPTSEAAKY (978 Da) and PMDYMVT (756 Da)	Hsu (2010)
Tuna dark muscle	Alcalase, α-chymotrypsin, neutrase, papain, pepsin, and trypsin	Antioxidant	in vitro	HLNLPTAVYMVT-OH (1222 Da)	Je et al. (2008)
Tuna dark muscle	Alcalase, neutrase, pepsin, papain, α-chymotrypsin, trypsin, and UF	Antihypertensive	in vitro in vivo	WPEAAELMMEVDP (1581 Da)	Qian et al. (2007)
Tuna dark muscle	Papain and Protease XXIII	Antiproliferative	in vitro	LPHVLTPEAGAT (1206 Da) and PTAEGGVYMVT (1124 Da)	Hsu et al. (2011)
Tuna processing waste biomass	Prolyve BS and fractionated by membranes process	Antioxidant	in vitro	YENGG (538.46 Da), EGYPWN (764.75 Da), YIVYPG (710.78 Da) and WGDAGGYY (887.85 Da)	Saidi et al. (2018)
Tuna backbone	Alcalase, α-chymotrypsin, neutrase, papain, pepsin, and trypsin	Antioxidant	in vitro	VKAGFAWTANQQLS (1519 Da)	Je et al. (2007)
Tuna cooking juice	Protease XXIII	Antioxidant	in vitro	PSHDAHPE (1010 Da), SHDAHPE (896 Da), VDHDHPE (953 Da), PKAVHE766, PAGY (457 Da), PHHADS (751 Da), and VDYP (544 Da)	Jao and Ko (2002)
Tuna cooking juice	Orientase	Antioxidant	in vitro	PVSHDHAPEY (1305 Da), PSDHDHE (938 Da) and VHDY (584 Da)	Hsu et al. (2009)

(Continued)

TABLE 5.1 (Continued)

Seafood	Production Method	Bioactivity Demonstrated	Study Type	Peptide Sequence or General Composition (Molecular Weight)	Reference
Tuna cooking juice	Orientase	Antihypertensive	in vitro	Constituents with molecular weight ranging from 240 to 565 Da	Hsu et al. (2007)
Tuna cooking juice	Protease XXIII and UF[d]	Antiproliferative	in vivo	KPEGMDPPI.SEPEDRRDGAAGPK (2449.292 Da) and KLPPLLLAKLL MSGKLLAEPCTGR (2562.405 Da)	Hung et al. (2014)
Scales of skipjack tuna (*Katsuwonus pelamis*)	Pepsin, papain, trypsin, Neutrase, Alcalase	Antioxidant	in vitro	HGP-Hyp-GEG (608.57 Da), DGPKGH (609.61 Da), and MLGPFGPS (804.92 Da)	Qiu et al. (2019)
Spanish mackerel (*Scomber morousniphonius*)	Pepsin	Antioxidant	in vitro	PFGPD (531.52 Da), PYGAKG, (591.69 Da), and YGPM (466.50 Da)	Zhang et al. (2019)
Skate (*Okamejei kenojei*) skin	Alcalase	Antihypertensive	in vitro	LGPLGHQ (720 Da) and MVGSAPGVL (829 Da)	Ngo et al. (2015)
Seabass (*Lates calcarifer*) skin	Alcalase	Antioxidant	in vitro	GLFGPR (646.3671 Da), GATGPQGPLGPR (1107.5905 Da), VLGPF (532.31130 Da), and QLGPLGPV (780.4614 Da)	Sae-Leaw et al. (2017)
Scales of large yellow croaker (*Pseudosciaena crocea*)	Alcalase	Antioxidant	in vitro	QRPPEPR (879.4 Da), EKVWKYCD (1070.4 Da), and VGLPGLSGPVG (952.5 Da)	Chen et al. (2020)
Atlantic mackerel by-products	Protamex	Antibacterial	in vitro	SIFIQRFTT, RKSGDPLGR, AKPGDGAGSGPR and GLPGPLGPAGPK	Ennaas et al. (2015)
Fish discarded (*Sardine pilchardus*)	Subtilisin-trypsin sequential hydrolysis	Antidiabetic	in vitro	NAPNPR (667.329 Da) and YACSVR (697.311 Da)	Rivero-Pino et al. (2020)

(*Continued*)

TABLE 5.1 (Continued)

Seafood	Production Method	Bioactivity Demonstrated	Study Type	Peptide Sequence or General Composition (Molecular Weight)	Reference
Salmon backbones	Corolase, Corolase PP, protamex, papain, bromelain, protex, sea zyme, and trypsin	Antioxidant, Antihypertensive, and Antidiabetic	in vitro	Hydrolysates with molecular weight > 2500, 1400–1600 and 650–700 Da	Slizyte et al. (2016)
Grass carp (*Ctenopharyngodon idella*) skin	Alcalase	Antioxidant	in vitro	PYSFK (640.74 Da), GFGPEL (618.89 Da), and VGGRP (484.56 Da)	Cai et al. (2015)
Bluefin leatherjacket heads (*Navodon septentrionalis*)	Papain and UF	Antioxidant	in vitro	WEGPK (615.69 Da), GPP (269.33 Da), and GVPLT (485.59 Da)	Chi et al. (2015)
Jumbo squid fins and arms	Trypsin and protease XIV	Antioxidant, Antiproliferative, and Antimutagenic	in vitro	Gly, Arg, Leu, Glu, Asp, and Pro	Suárez-Jiménez et al. (2015)
Squid (*Nototodarus sloanii*) pens	Trypsin, pepsin, HT (bacterial protease)	Antioxidant Antihypertensive	in vitro	High content of Cys	Shavandi et al. (2017)
Skate cartilage	Trypsin, alcalase, and UF	Antioxidant	in vitro	FIMGPY (726.90 Da), GPAGDY (578.58 Da), and IVAGPQ (583.69 Da)	Pan et al. (2016)
Skate cartilage	Alcalase	Antihypertensive	in vitro	Fractions (<1200 Da) with peptides: GIV, GAP*GF, GFP*GPA, SGNIGFP*GPK, and GIPGPIGPP*GPR.	Thuanthong et al. (2017)
Tilapia by-products	Alcalase and UF	Antihypertensive	in vitro	Hydrolysate rich in Glu, Gly, Lys, and Ala	Roslan et al. (2017)
Tilapia muscle co-products	Alcalase and UF	Antihypertensive	in vitro	TL, TI, IK, LR, LD, IQ, DI, AILE, ALLE, AlLE, and AIIE	Yesmine et al. (2017)

(*Continued*)

TABLE 5.1 (Continued)

Seafood	Production Method	Bioactivity Demonstrated	Study Type	Peptide Sequence or General Composition (Molecular Weight)	Reference
Turbot (*Scophthalmus maximus*)	Alcalase	Antioxidant Antihypertensive	in vitro	1000 Da and below 200 Da	Vazquez et al. (2020)
Wastes smooth-hound viscera	Neutrase, purafect, and UF	Antihypertensive	in vitro	GPAGPRGPAG, AVVPPSDKM, TTMYPGIA, and VKPLPQSG	Abdelhedi et al. (2018)
Abalone (*Haliotis discus hannai ino*) gonads	Alcalase followed by papain	Antihypertensive	in vitro	AMN	Wu et al. 2015
Giant grouper roe	Protease N and UF	Antiproliferative	in vitro	Hydrolysate with low molecular weight (<5000 Da)	Yang et al. (2016)

5.1a GI = Gastrointestinal digestion.
5.1b Antihypertensive (*in vitro*) = ACE inhibition activity; Antihypertensive (*in vivo*) = lowering blood pressure.
5.1c Antidiabetic = glucose lowering activity; DPP-IV inhibitory activity (*in vitro*).
5.1d UF = Ultrafiltration.

(Xu et al., 2019) and antidiabetic effects of sea cucumber (*Holothuria nobilis*) proteins hydrolyzed with an enzyme mixture of papain and Protamex (Wang et al., 2020). The bioactivity observed in the aforementioned studies, and in those listed in Table 5.1, comes not only from the raw material but also from the chemical-structural characteristics of the bioactive peptides and their sequence of amino acids. As previously mentioned, these characteristics are affected by the enzymatic hydrolysis process used and the degree of hydrolysis achieved, which increase the number of peptides of low molecular weight (Sandoval-Gallardo et al., 2020). Similarly, treatments applied prior to or after obtaining the peptides can have either positive or negative effects on bioactivity. For instance, cooking mackerel fillets using different methods (boiling, frying, roasting, or sterilized canned products) before subjecting them to enzymatic hydrolysis has been reported to change the bioactive characteristics of the hydrolysate, decreasing its antioxidant activity due to the loss of endogenous compounds and the oxidation of muscle components caused by the thermal processes applied (Korczek et al., 2020). In contrast, a combined ultrasound-microwave treatment has been observed to cause partial structural denaturation of myofibrillar proteins in the golden threadfin bream (*Nemipterus virgatus*), which increases their susceptibility to hydrolysis by exogenous enzymes, accelerating the hydrolytic process and increasing the concentration of peptides with antioxidant and antihypertensive activity (Li et al., 2020).

In general, most seafood peptides obtained through enzymatic hydrolysis exhibit antioxidant properties. This is mainly attributed to the exposure of amino acid residues, such as alanine (A), glycine (G), and proline (P); acidic amino acids, such as glutamic acid (E) and aspartic acid (D); basic amino acids, such as arginine (R), lysine (K), and histidine (H); and amino acids with aromatic residues, such as phenylalanine (F) and tyrosine (Y) (Zamora-Sillero, Gharsallaoui, and Prentice, 2018). These amino acid residues are encrypted in the protein and become exposed when peptide bonds are cleaved by the enzymes used in the hydrolysis process.

In addition to its antioxidant capacity, hydrolysis can produce other peptides with unique bioactivity, as reported by various authors that used the same protein source but varied the hydrolytic enzymes used. For example, low-molecular-weight (<1.5 kDa) peptides obtained from tuna fish dark muscle show differences in amino acid sequence and amino acid residues exposed, which confer upon them different biological activities (Table 5.1), such as *in vitro* antioxidant activity (Je et al., 2008; Hsu, 2010), antihypertensive activity (evaluated *in vitro* and *in vivo*) (Qian, Je, and Kim, 2007), and antiproliferative activity against the human breast cancer cell line MCF-7 (Hsu, Li-Chan, and Jao, 2011).

The type of protease used also affects the biological activity of the resulting peptides. For example, when protease XXIII was used to hydrolyze proteins recovered from tuna fish cooking juice, seven short-sequence (4–8 amino acids) antioxidant peptides of low molecular weight (<1 kDa) were produced (Jao and Ko, 2002). In another study where the same effluent was hydrolyzed with the same enzyme (protease XXIII), two peptides with high antiproliferative activity on MCF-7 cells (IC value = 1.39 mg/mL) were identified in the <2.5 kDa fraction (Hung et al., 2014) (Table 5.1). Two different studies that used orientase (instead of protease XXIII) to hydrolyze tuna fish cooking juice yielded three <1 kDa peptides with high *in vitro* antioxidant capacity (Hsu, Lu, and Jao 2009). Hsu (2007) reported small peptides (240-565 Da) that exhibited *in vitro* and *in vivo* ACE inhibitory capacity (Table 5.1).

A newly discovered biological activity of seafood peptides is their ability to inhibit the enzyme xanthine oxidase (XO), which is involved in the elevation of uric acid in blood (hyperuricemia). Yu et al. (2021) hydrolyzed the edible part (skeletal myosin) of tuna fish and found several peptides that inhibit XO; the peptide EEAK showed the highest XO inhibitory activity, with an IC_{50} of 173.00 ± 0.06 μM. This peptide thus represents a promising alternative for controlling gout and hyperuricemia (Table 5.1). Another recently studied bioactive property of seafood peptides is related to osteoporosis treatment. Two peptides (AWLNH and PHDL) produced from ark shell (*Scapharca subcrenata*) protein showed the capacity to restore femoral bone mineral density when injected at a dose of 0.2 mg/kg/day on ovariectomized (OVX) mice, improving the osteoporotic condition of the organisms studied (Oh et al., 2019).

Table 5.1 lists several studies on the production of bioactive peptides from edible parts and by-products of seafood. The table shows the broad range of peptide sequences obtained, which is related to the raw material used and the peptide production method.

Figure 5.3 shows the relationship between protein source, production method, and biological activity exhibited by seafood peptides. There is a broad range of bioactivity types, the most extensively studied being the antioxidant, antihypertensive, anti-inflammatory, antimicrobial, and antiproliferative activities. This range of biological activities suggests that seafood bioactive peptides may have therapeutic effects to support the treatment of various diseases and physiological conditions of high global public-health impact, such as obesity (Oh, Ahn, and Je, 2020), osteoporosis (Oh, Ahn, and Je, 2020), diabetes (Rivero-Pino, Espejo-Carpio, and Guadix, 2020), coronary disease (Nasri et al., 2018), and cancer (Hung et al., 2014). Thus, the nutraceutical application of seafood bioactive peptides is an actively growing area of research.

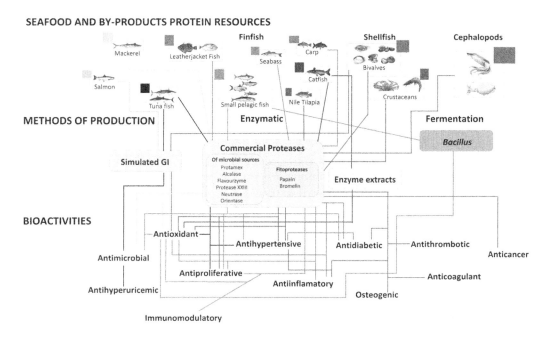

FIGURE 5.3 Biological activity of seafood peptides in relation to protein sources and production methods. (A color code was used to identify the relationships between protein source, production method, and bioactivity.)

TABLE 5.2 Commercial Products Derived from Seafood Bioactive Peptides

Seafood Source	Bioactivity	Manufacturer	Reference
White fish (*Molva molva*)	Antioxidant; low glycemic index; anti-stress	Fortidium Liquamen®	Guérard et al., (2010)
Bonito fish (*Sarda orientalis*)	Antihypertensive	PeptACE™	Guérard et al. (2019)
Bacalao (*Gadus morhua*)	Low glycemic index	Nutripeptin™	de la Fuente et al.(2020)
Sardine (*Sardinops sagax*)	Maintain healthy cardiovascular system	Valtyron®	Osajima et al. (2009)
Piked dogfish (*Squalus acanthias*)	Anticancer peptides	Neovastat®	Gingras et al. (2003)

Nutraceutical Application of Bioactive Seafood Peptides

Peptides obtained from seafood and its by-products can be used as functional ingredients and are also sources of high-quality nutrients. Thus, these peptides are being considered for the development of nutraceuticals or functional foods and are currently used in the treatment or prevention of diseases (Cheung et al., 2015). For instance, Table 5.2 lists some commercial products approved by the U.S. Food and Drug Administration that have been produced using seafood peptides as active ingredients and have demonstrated to yield health benefits due to their antioxidant, antihypertensive, antidiabetic, or anticancer properties. The latter property is of great interest in the pharmaceutical industry in their search for new sources of antitumor agents to treat this disease (de la Fuente et al., 2020).

REFERENCES

Abdelhedi, Ola, Rim Nasri, Letica Mora, Mourad Jridi, Fidel Toldrá, and Moncef Nasri. 2018. *In silico* analysis and molecular docking study of angiotensin I-converting enzyme inhibitory peptides from smooth-hound viscera protein hydrolysates fractionated by ultrafiltration. *Food Chemistry* 239:453–463.

Abejón, R, MP Belleville, J Sanchez-Marcano, A Garea, and A Irabien. 2018. Optimal design of industrial scale continuous process for fractionation by membrane technologies of protein hydrolysate derived from fish wastes. *Separation and Purification Technology* 197:137–146.

Al Khawli, Fadila, Mirian Pateiro, Rubén Domínguez, José M Lorenzo, Patricia Gullón, Katerina Kousoulaki, Emilia Ferrer, Houda Berrada, and Francisco J Barba. 2019. Innovative green technologies of intensification for valorization of seafood and their by-products. *Marine Drugs* 17 (12):689.

Alavi, Farhad, Majid Jamshidian, and Karamatollah Rezaei. 2019. Applying native proteases from melon to hydrolyze kilka fish proteins (*Clupeonella cultriventris caspia*) compared to commercial enzyme Alcalase. *Food Chemistry* 277:314–322.

Amado, Isabel Rodríguez, José Antonio Vázquez, Mª Pilar González, and Miguel Anxo Murado. 2013. Production of antihypertensive and antioxidant activities by enzymatic hydrolysis of pr otein concentrates recovered by ultrafiltration from cuttlefish processing wastewaters. *Biochem. Eng. J.* 76:43–54.

Amado, Isabel Rodríguez, José Antonio Vázquez, Pilar González, Diego Esteban-Fernández, Mónica Carrera, and Carmen Piñeiro. 2014. Identification of the major ACE-inhibitory peptides produced by enzymatic hydrolysis of a protein concentrate from cuttlefish wastewater. *Mar. Drugs* 12 (3):1390–1405.

Bashir, Khawaja Muhammad Imran, Jae Hak Sohn, Jin-Soo Kim, and Jae-Suk Choi. 2020. Identification and characterization of novel antioxidant peptides from mackerel (*Scomber japonicus*) muscle protein hydrolysates. *Food Chemistry* 323:126809.

Benvenuto, John A, Robert A Newman, Gary S Bignami, TJG Raybould, Martin N Raber, Laura Esparza, and Ronald S Walters. 1992. Phase II clinical and pharmacological study of didemnin B in patients with metastatic breast cancer. *Investigational New Drugs* 10 (2):113–117.

Bruno, Siewe Fabrice, Franck Junior Anta Akouan Ekorong, Sandesh S Karkal, MSB Cathrine, and Tanaji G Kudre. 2019. Green and innovative techniques for recovery of valuable compounds from seafood by-products and discards: A review. *Trends in Food Science & Technology* 85:10–22.

Bui, Xuan Dong, Cong Tuan Vo, Viet Cuong Bui, Thi My Pham, Thi Thu Hien Bui, Toan Nguyen-Sy, Thi Dong Phuong Nguyen, Kit Wayne Chew, MD Mukatova, and Pau Loke Show. 2020. Optimization of production parameters of fish protein hydrolysate from Sarda orientalis black muscle (by-product) using protease enzyme. *Clean Technologies and Environmental Policy* 23:31–40.

Cai, Luyun, Xiaosa Wu, Yuhao Zhang, Xiuxia Li, Shuai Ma, and Jianrong Li. 2015. Purification and characterization of three antioxidant peptides from protein hydrolysate of grass carp (*Ctenopharyngodon idella*) skin. *Journal of Functional Foods* 16:234–242.

Calabrò, V, and A Basile. 2011. Fundamental membrane processes, science and engineering. In *Advanced Membrane Science and Technology for Sustainable Energy and Environmental Applications*: Elsevier.

Caruso, Gabriella, Rosanna Floris, Claudio Serangeli, and Luisa Di Paola. 2020. Fishery Wastes as a yet undiscovered treasure from the sea: Biomolecules sources, extraction methods and valorization. *Marine Drugs* 18 (12):622.

Cerrato, Andrea, Anna Laura Capriotti, Federico Capuano, Chiara Cavaliere, Angela Michela Immacolata Montone, Carmela Maria Montone, Susy Piovesana, Riccardo Zenezini Chiozzi, and Aldo Laganà. 2020. Identification and antimicrobial activity of medium-sized and short peptides from yellowfin tuna (*Thunnus albacares*) simulated gastrointestinal digestion. *Foods* 9 (9):1185.

Chakrabarti, Subhadeep, Snigdha Guha, and Kaustav Majumder. 2018. Food-derived bioactive peptides in human health: Challenges and opportunities. *Nutrients* 10 (11):1738.

Chen, Shengyang, Qian Yang, Xuan Chen, Yongqi Tian, Zhiyu Liu, and Shaoyun Wang. 2020. Bioactive peptides derived from crimson snapper and *in vivo* anti-aging effects on fat diet-induced high fat *Drosophila melanogaster*. *Food & Function* 11 (1):524–533.

Chen, Wenwei, Yao Hong, Zhenbao Jia, and Guangrong Huang. 2020. Purification and identification of antioxidant peptides from hydrolysates of large yellow croaker (*Pseudosciaena crocea*) scales. *Transactions of the ASABE* 63 (2):289–294.

Chen, Yingyun, Xiang Gao, Yuxi Wei, Qi Liu, Yuhong Jiang, Ling Zhao, and Sadeeq Ulaah. 2018. Isolation, purification and the anti-hypertensive effect of a novel angiotensin I-converting enzyme (ACE) inhibitory peptide from *Ruditapes philippinarum* fermented with *Bacillus natto*. *Food & Function* 9 (10):5230–5237.

Chen, Yixuan, Jianchu Chen, Juan Chen, Huilin Yu, Yangfan Zheng, Jiawen Zhao, and Jiajin Zhu. 2020. Recent advances in seafood bioactive peptides and their potential for managing osteoporosis. *Critical Reviews in Food Science and Nutrition*:1–17. doi: 10.1080/10408398.2020.1836606

Cheung, Randy Chi Fai, Tzi Bun Ng, and Jack Ho Wong. 2015. Marine peptides: Bioactivities and applications. *Marine Drugs* 13 (7):4006–4043.

Chi, Chang-Feng, Fa-Yuan Hu, Bin Wang, Xi-Jie Ren, Shang-Gui Deng, and Chang-Wen Wu. 2015. Purification and characterization of three antioxidant peptides from protein hydrolyzate of croceine croaker (*Pseudosciaena crocea*) muscle. *Food Chemistry* 168:662–667.

da Silva, Cristiane Pereira, Ranilson Souza Bezerra, Ana Célia Oliveira dos Santos, Júlio Brando Messias, Claudio Renato Oliveira Beltrão de Castro, and Luiz Bezerra Carvalho Junior. 2017. Biological value of shrimp protein hydrolysate by-product produced by autolysis. *LWT* 80:456–461.

de la Fuente, Beatriz, Adrián Tornos, Andrea Prínceп, José M Lorenzo, Mirian Pateiro, Houda Berrada, Francisco J Barba, María-José Ruiz, and Francisco J Martí-Quijal. 2020. Scaling-up processes: Patents and commercial applications. In *Advances in Food and Nutrition Research*: Elsevier.

Defer, Diane, Florie Desriac, Joël Henry, Nathalie Bourgougnon, Michèle Baudy-Floc'H, Benjamin Brillet, Patrick Le Chevalier, and Yannick Fleury. 2013. Antimicrobial peptides in oyster hemolymph: The bacterial connection. *Fish & Shellfish Immunology* 34 (6):1439–1447.

Destoumieux-Garzón, Delphine, Rafael Diego Rosa, Paulina Schmitt, Cairé Barreto, Jeremie Vidal-Dupiol, Guillaume Mitta, Yannick Gueguen, and Evelyne Bachere. 2016. Antimicrobial peptides in marine invertebrate health and disease. *Philosophical Transactions of the Royal Society B: Biological Sciences* 371 (1695):20150300.

Djellouli, Mustapha, M Elvira López-Caballero, Salima Roudj, and Oscar Martínez-Álvarez. 2020. Hydrolysis of shrimp cooking juice waste for the production of antioxidant peptides and proteases by *Enterococcus faecalis* DM19. *Waste and Biomass Valorization* 13:37–3752. https://doi.org/10.1007/s12649-020-01263-3

dos Santos Aguilar, Jessika Gonçalves, and Hélia Harumi Sato. 2018. Microbial proteases: Production and application in obtaining protein hydrolysates. *Food Research International* 103:253–262.

Durand, Rachel, Geneviève Pellerin, Jacinthe Thibodeau, Erwann Fraboulet, André Marette, and Laurent Bazinet. 2020. Screening for metabolic syndrome application of a herring by-product hydrolysate after its separation by electrodialysis with ultrafiltration membrane and identification of novel anti-inflammatory peptides. *Separation and Purification Technology* 235:116205.

Ennaas, Nadia, Riadh Hammami, Lucie Beaulieu, and Ismail Fliss. 2015. Purification and characterization of four antibacterial peptides from protamex hydrolysate of Atlantic mackerel (*Scomber scombrus*) by-products. *Biochemical and Biophysical Research Communications* 462 (3):195–200.

FAO. 2020. *The State of World Fisheries and Aquaculture 2020. Sustainability in Action.* Food and Agriculture Organization of the United Nations.

Fernandes, Pedro. 2016. Enzymes in fish and seafood processing. *Frontiers in Bioengineering and Biotechnology* 4:59.

Gao, Xiujun, Peisheng Yan, Jianbing Wang, Xin Liu, and Jiajia Yu. 2016. Utilization of shrimp by-products by bioconversion with medical fungi for angiotensin I-converting enzyme inhibitor and antioxidant. *Journal of Aquatic Food Product Technology* 25 (5):694–707.

García-Santiago, Xela, Amaya Franco-Uría, Luis T Antelo, José A Vázquez, Ricardo Pérez-Martín, M Teresa Moreira, and Gumersindo Feijoo. 2020. Eco-efficiency of a marine biorefinery for valorization of cartilaginous fish biomass. *Journal of Industrial Ecology* 25:789–801. https://doi.org/10.1111/jiec.13066

Ghaly, AE, VV Ramakrishnan, MS Brooks, SM Budge, and D Dave. 2013. Fish processing wastes as a potential source of proteins. Amino acids and oils: A critical review. *J Microb Biochem Technol* 5 (4):107–129.

Gingras, Denis, Dominique Boivin, Christophe Deckers, Sébastien Gendron, Chantal Barthomeuf, and Richard Béliveau. 2003. Neovastat—A novel antiangiogenic drug for cancer therapy. *Anti-Cancer Drugs* 14 (2):91–96.

Gomez, Honey Lyn R, Jose P Peralta, Lhumen A Tejano, and Yu-Wei Chang. 2019. In silico and in vitro assessment of portuguese oyster (*Crassostrea angulata*) proteins as precursor of bioactive peptides. *International Journal of Molecular Sciences* 20 (20):5191.

Guénard, Frédéric, Hélène Jacques, Claudia Gagnon, André Marette, and Marie-Claude Vohl. 2019. Acute effects of single doses of Bonito fish peptides and vitamin D on whole blood gene expression levels: A randomized controlled trial. *International Journal of Molecular Sciences* 20 (8):1944.

Guérard, Fabienne, Nicolas Decourcelle, Claire Sabourin, Corinne Floch-Laizet, Laurent Le Grel, Pascal Le Floc'H, Florence Gourlay, Ronan Le Delezir, Pascal Jaouen, and Patrick Bourseau. 2010. Recent developments of marine ingredients for food and nutraceutical applications: A review. *Journal des Sciences Halieutique et Aquatique* 2:21–27.

Guo, Zixuan, Fujunzhu Zhao, Hui Chen, Maolin Tu, Shuaifei Tao, Zhenyu Wang, Chao Wu, Shudong He, and Ming Du. 2020. Heat treatments of peptides from oyster (*Crassostrea gigas*) and the impact on their digestibility and angiotensin I converting enzyme inhibitory activity. *Food Science and Biotechnology* 29 (7):961–967.

Haard, Norman F, Benjamin K Simpson, and Zdzisław E Sikorski. 1994. Biotechnological applications of seafood proteins and other nitrogenous compounds. In *Seafood proteins*: Springer.

Halim, NRA, HM Yusof, and NM Sarbon. 2016. Functional and bioactive properties of fish protein hydolysates and peptides: A comprehensive review. *Trends in Food Science & Technology* 51:24–33.

Hamzeh, Ali, Parinya Noisa, and Jirawat Yongsawatdigul. 2020. Characterization of the antioxidant and ACE-inhibitory activities of Thai fish sauce at different stages of fermentation. *Journal of Functional Foods* 64:103699.

Harnedy-Rothwell, Pádraigín A., Chris M. McLaughlin, Martina B. O'Keeffe, Aurélien V. Le Gouic, Philip J. Allsopp, Emeir M. McSorley, Shaun Sharkey, Jason Whooley, Brian McGovern, Finbarr P. M. O'Harte, and Richard J. FitzGerald. 2020. Identification and characterisation of peptides from a boarfish (*Capros aper*) protein hydrolysate displaying in vitro dipeptidyl peptidase-IV (DPP-IV) inhibitory and insulinotropic activity. *Food Research International* 131:108989.

Hsu, K.C.; Cheng, M.L.; & Hwang, J.S. 2007. Hydrolysates from tuna cooking juice as an anti-hypertensive agent. *J. Food Drug Anal.* 15 (2):169–173.

Hsu, Kuo-Chiang. 2010. Purification of antioxidative peptides prepared from enzymatic hydrolysates of tuna dark muscle by-product. *Food Chemistry* 122 (1):42–48.

Hsu, Kuo-Chiang, Eunice CY Li-Chan, and Chia-Ling Jao. 2011. Antiproliferative activity of peptides prepared from enzymatic hydrolysates of tuna dark muscle on human breast cancer cell line MCF-7. *Food Chemistry* 126 (2):617–622.

Hsu, Kuo-Chiang, Geng-Hwang Lu, and Chia-Ling Jao. 2009. Antioxidative properties of peptides prepared from tuna cooking juice hydrolysates with orientase (*Bacillus subtilis*). *Food Research International* 42 (5):647–652.

Hu, Xiao-Meng, Yu-Mei Wang, Yu-Qin Zhao, Chang-Feng Chi, and Bin Wang. 2020. Antioxidant peptides from the protein hydrolysate of monkfish (*Lophius litulon*) muscle: Purification, identification, and cytoprotective function on HepG2 cells damage by H2O2. *Marine Drugs* 18 (3):153.

Hung, Chuan-Chuan, Yu-Hsuan Yang, Pei-Feng Kuo, and Kuo-Chiang Hsu. 2014. Protein hydrolysates from tuna cooking juice inhibit cell growth and induce apoptosis of human breast cancer cell line MCF-7. *Journal of Functional Foods* 11:563–570.

Hyung, Jun-Ho, Chang-Bum Ahn, and Jae-Young Je. 2017. Ark shell protein hydrolysates inhibit adipogenesis in mouse mesenchymal stem cells through the down-regulation of transcriptional factors. *RSC Advances* 7 (11):6223–6228.

Idowu, Anthony Temitope, Soottawat Benjakul, Sittichoke Sinthusamran, Pornsatit Sookchoo, and Hideki Kishimura. 2019. Protein hydrolysate from salmon frames: Production, characteristics and antioxidative activity. *Journal of Food Biochemistry* 43 (2):e12734.

Ishak, NH, and NM Sarbon. 2018. A review of protein hydrolysates and bioactive peptides deriving from wastes generated by fish processing. *Food and Bioprocess Technology* 11 (1):2–16.

Jagonia, Rofel Vincent S, Rejemae G Dela Victoria, Lydia M Bajo, and Roger S Tan. 2019. Conus striatus venom exhibits non-hepatotoxic and non-nephrotoxic potent analgesic activity in mice. *Molecular Biology Reports* 46 (5):5479–5486.

Jakubczyk, Anna, Monika Karaś, Kamila Rybczyńska-Tkaczyk, Ewelina Zielińska, and Damian Zieliński. 2020. Current trends of bioactive peptides—New sources and therapeutic effect. *Foods* 9 (7):846.

Jao, Chia-Ling, and Wen-Ching Ko. 2002. 1,1-Diphenyl-2-picrylhydrazyl (DPPH) radical scavenging by protein hydrolyzates from tuna cooking juice. *Fisheries Science* 68 (2):430–435.

Je, Jae-Young, Zhong-Ji Qian, Hee-Guk Byun, and Se-Kwon Kim. 2007. Purification and characterization of an antioxidant peptide obtained from tuna backbone protein by enzymatic hydrolysis. *Process Biochemistry* 42 (5):840–846.

Je, Jae-Young, Zhong-Ji Qian, Sang-Hoon Lee, Hee-Guk Byun, and Se-Kwon Kim. 2008. Purification and antioxidant properties of bigeye tuna (*Thunnus obesus*) dark muscle peptide on free radical-mediated oxidative systems. *Journal of Medicinal Food* 11 (4):629–637.

Jemil, Ines, Ola Abdelhedi, Rim Nasri, Leticia Mora, Mourad Jridi, Maria-Concepción Aristoy, Fidel Toldrá, and Moncef Nasri. 2017. Novel bioactive peptides from enzymatic hydrolysate of Sardinelle (*Sardinella aurita*) muscle proteins hydrolysed by *Bacillus subtilis* A26 proteases. *Food Research International* 100:121–133.

Jin, Ai-Hua, Markus Muttenthaler, Sebastien Dutertre, SWA Himaya, Quentin Kaas, David J Craik, Richard J Lewis, and Paul F Alewood. 2019. Conotoxins: Chemistry and biology. *Chemical Reviews* 119 (21):11510–11549.

Jin, Ji-Eun, Chang-Bum Ahn, and Jae-Young Je. 2018. Purification and characterization of antioxidant peptides from enzymatically hydrolyzed ark shell (*Scapharca subcrenata*). *Process Biochemistry* 72:170–176.

Kim, Hyun-Soo, WonWoo Lee, Thilina U. Jayawardena, Nalae Kang, Min Cheol Kang, Seok-Chun Ko, Jeong Min Lee, Mi-Jin Yim, Dae-Sung Lee, and You-Jin Jeon. 2020. Potential precursor of Angiotensin-I Converting enzyme (ACE) inhibitory activity and structural properties of peptide from peptic hydrolysate of cutlassfish Muscle. *Journal of Aquatic Food Product Technology* 29 (6):544–552.

Kim, Se-Kwon, and Jayachandran Venkatesan. 2014. Introduction to seafood processing by-products. In *Seafood Processing By-Products*: Springer.

Kiran, N, G Siddiqui, AN Khan, K Ibrar, and P Tushar. 2014. Extraction and screening of bioactive compounds with antimicrobial properties from selected species of mollusk and crustacean. *J Clin Cell Immunol* 5 (1):1000189.

Kleekayai, Thanyaporn, Pádraigín A Harnedy, Martina B O'Keeffe, Alexey A Poyarkov, Adriana CunhaNeves, Worapot Suntornsuk, and Richard J FitzGerald. 2015. Extraction of antioxidant and ACE inhibitory peptides from Thai traditional fermented shrimp pastes. *Food Chemistry* 176:441–447.

Korczek, Klaudia Róża, Joanna Tkaczewska, Iwona Duda, and Władysław Migdał. 2020. Effect of heat treatment on the antioxidant and antihypertensive activity as well as *in vitro* digestion stability of mackerel (*Scomber scombrus*) protein hydrolysates. *Journal of Aquatic Food Product Technology* 29 (1):73–89.

Kucuk, Omer, Mary L Young, Thomas M Habermann, Barbara C Wolf, Jose Jimeno, and Peter A Cassileth. 2000. Phase II trial of didemnin B in previously treated non-Hodgkin's lymphoma: An eastern cooperative oncology group (ECOG) study. *American Journal of Clinical Oncology* 23 (3):273–277.

Kumari, Anjali, Shijin Ameri, Palavancha Ravikrishna, Arul Dhayalan, S Kamala-Kannan, T Selvankumar, and M Govarthanan. 2019. Isolation and characterization of conotoxin protein from *Conus inscriptus* and its potential anticancer activity against cervical cancer (HeLa-HPV 16 associated) cell lines. *International Journal of Peptide Research and Therapeutics* 26:1051–1059. https://doi.org/10.1007/s10989-019-09907-2

Latorres, JM, DG Rios, G Saggiomo, W Wasielesky, and C Prentice-Hernandez. 2018. Functional and antioxidant properties of protein hydrolysates obtained from white shrimp (*Litopenaeus vannamei*). *Journal of Food Science and Technology* 55 (2):721–729.

Lazcano-Pérez, Fernando, Sergio A Roman-Gonzalez, Nuria Sánchez-Puig, and Roberto Arreguin-Espinosa. 2012. Bioactive peptides from marine organisms: A short overview. *Protein and Peptide Letters* 19 (7):700–707.

Lebbe, Eline KM, Maarten GK Ghequire, Steve Peigneur, Bea G Mille, Prabha Devi, Samuthirapandian Ravichandran, Etienne Waelkens, Lisette D'Souza, René De Mot, and Jan Tytgat. 2016. Novel conopeptides of largely unexplored Indo Pacific *Conus* sp. *Marine drugs* 14 (11):199.

Lee, Jung Kwon, and Hee-Guk Byun. 2019. Characterization of antioxidative peptide purified from black eelpout (*Lycodes diapterus*) hydrolysate. *Fisheries and Aquatic Sciences* 22 (1):22.

Lee, Jung Kwon, Eunice CY Li-Chan, Joong-Kyun Jeon, and Hee-Guk Byun, eds. 2014. *Development of Functional Materials from Seafood By-Products by Membrane Separation Technology*, *Seafood Processing By-Products*: Springer.

Li-Chan, Eunice CY. 2015. Bioactive peptides and protein hydrolysates: Research trends and challenges for application as nutraceuticals and functional food ingredients. *Current Opinion in Food Science* 1:28–37.

Li, Zhiyu, Jianyi Wang, Baodong Zheng, and Zebin Guo. 2020. Impact of combined ultrasound-microwave treatment on structural and functional properties of golden threadfin bream (*Nemipterus virgatus*) myofibrillar proteins and hydrolysates. *Ultrasonics sonochemistry* 65:105063.

Lippe, Giovanna, Barbara Prandi, Tiziana Bongiorno, Francesca Mancuso, Emilio Tibaldi, Andrea Faccini, Stefano Sforza, and Mara Lucia Stecchini. 2021. The effect of pre-slaughter starvation on muscle protein degradation in sea bream (*Sparus aurata*): Formation of ACE inhibitory peptides and increased digestibility of fillet. *European Food Research and Technology* 247 (1):259–271.

Liu, Pengru, Xiongdiao Lan, Muhammad Yaseen, Shanguang Wu, Xuezhen Feng, Liqin Zhou, Jianhua Sun, Anping Liao, Dankui Liao, and Lixia Sun. 2019. Purification, characterization and evaluation of inhibitory mechanism of ACE inhibitory peptides from pearl oyster (*Pinctada fucata martensii*) meat protein hydrolysate. *Marine Drugs* 17 (8):463.

Marti-Quijal, Francisco J, Fabienne Remize, Giuseppe Meca, Emilia Ferrer, María-José Ruiz, and Francisco J Barba. 2020. Fermentation in fish and by-products processing: An overview of current research and future prospects. *Current Opinion in Food Science* 31:9–16.

Mason, A James, Philippe Bertani, Gilles Moulay, Arnaud Marquette, Barbara Perrone, Alex F Drake, Antoine Kichler, and Burkhard Bechinger. 2007. Membrane interaction of chrysophsin-1, a histidine-rich antimicrobial peptide from red sea bream. *Biochemistry* 46 (51):15175–15187.

Mazorra-Manzano, MA, JC Ramírez-Suárez, JM Moreno-Hernández, and R Pacheco-Aguilar. 2018. Seafood proteins. In *Proteins in Food Processing*: Elsevier.

Mazorra-Manzano, MA, JC Ramírez-Suarez, and RY Yada. 2018. Plant proteases for bioactive peptides release: A review. *Critical Reviews in Food Science and Nutrition* 58 (13):2147–2163.

Mirzapour-Kouhdasht, Armin, Marzieh Moosavi-Nasab, Young-Min Kim, and Jong-Bang Eun. 2021. Antioxidant mechanism, antibacterial activity, and functional characterization of peptide fractions obtained from barred mackerel gelatin with a focus on application in carbonated beverages. *Food Chemistry* 342:128339.

Murthy, Pushpa S, Amit Kumar Rai, and N Bhaskar. 2014. Fermentative recovery of lipids and proteins from freshwater fish head waste with reference to antimicrobial and antioxidant properties of protein hydrolysate. *Journal of Food Science and Technology* 51 (9):1884–1892.

Najafian, Leila, and Abdul Salam Babji. 2018. Fractionation and identification of novel antioxidant peptides from fermented fish (pekasam). *Journal of Food Measurement and Characterization* 12 (3):2174–2183.

Najafian, Leila, and Abdul Salam Babji. 2019. Purification and identification of antioxidant peptides from fermented fish sauce (Budu). *Journal of Aquatic Food Product Technology* 28 (1):14–24.

Nasri, Rim, Ola Abdelhedi, Ines Jemil, Ikram Ben Amor, Abdelfattah Elfeki, Jalel Gargouri, Ahmed Boualga, Maha Karra-Châabouni, and Moncef Nasri. 2018. Preventive effect of goby fish protein hydrolysates on hyperlipidemia and cardiovascular disease in Wistar rats fed a high-fat/fructose diet. *RSC Advances* 8 (17):9383–9393.

Navarro-Peraza, Rosa Stephanie, Idalia Osuna-Ruiz, María Elena Lugo-Sánchez, Ramón Pacheco-Aguilar, Juan Carlos Ramírez-Suárez, Armando Burgos-Hernández, Emmanuel Martínez-Montaño, and Jesús Aarón Salazar-Leyva. 2020. Structural and biological properties of protein hydrolysates from seafood by-products: A review focused on fishery effluents. *Food Science and Technology* 40:1–5.

Neshani, Alireza, Hosna Zare, Mohammad Reza Akbari Eidgahi, Azad Khaledi, and Kiarash Ghazvini. 2019. Epinecidin-1, a highly potent marine antimicrobial peptide with anticancer and immunomodulatory activities. *BMC Pharmacology and Toxicology* 20 (1):1–11.

Ngo, Dai-Hung, Kyong-Hwa Kang, BoMi Ryu, Thanh-Sang Vo, Won-Kyo Jung, Hee-Guk Byun, and Se-Kwon Kim. 2015. Angiotensin-I converting enzyme inhibitory peptides from antihypertensive skate (*Okamejei kenojei*) skin gelatin hydrolysate in spontaneously hypertensive rats. *Food Chemistry* 174:37–43.

Oh, Jae-Young, Jun-Geon Je, Hyo-Geun Lee, Eun- A. Kim, Sang I. Kang, Jung-Suck Lee, and You-Jin Jeon. 2020. Anti-hypertensive activity of novel peptides identified from olive flounder (*Paralichthys olivaceus*) surimi. *Foods* 9 (5):647. https://doi.org/10.3390/foods9050647

Oh, Yunok, Chang-Bum Ahn, Jun-Ho Hyung, and Jae-Young Je. 2019. Two novel peptides from ark shell protein stimulate osteoblast differentiation and rescue ovariectomy-induced bone loss. *Toxicology and Applied Pharmacology* 385:114779.

Oh, Yunok, Chang-Bum Ahn, and Jae-Young Je. 2020. Low molecular weight blue mussel hydrolysates inhibit adipogenesis in mouse mesenchymal stem cells through upregulating HO-1/Nrf2 pathway. *Food Research International* 136:109603.

Oh, Yunok, Chang-Bum Ahn, and Jae-Young Je. 2021. Cytoprotective role of edible seahorse (*Hippocampus abdominalis*)-derived peptides in H2O2-induced oxidative stress in human umbilical vein endothelial cells. *Marine Drugs* 19 (2):86. https://doi.org/10.3390/md19020086

Oh, Yunok, Chang-Bum Ahn, and Jae-Young Je. 2020. Ark shell protein-derived bioactive peptides promote osteoblastic differentiation through upregulation of the canonical Wnt/β-catenin signaling in human bone marrow-derived mesenchymal stem cells. *Journal of Food Biochemistry* 44 (10):e13440.

Olsen, Ragnar L, and Jogeir Toppe. 2017. Fish silage hydrolysates: Not only a feed nutrient, but also a useful feed additive. *Trends in Food Science & Technology* 66:93–97.

Osajima, Katsuhiro, Toshio Ninomiya, Melody Harwood, and Barbara Danielewska-Nikiel. 2009. Safety evaluation of a peptide product derived from sardine protein hydrolysates (valtyron). *International Journal of Toxicology* 28 (5):341–356.

Ozyurt, G, M Boga, Y Uçar, EK Boga, and A Polat. 2018. Chemical, bioactive properties and in vitro digestibility of spray-dried fish silages: Comparison of two discard fish (*Equulites klunzingeri* and *Carassius gibelio*) silages. *Aquaculture Nutrition* 24 (3):998–1005.

Özyurt, Gülsün, Ali Serhat Özkütük, Yılmaz Uçar, Mustafa Durmuş, and Yeşim Özoğul. 2018. Fatty acid composition and oxidative stability of oils recovered from acid silage and bacterial fermentation of fish (Sea bass–*Dicentrarchus labrax*) by-products. *International Journal of Food Science & Technology* 53 (5):1255–1261.

Pan, Xin, Yu-Qin Zhao, Fa-Yuan Hu, and Bin Wang. 2016. Preparation and identification of antioxidant peptides from protein hydrolysate of skate (*Raja porosa*) cartilage. *Journal of Functional Foods* 25:220–230.

Park, Soo Yeon, Young-Sang Kim, Chang-Bum Ahn, and Jae-Young Je. 2016. Partial purification and identification of three antioxidant peptides with hepatoprotective effects from blue mussel (*Mytilus edulis*) hydrolysate by peptic hydrolysis. *Journal of Functional Foods* 20:88–95.

Petrova, Inna, Ignat Tolstorebrov, and Trygve Magne Eikevik. 2018. Production of fish protein hydrolysates step by step: Technological aspects, equipment used, major energy costs and methods of their minimizing. *International Aquatic Research* 10 (3):223–241.

Pezeshk, Samaneh, Seyed Mahdi Ojagh, Masoud Rezaei, and Bahareh Shabanpour. 2019. Fractionation of protein hydrolysates of fish waste using membrane ultrafiltration: Investigation of antibacterial and antioxidant activities. *Probiotics and Antimicrobial Proteins* 11 (3):1015–1022.

Qara, Sepide, and Mohammad B. Habibi Najafi. 2018. Bioactive properties of Kilka (*Clupeonella cultriventris* caspi) fish protein hydrolysates. *Journal of Food Measurement and Characterization* 12 (4):2263–2270.

Qian, Zhong-Ji, Jae-Young Je, and Se-Kwon Kim. 2007. Antihypertensive effect of angiotensin I converting enzyme-inhibitory peptide from hydrolysates of bigeye tuna dark muscle, *Thunnus obesus*. *Journal of Agricultural and Food Chemistry* 55 (21):8398–8403.

Qiao, Meiling, Maolin Tu, Hui Chen, Fengjiao Mao, Cuiping Yu, and Ming Du. 2018a. Identification and *In silico* prediction of anticoagulant peptides from the enzymatic hydrolysates of *Mytilus edulis* proteins. *International Journal of Molecular Sciences* 19 (7):2100. https://doi.org/10.3390/ijms19072100

Qiao, Meiling, Maolin Tu, Zhenyu Wang, Fengjiao Mao, Hui Chen, Lei Qin, and Ming Du. 2018b. Identification and antithrombotic activity of peptides from blue mussel (*Mytilus edulis*) Protein. *International Journal of Molecular Sciences* 19 (1):138. https://doi.org/10.3390/ijms19010138

Qiu, Yi-Ting, Yu-Mei Wang, Xiu-Rong Yang, Yu-Qin Zhao, Chang-Feng Chi, and Bin Wang. 2019. Gelatin and antioxidant peptides from gelatin hydrolysate of skipjack tuna (*Katsuwonus pelamis*) scales: Preparation, identification and activity evaluation. *Marine drugs* 17 (10):565. https://doi.org/10.3390/md17100565

Rajabzadeh, Mahsa, Parastoo Pourashouri, Bahare Shabanpour, and Alireza Alishahi. 2018. Amino acid composition, antioxidant and functional properties of protein hydrolysates from the roe of rainbow trout (*Oncorhynchus mykiss*). *International Journal of Food Science & Technology* 53 (2):313–319.

Rangel, Marisa, Carlos José Correia de Santana, Andréia Pinheiro, Lilian Dos Anjos, Tania Barth, Osmindo Rodrigues Pires Júnior, Wagner Fontes, and Mariana S Castro. 2017. Marine depsipeptides as promising pharmacotherapeutic agents. *Current Protein and Peptide Science* 18 (1):72–91.

Rios-Herrera, Gissel Daniela, Idalia Osuna Ruiz, Crisantema Hernández, Angel Valdez-Ortiz, Jorge Manuel Sandoval-Gallardo, Emmanuel Martinez-Montano, Jorge Saúl Ramírez-Pérez, and Jesús Aarón Salazar-Leyva. 2019. Chihuil sea catfish *Bagre panamensis* viscera as a new source of serine proteases: Semi-purification, biochemical characterization and application for protein hydrolysates production. *Waste and Biomass Valorization* 11:5821–5833. https://doi.org/10.1007/s12649-019-00895-4

Rivero-Pino, Fernando, F. Javier Espejo-Carpio, and Emilia M. Guadix. 2020. Production and identification of dipeptidyl peptidase IV (DPP-IV) inhibitory peptides from discarded *Sardine pilchardus* protein. *Food Chemistry* 328:127096.

Robinson, Samuel D, Eivind AB Undheim, Beatrix Ueberheide, and Glenn F King. 2017. Venom peptides as therapeutics: Advances, challenges and the future of venom-peptide discovery. *Expert Review of Proteomics* 14 (10):931–939.

Romero-Garay, Martha Guillermina, Emmanuel Martínez-Montaño, Adrián Hernández-Mendoza, Belinda Vallejo-Cordoba, Aarón Fernando González-Córdova, Efigenia Montalvo-González, and María de Lourdes García-Magaña. 2020. *Bromelia karatas* and *Bromelia pinguin*: Sources of plant proteases used for obtaining antioxidant hydrolysates from chicken and fish by-products. *Applied Biological Chemistry* 63 (1):1–11.

Roslan, Jumardi, Siti Mazlina Mustapa Kamal, Khairul Faezah Md. Yunos, and Norhafizah Abdullah. 2017. Assessment on multilayer ultrafiltration membrane for fractionation of tilapia by-product protein hydrolysate with angiotensin I-converting enzyme (ACE) inhibitory activity. *Separation and Purification Technology* 173:250–257.

Sable, Rushikesh, Pravin Parajuli, and Seetharama Jois. 2017. Peptides, peptidomimetics, and polypeptides from marine sources: A wealth of natural sources for pharmaceutical applications. *Marine Drugs* 15 (4):124.

Sae-Leaw, Thanasak, Supatra Karnjanapratum, Yvonne C O'Callaghan, Martina B O'Keeffe, Richard J FitzGerald, Nora M O'Brien, and Soottawat Benjakul. 2017. Purification and identification of antioxidant peptides from gelatin hydrolysate of seabass skin. *Journal of food biochemistry* 41 (3):e12350.

Saidi, Sami, Mongi Saoudi, and Raja Ben Amar. 2018. Valorisation of tuna processing waste biomass: Isolation, purification and characterisation of four novel antioxidant peptides from tuna by-product hydrolysate. *Environmental Science and Pollution Research* 25 (18):17383–17392.

Salampessy, Junus, Narsimha Reddy, Michael Phillips, and Kasipathy Kailasapathy. 2017. Isolation and characterization of nutraceutically potential ACE-Inhibitory peptides from leatherjacket (*Meuchenia* sp.) protein hydrolysates. *LWT* 80:430–436.

Sandoval-Gallardo, Jorge Manuel, Idalia Osuna-Ruiz, Emmanuel Martínez-Montaño, Crisantema Hernández, Miguel Ángel Hurtado-Oliva, Ángel Valdez-Ortiz, Gissel Daniela Rios-Herrera, Jesús Aarón Salazar-Leyva, and Jorge Saúl Ramírez-Pérez. 2020. Influence of enzymatic hydrolysis conditions on biochemical and antioxidant properties of Pacific thread herring (*Ophistonema libertate*) hydrolysates. *CyTA - Journal of Food* 18 (1):392–400.

Shavandi, Amin, Zhihao Hu, SueSiang Teh, Jenny Zhao, Alan Carne, Adnan Bekhit, and Alaa El-Din A. Bekhit. 2017. Antioxidant and functional properties of protein hydrolysates obtained from squid pen chitosan extraction effluent. *Food Chemistry* 227:194–201.

Shin, Dong M, Paul Y Holoye, William K Murphy, Arthur Forman, and Sozos C Papasozomenos. 1991. Phase I/II clinical trial of didemnin B in non-small-cell lung cancer: Neuromuscular toxicity is dose-limiting. *Cancer Chemotherapy and Pharmacology* 29 (2):145–149.

Slizyte, Rasa, Katariina Rommi, Revilija Mozuraityte, Peter Eck, Kathrine Five, and Turid Rustad. 2016. Bioactivities of fish protein hydrolysates from defatted salmon backbones. *Biotechnology Reports* 11:99–109.

Song, Ru, Qing-qing Shi, Assane Gninguue, Rong-bian Wei, and Hong-yu Luo. 2017. Purification and identification of a novel peptide derived from by-products fermentation of spiny head croaker (*Collichthys lucidus*) with antifungal effects on phytopathogens. *Process Biochemistry* 62:184–192.

Souza, Andre LA, Paola Díaz-Dellavalle, Andrea Cabrera, Patricia Larrañaga, Marco Dalla-Rizza, and Salvatore G De-Simone. 2013. Antimicrobial activity of pleurocidin is retained in Plc-2, a C-terminal 12-amino acid fragment. *Peptides* 45:78–84.

Sperstad, Sigmund V, Tor Haug, Hans-Matti Blencke, Olaf B Styrvold, Chun Li, and Klara Stensvåg. 2011. Antimicrobial peptides from marine invertebrates: Challenges and perspectives in marine antimicrobial peptide discovery. *Biotechnology Advances* 29 (5):519–530.

Suarez-Jimenez, Guadalupe-Miroslava, Armando Burgos-Hernandez, and Josafat-Marina Ezquerra-Brauer. 2012. Bioactive peptides and depsipeptides with anticancer potential: Sources from marine animals. *Marine Drugs* 10 (5):963–986.

Suárez-Jiménez, Guadalupe Miroslava, Rosario Maribel Robles-Sánches, Glória Yépiz-Plascencia, Armando Burgos-Hernández, and Josafat Marina Ezquerra-Brauer. 2015. In vitro antioxidant, antimutagenic and antiproliferative activities of collagen hydrolysates of jumbo squid (*Dosidicus gigas*) by-products. *Food Science and Technology* 35 (3):421–427.

Sun, Lixia, Shanguang Wu, Liqin Zhou, Feng Wang, Xiongdiao Lan, Jianhua Sun, Zhangfa Tong, and Dankui Liao. 2017. Separation and characterization of angiotensin I converting enzyme (ACE) inhibitory peptides from *Saurida elongata* proteins hydrolysate by IMAC-Ni2+. *Marine Drugs* 15 (2):29. https://doi.org/10.3390/md15020029

Tacon, Albert GJ, and Marc Metian. 2018. Food matters: Fish, income, and food supply—A comparative analysis. *Reviews in Fisheries Science & Aquaculture* 26 (1):15–28.

Tao, Jing, Yu-Qin Zhao, Chang-Feng Chi, and Bin Wang. 2018. Bioactive peptides from cartilage protein hydrolysate of spotless smoothhound and their antioxidant activity in vitro. *Marine Drugs* 16 (4):100. https://doi.org/10.3390/md16040100

Thuanthong, Mantaka, Cristian De Gobba, Nualpun Sirinupong, Wirote Youravong, and Jeanette Otte. 2017. Purification and characterization of angiotensin-converting enzyme-inhibitory peptides from Nile tilapia (*Oreochromis niloticus*) skin gelatine produced by an enzymatic membrane reactor. *Journal of Functional Foods* 36:243–254.

Tkaczewska, Joanna, Justyna Borawska-Dziadkiewicz, Piotr Kulawik, Iwona Duda, Małgorzata Morawska, and Barbara Mickowska. 2020. The effects of hydrolysis condition on the antioxidant activity of protein hydrolysate from *Cyprinus carpio* skin gelatin. *LWT* 117:108616.

Toopcham, Tidarat, Sittiruk Roytrakul, and Jirawat Yongsawatdigul. 2015. Characterization and identification of angiotensin I-converting enzyme (ACE) inhibitory peptides derived from tilapia using *Virgibacillus halodenitrificans* SK1-3-7 proteinases. *Journal of Functional Foods* 14:435–444.

Ucak, Ilknur, Maliha Afreen, Domenico Montesano, Celia Carrillo, Igor Tomasevic, Jesus Simal-Gandara, and Francisco J Barba. 2021. Functional and bioactive properties of peptides derived from marine side streams. *Marine Drugs* 19 (2):71.

Unnikrishnan, Parvathy, Binsi Puthenveetil Kizhakkethil, Joshy Chalil George, Zynudheen Aliyamveetil Abubacker, George Ninan, and Ravishankar Chandragiri Nagarajarao. 2020. Antioxidant peptides from dark meat of yellowfin tuna (*Thunnus albacares*): Process optimization and characterization. *Waste and Biomass Valorization* 12:1845–1860. https://doi.org/10.1007/s12649-020-01129-8

Van'T Land, M, E Vanderperren, and Katleen Raes. 2017. The effect of raw material combination on the nutritional composition and stability of four types of autolyzed fish silage. *Animal Feed Science and Technology* 234:284–294.

Vázquez, José A., Isabel Rodríguez-Amado, Carmen G. Sotelo, Noelia Sanz, Ricardo I. Pérez-Martín, and Jesus Valcárcel. 2020. Production, characterization, and bioactivity of fish protein hydrolysates from aquaculture turbot (*Scophthalmus maximus*) Wastes. *Biomolecules* 10 (2):310. https://doi.org/10.3390/biom10020310

Vázquez, José Antonio, Araceli Meduíña, Ana I Durán, Margarita Nogueira, Andrea Fernández-Compás, Ricardo I Pérez-Martín, and Isabel Rodríguez-Amado. 2019. Production of valuable compounds and bioactive metabolites from by-products of fish discards using chemical processing, enzymatic hydrolysis, and bacterial fermentation. *Marine Drugs* 17 (3):139.

Verdes, Aida, Prachi Anand, Juliette Gorson, Stephen Jannetti, Patrick Kelly, Abba Leffler, Danny Simpson, Girish Ramrattan, and Mande Holford. 2016. From mollusks to medicine: A venomics approach for the discovery and characterization of therapeutics from Terebridae peptide toxins. *Toxins* 8 (4):117.

Villamil, Oscar, Henry Váquiro, and José F Solanilla. 2017. Fish viscera protein hydrolysates: Production, potential applications and functional and bioactive properties. *Food Chemistry* 224:160–171.

Wang, Kai, Xiaoxue Yang, Wenyong Lou, and Xuewu Zhang. 2020. Discovery of dipeptidyl peptidase 4 inhibitory peptides from Largemouth bass (*Micropterus salmoides*) by a comprehensive approach. *Bioorganic Chemistry* 105:104432.

Wang, Limei, Jiang Sun, Shuhui Ding, and Bin Qi. 2018. Isolation and identification of novel antioxidant and antimicrobial oligopeptides from enzymatically hydrolyzed anchovy fish meal. *Process Biochemistry* 74:148–155.

Wang, Tingting, Lin Zheng, Tiantian Zhao, Qi Zhang, Zhitong Liu, Xiaoling Liu, and Mouming Zhao. 2020. Anti-diabetic effects of sea cucumber (*Holothuria nobilis*) hydrolysates in streptozotocin and high-fat-diet induced diabetic rats via activating the PI3K/Akt pathway. *Journal of Functional Foods* 75:104224.

Wang, Xueqin, Huahua Yu, Ronge Xing, Song Liu, Xiaolin Chen, and Pengcheng Li. 2019. Preparation and identification of antioxidative peptides from Pacific herring (*Clupea pallasii*) protein. *Molecules* 24 (10):1946. https://doi.org/10.3390/molecules24101946

Wenno, Max Robinson, Eddy Suprayitno, and Hardoko Hardoko. 2016. The physicochemical characteristics and angiotensin converting enzyme (ACE) inhibitory activity of skipjack tuna (*Katsuwonus pelamis*) "Bakasang". *Jurnal Teknologi* 78 (4–2). https://doi.org/10.11113/jt.v78.8191

Wu, Di, Chao Wu, Maolin Tu, Cuiping Yu, and Ming Du. 2018. Identification and analysis of bioactive peptides from scallops (*Chlamys farreri*) protein by simulated gastrointestinal digestion. *Journal of Food Processing and Preservation* 42 (9):e13760.

Xu, Zhe, Fujunzhu Zhao, Hui Chen, Shiqi Xu, Fengjiao Fan, Pujie Shi, Maolin Tu, Ziye Wang, and Ming Du. 2019. Nutritional properties and osteogenic activity of enzymatic hydrolysates of proteins from the blue mussel (*Mytilus edulis*). *Food & Function* 10 (12):7745–7754.

Yang, Jing-Iong, Jen-Yang Tang, Ya-Sin Liu, Hui-Ru Wang, Sheng-Yang Lee, Ching-Yu Yen, and Hsueh-Wei Chang. 2016. Roe protein hydrolysates of giant grouper (*Epinephelus lanceolatus*) inhibit cell proliferation of oral cancer cells involving apoptosis and oxidative stress. *BioMed Research International* 2016. https://doi.org/10.1155/2016/8305073

Yang, Xiu-Rong, Yi-Ting Qiu, Yu-Qin Zhao, Chang-Feng Chi, and Bin Wang. 2019a. Purification and characterization of antioxidant peptides derived from protein hydrolysate of the marine bivalve mollusk *Tergillarca granosa*. *Marine Drugs* 17 (5):251. https://doi.org/10.3390/md17050251

Yang, Xiu-Rong, Lun Zhang, Dong-Ge Ding, Chang-Feng Chi, Bin Wang, and Jian-Cong Huo. 2019b. Preparation, identification, and activity evaluation of eight antioxidant peptides from protein hydrolysate of Hairtail (*Trichiurus japonicas*) muscle. *Marine Drugs* 17 (1):78. https://doi.org/10.3390/md17020078

Yang, Xiu-Rong, Yu-Qin Zhao, Yi-Ting Qiu, Chang-Feng Chi, and Bin Wang. 2019. Preparation and characterization of gelatin and antioxidant peptides from gelatin hydrolysate of skipjack tuna (*Katsuwonus pelamis*) bone stimulated by *in vitro* gastrointestinal digestion. *Marine Drugs* 17 (2):78.

Yesmine, Ben Henda, Bonnet Antoine, Nunes Gonzalez da Silva Ortência Leocádia, Boscolo Wilson Rogério, Arnaudin Ingrid, Bridiau Nicolas, Maugard Thierry, Piot Jean-Marie, Sannier Frédéric, and Bordenave-Juchereau Stéphanie. 2017. Identification of ace inhibitory cryptides in Tilapia protein hydrolysate by UPLC–MS/MS coupled to database analysis. *Journal of Chromatography B* 1052:43–50.

Yu, Dingyi, Cong Wang, Yufeng Song, Junxiang Zhu, and Xiaojun Zhang. 2019. Discovery of novel angiotensin-converting enzyme inhibitory peptides from *Todarodes pacificus* and their inhibitory mechanism: *In silico* and *in vitro* studies. *International Journal of Molecular Sciences* 20 (17):4159. https://doi.org/10.3390/ijms20174159

Yu, Zhipeng, Ruotong Kan, Sijia Wu, Hui Guo, Wenzhu Zhao, Long Ding, Fuping Zheng, and Jingbo Liu. 2021. Xanthine oxidase inhibitory peptides derived from tuna protein: Virtual screening, inhibitory activity, and molecular mechanisms. *Journal of the Science of Food and Agriculture* 101 (4):1349–1354.

Yu, Zhipeng, Sijia Wu, Wenzhu Zhao, Geng Mi, Long Ding, Jianrong Li, and Jingbo Liu. 2020. Identification of novel angiotensin I-converting enzyme inhibitory peptide from collagen hydrolysates and its molecular inhibitory mechanism. *International Journal of Food Science & Technology* 55 (9):3145–3152.

Zamora-Sillero, Juan, Adem Gharsallaoui, and Carlos Prentice. 2018. Peptides from fish by-product protein hydrolysates and its functional properties: An overview. *Marine Biotechnology* 20 (2):118–130.

Zhang, Jing-Bo, Yu-Qin Zhao, Yu-Mei Wang, Chang-Feng Chi, and Bin Wang. 2019. Eight collagen peptides from hydrolysate fraction of spanish mackerel skins: Isolation, identification, and *in vitro* antioxidant activity evaluation. *Marine Drugs* 17 (4):224. https://doi.org/10.3390/md17040224

Zhang, Lun, Guo-Xu Zhao, Yu-Qin Zhao, Yi-Ting Qiu, Chang-Feng Chi, and Bin Wang. 2019. Identification and active evaluation of antioxidant peptides from protein hydrolysates of skipjack tuna (*Katsuwonus pelamis*) head. *Antioxidants* 8 (8):318.

Zhang, Ningning, Chong Zhang, Yuanyuan Chen, and Baodong Zheng. 2017. Purification and characterization of antioxidant peptides of *Pseudosciaena crocea* protein hydrolysates. *Molecules* 22 (1):57. https://doi.org/10.3390/molecules22010057

Zhang, Wanwan, Yifang Wei, Xiaoxiao Cao, Kaixin Guo, Qiangqiang Wang, Xiaochun Xiao, Xufeng Zhai, Dingding Wang, and Zebo Huang. 2021. Enzymatic preparation of *Crassostrea oyster* peptides and their promoting effect on male hormone production. *Journal of Ethnopharmacology* 264:113382.

Zhang, Yuqi, Huaigao Liu, Hui Hong, and Yongkang Luo. 2019. Purification and identification of dipeptidyl peptidase IV and angiotensin-converting enzyme inhibitory peptides from silver carp (*Hypophthalmichthys molitrix*) muscle hydrolysate. *European Food Research and Technology* 245 (1):243–255.

CHAPTER 6

Cereals

Hitomi Kumagai

CONTENTS

Introduction	129
Cereal Proteins	129
Antidiabetic Activity	130
Antihypertensive Activity	132
Immunomodulatory Activity	133
Opioid Activity	133
Antioxidant Activity	134
Conclusions	135
References	135

INTRODUCTION

Cereals, such as wheat and rice, are staple foods in many countries. Although the protein content in wheat and rice is only 8–13% and 6–7%, the daily protein intake from wheat and rice products is 7 g and 8 g, respectively, according to the National Health and Nutrition Survey in Japan [1]. As the daily protein intake from fish and meat products is 13 g and 17 g, respectively, people obtain about one-third of their protein intake from cereal sources, which is not inconsequential. When protein in food is taken into the body, it is hydrolyzed by proteases such as pepsin, trypsin, and chymotrypsin in the gut. The produced amino acids and small peptides are absorbed from the small intestine to the portal vein and transported to the liver. Therefore, bioactive proteins and peptides mainly exert their effects in the gut; however, small peptides can exert functional activities in the blood and organs if they are absorbed from the small intestine and transported to various parts of the body. This chapter summarizes the physiological functions of proteins and peptides from cereals.

CEREAL PROTEINS

Proteins are classified into four groups according to their solubility in solvents as follows: water-soluble (albumin), salt-soluble (globulin), 60–70% aqueous alcohol-soluble (prolamin), and dilute acid/alkali-soluble (glutelin). The contents of albumin, globulin, prolamin (gliadin), and glutelin (glutenin) in wheat protein are 5%, 9%, 40%, and 46%, respectively, and those of albumin, globulin, prolamin (zein), and glutelin in corn protein are 3–4%, 2–3%, 51–55%, and 33–39%, respectively. On the other hand, the contents of albumin, globulin, prolamin, and glutelin in rice endosperm are 4–22%, 5–13%, 1–5%,

DOI: 10.1201/9781003106524-8

and 70–80%, while those in rice bran are 24–40%, 15–35%, 1–6%, and 11–38%, respectively [2–8].

Cereals are rich in glutamine residues. Wheat gliadin contains 40–55% glutamine and 20–30% proline, and glutenin contains about 29% glutamine and 12% proline [9–11]. On the other hand, rice prolamin contains about 20% glutamine, and glutelin contains about 17% glutamine and 9% asparagine [12, 13].

ANTIDIABETIC ACTIVITY

Diabetes mellitus is a group of disorders characterized by hyperglycemia and is a serious disease that can lead to various complications such as arteriosclerosis obliterans, diabetic neuropathy, diabetic retinopathy, diabetic nephropathy, ischemia disease, cerebral infarction, and immune deficiency. The number of patients suffering from diabetes mellitus was estimated to be 422 million worldwide in 2014 [14], and about 1.3 million deaths are attributed to the disease. Approximately 95% of all diabetes cases are categorized as type 2, which is characterized by insulin resistance and/or relative deficiency. In order to prevent type 2 diabetes, it is important to control postprandial blood glucose levels in our daily life by appropriate food intake.

When starch is consumed, it is partially hydrolyzed by α-amylase in saliva, and the hydrolysate is transported to the small intestine via the stomach. There it is further hydrolyzed to glucose by α-amylase and maltase. Incretin peptide hormones such as glucagon-like peptide-1 (GLP-1) and glucose-dependent insulinotropic polypeptide (GIP) are secreted from intestinal mucosal cells in response to nutrient ingestion, which sends a signal for insulin to be released from the pancreas to the blood. Then, insulin promotes the uptake of glucose into tissues such as skeletal muscle and adipose tissue, which reduces the blood glucose level. Meanwhile, dipeptidyl peptidase 4 (DPP-4) functionally inactivates incretin by cleaving it, which reduces insulin secretion. Therefore, inhibition of digestive enzymes, such as α-amylase and maltase, prevention of glucose absorption from the small intestine, inhibition of DPP-4 activity, and promotion of insulin secretion would be effective to suppress elevation of the blood glucose level. However, excessive insulin secretion poses a risk, as it may lead to hypoglycemia (Figure 6.1).

Cereals commonly possess α-amylase inhibitors, likely because of their protective activity against insects [15–20]. Wheat α-amylase inhibitors have been identified in the albumin fraction [21, 22]. The molecular masses of α-amylase inhibitors in wheat albumin are reported as 12, 24, and 60 kDa. Members of the 12-kDa family (0.28, 0.32, 0.35, 0.39, and 0.48 inhibitors) are monomers, while those of the 24-kDa family (0.19, 0.36, 0.38, and 0.53 inhibitors) are dimers of the 12-kDa subunits with 10 cysteine residues in each subunit [8, 16, 22–25]. The molecular masses of α-amylase inhibitors in rice albumin are found to be 14, 16, and 25-kDa. PIs of 14-kDa proteins are between 6 and 9 and that of 21-kDa protein is 9 [26–28].

Wheat albumin has strong inhibitory activity against α-amylase from both insects and mammals, and suppresses the increase in blood glucose level in rats following oral administration of starch [28]. Wheat albumin suppresses postprandial hyperglycemia when administered to dogs together with canned dog food [29] and to human subjects together with a rice-containing meal [30]. On the other hand, rice albumin inhibits α-amylase from insects but not from mammals [28]. However, the oral administration of rice albumin together with starch suppresses the increase in blood glucose level. In addition, rice albumin suppresses postprandial hyperglycemia even upon glucose administration [28].

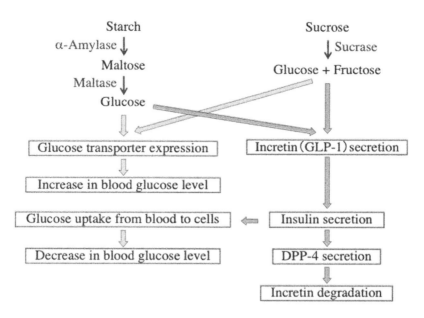

FIGURE 6.1 Blood glucose regulation by incretin, insulin, and DPP-4.

Rice albumin of 16 kDa is hydrolyzed to indigestible high-molecular-weight peptides of 14 kDa and low-molecular-weight peptides of less than 2 kDa by trypsin; the high-molecular-weight peptides adsorb glucose and promote its excretion like dietary fiber, while the low-molecular-weight peptides suppress the expression of sodium-dependent glucose transporter-1 [31]. Therefore, rice albumin suppresses the postprandial elevation of blood glucose level by dual functions. Buckwheat albumin inhibits α-amylase from insects and mammalian pancreas but not from human saliva [32]. Different from rice albumin, buckwheat albumin is hydrolyzed by pepsin and trypsin but retains high inhibitory activity against α-amylase even after digestion by pepsin for 2 h and trypsin for 6 h. Wheat albumin inhibits α-amylase in a noncompetitive manner, while buckwheat albumin inhibits it in a competitive manner [32]. Therefore, the bioactive peptides from buckwheat albumin may be glycosylated.

A rice protein hydrolysate prepared by simulated gastrointestinal digestion shows inhibitory activity against DPP-4 [33], and dipeptides, such as Ile-Pro, Met-Pro, Val-Pro, and Leu-Pro, are identified [34]. Some peptides stimulate GLP-1 secretion and increase the plasma insulin level, which results in a reduction in blood glucose level. Oral administration of a hydrolysate prepared from rice-endosperm protein or rice-bran protein by papain to rats increases total GLP-1 concentration in the portal vein [35]. The oral glucose tolerance test shows that oral administration of the hydrolysate retards the elevation of blood glucose concentration, but there is no difference in area under the curve. Intraperitoneal injection of glucose after oral administration of the hydrolysate from rice-endosperm protein or rice-bran increases total GLP-1 and insulin levels significantly. Ileal administration of the hydrolysate decreases DPP-4 activity and increases active GLP-1 level. Although there is no detailed information about the rice protein used for the preparation of these hydrolysates, it would likely not be from the albumin fraction, as the oral administration of rice albumin together with glucose has suppressive effects on plasma insulin levels [28].

A hydrolysate from corn zein stimulates GLP-1 secretion [36, 37]. Oat, buckwheat, and barley hydrolysates show inhibitory activity against DPP-4,

Glu-Phe-Leu-Leu-Ala-Gly-Asn-Asn-Lys and Leu-Gln-Ala-Phe-Glu-Pro-Leu-Arg being identified as bioactive peptides from oat [38].

ANTIHYPERTENSIVE ACTIVITY

Blood pressure is primarily regulated by the renin-angiotensin and kallikrein-kinin systems. Angiotensin-converting enzyme (ACE) converts angiotensin I to angiotensin II, angiotensin I (1-9) to angiotensin (1-7), and bradykinin to inactive bradykinin 1-5. When angiotensin II binds to type 1 angiotensin-II receptors (AT1) on smooth muscle cells in blood vessels and heart tissues, vasoconstriction occurs which leads to an increase in blood pressure. In addition, as bradykinin has a vasodilatory action, ACE inhibition is considered to suppress the increase in blood pressure (Figure 6.2).

Some peptides of food origin that are essentially composed of hydrophobic and/or aromatic amino-acid residues [39] and/or Pro-Pro sequences [40, 41] are known to inhibit ACE. Various peptides with ACE inhibitory activity are found in the hydrolysates of food proteins.

In silico analysis suggests that rice-bran protein has the potential to produce peptides with ACE inhibitory activity [42]. Rice-bran albumin hydrolysates produced by proteases such as Alcalase and Protamax show high ACE inhibitory activity [43]. Peptides from rice-bran protein digested by thermolysin show antihypertensive effects and reduce ACE activity in a spontaneously hypertensive rat (SHR) model, Leu-Arg-Ala and Tyr-Tyr being identified as the active peptides [44]. A rice-bran protein hydrolysate of the Thai jasmine variety also reduces blood pressure and ACE activity in a rat model of renovascular hypertension [45], Gly-Ser-Gly-Tyr-Phe being found to be the active peptide [46]. Red-mold rice prepared using *Monascus purpureus* shows ACE inhibitory activity, with Ile-Val, Val-Val-Tyr, Val-Phe, and Val-Trp identified as active peptides [47].

Wheat-germ protein, a by-product of flour milling, produces various ACE inhibitory peptides. A wheat-germ hydrolysate prepared using *Bacillus licheniformis* alkaline protease shows potent ACE inhibitory activity, Ile-Val-Tyr being identified as an active peptide [48]. The administration of Ile-Val-Tyr to SHRs reduces blood pressure due to the combined effect of itself and Val-Tyr produced from Ile-Val-Tyr by the action of aminopeptidases in plasma [49]. A hydrolysate of wheat-germ protein prepared by ultrasound before

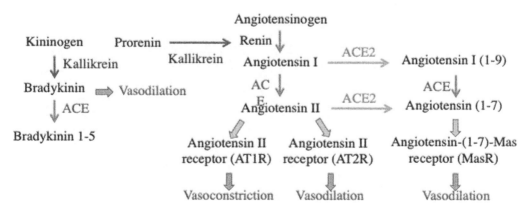

FIGURE 6.2 Blood pressure regulation by rennin-angiotensin and kallikrein-kinin systems.

enzymatic treatment exhibits ACE inhibitory activity [50]. Ser-Gly-Gly-Ser-Tyr-Ala-Asp-Glu-Leu-Val-Ser-Thr-Ala-Lys in the hydrolysate of wheat-germ protein prepared with proteinase K shows ACE inhibitory activity [51], while Leu-Gln-Pro, Ile-Gln-Pro, Leu-Arg-Pro, Val-Tyr, Ile-Tyr, and Thr-Phe from the autolysate of the by-product fraction of milled parts of wheat seeds exhibit ACE inhibitory activity [52]. Among peptides of wheat-bran protein prepared by Alcalase, those <1 kDa in size inhibit ACE and renin activities *in vitro* [53]. In addition, oral administration of the <1 kDa fraction to SHRs resulted in a decrease in systolic blood pressure, Asn-Leu, Gln-Leu, Phe-Leu, His-Ala-Leu, Ala-Ala-Val-Leu, Ala-Lys-Thr-Val-Phe, and Thr-Pro-Leu-Thr-Arg being identified in the fraction. A gliadin hydrolysate digested by trypsin as well as other proteases such as thermolysin, Clarex, Alcalase, and Esperase shows high ACE inhibitory activity, which produces fractions enriched in histidine and hydrophobic amino acids such as Pro, Val, Ile, Leu, and Phe [54]. Ile-Ala-Pro identified in a gliadin hydrolysate prepared with an acid protease has potent ACE inhibitory activity, and its intravenous administration to SHRs decreases blood pressure [55]. Wheat gluten hydrolyzed by Alcalase plus PaproA produces anti-hypertensive peptides containing tryptophan at the carboxyl-end, Ser-Ala-Gly-Gly-Tyr-Ile-Trp and Ala-Pro-Ala-Thr-Pro-Ser-Phe-Trp [56]. The intake of wholegrain diets (whole wheat/brown rice, barley, and half barley/half whole wheat-brown rice) reduces diastolic and mean arterial pressures and that of wheat/rice and half-and-half reduces systolic pressure in human subjects [57].

Gly-Pro-Pro purified from a buckwheat extract shows ACE inhibitory activity [58]. A hydrolysate prepared from buckwheat powder using pepsin, followed by chymotrypsin and trypsin, inhibits ACE activity, and its oral administration to SHRs decreases systolic blood pressure [59]. Tyr-Gln-Tyr, Val-Lys, Pro-Ser-Tyr, Phe-Tyr, and Leu-Gly-Ile show potent ACE inhibitory activity among the 11 peptides identified. The oral administration of buckwheat sprouts fermented using *Lactobacillus plantarum* KT to SHRs decreases both systolic and diastolic blood pressures, but that of buckwheat sprouts themselves does not have such a hypotensive effect [60]. Asp-Val-Trp-Tyr, Phe-Gln, Val-Val-Gly, Trp-Thr-Phe-Arg, Phe-Asp-Ala-Arg-Thr, and Val-Ala-Glu identified in the lactic acid-fermented buckwheat sprouts show blood pressure lowering effects when orally administered to SHRs.

An α-zein hydrolysate prepared using thermolysin inhibits ACE activity, Leu-Arg-Pro, Leu-Ser-Pro, and Leu-Gln-Pro being identified as active peptides [41]. Oral administration of Leu-Arg-Pro to SHRs decreases blood pressure.

IMMUNOMODULATORY ACTIVITY

Oryzatensin, Gly-Tyr-Pro-Met-Tyr-Pro-Leu-Pro-Arg, obtained by tryptic hydrolysis of rice protein exhibits immunomodulatory activity showing a biphasic ileum contraction [61]. The contractile profile of oryzatensin is similar to that of human complement C3a (70-77) and acts through complement C3a receptors [62].

OPIOID ACTIVITY

A pepsin-derived hydrolysate of wheat gluten shows opioid activity [63], and the bioactive peptides are Gly-Tyr-Tyr-Pro-Thr, Gly-Tyr-Tyr-Pro, Tyr-Gly-Gly-Trp-Leu, and Tyr-Gly-Gly-Trp [64]. Intracerebroventricular administration of gluten exorphin A5,

Gly-Tyr-Tyr-Pro-Thr, shows an antinociceptive effect, while its oral administration has no such effect. On the other hand, oral administration of gluten exorphin A5 suppresses the endogenous pain-inhibitory system [65].

ANTIOXIDANT ACTIVITY

Both rice protein and rice-bran protein hydrolysates show antioxidant activity as evaluated by the Hydrophilic Oxygen Radical Absorbance Capacity (H-ORAC) assay [34]. Among albumin, globulin, and glutelin fractionated from rice-bran protein, the antioxidant activity is higher in the order of albumin > globulin > glutelin by the Oxygen Radical Absorbance Capacity (ORAC) assay [66]. In addition, the denatured protein shows higher antioxidant activity than the native one. The molecular masses of the peptides identified are 800–2100 Da consisting of 6–21 amino-acid residues. Typical antioxidant peptides have hydrophobic amino acids at the N-terminus or in the third position adjacent to the C-terminus. Rice protein hydrolyzed using *Bacillus pumilus* AG1 shows high antioxidant activity by the 2,2'-azino-bis(3-ethylbenzothiazolin-6-sulfonic acid) (ABTS) method [67]. Rice-bran protein hydrolyzed by pepsin shows a low degree of hydrolysis and high antioxidant activity by the ABTS method [68]. The molecular masses of the 19 peptides identified are 630–3611 Da consisting of 6–30 amino-acid residues.

In addition to anti-hypertensive activity, peptides (<1 kDa) from wheat-bran protein prepared by Alcalase exhibit antioxidant activity by the ORAC assay [53]. A gluten hydrolysate fractionated by autofocusing also shows antioxidant activity [69]. The acidic fractions show strong DPPH radical scavenging activity, while the basic fractions suppress 2,2'-azobis(2-amidinopropane) dihydrochloride-induced oxidation of linoleic acid. An Alcalase-produced hydrolysate from purple wheat bran is revealed to have Trolox equivalent antioxidant capacity (TEAC), Cys-Gly-Phe-Pro-Gly-His-Cys, Arg-Asn-Phe, Gln-Ala-Cys, Ser-Ser-Cys, and Trp-Phe being identified as active peptides. Among them, Gln-Ala-Cys and Ser-Ser-Cys have stronger antioxidant capacity than the others [70].

A hydrolysate from a buckwheat protein isolate prepared by pepsin and pancreatin exhibits antioxidant activity as evaluated by scavenging assays for 2,2'-azino-bis(3-ethylbenzothiazolin-6-sulfonic acid) and hydroxyl radicals, Trp-Pro-Leu, Val-Pro-Trp, Val-Phe-Pro-Trp, and Pro-Trp being identified as active peptides [58]. Tartary buckwheat albumin hydrolyzed by alkaline protease shows hydroxyl-radical (OH·) scavenging activity [71]. Specifically, Gly-Glu-Val-Pro-Try, Try-Met-Glu-Asn-Phe, and Ala-Phe-Tyr-Arg-Trp (AFYRW) are identified as active peptides, AFYRW showing the highest scavenging activity of OH· and DPPH·(1,1-diphenyl-2-picrylhydrazyl) and inhibitory activity against lipid peroxidation.

Corn gluten meal hydrolyzed by alkaline protease and Flavourzyme shows antioxidant activity as evaluated by free radical scavenging capacity (1,1-diphenyl-2-picrylhydrazyl/2,2-azino-bis(3-ethylbenzothiazolin-6-sulphonic acid) diammonium salt/hydroxyl radical/superoxide radical anion), metal ion (Fe^{2+}/Cu^{2+}) chelating activity and lipid peroxidation inhibitory capacity, Leu-Pro-Phe, Leu-Leu-Pro-Phe, and Phe-Leu-Pro-Phe being identified as active peptides [72]. Similarly, a hydrolysate from corn gluten meal prepared using Alcalase and Flavourzyme is evaluated for DPPH radical scavenging activity, Fe^{2+} chelating activity, reducing power, hydroxyl radical scavenging activity, and superoxide anion radical scavenging activity, Cys-Ser-Gln-Ala-Pro-Leu-Ala, Tyr-Pro-Lys-Leu-Ala-Pro-Asn-Glu, and Tyr-Pro-Gln-Leu-Leu-Pro-Asn-Glu being identified as active peptides [73].

CONCLUSIONS

Cereal proteins produce bioactive peptides with various functions such as anti-diabetic, anti-hypertensive, and antioxidant activities. Most cereal plants possess α-amylase inhibitors in the albumin fraction as a natural defense mechanism against insects, and those in wheat and buckwheat inhibit the activity of α-amylase even from mammals. In addition, although not inhibiting the activity of mammalian α-amylase, rice albumin suppresses glucose absorption from the small intestine by promoting the excretion of glucose and decreasing the expression of a glucose transporter. Therefore, rice albumin suppresses postprandial hyperglycemia following the consumption of starch-based, sugar-based, and even glucose-based foods. As these cereal proteins exert their functions in the gut and are tasteless and odorless, they can be used as active components in various functional foods. On the other hand, in order to inhibit DPP-4 and ACE activities and suppress oxidative stress in the liver, bioactive components need to be absorbed from the small intestine and transported to the blood and organs, indicating that active dipeptides or tripeptides themselves should be included in functional foods. Many peptides have a bitter taste, which poses challenges for their addition to foods in large quantities. Therefore, in order to apply scientific findings to functional foods, the development of food-processing techniques to produce functional foods with favorable palatability is also required.

REFERENCES

1. Ministry of Health, Labour and Welfare. *The National Health and Nutrition Survey in Japan* 2018.
2. Chavan J, Duggal S. Studies on the essential amino acid composition, protein fractions and biological value (BV) of some new varieties of rice. *J Sci Food Agric* 1978; 29(3):225–229.
3. Landry J, Moureaux T. Distribution and amino acid composition of protein groups located in different histological parts of maize grain. *J Agric Food Chem* 1980; 28(6):1186–1191.
4. Hamada JS. Characterization of protein fractions of rice bran to devise effective methods of protein solubilization. *Cereal Chem* 1997; 74(5):662–668.
5. Ju ZY, Rath HN. Extraction, denaturation and hydrophobic properties of rice flour proteins. *J Food Sci* 2001; 66(2):229–232.
6. Fabian C, Ju YH. A review on rice bran protein: Its properties and extraction methods. *Crit Rev Food Sci Nutr* 2011; 51(9):816–827.
7. Kim J-W, Kim B-C, Lee J-H, Lee D-R, Rehman S, Yun SJ. Protein content and composition of waxy rice grains. *Pak J Bot* 2013; 45(1):151–156.
8. Wang C, Li D, Xu F, Hao T, Zhang M. Comparison of two methods for the extraction of fractionated rice bran protein. *J Chem* 2014; 2014, 546345.
9. Kasarda DD, Autran JC, Lew EJL, Nimmo CC, Shewry PR. N-terminal amino acid sequences of ω-gliadins and ω-secalins. Implications for the evolution of prolamin genes. *Biochim Biophys Acta* 1983; 747(1–2):138–150.
10. Shewry PR, Tatham AS, Forde J, Kreis M, Miflin BJ. The classification and nomenclature of wheat gluten proteins: A reassessment. *J Cereal Sci* 1986; 4(2):97–106.
11. DuPont FM, Vensel WH, Chan R, Kasarda DD. Characterization of the 1B-type ω-gliadins from *Triticum aestivum* cultivar butte. *Cereal Chem* 2000; 77(5):607–614.

12. Padhye VW, Salunkhe DK. Extraction and characterization of rice proteins. *Cereal Chem* 1979; 56(5):389–393.
13. Hibino T, Kidzu K, Masumura T, Ohtsuki K, Tanaka K, Kawabata M. Amino acid compoisiton of rice prolamin polypeptides. *Agric Biol Chem* 1989; 53(2):513–518.
14. World Health Organization. *Global Report on Diabetes* 2016.
15. Shainkin R, Birk Y. α-Amylase inhibitors from wheat. Isolation and characterization. *Biochim Biophys Acta* 1970; 221(3):502–513.
16. Silano V, Pocchiari F, Kasarda DD. Physical characterization of α-amylase inhibitors from wheat. *Biochim Biophys Acta Protein Struct* 1973; 317(1):139–148.
17. Buonocore V, Petrucci T, Silano V. Wheat protein inhibitors of α-amylase. *PhytoChemistry* 1977; 16(7):811–820.
18. Blanco-Labra A, Iturbe-Chiñas FA. Purification and characterization of an alpha-amylase inhibitor from maize (Zea maize). *J Food Biochemistry* 1981; 5(1):1–17.
19. Weselake RJ, MacGregor AW, Hill RD, Duckworth HW. Purification and characteristics of an endogenous α-amylase inhibitor from barley kernels. *Plant Physiol* 1983; 73(4):1008–1012.
20. Feng GH, Chen M, Kramer KJ, Reeck GR. Amylase inhibitors from rice: Fractionation and selectivity toward insect, mammalian, and bacterial α-amylases. *Cereal Chem* 1991; 68:516–526.
21. Shewry PR, Lafiandra D, Salcedo G, Aragoncillo C, Garcia-Olmedo F, Lew EJL, Dietler MD, Kasarda DD. N-terminal amino acid sequences of chloroform/methanol-soluble proteins and albumins from endosperms of wheat, barley and related species. Homology with inhibitors of α-amylase and trypsin and with 2S storage globulins. *FEBS Lett* 1984; 175(2):359–363.
22. Buonocore V, De Biasi M-G, Giardina P, Poerio E, Silano V. Purification and properties of an α-amylase tetrameric inhibitor from wheat kernel. *Biochim Biophys Acta* 1985; 831(1):40–48.
23. Deponte R, Parlamenti R, Petrucci T, Silano V, Tomasi M. Albumin α-amylase inhibitor families from wheat flour. *Cereal Chem* 1976; 53(5):805–820.
24. Choudhury A, Maeda K, Murayama R, Dimagno EP. Character of a wheat amylase inhibitor preparation and effects on fasting human pancreaticobiliary secretions and hormones. *Gastroenterology* 1996; 111(5):1313–1320.
25. Wang J-R, Yan Z-H, Wei Y-M, Nevo E, Baum BR, Zheng Y-L. Molecular characterization of dimeric alpha-amylase inhibitor genes in wheat and development of genome allele-specific primers for the genes located on chromosome 3BS and 3DS. *J Cereal Sci* 2006; 43(3):360–368.
26. Feng G-H, Chen M, Kramer KJ, Reeck GR. α-amylase inhibitors from rice: Fractionation and selectivity toward insect, mammalian, and bacterial α-amyases. *Cereal Chem* 1991; 68(5):516–521.
27. Yamagata H, Kunimatsu K, Kamasaka H, Kuramoto T, Iwasaki T. Rice bifunctional α-amylase/ subtilisin inhibitor: Characterization, localization, and changes in developing and germinating seeds. *Biosci Biotechnol Biochem* 1998; 62(5):978–985.
28. Ina S, Ninomiya K, Mogi T, Hase A, Ando T, Matsukaze N, Ogihara J, Akao M, Kumagai H, Kumagai H. Rice (Oryza sativa japonica) albumin suppresses the elevation of blood glucose and plasma insulin levels after oral glucose loading. *J Agric Food Chem* 2016; 64(24):4882–4890.
29. Koike D, Yamadera K, DiMagno EP. Effect of a wheat amylase inhibitor on canine carbohydrate digestion, gastrointestinal function, and pancreatic growth. *Gastroenterology* 1995; 108(4):1221–1229.

30. Kodama T, Miyazaki T, Kitamura I, Suzuki Y, Namba Y, Sakurai J, Torikai Y, Inoue S. Effects of single and long-term administration of wheat albumin on blood glucose control: Randomized controlled clinical trials. *Eur J Clin Nutr* 2005; 59(3):384–392.
31. Ina S, Hamada A, Nakamura H, Yamaguchi Y, Kumagai H, Kumagai H. Rice (Oryza sativa japonica) albumin hydrolysates suppress postprandial blood glucose elevation by adsorbing glucose and inhibiting Na$^+$-D-glucose cotransporter SGLT1 expression. *J Funct Foods* 2020; 64, 103603.
32. Ninomiya K, Ina S, Hamada A, Yamaguchi Y, Akao M, Shinmachi F, Kumagai H, Kumagai H. Suppressive effect of the α-amylase inhibitor albumin from buckwheat (Fagopyrum esculentum Moench) on postprandial hyperglycaemia. *Nutrients* 2018; 10(10):1503.
33. Nongonierma AB, FitzGerald RJ. Investigation of the potential of hemp, pea, rice and soy protein hydrolysates as a source of dipeptidyl peptidase IV (DPP-IV) inhibitory peptides. *Food Dig: Res Curr Opin* 2015; 6:19–29.
34. Hatanaka T, Uraji M, Fujita A, Kawakami K. Anti-oxidation activities of rice-derived peptides and their inhibitory effects on dipeptidylpeptidase-IV. *Int J Pept Res Ther* 2015; 21(4):479–485.
35. Ishikawa Y, Hira T, Inoue D, Harada Y, Hashimoto H, Fujii M, Kadowaki M, Hara H. Rice protein hydrolysates stimulate GLP-1 secretion, reduce GLP-1 degradation, and lower the glycemic response in rats. *Food Funct* 2015; 6(8):2525–2534.
36. Hira T, Mochida T, Miyashita K, Hara H. GLP-1 secretion is enhanced directly in the ileum but indirectly in the duodenum by a newly identified potent stimulator, zein hydrolysate, in rats. *Am J Physiol Gastrointest Liver Physiol* 2009; 297(4):G663–G671.
37. Higuchi N, Hira T, Yamada N, Hara H. Oral administration of corn zein hydrolysate stimulates GLP-1 and GIP secretion and improves glucose tolerance in male normal rats and Goto-Kakizaki rats. *Endocrinologist* 2013; 154(9):3089–3098.
38. Wang F, Yu G, Zhang Y, Zhang B, Fan J. Dipeptidyl peptidase IV inhibitory peptides derived from oat (Avena sativa L.), buckwheat (Fagopyrum esculentum), and highland barley (Hordeum vulgare trifurcatum (L.) Trofim) proteins. *J Agric Food Chem* 2015; 63(43):9543–9549.
39. Matsui T, Matsufuji H, Seki E, Osajima K, Nakashima M, Osajima Y. Inhibition of angiotensin I-converting enzyme by Bacillus licheniformis alkaline protease hydrolyzates derived from sardine muscle. *Biosci Biotechnol Biochem* 1993; 57(6):922–925.
40. Maruyama S, Miyoshi S, Kaneko T, Tanaka H. Angiotensin I-converting enzyme inhibitory activities of synthetic peptides related to the tandem repeated sequence of a maize endosperm protein. *Agric Biol Chem* 1989; 53(4):1077–1081.
41. Miyoshi W, Ishikawa H, Kaneko T, Fukui F, Tanaka H, Maruyama S. Structures and activity of angiotensin-converting enzyme inhibitors in an α-zein hydrolysate. *Agric Biol Chem* 1991; 55(5):1313–1318.
42. Pooja K, Rani S, Prakash B. In silico approaches towards the exploration of rice bran proteins-derived angiotensin-I-converting enzyme inhibitory peptides. *Int J Food Prop* 2017; 20(52):S2178–S2191.
43. Uraipong C, Zhao J. Identification and functional characterisation of bioactive peptides in rice bran albumin hydrolysates. *Int J Food Sci Technol* 2016; 51(10):2201–2208.
44. Shobako N, Ogawa Y, Ishikado A, Harada K, Kobayashi E, Suido H, Kusakari T, Maeda M, Suwa M, Matsumoto M, Kanamoto R, Ohinata K. A novel antihypertensive peptide identified in thermolysin-digested rice bran. *Mol Nutr Food Res* 2018; 62(4), 1700732.

45. Boonla O, Kukongviriyapan U, Pakdeechote P, Kukongviriyapan V, Pannangpetch P, Thawornchinsombut S. Peptides-derived from Thai rice bran improves endothelial function in 2K-1C renovascular hypertensive rats. *Nutrients* 2015; 7(7):5783–5799.
46. Suwannapan O, Wachirattanapongmetee K, Thawornchinsombut S, Katekaew S. Angiotensin-I-converting enzyme (ACE)-inhibitory peptides from Thai jasmine rice bran protein hydrolysates. *Int J Food Sci Technol* 2020; 55(6):2441–2450.
47. Kuba M, Tanaka K, Sesoko M, Inoue F, Yasuda M. Angiotensin I-converting enzyme inhibitory peptides in red-mold rice made by Monascus purpureus. *Process Biochem* 2009; 44(10):1139–1143.
48. Matsui T, Li C-H, Osajima Y. Preparation and characterization of novel bioactive peptides responsible for angiotensin I-converting enzyme inhibition from wheat germ. *J Pept Sci* 1999; 5(7):289–297.
49. Matsui T, Li C-H, Tanaka T, Maki T, Osajima Y, Matsumoto K. Depressor effect of wheat germ hydrolysate and its novel angiotensin I-converting enzyme inhibitory peptide, *Ile-Val-Tyr, and the Metabolism in Rat and Human Plasma. Biol Pharm Bull* 2000; 23(4):427–431.
50. Huang L, Liu B, Ma H, Zhang X. Combined effect of ultrasound and enzymatic treatments on production of ACE inhibitory peptides from wheat germ protein. *J Food Process Preserv* 2014; 38(4):1632–1640.
51. Karami Z, Peighambardoust SH, Hesari J, Akbari-Adergani B, Andreu D. Antioxidant, anticancer and ACE-inhibitory activities of bioactive peptides from wheat germ protein hydrolysates. *Food Biosci* 2019; 32, 100450.
52. Nogata Y, Nagamine T, Yanaka M, Ohta H. Angiotensin I converting enzyme inhibitory peptides produced by autolysis reactions from wheat bran. *J Agric Food Chem* 2009; 57(15):6618–6622.
53. Zou Z, Wang M, Wang Z, Aluko RE, He R. Antihypertensive and antioxidant activities of enzymatic wheat bran protein hydrolysates. *J Food Biochem* 2020; 44(1):e13090.
54. Thewissen BG, Pauly A, Celus I, Brijs K, Delcour JA. Inhibition of angiotensin I-converting enzyme by wheat gliadin hydrolysates. *Food Chem* 2011; 127(4):1653–1658.
55. Motoi H, Kodama T. Isolation and characterization of angiotensin I-converting enzyme inhibitory peptides from wheat gliadin hydrolysate. *Nahrung* 2003; 47(5):354–358.
56. Zhang P, Chang C, Liu H, Li B, Yan Q, Jiang Z. Identification of novel angiotensin I-converting enzyme (ACE) inhibitory peptides from wheat gluten hydrolysate by the protease of Pseudomonas aeruginosa. *J Funct Foods* 2020; 65, 103751.
57. Behall KM, Scholfield DJ, Hallfrisch J. Whole-grain diets reduce blood pressure in mildly hypercholesterolemic men and women. *J Am Diet Assoc* 2006; 106(9):1445–1449.
58. Ma Y, Xiong YL, Zhai J, Zhu H, Dziubla T. Fractionation and evaluation of radical scavenging peptides from in vitro digests of buckwheat protein. *Food Chem* 2010; 118(3):582–588.
59. Li C-H, Matsui T, Matsumoto K, Yamasaki R, Kawasaki T. Latent production of angiotensin I-converting enzyme inhibitors from buckwheat protein. *J Pept Sci* 2002; 8(6):267–274.
60. Koyama M, Naramoto K, Nakajima T, Aoyama T, Watanabe M, Nakamura K. Purification and identification of antihypertensive peptides from fermented buckwheat sprouts. *J Agric Food Chem* 2013; 61(12):3013–3021.

61. Takahashi M, Moriguchi S, Yoshikawa M, Sasaki R. Isolation and characterization of oryzatensin: A novel bioactive peptide with ileum-contracting and immunomodulating activities derived from rice albumin. *Biochem Mol Biol Int* 1994; 33(6):1151–1158.
62. Takahashi M, Moriguchi S, Ikeno M, Kono S, Ohata K, Usui H, Kurahashi K, Sasaki R, Yoshikawa M. Studies on the ileum-contracting mechanisms and identification as a complement C3a receptor agonist of oryzatensin, a bioactive peptide derived from rice albumin. *Peptides* 1996; 17(1):5–12.
63. Zioudrou C, Streaty RA, Klee WA. Opioid peptides derived from food proteins. The exorphins. *J Biol Chem* 1979; 254(7):2446–2449.
64. Fukudome S, Yoshikawa W. Opioid peptides derived from wheat gluten: Their isolation and characterization. *FEBS Lett* 1992; 296(1):107–111.
65. Takahashi M, Fukunaga H, Kaneto H, Fukudome S, Yoshikawa M. Behavioral and pharmacological studies on gluten exorphin A5, a newly isolated bioactive food protein fragment, in mice. *Jpn J Pharmacol* 2000; 84(3):259–265.
66. Wattanasiritham L, Theerakulkait C, Wickramasekara S, Maier CS, Stevens JF. Isolation and identification of antioxidant peptides from enzymatically hydrolyzed rice bran protein. *Food Chem* 2016; 192:156–162.
67. Piu LD, Tassoni A, Serrazanetti DI, Ferri M, Babini E, Tagliazucchi D, Gianotti A. Exploitation of starch industry liquid by-product to produce bioactive peptides from rice hydrolyzed proteins. *Food Chem* 2014; 155:199–206.
68. Adebiyi AP, Adebiyi AO, Yamashita J, Ogawa T, Muramoto K. Purification and characterization of antioxidative peptides derived from rice bran protein hydrolysates. *Eur Food Res Technol* 2009; 228(4):553–563.
69. Park EY, Morimae M, Matsumura Y, Nakamura Y, Sato K. Antioxidant activity of some protein hydrolysates and their fractions with different isoelectric points. *J Agric Food Chem* 2008; 23(19):9246–9251.
70. Zhao Y, Zhao Q, Lu Q. Purification, structural analysis, and stability of antioxidant peptides from purple wheat bran. *BMC Chem* 2020; 14(1):58.
71. Luo X, Fei Y, Xu Q, Lei T, Mo X, Wang Z, Zhang L, Mou X, Li H. Isolation and identification of antioxidant peptides from Tartary buckwheat albumin (Fagopyrum tataricum Gaertn.) and their antioxidant activities. *J Food Sci* 2020; 85(3):611–617.
72. Zhuang H, Tang N, Yuan Y. Purification and identification of antioxidant peptides from corn gluten meal. *J Funct Foods* 2013; 5(4):1810–1821.
73. Jin D-X, Liu X-L, Zheng Z-Q, Wang X-J, He J-F. Preparation of antioxidative corn protein hydrolysates, purification and evaluation of three novel corn antioxidant peptides. *Food Chem* 2016; 204:427–436.

CHAPTER 7

Soybean

Toshihiro Nakamori

CONTENTS

References 149

Soybeans are an important plant resource that are grown worldwide. In fiscal year 2019, worldwide production of soybeans amounted to approximately 340 million tons (Figure 7.1), making it one of the world's most heavily produced oilseeds[1]. Most soybeans are grown as feed. Outside of Asian regions, soybean production began in earnest in the United States after 1930, and production increased dramatically in a short period of time, resulting in soybeans being called a "miracle crop"[2]. Soybeans have outstanding productivity and nutritional value. However, only approximately 30% of the volume of soybeans produced worldwide are used for human consumption. Soybean protein has high nutritional value and significant economic benefits over raw dairy products, for example. Raising the proportion of soybeans used for food is important from the perspective of reducing environmental burdens and securing food resources.

The soybean growing regions are very broad, extending from tropical regions at the equator to southern parts of Sweden and Canada in the north and southern parts of Argentina and Australia in the south. In addition, there are a large number of soybean varieties, in accordance with the cultivation environment and usage applications in each region. Soybeans are generally composed of 12% water, 35% protein, 19% lipids, 29% carbohydrates, and 5% ash (Figure 7.2).[3] Soybean protein is mainly composed of glycinin and a type of glycinin called β-conglycinin. In addition, it also contains phospholipid-associated proteins,[4,5] enzymes such as albumin, trypsin inhibitors, amylase, various types of peptides,[6] and trace amounts of amino acids. Soybeans contain approximately 30% fine-quality protein that constitutes the best amino acid-balanced protein,[7] and are a source of vegetable protein with a nutritional value so high and comparable to meats as to be nicknamed "field-grown meat". In Japan, where there are many types of traditionally processed soybean-based foods, such as soy milk, tofu, soy sauce, and miso (Figure 7.3). Nongenetically modified organisms (Non-GMOs) are mainly used for food. Soybean protein and soybean peptides are produced from defatted, oil-extracted soybeans (Figure 7.4). Combining functional research and the technology needed to manufacture peptides by breaking down soybean protein with enzymes has the potential to create functional foods to prevent lifestyle-related diseases, such as obesity, hyperlipidemia (dyslipidemia), and hypercholesterolemia.

Carroll et al. reported in 1975 that soybean protein acts to reduce plasma low-density lipoprotein (LDL) concentration.[8] Later, Sugano et al. elucidated the mechanism by which the digestion-resistant high molecular fraction (HMF) combines with excessive bile acid inside the intestinal tract and excretes it from the body (Figure 7.5).[9-12] The U.S.

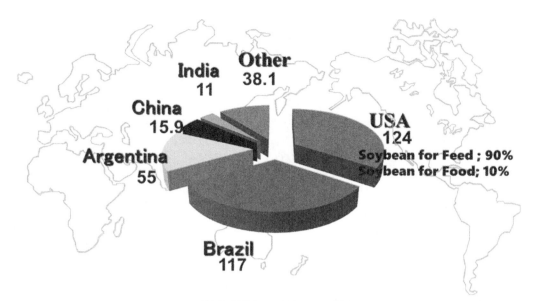

FIGURE 7.1 World soybean production.

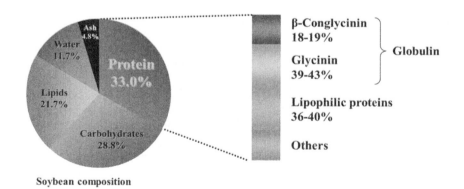

FIGURE 7.2 Various proteins comprise about one-third of each soybean.

Food and Drug Administration (FDA) has taken note of the cholesterol-reducing effects of soybean protein and has approved the display of a label stating, "A food that reduces the risk of coronary heart disease" on items that can contribute toward a daily recommended allowance of 25 g per day of soybean protein.[13] Soybean protein has become a functional component of foods for specified health uses (FOSHU) in Japan, and several related products are already being sold.[14–16]

Soybean protein is known to reduce the concentration of triglycerides (TGs) in the serum and liver more effectively than casein.[17–19] Moreover, since it reduces adipose tissue mass through vesiculation of white fat,[20] it is likely to reduce obesity. In terms of the mechanism by which this occurs, researchers have reported that blood insulin concentration decreases upon ingestion of soybean protein. It appears to suppress the expression of the transcription factor SREBP-1, which controls fatty acid synthesis in the liver, which

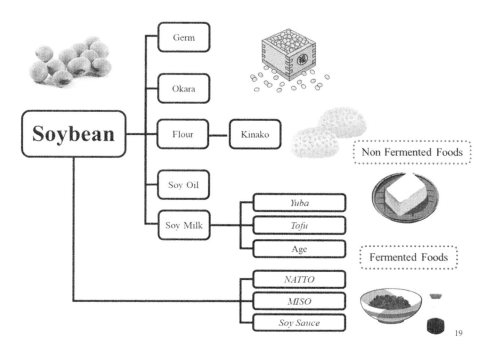

FIGURE 7.3 Traditional soybean products.

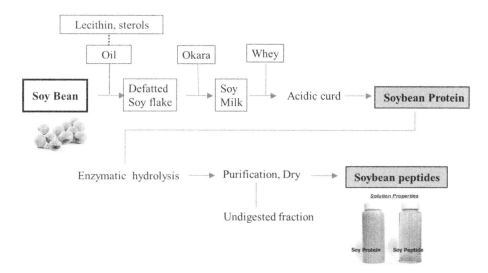

FIGURE 7.4 Soybean protein and soybean peptides manufacturing process.

is induced by high insulin concentrations and, as a result, suppresses the ability to synthesize fatty acids. This results in the prevention of liver fat accumulation.[21,22] However, it has not been clarified experimentally whether the suppression of fatty acid synthesis in the liver is due to peptides derived from soybean protein, or to trace ingredients such as isoflavones that are also present in soybean protein.[23–25] Komatsu et al. compared the type of protein vis-à-vis diet-induced heat generation in obese patients. They showed

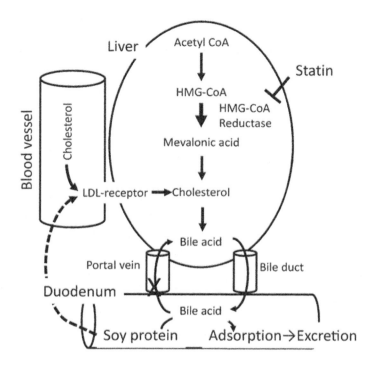

FIGURE 7.5 The mechanism by which soybean protein and statins lower cholesterol. Statin inhibits 3-hydroxy-3-methyl-glutaryl-coenzyme A (HMG-CoA) reductase, and soybean protein removes cholesterol-derived bile acids from the body in the feces.[12]

that diet-induced thermogenesis tended to be higher with soybean peptides than with either soybean protein or lactoalbumin, and reported that this was not only the result of regulation of the sympathetic nervous system but also endocrine secretion.[26,27] Kwark et al. investigated the impact on body fat of proteins ingested after physical exercise and reported that the rate of subcutaneous fat consumption was the greatest when soybean peptides had been ingested. However, the mechanism of action was not discussed.[28]

To treat and prevent obesity, it is necessary to adjust the size of meals and reduce the energy intake. Appetite becomes a problem in this case, making it difficult for subjects to continually control their food intake. Although surgery is available to reduce the size of the stomach, it entails substantial physical and economic burdens. Appetite is regulated at the hypothalamus and brainstem, but it is known that several types of gastrointestinal hormones also control appetite via the vagus nerve. One such hormone is cholecystokinin (CCK), which is known to bind to CCK receptors and stimulate the satiety center via the vagus nerve. Hara et al. identified a peptide produced from soybean protein decomposition that promotes CCK secretion and discovered that, in animal tests, it reduced food intake.[29,30] Although we cannot state unequivocally that the satiety center is stimulated by CCK alone, it might be used in a complementary fashion whenever the amount of food intake needs to be controlled, such as during obesity treatment, and to create a feeling of satiety. Although its effects in animals have been confirmed, whether CCK is also effective in humans is unknown. After meal-derived proteins are broken down by gastric fluids and proteases contained in pancreatic fluid, amino acids are absorbed through the action of amino acid transporters. Peptides are rapidly absorbed into the body through the action of a peptide transporter, PepT1.[31,32] Absorption of food-derived peptides by

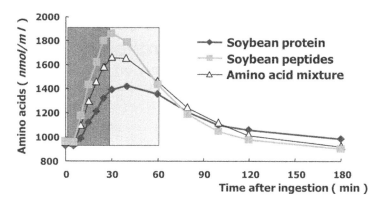

Serum essential amino acid levels in peripheral blood vessel after ingestion of the test beverages

FIGURE 7.6 Serial changes in serum concentrations of essential amino acids following intake of soybean protein, soybean peptides, equivalent amino acid mixture, or placebo.[36] Data are mean±SEM values of 12 subjects.

the body is dependent on their chain lengths, and di- and tripeptides are believed to be the absorption limit through this PepT1.[33–35] Upon confirming the rate at which soybean peptides enter the bloodstream, we noted that soybean peptides were almost fully absorbed within 60 min using an absorption test with human subjects (Figure 7.6), and that the concentration of free amino acids in the blood had increased.[36] PepT1 has low substrate specificity, but is only capable of transporting di- or tripeptides.[37] We therefore believe that peptides that have not been broken down and instead remain inside the body as absorbed di- and tripeptides manifest physiological functions, making them bioactive peptides. Table 7.1 shows the bioactive di- and tripeptides that have been isolated and identified from soybean proteins and processed soybean products. Val-Pro-Tyr and Val-Ile-Lys have been reported to show anti-inflammatory effects.[38,39] Trp-Leu, Val-Phe, Leu-Leu-Phe, Leu-Asn-Phe, Leu-Ser-Trp, and Lue-Glu-Phe, which are derived from soybean protein, show antihypertensive effects.[40,41] Antihypertensive peptides have been also isolated from fermented soybean products.[42–44]

As obesity progresses, TG synthesis in the liver is stepped up by the inflow of free fatty acids, especially from visceral fat. A reduction in insulin sensitivity in line with the

TABLE 7.1 Bioactive Di- and Tripeptides Derived from Soybeans

Source	Peptide Sequence	Activity	Reference
Soybean protein	Val-Pro-Tyr	anti-inflammation	34
Soybean milk	Val-Ile-Lys	anti-inflammation	35
Soybean protein	Trp-Leu	antihypertensive	36
Soybean protein	Val-Phe, Leu-Leu-Phe, Leu-Asn-Phe, Leu-Ser-Trp, Lue-Glu-Phe	antihypertensive	37
Korean fermented soybean paste	His-His-Leu	antihypertensive	38
Fermented soybean seasoning	Ser-Tyr Gly-Tyr	antihypertensive	39 40

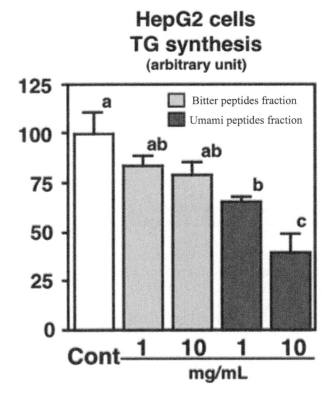

FIGURE 7.7 Effects of soy peptides on triglyceride (TG) synthesis in HepG2 cells (1 or 10 mg/mL). Values are expressed as mean ± standard error of 5 samples *in vitro*.[49] Different letters indicate a significant difference at $P < 0.05$.

hypertrophy of visceral fat acts on the liver and further exacerbates fatty acid synthesis. Due to this supply of fatty acids in the liver, TGs, which are the major lipids constituting very low-density lipoproteins (VLDLs), are synthesized in excess. This, in turn, promotes the synthesis and secretion of VLDLs.[45] If the level of TGs in the blood is high, it is liable to progress to arteriosclerosis, so there is a need to develop drugs and foods to prevent this. Using rats, Aoyama et al. reported that soybean peptides had greater TG reduction in the liver than soybean protein and suggested that some soybean peptides may be bioactive.[46] However, they were unable to find any studies reporting isolation of the main active component.[47] Tamaru et al. also reported that the low-molecular-weight fraction of soybean peptides acts strongly to reduce TG concentration in rat serum and liver. A liver perfusion test indicated that a rise in serum TG concentration was suppressed by reducing the amount of TG excreted from the liver through the promotion of β-oxidation and suppression of TG synthesis.[48] However, to date, no studies have reported the isolation and identification of TG synthesis-suppressing peptides originating from soybean protein.

To isolate and identify TG-reducing peptides, we first took note of the characteristics of the amino acid sequences of soybean proteins.[49] We then used a hydrophobic resin to separate bitter peptides fraction with a high hydrophobic amino acid content, and umami *peptides* fraction with a high hydrophilic amino acid content, and investigated them using a HepG2 cell culture system. The *umami peptides* fraction showed powerful suppression of TG synthesis, so we used this fraction as our starting material and succeeded in isolating

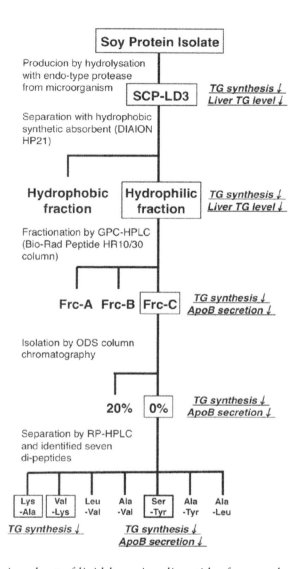

FIGURE 7.8 Screening chart of lipid-lowering dipeptides from soybean protein.[49]

and identifying a dipeptide that reduced TG production (Figure 7.7). Because of this, bioactive peptides that are taken into the body reach the liver via the portal artery and demonstrate TG-reducing effects. They appear to be either di- or tripeptides. To explore the bioactive peptides contained in the *umami* peptides fraction, which show the ability to reduce TG, we used gel filtration chromatography (GPC-HPLC) to fractionate them into di- and tripeptides. We ultimately isolated and identified seven peptides using reversed-phase chromatography: Lys-Ala, Val-Lys, Leu-Val, Ala-Val, Ser-Tyr, Ala-Tyr, and Ala-Leu (Figure 7.8). Next, we created synthetic versions of these dipeptides and confirmed their TG-synthetic activity (Figure 7.9), and ApoB100 secretion suppression activity (Figure 7.9) in HepG2 cells. The results showed a significantly greater decrease in TG synthesis for Ser-Tyr, Val-Lys, and Lys-Ala compared to the non-peptide-added group (Figure 7.9). Significant ApoB100 suppression activity was observed for Ser-Tyr (Figure 7.9).

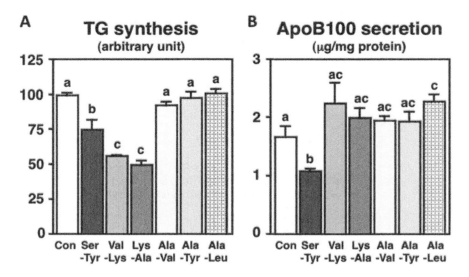

FIGURE 7.9 Effects of synthesized dipeptides (5 mg/mL) on triglyceride (TG) synthesis (A) and apolipoprotein (Apo) B100 secretion (B) in HepG2 cells [49]. Values are expressed as mean ± standard error of 5 samples. Different letters indicate a significant difference at $P < 0.05$.

TABLE 7.2 Amino Acid Sequence of Major Components in Soybean Protein[49]

Glycinin [Glycine max (soybean)]/CAA37044

MGKPFTLSLSSLCLLLLSSACFAISSSKLNECQLNNLALEPDHRVEFEGGLIQTWNSQHPELKCAGVTVSKLTLNRNGLHLP**SY**SPYPRMIIIAQGKGALQCKPGCPETFE EPQEQSNRRGSRSQKQQLQDSHQKIRHFNEGDVLVIPPGVPYWTYNTGDEPVVAISLLDTSNFNNQLDQTPRVFYLAGNPDIEYPETMQQQQQQKSHGGRKQGQHQ QEEEEEGGSVLSGFSKHFLAQSFNTNEDIAEKLQSPDDERKQIVTVEGGLSVISPKWQEQQDEDEDEDEDEDEQIPSHPPRRPSHGKREQDEDEDEDEDKPRPSRPSQ GKREQDQDQDEDEDEDEDQPRKSREWRSKKTQPRRPRQEEPRERGCETRNGVEENICTLKLHENIARPSRADFYNP**KA**GRISTLNSLTLPALRQFQLSAQYVVLYKN GIYSPHWNLNANSVIYVTRGGQKVRVVNCQGNAVFDGELRRGQLLVVPQNFWAEQAGEQGFEYIVFKTHHNAVT**SYL**KDVFRAIPSEVLAHS**YN**LRQSQVSELKYEGN WGPLVNPESQQGSPRV**K**VA

Beta-conglycinin alpha subunit [Glycine max (soybean)]/BAE46788

MMRARFPLLLLGLVFLASVSVSFGIAYWEKENPKHNKCLQSCNSERD**SY**RNQACHARCNLLKVEKEECEEGEIPRPRPRPQHPEREPQQPGEKEEDEDEQPRPIPFPRPQ PRQEEEHEQREEQEWPRKEEKRGEKGSEEEDEDEDEEQDERQFPFPRPPHQKEERKQEEDEDEEQQRESEESEDSELRRHKNKNPFLFGSNRFETLFKNQYGRIRVLQRF NQRSPQLQNLRDYRILEFNSKPNTLLLPNHADADYLIVILNGTAILSLVNNDDRD**SY**RLQSGDALRVPSGTTYYVVNPDNNENLRLITLAIPVNKPGRFESFFLSSTEAQQ **SYL**QGFSRNILEAS**Y**DTKFEEINKVLFSREEGQQQGEQRLQESVIVEISKEQIRALSKRAKSSSRKTISSEDKPFNLRSRDPIYSNKLGKFFEITPEKNPQLRDLDIFLSIVDMNE GALLLPHFNS**KA**IVLVINEGDANIELVGLKEQQQEQQQEEQPLEVRKYRAELSEQDIFVIPAGYPVVVNATSNLNFFAIGINAENNQRNFLAGSQDNVISQIPSQVQELA FPGSAQAVEKLLKNQRE**SY**FVDAQPKKKEEGNKGRKGPLSSILRAFY

Beta-conglycinin alpha prime subunit [Glycine max (soybean)]/BAB64303

MMRARFPLLLLGVVFLASVSVSFGIAYWEKQNPSHNKCLRSCNSEKD**SY**RNQACHARCNLLKVEEEEECEEGQIPRPRPQHPERERQQHGEKEEDEGEQPRPFPFPRP RQPREQGEHEQKEEHEWHRKEEKHGGKGSEEEQDGREHPRPHQPHQKEEEKHEWQHKQEKHQGKESEEEEEDQDEDEEQDKESQESEGSESQREPRRHKNKNPFHF NSKRFQTLFKNQYGHVRVLQRFNKRSQQLQNLRDYRILEFNSKPNTLLLPPHHADADYLIVILNGTAILTLVNNDDRD**SY**NLQSGDALRVPAGTTYYVVNPDNDENLRM ITLAIPVNKPGRFESFFLSSTQAQQ**SY**LQGFSKNILEA**SY**DTKFEEINKVLFGREEGQQQGEERLQESVIVEISKKQIRELSKRAKSSSRKTISSEDKPFNLRSRDPIYSNKLGKL FEITPEKNPQLRDLDVFLSVVDMNEGALFLPHFNS**KA**IVLVINEGEANIELVGIKEQQQRQQQEEQPLEVRKYRAELSEQDIFVIPAGYPVVVNATSDLNFFAFGINAEN NQRNFLAGSKDNVISQIPSQVQELAFLGSAKDIENLIKSQSE**SY**FVDAQPQQKEEGNKGRKGPLSSILRAFY

Beta-conglycinin beta subunit [Glycine max (soybean)]/BAB64306

MMRVRFPLLVLLGTVFLASVCVSLKVREDENNPFYFRSSNSFQTLFENQNGRIRLLQRFNKRSPQLENLRDYRIVQFQSKPNTILLPHHADADFLLFVLSGRAILTLVN NDDRD**SY**NLHPGDAQRIPAGTTYYLVNPHDHQNLKIIKLAIPVNKPSRYDDFFLSSTQAQQ**SYL**QGFSHNILETSFHSEFEEINRVLFGEEEEQRQQEGVIVELSKEQIRQLS RRAKSSSRKTISSEDEPFNLRSRNPIYSNNFGKFFEITPEKNPQRDLDIFLSSVDINEGALLLPHFNS**KA**IVLVINEGDANIELVGIKEQQQKQKQEEEPLEVQRYRAELSE DDVFVIPAAYPFVVNATSNLNFLAFGINAENNQRNFLAGEKDNVVRQIERQVQELAFPGSAQDVERLLKKQRE**SY**FVDAQPQQKEEGSKGRKGPFPSILGALY

Trypsin inhibitor [Glycine max (soybean)]/AAF87095

MPSTWGAAGGGLKLGRTGNSNCPVTVLQDYSEIFRGTP**VK**FSIPGISPGIIFTGTPLEIEFAEKPYCAESSKWVAFVDNEIQ**KA**CVGIGGPEGHPGQQTFSGTFSIQKY KFGYKLVFCITGSGTCLDIGRFDAKNGEGGRRLNLTEHEAFDIVFIEASKVDGIIKSVV

Lipoxygenase [Glycine max (soybean)]/CAA39604

MFGIFDKGQKIKGTVVLMPKNVLDFNAITSIGKGGVGIDTATGILGQGVSLVGGVIDTATSFLGRNISMQLISATQTDGSGNGKVGKEVYLEKHLPTLPTLGARQDAFSIF FEWDASFGIPGAFYIKNFMTDEFFLVS**VK**LEDIPNHGTIEFVCNSWVYNFR**SY**KKNRIFFVNDTYLPSATPAPLLKYRKEELEVLRGDGTGKRKDFDRIYDYDVYNDLGN PDGGDPRPILGGSSIYPYPRRVRTGRERTRTDPNSEKPGEVYVPRDENFGHLKSSDFLTYGIKSLSHDVIPLFKSAIFQLRVTSSEFESFEDVRSLYEGGIKLPTDILSQISPL PALKEIFRTDGENVLQFPPPHVAKVSKSGWMTDEEFAREVIAGVNPNVIRRLQEFPPKSTLDPTLYGDQTSTITKEQLEINMGGVVTVEEALSTQRLFILDYQDAFIPYLTR INSLPTA**KA**YATRTILFLKDDGTLKPLAIELSKPHPDGDNLGPESIVVLPATEGVDSTIWLLA**KA**HVIVNDSGYHQLVSHWLNTHAVMEPFAIATNRHLSVLHPIYKLLYPHYR DTININGLARQSLINADGIIEKSFLPGKYSIEMSSSVYKNWVFTDQALPADLVKRGLAIEDPSAPHGLRLVIEDYPYAVDGLEIWDAIKTWVHEYYSLYYPTDAAV QQDTELQAWWKEAVEKGHGDLKEKPWWPKMQTTEDLIQSCSIIVWTAS**AL**HAAVNFGQYPYGGLILNRPTLARRFIPAEGTPEYDEM**VK**NPQ**KA**YLRTITPKFETL IDLSVIEILSRHASDEIYLGERETPNWTTDK**KA**LEAFKRFGSKLTGIEGKINARNSDPSLRNRTGPVQLPYTLLHRSSEEGLTFKGIPNSISI

Lys-Ala, KA; Ser-Tyr, SY; Val-Lys, VK.

Since Ser-Tyr suppressed TG synthesis, it appears to be the peptide with the strongest TG-reduction effects. Two other dipeptides, Val-Lys and Lys-Ala, showed strong suppression of TG synthesis. Excessive TG synthesis in liver cells induces secretion of TG-rich VLDL, which is larger in size than normal VLDL. This results in the production of small-dense LDL, which can be taken incorporated into the vascular wall cells and is believed to cause progression of arteriosclerosis.[50,51] Val-Lys and Lys-Ala may contribute to reducing the size of large VLDL molecules. In terms of the primary structure of soybean protein, Ser-Tyr and Lys-Ala are widely distributed, including in glycinin; the α, α', and β subunits of β-conglycinin; and lipoxygenase, so they appear to be dipeptides that are relatively easy to produce via enzymatic breakdown (Table 7.2). Ser-Tyr, however, is not present in casein; it is a peptide unique to soybean protein. Therefore, it is believed to be one of the major components of the TG-reducing peptides. Suppression of fatty acid synthesis in the liver has been noted to act on peptides derived from soybean protein, and we were able to prove that these peptides exert this precise action.

These peptides isolated from the umami peptides fraction are derived from the hydrophilic-rich sequences of soybean protein. Since these sequences are common to all soybean-processed foods, it is highly significant that peptides that play a part in producing umami peptides also have physiological functions.

Proteins are the main nutritional source of amino acids. However, it is possible that certain soybean protein amino acid sequences contain bioactive peptides that demonstrate a variety of bioactivities. It is therefore possible to regard proteins not only as a source of nutrition but also as a source of peptides that manifest a variety of physiological functions. We hope that soybean protein and its processed products will continue to contribute to human health.

REFERENCES

1. USDA (United States Department of Agriculture). *World Agricultural Production Report*.
2. Manu SM, Halagalimath SP, Chandranath HT, Biradar BD. Effect of nutrient levels and plant growth regulators on harvest index and economics of soybean (*Glycine max*). *Int. J. Curr. Microbiol. App. Sci.* 2020; 9(3): 890–897.
3. Perkins EG. Chapter 2 – Composition of soybeans and soybean products. In *Practical Handbook of Soybean Processing and Utilization*, edited by D.R. Erickson. AOCS Press, 1995; 9–28.
4. Samoto M, Maebuchi M, Miyazaki C, Kugitani H, Kohno M, Hirotsuka M, Kito M. Abundant proteins associated with lecithin in soy protein isolate. *Food Chem.* 2007; 102(1): 317–322.
5. Sirison J, Matsumiya K, Samoto M, Hidaka H, Kouno M, Matsumura Y. Solubility of soy lipophilic proteins: Comparison with other soy protein fractions. *Biosci. Biotechnol. Biochem.* 2017; 81(4): 790–802.
6. Jeong HJ, Park JH, Lam Y, de Lumen BO. Characterization of lunasin Isolated from soybean. *J. Agric. Food Chem.* 2003; 51(27): 7901–7906.
7. Lusas EW, Riaz MN. Soy protein products: Processing and use. *J. Nutr.* 1995; 125(3) Supplement: 573S–580S.
8. Carroll KK, Hamilton RMG. Effects of dietary protein and carbohydrate on plasma cholesterol levels in relation to atherosclerosis. *J. Food Sci.* 1975; 40(1): 18–23.

9. Sugano M, Goto S, Yamada Y, Yoshida K, Hashimoto Y, Matsuo T, Kimoto M. Cholesterol-lowering activity of various undigested fractions of soybean protein in rats. *J. Nutr.* 1990; 120(9): 977–985.
10. Ogawa T, Gatchalian-Yee M, Sugano M, Hashimoto Y, Matsuo T, Kimoto M. Hypocholesterolemic effect of undigested fraction of soybean protein in rats fed no cholesterol. *Biosci. Biotechnol. Biochem.* 1992; 56(11): 1845–1848.
11. Gatchalian-Yee M, Arimura Y, Ochiai E, Yamada K, Sugano M. Soybean protein lowers serum cholesterol levels in hamsters: Effect of debittered undigested fraction. *Nutrition* 1997; 13(7–8): 633–639.
12. Kohno M. Soybean protein and peptide as complementation medical food materials for treatment of dyslipidemia and inflammatory disorders. *Food Sci. Technol. Res.* 2017; 23(6): 773–782.
13. U.S. FOOD & DRUG. Code of federal regulations, title 21, volume 2, Revised as of April 1, 2020; CITE: 21CFR101.82.
14. Takamatsu K, Tachibana N, Matsumoto I, Abe K. Soy protein functionality and nutrigenomic analysis. *BioFactors* 2004; 21(1–4): 49–53.
15. Yamamoto T. *Soybean Components and Food for Specified Health Uses (FOSHU). SOY in Health and Disease Prevention.* 1st edition. CRC Press; 2005.
16. Iwatani S, Yamamoto N. Functional food products in Japan: A review. *Food Sci. Hum. Wellness* 2019; 8(2): 96–101.
17. Sirtori CR, Lovati MR. Soy proteins and cardiovascular disease. *Curr. Atheroscler. Rep.* 2001; 3(1): 47–53.
18. Aoyama T, Fukui K, Nakamori T, Hashimoto Y, Yamamoto T, Takamatsu K, Sugano M. Effect of soy and milk whey protein isolates and their hydrolysates on weight reduction in genetically obese mice. *Biosci. Biotechnol. Biochem.* 2000; 64(12): 2594–2600.
19. Aoyama T, Fukui K, Takamatsu K, Hashimoto Y, Yamamoto T. Soy protein isolate and its hydrolysate reduce body fat of dietary obese rats and genetically obese mice (yellow KK). *Nutrition* 2000; 16(5): 349–354.
20. Vázquez-Vela MEF, Torres N, Tovar AR. White adipose tissue as endocrine organ and its role in obesity. *Arch. Med. Res.* 2008; 39(8): 715–728.
21. Tovar AR, Torre-Villalvazo I, Ochoa M, Elias AL, Ortiz V, Aguilar-Salinas CA, Torres N. Soy protein reduces hepatic lipotoxicity in hyperinsulinemic obese Zucker fa/fa rats. *J. Lipid Res.* 2005; 46(9): 1823–1832.
22. Ascencio C, Torres N, Isoard-Acosta F, Gomez-Perez FJ, Hernandez-Pando R, Tovar AR. Soy protein affects serum insulin and hepatic SREBP-1 mRNA and reduces fatty liver in rats. *J. Nutr.* 2004; 134(3): 522–529.
23. Borradaile NM, de Dreu LE, Wilcox LJ, Edwards JY, Huff MW. Soya phytoestrogens, genistein and daidzein, decrease apolipoprotein B secretion from HepG2 cells through multiple mechanisms. *Biochem. J.* 2002; 366(2): 531–539.
24. Takahashi Y, Ide T. Effects of soy protein and isoflavone on hepatic fatty acid synthesis and oxidation and mRNA expression of uncoupling proteins and peroxisome proliferator-activated receptor γ in adipose tissues of rats. *J. Nutr. Biochem.* 2008; 19(10): 682–693.
25. Naaz A, Yellayi S, Zakroczymski MA, Bunick D, Doerge DR, Lubahn DB, Helferich WG, Cooke PS. The soy isoflavone genistein decreases adipose deposition in mice. *Endocrinology* 2003; 144(8): 3315–3320.

26. Komatsu T, Komatsu K, Matsuo M, Nagata M, Yamagishi M. Comparison between effects of energy restricted diets supplemented with soybean peptide and lactalbumin on energy, protein and lipid metabolisms in treatment of obese children. *Nutr. Sci. Soy Protein Jpn.* 1990; 11: 98–103.
27. Komatsu T, Komatsu K, Shiraishi M, Nagata M, Yamagishi M. Effects of diet containing soybean peptide or lactalbumin on basal energy expenditure of obese children reducing body weight and on thermogenesis after meal. *Nutr. Sci. Soy Protein Jpn.* 1991; 21: 80–84.
28. Kwak JH, Ahn CW, Park SH, Jung SU, Min BJ, Kim OY, Lee JH. Weight reduction effects of a black soy peptide supplement in overweight and obese subjects: Double blind, randomized, controlled study. *Food Funct.* 2012; 3(10): 1019–1024.
29. Nishi T, Hara H, Asano K, Tomita F. The soybean beta-conglycinin beta 51–63 fragment suppresses appetite by stimulating cholecystokinin release in rats. *J. Nutr.* 2003; 133(8): 2537–2542.
30. Hira T, Maekawa T, Asano K, Hara H. Cholecystokinin secretion induced by beta-conglycinin peptone depends on Galpaq-mediated pathway in enteroendocine cells. *Eur. J. Nutr.* 2009; 48(2): 124–127.
31. Adibi SA, Morse EL. The number of glycine residues which limits intact absorption of glycine oligopeptides in human jejunum. *J. Clin. Invest.* 1977; 60(5): 1008–1016.
32. Brodin B, Nielsen CU, Steffansen B, Frøkjaer S. Transport of peptidomimetic drugs by the intestinal di/tri-peptide transporter, PEPT1. *Pharmacol. Toxicol.* 2002; 90(6): 285–296.
33. Brandsch M, Knütter I, Leibach FH. The intestinal H+/peptide symporter PepT1: Structure-affinity relationships. *Eur. J. Pharm. Sci.* 2004; 21(1): 53–60.
34. Vig BS, Stouch TR, Timoszyk JK, Quan Y, Wall DA, Smith RL, Faria TN. Human PepT1 pharmacophore distinguishes between dipeptide transport and binding. *J. Med. Chem.* 2006; 49(12): 3636–3644.
35. Kovacs-Nolana J, Zhanga H, Ibuki M, Nakamori T, Yoshiura K, Turnere PV, Matsui T, Mine Y. The PepT1-transportable soy tripeptide VPY reduces intestinal inflammation. *Biochim. Biophys. Acta* 2012; 1820(11): 1753–1763.
36. Maebuchi M, Samoto M, Kohno M, Ito R, Koileda T, Hirotsuka M, Nakabou Y. Improvement in the intestinal absorption of soy protein by enzymatic digestion to oligopeptide in healthy adult men. *Food Sci. Technol. Res.* 2007; 13(1): 45–53.
37. Ito K, Hikida A, Kawai S, Tuyet Lan VT, Motoyama T, Kitagawa S, Yoshikawa Y, Kato R, Kawarasaki Y. Analysing the substrate multispecificity of a proton-coupled oligopeptide transporter using a dipeptide library. *Nat. Commun.* 2013; 4(2502). https://doi.org/10.1038/ncomms3502
38. Kovacs-Nolana J, Zhanga H, Ibuki M, Nakamori T, Yoshiura K, Turnere PV, Matsui T, Mine Y. The PepT1-transportable soy tripeptide VPY reduces intestinal inflammation. *Biochim. Biophys. Acta* 2012; 1820(11): 1753–1763.
39. Dia VP, Bringe NA, de Mejia EG. Peptides in pepsin-pancreatin hydrolysates from commercially available soy products that inhibit lipopolysaccharide-induced inflammation in macrophages. *Food Chem.* 2014; 152: 423–431.
40. Wang W, Dia VP, Vasconez M, de Mejia EG, Nelson RL. Analysis of soybean protein-derived peptides and the effect of cultivar, environmental conditions, and processing on lunasin concentration in soybean and soy products. *J. AOAC Int.* 2008; 91(4): 936–946.

41. Gu Y, Wu J. LC-MS/MS coupled with QSAR modeling in characterising of angiotensin I-converting enzyme inhibitory peptides from soybean proteins. *Food Chem.* 2013; 141(3): 2682–2690.
42. Shin ZI, Yu R, Park SA, Chung DK, Ahn CW, Nam HS, Kim KS, Lee HJ. His-His-Leu, an angiotensin I converting enzyme inhibitory peptide derived from Korean soybean paste, exerts antihypertensive activity in vivo. *J. Agric. Food Chem.* 2001; 49(6): 3004–3009.
43. Nakahara T, Sano A, Yamaguchi H, Sugimoto K, Chikata H, Kinoshita E, Uchida R. Antihypertensive effect of peptide-enriched soy sauce-like seasoning and identification of its angiotensin I-converting enzyme inhibitory substances. *J. Agric. Food Chem.* 2010; 58(2): 821–827.
44. Nakahara T, Sugimoto K, Sano A, Yamaguchi H, Katayama H, Uchida R. Antihypertensive mechanism of a peptide-enriched soy sauce-like seasoning: The active constituents and its suppressive effect on renin-angiotensin-aldosterone system. *J. Food Sci.* 2011; 76(8): H201–H206.
45. Sirtori CR, Lovati MR. Soy proteins and cardiovascular disease. *Curr. Atheroscler. Rep.* 2001; 3(1): 47–53.
46. Aoyama T, Fukui K, Nakamori T, Hashimoto Y, Yamamoto T, Takamatsu K, Sugano M. Effect of soy and milk whey protein isolates and their hydrolysates on weight reduction in genetically obese mice. *Biosci. Biotechnol. Biochem.* 2000; 64(12): 2594–2600.
47. Aoyama T, Fukui K, Takamatsu K, Hashimoto Y, Yamamoto T. Soy protein isolate and its hydrolysate reduce body fat of dietary obese rats and genetically obese mice (yellow KK). *Nutrition* 2000; 16(5): 349–354.
48. Tamaru S, Kurayama T, Sakono M, Fukuda N, Nakamori T, Furuta H, Tanaka K, Sugano M. Effects of dietary soybean peptides on hepatic production of ketone bodies and secretion of triglyceride by perfused rad liver. *Biosci. Biotechnol. Biochem.* 2007; 71(10): 2451–2457.
49. Inoue N, Nagao K, Sakata K, Yamano N, Ranawakage Gunawardena PE, Han SY, Matsui T, Nakamori T, Furuta H, Takamatsu K, Yanagita T. Screening of soy protein-derived hypotriglyceridemic dipeptides in vitro and in vivo. *Lipids Health Dis.* 2011; 10(85). https://doi.org/10.1186/1476-511X-10-85
50. Inoue N, Yamano N, Sakata K, Nagao K, Hama Y, Yanagita T. The sulfated polysaccharide porphyran reduces apolipoprotein B100 secretion and lipid synthesis in HepG2 cells. *Biosci. Biotechnol. Biochem.* 2009; 73(2): 447–449.
51. Inoue N, Yamano N, Sakata K, Arao K, Kobayashi T, Nagao T, Shimada Y, Nagao K, Yanagita T. Linoleic acid-menthyl ester reduces the secretion of apolipoprotein B100 in HepG2 cells. *J. Oleo Sci.* 2009; 58(4): 171–175.

CHAPTER 8

Bioactive Peptides in Pulses

*Anne Pihlanto, Elena Maestri, Markus Nurmi,
Nelson Marmiroli, and Minna Kahala*

CONTENTS

Introduction	153
Pulse Proteins	155
Bioactive Peptides	157
In Silico Approach	157
Production of Bioactive Peptides	165
Antihypertensive Pulse Peptides	165
Antioxidative	167
Anticholesterolemic	169
Antimicrobial	169
Other Activities and Multifunctional Peptides	170
Health Benefits	171
Conclusions	172
References	172

INTRODUCTION

Pulse crops belong to the family of cool season, annually grown leguminous crops and are harvested for their seed only which are distinguished from leguminous oil seeds by their low-fat content. According to the Codex Alimentarius (http://www.fao.org/fao-who-codexalimentarius/sh-proxy/en/?lnk=1&url=https%253A%252F%252Fworkspace.fao.org%252Fsites%252Fcodex%252FStandards%252FCXS%2B171-1989%252FCXS_171e.pdf), pulses comprise the following:

- Beans of *Phaseolus* spp. (except *Phaseolus mungo* L. syn. *Vigna mungo* (L.) Hepper and *Phaseolus aureus* Roxb. syn. *Phaseolus radiatur* L., *Vigna radiata* (L.) Wilczek)
- Lentils of *Lens culinaris* Medic. Syn. *Lens esculenta* Moench.
- Peas of *Pisum sativum* L.
- Chickpeas of *Cicer arietinum* L.
- Field beans of *Vicia faba* L.
- Cow peas of *Vigna unguiculata* (L.) Walp., syn. *Vigna sesquipedalis* Fruhw., *Vigna sinensis* (L.) Savi exd Hassk

These crops are produced on many continents worldwide. North America, specifically Canada, and areas within Asia and the Middle East are responsible for the majority of pulse crop production and exportation. Importation of pulses occurs most frequently in populated countries such as India and Egypt, where pulses are a staple of the diet.

Pulse crops are an excellent source of protein, carbohydrates, and fiber, and provide many essential vitamins and minerals. Their highly nutritional properties have been associated with many beneficial health-promoting properties, such as managing high cholesterol and type-2 diabetes and preventing various forms of cancer. However, pulse crops and other leguminous crops also contain many antinutritional proteins, such as lectins, protease inhibitors. Various deleterious effects may occur following the ingestion of raw pulse seeds or flours, such as hemagglutination, bloating, vomiting, and pancreatic enlargement, due to the activity of the antinutritional compounds inside the host. Conversely, antinutritional compounds in pulses may have many beneficial properties in the treatment and/or prevention of disease when properly processed.

Generally, enzymatic hydrolysis is widely applied to upgrade functional features (such as emulsifying properties of hydrolysed protein) and nutritional properties of proteins. It has been reported that additional advantage of hydrolysis can be the development of hydrophobicity since proteolysis unfolds the protein chains. The cleavage of peptide bonds enhances levels of free amino and carboxyl groups resulting in enhanced solubility. Therefore, hydrolysis can increase or decrease the hydrophobicity, which mostly depends on the nature of the precursor protein and molecular weight of the generated peptides. Moreover, hydrolysis leads to production of small bioactive peptides, and bitterness of peptides below 1000 Da is much less than fractions with a higher molecular mass. Some factors to consider in producing bioactive peptides include hydrolysis time, degree of hydrolysis of the proteins, enzyme–substrate ratio, and pretreatment of the protein prior to hydrolysis (Pihlanto and Mäkinen, 2013).

Pulse-based fermented products are frequently consumed globally. Fermentation can be used to affect the taste, texture, and shelf life of the product and to enhance the nutritional value and to improve the bioactivity of plant materials. It has been shown to degrade antinutritional compounds, like phytates, oligosaccharides, tannins, vicines, protease inhibitors, lectins but also to improve protein digestibility, release bioactive peptides from the native protein, and increase the level of soluble phenolic compounds (Adebo et al., 2017; Di Stefano, et al. (2019)). Bioactive peptides which can exhibit bioactivities with beneficial health effects, such as antihypertensive, antimicrobial, antioxidant, antidiabetic, immunomodulatory, and anticancer activities, have been reported in fermented foods (Maleki and Razavi, 2020; Sanjukta and Rai, 2016; Martinez-Villaluenga, Penas, and Frias, 2017; Matemu, Nakamura, and Katayama, 2021). During fermentation, peptides are released by the action of proteolytic enzymes produced by the microorganisms involved in fermentation. Thus, proteinase activity and presence of peptidases are prerequisites for selecting the microbial starters for producing bioactive peptides (Sanjukta and Rai, 2016; Rizzello et al., 2012; 2015). The liberation of bioactive peptides during fermentation depends on the microbial starter culture and applied parameters, like fermentation time and the raw material protein content and composition (Jakubczyk et al., 2013, 2017; Maleki and Razavi, 2020; Sanjukta, and Rai, 2016). Both liquid and solid-state fermentation has been used (Sanjukta and Rai, 2016; Torino et al., 2013) with several bacterial genera, such as *Bacillus*, *Lactococcus* and *Lactobacillus,* and fungi, such as *Rhizopus oligosporus* (Maleki and Razavi, 2020; Rochín-Medina et al., 2015).

Germination or sprouting is economical and effective processing method to improve seed protein quality and to produce bioactive compounds. During the germination process, seed storage proteins are digested by the specific enzymes to smaller fragments and amino acids and finally used to synthetize new proteins. Starch and lipids are also processed further to meet the needs of developing plant and this can have effect to the fermentation when these two bioprocessing methods are used in parallel. Germination conditions have a great effect to the end-product and to the bioactivity of the seeds. These parameters can be temperature, soaking, germination time, humidity, elicitors, and light. Especially germination time and temperature are critical parameters and vary between the species (De Souza et al., 2014; Maleki and Razavi, 2020).

In this review, we give an overview on the release of encrypted bioactive peptides from a range of pulses by enzymatic, fermentation or germination, as well as the *in silico* approach to predict the potential bioactive sequences within parent proteins.

PULSE PROTEINS

The protein content of pulses ranges from 20% to 40% dry weight; within these percentages, the most abundant are seed storage proteins (SSPs). Depending on the source, seeds usually present 1 or 2 predominant types of SSP (Table 8.1). The rest are minor or housekeeping proteins, which include enzymes, protease, and amylase inhibitors; lectins, lipooxygenase, defense proteins, and others (Duranti, 2006; Roy et al., 2010).

Pulse seeds accumulate protein throughout their development, hence mature pulse seeds are normally high in protein. Chickpea, lentil, and dry pea contain approximately 22%, 28.6%, and 23.3% protein, respectively, on a dry weight basis (DW) (Pulse Canada, 2004; Sotelo and Adsule, 1996). However, these percentages may vary slightly depending on plant species, variety, maturity and growing conditions. The majority of the protein found within pulse seeds is in the form of storage proteins, which are classified as albumins, globulins, and glutelins based on their solubility properties. Globulins, soluble in salt-water solutions, represent approximately 70% of the total protein found in pulses. Pulse globulins are generally categorized as 7S vicilin-type and 11S legumin-type according to their sedimentation coefficients. Globulins commonly found in legumes are legumin, vicilin, and convicilin (Duranti, 2006; Marambe and Wanasundara, 2012). Albumins, soluble in water, account for 10–20% of the total protein in pulses. Finally, glutelins, soluble in dilute acid and base, account for 10–20% of the total protein found in the seeds of pulses (Duranti, 2006; Roy, Boye, and Simpson, 2010). Nutritionally, pulse storage proteins are relatively low in sulfur-containing amino acids, such as methionine, cysteine, and tryptophan. However, the lysine content is relatively high compared to cereal crops (Mattila et al., 2018). There are many other types of protein found in legumes including various enzymes, protease and amylase inhibitors, defense proteins and lectins, which are collectively known as antinutritional compounds (ANCs). The majority of these proteins are within the water-soluble albumin class of legume proteins (Roy et al., 2010). They are distinct from other storage proteins with respect to functionality, as they have evolved within the seed as a protective mechanism. Due to the harmful effects of these compounds, they can be partly or totally inactivated after cooking or processes like fermentation, germination, and dehulling (Kahala et al., in press). Once inactivated, lectins or protease inhibitors may present potential health benefits. Protease inhibitors are prospective anti-inflammatory and anticancer agents. Whereas, lectins have demonstrated to play a key role in preventing certain cancers and the activation of certain innate defense mechanisms. Besides, lectins have also been proposed as therapeutic agents for preventing or controlling obesity (Roy et al., 2010).

TABLE 8.1 Storage Proteins on Different Pulses

Pulse (Common Names)	Storage Proteins	Total Protein Content (%)	Reference
Phaseolus spp. (dry beans)	Globulins 45–70% of total proteins (40–50% storage protein is Phaseolin, (7)S Vicilin- like, smaller amount of legumin-like (11S) Albumins 10–30% of total storage proteins	17.5–28.7	Sathe, 2002
Vicia faba (dry broad beans)	70% globulins: Vicilin (7S) and Legumin (11S) like	avg. 30	Warsame et al., 2018
Vigna radiata (mung bean)	Globulins (60% of total proteins): basic type (7S), vicilin (8S) and legumin (11S) (Legumin: Vicilin 1:2) Albumins 25% of total proteins	21–31	Yi-Shen, Shuai, and FitzGerald, 2018
Pisum spp. (dry peas)	Major storage proteins: 11S, 7S and albumins. Globulins 49.2–81.8% of total proteins (legumins 5.9–24.5% and vicilin 26.3–50.2%) Albumins: albumin 2, defensins 1 and 2 and Bowman–Birk inhibitors	23–31	Tzitzikas et al., 2006 Rubio et al., 2014
Cicer arietinum (Chickpea)	Globulin 62% of total proteins: Legumin and vicilin 6:1 Albumin: Globulin 1:4	20–28	Singh et al., 1988
Vigna unguiculata spp. Dekindtiana (dry cowpea, blackeye pea, blackeye bean)	Globulins 51% of the total seed protein, and albumins composing approximately 45%. Globulins: α-vignin (16.5S, major nonglycosylated, β-vignin (13S, major, glycosylated) and, γ-vignin (minor)	21–38	Freitas et al., 2004
Lens culinaris (lentil)	79% of total proteins are storage proteins: 72% Vicilin (7S), 21% legumin (11S)	24–30	Scippa et al., 2010
Lupinus spp. (lupins)	Globulins: α –Conglutin (legumin, 11 S) 35–37%; β –Conglutin (vicilin 7S) 44–45% of total storage protein and g-Conglutin 5% of total protein	33–40	Cabello-Hurtado et al., 2016

BIOACTIVE PEPTIDES

In Silico Approach

Correctly documented bioactive peptides can be forecasted from a known protein amino acid sequence. Computer-assisted databases are available to predict bioactive peptides located within a parent protein. Other databases predict the precursor protein of a bioactive peptide from a known amino acid sequence (Marambe and Wanasundara, 2012). BIOPEP is a peptide sequences database integrated with a program that allows classifying food proteins as potential sources of BPs. Using this database and common bean (*Phaseolus vulgaris*) proteins sequences from UniPort database, Carrasco-Castilla et al. (2012) identified sequences with 12 different biological activities corresponding to 15 seed proteins. Mojica and de Mejia (2019) have applied *in silico* digestion tools to the main storage proteins in different legumes (bean, lentil, peanut, chickpea, and pea), starting from sequences of the protein database UniProtKB. Digestion was simulated with gastrointestinal proteases with the tool PeptideCutter [https://web.expasy.org/peptide_cutter/]. According to the potential bioactivity of the obtained peptides, the majority are DPP-IV inhibitors or ACE inhibitors followed by antioxidative peptides.

Tables 8.2 and 8.3 list some of the best-known bioactive peptides from pulses, which are further described in the following subchapters. The sequences we have chosen are derived from pulses or from other similar legumes, such as mung bean or lupin. The protein originating the bioactive peptides is sometimes not mentioned in the original literature, and therefore we applied two types of database searches to identify the original protein: search for homology in non-redundant sequence databases, and search for the exact sequence. Additionally, we obtained the information about the possible presence of the same bioactive peptide in other pulses.

More specifically, searches were performed in April 2021 with the following tools:

1) PSI-BLAST (Position-Specific Iterated BLAST) on non-redundant protein sequences [https://blast.ncbi.nlm.nih.gov/Blast.cgi?PAGE_TYPE=BlastSearch&PROGRAM=blastp&BLAST_PROGRAMS=psiBlast]
2) PeptideMatch at the Protein Information Resources on the UniProtKB release 2021_02 [https://research.bioinformatics.udel.edu/peptidematch/index.jsp]

Table 8.2 is not complete with information about all peptides for several reasons: (1) peptides with two residues cannot be used for searches with these tools; (2) peptides with three residues cannot be used in PSI-BLAST; (3) short peptides of up to five or six residues might provide a large number of positive hits which cannot be sorted and analyzed; (4) genomic information for legumes and pulses is not complete. Due to this the di-, tri- and tetrapeptides are in Table 8.3.

In most cases, one or both of the methods tested can identify the protein of origin of the bioactive peptides, which is often a storage protein such as legumin, convicilin, or phaseolin. On the other hand, in several cases BLAST fails to identify homologous sequences in the same plant species or in other plant species. There are examples of peptides obtained from mung bean through fermentation (Lapsongphon and Yongsawatdigul, 2013) which are homologous or identical to sequences in animals or bacteria, but show no homology with plant sequences. In some cases, such as AVKPEPAR from mung bean, homologous sequences can be found in other legumes but not among pulses.

TABLE 8.2 Examples of Pulse Derived Bioactive Peptides with the Potential Precursor Proteins

Peptide Sequence	Technique for Peptide Production	Activity	Original Paper	https://research.bioinformatics.udel.edu/ peptidematch Homology search NCBI BLASTP	Comments (hits to pulse proteins)	Reference
Scientific name: *Cicer arietinum* **Common name:** Chickpea						
MDFLI MDLA MFDL		ACE inhibition 11 μg/ml 13 μg/ml 13 μg/ml		518 matched proteins in plants – Transmembrane 9 superfamily member 29,857 matched proteins in plants – 135 in Cicer 17,473 matched proteins in plants – 82 hits in Cicer	P. vulgaris Uncharacterized protein P. sativum: Apyrase P. vulgaris 107 hits, no storage proteins P. sativum: IUDP-glucose 4-epimerase and UDP-xylose 4-epimerase 1 V. unguiculata: 110 hits including lectin P. vulgaris: 75 hits, no storage proteins V. unguiculata: 56 hits, no storage proteins	Yust et al., 2003
VGDI	Alcalase	Antioxidant DPPH-scavenging (67.32%)		60,071 matched proteins in plants – 203 hits in Cicer, no storage proteins	P. vulgaris: 214 hits, no storage proteins P. sativum: 16 hits Indole-3-acetyl-amido synthetase; ABI3-like factor and others V. faba: Auxin-responsive GH3 family protein; Cellulose synthase V. unguiculata: 201 hits, no storage proteins	Ghribi et al., 2015
RQSHFANAQP	Alcalase, Flavourzyme	Antioxidant		3 proteins matched – vicilin-like and provicilin (C. arietinum: vicilin-like)	L. culinaris: allergen Len c 1.0101, partial P. sativum: vicilin V. faba: vicillin	Kou et al., 2013
ALEPDHR SAEHGSLH TETWNPNHPEL	Pepsin and pancreatin	Antioxidant		115 proteins matched – legumin J-like – C. arietium 14 proteins matched – legumin A-like – C. arietium 16 proteins matched – legumin J-like – C. arietium 10,470 in plants	P. sativum: legumin J V. faba: legumin type B precursor P. sativum: legumin A V. faba: legumin P. sativum: legumin J V. faba: legumin type B precursor	Torres-Fuentes et al., 2015

(*Continued*)

TABLE 8.2 (Continued)

Peptide Sequence	Technique for Peptide Production	Activity	Original Paper	https://research.bioinformatics.udel.edu/peptidematch Homology search NCBI BLASTP	Comments (hits to pulse proteins)	Reference
LTEIIP	Germination and Neutrase	Antioxidant		133 proteins matched in plants – uncharacterized protein LOC101512176		Wali et al., 2020
NRYHE	Alcalase	Antioxidant		8,254 proteins matched in plants, no pulses		Zhang et al., 2011
RIKTVTSFDLPALRFLKL RIKTVTSFDLPALRWLKL	Chymotrypsin	Antimicrobial	Legumin	1 hit – Legumin-like (C. arietinum) 1 hit – Legumin A-like (C. arietinum)	P. sativum: legumin V. faba: legumin A P. sativum: legumin V. faba: legumin A2	Heymich et al., 2021
ARCENFADSYRQPPIS SSQT GVGYKVVVTTTAAADD DDVV		Antimicrobial		1 hit – Cicerin (Fragment) 1 hit – Arietin (Fragment)		Ng, 2004
Lens culinaris; Lentils						
SDQENPFIFK	*L. plantarum* CECT 748 combined with Savinase	ACE inhibitors, possible antioxidant, oxygen radical-scavenging, hypoglycemic	Allergen Len c 1.0102 Convicilin ADP-glucose pyrophosphorylase	12 proteins matched - Allergen Len c 1.0102 (Fragment) L. culinaris: vicilin C	P. sativum: vicilin 47k C. arietinum: vicilin-like V. faba: vicilin	Bautista-Expósito et al., 2018
LLSGTQNQPSLSGF NSLTLPILRYL TLEPNSVFLPVLLH	Savinase and ultrafiltration (3 kDa permeates)	Antioxidant (0.013 Trolox equivalent) ACE inhibition (119.75 µM) Antioxidant (1.432, Trolox equivalent); ACE inhibition (77.14 µM) Antioxidant (0.139 Trolox equivalent); ACE inhibition (117.81 µM)	Globulin Globulin Globulin	2 proteins matched – Allergen Len c 1.0102/1 (Fragment) 5 proteins matched – no in Lens 35 proteins matched – no in Lens	P. vulgaris: Hypothetical protein PHAVU_009G112500g P. sativum: vicilin 47k V. faba: vicillin V. unguiculata: ATP-dependent 6-phosphofructokinase 6-like P. sativum: legumin J K C. arietinum: legumin J-like V. faba: legumin type B Legumin type B precursor P. vulgaris: hypothetical protein PHAVU_002G027900g C. arietinum: vicilin-like seed storage protein At2g18540 V. unguiculata: vicilin-like	Garcia-Mora et al., 2017

(Continued)

TABLE 8.2 (Continued)

Peptide Sequence	Technique for Peptide Production	Activity	Original Paper	https://research.bioinformatics.udel.edu/peptidematch Homology search NCBI BLASTP	Comments (hits to pulse proteins)	Reference
Phaseolus vulgaris Common bean						
KTCENLADTYKGPCF TTGSCDDHCK		Antimicrobial		Defensin D1, Knot1 domain-containing protein (antifungal plant defensin PvD1: P. vulgaris,	L. culinaris: defensin P. sativum: defensin C. arietinum: defensin V. faba: defensin-like V. unguiculata: defensin	Chan et al., 2012
GEGSGA GLTSK LSGNK MPACGSS MTEEY	Cooking, pepsin & porcine pancreatic α-amylase treatment	ACE inhibition, antiproliferative (colon cancer cells) ACE inhibition, antiproliferative ACE inhibition ACE inhibition, antioxidant ACE inhibition		636 proteins matched in plants – Glycine-rich cell wall structural protein 1.0 precursor – Protein kinase domain-containing protein 11,157 proteins matched in plants 8,903 proteins matched in plants 19 proteins matched in plants, no in Phaseolus 10,566 proteins matched in plants	P. sativum: Class II knotted-1-like 4 homeobox transcription factor P. sativum: homeobox protein HD1 V. unguiculata: Uncharacterized protein L. culinaris: NADH-plastoquinone oxidoreductase subunit I V. faba NADH-plastoquinone oxidoreductase subunit I	Luna Vital et al., 2015a,b; Luna-Vital, Mejia & Loarca-Pina, 2016 Vital et al, 2014
TTGGKGGK CPGNK GHVPP KMARPV MPHLK KTYGL	Alcalase/ in silico Pepsin/ pancreatin In vitro digestion	Antidiabetic ACE and DPPIV inhibition	Arcelin Peroxiredoxin Cylophilin Phenylalanine ammonia-lyase class III Putative resistance protein TIR 34	37 proteins matched in plants – 0 in pulses 380 proteins matched in plants – Hexosyltransferase, AAI domain-containing protein 21,604 proteins matched in plants 30 proteins matched in plants – no in pulses (soybean hypothetical) 814 proteins matched in plants – Thioredoxin domain-containing protein (fragment) 1,491 proteins matched in plants	C. arietinum: uncharacterized protein LOC101496282 C. arietinum: receptor protein kinase TMK1-like V. unguiculate: GYF domain-containing protein	Mojica & De Mejia, 2016; Mojica, Luna-Vital & de Mejia, 2017
PLPLHMLP PLPPHHMLP PPHMLP		Antioxidant		5 proteins matched – 0 in plants (hypothetical protein PHAVU_003G206300g P. vulgaris – serine/threonine protein kinase from BLAST) No matches 142 proteins matched in plants – no in pulses	C. arietinum: serine/ threonine-protein kinase AFC2-like isoform X1 V. unguiculate: thaumatin	Ngoh & Gan, 2016

(Continued)

TABLE 8.2 (Continued)

Peptide Sequence	Technique for Peptide Production	Activity	Original Paper	https://research.bioinformatics.udel.edu/peptidematch Homology search NCBI BLASTP	Comments (hits to pulse proteins)	Reference
PVNNPQIH	Alcalase/papain /pepsin/trypsin-α-chymotrypsin	ACE inhibition (206.7 μM)		18 proteins matched – Phaseolin (P. vulgaris)		Rui et al. (2015)
FVVAEQAGNEEGFE GDTVTVEFDTFLSR INEGSLLLPH	L. plantarum 299v	ACE and α-amylase inhibition	Described in vicilin, described in legumin J-like	1 protein matched – no hit in Phaseolus 38 proteins matched, all Phaseolus – Alpha amylase inhibitor-1 6 proteins matched – no in Phaseolus (Hypothetical protein PHAVU_007G059800g [P. vulgaris] 7S globulin, beta-conglycinin from BLAST)	P. sativum: legumin C. arietinum: legumin J-like V. faba: legumin V. unguiculate: 11S seed storage C. arietinum: lectin trimannoside complex L. culinaris: convicilin P. sativum: vicilin C. arietinum: vicilin-like, provicilin V. faba: vicilin V. unguiculate: beta-conglycinin beta subunit 1-like	Jakubczyk et al., 2017
Pisum sativum, Pea						
KEDDEEEQGEEE		ACE Inhibition (64.04 μg/ml)		1 protein matched – Vicilin precursor (P. sativum)		Jakubczyk et al., 2013
Vicia faba, Faba bean						
TDALEPDNRIESEGGLIET WNPNNRQ	L. plantarum 299v	Potential activity of peptides according to BIOPEP database ACE inhibition, Glucose uptake stimulating, vasoactive substance release, Antioxidative, Hypotensive, DPPIV inhibition		no hits (legumin A2 primary translation product [Vicia faba var. minor]	P. vulgaris: legumin P. sativum: legumin A-like V. unguiculata: 11S seed storage	Jakubczyk et al., 2019

(*Continued*)

TABLE 8.2 (Continued)

Peptide Sequence	Technique for Peptide Production	Activity	Original Paper	https://research.bioinformatics.udel.edu/peptidematch Homology search NCBI BLASTP	Comments (hits to pulse proteins)	Reference
Vigna radiata, Mung bean						
KDYRL VTPALR		ACE inhibition (17.9 µg/ml)		15,821 proteins matched in plants 133 in plants – no in Vigna	P. vulgaris: Uncharacterized protein P. vulgaris: Uncharacterized protein	Li et al., 2006
AVKPEPAR FLGSFLYEYSR HNVAMER LGSFLYEYSR LLPHLR		Antioxidant	Virgibacillus	14 proteins matched – no in plants 23 proteins matched – no in plants 19 proteins matched – no in pulses 23 proteins matched – 0 in plants 425 proteins matched in plants – lysine-specific demethylase JMJ25 isoform X1	P. sativum: ethylene-insensitive protein 2 C. arietinum: ethylene-insensitive protein 2 V. unguiculate: flowering time control protein FPA-like V. unguiculate: Transcription factor TCP subgroup	Lapsongphon & Yongsawatdigul, 2013
HLNVVHEN LPRL LRLESF YADLVE	Bromelain	ACE inhibition (50.88 µµM); Renin inhibition (34.34%) ACE inhibition (1912 µM); Renin inhibition (46.32%) ACE inhibition (5.39 µM); Renin inhibition (29.99%) Renin inhibition (97.06%)		no hit (hypothetical soybean) 121,937 proteins matched in plants 706 proteins matched in plants – eukaryotic translation initiation factor 4G isoform X1/X2 274 proteins matched in plants – T-complex protein 1 subunit delta	C. arietinum: low-temperature-induced cysteine proteinase-like-eukaryotic translation initiation factor 4G-like isoform X3 V. unguiculata: Protein phosphatase 1 regulatory subunit 12A n P. vulgaris: T-complex protein 1 subunit delta C. arietinum: Prolyl-tRNA synthetase; T-complex protein 1 subunit delta V. unguiculata: T-complex protein 1 subunit delta; S-adenosyl-L-methionine-dependent methyltransferase	Sonklin et al., 2020

TABLE 8.3 Di-, Tri-, and Tetra- Bioactive Peptides from Pulses

Source	Peptide Sequence	Technique for Peptide Production	Activity	Protein Matches in Plants https://research.bioinformatics.udel.edu/peptidematch	Reference
Chickpea	DHG	Alcalase	DPPH scavenging (51.66%)	328,111	Ghribi et al., 2015
	FVPH	Pepsin & Pancreatin	Antioxidant, hypotensive *in vivo*	10,470	Torres-Fuentes et al., 2015
Pea	IR	Alcalase	ACE inhibition (2.25 mM) Renin inhibition (9.2 mM)		Li and Aluko, 2010
	KF		ACE inhibition (7.23 mM) Renin inhibition (17.84 mM)		
	EF		ACE inhibition (2.98 mM) Renin inhibition (22.66 mM)		
Common bean	GHVP	Pepsin/pancreatin	ACE inhibition	21,604	Mojica, Luna-Vital, and Mejia, 2017
Lupin	FVPY		Antioxidant (ABTS, DPPH and superoxide anion scavenging)	13,858	Babini et al., 2017

The search with PeptideMatch does not recover homologous sequences, but rather proteins containing the same exact sequence. It is a useful tool to recover other potential sources of an interesting peptide. Just to give an example, the antioxidant peptide RQSHFANAQP deriving from the vicilin-like protein of chickpea can also be found in vicilins from lentil, pea, and faba beans, but it is not present in proteins from bean and cowpea. The database search indicated that some of the identified peptides are derived from other sources, like microbes, enzymes or are contaminated with other proteins.

Based on our results, we counted the number of peptides isolated from one pulse which could be found also in proteins from other pulses and we have built the chord diagram shown in Figure 8.1. Peptides from chickpea are often found in pea and faba bean and vice versa; peptides from bean mainly find correspondences in cowpea but also chickpea and pea; peptides from lentil find corresponding peptides in all pulses at the same frequency. Taxonomically, *Cicer*, *Lens*, *Pisum*, and *Vicia* belong to the Vicioid clade, whereas *Phaseolus* and *Vigna* belong to the Millettioids clade (Wojciechowski et al., 2004). A higher degree of similarity within clades is to be expected.

Computational modeling has indicated that among the bioactive peptides identified in the mixtures of cowpeas obtained after simulated gastrointestinal digestion, TTAGLLE and KVSVVAL peptides may be good DPP-IV inhibitors due to their interactions with the active site or catalytic region of the enzyme (de Souza Rocha et al., 2014). Pulses such as cowpea may provide peptides with potential to contribute to reduce the risk of type II diabetes. *In silico* binding affinities and interactions between the N-terminal domain of Niemann-Pick C1 like-1 (NPC1L1 mediates cholesterol absorption at the apical membrane of enterocytes) and 14 pulse peptides (5≥ amino acids) derived through

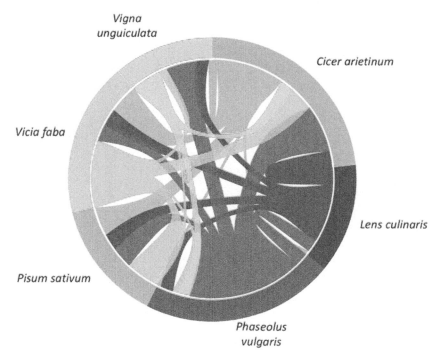

FIGURE 8.1 Chord diagram showing the relationship of precursor proteins of selected pulse bioactive peptides.

pepsin-pancreatin digestion have been investigated. The common bean peptide Tyr-Ala-Ala-Ala-Thr (YAAAT) found from black or navy bean proteins, had the highest binding affinity, even higher affinities than the control (ezetimibe). Due to their high affinity for the N-terminal domain of NPC1L1, black and cowpea bean peptides produced in the digestive track have the potential to disrupt interactions between NPC1L1 and membrane proteins that lead to cholesterol absorption. Pigeon pea (*Cajanus cajan*) was fermented with *Aspergillus niger*, and an ACE-inhibitory octapeptide, Val-Val-Ser-Leu-Ser-Ile-Pro-Arg (VVSLSIPR), was identified. Molecular docking and simulation studies showed a strong and stable interaction of the peptide with ACE (Hernandez and Mejia, 2017:Nawaz et al., 2017).

Production of Bioactive Peptides

Antihypertensive Pulse Peptides

Plant anti-hypertensive peptides have been widely studied due to the incidence of hypertension and cardiovascular diseases. Hypertension is related to the activity of the renin-angiotensin aldosterone system (RAS) and, more specifically, to the ACE. This enzyme removes two C-terminal residues from the biologically inactive angiotensin I, converting it into angiotensin II, a potent vasoconstrictor, with effects on arteries and blood vessels, namely by increasing the blood pressure. The second strategy involves the inhibition of renin directly, which could potentially result in better inhibition of RAS since renin catalyzes the rate-degerming step in the cascade of reactions in RAS (Aluko, 2012). In general, most ACE inhibitory peptides from all kinds of food sources are relatively short sequences containing from 2 to 12 amino acids. Also, several publications have identified structural features from the C-terminal tripeptide residue that play a predominant role in competitive binding to the active site of ACE. Peptides that inhibit ACE or renin have been produced from the digestion of several plant proteins, including pulses (Pihlanto and Mäkinen, 2013; 2017). Several scientific works report the ACE-inhibition of pulse hydrolysates from using diverse enzymes (pepsin, pancreatin, trypsin, Alcalase, papain) and treatments (pH, temperature) for enzymatic digestion. The hydrolysis time is a very important factor related to the peptide size; in most cases, the hydrolysates' ACE inhibitory potency increases with the hydrolysis time, implying thus that the smaller peptides are more active. Renin-inhibiting peptides or hydrolysates have been discovered from hydrolysates of yellow field pea (Aluko et al., 2015; Girgih et al., 2016), lima bean (Ciau-Solís, Acevedo-Fernández, and Betancur-Ancona, 2018), kidney bean (Mundi and Aluko, 2014) and mung bean (Sonkin et al., 2020).

Hydrolysates of chickpea and mung bean obtained by Alcalase treatments are good sources of ACE-inhibitory peptides (Li et al., 2006a; Yust et al., 2003). In a report by Li et al., (2005) mung bean hydrolysate produced by Alcalase exhibited high ACE inhibition (IC50 value 0.64 mg prot/mL) and hydrolysates using trypsin and chymotrypsin had ACE inhibition of 83.95% and 93.69%, respectively. Pedroche et al., (2002) hydrolysed chickpea protein isolate with Alcalase to produce a bioactive hydrolysate having ACE-inhibitory properties with an IC_{50} value of 0.190 mg/mL. Four peptide-fractions with average molecular weight of 900 Da, representing peptides with six to eight amino acid residues were isolated with IC_{50} values of 0.103–0.117 mg/mL. Six ACE-inhibitory peptides with IC_{50} values ranging from 0.011 to 0.012 mg/mL have been isolated from chickpea hydrolysed by Alcalase. All identified peptides contained Met and were rich in other hydrophobic amino acids (Yust et al., 2003) The protein hydrolysates obtained by digestion (with pepsin) of defatted flour from common bean, pea, chickpea, lentil, and lupin, showed anti-hypertensive activity, the lupin peptide being the most efficient with an IC_{50} value of 0.226 mg/mL (mostly resulting from α- and β-conglutin) (Boschin et al., 2014). *In vitro* gastrointestinal digestion increased the ACE-inhibitory potencies of common dry beans, dry pinto beans, and green lentils, with IC_{50} values of 0.78–0.83, 0.15–0.69 and 0.008–0.89 mg protein/mL, respectively (Akıllıoglu and Karakaya, 2009). The ACE inhibitory property of the lentil tryptic hydrolysates varied as a function of the protein fraction with the total lentil protein hydrolysate having the lowest IC_{50} (0.440 ± 0.004 mg/mL). This indicates that lentil varieties having higher amounts of legumin and albumin proteins may have higher ACE-inhibitory properties (Boye et al., 2010).

Three dipeptides, Ile-Arg (IR), Lys-Phe (KF), and Glu-Phe (EF), were isolated and identified from Alcalase hydrolysate of pea protein isolate (Table 8.3). The peptides showed strong inhibitions (IC_{50} values <25 mM) of ACE and renin (Li and Aluko, 2010). Sonklin et al. (2020) identified five peptides from mung bean Bromelain digests with ACE and renin inhibition. The results showed that Leu-Arg-Leu-Glu-Ser-Phe (LRLESF) was the most potent inhibitor of ACE (IC50=5.4 μM), and Tyr-Ala-Asp-Leu-Val-Glu (YADLVE) was the strongest renin inhibitor with 97% inhibition, while LRLESF was the weakest renin inhibitor (~30%).

Fermentation parameters, as well as the microbial strains selected, have an effect on the level of protein content and released peptides. In a study conducted by Torino et al. (2013), spontaneous fermentation of lentils and fermentation with *L. plantarum* improved ACE inhibitory activity in lentil extracts, up to inhibition values of 92% and 93%, and with IC50 values of 0.18 and 0.20 mg protein/mL, respectively. Results from the same study revealed that in solid-state fermentation with *B. subtilis*, lower ACE inhibitory activity, up to 39% after 96 h fermentation, was observed in lentils, possibly due to variations in released peptides (Torino et al., 2013). Rui et al. (2015) observed that fermentations of navy bean milk with *L. bulgaricus* and two *L. plantarum* strains increased ACE inhibitory activity IC50 values ranging from 101 to 109 μg protein/mL. Hydrolysis with Savinase, fermentation with *L. plantarum*, and the combination of both treatments were shown to have effect on the peptide concentration of lentil soluble fraction (Bautista-Exposito et al., 2018). Processing enhanced the ACE inhibitory activity of lentil soluble fraction from 81% in nonprocessed lentil to 93–95% in treated lentil samples and peptides having amino acid sequences reported as ACE inhibitors, possible antioxidant, oxygen radical-scavenging and hypoglycemic activity were found (Table 8.2).

Jakubczyk et al. (2017) demonstrated the influence of fermentation by *L. plantarum* 299v on the release of inhibitory peptides during *in vitro* digestion and reported the highest level of lipase and ACE inhibitory activity in beans in 30°C for 3 days, with IC50 of 1.19 and 0.28 mg/mL, respectively. The fractions with the highest inhibitory activity were further studied and the peptides with a molecular weight of 3.5–7 kDa and the highest ACE and α-amylase inhibitory activity, obtained from seeds after fermentation at 22 °C for 3 h and at 30 °C for 3 days were identified and sequences determined (Table 8.2). Xiao et al. (2018) demonstrated the generation of low-molecular-weight peptides and ACE inhibitory activity with IC50 value of 0,63 mg protein/ml in red beans (*Phaseolus angularis*) during solid-state fermentation with *Cordyceps militaris*, while Rochín-Medina et al. (2015) reported black bean flour fermentation by *R. oligoporus* with IC50 of 0.0321 μg/mL.

Jakubczyk et al. (2013) reported that fermentation of pea seeds by *L. plantarum* 299v enables the release of ACE inhibitory peptides during gastrointestinal digestion. The peptide present in the hydrolysate fraction with highest ACE inhibitory activity

(IC50 64.04 μg/mL) was identified as KEDDEEEEQGEEE. Vermeirssen et al. (2003) studied the fermentation of pea with yeast and different lactobacilli, both in monoculture and in combination, and observed an increase in ACE inhibitory activity. However, after subsequent *in vitro* digestion both nonfermented and fermented samples reached maximum ACE inhibitory activity. The authors suggested that *in vitro* gastrointestinal digestion was the predominant factor controlling the formation of ACE inhibitory activity. Release of bioactive peptides and ACE inhibitory activity of 67.5% after 5 h fermentation of mung bean with *L. plantarum* B1-6 was evidenced in the study of Wu et al. (2015).

Bamdad et al. (2009) showed that germination at 20 °C in darkness improved the ACE inhibitory activity on lenses, and the activity increased until the fifth day of germination, when it reached nearly 90% of the ACE inhibition. Further, elicitors can be used to enhance lentil germination rate with no or minor decrease of ACE-inhibitory activity (Peñas et al., 2015). Surprisingly, Jamdar, Deshpande, and Marathe (2017) reported that no ACE inhibitory activity on germinated lentils was detected. This could be due to the different germination conditions since temperature is one of the key factors for bioactive peptide formation. When red lentils were germinated at 30 °C, ACE inhibitory activity was decreased but increased at 40 °C (Mamilla and Mishra, 2017). Similar variation was seen between the legume species. Germination at 30 °C improved mung bean and kidney bean ACE inhibitory activity significantly, but the improvement was less when germinated at 40 °C (Mamilla and Mishra, 2017). On the other hand, chickpea had higher activity at 40 °C. Bioactive peptides are produced also by the enzymatic treatment, and this processing method can be combined with the germination to improve the process. For example, less enzyme can be used when seeds of lima bean are first germinated and then hydrolyzed with pepsin-pancreatin to achieve sufficient ACE inhibitory effect (Magana et al., 2015).

Antioxidative

Free radicals are generated daily as the result of body metabolism. However, an imbalance in reactive species production and endogenous/exogenous antioxidant defenses can lead to a condition known as oxidative stress, an underlying mechanism of various human pathologies. Peptides can contribute to the antioxidant defense in the body, being able to rapidly scavenge reactive oxygen species before cellular damage, thereby inactivating them. The antioxidant activity measured in protein hydrolysates and peptides from pulse extracts has been evaluated by different methods: 2,2′-azino-bis-3-ethyl-benzthiazoline-6-sulphonic acid (ABTS) and 2,2-diphenyl-1-picrylhy-drazyl (DPPH) radical scavenging action, trolox equivalent antioxidant capacity (TEAC), reducing power, or oxygen radical absorbance capacity (ORAC), validating its correlation with the proteolytic treatment type and hydrolysis duration. Even though the meaning of the results obtained with these assays is limited as they use no physiological radicals, it is still possible to achieve representative data evaluating antioxidant activity.

Regarding antioxidant activity of chickpea proteins, hydrolysis with alcalase or flavourzyme showed better reducing power and DPPH scavenging effect than chickpea protein isolate (Yust et al., 2012). Pea protein hydrolysates by thermolysin contained low molecular weight (<3 kDa) peptides with various antioxidant activities that were dependent on the amounts of hydrophobic and aromatic amino acid constituents (Pownall, Udenigwe, and Aluko, 2010). Peptide fractions with the least cationic property had the significantly strongest scavenging activity against DPPH and H_2O_2. Generally, the scavenging of O_2^- and H_2O_2 was negatively related with the cationic property of the peptide fractions (Pownall, Udenigwe, and Aluko, 2011). Chickpea hydrolysate with antioxidant activities was prepared from chickpea protein isolates by Alcalase. This hydrolysate was separated with Sephadex G-25 to four fractions and fraction with 200–3000 Da had the highest antioxidant activities assayed by free radical scavenging effects (Li et al., 2008.) The active peptide was identified as Asn-Arg-Tyr-His-Glu (NRYHE). This peptide quenched the free radical sources DPPH, hydroxyl, and superoxide free radicals. Furthermore, the inhibition of the peptide on lipid peroxidation was greater than that of α-tocopherol (Zhang et al.,2011). Torres-Fuentes et al. (2015) have investigated the reducing power, free radical scavenging and cellular antioxidant activities of chickpea hydrolysates obtained with enzymes of the gastrointestinal digestion (pepsin and pancreatin). The identified peptides (ALEPDHR, TETWNPNHPEL, FVPH, and SAEHGSLH) corresponded with legumin and contained the antioxidant His in the sequence.

Lentil protein hydrolysate generated with pepsin+pancreatin showed higher potential to scavenge the radical ABTS• compared to the lentil protein hydrolysate generated using Alcalase (Moreno et al., 2020). Ngoh and Can (2016) determined the antioxidant activity present in pinto bean peptides; in this study, peptide fractions < 3 kDa exhibited 42.2% inhibition of ABTS•+ scavenging activity. The results for DPPH• suggest that lower degree of hydrolysis in lentil and black bean Alcalase hydrolysates may enhance the potential to scavenge the radical DPPH• (do Evangelho et al., 2017; Moreno et al., 2020). Likewise, similar findings were observed in a study performed by Kou et al. (2013) showing that chickpea peptides obtained by Alcalase exhibited 41.3% DPPH• radical scavenging activity.

Another free radical scavenging evaluation includes NO, which plays an important role in inflammatory processes; high levels of NO and its oxidized derivatives are known to be toxic, resulting in vascular damage and other ailments. Legume protein hydrolysates showed the potential to scavenge NO with important results. In the case of the NO scavenging capacity, the most potent legume hydrolysates were the ones hydrolysed with pepsin/pancreatin. Pepsin/pancreatin protein hydrolysates presented a more extensive degree of hydrolysis, leading to more diversity of peptides in the protein hydrolysate. In a study performed by Oseguera-Toledo et al. (2015), potent hydrolysate fractions of 5–10 kDa were obtained with the enzyme Alcalase. These hydrolysates from black bean demonstrated the capacity to scavenge the radical NO with a range of inhibition from 57.46% to 68.26%.

Effect of fermentation on antioxidative properties of pulses has been established in several studies (Wu et al., 2015). Pulses contain soluble phenolic compounds which are known to affect antioxidant activity and may be the compounds mainly contributing to the antioxidant activity (Limón et al., 2015). Phenolic and peptide fractions from bean milk and yogurt prepared from common beans, navy beans (cultivar Bolt), and light red kidney beans (cultivar Inferno) showed strong antioxidant and anti-inflammatory effects in cell-based assays after simulated *in vitro* gastric digestion (Chen et al., 2019). Potential contribution of two γ-glutamyl dipeptides on the antioxidant and anti-inflammatory effects was reported. Released amino acids and peptides were suggested to contribute in the increase in antioxidant activity observed in chickpea fermented by *B. subtilis* (Li and Wang, 2021). Pea protein-derived peptides produced in fermentation by *L. rhamnosus* BGT10 exhibited high antioxidant activity in fraction with MW <10 kDa (Stanisavljević et al., 2015). In a study of Jakubczyk et al. (2019), one of the peptide fractions obtained from seeds of faba bean after fermentation (*L. plantarum)* and *in vitro* digestion was shown to have antiradical activity (ABTS•+) with IC50 of 0.99 mg/mL. Antioxidant activity shown in bitter beans (*Parkia speciosa*) fermented with *L. fermentum* ATCC9338 was proposed to be due to the presence of low-molecular-weight peptides (Muhialdin, Rani, and Hussin, 2020). Peptides PVNNNAWAYATNFVPGK and EAKPSFYLK, identified from the processed raw material, were proven to have strong antioxidant activity by DPPH assay.

Germination can either improve (De Souza et al., 2014) or decrease (López-Barrios et al., 2014) the antioxidative activity of the pulses depending on species. In the case of black bean, antioxidative activity was compromised in germinated seeds but improved anti-inflammatory properties (López-Barrios et al., 2016).

Anticholesterolemic

Cholesterol absorption by humans requires its solubilization in micelles formed by bile salts, and it has been reported that peptides derived from digestion of food proteins may decrease cholesterol solubility by competing with it for the micellar composition. The key enzyme in cholesterol metabolism is 3-hydroxy-3methyl-glutaryl-coenzyme A (HMGCoAR). HMGCoAR inhibition leads to lowered blood cholesterol levels and is the target of statins. HMGCoAR activity is regulated by phosphorylation via the AMPK pathway. Hydrolysis of chickpea protein with Alcalase or Flavourzyme resulted in better hypocholesterolemic activity when compared with chickpea protein isolate. The highest cholesterol micellar solubility inhibition (50%) was found after 60 min of treatment with Alcalase followed by 30 min of hydrolysis with Flavourzyme (Yust et al., 2012). Marques et al. (2015) observed that after simulation of human digestion, peptides liberated from raw and cooked cowpea protein inhibit routes that interfere in the *in vitro* lipid metabolism. Protein and peptides of cowpea were able to inhibit the enzyme HMGCoAR and reduced cholesterol micellar solubilization *in vitro*. Ashraf et al. (2020) showed that thermally processed proteins from adzuki bean (*Vigna angularis*) and faba bean contained low-molecular-weight peptides (< 3 kDa) and the faba bean peptides showed significantly increased inhibition of cholesterol solubilization into micelles (45.1 ± 1.6%). Whereas peptides from the native faba bean protein exhibited significantly increased inhibition of HMG Co-AR catalytic activity (84.1 ± 2.7%) compared with the adzuki bean and thermally processed peptides.

Antimicrobial

Pathogenic fungi can be dangerous for humans and plants. Fungal infections might destroy plants and significantly reduce crop yields. As a defense system, plants produce compounds, including proteins and peptides, with fungal growth inhibitory activity (Luna-Vital et al., 2015a; Pina-Pérez and Pérez, 2018). Natural compounds with inhibitory activity against pathogens and microbes are nowadays potential candidates as alternative antimicrobial molecules in food and health (Pina-Pérez and Pérez, 2018). Recent studies reported that the antifungal properties of common beans are expressed in plant (e.g., inhibiting the growth of *Rhizoctonia solani*, *Mycosphaerella arachidicola*, *Fusarium oxysporum*, *Verticillium dahlia*, and/or *Setosphaeria turcica*) and human (e.g., *Helminthosporium maydis* or *Candida albicans*) pathogens (Luna-Vital et al., 2015b). An antifungal peptide (lectin-like peptide) was isolated from dry red lentil seeds, with *M. arachidicola* growth inhibition, suggesting the potential use of the lentil antifungal peptide in transgenic plants against banana (and also cotton) pathogens (Wang and Ng, 2007). Lupin protein hydrolysates have shown antibacterial activity against Gram-positive and Gram-negative species in different food systems, suggesting their potential use as a food biopreserver (Osman, El-Araby, and Taha, 2016). In this way, several pulses proteins/peptides have been studied due to their positive antimicrobial potential against several foodborne pathogens, such as *Listeria monocytogenes* (chickpea methylated proteins);

Salmonella spp. (chickpea protein and pea purified peptide seed); *Escherichia coli* (lectins from lentils, common pea peptide fractions); *Staphylococcus aureus* (common pea peptide fractions and purified lentils lectins); and also pathogens, such as *Pseudomonas aeruginosa* (*V. faba* peptides) (Pina-Pérez and Pérez, 2018).

Boiling and fermentation processes clearly increased the antibacterial activity of bitter beans (*Parkia speciosa*) towards four selected pathogenic microorganisms (Muhialdin et al., 2020). When the peptides, released in the fermentation process by *L. fermentum* ATCC9338, were fractionated and identified, three of the peptides EAKPSFYLK, PVNNNAWAYATNFVPGK, and AIGIFVKPDTAV, were demonstrated to have antibacterial activity (Muhialdin et al., 2020).

Other Activities and Multifunctional Peptides

Lentil hydrolyzed by Savinase® produced multiactive hydrolysates (ACE-inhibitory, antioxidant and anti-inflammatory) (Garcia-Mora et al., 2014). In further studies, three peptides (LLSGTQNQPSFLSGF, NSLTLPILRYL, TLEPNSVFLPVLLH) showed the highest antioxidant (0.013–1.432 µmol Trolox eq./µmol peptide) and ACE inhibitory activities (IC_{50} = 44–120 µM). In addition, gastrointestinal digestion of peptides improved their dual activity (10–14 µmol Trolox eq./µmol peptide; IC_{50} = 11–21 µM) (Carcia-Mora et al., 2017). Xie et al. (2019) hydrolysed mung bean protein with alcalase, neutrase, papain, and protamex and fractionated the hydrolysate according to the molecular mass into three fractions to study antioxidant and ACE-inhbitory properties. Fractions with <3kDa showed the best scavenge DPPH, hydroxyl radicals, superoxide radicals, Fe^{2+} chelating activities, and the best ACE inhibitory activity (IC_{50} = 4.66 µg/mL) compared to hydrolysates and other fractions (3–10 kDa and <10 kDa).

The fermentation of faba bean seeds at 30 °C for 3 days with *L. plantarum* 299v released peptide fraction with molecular mass under 3.0 kDa with the highest potential for inhibition of metabolic syndrome, the IC50 values of ACE, LOX, and pancreatic lipase activity were 0.05, 0.10, and 0.46 mg/mL, respectively. The identified peptides had more than 20 amino acids (TDALEPDNRIESEGGLIETWNPNNRQ, FEEPQQSEQGEGR, GSRQEEDEDEDE, WMYNDQDIPVINNQLDQMPR, RGEDEDDKEKRHSQKGES, and RLNIGSSSSPDIYNPQAGR) derived from an N-terminal incomplete legumin A1 pre-pro-polypeptide partial (*Vicia faba* var. minor) were identified. (Jakubczyk et al.,2019).

An important biological activity is the dipeptidyl peptidase-IV (DPP-IV) inhibition due to its role in current type II diabetes treatment. DPP-IV degrades glucagon-like peptide-1 (GLP-1), an incretin hormone responsible for the regulation of postprandial blood glucose levels stimulating insulin secretion (Kim et al., 2008). Germination process (25 °C, up to 48) of cowpea and the alcalase hydrolysis (4 h) was used to produce ingredients with high antioxidant capacity. On the other hand, the non-germinated and 1 h Alcalase hydrolysates showed the highest DPP-IV inhibition (IC_{50} = 0.58 mg protein/mL), after *in vitro* simulated gastrointestinal digestion. Low molecular mass peptides produced, especially after short-time germination, inhibited DPP-IV activity (de Souza Rocha et al., 2014). Based on the sequence of the peptide obtained in the fermentation of faba bean, Jakubczyk et al. (2019) suggested potential anti-diabetic activity such as DPP-IV inhibition or glucose uptake stimulation. In the study of Di Stefano et al. (2019), germination and solid-state fermentation with *L. plantarum* were shown to alter the phenolic and protein profiles of yellow pea and green lentil, with improvement of their DPP-IV and α-glucosidase inhibitory activities.

Lunasin is a widely studied peptide for its anticancer activities and is reported especially in soybean products (Rizzello et al., 2016; Sanjukta and Rai, 2016). An increase in the concentration of lunasin has been reported also during fermentation of cereals, pseudocereals, and soybean flours with sourdough lactic acid bacteria (Rizzello et al., 2012; 2016). Lunasin-like immunoreactive polypeptides were found to be released from native legume proteins during fermentation of Italian legumes (*P. vulgaris, Cicer arietinum, Lathyrus sativus, Lens culinaris and Pisum sativum*) by *L. plantarum* and *L. brevis* (Rizzello et al., 2015). Furthermore extracts from legume sourdoughs showed noticeable inhibitory effect on proliferation of human adenocarcinoma Caco-2 cells.

Health Benefits

Different potential health benefits of pulses and their constituents have been investigated, particularly in the area of cardiovascular disease and type 2 diabetes. The health benefits derive mainly from the concentration and properties of starch, protein, fiber, vitamins, minerals, and phytochemicals in pulses (e.g., Hou et al., 2019; Mudryj, Yu, and Aukema, 2014; Roy et al., 2010). Fiber from the seed coat and the cell walls of the cotyledon contributes to gastrointestinal function and health and reduces the digestibility of starch in pulses. The intermediate amylose content of pea starch also contributes to its lower glycemic index and reduced starch digestibility.

Several animal studies using different models have been conducted to investigate the favorable effects of pulse hydrolysates or peptides on high blood pressure. Pea and mung bean protein digests were shown to have antihypertensive activity in SHR (Li et al., 2006b; Hsu et al., 2011). Mung bean protein hydrolysate prepared with Alcalase decreased significantly SBP (−30.8 mmHg) of SHR 6 h after a single oral administration at a dose of 600 mg/mL. The blood pressure-lowering effect continued for at least 8 h, and the blood pressure returned to initial levels at 12 h after administration (Li et al., 2006b). Single administration of mung bean raw sprout extract (at dose of 600 mg/kg) significantly reduced SBP (−40 mmHg) 6 h after administration. Plasma ACE activities in the treated rats also decreased (0.007 Unit/mL). A long-term intervention (1 month) test showed that blood pressure in the treated animals fluctuated according to the treatments. While raw sprout extract showed effective results after 1 week of intervention, dried sprout extracts did not have significant effects until 2 weeks (Hsu et al., 2011). Pea protein hydrolysate was made by thermolysin action followed by membrane filtration. Oral administration of the pea protein hydrolysate, containing <3 kDa peptides, to SHR at doses of 100 and 200 mg/kg body weight led to a lowering of SBP, with a maximum reduction of 19 mmHg at 4 h. In contrast, orally administered unhydrolysed pea protein isolate had no blood pressure reducing effect in SHR, suggesting that thermolysin hydrolysis may have been responsible for releasing bioactive peptides from the native protein (Li et al., 2011). The effect of renin and ACE inhibitory hydrolysates and peptides derived from mung meal bromelain hydrolysate after oral administration to SHR was studied. Blood pressure was reduced with hydrolysates and peptides, however, the peptides (LPRL, YADLVE, LRLESF, HLNVVHEN, and PGSGCAGTDL) had significantly ($p < 0.05$) stronger SBP reductions (up to 36 mmHg) than the hydrolysate (up to 15 mmHg). In particular, peptide YADLVE had long-lasting effect as evident in the −27 mmHg value at 24 h when compared to <18 mmHg for hydrolysate and other peptides. The results suggest that YADLVE may be more resistant to endogenous enzyme breakdown or binds more tightly to ACE and renin (Sonkin et al., 2020).

CONCLUSIONS

The abundance and availability of pulse crops worldwide, coupled with their high nutritional value, have resulted in pulse crops being one of the most widely produced and consumed agricultural commodities. Pulse crops have long been known for their nutritional and health-promoting properties, such as being an excellent source of protein, fiber, carbohydrates, and for their role in decreasing the risk of certain cancers, managing obesity, lowering cholesterol and type-2 diabetes. Recently, the bioactive properties of peptides derived from pulse seeds have gained increased recognition in the areas of food science and nutrition for their potential benefits in treating and/or reducing the onset of disease. Pulse seeds may, therefore, be potentially excellent sources of beneficial bioactive peptides, and techniques for the efficient extraction and fractionation of these peptides are needed. Further research is needed to improve our understanding of the mechanisms involved in the absorption into the blood stream, target sites and activity in various tissues of biologically active compounds derived from pulses. Another area requiring research is comparative studies of the bioactivities of various pulse cultivars of the same species and between different pulse crops, in order to determine the cultivar of a specific pulse that has the greatest concentration and activity of a desired biologically active compound.

REFERENCES

Adebo, O. A., Njobeh, P. B., Adebiyi, J. A., Gbashi, S., Phoku, J. Z., & Kayitesi, E. (2017). Fermented pulse-based food products in developing nations as functional foods and ingredients. In Hueda, M. C. (Ed.), *Functional food—Improve health through adequate food* (pp. 77–109). Croatia: IntechOpen.

Akıllıoğlu, H. G., & Karakaya, S. (2009). Effects of heat treatment and in vitro digestion on the angiotensin converting enzyme inhibitory activity of some legume species. *European Food Research and Technology*, 229(6), 915–921.

Aluko, R. (2012). Bioactive peptides. In Aluko, R. (Ed.), *Functional foods and nutraceuticals* (pp. 37–61). New York: Springer.

Aluko, R. E., Girgih, A. T., He, R., Malomo, S., Li, H., Offengenden, M., & Wu, J. (2015). Structural and functional characterization of yellow field pea seed (*Pisum sativum* L.) protein-derived antihypertensive peptides. *Food Research International*, 77, 10–16.

Ashraf, J., Awais, M., Liu, L., Khan, M. I., Tong, L. T., Ma, Y., ... & Zhou, S. (2020). Effect of thermal processing on cholesterol synthesis, solubilisation into micelles and antioxidant activities using peptides of *Vigna angularis* and *Vicia faba*. *LWT - Food Science and Technology*, 129, 109504. https://doi.org/10.1016/j.lwt.2020.109504.

Babini, E. Davide, T., Martini, S., Dei Più, L., & Gianotti, A. LC-ESI-QTOF-MS identification of novel antioxidant peptides obtained by enzymatic and microbial hydrolysis of vegetable proteins. *Food Chemistry*, 228, 186–196.

Bamdad, F., Dokhani, S., Keramat, J., & Zareie, R. (2009). The impact of germination and in vitro digestion on the formation of angiotensin converting enzyme (ACE) inhibitory peptides from lentil proteins compared to whey proteins. *Heart Failure*, 3, 4.

Bautista-Expósito, S., Martínez-Villaluenga, C., Dueñas, M., Silván, J. M., Frias, J., & Peñas, E. (2018). Combination of pH-controlled fermentation in mild acidic conditions and enzymatic hydrolysis by savinase to improve metabolic health-promoting properties of lentil. *Journal of Functional Foods*, 48, 9–18.

Boschin, G., Scigliuolo, G. M., Resta, D., & Arnoldi, A. (2014). ACE-inhibitory activity of enzymatic protein hydrolysates from lupin and other legumes. *Food Chemistry*, *145*, 34–40.

Boye, J. I., Roufik, S., Pesta, N., & Barbana, C. (2010). Angiotensin I-converting enzyme inhibitory properties and SDS-PAGE of red lentil protein hydrolysates. *LWT - Food Science and Technology*, *43*(6), 987–991.

Cabello-Hurtado, F., Keller, J., Ley, J., Sanchez-Lucas, R., Jorrín-Novo, J. V., & Aïnouche, A. (2016). Proteomics for exploiting diversity of lupin seed storage proteins and their use as nutraceuticals for health and welfare. *Journal of Proteomics*, *143*, 57–68.

Carrasco-Castilla, J., Hernández-Álvarez, A. J., Jiménez-Martínez, C., Jacinto-Hernández, C., Alaiz, M., Girón-Calle, J., ... & Dávila-Ortiz, G. (2012). Antioxidant and metal chelating activities of Phaseolus vulgaris L. var. Jamapa protein isolates, phaseolin and lectin hydrolysates. *Food Chemistry*, *131*(4), 1157–1164.

Chan, Y. S., Wong, J. H., Fang, E. F., Pan, W. L., & Ng, T. B. (2012). An antifungal peptide from *Phaseolus vulgaris* cv. brown kidney bean. *Acta Biochimica et Biophysica Sinica*, *44*(4), 307–315.

Chen, Y., Zhang, H., Liu, R., Mats, L., Zhu, H., Pauls, K. P., ... & Tsao, R. (2019). Antioxidant and anti-inflammatory polyphenols and peptides of common bean (*Phaseolus vulga* L.) milk and yogurt in Caco-2 and HT-29 cell models. *Journal of Functional Foods*, *53*, 125–135.

Ciau-Solís, N. A., Acevedo-Fernández, J. J., & Betancur-Ancona, D. (2018). In vitro renin–angiotensin system inhibition and in vivo antihypertensive activity of peptide fractions from lima bean (*Phaseolus lunatus* L.). *Journal of the Science of Food and Agriculture*, *98*(2), 781–786.

de Souza Rocha, T., Hernandez, L. M. R., Chang, Y. K., & de Mejía, E. G. (2014). Impact of germination and enzymatic hydrolysis of cowpea bean (*Vigna unguiculata*) on the generation of peptides capable of inhibiting dipeptidyl peptidase IV. *Food Research International*, *64*, 799–809.

Di Stefano, E., Tsopmo, A., Oliviero, T., Fogliano, V., & Udenigwe, C. C. (2019). Bioprocessing of common pulses changed seed microstructures, and improved dipeptidyl peptidase-IV and α-glucosidase inhibitory activities. *Scientific Reports*, *9*(1), 1–13.

do Evangelho, J. A., Vanier, N. L., Pinto, V. Z., De Berrios, J. J., Dias, A. R. G., & da Rosa Zavareze, E. (2017). Black bean (*Phaseolus vulgaris* L.) protein hydrolysates: Physicochemical and functional properties. *Food Chemistry*, *214*, 460–467.

Duranti, M. (2006). Grain legume proteins and nutraceutical properties. *Fitoterapia*, *77*(2), 67–82.

Freitas, R. L., Teixeira, A. R., & Ferreira, R. B. (2004). Characterization of the proteins from Vigna unguiculata seeds. *Journal of Agricultural and Food Chemistry*, *52*(6), 1682–1687.

García-Mora, P., Martín-Martínez, M., Bonache, M. A., González-Múniz, R., Peñas, E., Frias, J., & Martinez-Villaluenga, C. (2017). Identification, functional gastrointestinal stability and molecular docking studies of lentil peptides with dual antioxidant and angiotensin I converting enzyme inhibitory activities. *Food Chemistry*, *221*, 464–472.

Garcia-Mora, P., Peñas, E., Frias, J., & Martínez-Villaluenga, C. (2014). Savinase, the most suitable enzyme for releasing peptides from lentil (Lens culinaris var. Castellana) protein concentrates with multifunctional properties. *Journal of Agricultural and Food Chemistry*, *62*(18), 4166–4174.

Ghribi, A. M., Sila, A., Przybylski, R., Nedjar-Arroume, N., Makhlouf, I., Blecker, C., ... & Besbes, S. (2015). Purification and identification of novel antioxidant peptides from enzymatic hydrolysate of chickpea (*Cicer arietinum* L.) protein concentrate. *Journal of Functional Foods*, 12, 516–525.

Girgih, A. T., Nwachukwu, I. D., Onuh, J. O., Malomo, S. A., & Aluko, R. E. (2016). Antihypertensive properties of a pea protein hydrolysate during short-and long-term oral administration to spontaneously hypertensive rats. *Journal of Food Science*, 81(5), H1281–H1287.

Hernandez, L. M. R., & de Mejia, E. G. (2017). Bean peptides have higher in silico binding affinities than ezetimibe for the N-terminal domain of cholesterol receptor Niemann-Pick C1 Like-1. *Peptides*, 90, 83–89.

Heymich, M. L., Friedlein, U., Trollmann, M., Schwaiger, K., Böckmann, R. A., & Pischetsrieder, M. (2021). Generation of antimicrobial peptides Leg1 and Leg2 from chickpea storage protein, active against food spoilage bacteria and foodborne pathogens. *Food Chemistry*, 347, 128917. https://doi.org/10.1016/j.foodchem.2020.128917.

Hou, D., Yousaf, L., Xue, Y., Hu, J., Wu, J., Hu, X., ... & Shen, Q. (2019). Mung bean (*Vigna radiata* L.): Bioactive polyphenols, polysaccharides, peptides, and health benefits. *Nutrients*, 11(6), 1238.

Hsu, G. S. W., Lu, Y. F., Chang, S. H., & Hsu, S. Y. (2011). Antihypertensive effect of mung bean sprout extracts in spontaneously hypertensive rats. *Journal of Food Biochemistry*, 35(1), 278–288.

Jakubczyk, A., Karaś, M., Baraniak, B., & Pietrzak, M. (2013). The impact of fermentation and in vitro digestion on formation angiotensin converting enzyme (ACE) inhibitory peptides from pea proteins. *Food Chemistry*, 141(4), 3774–3780.

Jakubczyk, A., Karaś, M., Złotek, U., & Szymanowska, U. (2017). Identification of potential inhibitory peptides of enzymes involved in the metabolic syndrome obtained by simulated gastrointestinal digestion of fermented bean (*Phaseolus vulgaris* L.) seeds. *Food Research International*, 100(1), 489–496.

Jakubczyk, A., Karaś, M., Złotek, U., Szymanowska, U., Baraniak, B., & Bochnak, J. (2019). Peptides obtained from fermented faba bean seeds (*Vicia faba*) as potential inhibitors of an enzyme involved in the pathogenesis of metabolic syndrome. *LWT - Food Science and Technology*, 105, 306–313.

Jamdar, S. N., Deshpande, R., & Marathe, S. A. (2017). Effect of processing conditions and in vitro protein digestion on bioactive potentials of commonly consumed legumes. *Food Bioscience*, 20, 1–11.

Kahala, M., Mäkinen, S., & Pihlanto, A. (in press, 2021). Impact of fermentation on antinutritional factors. In Rai, A. K. & Anu Appaiah, K. A. (Eds.), *Bioactive compounds in fermented foods, health aspects*. Boca Raton, FL: CRC Press.

Kim, S. J., Nian, C., Doudet, D. J., & McIntosh, C. H. (2008). Inhibition of dipeptidyl peptidase IV with sitagliptin (MK0431) prolongs islet graft survival in streptozotocin-induced diabetic mice. *Diabetes*, 57(5), 1331–1339.

Kou, X., Gao, J., Xue, Z., Zhang, Z., Wang, H., & Wang, X. (2013). Purification and identification of antioxidant peptides from chickpea (*Cicer arietinum* L.) albumin hydrolysates. *LWT - Food Science and Technology*, 50(2), 591–598.

Lapsongphon, N., & Yongsawatdigul, J. (2013). Production and purification of antioxidant peptides from a Mung bean meal hydrolysate by *Virgibacillus* sp. SK37 proteinase. *Food Chemistry*, 141(2), 992–999.

Li, G. H., Le, G. W., Liu, H., & Shi, Y. H. (2005). Mung-bean protein hydrolysates obtained with alcalase exhibit angiotensin I-converting enzyme inhibitory activity. *Food Science and Technology International, 11*(4), 281–287.

Li, G. H., Shi, Y. H., Liu, H., & Le, G. W. (2006b). Antihypertensive effect of alcalase generated mung bean protein hydrolysates in spontaneously hypertensive rats. *European Food Research and Technology, 222*(5), 733–736.

Li, G. H., Wan, J. Z., Le, G. W., & Shi, Y. H. (2006a). Novel angiotensin I-converting enzyme inhibitory peptides isolated from alcalase hydrolysate of mung bean protein. *Journal of Peptide Science, 12*(8), 509–514.

Li, H., & Aluko, R. E. (2010). Identification and inhibitory properties of multifunctional peptides from pea protein hydrolysate. *Journal of Agricultural and Food Chemistry, 58*(21), 11471–11476.

Li, H., Prairie, N., Udenigwe, C. C., Adebiyi, A. P., Tappia, P. S., Aukema, H. M., ... & Aluko, R. E. (2011). Blood pressure lowering effect of a pea protein hydrolysate in hypertensive rats and humans. *Journal of Agricultural and Food Chemistry, 59*(18), 9854–9860.

Li, W., & Wang, T. (2021). Effect of solid-state fermentation with Bacillus subtilis lwo on the proteolysis and the antioxidative properties of chickpeas. *International Journal of Food Microbiology, 338*, 108988. https://doi.org/10.1016/j.ijfoodmicro.2020.108988.

Li, Y., Jiang, B., Zhang, T., Mu, W., & Liu, J. (2008). Antioxidant and free radical-scavenging activities of chickpea protein hydrolysate (CPH). *Food Chemistry, 106*(2), 444–450.

Limón, R. I., Peñas, E., Torino, M. I., Martínez-Villaluenga, C., Dueñas, M., & Frias, J. (2015). Fermentation enhances the content of bioactive compounds in kidney bean extracts. *Food Chemistry, 172*, 343–352.

López-Barrios, L., Gutiérrez-Uribe, J. A., & Serna-Saldívar, S. O. (2014). Bioactive peptides and hydrolysates from pulses and their potential use as functional ingredients. *Journal of Food Science, 79*(3), R273–R283.

Luna-Vital, D., & de Mejía, E. G. (2018). Peptides from legumes with antigastrointestinal cancer potential: Current evidence for their molecular mechanisms. *Current Opinion in Food Science, 20*, 13–18.

Luna-Vital, D. A., De Mejía, E. G., & Loarca-Piña, G. (2016). Selective mechanism of action of dietary peptides from common bean on HCT116 human colorectal cancer cells through loss of mitochondrial membrane potential and DNA damage. *Journal of Functional Foods, 23*, 24–39.

Luna-Vital, D. A., De Mejía, E. G., Mendoza, S., & Loarca-Piña, G. (2015a). Peptides present in the non-digestible fraction of common beans (Phaseolus vulgaris L.) inhibit the angiotensin-I converting enzyme by interacting with its catalytic cavity independent of their antioxidant capacity. *Food and Function, 6*(5), 1470–1479.

Luna-Vital, D. A., Mojica, L., de Mejía, E. G., Mendoza, S., & Loarca-Piña, G. (2015b). Biological potential of protein hydrolysates and peptides from common bean (*Phaseolus vulgaris* L.): A review. *Food Research International, 76*, 39–50.

Magaña, M. D., Segura-Campos, M., Dávila-Ortiz, G., Betancur-Ancona, D., & Chel-Guerrero, L. (2015). ACE-I inhibitory properties of hydrolysates from germinated and ungerminated Phaseolus lunatus proteins. *Food Science and Technology, 35*(1), 167–174.

Maleki, S., & Razavi, S. H. (2020). Pulses' germination and fermentation: Two bioprocessing against hypertension by releasing ACE inhibitory peptides. *Critical Reviews in Food Science and Nutrition, 61*(7), 2876–2893. doi: 10.1080/10408398.2020.1789551

Mamilla, R. K., & Mishra, V. K. (2017). Effect of germination on antioxidant and ACE inhibitory activities of legumes. *LWT - Food Science and Technology, 75*, 51–58.

Marambe, P. W. M. L. H. K., & Wanasundara, J. P. D. (2012). Seed storage proteins as sources of bioactive peptides. In Florence, E. & Uruakpa, O. (Eds.), *Bioactive molecules in plant foods*. New York: Nova Science Publishers.

Marques, M. R., Freitas, R. A. M. S., Carlos, A. C. C., Siguemoto, É. S., Fontanari, G. G., & Arêas, J. A. G. (2015). Peptides from cowpea present antioxidant activity, inhibit cholesterol synthesis and its solubilisation into micelles. *Food Chemistry, 168*, 288–293.

Martinez-Villaluenga, C., Penas, E., & Frias, J. (2017). Bioactive peptides in fermented foods: Production and evidence for health effects. In Frias, J., Martinez-Villaluenga, & Peñas, J. (Eds.), *Fermented foods in health and disease prevention* (pp. 23–47). Cambridge: Academic Press.

Matemu, A., Nakamura, S., & Katayama, S. (2021). Health benefits of antioxidative peptides derived from legume proteins with a high amino acid score. *Antioxidants, 10*(2), 316.

Mattila, P., Mäkinen, S., Eurola, M., Jalava, T., Pihlava, J. M., Hellström, J., & Pihlanto, A. (2018). Nutritional value of commercial protein-rich plant products. *Plant Foods for Human Nutrition, 73*(2), 108–115.

Mittal, R., Nagi, H. P. S., Sharma, P., & Sharma, S. (2012). Effect of processing on chemical composition and antinutritional factors in chickpea flour. *Journal of Food Science and Engineering, 2*(3), 180–186.

Mojica, L., & De Mejía, E. G. (2015). Characterization and comparison of protein and peptide profiles and their biological activities of improved common bean cultivars (*Phaseolus vulgaris* L.) from Mexico and Brazil. *Plant Foods for Human Nutrition, 70*(2), 105–112.

Mojica, L., & De Mejía, E. G. (2016). Optimization of enzymatic production of antidiabetic peptides from black bean (*Phaseolus vulgaris* L.) proteins, their characterization and biological potential. *Food and Function, 7*(2), 713–727.

Mojica, L., & González de Mejía, E. (2019). Legume bioactive peptides. In Martin-Cabrejas, M. A. (Ed.), *Legumes: Nutritional quality, processing and potential health benefits. Food Chemistry, function and analysis No. 8* (pp. 106–138). London: Royal Society of Chemistry.

Mojica, L., Luna-Vital, D. A., & González de Mejía, E. (2017). Characterization of peptides from common bean protein isolates and their potential to inhibit markers of type-2 diabetes, hypertension and oxidative stress. *Journal of the Science of Food and Agriculture, 97*(8), 2401–2410.

Moreno, C., Mojica, L., González de Mejía, E., Camacho Ruiz, R. M., & Luna-Vital, D. A. (2020). Combinations of legume protein hydrolysates synergistically inhibit biological markers associated with adipogenesis. *Foods, 9*(11), 1678. https://doi.org/10.3390/foods9111678.

Mudryj, A. N., Yu, N., & Aukema, H. M. (2014). Nutritional and health benefits of pulses. *Applied Physiology, Nutrition, and Metabolism, 39*(11), 1197–1204.

Muhialdin, B. J., Rani, N. F. A., & Hussin, A. S. M. (2020). Identification of antioxidant and antibacterial activities for the bioactive peptides generated from bitter beans (*Parkia speciosa*) via boiling and fermentation processes. *LWT, 131*, 109776. https://doi.org/10.1016/j.lwt.2020.109776.

Mundi, S., & Aluko, R. E. (2014). Inhibitory properties of kidney bean protein hydrolysate and its membrane fractions against renin, angiotensin converting enzyme, and free radicals. *Austin Journal of Nutrition and Food, 2*(1), 1008–1018.

Nawaz, K. A., David, S. M., Murugesh, E., Thandeeswaran, M., Kiran, K. G., Mahendran, R., ... & Angayarkanni, J. (2017). Identification and in silico characterization of a novel peptide inhibitor of angiotensin converting enzyme from pigeon pea (*Cajanus cajan*). *Phytomedicine*, 36, 1–7.

Ng, T. B. (2004). Antifungal proteins and peptides of leguminous and non-leguminous origins. *Peptides*, 25(7), 1215–1222.

Ngoh, Y. Y., & Gan, C. Y. (2016). Enzyme-assisted extraction and identification of antioxidative and α-amylase inhibitory peptides from Pinto beans (*Phaseolus vulgaris* cv. Pinto). *Food Chemistry*, 190, 331–337.

Oseguera-Toledo, M. E., de Mejia, E. G., & Amaya-Llano, S. L. (2015). Hard-to-cook bean (*Phaseolus vulgaris* L.) proteins hydrolyzed by alcalase and bromelain produced bioactive peptide fractions that inhibit targets of type-2 diabetes and oxidative stress. *Food Research International*, 76(3), 839–851.

Osman, A., El-Araby, G. M., & Taha, H. (2016). Potential use as a bio-preservative from lupin protein hydrolysate generated by alcalase in food system. *Journal of Applied Biology and Biotechnology*, 4, 076–081.

Pedroche, J., Yust, M. M., Girón-Calle, J., Alaiz, M., Millán, F., & Vioque, J. (2002). Utilisation of chickpea protein isolates for production of peptides with angiotensin I-converting enzyme (ACE)-inhibitory activity. *Journal of the Science of Food and Agriculture*, 82(9), 960–965.

Peñas, E., Limón, R. I., Martínez-Villaluenga, C., Restani, P., Pihlanto, A., & Frias, J. (2015). Impact of elicitation on antioxidant and potential antihypertensive properties of lentil sprouts. *Plant Foods for Human Nutrition*, 70(4), 401–407.

Pihlanto, A., & Mäkinen, S. (2013). *Antihypertensive properties of plant protein derived peptides* (pp. 145–182). New York: Intech Publishers.

Pihlanto, A., & Mäkinen, S. (2017). The function of renin and the role of food-derived peptides as direct renin inhibitors. In Tolekova, A.N. (Ed.), *Renin-angiotensin system-past, present and future* (pp. 241–258). Croatia: IntechOpen.

Pina-Pérez, M. C., & Pérez, M. F. (2018). Antimicrobial potential of legume extracts against foodborne pathogens: A review. *Trends in Food Science and Technology*, 72, 114–124.

Pownall, T. L., Udenigwe, C. C., & Aluko, R. E. (2010). Amino acid composition and antioxidant properties of pea seed (*Pisum sativum* L.) enzymatic protein hydrolysate fractions. *Journal of Agricultural and Food Chemistry*, 58(8), 4712–4718.

Pownall, T. L., Udenigwe, C. C., & Aluko, R. E. (2011). Effects of cationic property on the in vitro antioxidant activities of pea protein hydrolysate fractions. *Food Research International*, 44(4), 1069–1074.

Pulse Canada. (2004). Canadian dry peas. Retrieved from http://www.pulsecanada.com (accessed 21.08.08).

Rizzello, C. G., Hernández-Ledesma, B., Fernández-Tomé, S., Curiel, J. A., Pinto, D., Marzani, B., ... & Gobbetti, M. (2015). Italian legumes: Effect of sourdough fermentation on lunasin-like polypeptides. *Microbial Cell Factories*, 14(1), 1–20.

Rizzello, C. G., Nionelli, L., Coda, R., & Gobbetti, M. (2012). Synthesis of the cancer preventive peptide lunasin by lactic acid bacteria during sourdough fermentation. *Nutrition and Cancer*, 64(1), 111–120.

Rizzello, C. G., Tagliazucchi, D., Babini, E., Rutella, G. S., Saa, D. L. T., & Gianotti, A. (2016). Bioactive peptides from vegetable food matrices: Research trends and novel biotechnologies for synthesis and recovery. *Journal of Functional Foods*, 27, 549–569.

Rochín-Medina, J. J., Gutiérrez-Dorado, R., Sánchez-Magaña, L. M., Milán-Carrillo, J., Cuevas-Rodríguez, E. O., Mora-Rochín, S., ... & Reyes-Moreno, C. (2015). Enhancement of nutritional properties, and antioxidant and antihypertensive potential of black common bean seeds by optimizing the solid state bioconversion process. *International Journal of Food Sciences and Nutrition*, 66(5), 498–504.

Roy, F., Boye, J. I., & Simpson, B. K. (2010). Bioactive proteins and peptides in pulse crops: Pea, chickpea and lentil. *Food Research International*, 43(2), 432–442.

Rubio, L. A., Pérez, A., Ruiz, R., Guzmán, M. Á., Aranda-Olmedo, I., & Clemente, A. (2014). Characterization of pea (*Pisum sativum*) seed protein fractions. *Journal of the Science of Food and Agriculture*, 94(2), 280–287.

Rui, X., Wen, D., Li, W., Chen, X., Jiang, M., & Dong, M. (2015). Enrichment of ACE inhibitory peptides in navy bean (*Phaseolus vulgaris*) using lactic acid bacteria. *Food and Function*, 6(2), 622–629.

Sanjukta, S., & Rai, A. K. (2016). Production of bioactive peptides during soybean fermentation and their potential health benefits. *Trends in Food Science and Technology*, 50, 1–10.

Sathe, S. K. (2002). Dry bean protein functionality. *Critical Reviews in Biotechnology*, 22(2), 175–223.

Scippa, G. S., Rocco, M., Ialicicco, M., Trupiano, D., Viscosi, V., Di Michele, M., ... & Scaloni, A. (2010). The proteome of lentil (*Lens culinaris* Medik.) seeds: Discriminating between landraces. *Electrophoresis*, 31(3), 497–506.

Singh, D. K., Rao, A. S., Singh, R., & Jambunathan, R. (1988). Amino acid composition of storage proteins of a promising chickpea (*Cicer arietinum* L) cultivar. *Journal of the Science of Food and Agriculture*, 43(4), 373–379.

Sonklin, C., Alashi, M. A., Laohakunjit, N., Kerdchoechuen, O., & Aluko, R. E. (2020). Identification of antihypertensive peptides from mung bean protein hydrolysate and their effects in spontaneously hypertensive rats. *Journal of Functional Foods*, 64, 103635.

Sotelo, A., & Adsule, R. N. (1996). Chickpea (*Cicer arietinum* L.). In Smartt, J. & Nwokolo, E. (Eds.), *Food and feed from legumes and oilseeds* (pp. 82–89). Boston, MA: Springer.

Stanisavljević, N. S., Vukotić, G. N., Pastor, F., Suznjevic, D., Jovanović, Ž. S., Strahinic, I. D., ... & Radovic, S. S. (2015). Antioxidant activity of pea protein hydrolysates produced by batch fermentation with lactic acid bacteria. *Archives of Biological Sciences*, 67(3), 1033–1042.

Torino, M. I., Limón, R. I., Martínez-Villaluenga, C., Mäkinen, S., Pihlanto, A., Vidal-Valverde, C., & Frias, J. (2013). Antioxidant and antihypertensive properties of liquid and solid state fermented lentils. *Food Chemistry*, 136(2), 1030–1037.

Torres-Fuentes, C., del Mar Contreras, M., Recio, I., Alaiz, M., & Vioque, J. (2015). Identification and characterization of antioxidant peptides from chickpea protein hydrolysates. *Food Chemistry*, 180, 194–202.

Tzitzikas, E. N., Vincken, J. P., de Groot, J., Gruppen, H., & Visser, R. G. (2006). Genetic variation in pea seed globulin composition. *Journal of Agricultural and Food Chemistry*, 54(2), 425–433.

Vermeirssen, V., Van Camp, J., Decroos, K., Van Wijmelbeke, L., & Verstraete, W. (2003). The impact of fermentation and in vitro digestion on the formation of angiotensin-I-converting enzyme inhibitory activity from pea and whey protein. *Journal of Dairy Science*, 86(2), 429–438.

Vital, D. A. L., De Mejía, E. G., Dia, V. P., & Loarca-Piña, G. (2014). Peptides in common bean fractions inhibit human colorectal cancer cells. *Food Chemistry*, *157*, 347–355.

Wali, A., Mijiti, Y., Yanhua, G., Yili, A., Aisa, H. A., & Kawuli, A. (2020). Isolation and identification of a novel antioxidant peptide from chickpea (*Cicer arietinum* L.) sprout protein hydrolysates. *International Journal of Peptide Research and Therapeutics*, *27*, 1–9. https://doi.org/10.1007/s10989-020-10070-2

Wang, H. X., & Ng, T. B. (2007). An antifungal peptide from red lentil seeds. *Peptides*, *28*(3), 547–552.

Warsame, A. O., O'Sullivan, D. M., & Tosi, P. (2018). Seed storage proteins of faba bean (*Vicia faba* L): Current status and prospects for genetic improvement. *Journal of Agricultural and Food Chemistry*, *66*(48), 12617–12626.

Wojciechowski, M. F., Lavin, M., & Sanderson, M. J. (2004). A phylogeny of legumes (*Leguminosae*) based on analysis of the plastid *matK* gene resolves many well-supported subclades within the family. *American Journal of Botany*, *91*(11), 1846–1862.

Wu, H., Rui, X., Li, W., Chen, X., Jiang, M., & Dong, M. (2015). Mung bean (*Vigna radiata*) as probiotic food through fermentation with *Lactobacillus plantarum* B1-6. *LWT - Food Science and Technology*, *63*(1), 445–451.

Xiao, Y., Sun, M., Zhang, Q., Chen, Y., Miao, J., Rui, X., & Dong, M. (2018). Effects of *Cordyceps militaris* (L.) Fr. fermentation on the nutritional, physicochemical, functional properties and angiotensin I converting enzyme inhibitory activity of red bean (*Phaseolus angularis* [Willd.] WF Wight.) flour. *Journal of Food Science and Technology*, *55*(4), 1244–1255.

Xie, J., Du, M., Shen, M., Wu, T., & Lin, L. (2019). Physico-chemical properties, antioxidant activities and angiotensin-I converting enzyme inhibitory of protein hydrolysates from Mung bean (Vigna radiate). *Food Chemistry*, *270*, 243–250.

Yi-Shen, Z., Shuai, S., & FitzGerald, R. (2018). Mung bean proteins and peptides: Nutritional, functional and bioactive properties. *Food and Nutrition Research*, *62*. https://doi.org/10.29219/fnr.v62.1290.

Yust, M. D. M., Millán-Linares, M. D. C., Alcaide-Hidalgo, J. M., Millán, F., & Pedroche, J. (2012). Hypocholesterolaemic and antioxidant activities of chickpea (*Cicer arietinum* L.) protein hydrolysates. *Journal of the Science of Food and Agriculture*, *92*(9), 1994–2001.

Yust, M. M., Pedroche, J., Giron-Calle, J., Alaiz, M., Millán, F., & Vioque, J. (2003). Production of ACE inhibitory peptides by digestion of chickpea legumin with alcalase. *Food Chemistry*, *81*(3), 363–369.

Zhang, T., Li, Y., Miao, M., & Jiang, B. (2011). Purification and characterisation of a new antioxidant peptide from chickpea (*Cicer arietium* L.) protein hydrolysates. *Food Chemistry*, *128*(1), 28–33.

CHAPTER 9

Other Nonconventional Sources

Semih Ötleş and Vasfiye Hazal Özyurt

CONTENTS

Introduction	181
Plant-Based By-Products	181
Animal-Based By-Products	183
Conclusion	184
References	184

INTRODUCTION

Food by-products (or waste) are defined as materials generated by a production chain, and it can be any food or inedible parts of food. Food processing produces a great amount of waste during different production activities (agro-food industry, slaughtering, cutting, processing, storage, etc.) (Esparza, Jiménez-Moreno, Bimbela, Ancín-Azpilicueta, and Gandía, 2020). A problematic aspect of food by-products is its increasing volume, which became detrimental to the environment with industrialization and expansion of the manufacture of food products in the 20th century (Imbert, 2017). By-products are discarded in huge amounts every day. Food waste is a valuable source of carbohydrates, proteins, lipids, and other potentially useful nutrients, making it a very interesting material for possible reuses in food (Prandi et al., 2019).

Bioactive peptides are known as the small fragments of 2–20 amino acids. They are obtained from a host protein. However, bioactive peptides are inactive inside of their host protein and are activated only after being released from their host. This happens either during food aging/digestion or enzymatic hydrolysis/artificial synthesis. Peptides exhibit interesting bioactivities such as antioxidants, blood pressure-raising agents, or anticancer and antimicrobial agents (Bechaux, Gatellier, Page, Drillet, and Sante-lhoutellier, 2019). Bioactive peptides can have one or more specific activities depending on their amino acid composition and secondary structure. Food wastes, which have a high amount of protein, may be an important source for bioactive peptides.

This chapter will highlight the preparation of bioactive peptides from plant-based and animal-based by-products, their biological activity, and their sequences.

PLANT-BASED BY-PRODUCTS

The sources of plant-based by-products are cereals, roots and tubers, oil crops and pulses, and fruits and vegetables. The utilization of plant-based by-products to obtain value-added products reduces associated costs and improves accessibility to food. With increasing demand for products with highly appreciated properties, plant-based by-products produce the potential due to their valuable composition (Esparza, Jim, Bimbela, Ancín-azpilicueta, and Gandía, 2020). The following paragraphs summarize some results for peptides obtained from plant-based by-products.

Puchalska et al. 2012 investigated three antihypertensive peptides (LQP, LSP, and LRP) in different maize crops. They found that the content of LRP peptide was very low regardless of the maize variety, while the content of LQP and LSP significantly varied among studied maize lines (Puchalska, Marina, and García, 2012).

He et al. 2013 hydrolyzed protein isolates from rapeseed and purified bioactive peptides (LY, TF, and RALP) using Alcalase. Their bioactivities were attributed to the number of hydrophobic amino acid residues (He et al., 2013).

Hong et al. 2014 purified Gly-Ser-Gln with antioxidant activity from Chinese leek seeds (Hong, Chen, Hu, Yang, and Wang, 2014).

Alashi et al. 2014 prepared canola protein hydrolysates with antioxidant activity using different proteases and ultrafiltration membranes. Scavenging of superoxide radicals was generally weak, while canola protein hydrolysates inhibited linoleic acid oxidation with greater efficiency (Alashi et al., 2014).

Gonzalez-Garcia et al. (2014) extracted proteins from plum seed to evaluate their potential as a bioactive peptide source. The bioactive peptides were produced using different enzymes like Alcalase, Thermolysin, Flavourzyme, and Protease P. Bioactive peptide prepared by Alcalase was found to have higher ABTS radical-scavenging capacity, lipid peroxidation inhibition capacity, hydroxyl radical-scavenging capacity, and inhibition of ACE enzyme than other bioactive peptides (González-garcía, Marina, and García, 2014).

The antihypertensive and antioxidant activities of protein hydrolysates from cauliflower by-products were investigated in cell-free systems (Chiozzi et al., 2016). For that reason, Chiozzi et al. 2016 developed an analytical strategy for the production of the bioactive peptides from cauliflower waste proteins testing different extraction protocols and different enzymes. In this way, three novel ACE-inhibitory peptides were successfully identified and validated from cauliflower waste hydrolysate.

Xu et al. 2016 purified an angiotensin I-converting enzyme (ACE) inhibitory peptide from cauliflower by-products, protein hydrolysate-testing different enzymes such as pepsin, pancreatin, and Alcalase 2.4L. Alcalase 2.4L was found to have the highest ACE inhibitory activity, and this bioactive peptide was identified to be a dipeptide (Val-Trp) (Xu et al., 2016).

Caliceti et al. (2019) characterized peptide hydrolysates from cauliflower leaves and evaluated their biological activities. One fraction presented the inhibition of intracellular xanthine oxidase activity, another fraction induced the antioxidant enzyme superoxide dismutase 1 (Caliceti et al., 2019).

Famuwagun et al. (2020) produced protein isolates from Amaranth (ALI), eggplant (ELI), and fluted pumpkin (FLI) leaves and studied their conformational structures (Famuwagun et al., 2020).

Han et al. (2021) hydrolyzed the different oilseed proteins (flaxseed, rapeseed, sunflower, sesame, and soybean) using alcalase and pepsin. The protein hydrolysates and their low Mw fractions were evaluated for their *in vitro* antioxidant, antihypertensive, and antidiabetic capabilities. The low Mw fractions acted as the dipeptidyl-peptidase IV inhibitors, while alcalase-treated soybean peptides were found to have antioxidant activities. Moreover, soybean protein hydrolysates showed a higher α-glucosidase inhibitory activity (Han, Álvarez, Maycock, Murray, and Boesch, 2021).

ANIMAL-BASED BY-PRODUCTS

The valorization of animal-based by-products may both reduce the carbon footprint and support food manufacturing to create benefit. Moreover, the use of animal-based by-products to produce bioactive peptides would be a good alternative since they are high-grade products. It is both an increase of economic value and a repurposing of animal-based by-products that often-become waste. The following part summarizes the production of bioactive peptides from various animal-based by-products.

Whey proteins and their derived peptides from the hydrolysis of proteins and direct fermentation of whey have garnered significant attention (Dullius, Inês, Fernanda, and Souza, 2018; Fitzgerald, 1998; Mazorra-Manzano, Robles-Porchas, and Gonz, 2020; Miralles, Krause, Ramos, and Amigo, 2006). There are many studies showing their biological activities such as antioxidant, antimicrobial (Sah, Vasiljevic, Mckechnie, and Donkor, 2016), antihypertensive (Bhat, Kumar, and Bhat, 2015), anticancer (Hernández-ledesma and Hsieh, 2015), opioid (Garg, Nurgali, and Mishra, 2016), and immunomodulatory functions (Santiago-López, Hernández-Mendoza, Mata-Haro, and González-Córdova, 2016). Moreover, the production of whey bioactive peptide-containing foods on an industrial scale has resulted in several products on the international market. The BioZate, BiopureGMP, Vivinal ALPHA, Praventin, Dermylex, Hilmar 8390, and NOP-47 product lines are commercially sold as functional ingredients (Dullius et al., 2018). Detailed knowledge about bioactive peptides derived from whey proteins are found in Dullius et al., 2018.

Yu et al. (2006) isolated and characterized angiotensin I-converting enzyme (ACE) inhibitory peptides, LGFPTTKTYFPHF and VVYPWT, derived from porcine hemoglobin (Yu et al., 2006).

Saiga et al. (2008) hydrolyzed collagen from chicken legs, using various enzymes, and determined their angiotensin I-converting enzyme (ACE)-inhibitory activity. The peptides with a Gly-Ala-Hyp-Gly-Leu-Hyp-Gly-Pro sequence had high activity (Saiga et al., 2008).

Nakade et al. 2008 isolated and purified a peptide, YYRA (Tyr-Tyr-Arg-Ala), having angiotensin-converting enzyme (ACE) inhibitory activity from chicken bone extract using enzymes (Nakade et al., 2008).

Castro et al. 2009 investigated the inhibition effect of bovine collagen hydrolysate on melanoma cell (B16F10) proliferation. Apoptosis was found extremely high even at low concentrations (Castro, Maria, Bouhallab, and Sgarbieri, 2009).

Lee et al (2011) obtained skate skin protein hydrolysates and its corresponding fraction hydrolysis using alcalase, α-chymotrypsin, neutrase, pepsin, papain, and trypsin. They evaluated the angiotensin-I converting enzyme (ACE) inhibitory activity. The purified peptides were identified as Pro–Gly–Pro–Leu–Gly–Leu–Thr–Gly–Pro (975.38 Dalton (Da)), and Gln–Leu–Gly–Phe–Leu–Gly–Pro–Arg. Their IC_{50} values were found between 95 µM and 148µM (Lee, Jeon, and Byun, 2011).

Benarjee and Shanthi (2012) identified a bioactive peptide, AKGANGAPGIAGAPG FPGARGPSGPQGPSGPP, from the collogen of bovine tendon and displayed its angiotensin II-converting enzyme (ACE) inhibitory activity (Banerjee and Shanthi, 2012).

Ngo et al. (2015) investigated the antihypertensive effect of bioactive peptides, GASSGMPG and LAYA, from skate (*Okamejei kenojei*) skin gelatin. Bioactive peptides isolated from skate skin gelatin have ACE activity (Ngo et al., 2015).

Thuanthong et al. 2017 identified five bioactive peptides from Nile tilapia. Their IC_{50} values of ACE activities were found to be between 760 and 1490 µM (Thuanthong, Gobba, Sirinupong, Youravong, and Otte, 2017).

CONCLUSION

Large amounts of waste and by-products are generated every year from industrial food manufacturing processes. Scientific studies reveal that many food wastes have peptide content showing significant numbers of bioactive properties. As summarized in this chapter, plant-based and animal-based food by-products have potential as bioactive peptide sources. Bioactive peptides obtained from protein-rich plant and animal wastes can show different biological activities, such as antihypertensive, antioxidant, antimicrobial, neuroprotection, and antihyperglycemic, etc. on metabolism for different individuals. It is very important to increase *in vitro* and *in vivo* studies to determine and verify the action mechanisms of bioactive peptides. As a result, more scientific studies are needed to address the mechanisms of action of bioactive peptides derived from food waste and by-products on human metabolism, the preservation of their biological effects in the gastrointestinal system, as well as the usage of them as sources for functional foods.

REFERENCES

Alashi, A. M., Blanchard, C. L., Mailer, R. J., Agboola, S. O., Mawson, A. J., He, R., ... Aluko, R. E. (2014). Antioxidant properties of Australian canola meal protein hydrolysates. *Food Chemistry*, 146, 500–506. https://doi.org/10.1016/j.foodchem.2013.09.081

Banerjee, P., & Shanthi, C. (2012). Isolation of novel bioactive regions from bovine Achilles tendon collagen having angiotensin I-converting enzyme-inhibitory properties. *Process Biochemistry*, 47(12), 2335–2346. https://doi.org/10.1016/j.procbio.2012.09.012

Bechaux, J., Gatellier, P., Page, J. Le, Drillet, Y., & Sante-Lhoutellier, V. (2019). A comprehensive review of bioactive peptides obtained from animal by-products and their applications. *Food and Function*, 10(10), 6244–6266. https://doi.org/10.1039/c9fo01546a

Bhat, Z. F., Kumar, S., & Bhat, H. F. (2015). Antihypertensive peptides of animal origin : A review. *Critical Reviews in Food Science and Nutrition*, 37–41. https://doi.org/10.1080/10408398.2014.898241

Caliceti, C., Capriotti, A. L., Calabria, D., Bonvicini, F., Chiozzi, R. Z., Montone, C. M., ... Roda, A. (2019). Peptides from cauliflower by-products, obtained by an efficient, ecosustainable, and semi-industrial method, exert protective effects on endothelial function. *Hindawi Oxidative Medicine and Cellular Longevity*, 2019, Article ID 1046504, 13. https://doi.org/10.1155/2019/1046504

Castro, G. A., Maria, D. A., Bouhallab, S., & Sgarbieri, V. C. (2009). In vitro impact of a whey protein isolate (WPI) and collagen hydrolysates (CHs) on B16F10 melanoma cells proliferation, 56(1), 51–57. https://doi.org/10.1016/j.jdermsci.2009.06.016

Chiozzi, R. Z., Capriotti, A. L., Cavaliere, C., Barbera, G. La, Piovesana, S., & Laganà, A. (2016). Identification of three novel angiotensin-converting enzyme inhibitory peptides derived from cauliflower by-products by multidimensional liquid chromatography and bioinformatics. *Journal of Functional Foods*, 27, 262–273. https://doi.org/10.1016/j.jff.2016.09.010

Dullius, A., Inês, M., Fernanda, C., & Souza, V. De (2018). Whey protein hydrolysates as a source of bioactive peptides for functional foods – Biotechnological facilitation of industrial scale-up. *Journal of Functional Foods*, 42, 58–74. https://doi.org/10.1016/j.jff.2017.12.063

Esparza, I., Jim, N., Bimbela, F., Ancín-Azpilicueta, C., & Gandía, L. M. (2020). Fruit and vegetable waste management : Conventional and emerging approaches, 265(March). https://doi.org/10.1016/j.jenvman.2020.110510

Esparza, I., Jiménez-Moreno, N., Bimbela, F., Ancín-Azpilicueta, C., & Gandía, L. M. (2020). Fruit and vegetable waste management: Conventional and emerging approaches. *Journal of Environmental Management*, 265(February). https://doi.org/10.1016/j.jenvman.2020.110510

Famuwagun, A. A., Alashi, A. M., Gbadamosi, S. O., Taiwo, K. A., Oyedele, D. J., Adebooye, O. C., & Aluko, R. E. (2020). Comparative study of the structural and functional properties of protein isolates prepared from edible vegetable leaves. *International Journal of Food Properties*, 23(1), 955–970. https://doi.org/10.1080/10942912.2020.1772285

Fitzgerald, R. J. (1998). Potential uses of caseinophosphopeptides. *International Dairy Journal*, 8(5–6), 451–457.

Garg, S., Nurgali, K., & Mishra, V. K. (2016). Food proteins as source of opioid peptides—A review. *Current Medical Chemistry*, 23(9), 893–910.

González-García, E., Marina, M. L., & García, M. C. (2014). Plum (*Prunus domestica* L.) by-product as a new and cheap source of bioactive peptides: Extraction method and peptides characterization. *Journal of Functional Foods*, 11, 428–437. https://doi.org/10.1016/j.jff.2014.10.020

Han, R., Álvarez, A. J. H., Maycock, J., Murray, B. S., & Boesch, C. (2021). Comparison of alcalase- and pepsin-treated oilseed protein hydrolysates – Experimental validation of predicted antioxidant, antihypertensive and antidiabetic properties. *Current Research in Food Science*. https://doi.org/10.1016/j.crfs.2021.03.001

He, R., Malomo, S. A., Alashi, A., Girgih, A. T., Ju, X., & Aluko, R. E. (2013). Purification and hypotensive activity of rapeseed protein-derived renin and angiotensin converting enzyme inhibitory peptides. *Journal of Functional Foods*, 5(2), 781–789. https://doi.org/10.1016/j.jff.2013.01.024

Hernández-Ledesma, B., & Hsieh, C. (2015). Chemopreventive role of food-derived proteins and peptides: A review. *Critical Reviews in Food Science and Nutrition*, 8398(November). https://doi.org/10.1080/10408398.2015.1057632

Hong, J., Chen, T., Hu, P., Yang, J., & Wang, S. (2014). Purification and characterization of an antioxidant peptide (GSQ) from Chinese leek (*Allium tuberosum* Rottler) seeds. *Journal of Functional Foods*, 10, 144–153. https://doi.org/10.1016/j.jff.2014.05.014

Imbert, E. (2017). Food waste valorization options: Opportunities from the bioeconomy. *Open Agriculture*, 2(1), 195–204. https://doi.org/10.1515/opag-2017-0020

Lee, J. K., Jeon, J., & Byun, H. (2011). Effect of angiotensin I converting enzyme inhibitory peptide purified from skate skin hydrolysate. *Food Chemistry*, 125(2), 495–499. https://doi.org/10.1016/j.foodchem.2010.09.039

Mazorra-Manzano, M. A., Robles-Porchas, G. R., González-Velázquez, D. A., Torres-Llanez, M. J., Martínez-Porchas, M., García-Sifuentes, C. O., González-Córdova, A. F., & Vallejo-Córdoba, B. (2020). Cheese whey fermentation by its native microbiota: Proteolysis and bioactive peptides release with ACE-inhibitory activity. *Fermentation*, 6, 19. https://doi.org/10.3390/fermentation6010019

Miralles, B., Krause, I., Ramos, M., & Amigo, L. (2006). Comparison of capillary electrophoresis and isoelectric focusing for analysis of casein/caseinate addition in processed cheeses, 16(12), 1448–1453. https://doi.org/10.1016/j.idairyj.2005.10.025

Nakade, K., Kamishima, R., Inoue, Y., Ahhmed, A., Kawahara, S., Nakayama, T., ... Aoki, T. (2008). Identification of an antihypertensive peptide derived from chicken bone extract. (December 2007), 710–715. https://doi.org/10.1111/j.1740-0929.2008.00584.x

Ngo, D., Kang, K., Ryu, B., Vo, T., Jung, W., Byun, H., & Kim, S. (2015). Angiotensin-I converting enzyme inhibitory peptides from antihypertensive skate (*Okamejei kenojei*) skin gelatin hydrolysate in spontaneously hypertensive rats. *Food Chemistry*, *174*, 37–43. https://doi.org/10.1016/j.foodchem.2014.11.013

Prandi, B., Faccini, A., Lambertini, F., Bencivenni, M., Jorba, M., Van Droogenbroek, B., ... Sforza, S. (2019). Food wastes from agrifood industry as possible sources of proteins: A detailed molecular view on the composition of the nitrogen fraction, amino acid profile and racemisation degree of 39 food waste streams. *Food Chemistry*, *286*(February), 567–575. https://doi.org/10.1016/j.foodchem.2019.01.166

Puchalska, P., Marina, M. L., & García, M. C. (2012). Development of a reversed-phase high-performance liquid chromatography analytical methodology for the determination of antihypertensive peptides in maize crops. *Journal of Chromatography. Part A*, *1234*, 64–71. https://doi.org/10.1016/j.chroma.2011.12.079

Sah, B. N. P., Vasiljevic, T., Mckechnie, S., & Donkor, O. N. (2016). Antioxidative and antibacterial peptides derived from bovine milk proteins. *8398*(August). https://doi.org/10.1080/10408398.2016.1217825

Saiga, A., Iwai, K. W., Hayakawa, T., Takahata, Y., Kitamura, S., Nishimura, T., & Morimatsu, F. (2008). Angiotensin I-converting enzyme-inhibitory peptides obtained from chicken collagen hydrolysate. *Journal of Aagricultural and Food Chemistry*, *56*(20), 9586–9591.

Santiago-López, L., Hernández-Mendoza, A., Mata-Haro, V., & González-Córdova, A. F. (2016). Food-derived immunomodulatory peptides. (December 2015). https://doi.org/10.1002/jsfa.7697

Thuanthong, M., Gobba, C. De, Sirinupong, N., Youravong, W., & Otte, J. (2017). Purification and characterization of angiotensin-converting enzyme-inhibitory peptides from Nile tilapia (*Oreochromis niloticus*) skin gelatine produced by an enzymatic membrane reactor. *Journal of Functional Foods*, *36*, 243–254. https://doi.org/10.1016/j.jff.2017.07.011

Xu, Y., Bao, T., Han, W., Chen, W., Zheng, X., & Wang, J. (2016). Purification and identification of an angiotensin I-converting enzyme inhibitory peptide from cauliflower by-products protein hydrolysate. *Process Biochemistry*, *51*(9), 1299–1305. https://doi.org/10.1016/j.procbio.2016.05.023

Yu, Y., Hu, J., Miyaguchi, Y., Bai, X., Du, Y., & Lin, B. (2006). Isolation and characterization of angiotensin I-converting enzyme inhibitory peptides derived from porcine hemoglobin. *Peptides*, *27*(11), 2950–2956. https://doi.org/10.1016/j.peptides.2006.05.025

SECTION 3

From Proteins to Peptides

CHAPTER 10

Enzymatic Hydrolysis of Proteins

Mohammad Zarei, Belal J. Muhialdin, Kambiz Hassanzadeh, Chay Shyan Yea, and Raman Ahmadi

CONTENTS

An Overview of Protein Hydrolysis	189
Enzymatic Protein Hydrolysis	190
Degree of Hydrolysis	194
Proteases Used in Bioactive Peptides Generation	195
Protein Sources of Bioactive Peptides	199
References	199

AN OVERVIEW OF PROTEIN HYDROLYSIS

Bioactive peptides can be produced using different approaches, such as chemical, enzymatic, and fermentation (microbial) methods (Kim and Wijesekara, 2010). In chemical hydrolysis, an acid or a base is used to hydrolyze the proteins. Complete hydrolysis of protein using this method is performed at a high temperature usually up to 110°C for 20 h or more but not more than 4 days. HCl 6 M is the most common acid used in this process since it degrades the peptide bonds faster than sulfuric acid even at the same concentration. The optimal hydrolysis time for complete hydrolysis using HCl is 24 h (Ashaolu, 2020). Since the protein is almost completely hydrolyzed to amino acids during this time and concentration, it is not suitable to be used to generate the peptides. It is used typically for quantification of the amino acids in a sample; however, all amino acids cannot be completely recovered. Some amino acids, such as asparagine and glutamine, are hydrolyzed to their acid form. While tryptophan is destroyed, sulfur-containing amino acids (e.g., cysteine, methionine) are partially destroyed and some amino acids can be preserved if a pretreatment is performed. Peptides are normally generated by partial hydrolysis, which is carried out at much shorter times (e.g., 2–6 h) than complete hydrolysis (Ashaolu, 2020; Hou, Wu, Dai, Wang, and Wu, 2017). Complete hydrolysis of proteins using alkaline hydrolysis is carried out using a high temperature (e.g., 105°C) for 20 h using calcium, sodium, or potassium hydroxide. Shorter times (4–6 h) and lower temperatures (e.g., 27°C–55°C) are typically used for partial hydrolysis to produce peptides in the food industry (Hou et al., 2017; Pasupuleti, Holmes, and Demain, 2008). Since this method results in almost complete destruction of amino acids, it is not widely used in the food industry.

Although chemical hydrolysis is economical and fast and its protein yield is higher, there is, however, limited control of the consistency of the hydrolysate generated and significant variations in the amino-acid profile due to nonspecific hydrolysis (Siddik,

Howieson, Fotedar, and Partridge, 2021). Chemical hydrolysis particularly damages the food protein by converting the L-form amino acids to D-form amino acids and potentially yields toxic substances, such as lysinoalanine (Clemente, 2000). The destruction of some amino acids and peptides during chemical hydrolysis also leads to hydrolysis products with reduced nutritional properties (Neklyudov, Ivankin, and Berdutina, 2000). Recently there has been significant concern expressed in many countries about the use of chemical hydrolysis in which 3-chloro-1, 2-propanediol (MCPD) and 1, 3-dichloro-2-propanol (DCP)] are produced. Therefore, manufacturers prefer to use the enzymatic hydrolysis as a mild and controllable method to hydrolyze protein (Ashaolu, 2020).

The milder conditions of enzyme-catalyzed hydrolysis are less damaging to the protein substrate and are an ideal means of generating food-grade protein hydrolysates. The overall amino acid compositions of the enzymatic hydrolysis are similar to that of the original protein source (Clemente, 2000). In this method, proteins are exposed to one or more proteases at the optimum respective pH and temperature under controlled conditions using a specific buffer for each protease in the buffering system or by adding alkali or acid in the thermostated instrument (Kaur, Kehinde, Sharma, Sharma, and Kaur, 2020). The protein hydrolysates obtained in this method show better technological properties such as heat stability, solubility, and relatively high resistance to precipitation by many agents (Peng, Kong, Xia, and Liu, 2010). Thus, enzymatic hydrolysis of whole protein molecules is a widely used method for producing bioactive peptides. In most cases, the peptides released have shown a better bioactivity compared with the parent proteins.

The other method to produce bioactive peptides is microbial protein fermentation, in which the endogenous proteases of microorganisms degrade the protein to generate bioactive peptides. Numerous protein hydrolysates and bioactive peptides with various bioactivities have been produced from different protein sources (Aguilar-Toalá et al., 2017; Al-Sahlany et al., 2020; Arulrajah, Muhialdin, Zarei, Hasan, and Saari, 2020; Moreno-Montoro et al., 2017; Muhialdin and Algboory, 2018; Ricci et al., 2019; Zanutto-Elgui et al., 2019).

Bioactive peptides are specific protein fragments that have a positive impact on the body function or conditions and ultimately may influence the body systems and health including the antioxidant, opioid, immunomodulatory, antibacterial, antithrombotic, and antihypertensive activities (Kitts and Weiler, 2003). They are products of hydrolysis of various protein sources. These peptides usually consist of 3–20 amino acids and are released from the original protein after degradation (Möller, Scholz-Ahrens, Roos, and Schrezenmeir, 2008). Since the use of the enzymatic hydrolysis of proteins is emerging, our focus in this chapter will be on different aspects of enzymatic protein hydrolysis, its mechanism of action, and the protein hydrolysates and bioactive peptides obtained from different sources (Figure 10.1).

ENZYMATIC PROTEIN HYDROLYSIS

Hydrolysis is a chemical reaction, in which the peptide bonds are degraded into short peptides or free amino acids. Several factors can affect the protein hydrolysis and the bioactive peptides-generated properties. The enzymes used for hydrolysis, protein hydrolysis conditions including pH, temperature, enzyme–substrate ratio, protein hydrolysis time, mixing type methods, consequently the size of peptides and pretreatments of protein before digestion, are important factors which influence the bioactivity of peptides.

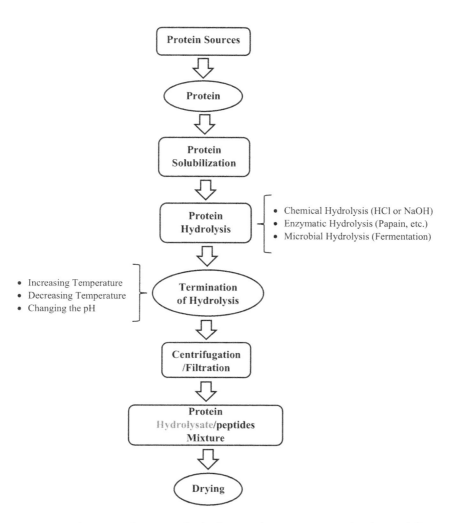

FIGURE 10.1 Production of protein hydrolysates/bioactive peptides from different protein sources. Protein hydrolysates can be obtained from animal, plant, or marine protein sources using different protein hydrolysis approaches including chemical, enzymatic, or microbial methods.

To maintain the optimum pH of the medium for enzymatic protein hydrolysis, a suitable buffer must be used, adjusted to the desired pH and compatible with the enzyme used. Since pH and temperature of the protein solution are correlated, the temperature should be adjusted to the optimal temperature before adjusting the pH (optimum). Then the enzyme could be added to the solution after dissolving the protein according to the desired enzyme/substrate ratio (Aluko, 2018).

Enzyme or enzymes addition is a crucial step in bioactive peptides production, which could be performed based on two different approaches of single enzyme or multiple enzymes digestion systems. In the single enzyme digestion system (Figure 10.2) only one specific proteolytic enzyme is used to hydrolyze the protein and generate the peptides. As said before, the condition should be adjusted according to the optimal condition for each specific enzyme including hydrolysis temperature, time, pH, enzyme/substrate ratio, etc.

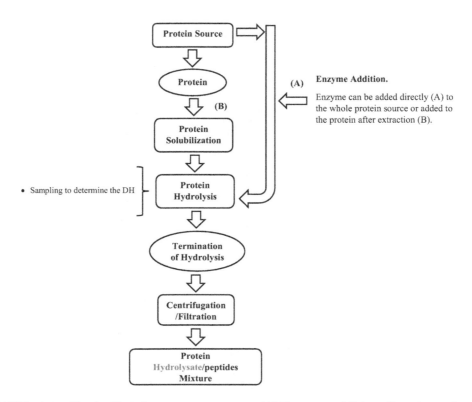

FIGURE 10.2 Single digestion enzyme system; (A) Enzyme addition directly to the protein source; (B) Enzyme addition to the protein after extraction.

The progress of hydrolysis and resulting protein hydrolysate could be monitored and controlled by determination of degree of hydrolysis. To do so, the aliquots of sample can be taken in specific time periods (time interval) to determine the degree of hydrolysis (DH) and set up the hydrolysis endpoint, where the protease is not active anymore, and the DH reaches the plateau in the DH-time chart.

In the multiple enzyme digestion system, proteases can be added simultaneously – but only if they have optimal catalytic activities at similar pH and temperature. However, if the enzymes have different optimum catalytic conditions, then sequential addition is required (Figure 10.3) (Aluko, 2018). As shown, in sequential digestion the first enzyme is added for a period of time and then the second enzyme will be added, while in the simultaneous enzyme addition, the enzymes are added at the same time. It is important to remember that the set of conditions of hydrolysis for the protease addition in the simultaneous enzyme method must be the same.

Protein hydrolysates are classified into three main groups based on their degree of hydrolysis (DH) that determine their application: (1) protein hydrolysates with low DH and improved functional features; (2) protein hydrolysates with various DHs (generally used as flavorings); and (3) protein hydrolysates with broad DH, which are mostly used as nutritional supplements and in special medical diets (Sarmadi and Ismail, 2010).

Most of the peptides reported by previous researchers, have been produced by enzymatic hydrolysis or released through fermentation of various foods (Byun and Kim, 2001; X. Gu, Hou, Li, Wang, and Wang, 2013; He, Malomo, Girgih, Ju, and Aluko, 2013; Himaya, Ngo, Ryu, and Kim, 2012; Je, Park, and Kim, 2005; Rho, Lee, Chung, Kim, and

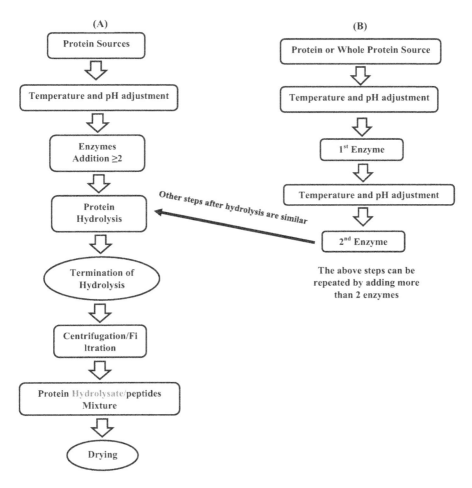

FIGURE 10.3 Multiple enzymes digestion system; (A) Simultaneous enzyme addition; (B) Consecutive enzyme addition.

Lee, 2009; Samarakoon et al., 2013; Yea et al., 2014; Zhuang, Tang, Dong, Sun, and Liu, 2013). Selecting a protein source is an important parameter for generating the bioactive peptides with high activity. After selecting the protein food source, enzymatic hydrolysis is performed by a protease or a combination of specific or nonspecific proteases.

Production of protein hydrolysate with high yield and high bioactivity is an important challenge in bioactive peptide releasing and generation. Thus, in most cases the peptide products are further processed to enhance the bioactivity. The bioactive peptide properties which are important to increase the functionality of the product are the peptide size, hydrophobicity, and net charge. Size-exclusion chromatography, gel filtration, and membrane ultra-filtration can be used to fractionate the bioactive peptides based on a proper molecular weight and size range (Bamdad, Wu, and Chen, 2011; Chabanon, Chevalot, Framboisier, Chenu, and Marc, 2007; H. M. Chen, Muramoto, and Yamauchi, 1995; Tan, Ayob, and Wan Yaacob, 2013; Wu, Aluko, and Muir, 2008; Xia, Bamdad, Gänzle, and Chen, 2012; Zarei et al., 2014; M. Zhang, Mu, and Sun, 2014). Moreover, reversed-phase high performance liquid chromatography (RP-HPLC) on a hydrophobic column can be used to fractionate the protein hydrolysate based on hydrophobic–hydrophilic properties (Ghassem et al., 2014; Kang, Lee, Kim, Lee, and Lee, 2003; McCann et al., 2006; Memarpoor-Yazdi, Asoodeh, and Chamani, 2012; Pan, Cao, Guo, and Zhao, 2012).

DEGREE OF HYDROLYSIS

The degree of hydrolysis (DH) is defined as the percentage of hydrolyzed peptide bonds. The techniques described here are based on the assumption that a free amino group and a free carboxyl group are released every time a peptide bond is hydrolyzed. They quantify the increase in the concentration of such groups as a variable to evaluate the progress of hydrolysis (Navarrete del Toro and García-Carreño, 2002).

According to the literature, several methods have been used to monitor the protein hydrolysis progression. The trinitro-benzene-sulfonic acid (TNBS), trichloroacetic acid soluble nitrogen (SN-TCA), formol titration, pH-stat, osmometry, ninhydrin reaction, soluble nitrogen content, O-phthaldialdehyde (OPA) are widely used for the determination of degree of hydrolysis.

The pH-stat technique is based on adding a base to keep the pH constant during protein hydrolysis. Degree of hydrolysis is proportional to the amount of base used in the hydrolysis. Practically, the pH-stat method is limited to the pH higher than 7. Using the pH-stat to reach the degree of hydrolysis higher than 30% is not economically feasible for a single enzyme. Thus, to obtain a high degree of hydrolysis, a combination of enzymes is required that is not in the range of pH-stat control. Another problem in using the pH-stat technique is the addition of a base that is not desirable in the end product (Jacobsen, Leonis, Linderstrøm-Lang, and Ottesen, 2006; Nielsen, Petersen, and Dambmann, 2001).

Changing the freezing point of a mixture is an important parameter that can be correlated to the degree of hydrolysis, which is measured using an osmometer (cryoscope). This method cannot be used in highly viscous solutions. Moreover, the content of nonprotein compounds is another limitation due to changing the freezing point, which cannot be correlated to the degree of hydrolysis of protein alone (Nielsen et al., 2001).

Determination of soluble nitrogen in trichloroacetic acid (SN-TCA) is another method to evaluate the DH. This technique is based on the correlation between SN-TCA and base consumption in the pH-stat. The limitation is that it cannot show a good performance when it is used in a system with high exopeptidase activity because a protein mixture containing a exopeptidase and endopeptidase will not result in the same increase in solubility (Margot, Flaschel, and Renken, 1994; Silvestre, 1997).

Trinitro-benzene-sulfonic acid (TNBS) reacts with primary amines under slightly alkaline conditions, and lowering the pH stops the reaction. However, TNBS does not measure the presence of proline and hydroxyproline (Lemieux, Puchades, and Simard, 1990). Furthermore, TNBS is laborious, unstable, toxic, and the solid TNBS must be handled carefully due to risk of explosion (Nielsen et al., 2001; Silvestre, 1997).

The ninhydrin assay principle is based on the reaction between ninhydrin and an α-amino group (Lemieux et al., 1990). It is sensitive to a free amino acid concentration of 2 µM but has the disadvantages of requiring heating and cooling steps and that the ninhydrin reagent will also react with ammonia and environmental oxygen (Spellman, McEvoy, O'cuinn, and FitzGerald, 2003).

The OPA method is based on the specific reaction between o-phthaldialdehyde (OPA) and an α-amino group of the peptide, in the presence of the β-mercaptoethanol in an alkaline environment to form 1-alkylthio-2-alkyl-substituted isoindoles. The isoindoles formed can be quantified spectrophotometrically at 340 nm or fluorometrically at 455 nm (Church, Swaisgood, Porter, and Catignani, 1983; Spellman et al., 2003). The o-phthaldialdehyde assay is widely used to determine the degree of hydrolysis of food protein hydrolysates (Robinson, 2010).

The assay is more accurate, easier and faster to perform, has a broader application range, and is environmentally safer than the other assays that are used to determine the degree of hydrolysis (Vigo, Malec, Gomez, and Llosa, 1992). Moreover, it can be used at room temperature and has a good resolution down to a 7 µM concentration of α-amino groups in aqueous media (Lemieux et al., 1990). The incubation time of reaction of OPA with the primary amines is around 2 min at room temperature. Furthermore, the peptide content can be linearly related to a peptide or amino acid solution of known concentration. One limitation of the OPA assay is the low response of Cys, Lys, and hydroxylysine and the lack of reaction between OPA with proline and hydroxyproline (Zumwalt and Gehrke, 1988).

Nielsen et al. (2001) compared the TNBS method with the OPA technique, and they concluded that the OPA method of analyzing the DH of protein is more accurate. They also demonstrated that this method is easier and faster to carry out, has a broader application range, and is environmentally safer than the TNBS method.

PROTEASES USED IN BIOACTIVE PEPTIDES GENERATION

Proteolytic enzymes catalyze peptide bond cleavage in proteins. They are degradative enzymes which catalyze the total hydrolysis of proteins (Rao, Tanksale, Ghatge, and Deshpande, 1998). According to the literature, different enzymes have been used to produce protein hydrolysates and bioactive peptides from food protein sources. Exogenous enzymes are preferable to the autolytic process because a shorter time is required to reach a similar degree of hydrolysis. Moreover, hydrolysis reaction can be controlled better when exogenous enzymes are used (Samaranayaka and Li-Chan, 2011).

The proteolytic enzymes, such as microbial enzymes (alcalase, flavourzyme, and protamex), enzymes from plant sources (papain and bromelain) and some enzymes of animal origin (pepsin, trypsin and chymotrypsin) have been widely used in the generation of bioactive peptides (Ahn, Kim, and Je, 2014; Bamdad et al., 2011; Baratzadeh, Asoodeh, and Chamani, 2013; Barbana and Boye, 2011; Forghani et al., 2012; Girgih, Udenigwe, Hasan, Gill, and Aluko, 2013; Torres-Fuentes, Alaiz, and Vioque, 2011; Udenigwe, Lin, Hou, and Aluko, 2009; J. Wang, Zhao, Zhao, and Jiang, 2007; Zarei et al., 2014).

The performance of papain depends on the plant source of protein, the climatic conditions for growth, and the methods used for its extraction and purification. Papain involves a polypeptide chain with three disulfide bridges and a sulfhydryl group necessary for the activity of the enzyme. Papain digests most protein substrates more significantly than pancreatic proteases. Papain reveals a broad specificity, degrading peptide bonds of basic amino acids, leucine, or glycine. The pH range of enzyme activity is between pH 5 and 9 with an optimum of pH = 6 – 7 and optimum temperature of 65 °C. However, it is stable up to 80 or 90 °C in the presence of substrates. It is a protease suitable for preparing flavored protein hydrolysates as well as highly soluble proteins.

Bromelain is characterized as a cysteine protease with activity pH between 5 and 9. The inactivation temperature of bromelain is 70 °C, which is lower than papain's inactivation temperature. The optimum pH range for bromelain is around 4.5 – 7.5, and its optimum temperature range is 35 – 45 °C. Two forms of bromelain (A and B) with similar specificity have been isolated from the pineapple stem. The preferential cleavage site for bromelain to hydrolyze the proteins, peptides etc., is the carbonyl end of lysine, alanine, tyrosine, and glycine (Arbige and Pitcher, 1989).

Trypsin is the main intestinal digestive enzyme responsible for the hydrolysis of food proteins. It is a serine protease and hydrolyzes peptide bonds. The carboxyl groups are contributed by the lysine and arginine residues. Trypsin cleaves the peptides on the C-terminal side of lysine and arginine amino acid residues. The rate of hydrolysis is slower if an acidic residue is on the either side of the cleavage site, and no cleavage occurs if a proline residue is on the carboxyl side of the cleavage site. The optimum pH of trypsin is 7 – 9, and the optimum temperature of activity is 37 °C.

Chymotrypsin is found in animal pancreatic extract. Pure chymotrypsin is an expensive enzyme which is used only for diagnostic and analytical applications. It is specific for the hydrolysis of peptide bonds in which the carboxyl groups are provided by one of the three aromatic amino acids (i.e., phenylalanine, tyrosine, or tryptophan). It is used extensively in the de-allergenizing of milk protein hydrolysates (Burrell, 1993). The optimum pH and temperature of chymotrypsin are 7.8 and 50 °C, respectively.

TABLE 10.1 Animal Protein Sources Used to Generate the Protein Hydrolysates and Bioactive Peptides

Animal Protein Source	Protease Used	Bioactivity	Reference
Whey protein	Chymotrypsin	Antioxidant	Bustamante, González, Sforza, and Tedeschi (2021)
Human milk	Pepsin	Antioxidant	Hernandez, Quiros, Amigo, and Recio (2007)
Meat muscle protein	Eight proteases	ACE inhibition	Arihara, Nakashima, Mukai, Ishikawa, and Itoh (2001); Suwaluk, Chansuwan, Sirinupong, and Chinachoti (2021)
Egg white protein ovotransferrin	Pepsin, Thermolysin	Antioxidant	Huang, Majumder, and Wu (2010); Rathnapala, Ahn, and Abeyrathne (2021)
Milk protein	Trypsin	Antioxidant	Ballatore et al. (2021); Pan et al. (2012)
Porcine blood plasma	Alcalase	Antioxidant	Q. Liu, Kong, Xiong, and Xia (2010); López-Pedrouso, Borrajo, Amarowicz, Lorenzo, and Franco (2021)
Casein	Pepsin	Antioxidant	Irshad, Kanekanian, Peters, and Masud (2013); Suetsuna, Ukeda, and Ochi (2000)

Pepsin is an acidic protease which is found in the stomachs of almost all vertebrates. The active enzyme is liberated from its zymogen, (pepsinogen), by autocatalysis in the presence of hydrochloric acid. Pepsin, unlike some other endopeptidases, hydrolyzes only peptide bonds. It does not hydrolyze non-peptide amide or ester linkages. Pepsin exhibits preferential cleavage of hydrophobic residues, preferably aromatic residues. Increased susceptibility to hydrolysis occurs if there is a sulfur-containing amino acid close to the peptide bond, which has an aromatic amino acid. The other typical protease is alcalase which is an endo-protease of the serine type. It has a very broad substrate specificity with a preference for a large uncharged residue in the P1 position. It hydrolyzes native and

TABLE 10.2 Plant Protein Sources Used to Generate Protein Hydrolysates and Bioactive Peptides

Plant Protein Source	Protease Used	Bioactivity	Reference
Soy protein	Alcalase	ACE inhibition	Wu and Ding (2001); Xu et al. (2021)
Wheat gliadin	Alcalase and Esperase	ACE inhibition	Thewissen, Pauly, Celus, Brijs, and Delcour (2011)
Rice Endosperm	Different proteases	ACE inhibition	Uraipong and Zhao (2016); J. Zhang et al. (2009)
Chickpea	Alcalase	Antioxidant	Ghribi et al. (2015); Y. Li, Jiang, Zhang, Mu, and Liu (2008)
Flaxseed	Different proteases	ACE inhibition	Ji, Xu, Udenigwe, and Agyei (2020); Udenigwe et al. (2009)
Canola	Flavourzyme	Antioxidant	Cumby, Zhong, Naczk, and Shahidi (2008)
Walnut	Pepsin, neutrase, and alcalase	Antioxidant	N. Chen, Yang, Sun, Niu, and Liu (2012); M.-C. Liu et al. (2016)
Hemp seed	Pepsin and pancreatin	Antioxidant	Girgih, Udenigwe, and Aluko (2011); Lin, Pangloli, Meng, and Dia (2020)
Corn	Alcalase	Antioxidant	Jin, Liu, Zheng, Wang, and He (2016); X. Li, Han, and Chen (2008)
Lentil	Trypsin	ACE inhibition	Boye, Roufik, Pesta, and Barbana (2010)
Pea	Thermolysin	Antioxidant	Pownall, Udenigwe, and Aluko (2010)
Mung bean	Alcalase	ACE inhibition	G. H. Li, Wan, Le, and Shi (2006)
Winged seed bean	Papain	Antioxidant and ACE inhibition	Yea et al. (2014)
Apricot almond	Neutrase	ACE inhibition	Chunyan Wang, Tian, and Wang (2011)
Rapeseed protein	Subtilisin	Antihypertension	Marczak et al. (2003)
Peanut	Alcalase, neutrase	ACE inhibition	H. Liu, Li, Shi, and Le (2005)
Sesame peptide powder	Papain, alcalase, trypsin	Antioxidant	Chan Wang, Li, and Ao (2012)
Sunflower	Flavourzyme	Antioxidant	Ren, Zheng, Liu, and Liu (2010)

denatured proteins and is active under alkaline conditions. In other words, it can hydrolyze most peptide bonds within a protein molecule. Alcalase is active between pH 6.5 and 8.5. It functions between 45 and 65 °C with a maximum activity at about 60 °C, above which the activity falls rapidly. Both pure and crude enzymes can be used to produce bioactive peptides. However, to reduce the production cost, crude enzyme mixtures are preferred (Mine, 2010).

TABLE 10.3 Marine Protein Sources Used to Generate the Protein Hydrolysates and Bioactive Peptides

Marine Protein Source	Protease Used	Bioactivity	Reference
Sardine (muscle)	Pepsin	ACE inhibition	Martínez-Alvarez, Batista, Ramos, and Montero (2016); Osajima (1993)
Yellowfin sole protein	Six proteases	Antioxidant	Jun, Park, Jung, and Kim (2004)
Alaska Pollack skin gelatin	Alcalase	Antioxidant	S.K. Kim et al. (2001)
Tilapia protein	Flavourzyme, Neutrase	ACE inhibition	Raghavan and Kristinsson (2009)
Yellowfin tuna	Alcalase	-	Guerard, Dufosse, De La Broise, and Binet (2001); Han, Byun, Park, and Kim (2015)
Round scad (muscle)	Alcalase and Flavourzyme	Antioxidant	Thiansilakul, Benjakul, and Shahidi (2007)
Conger eel (muscle)	Trypsin	Antioxidant	Ranathunga, Rajapakse, and Kim (2006)
Yellow stripe trevally	Alcalase and Flavourzyme	Antioxidant	Klompong, Benjakul, Kantachote, and Shahidi (2007)
Hoki skin gelatin	Trypsin	Antioxidant	Mendis, Rajapakse, and Kim (2005)
Herring	Alcalase	Antioxidant	Sathivel et al. (2003)
Pollack	Three proteases	ACE inhibition	Byun and Kim (2001)
Chum salmon	Complex proteases	Immune stimulant	Yang et al. (2009)
Pacific oysters	Crude protease solution from *Bacillus*	Multifunctional properties	Y. Wang et al. (2010)
Jellyfish collagen	Protamex	Antioxidant	Ding et al. (2011)
Rockfish	Alcalase and Flavourzyme	Multifunctional properties	H. J. Kim et al. (2011)
Atlantic salmon	Alcalase and papain	ACE inhibition	R. Z. Gu, Li, Liu, Yi, and Cai (2011)
Tuna dark muscle	Papain	Anticancer	Hsu, Li-Chan, and Jao (2011)
Purple sea urchin	Four proteases	Antioxidant	Qin et al. (2011)

PROTEIN SOURCES OF BIOACTIVE PEPTIDES

As shown in Table 10.1, various plants, animal, and marine food proteins have been widely used to generate the protein hydrolysates and bioactive peptides. Some previous studies have been done on milk, cheese, casein, whey, meat, and egg as protein sources to produce the protein hydrolysates and bioactive peptides (Table 10.1). Moreover, several researchers have used marine products such as fish, salmon, oyster, macroalgae, squid, sea urchin, shrimp, snow crab, and seahorse as protein sources (Table 10.3). Soy, lentil, chickpea, pea, beans, oat, wheat, hemp seed, canola, walnut, winged bean seed, barley, and flaxseed are the typical plant food used to generate the bioactive peptides and protein hydrolysates (Table 10.2).

Food protein sources are selected as a source of protein hydrolysate and bioactive peptides based on two criteria (Udenigwe and Aluko, 2012):

1. They are by-products of the food industry with high protein content or increasing the added-value of underutilized proteins
2. They contain a specific peptide or amino acid with pharmacological interest.

Although one criterion is sufficient to select the best protein sources with which to generate a protein hydrolysate and produce the bioactive peptides, a combination of criteria will achieve better results.

Plant proteins have been used to formulate a number of foods as an alternative of animal protein sources. Many plant protein sources have been investigated to produce protein hydrolysates in medical foods based on the nutritional and economic criteria. Soy bean is the most important plant source, which has been applied in some nutritional and functional foods (Clemente, 2000).

Other plant protein sources such as chickpeas, peas, and winged bean seed as well as some oil industry by-products (sunflower and rapeseed) have been recently reported (Barbana and Boye, 2010; He, et al., 2013; Ren et al., 2010; Yea et al., 2014; Zhang, Jiang, Miao, Mu, and Li, 2012). The main drawback of plant protein sources compared to casein or whey protein is the low level of some essential amino acids, such as sulfur-containing amino acids in legume protein hydrolysates, which must be added to the formulation to reach the meet standards.

REFERENCES

Aguilar-Toalá, J., Santiago-López, L., Peres, C., Peres, C., Garcia, H., Vallejo-Cordoba, B., ... Hernández-Mendoza, A. (2017). Assessment of multifunctional activity of bioactive peptides derived from fermented milk by specific *Lactobacillus plantarum* strains. *Journal of Dairy Science*, 100(1), 65–75.

Ahn, C. B., Kim, J. G., & Je, J. Y. (2014). Purification and antioxidant properties of octapeptide from salmon byproduct protein hydrolysate by gastrointestinal digestion. *Food Chemistry*, 147, 78–83.

Al-Sahlany, S. T. G., Altemimi, A. B., Abd Al-Manhel, A. J., Niamah, A. K., Lakhssassi, N., & Ibrahim, S. A. (2020). Purification of bioactive peptide with antimicrobial properties produced by Saccharomyces cerevisiae. *Foods*, 9(3), 324. https://doi.org/10.3390/foods9030324

Aluko, R. (2018). Food protein-derived peptides: Production, isolation, and purification. In Yada, R.Y. (ed.), *Proteins in food processing* (pp. 389–412). Elsevier.

Arbige, M. V., & Pitcher, W. H. (1989). Industrial enzymology: A look towards the future. *Trends in Biotechnology*, 7(12), 330–335.

Arihara, K., Nakashima, Y., Mukai, T., Ishikawa, S., & Itoh, M. (2001). Peptide inhibitors for angiotensin I-converting enzyme from enzymatic hydrolysates of porcine skeletal muscle proteins. *Meat Science*, 57(3), 319–324.

Arulrajah, B., Muhialdin, B. J., Zarei, M., Hasan, H., & Saari, N. (2020). Lacto-fermented Kenaf (*Hibiscus cannabinus* L.) seed protein as a source of bioactive peptides and their applications as natural preservatives. *Food Control*, 110, 106969.

Ashaolu, T. J. (2020). Health applications of soy protein hydrolysates. *International Journal of Peptide Research and Therapeutics*, 26, 2333–2343. https://doi.org/10.1007/s10989-020-10018-6

Ballatore, M. B., Escobar, F. M., Rossi, Y. E., del Rosario Bettiol, M., Torres, C. V., Montenegro, M. A., & Cavaglieri, L. R. (2021). Cytotoxic activity and genotoxicity of antioxidant WPC-hydrolysates and their probiotics compatibility as potential functional feed additive. *Food Bioscience*, 41, 100922.

Bamdad, F., Wu, J., & Chen, L. (2011). Effects of enzymatic hydrolysis on molecular structure and antioxidant activity of barley hordein. *Journal of Cereal Science*, 54(1), 20–28.

Baratzadeh, M. H., Asoodeh, A., & Chamani, J. (2013). Antioxidant peptides obtained from goose egg white proteins by enzymatic hydrolysis. *International Journal of Food Science & Technology*, 48(8), 1603–1609.

Barbana, C., & Boye, J. I. (2010). Angiotensin I-converting enzyme inhibitory activity of chickpea and pea protein hydrolysates. *Food Research International*, 43(6), 1642–1649.

Barbana, C., & Boye, J. I. (2011). Angiotensin I-converting enzyme inhibitory properties of lentil protein hydrolysates: Determination of the kinetics of inhibition. *Food Chemistry*, 127(1), 94–101. http://dx.doi.org/10.1016/j.foodchem.2010.12.093

Boye, J. I., Roufik, S., Pesta, N., & Barbana, C. (2010). Angiotensin I-converting enzyme inhibitory properties and SDS-PAGE of red lentil protein hydrolysates. *LWT-Food Science and Technology*, 43(6), 987–991.

Burrell, M. M. (1993). *Enzymes of molecular biology*. Humana Press.

Bustamante, S. Z., González, J. G., Sforza, S., & Tedeschi, T. (2021). Bioactivity and peptide profile of whey protein hydrolysates obtained from Colombian double-cream cheese production and their products after gastrointestinal digestion. *LWT*, 145, 111334.

Byun, H. G., & Kim, S. K. (2001). Purification and characterization of angiotensin I converting enzyme (ACE) inhibitory peptides from Alaska pollack (*Theragra chalcogramma*) skin. *Process Biochemistry*, 36(12), 1155–1162.

Chabanon, G., Chevalot, I., Framboisier, X., Chenu, S., & Marc, I. (2007). Hydrolysis of rapeseed protein isolates: Kinetics, characterization and functional properties of hydrolysates. *Process Biochemistry*, 42(10), 1419–1428.

Chen, H. M., Muramoto, K., & Yamauchi, F. (1995). Structural analysis of antioxidative peptides from soybean. beta.-conglycinin. *Journal of Agricultural and Food Chemistry*, 43(3), 574–578.

Chen, N., Yang, H., Sun, Y., Niu, J., & Liu, S. (2012). Purification and identification of antioxidant peptides from walnut (*Juglans regia* L.) protein hydrolysates. *Peptides*, 38(2), 344–349. http://dx.doi.org/10.1016/j.peptides.2012.09.017

Church, F. C., Swaisgood, H. E., Porter, D. H., & Catignani, G. L. (1983). Spectrophotometric assay using o-phthaldialdehyde for determination of proteolysis in milk and isolated milk proteins 1. *Journal of Dairy Science*, 66(6), 1219–1227.

Clemente, A. (2000). Enzymatic protein hydrolysates in human nutrition. *Trends in Food Science & Technology*, 11(7), 254–262.

Cumby, N., Zhong, Y., Naczk, M., & Shahidi, F. (2008). Antioxidant activity and water-holding capacity of canola protein hydrolysates. *Food Chemistry*, 109(1), 144–148.

Ding, J. F., Li, Y. Y., Xu, J. J., Su, X. R., Gao, X., & Yue, F. P. (2011). Study on effect of jellyfish collagen hydrolysate on anti-fatigue and anti-oxidation. *Food Hydrocolloids*, 25(5), 1350–1353.

Forghani, B., Ebrahimpour, A., Bakar, J., Abdul Hamid, A., Hassan, Z., & Saari, N. (2012). Enzyme hydrolysates from *Stichopus horrens* as a new source for angiotensin-converting enzyme inhibitory peptides. *Evidence-Based Complementary and Alternative Medicine*, 2012. doi:10.1155/2012/236384

Ghassem, M., Fern, S., Said, M., Ali, Z., Ibrahim, S., & Babji, A. (2014). Kinetic characterization of *Channa striatus* muscle sarcoplasmic and myofibrillar protein hydrolysates. *Journal of Food Science and Technology*, 51(3), 467–475. 10.1007/s13197-011-0526-6

Ghribi, A. M., Sila, A., Przybylski, R., Nedjar-Arroume, N., Makhlouf, I., Blecker, C., ... Besbes, S. (2015). Purification and identification of novel antioxidant peptides from enzymatic hydrolysate of chickpea (*Cicer arietinum* L.) protein concentrate. *Journal of Functional Foods*, 12, 516–525.

Girgih, A. T., Udenigwe, C. C., & Aluko, R. E. (2011). *In vitro* antioxidant properties of hemp seed (*Cannabis sativa* L.) protein hydrolysate fractions. *Journal of the American Oil Chemists' Society*, 88(3), 381–389.

Girgih, A. T., Udenigwe, C. C., Hasan, F. M., Gill, T. A., & Aluko, R. E. (2013). Antioxidant properties of Salmon (*Salmo salar*) protein hydrolysate and peptide fractions isolated by reverse-phase HPLC. *Food Research International*, 52(1), 315–322.

Gu, R. Z., Li, C. Y., Liu, W. Y., Yi, W. X., & Cai, M. Y. (2011). Angiotensin I-converting enzyme inhibitory activity of low-molecular-weight peptides from Atlantic salmon (*Salmo salar* L.) skin. *Food Research International*, 44(5), 1536–1540. http://dx.doi.org/10.1016/j.foodres.2011.04.006

Gu, X., Hou, Y.-K., Li, D., Wang, J.-Z., & Wang, F.-J. (2013). Separation, purification and identification of angiotensin I–converting enzyme inhibitory peptides from walnut (*Juglans regia* L.) hydrolyzate. *International Journal of Food Properties*, 18(2), 266–276.

Guerard, F., Dufosse, L., De La Broise, D., & Binet, A. (2001). Enzymatic hydrolysis of proteins from yellowfin tuna (*Thunnus albacares*) wastes using alcalase. *Journal of Molecular Catalysis B: Enzymatic*, 11(4), 1051–1059.

Han, Y., Byun, S. H., Park, J. H., & Kim, S. B. (2015). Bioactive properties of enzymatic hydrolysates from abdominal skin gelatin of yellowfin tuna (T hunnus albacares). *International Journal of Food Science & Technology*, 50(9), 1996–2003.

He, R., Malomo, S. A., Alashi, A., Girgih, A. T., Ju, X., & Aluko, R. E. (2013). Purification and hypotensive activity of rapeseed protein-derived renin and angiotensin converting enzyme inhibitory peptides. *Journal of Functional Foods*, 5(2), 781–789. http://dx.doi.org/10.1016/j.jff.2013.01.024

He, R., Malomo, S. A., Girgih, A. T., Ju, X., & Aluko, R. E. (2013). Glycinyl-histidinyl-serine (GHS), a novel rapeseed protein-derived peptide has blood pressure-lowering effect in spontaneously hypertensive rats. *Journal of Agricultural and Food Chemistry, 61*(35), 8396–8402.

Hernandez, B., Quiros, A., Amigo, L., & Recio, I. (2007). Identification of bioactive peptides after digestion of human milk and infant formula with pepsin and pancreatin. *International Dairy Journal, 17*(1), 42–49.

Himaya, S. W. A., Ngo, D.-H., Ryu, B., & Kim, S.-K. (2012). An active peptide purified from gastrointestinal enzyme hydrolysate of Pacific cod skin gelatin attenuates angiotensin-1 converting enzyme (ACE) activity and cellular oxidative stress. *Food Chemistry, 132*(4), 1872–1882. http://dx.doi.org/10.1016/j.foodchem.2011.12.020

Hou, Y., Wu, Z., Dai, Z., Wang, G., & Wu, G. (2017). Protein hydrolysates in animal nutrition: Industrial production, bioactive peptides, and functional significance. *Journal of Animal Science and Biotechnology, 8*(1), 1–13.

Hsu, K. C., Li-Chan, E. C. Y., & Jao, C. L. (2011). Antiproliferative activity of peptides prepared from enzymatic hydrolysates of tuna dark muscle on human breast cancer cell line MCF-7. *Food Chemistry, 126*(2), 617–622. http://dx.doi.org/10.1016/j.foodchem.2010.11.066

Huang, W. Y., Majumder, K., & Wu, J. (2010). Oxygen radical absorbance capacity of peptides from egg white protein ovotransferrin and their interaction with phytochemicals. *Food Chemistry, 123*(3), 635–641. http://dx.doi.org/10.1016/j.foodchem.2010.04.083

Irshad, I., Kanekanian, A., Peters, A., & Masud, T. (2013). Antioxidant activity of bioactive peptides derived from bovine casein hydrolysate fractions. *Journal of Food Science and Technology, 52*, 231–239. https://doi.org/10.1007/s13197-012-0920-8

Jacobsen, C. F., Leonis, J., Linderstrøm-Lang, K., & Ottesen, M. (2006). The pH-stat and its use in biochemistry. *Methods of Biochemical Analysis, 4*, 171–210.

Je, J. Y., Park, P. J., & Kim, S. K. (2005). Antioxidant activity of a peptide isolated from Alaska pollack (*Theragra chalcogramma*) frame protein hydrolysate. *Food Research International, 38*(1), 45–50.

Ji, D., Xu, M., Udenigwe, C. C., & Agyei, D. (2020). Physicochemical characterisation, molecular docking, and drug-likeness evaluation of hypotensive peptides encrypted in flaxseed proteome. *Current Research in Food Science, 3*, 41–50.

Jin, D.-x., Liu, X.-l., Zheng, X.-q., Wang, X.-j., & He, J.-f. (2016). Preparation of antioxidative corn protein hydrolysates, purification and evaluation of three novel corn antioxidant peptides. *Food Chemistry, 204*, 427–436.

Jun, S. Y., Park, P. J., Jung, W. K., & Kim, S. K. (2004). Purification and characterization of an antioxidative peptide from enzymatic hydrolysate of yellowfin sole (*Limanda aspera*) frame protein. *European Food Research and Technology, 219*(1), 20–26.

Kang, D. G., Lee, Y. S., Kim, H. J., Lee, Y. M., & Lee, H. S. (2003). Angiotensin converting enzyme inhibitory phenylpropanoid glycosides from *Clerodendron trichotomum*. *Journal of Ethnopharmacology, 89*(1), 151–154.

Kaur, A., Kehinde, B. A., Sharma, P., Sharma, D., & Kaur, S. (2020). Recently isolated food-derived antihypertensive hydrolysates and peptides: A review. *Food Chemistry, 346*, 128719. https://doi.org/10.1016/j.foodchem.2020.128719

Kim, H. J., Park, K. H., Shin, J. H., Lee, J. S., Heu, M. S., Lee, D. H., & Kim, J.-S. (2011). Antioxidant and ACE inhibiting activities of the rockfish sebastes hubbsi skin gelatin hydrolysates produced by sequential two-step enzymatic hydrolysis. *Korean Journal of Fisheries and Aquatic Sciences, 14*(1), 1–10.

Kim, S. K., Kim, Y., Byun, H. G., Nam, K. S., Joo, D. S., & Shahidi, F. (2001). Isolation and characterization of antioxidative peptides from gelatin hydrolysate of Alaska pollack skin. *Journal of Agricultural and Food Chemistry, 49*(4), 1984–1989.

Kim, S. K., & Wijesekara, I. (2010). Development and biological activities of marine-derived bioactive peptides: A review. *Journal of Functional Foods, 2*(1), 1–9.

Kitts, D. D., & Weiler, K. (2003). Bioactive proteins and peptides from food sources. Applications of bioprocesses used in isolation and recovery. *Current Pharmaceutical Design, 9*(16), 1309–1323.

Klompong, V., Benjakul, S., Kantachote, D., & Shahidi, F. (2007). Antioxidative activity and functional properties of protein hydrolysate of yellow stripe trevally (*Selaroides leptolepis*) as influenced by the degree of hydrolysis and enzyme type. *Food Chemistry, 102*(4), 1317–1327.

Lemieux, L., Puchades, R., & Simard, R. (1990). Free amino acids in cheddar cheese: Comparison of quantitation methods. *Journal of Food Science, 55*(6), 1552–1554.

Li, G. H., Wan, J. Z., Le, G. W., & Shi, Y. H. (2006). Novel angiotensin I-converting enzyme inhibitory peptides isolated from Alcalase hydrolysate of mung bean protein. *Journal of Peptide Science, 12*(8), 509–514.

Li, X., Han, L., & Chen, L. (2008). In vitro antioxidant activity of protein hydrolysates prepared from corn gluten meal. *Journal of the Science of Food and Agriculture, 88*(9), 1660–1666.

Li, Y., Jiang, B., Zhang, T., Mu, W., & Liu, J. (2008). Antioxidant and free radical-scavenging activities of chickpea protein hydrolysate (CPH). *Food Chemistry, 106*(2), 444–450.

Lin, Y., Pangloli, P., Meng, X., & Dia, V. P. (2020). Effect of heating on the digestibility of isolated hempseed (Cannabis sativa L.) protein and bioactivity of its pepsin-pancreatin digests. *Food Chemistry, 314*, 126198.

Liu, H., Li, G. H., Shi, Y. H., & Le, G. W. (2005). Angiotensin I-converting enzyme inhibitory activity of peanut protein hydrolysates prepared with alcalase [J]. *Journal of Peanut Science, 1*, 002.

Liu, M.-C., Yang, S.-J., Hong, D., Yang, J.-P., Liu, M., Lin, Y., … Wang, C.-J. (2016). A simple and convenient method for the preparation of antioxidant peptides from walnut (Juglans regia L.) protein hydrolysates. *Chemistry Central Journal, 10*(1), 1–11.

Liu, Q., Kong, B., Xiong, Y. L., & Xia, X. (2010). Antioxidant activity and functional properties of porcine plasma protein hydrolysate as influenced by the degree of hydrolysis. *Food Chemistry, 118*(2), 403–410. http://dx.doi.org/10.1016/j.foodchem.2009.05.013

López-Pedrouso, M., Borrajo, P., Amarowicz, R., Lorenzo, J. M., & Franco, D. (2021). Peptidomic analysis of antioxidant peptides from porcine liver hydrolysates using SWATH-MS. *Journal of Proteomics, 232*, 104037.

Marczak, E. D., Usui, H., Fujita, H., Yang, Y., Yokoo, M., Lipkowski, A. W., & Yoshikawa, M. (2003). New antihypertensive peptides isolated from rapeseed. *Peptides, 24*(6), 791–798.

Margot, A., Flaschel, E., & Renken, A. (1994). Continuous monitoring of enzymatic whey protein hydrolysis. Correlation of base consumption with soluble nitrogen content. *Process Biochemistry, 29*(4), 257–262.

Martínez-Alvarez, O., Batista, I., Ramos, C., & Montero, P. (2016). Enhancement of ACE and prolyl oligopeptidase inhibitory potency of protein hydrolysates from sardine and tuna by-products by simulated gastrointestinal digestion. *Food & Function, 7*(4), 2066–2073.

McCann, K., Shiell, B., Michalski, W., Lee, A., Wan, J., Roginski, H., & Coventry, M. (2006). Isolation and characterisation of a novel antibacterial peptide from bovine α$_{s1}$-casein. *International Dairy Journal, 16*(4), 316–323.

Memarpoor-Yazdi, M., Asoodeh, A., & Chamani, J. (2012). A novel antioxidant and antimicrobial peptide from hen egg white lysozyme hydrolysates. *Journal of Functional Foods, 4*(1), 278–286.

Mendis, E., Rajapakse, N., & Kim, S. K. (2005). Antioxidant properties of a radical-scavenging peptide purified from enzymatically prepared fish skin gelatin hydrolysate. *Journal of Agricultural and Food Chemistry, 53*(3), 581–587.

Mine, Y. (2010). *Bioactive proteins and peptides as functional foods and nutraceuticals* (Vol. 29). Wiley-Blackwell.

Möller, N. P., Scholz-Ahrens, K. E., Roos, N., & Schrezenmeir, J. (2008). Bioactive peptides and proteins from foods: Indication for health effects. *European Journal of Nutrition, 47*(4), 171–182.

Moreno-Montoro, M., Olalla-Herrera, M., Rufián-Henares, J. Á., Martínez, R. G., Miralles, B., Bergillos, T., ... Jauregi, P. (2017). Antioxidant, ACE-inhibitory and antimicrobial activity of fermented goat milk: Activity and physicochemical property relationship of the peptide components. *Food & Function, 8*(8), 2783–2791.

Muhialdin, B. J., & Algboory, H. L. (2018). Identification of low molecular weight antimicrobial peptides from Iraqi camel milk fermented with *Lactobacillus plantarum*. *PharmaNutrition, 6*(2), 69–73.

Navarrete del Toro, M., & García-Carreño, F. L. (2002). Evaluation of the progress of protein hydrolysis. *Current Protocols in Food Analytical Chemistry*. http://www.bashanfoundation.org/contributions/Garcia-F/carrenodigestibility.pdf

Neklyudov, A., Ivankin, A., & Berdutina, A. (2000). Production and purification of protein hydrolysates (review). *Applied Biochemistry and Microbiology, 36*(4), 317–324.

Nielsen, P., Petersen, D., & Dambmann, C. (2001). Improved method for determining food protein degree of hydrolysis. *Journal of Food Science-Chicago, 66*(5), 642–646.

Osajima, Y. (1993). Inhibition of angioteusin I-converting enzyme by *Bacillus licheniformis* alkaline protease hydrolyzates derived from sardine muscle. *Bioscience, Biotechnology, and Biochemistry, 57*(6), 922–925.

Pan, D., Cao, J., Guo, H., & Zhao, B. (2012). Studies on purification and the molecular mechanism of a novel ACE inhibitory peptide from whey protein hydrolysate. *Food Chemistry, 130*(1), 121–126.

Pasupuleti, V. K., Holmes, C., & Demain, A. L. (2008). Applications of protein hydrolysates in biotechnology. In Pasupuleti, V., & Demain, A. (eds), *Protein hydrolysates in biotechnology* (pp. 1–9). Springer, Dordrecht. https://doi.org/10.1007/978-1-4020-6674-0_1

Peng, X., Kong, B., Xia, X., & Liu, Q. (2010). Reducing and radical-scavenging activities of whey protein hydrolysates prepared with alcalase. *International Dairy Journal, 20*(5), 360–365.

Pownall, T. L., Udenigwe, C. C., & Aluko, R. E. (2010). Amino acid composition and antioxidant properties of pea seed (Pisum sativum L.) enzymatic protein hydrolysate fractions. *Journal of Agricultural and Food Chemistry, 58*(8), 4712–4718.

Qin, L., Zhu, B.-W., Zhou, D.-Y., Wu, H.-T., Tan, H., Yang, J.-F., ... Murata, Y. (2011). Preparation and antioxidant activity of enzymatic hydrolysates from purple sea urchin (Strongylocentrotus nudus) gonad. *LWT - Food Science and Technology, 44*(4), 1113–1118. http://dx.doi.org/10.1016/j.lwt.2010.10.013

Raghavan, S., & Kristinsson, H. G. (2009). ACE-inhibitory activity of tilapia protein hydrolysates. *Food Chemistry*, *117*(4), 582–588. http://dx.doi.org/10.1016/j.foodchem.2009.04.058

Ranathunga, S., Rajapakse, N., & Kim, S.-K. (2006). Purification and characterization of antioxidative peptide derived from muscle of conger eel (Conger myriaster). *European Food Research and Technology*, *222*(3–4), 310–315.

Rao, M. B., Tanksale, A. M., Ghatge, M. S., & Deshpande, V. V. (1998). Molecular and biotechnological aspects of microbial proteases. *Microbiology and Molecular Biology Reviews*, *62*(3), 597–635.

Rathnapala, C., Ahn, D. U., & Abeyrathne, S. (2021). Enzymatic hydrolysis of ovotransferrin and the functional properties of its hydrolysates. *Food Science of Animal Resources*, *41*(4), 608–622. doi:10.5851/kosfa.2021.e19

Ren, J., Zheng, X. Q., Liu, X. L., & Liu, H. (2010). Purification and characterization of antioxidant peptide from sunflower protein hydrolysate. *Food Technology and Biotechnology*, *48*(4), 519.

Rho, S. J., Lee, J. S., Chung, Y. I., Kim, Y. W., & Lee, H. G. (2009). Purification and identification of an angiotensin I-converting enzyme inhibitory peptide from fermented soybean extract. *Process Biochemistry*, *44*(4), 490–493.

Ricci, A., Bernini, V., Maoloni, A., Cirlini, M., Galaverna, G., Neviani, E., & Lazzi, C. (2019). Vegetable by-product lacto-fermentation as a new source of antimicrobial compounds. *Microorganisms*, *7*(12), 607.

Robinson, M. A. (2010). *Production, fractionation, and evaluation of antioxidant potential of peptides derived from soy protein digests*. UWSpace. http://hdl.handle.net/10012/5257

Samarakoon, K. W., Kwon, O. N., Ko, J. Y., Lee, J. H., Kang, M. C., Kim, D., ... Jeon, Y. J. (2013). Purification and identification of novel angiotensin-I converting enzyme (ACE) inhibitory peptides from cultured marine microalgae (Nannochloropsis oculata) protein hydrolysate. *Journal of Applied Phycology*, *25*, 1595–1606. https://doi.org/10.1007/s10811-013-9994-6

Samaranayaka, A. G. P., & Li-Chan, E. C. Y. (2011). Food-derived peptidic antioxidants: A review of their production, assessment, and potential applications. *Journal of Functional Foods*, *3*(3), 229–254.

Sarmadi, B. H., & Ismail, A. (2010). Antioxidative peptides from food proteins: A review. *Peptides*, *31*(10), 1949–1956.

Sathivel, S., Bechtel, P., Babbitt, J., Smiley, S., Crapo, C., Reppond, K., & Prinyawiwatkul, W. (2003). Biochemical and functional properties of herring (Clupea harengus) byproduct hydrolysates. *Journal of Food Science*, *68*(7), 2196–2200.

Siddik, M. A., Howieson, J., Fotedar, R., & Partridge, G. J. (2021). Enzymatic fish protein hydrolysates in finfish aquaculture: A review. *Reviews in Aquaculture*, *13*(1), 406–430.

Silvestre, M. (1997). Review of methods for the analysis of protein hydrolysates. *Food Chemistry*, *60*(2), 263–271.

Spellman, D., McEvoy, E., O'cuinn, G., & FitzGerald, R. (2003). Proteinase and exopeptidase hydrolysis of whey protein: Comparison of the TNBS, OPA and pH stat methods for quantification of degree of hydrolysis. *International Dairy Journal*, *13*(6), 447–453.

Suetsuna, K., Ukeda, H., & Ochi, H. (2000). Isolation and characterization of free radical scavenging activities peptides derived from casein. *The Journal of Nutritional Biochemistry*, *11*(3), 128–131.

Suwaluk, R., Chansuwan, W., Sirinupong, N., & Chinachoti, P. (2021). Biological properties of peptide released by in-vitro stimulated digestion of cooked eats. *Journal of Food and Nutrition Research*, 9(2), 87–95.

Tan, Y. N., Ayob, M. K., & Wan Yaacob, W. A. (2013). Purification and characterisation of antibacterial peptide-containing compound derived from palm kernel cake. *Food Chemistry*, 136(1), 279–284. Retrieved from http://www.sciencedirect.com/science/article/pii/S0308814612012952

Thewissen, B. G., Pauly, A., Celus, I., Brijs, K., & Delcour, J. A. (2011). Inhibition of angiotensin I-converting enzyme by wheat gliadin hydrolysates. *Food Chemistry*, 127(4), 1653–1658. http://dx.doi.org/10.1016/j.foodchem.2010.11.171

Thiansilakul, Y., Benjakul, S., & Shahidi, F. (2007). Antioxidative activity of protein hydrolysate from round scad muscle using alcalase and flavourzyme. *Journal of Food Biochemistry*, 31(2), 266–287.

Torres-Fuentes, C., Alaiz, M., & Vioque, J. (2011). Affinity purification and characterisation of chelating peptides from chickpea protein hydrolysates. *Food Chemistry*, 129(2), 485–490. http://dx.doi.org/10.1016/j.foodchem.2011.04.103

Udenigwe, C. C., & Aluko, R. E. (2012). Food protein-derived bioactive peptides: Production, processing, and potential health benefits. *Journal of Food Science*, 77(1), R11–R24.

Udenigwe, C. C., Lin, Y.-S., Hou, W.-C., & Aluko, R. E. (2009). Kinetics of the inhibition of renin and angiotensin I-converting enzyme by flaxseed protein hydrolysate fractions. *Journal of Functional Foods*, 1(2), 199–207.

Uraipong, C., & Zhao, J. (2016). Rice bran protein hydrolysates exhibit strong in vitro α-amylase, β-glucosidase and ACE-inhibition activities. *Journal of the Science of Food and Agriculture*, 96(4), 1101–1110. http://dx.doi.org/10.1002/jsfa.7182

Vigo, M., Malec, L., Gomez, R., & Llosa, R. (1992). Spectrophotometric assay using o-phthaldialdehyde for determination of reactive lysine in dairy products. *Food Chemistry*, 44(5), 363–365.

Wang, C., Li, B., & Ao, J. (2012). Separation and identification of zinc-chelating peptides from sesame protein hydrolysate using IMAC-Zn^{2+} and LC–MS/MS. *Food Chemistry*, 134(2), 1231–1238. http://dx.doi.org/10.1016/j.foodchem.2012.02.204

Wang, C., Tian, J., & Wang, Q. (2011). ACE inhibitory and antihypertensive properties of apricot almond meal hydrolysate. *European Food Research and Technology*, 232(3), 549–556.

Wang, J., Zhao, M., Zhao, Q., & Jiang, Y. (2007). Antioxidant properties of papain hydrolysates of wheat gluten in different oxidation systems. *Food Chemistry*, 101(4), 1658–1663.

Wang, Y., He, H., Wang, G., Wu, H., Zhou, B., Chen, X., & Zhang, Y. (2010). Oyster (Crassostrea gigas) hydrolysates produced on a plant scale have antitumor activity and immunostimulating effects in BALB/c mice. *Marine Drugs*, 8(2), 255–268.

Wu, J., Aluko, R. E., & Muir, A. D. (2008). Purification of angiotensin I-converting enzyme-inhibitory peptides from the enzymatic hydrolysate of defatted canola meal. *Food Chemistry*, 111(4), 942–950. http://dx.doi.org/10.1016/j.foodchem.2008.05.009

Wu, J., & Ding, X. (2001). Hypotensive and physiological effect of angiotensin converting enzyme inhibitory peptides derived from soy protein on spontaneously hypertensive rats. *Journal of Agricultural and Food Chemistry*, 49(1), 501–506.

Xia, Y., Bamdad, F., Gänzle, M., & Chen, L. (2012). Fractionation and characterization of antioxidant peptides derived from barley glutelin by enzymatic hydrolysis. *Food Chemistry*, *134*(3), 1509–1518. http://dx.doi.org/10.1016/j.foodchem.2012.03.063

Xu, Z., Wu, C., Sun-Waterhouse, D., Zhao, T., Waterhouse, G. I., Zhao, M., & Su, G. (2021). Identification of post-digestion angiotensin-I converting enzyme (ACE) inhibitory peptides from soybean protein Isolate: Their production conditions and in silico molecular docking with ACE. *Food Chemistry*, *345*, 128855.

Yang, R., Zhang, Z., Pei, X., Han, X., Wang, J., Wang, L., … Li, Y. (2009). Immunomodulatory effects of marine oligopeptide preparation from Chum Salmon (*Oncorhynchus keta*) in mice. *Food Chemistry*, *113*(2), 464–470.

Yea, C. S., Ebrahimpour, A., Hamid, A. A., Bakar, J., Muhammad, K., & Saari, N. (2014). Winged bean [Psophorcarpus tetragonolobus (L.) DC] seeds as an underutilised plant source of bifunctional proteolysate and biopeptides. *Food & Function*, *5*(5), 1007–1016.

Zanutto-Elgui, M. R., Vieira, J. C. S., do Prado, D. Z., Buzalaf, M. A. R., de Magalhães Padilha, P., de Oliveira, D. E., & Fleuri, L. F. (2019). Production of milk peptides with antimicrobial and antioxidant properties through fungal proteases. *Food Chemistry*, *278*, 823–831.

Zarei, M., Ebrahimpour, A., Abdul-Hamid, A., Anwar, F., Bakar, F. A., Philip, R., & Saari, N. (2014). Identification and characterization of papain-generated antioxidant peptides from palm kernel cake proteins. *Food Research International*, *62*, 726–734. http://dx.doi.org/10.1016/j.foodres.2014.04.041

Zhang, J., Zhang, H., Wang, L., Guo, X., Wang, X., & Yao, H. (2009). Antioxidant activities of the rice endosperm protein hydrolysate: Identification of the active peptide. *European Food Research and Technology*, *229*(4), 709–719.

Zhang, M., Mu, T.-H., & Sun, M.-J. (2014). Purification and identification of antioxidant peptides from sweet potato protein hydrolysates by Alcalase. *Journal of Functional Foods*, *7*, 191–200.

Zhang, T., Jiang, B., Miao, M., Mu, W., & Li, Y. (2012). Combined effects of high-pressure and enzymatic treatments on the hydrolysis of chickpea protein isolates and antioxidant activity of the hydrolysates. *Food Chemistry*, *135*(3), 904–912. http://dx.doi.org/10.1016/j.foodchem.2012.05.097

Zhuang, H., Tang, N., Dong, S. t., Sun, B., & Liu, J. b. (2013). Optimisation of antioxidant peptide preparation from corn gluten meal. *Journal of the Science of Food and Agriculture*, *93*(13), 3264–3270.

Zumwalt, R. W., & Gehrke, C. W. (1988). Amino acid analysis: A survey of current techniques. In Cherry, J.P., & Barford, R.A., (eds), *Methods for Protein Analysis* (pp. 13–35). American Oil Chemists' Society (AOCS), Champaign, IL.

CHAPTER 11

Protein Hydrolysates in Animal Nutrition
Industrial Production, Bioactive Peptides, and Functional Significance

Yongqing Hou, Zhenlong Wu, Zhaolai Dai, Genhu Wang, and Guoyao Wu

CONTENTS

Background	210
Definitions of Amino Acids, Peptides, and Protein	210
Industrial Production of Protein Hydrolysates	212
General Considerations of Protein Hydrolysis	212
Degree of Hydrolysis	212
Methods for Protein Hydrolysis	214
Acid Hydrolysis of Proteins	214
Alkaline Hydrolysis of Proteins	214
Cell-Free Proteases	214
Microbial Hydrolysis of Protein	217
Bioactive Peptides in Protein Hydrolysates	218
Definition	218
Transport of Small Peptides in the Small Intestine	218
ACE-Inhibitory Peptides	218
Antioxidative and Antimicrobial Peptides	218
Opioid Peptides	220
Applications of Plant and Animal Protein Hydrolysates in Animal Nutrition	220
General Consideration	220
Plant Peptides	222
Animal Peptides	224
Potential scale and economic value for The global use of animal and plant protein Hydrolysates in animal feeding	225
Future Research Directions	226
Conclusion	227
Abbreviations	227
Acknowledgments	227
Funding	227
Availability of Data and Materials	228

DOI: 10.1201/9781003106524-14

Authors' Contributions	228
Competing Interests	228
Consent for Publication	228
Ethics Approval and Consent to Participate	228
References	228

BACKGROUND

A protein is a macromolecule usually consisting of 20 different amino acids (AAs) linked via peptide bonds. Selenoproteins contain selenocysteine as a rare AA, but no free selenocysteine is present in animal cells. Protein is a major component of animal tissues (e.g., skeletal muscle, mammary glands, liver, and the small intestine) and products (e.g., meat, milk, egg, and wool). For example, the protein content in the skeletal muscle of growing beef cattle or pigs is approximately 70% on a dry-matter basis [1]. Thus, adequate intake of dietary protein is essential for maximum growth, production performance, and feed efficiency in livestock, poultry, and fish. After being consumed in a meal by animals, the proteins in feed ingredients (e.g., blood meal, meat and bone meal, intestinal mucosa powder, fish meal, soybean meal, peanut meal, and cottonseed meal) are hydrolyzed into small peptides (di- and tripeptides) and free AAs by proteases and oligopeptidases in the small intestine [2]; however, the types of resultant peptides can vary greatly with the physiological conditions of the animals and the composition of their diets. To consistently manufacture peptides from the proteins of animal and plant sources, robust chemical, enzymatic, or microbial methods have been used before feeding to improve their nutritional quality and reduce any associated antinutritional factors [3, 4]. The last two methods (enzymatic and microbial) can also improve the solubility, viscosity, emulsification, and gelation of peptides.

In animal production, high-quality protein is not hydrolyzed as feed additives. Only animal by-products, brewer's by-products, and plant ingredients containing antinutritional factors are hydrolyzed to produce peptides for animal feeds. Proteases isolated from various sources (including bacteria, plants, and yeast) are used for the enzymatic method, whereas intact microorganisms are employed for culture in the microbial approach. To date, protein hydrolysates have been applied to such diverse fields as medicine, nutrition (including animal nutrition), and biotechnology [5]. The major objectives of this article are to highlight enzyme- and fermentation-based techniques for the industrial preparation of protein hydrolysates and to discuss the nutritional and functional significance of their bioactive peptides in animal feeding.

DEFINITIONS OF AMINO ACIDS, PEPTIDES, AND PROTEIN

Amino acids are organic substances that contain both amino and acid groups. All proteinogenic AAs have an α-amino group and, except for glycine, occur as L-isomers in animals and feedstuffs. A peptide is defined as an organic molecule consisting of two or more AA residues linked by peptide bonds [2]. The formation of one peptide bond results in the removal of one water molecule. In most peptides, the typical peptide bonds are formed from the α-amino and α-carboxyl groups of adjacent AAs. Peptides can be classified according to the number of AA residues. An oligopeptide is comprised of 2 to 20 AA residues. Those oligopeptides containing ≤ 10 AA residues are called "small oligopeptides" (or small peptides), whereas those oligopeptides containing 10 to 20 AA residues are called "large oligopeptides" (or large peptides). A peptide, which contains ≥ 21 AA residues and does not have a 3-dimensional structure, is termed a "polypeptide" [6]. A protein consists of one or more high-molecular-weight polypeptides.

The dividing line between proteins and polypeptides is usually their molecular weight. Generally speaking, polypeptides with a molecular weight of ≥ 8,000 Daltons (Da) (i.e., ≥ 72 AA residues) are referred to as "proteins" [6]. For example, ubiquitin (a single chain of 72 AA residues) and casein α-S1 (200 AA residues) are proteins, but glucagon (29 AA residues) and oxytocin (9 AA residues) are peptides. However, the division between proteins and peptides simply on the basis of their molecular weights is not absolute. For example, insulin [51 AA residues (20 in chain A and 31 in chain B)] is well recognized as a protein because it has the defined 3-dimensional structure exhibited by proteins. In contrast, PEC-60 (a single chain of 60 AA residues) [7] and dopuin (a single chain of 62 AA residues) [8], which are isolated from the pig small-intestinal mucosae, are called "polypeptides". Figure 11.1 illustrates the four orders of protein structures (1): primary structure (the sequence of AAs along the polypeptide chain); (2) secondary structure (the conformation of the polypeptide backbone); (3) tertiary structure (the three-dimensional arrangement of proteins); and (4) quaternary structure (the spatial arrangement of polypeptide subunits). The primary sequence of AAs in a protein determines its secondary,

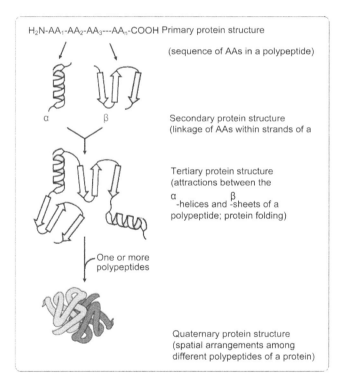

FIGURE 11.1 The four orders of protein structures. A protein has (1): a primary structure (the sequence of AAs along the polypeptide chain; (2) a secondary structure (the conformation of the polypeptide backbone); (3) a tertiary structure (the three-dimensional arrangement of protein); and (4) a quaternary structure (the spatial arrangement of polypeptide subunits). The primary sequence of AAs in a protein determines its secondary, tertiary, and quaternary structures, as well as its biological functions.

tertiary, and quaternary structures, as well as its biological functions. The forces stabilizing polypeptide aggregates are hydrogen and electrostatic bonds between AA residues. Trichloroacetic acid (TCA; the final concentration of 5%) or perchloric acid (PCA; the final concentration of 0.2 mol/L) can fully precipitate proteins, but not peptides, from animal tissues, cells, plasma, and other physiological fluids (e.g., rumen, allantoic, amniotic, intestinal-lumen fluids, and digesta) [9, 10]. Ethanol (the final concentration of 80%) can effectively precipitate both proteins and nucleic acids from aqueous solutions [11]. This method may be useful to remove water-soluble inorganic compounds (e.g., aluminum) from protein hydrolysates. Of note, 1% tungstic acid can precipitate both proteins and peptides with ≥ 4 AA residues [10]. Thus, PCA or TCA can be used along with tungstic acid to distinguish small and large peptides.

INDUSTRIAL PRODUCTION OF PROTEIN HYDROLYSATES

General Considerations of Protein Hydrolysis

The method of choice for the hydrolysis of proteins depends on their sources. For example, proteins from feathers, bristles, horns, beaks, and wool contain the keratin structure and, therefore, are usually hydrolyzed by acidic or alkaline treatment, or by bacterial keratinases [3]. In contrast, animal products (e.g., casein, whey, intestine, and meat) and plant ingredients (e.g., soy, wheat, rice, pea, and cottonseed proteins) are often subject to general enzymatic or microbial hydrolysis [4, 5]. The hydrolysis of proteins by cell-free proteases, microorganisms, acids, or bases results in the production of protein hydrolysates. The general procedures are outlined in Figure 11.2. Depending on the method used, the hydrolysis times may range from 4 to 48 h. In cases where bacteriostatic or bactericidal preservatives (e.g., benzoic acid) are used in the prolonged hydrolysis of proteins by enzymes or microorganisms, the hydrolysis is usually terminated by heating to deactivate the enzyme or enzyme systems. After hydrolysis, the insoluble fractions are separated from the protein hydrolysates with the use of a centrifuge, a filter (e.g., with a 10,000 Dalton molecular weight cut-off), or a micro filtration system [3]. The filtration process is often repeated several times to obtain a desirable color and clarity of the solution. Charcoal powder is commonly used to decolorize and remove haze-forming components. If very low concentrations of salts are desired, the filtrate may be subjected to exchange chromatography to remove the salts. After filtration, the protein hydrolysate product is heat-treated (pasteurized) to kill or reduce the microorganisms. Finally, the product is dried and packaged.

Degree of Hydrolysis

The protein hydrolysates include free AAs, small peptides, and large peptides. The proportions of these products vary with the sources of proteins, the quality of water, the type of proteases, and the species of microbes. The degree of hydrolysis (i.e., the extent to which the protein is hydrolyzed) is measured by the number of peptide bonds cleaved, divided by the total number of peptide bonds in a protein and multiplied by 100 [3]. The number of peptide bonds cleaved is measured by the moles of free AAs plus the moles of TCA- or PCA-soluble peptides. Due to the lack of standards for all the peptides generated from protein hydrolysis, it is technically challenging to quantify peptides released from

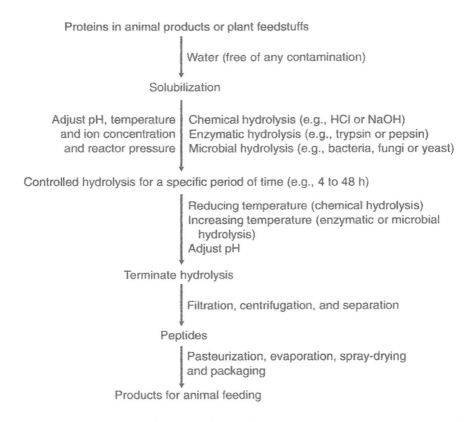

FIGURE 11.2 General procedures for the production of peptides from animal and plant proteins. Peptides (including bioactive peptides) can be produced from proteins present in animal products (including by-products) or plant-source feedstuffs material (e.g., soybeans and wheat) through chemical, enzymatic, or microbial hydrolysis. These general procedures may need to be modified for peptide production, depending on protein sources and product specifications calculated as (total AAs in protein-free AAs)/total AAs in protein x 100%. High-performance liquid chromatography (HPLC) is widely used to determine free AAs [12]. HPLC and other analytical techniques (e.g., nuclear magnetic resonance spectroscopy, matrix assisted laser desorption ionization–time of flight mass spectrometry, peptide mapping, and ion-exchange chromatography) are often employed to characterize peptides in protein hydrolysates [13, 14]. When standards are available, HPLC can be used to analyze peptides.

animal, plant, or microbial sources of proteins. The percentage of AAs in the free form or the peptide form is calculated as follows:

Percentage of AAs in the free form %
  = (Total free AAs/Total AAs in protein) × 100 %
Percentage of AAs in peptides %
  = (Total AAs in peptides/Total AAs in protein) × 100%

When the catabolism of AAs is limited (as in emzymatic hydrolysis), the percentage of AAs in peptides is calculated as (total AAs in protein – free AAs)/total AAs in protein ×

100%. High-performance liquid chromatography (HPLC) is widely used to determine free AAs [12]. HPLC and other analytical techniques (e.g., nuclear magnetic resonance spectroscopy, matrix assisted laser desorption ionization-time of flight mass spectrometry, peptide mapping, and ion-exchange chromatography) are often employed to characterize peptides in protein hydrolysates [13, 14]. When standards are available, HPLC can be used to analyze peptides.

When the catabolism of AAs is limited (as in enzymatic hydrolysis), the percentage of AAs in peptides is

METHODS FOR PROTEIN HYDROLYSIS

Acid Hydrolysis of Proteins

Acid hydrolysis of a protein (gelatin) at a high temperature was first reported by the French chemist H. Braconnot in 1920. It is now established that the complete hydrolysis of protein in 6 mol/L HCl occurs at 110 °C for 24 h [12]. A much shorter period of time (e.g., 2–6 h) is used to produce peptides [3]. After the hydrolysis, the product is evaporated, pasteurized, and spray dried. The majority of acid protein hydrolysates are used as flavor enhancers (e.g., hydrolyzed vegetable protein) [5]. The method of acid hydrolysis of a protein offers the advantage of low cost. However, this process results in the complete destruction of tryptophan, a partial loss of methionine, and the conversion of glutamine into glutamate and of asparagine into aspartate [5].

Alkaline Hydrolysis of Proteins

Alkaline agents, such as calcium, sodium, or potassium hydroxide (e.g., 4 mol/L), can be used at a high temperature (e.g., 105 °C) for 20 h to completely hydrolyze protein [12, 15]. Lower temperatures (e.g., 27–55 °C) and a shorter period of the hydrolysis time (e.g., 4–8 h) are often desirable for the generation of peptides in the food industry [5]. After hydrolysis, the product is evaporated, pasteurized, and spray dried. Like acid hydrolysis of proteins, alkaline hydrolysis of proteins offers the advantage of low cost, but it also can have a 100% recovery rate of tryptophan [12]. However, this process results in the complete destruction of most AAs (e.g., 100% loss). Thus, although alkaline hydrolysis is often used for the production of foaming agents (e.g., substitutes for egg proteins) and fire extinguisher foams, it is not widely used in the food industry.

Cell-Free Proteases

The peptide bonds of proteins can be broken down by many different kinds of proteases, which can be classified as exopeptidases and endopeptidases based on the type of reaction, namely, hydrolysis of a peptide bond in the terminal region (an exopeptidase) or within an internal region (an endopeptidase) of a protein [2]. Some proteases hydrolyze dipeptides (dipeptidases), whereas others remove terminal AA residues that are substituted, cyclized, or linked by isopeptide bonds (namely, peptide linkages other than those of α-carboxyl to α-amino groups; e.g., ω-peptidases). When a protease exhibits a marked preference for a peptide bond formed from a particular AA residue, the name of this

AA is used to form a qualifier (e.g., "leucine" aminopeptidase and "pro-line" endopeptidase). In contrast, for enzymes with very complex or broad specificity, alphabetical or numerical serial names (e.g., peptidyl-dipeptidase A, peptidyl-dipeptidase B, dipeptidyl-peptidase I, and dipeptidyl-peptidase II) are employed for protein hydrolysis. Some proteases may have both exopeptidase and endopeptidase properties (e.g., cathepsins B and H). Enzymatic hydrolysis takes place under mild conditions (e.g., pH 6–8 and 30–60 °C) and minimizes side reactions.

Most of the cell-free enzymes for producing protein hydrolysates are obtained from animal, plant, and microbial sources (Table 11.1). Enzymes of animal sources (particularly pigs) for protein hydrolysis are pancreatin, trypsin, pepsin, carboxypeptidases and aminopeptidases; enzymes of plant sources are papain and bromelain; and enzymes of bacterial and fungal sources are many kinds of proteases with a broad spectrum of optimal temperatures, pH, and ion concentrations [16, 17]. The enzymes from commercial sources may be purified, semi-purified, or crude from the biological sources. The hydrolysis of proteins can be achieved by a single enzyme (e.g., trypsin) or multiple enzymes (e.g., a mixture of proteases known as Pronase, pepsin, and prolidase). The choice of enzymes depends on the protein source and the degree of hydrolysis. For example, if the protein has a high content of hydrophobic AAs, the enzyme of choice would be the one that preferentially breaks downs the peptide bonds formed from these AAs. Fractionation of protein hydrolysates is often performed to isolate specific peptides or remove undesired peptides. It is noteworthy that the hydrolysis of some proteins (e.g., soy proteins and casein with papain for 18 h) can generate hydrophobic peptides and AAs with bitterness [18]. The addition of porcine kidney cortex homogenate or activated carbon to protein hydrolysates can reduce the bitterness of the peptide product [3]. Compared to acid and alkaline hydrolysis of proteins, the main advantages of enzyme hydrolysis of proteins are: (1) the hydrolysis conditions (e.g., temperature and pH) are mild and do not result in any loss of AAs; (2) the proteases are more specific and precise to control the degree of peptide-bond hydrolysis; and (3) the small amounts of enzymes can be easily deactivated after hydrolysis (e.g., 85 °C for 3 min) to facilitate the isolation of the protein hydrolysates. The disadvantages of enzymatic hydrolysis of protein include the relatively high cost and the potential presence of enzyme inhibitors in the raw protein materials.

The efficiency and specificity of protein hydrolysis differ between microbial- and animal-source proteases [19], as reported for their lipase activity [20, 21]. For example, the hydrolysis of 18 mg casein by 40 μg pancreatin (pancreatic enzymes from the porcine pancreas) for 1–2 h in a buffer solution (43 mmol/L NaCl, 7.3 mmol/L disodium tetraborate, 171 mmol/L boric acid, and 1 mmol/L $CaCl_2$, pH 7.4) yields the numbers and sequences of peptides differently than NS 4 proteases (from *Nocardiopsis prasina*) and NS 5 proteases (from *Bacillus subtilis*) [19]. The pancreatin exhibited the activities of trypsin (cleavage of peptide bonds from Arg and Lys sites), chymotrypsin (cleavage of peptide bonds from Phe, Trp, Tyr, and Leu), and elastase (cleavage of peptide bonds from (Ala and other aliphatic AAs). In contrast, the microbial proteases were characterized by relatively low trypsin, carboxypeptidase and elastase activities, but high chymotrypsin activity. The time course of hydrolysis is similar among the microbial and pancreatic enzymes when caseins are the substrates. However, the rates of peptide generation are higher for pancreatin than the microbial enzymes when soya protein is the substrate. The efficiency of production of peptides with a molecular weight less than 3 kDa is higher for pancreatin than the microbial enzymes when caseins are substrates, but it is similar among the microbial and pancreatic enzymes when soya protein is the substrate. In contrast, the efficiency of production of peptides with a molecular weight between 3 and 10

TABLE 11.1 Proteases Commonly Used for Protein Hydrolysis

Class of Enzyme	Name of Enzyme	EC Number	Specific Cleavage
Endopeptidases			
Aspartate protease	Chymosin (rennin; pH 1.8–2)	3.4.23.4	the Phe-Met bond, clotting of milk
	Pepsin A (pH 1.8–2)	3.4.23.1	Aromatic AAs, hydrophobic AAs
Cysteine protease	Bromelain (from pineapples)	3.4.22.4	Ala, Gly, Lys, Phe, Tyr
	Cathepsin B	3.4.22.1	Arg, Lys, Phe
	Ficain (ficin; from fig tree)	3.4.22.3	Ala, Asn, Gly, Leu, Lys, Tyr, Val
	Papain (from papaya)	3.4.22.2	Arg, Lys, Phe
Metallo protease	Bacillolysin (*Bacillus* bacteria)	3.4.24.28	Aromatic AAs, Ile, Leu, Val
	Thermolysin (*Bacillus* bacteria)	3.4.24.27	Aromatic AAs, Ile, Leu, Val
Serine protease	Chymotrypsin (pH 8–9)	3.4.21.1	Aromatic AAs, Leu
	Subtilisin (from *Bacillus* bacteria)	3.4.21.14	Mainly hydrophobic AAs
	Trypsin (pH 8–9)	3.4.21.4	Arg, Lys
Exopeptidases			
Aminopeptidases	Aminopeptidase[a]	3.4.11.1	AA at the N-terminus of protein/peptide
	Aminopeptidase Y[b]	3.4.11.15	Lys at the N-terminus of protein/peptide
Carboxypeptidase	Carboxypeptidase[c]	3.4.16.1	Acidic, neutral, and basic AAs
	Glycine carboxypeptidase[d]	3.4.17.4	Gly at the C-terminus of protein/peptide
	Alanine carboxypeptidase[e]	3.4.17.8	D-Ala at the C-terminus of peptide
	Carboxypeptidase S[f]	3.4.17.9	Gly at the C-terminus of protein/peptide
Dipeptidase	Dipeptidase 1[f]	3.4.13.11	A wide range of dipeptides
	Proline dipeptidase (prolidase)[a]	3.4.13.9	AA-Pro or -hydroxyproline at the C-terminus (not Pro-Pro)
	Prolyl dipeptidase[a]	3.3.13.8	Pro-AA or Hydroxyproline-AA
Endo- and exo-peptidases			
Pronase	A mixture of proteases[a] (from *Streptomyces griseus*)	3.4.24	Acidic, neutral, and basic AAs
Other peptidases	Dipeptidyl-peptide III[f]	3.4.14.4	Release of an N-terminal dipeptide from a peptide comprising four or more AA residues, with broad specificity

(*Continued*)

TABLE 11.1 (Continued)

Class of Enzyme	Name of Enzyme	EC Number	Specific Cleavage
	Dipeptidyl-peptidase IV[g]	3.4.14.5	Release of an N-terminal dipeptide from a peptide consisting of proline[h]

Adapted from Kunst [16] and Dixon and Webb [17]. *AA* amino acid
[a] Metallopeptidase (requiring Mn^{2+}, Mg^{2+} or Zn^{2+} for activation)
[b] Metallopeptidase (requiring Co^{2+} for activation; inhibited by Zn^{2+} and Mn^{2+})
[c] Serine carboxypeptidase
[d] Strongly inhibited by Ag^+ and Cu^{2+}
[e] Metallopeptidase (requiring Mn^{2+}, Mg^{2+}, Zn^{2+}, Ca^{2+} or Co^{2+} for activation)
[f] Metallopeptidase (requiring Zn^{2+} for activation)
[g] Serie protease
[h] AA_1-Pro-AA_2, where AA_2 is neither proline nor hydroxyproline

kDa is similar among the microbial and pancreatic enzymes when caseins are substrates or when soya protein is the substrate within 1 h incubation, but it is higher for the microbial enzymes than the animal-source pancreatin.

Microbial Hydrolysis of Protein

Microorganisms release proteases to hydrolyze extracellular proteins into large peptides, small peptides, and free AAs. Small peptides can be taken up by the microbes to undergo intracellular hydrolysis, yielding free AAs. Microorganisms also produce enzymes other than proteases to degrade complex carbohydrates and lipids [22]. Protein fermentation is classified into a liquid- or solid-state type. Liquid-state fermentation is performed with protein substrates under high-moisture fermentation conditions, whereas the solid-state fermentation is carried out under low-moisture fermentation conditions. The low moisture level of the solid-state fermentation can help to reduce the drying time for protein hydrolysates.

Soy sauce (also called soya sauce), which originated in China in the 2nd century AD, was perhaps the earliest product of protein fermentation by microorganisms [3]. The raw materials were boiled soybeans, roasted grain, brine, and *Aspergillus oryzae* or *Aspergillus sojae* (a genus of fungus). In Koji culturing, an equal amount of boiled soybeans and roasted wheat is cultured with *Aspergillus oryzae*, *A. sojae*, and *A. tamari*, *Saccharomyces cerevisiae* (yeasts), and bacteria, such as *Bacillus* and *Lactobacillus* species. Over the past two decades, various microorganisms have been used to hydrolyze plant-source proteins, such as *Lactobacillus rhamnosus* BGT10 and *Lactobacillus zeae* LMG17315 for pea proteins, *Bacillus natto* or *B. subtilis* for soybean, and fungi *A. oryzae* or *R. oryzae* for soybean [23–25]. Lactic acid bacteria, such as *Lactobacillus* and *Lactococcus* species, are commonly used to ferment milk products. The major advantages of fermentation are that the appropriately used microorganisms can not only break down proteins into peptides and free AAs, but can also remove hyper-allergic or antinutritional factors present in the matrix of the ingredients (e.g., trypsin inhibitors, glycinin, β-conglycinin, phytate, oligosaccharides raffinose and stachyose, saponins in soybeans). The disadvantages of the microbial hydrolysis of protein are the relatively high cost, as well as changes in microbial activity under various conditions and, therefore, inconsistency in the production of peptides and free AAs.

BIOACTIVE PEPTIDES IN PROTEIN HYDROLYSATES

Definition

Bioactive peptides are defined as the fragments of AA sequences in a protein that confer biological functions beyond their nutritional value [25]. They have antimicrobial, antioxidant, antihypertensive, and immunomodulatory activities. These bioactive peptides are usually 2–20 AA residues in length, but some may consist of >20 AA residues [23]. Many of them exhibit common structural properties, such as a relatively small number of AAs, a high abundance of hydrophobic AA residues, and the presence of Arg, Lys, and Pro residues [24]. In animals, endogenous peptides fulfil crucial physiological or regulatory functions. For example, PEC-60 activates Na/K ATPase in the small intestine and other tissues [26]. Additionally, many intestinal peptides (secreted by Paneth cells) have an antimicrobial function [27]. Furthermore, the brain releases numerous peptides to regulate endocrine status, food intake, and behavior in animals [28].

Transport of Small Peptides in the Small Intestine

In the small intestine, peptide transporter 1 (PepT1) is responsible for the proton-driven transport of extracellular di- and tripeptides through the apical membrane of the enterocyte into the cell [29]. However, due to the high activity of intracellular peptidases in the small intestine [2], it is unlikely that a nutritionally significant quantity of peptides in the lumen of the gut can enter the portal vein or the lymphatic circulation. It is possible that a limited, but physiologically significant, number of peptides (particularly those containing an amino acid) may be absorbed intact from the luminal content to the bloodstream through M cells, exosomes, and enterocytes via transepithelial cell transport [30, 31]. Diet-derived peptides can exert their bioactive (e.g., physiological and regulatory) actions at the level of the small intestine, and the intestinally generated signals can be transmitted to the brain, the endocrine system, and the immune system of the body to beneficially impact the whole body.

ACE-Inhibitory Peptides

The first food-derived bioactive peptide, which enhanced vitamin D-independent bone calcification in rachitic infants, was produced from casein [32]. To date, many angiotensin-I converting enzyme (ACE)-inhibitory peptides have been generated from milk or meat (Table 11.2). ACE removes the C-terminal dipeptide His-Leu in angiotensin I (Ang I) to form Ang II (a potent vasoconstrictive peptide), thereby conferring their antihypertensive effects [33]. The best examples for ACE-inhibitory peptides are Ile-Pro-Pro (IPP) and Val-Pro-Pro (VPP), both of which are derived from milk protein through the hydrolysis of neutral protease, alkaline protease, or papain [34]. There is evidence that these two proline-rich peptides may partially escape gastrointestinal hydrolysis and be transported across the intestinal epithelium into the blood circulation [35]. Similarly, the hydrolysis of proteins from meat [36] and egg yolk [37] also generates potent ACE inhibitors.

Antioxidative and Antimicrobial Peptides

Many small peptides from animal products (e.g., fish and meat) (Table 11.3) and plant-source feedstuffs [25] have antioxidative functions by scavenging free radicals and/or

TABLE 11.2 Antihypertensive Peptides Generated from the Hydrolysis of Animal Products

Source	Protease(s)	Amino Acid Sequence	IC$_{50}$, µmol/L[a]
Pig muscle myosin	Thermolysin	Ile-Thr-Thr-Asn-Pro	549
Pig muscle myosin	Pepsin	Lys-Arg-Val-Ile-Thr-Tyr	6.1
Pig muscle actin	Pepsin	Val-Lys-Arg-Gly-Phe	20.3
Pig muscle troponin	Pepsin	Lys-Arg-Gln-Lys-Tyr-Asp-Ile	26.2
Pig muscle	Pepsin + Pancreatin	Lys-Leu-Pro	500
Pig muscle	Pepsin + Pancreatin	Arg-Pro-Arg	382
Chicken muscle	Thermolysin	Leu-Ala-Pro	3.2
Chicken muscle myosin	Thermolysin	Phe-Gln-Lys-Pro-Lys-Arg	14
Chicken muscle	Thermolysin	Ile-Lys-Trp	0.21
Chicken collagen	*Aspergillus* proteases + Proteases FP, A, G and N	Gly-Ala-X-Gly-Leu-X-Gly-Pro	29.4
Cow muscle	Thermolysin + Proteinase A	Val-Leu-Ala-Gln-Tyr-Lys	32.1
Cow muscle	Thermolysin + Proteinase A	Phe-His-Gly	52.9
Cow muscle	Proteinase K	Gly-Phe-His-Ile	64.3
Cow skin gelatin	Alcalase + Pronase E + Collagenase	Gly-Pro-Val	4.67
Cow skin gelatin	Alcalase + Pronase E + Collagenase	Gly-Pro-Leu	2.55
Bonito (fish) muscle	Thermolysin	Leu-Lys-Pro-Asn-Met	2.4
Bonito (fish) muscle	Thermolysin	Leu-Lys-Pro	0.32
Bonito (fish) muscle	Thermolysin	Ile-Lys-Pro	6.9
Salmon muscle	Thermolysin	Val-Trp	2.5
Salmon muscle	Thermolysin	Met-Trp	9.9
Salmon muscle	Thermolysin	Ile-Trp	4.7
Sardine muscle	Alcalase	Ile-Tyr	10.5
Sardine muscle	Alcalase	Ala-Lys-Lys	3.13
Sardine muscle	Alcalase	Gly-Trp-Ala-Pro	3.86
Sardine muscle	Alcalase	Lys-Tyr	1.63
Alaska pollack skin	Alcalase + Pronase + Collagenase	Gly-Pro-Leu	2.65
Alaska pollack skin	Alcalase + Pronase + Collagenase	Gly-Pro-Met	17.1
Shark muscle	Protease SM98011	Glu-Tyr	1.98
Shark muscle	Protease SM98012	Phe-Glu	2.68

(*Continued*)

TABLE 11.2 (Continued)

Source	Protease(s)	Amino Acid Sequence	IC$_{50}$, µmol/L[a]
Shark muscle	Protease SM98013	Cys-Phe	1.45
Egg yolk	Pepsin	Tyr-Ile-Glu-Ala-Val-Asn-Lys-Val-Ser-Pro-Arg-Ala-Gly-Gln-Phe	9.4[b]
Egg yolk	Pepsin	Tyr-Ile-Asn-Gln-Met-Pro-Gln-Lys-Ser-Arg-Glu	10.1[b]

Adapted from Ryan JT et al. [28], Ryder et al. [33], and Zambrowicz et al. [34]
"X" hydroxyproline

[a] Inhibition of angiotensin-I converting enzyme (ACE) activity. All values are expressed as µM, except for egg yolk-derived peptides (µg/mL) as indicated by a superscript "b"

inhibiting the production of oxidants and pro-inflammatory cytokines [38–41]. These small peptides can reduce the production of oxidants by the small intestine, while enhancing the removal of the oxidants, resulting in a decrease in their intracellular concentrations and alleviating oxidative stress (Figure 11.3). Many of the bioactive peptides have both ACE-inhibitory and antioxidative effects [36, 37]. Additionally, some peptides from animal (Table 11.4) and plant protein hydrolysates [25] also have antimicrobial effects, as reported for certain endogenous peptides in the small intestine [27]. These antimicrobial peptides exert their actions by damaging the cell membrane of bacteria, interfering with the functions of their intracellular proteins, inducing the aggregation of cytoplasmic proteins, and affecting the metabolism of bacteria [42–44], but the underlying mechanisms remain largely unknown [27].

Opioid Peptides

The hydrolysis of certain proteins [e.g., casein, gluten (present in wheat, rye and barley), and soybeans] in the gastrointestinal tract can generate opioid peptides [45]. This can be performed *in vitro* by using digestive enzymes from the small intestine of mammals (e.g., pigs). Opioid peptides are oligopeptides (typically 4–8 AA residues in length) that bind to opioid receptors in the brain to affect the gut function [46, 47], as well as the behavior and food intake of animals (Table 11.5). Furthermore, the protein hydrolysates containing opioid-like peptides may be used as feed additives to alleviate stress, control pain and sleep, and modulate satiety in animals.

APPLICATIONS OF PLANT AND ANIMAL PROTEIN HYDROLYSATES IN ANIMAL NUTRITION

General Consideration

A major goal for animal agriculture is to enhance the efficiency of feed utilization for milk, meat, and egg production [48]. This approach requires optimal nutrition to support the function of the small intestine as the terminal site for the digestion and absorption of dietary nutrients [49]. To date, peptides generated from the hydrolysis of plant and animal proteins are included in the diets for feeding pigs, poultry, fish, and companion

TABLE 11.3 Antioxidative Peptides Generated from the Hydrolysis of Animal Proteins

Source	Protease(s)	Amino acid sequence
Pig muscle actin	Papain + Actinase E	Asp-Ser-Gly-Val-Thr
Pig muscle	Papain + Actinase E	Ile-Glu-Ala-Glu-Gly-Glu
Pig muscle tropomyosin	Papain + Actinase E	Asp-Ala-Gln-Glu-Lys-Leu-Glu
Pig muscle tropomyosin	Papain + Actinase E	Glu-Glu-Leu-Asp-Asn-Ala-Leu-Asn
Pig muscle myosin	Papain + Actinase E	Val-Pro-Ser-Ile-Asp-Asp-Gln-Glu-Glu-Leu-Met
Pig collagen	Pepsin + Papain + others[a]	Gln-Gly-Ala-Arg
Pig blood plasma	Alcalase	His-Asn-Gly-Asn
Chicken muscle	—	His-Val-Thr-Glu-Glu
Chicken muscle	—	Pro-Val-Pro-Val-Glu-Gly-Val
Deer muscle	Papain	Met-Gln-Ile-Phe-Val-Lys-Thr-Leu-Thr-Gly
Deer muscle	Papain	Asp-Leu-Ser-Asp-Gly-Glu-Gln-Gly-Val-Leu
Bovine milk casein	Pepsin, pH 2, 24 h	Tyr-Phe-Tyr-Pro-Glu-Leu
Bovine milk casein	Pepsin, pH 2, 24 h	Phe-Tyr-Pro-Glu-Leu
Bovine milk casein	Pepsin, pH 2, 24 h	Tyr-Pro-Glu-Leu
Bovine milk casein	Pepsin, pH 2, 24 h	Pro-Glu-Leu
Bovine milk casein	Pepsin, pH 2, 24 h	Glu-Leu
Bovine milk casein	Trypsin, pH 7.8, 24–28 h	Val-Lys-Glu-Ala-Met-Pro-Lys
Bovine milk casein	Trypsin, pH 7.8, 24–28 h	Ala-Val-Pro-Tyr-Pro-Gln-Arg
Bovine milk casein	Trypsin, pH 7.8, 24–28 h	Lys-Val-Leu-Pro-Val-Pro-Glu-Lys
Bovine milk casein	Trypsin, pH 7.8, 24–28 h	Val-Leu-Pro-Val-Pro-Glu-Lys
Bovine whey protein	Thermolysin, 80 °C, 8 h	Leu-Gln-Lys-Trp
Bovine whey protein	Thermolysin, 80 °C, 8 h	Leu-Asp-Thr-Asp-Tyr-Lys-Lys
Bovine β-Lactoglobulin	Corolase PP, 37 °C, 24 h	Trp-Tyr-Ser-Leu-Ala-Met-Ala-Ala-Ser-Asp-Ile
Bovine β-Lactoglobulin	Corolase PP, 37 °C, 24 h	Met-His-Ile-Arg-Leu
Bovine β-Lactoglobulin	Corolase PP, 37 °C, 24 h	Try-Val-Glu-Glu-Leu
Egg yolk	Pepsin	Tyr-Ile-Glu-Ala-Val-Asn-Lys-Val-Ser-Pro-Arg-Ala-Gly-Gln-Phe
Egg yolk	Pepsin	Tyr-Ile-Asn-Gln-Met-Pro-Gln-Lys-Ser-Arg-Glu

Adapted from Ryder et al. [33], Zambrowicz et al. [34], Shimizu and Son [35], Bah et al. [36], Memarpoor-Yazdia et al. [37], and Power et al. [38]

[a] Bovine pancreatic proteases plus bacterial proteases from *Streptomycest bacillus*

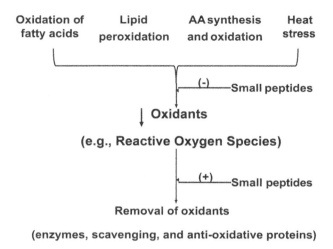

FIGURE 11.3 Inhibition of cellular oxidative stress by dietary small peptides in the small intestine. The small peptides, which are supplemented to the diets of animals (particularly young animals), can reduce the production of oxidants by the small intestine and enhance the removal of the oxidants, leading to a decrease in their intracellular concentrations and alleviating oxidative stress. (−), inhibition; (+), activation; ↓, decrease.

animals. The outcomes are positive and cost-effective for the improvement of intestinal health, growth, and production performance [50]. The underlying mechanisms may be that: (1) the rate of absorption of small peptides is greater than that of an equivalent amount of free AAs; (2) the rate of catabolism of small peptides by the bacteria of the small intestine is lower than that of an equivalent amount of free AAs; (3) the composition of AAs entering the portal vein is more balanced with the intestinal transport of small peptides than that of individual AAs; (4) the provision of functional AAs (e.g., glycine, arginine, glutamine, glutamate, proline, and taurine) enhances antioxidative reactions and muscle protein synthesis [51, 52]; and (5) specific peptides can improve the morphology, motility, and function of the gastrointestinal tract (e.g., secretion, motility, and anti-inflammatory reactions), endocrine status in favor of anabolism, and feed intake, compared with an equivalent amount of free AAs. In swine nutrition research, most of the studies involving the addition of peptides to diets have been conducted with postweaning pigs to improve palatability, growth, health, and feed efficiency [53–58]. This is primarily because young animals have immature digestive and immune systems and weanling pigs suffer from reduced feed intake, gut atrophy, diarrhea, and impaired growth. Moreover, peptide products have been supplemented to the diets of calves [59], poultry [60, 61], fish [62, 63], and companion animals [64] to improve their nutrition status, gut function, and infectious disease resistance.

Plant Peptides

As noted previously, plant-source protein ingredients often contain allergenic proteins and other antinutritional factors which can limit their practical use, particularly in the diets of young animals [50] and companion animals [64]. For example, soybeans can be processed to manufacture soybean meal and soybean protein concentrates for the

TABLE 11.4 Antimicrobial Peptides Generated from the Hydrolysis of Animal Proteins or Synthesized by Intestinal Mucosal Cells

Source	Amino Acid Sequence	Gram-Positive Bacteria	Gram-Negative Bacteria
Bovine meat	Gly-Leu-Ser-Asp-Gly-Glu-Trp-Gln	*Bacillus cereus* *Listeria monocytogenes*	*Salmonella typhimurium* *Escherichia coli*
	Gly-Phe-His-Ile	No effect	*Pseudomonas aeruginosa*
	Phe-His-Gly	No effect	*Pseudomonas aeruginosa*
Bovine collagen	Peptides < 2 kDa (by collagenase)[a]	*Staphylococcus aureus*	*Escherichia coli*
Goat whey	GWH (730 Da) and SEC-F3 (1,183 Da) (hydrolysis by Alcalase)	*Bacillus cereus* *Staphylococcus aureus*	*Salmonella typhimurium* *Escherichia coli*
Red blood cells	Various peptides (24-h hydrolysis by fugal proteases)	*Staphylococcus aureus*	*Escherichia coli* *Pseudomonas aeruginosa*
Hen egg white lysozyme	Asn-Thr-Asp-Gly-Ser-Thr-Asp-Tyr-Gly- Ile-Leu-Gln-Ile-Asn-Ser-Arg (hydrolysis by papain and trypsin)[b]	*Leuconostoc-mesenteroides*	*Escherichia coli*
Trout by-products	Various peptides (20–30% of hydrolysis) (hydrolysis by trout pepsin)	*Renibacterium-salmoninarum*	*Flavobacterium psychrophilum*
Small intestine (Paneth cells)	α-Defensins, lysozyme C, angiogenin-4 and cryptdin-related sequence peptides	Gram-positive bacteria broad-spectrum)	Gram-negative bacteria (broad-spectrum)
	Phospholipid-*sn*-2 esterase and C-type lectin	Gram-positive bacteria (broad-spectrum)	No effect

Adapted from Lima et al. [39], Osman et al. [40], and Wald et al. [41]
[a] minimal inhibition concentrations = 0.6 – 5 mg/mL
[b] minimal inhibition concentrations = 0.36 – 0.44 μg/mL

elimination of some anti-nutritional substances. However, the soy products still contain considerable amounts of protein-type allergens (e.g., glycinin and β-conglycinin) and significant quantities of trypsin inhibitors, lectins (hemagglutinins), phytic acid, soy oligosaccharides (raffinose and stachyose), and steroid glycosides (soy saponins) [18, 24, 25]. Fermentation of soybean by the commonly used microorganisms (e.g., *Aspergillus* species, *Bacillus* species, and *Lactobacillus* species) has been reported to improve growth performance on feed efficiently in weanling pigs [50]. Thus, 3- to 7-week-old pigs can be fed a corn- and soybean meal-based diet without affecting growth performance or feed efficiency [54]. Similar results were obtained for the Atlantic salmon fed a diet containing 40% protein from fermented soy white flakes [60]. Of interest, 50% of fish meal in the diet of juvenile red sea bream can be replaced by the same percentage of soybean protein hydrolysate [63]. The inclusion of plant-protein hydrolysate in diets is important in

TABLE 11.5 Opioid Peptides Generated from the Enzymatic Hydrolysis of Animal and Plant Proteins in the Gastrointestinal Tract

Source	Name of Opioid Peptide	Amino Acid Sequence
Mil casein	Bovine β-casomorphin 1–3	Tyr-Pro-Phe-OH
	Bovine β-casomorphin 1–4	Tyr-Pro-Phe-Pro-OH
	Bovine β-casomorphin 1–4 Amide	Tyr-Pro-Phe-Pro-NH$_2$
	Bovine β-casomorphin 5	Tyr-Pro-Phe-Pro-Gly-OH
	Bovine β-casomorphin 7	Tyr-Pro-Phe-Pro-Gly-Pro-Ile-OH
	Bovine β-casomorphin 8	Tyr-Pro-Phe-Pro-Gly-Pro-Ile-Pro-OH[a]
Gluten protein	Gluten exorphin A5	Gly-Tyr-Tyr-Pro-Thr-OH
	Gluten exorphin B4	Tyr-Gly-Gly-Trp-OH
	Gluten exorphin C	Tyr-Pro-Ile-Ser-Leu-OH
	Gliadorphin	Tyr-Pro-Gln-Pro-Gln-Pro-Phe-OH
Soybean protein	Soymorphin-5[b]	Tyr-Pro-Phe-Val-Val-OH
	Soymorphin-5, amide	Tyr-Pro-Phe-Val-Val-NH$_2$
	Soymorphin-6	Tyr-Pro-Phe-Val-Val-Asn-OH
	Soymorphin-7	Tyr-Pro-Phe-Val-Val-Asn-Ala-OH
Spinach protein	Rubiscolin-5	Gly-Tyr-Tyr-Pro-OH
	Rubiscolin-6	Gly-Tyr-Tyr-Pro-Thr-OH

Adapted from Li-Chan [21], López-Barrios et al. [22], Shimizu and Son [35], Bah et al. [36], and Froetschel [42]

[a] Another form of bovine β-casomorphin 8 has histidine instead of proline in position 8, depending on whether the peptide is derived from A1 or A2 beta-casein

[b] Derived from β-conglycinin β-subunit

aquaculture because fish meal is becoming scarce worldwide. Furthermore, as a replacement of the expensive skim milk powder, the hydrolysate of soy protein isolate (19.7% in diet) can be used to sustain high growth-performance in calve [59]. Finally, acidic hydrolysates of plant proteins (e.g., wheat gluten which contains a high amount of glutamine plus glutamate), often called "hydrolyzed vegetable proteins", can be included at a 1–2% level in the diets of companion animals to provide savory flavors due to the high abundance of glutamate in the products [64].

Animal Peptides

Postweaning piglets fed a diet containing 6% spray-dried porcine intestine hydrolysate (SDPI; the coproduct of heparin production) for 2 wk had better growth performance than those fed the control diet, the basal diet containing spray-dried plasma, or the basal diet containing dried whey [55, 56]. There was a carryover effect on enhancing growth performance during weeks 3–5 postweaning in piglets that were previously fed the SDPI [56], which was likely due to an increased area of the intestinal villus as well as improved digestion and absorption of dietary nutrients [57]. Similarly, [58] reported that piglets (weaned at 20 days of age) were fed a soybean meal and soybean protein concentrates for the elimination of some antinutritional substances. However, the soy products still contain considerable amounts of protein-type allergens (e.g., glycinin and β-conglycinin) and significant quantities of trypsin inhibitors, lectins (hemagglutinins), phytic acid, soy oligosaccharides (raffinose and stachyose), and steroid glycosides (soy saponins) [18, 24, 25]. Fermentation of soybeans by the commonly used microorganism (e.g., *Aspergillus* species, *Bacillus* species, and *Lactobacillus* species) has been reported to improve growth performance and feed efficiency in weanling pigs [50]. Thus, 3–7-week-old pigs fed a corn and soybean meal-based diet containing 3% or 6% fermented soybean meal grew at a rate comparable to that of the same percentage of dried skim milk [54]. Likewise, 4.9% fermented soybean meal could replace 3.7% spray-dried plasma protein in the diets of 3–7-week-old pigs fed a corn and soybean meal-based diet without affecting growth performance or feed efficiency [54]. Similar results were obtained for the Atlantic salmon fed a diet containing 40% protein from fermented soy white flakes [60]. Of interest, 50% of fish meal in the diet of juvenile red sea bream can be replaced by the same percentage of soybean protein hydrolysate [63]. The inclusion of plant-protein hydrolysate in diets is important in aquaculture because fish meal is becoming scarce worldwide. Furthermore, as a weanling diet containing 1.5, 3, or 4.5% SDPI had better growth performance and greater feed efficiency in comparison to piglets consuming the same amount of a fish meal-supplemented diet. Of note, these effects of the SDPI supplementation were dose-dependent. In addition, postweaning piglets fed a corn/soybean meal-based and dried whey-based diet containing 6% enzymatically hydrolyzed proteins (from a blend of swine blood and selected poultry tissues) exhibited a growth rate and a feed efficiency that were comparable to those for piglets fed a diet containing the same percentage of spray-dried blood cells [53]. Likewise, the inclusion of 2.5, 5, or 7.5% hydrolyzed porcine mucosa in a corn and soybean meal-based diet enhanced daily weight gain and nutrient retention in growing chicks [61]. Furthermore, broilers fed a diet containing 5% Atlantic salmon protein hydrolysates (from the viscera) had better growth performance than those fed a diet with or without 4% fish meal [60]. Finally, addition of the protein hydrolysate of fish by-products to the diet (at a 10% inclusion level) improved intestinal development, growth, immunological status, and survival in European sea bass larvae challenged with *Vibrio anguillarum* (a Gram-negative bacterium) [65]. Thus, SDPI or other hydrolysates of animal proteins hold promise for animal production.

POTENTIAL SCALE AND ECONOMIC VALUE FOR THE GLOBAL USE OF ANIMAL AND PLANT PROTEIN HYDROLYSATES IN ANIMAL FEEDING

Industrial processing of domestic farm animals generates large amounts of tissue (30–40% of body weight) not consumed by humans, including viscera, carcass-trimmings, bone (20–30% of body weight), fat, skin, feet, small-intestinal tissues (2% of body weight), feather (up to 10% of body weight), and collectible blood (5% body weight), with the global quantity of human-inedible livestock and poultry by-products being ~54 billion kg/yr [66–68]. Likewise, fish processing industries produce large amounts of waste (up to 55% of body weight), such as muscle-trimmings (15–20%), skin and fins (1–3%), bones (9–15%), heads (9–12%), viscera (12–18%), and scales, with the global quantity of fish by-products being ~6 billion kg/yr [66–69]. Thus, the global annual volume of total animal by-products generated by the processing industries is approximately 60 billion kg

TABLE 11.6 Potential Scale and Economic Values for the Global Use of Animal and Plant Protein Hydrolysates (PH) in Animal Feeding

Type	Annual Global Production[a] Billion kg/yr	Annual Use for Animal Feeding[a] Billion kg/yr	Amount Used for Production of PH[b] Billion kg/yr	Current Price[c] US $/kg	Total Value for PH Billion US $/yr
ABP	172	60	3	1.5	4.50
Soybean	180	135	6.75	0.575	3.88
Wheat	750	255	12.75	1.57	20.02

ABP, animal by-products (including livestock, poultry, and fish)
[a] Food and Agriculture Organization [69]
[b] Assuming that 5% of the ABP or plant products for animal feeds are used to produce protein hydrolysates
[c] The prices for peptone (a representative of animal protein hydrolysates), fermented soybean, and hydrolyzed wheat protein [70]

annually. Assuming that only 5% of the animal by-products and plant products for feed are used for protein hydrolysis, and based on the current average prices of animal, soybean, and wheat protein hydrolysates [70], their yields are 3, 6.75, and 12.75 billion kg/yr, respectively, and their economic values are 4.5, 3.88, and 20.02 billion US $/yr (Table 11.6). Thus, protein hydrolysates from the by-products of pigs or poultry and from plant ingredients hold great promise in sustaining animal agriculture and managing companion animals worldwide.

FUTURE RESEARCH DIRECTIONS

The nutritional value of protein hydrolysates as flavor enhancers, functional ingredients, and precursors for protein synthesis depends on the composition of free AAs, small peptides, and large peptides in the products, as well as their batch-to-batch consistency. At present, such data are not available for the commercially available products of animal or plant hydrolysates and should be obtained with the use of HPLC and mass spectrometry. Only when the composition of protein hydrolysates is known, can we fully understand their functionally active components and the mechanisms of their actions. In addition, the net rates of the transport of small peptides across the small intestine are not known for all the protein hydrolysates currently used in animal feeding. This issue can be readily addressed with the use of Ussing chambers [71]. There is also concern that some animal protein hydrolysates, which contain a high proportion of oligopeptides with a high abundance of basic AAs, have a low palatability for animals (particularly weanling piglets), and, therefore, the inclusion of the protein hydrolysates in animal feeds may be limited. Such a potential problem may be substantially alleviated through: (1) the addition of exopeptidases and a longer period of hydrolysis to remove basic and aliphatic AAs from the C- and N-terminals of the polypeptides; and (2) appropriate supplementation with glycine, monosodium glutamate, and inosine. Furthermore, the role of animal and plant protein hydrolysates in the signaling of intestinal epithelial cells and bacteria and metabolic regulation in these cells should be investigated to better understand how these beneficial products improve gut integrity, immunity, and health. Finally, the potential of protein hydrolysates as alternatives to dietary antibiotics should be explored along with studies to elucidate the underlying mechanisms. All these new lines of research will be particularly important for animals with compromised intestinal structure and function (e.g., neonates with intrauterine growth restriction and early-weaned mammals) and/or raised under adverse environmental conditions (e.g., high or low ambient temperatures).

CONCLUSION

Plant- and animal-protein hydrolysates provide highly digestible peptides and bioactive peptides, as well as specific AAs (e.g., glutamate) to confer nutritional and physiological or regulatory functions in animals. The industrial production of these protein hydrolysates involves: (1) strong acidic or alkaline conditions, (2) mild enzymatic methods, or (3) fermentation by microorganisms. The degree of hydrolysis is assessed by the number of peptide bonds cleaved divided by the total number of peptide bonds in a protein. Chemical hydrolysis is often employed to generate savory flavors, whereas microbial fermentation not only produces peptides but also removes antinutritional factors in protein ingredients. In addition to their nutritional value to supply AAs, bioactive peptides (usually 2–20 AA residues in length) have antimicrobial, antioxidant, antihypertensive, and immunomodulatory roles. These peptides exert beneficial effects on improving intestinal morphology, function, and resistance to infectious diseases in animals (including pigs, calves, chickens, companion animals, and fish), thereby enhancing their health and well-being, as well as growth performance and feed efficiency. This provides a cost-effective approach to converting animal by-products, brewer's by-products, or plant feedstuffs into high-quality protein-hydrolysate ingredients to feed livestock, poultry, fish, and companion animals.

ABBREVIATIONS

AA	Amino acid
ACE	Angiotensin-I converting enzyme
HPLC	High-performance liquid chromatography
PCA	Perchloric acid
PH	Protein hydrolysates
SDPI	Spray-dried porcine intestine hydrolysate
TCA	Trichloroacetic acid

ACKNOWLEDGMENTS

We thank our colleagues for collaboration on animal nutrition research.

FUNDING

Work in our laboratories was supported by the National Natural Science Foundation of China (31572416, 31372319, 31330075 and 31110103909), Hubei Provincial Key Project for Scientific and Technical Innovation (2014ABA022), Hubei Hundred Talent program, Natural Science Foundation of Hubei Province (2013CFA097), Agriculture and Food Research

Initiative Competitive Grants (2014-67015-21770 and 2015-67015-23276) from the USDA National Institute of Food and Agriculture, and Texas A&M AgriLife Research (H-8200).

AVAILABILITY OF DATA AND MATERIALS

Not applicable.

AUTHORS' CONTRIBUTIONS

GW conceived this project. YQH and GW wrote the manuscript. ZLW, ZLD, and GHW contributed to the discussion and revision of the article. GW had the primary responsibility for the content of the paper. All authors read and approved this manuscript.

COMPETING INTERESTS

None of the authors have any competing interests in the manuscript.

CONSENT FOR PUBLICATION

All authors read and approved the final manuscript.

ETHICS APPROVAL AND CONSENT TO PARTICIPATE

This article reviews published studies and does not require the approval of animal use or consent to participate.

REFERENCES

1. Wu G, Cross HR, Gehring KB, Savell JW, Arnold AN, McNeill SH. Composition of free and peptide-bound amino acids in beef chuck, loin, and round cuts. *J Anim Sci.* 2016;94:2603–13.
2. Wu G. *Amino acids: biochemistry and nutrition*. Boca Raton: CRC Press; 2013.
3. Pasupuleki VK, Braun S. State of the art manufacturing of protein hydrolysates. In: Pasupuleki VK, Demain AL, editors. *Protein hydrolysates in biotechnology*. New York: Springer Science; 2010. p. 11–32.
4. Dieterich F, Rogerio W, Bertoldo MT, da Silva VSN, Gonçalves GS, Vidotti RM. Development and characterization of protein hydrolysates originated from animal agro industrial by-products. *J Dairy Vet Anim Res.* 2014;1:00012.
5. Pasupuleki VK, Holmes C, Demain AL. Applications of protein hydrolysates in biotechnology. In: Pasupuleki VK, Demain AL, editors. *Protein hydrolysates in biotechnology*. New York: Springer Science; 2010. p. 1–9.
6. Kyte J. *Structure in protein chemistry*. 2nd ed. New York: Garland Science; 2006. p. 832.

7. Agerberth B, Söderling-Barros J, Jörnvall H, Chen ZW, Ostenson CG, Efendić S, et al. Isolation and characterization of a 60-residue intestinal peptide structurally related to the pancreatic secretory type of trypsin inhibitor: influence on insulin secretion. *Proc Natl Acad Sci U S A*. 1989;86:8590–4.
8. Chen ZW, Bergman T, Ostenson CG, Efendic S, Mutt V, Jörnvall H. Characterization of dopuin, a polypeptide with special residue distributions. *Eur J Biochem*. 1997;249:518–22.
9. Rajalingam D, Loftis C, Xu JJ, Kumar TKS. Trichloroacetic acid-induced protein precipitation involves the reversible association of a stable partially structured intermediate. *Protein Sci*. 2009;18:980–93.
10. Moughan PJ, Darragh AJ, Smith WC, Butts CA. Perchloric and trichloroacetic acids as precipitants of protein in endogenous ileal digesta from the rat. *J Sci Food Agric*. 1990;52:13–21.
11. Wilcockson J. The differential precipitation of nucleic acids and proteins from aqueous solutions by ethanol. *Anal Biochem*. 1975;66:64–8.
12. Dai ZL, Wu ZL, Jia SC, Wu G. Analysis of amino acid composition in proteins of animal tissues and foods as pre-column o-phthaldialdehyde derivatives by HPLC with fluorescence detection. *J Chromatogr B*. 2014;964:116–27.
13. Sapan CV, Lundblad RL. Review of methods for determination of total protein and peptide concentration in biological samples. *Proteomics Clin Appl*. 2015;9:268–76.
14. Larive CK, Lunte SM, Zhong M, Perkins MD, Wilson GS, Gokulrangan G, et al. Separation and analysis of peptides and proteins. *Anal Chem*. 1999;71:389R–423R.
15. McGrath R. Protein measurement by ninhydrin determination of amino acids released by alkaline hydrolysis. *Anal Biochem*. 1972;49:95–102.
16. Kunst T. Protein modification in optimize functionality: protein hydrolysates. In: Whitaker J, Voragen A, Wong D, editor. *Handbook of food enzymology*. New York: Marcel Dekker; 2003. p. 222–36.
17. Dixon MM, Webb EC. *Enzymes*. 3rd ed. New York: Academic; 1979.
18. Kim MR, Kawamura Y, Lee CH. Isolation and identification of bitter peptides of tryptic hydrolysate of soybean 11S glycinin by reverse-phase high-performance liquid chromatography. *J Food Sci*. 2003;68:2416–22.
19. Andriamihaja M, Guillot A, Svendsen A, Hagedorn J, Rakotondratohanina S, Tome' D, et al. Comparative efficiency of microbial enzyme preparations versus pancreatin for *in vitro* alimentary protein digestion. *Amino Acids*. 2013;44:563–72.
20. Layer P, Keller J. Lipase supplementation therapy: standards, alternatives, and perspectives. *Pancreas*. 2003;26:1–7.
21. Sikkens EC, Cahen DL, Kuipers EJ, Bruno MJ. Pancreatic enzyme replacement therapy in chronic pancreatitis. *Best Pract Res Clin Gastroenterol*. 2010;24:337–47.
22. Smid EJ, Lacroix C. Microbe-microbe interactions in mixed culture food fermentations. *Curr Opin Biotechnol*. 2013;24:148–54.
23. Bah CS, Carne A, McConnell MA, Mros S, Bekhit A-D. Production of bioactive peptide hydrolysates from deer, sheep, pig and cattle red blood cell fractions using plant and fungal protease preparations. *Food Chem*. 2016;202:458–66.
24. Li-Chan ECY. Bioactive peptides and protein hydrolysates: research trends and challenges for application as nutraceuticals and functional food ingredients. *Curr Opin Food Sci*. 2015;1:28–37.

25. López-Barrios L, Gutiérrez-Uribe JA, Serna-Saldívar SO. Bioactive peptides and hydrolysates from pulses and their potential use as functional ingredients. *J Food Sci.* 2014;79:R273–83.
26. Kairane C, Zilmer M, Mutt V, Sillard R. Activation of Na, K-ATPase by an endogenous peptide, PEC-60. *FEBS Lett.* 1994;345:1–4.
27. Bevins CL, Salzman NH. Paneth cells, antimicrobial peptides and maintenance of intestinal homeostasis. *Nature Rev Microbiol.* 2011;9:356–68.
28. Engel JA, Jerlhag E. Role of appetite-regulating peptides in the pathophysiology of addiction: implications for pharmacotherapy. *CNS Drugs.* 2014;28:875–86.
29. Zhanghi BM, Matthews JC. Physiological importance and mechanisms of protein hydrolysate absorption. In: Pasupuleki VK, Demain AL, editors. *Protein hydrolysates in biotechnology.* New York: Springer Science; 2010. p. 135–77.
30. Gardner ML. Absorption of intact peptides: studies on transport of protein digests and dipeptides across rat small intestine *in vitro*. *Q J Exp Physiol.* 1982;67:629–37.
31. Gardner ML, Wood D. Transport of peptides across the gastrointestinal tract. *Biochem Soc Trans.* 1989;17:934–7.
32. Mellander O. The physiological importance of the casein phosphopeptide calcium salts. II. Peroral calcium dosage of infants. *Acta Soc Med Ups.* 1950; 55:247–55.
33. Ryan JT, Ross RP, Bolton D, Fitzgerald GF, Stanton C. Bioactive peptides from muscle sources: meat and fish. *Nutrients.* 2011;3:765–91.
34. Power O, Jakeman P, FitzGerald RJ. Antioxidative peptides: enzymatic production, in4vitro and *in vivo* antioxidant activity and potential applications of milk-derived antioxidative peptides. *Amino Acids.* 2013;44:797–820.
35. Martínez-Augustin O, Rivero-Gutiérrez B, Mascaraque C, de Medina FS. Food-derived bioactive peptides and intestinal barrier function. *Int J Mol Sci.* 2014;15:22857–73.
36. Ryder K, Ael-D B, McConnell M, Carne A. Towards generation of bioactive peptides from meat industry waste proteins: generation of peptides using commercial microbial proteases. *Food Chem.* 2016;208:42–50.
37. Zambrowicz A, Pokora M, Setner B, Dąbrowska A, Szołtysik M, Babij K, et al. Multifunctional peptides derived from an egg yolk protein hydrolysate: isolation and characterization. *Amino Acids.* 2015;47:369–80.
38. Shimizu M, Son DO. Food-derived peptides and intestinal functions. *Curr Pharm Des.* 2007;13:885–95.
39. Bah CS, Bekhit A-D, McConnell MA, Carne A. Generation of bioactive peptide hydrolysates from cattle plasma using plant and fungal proteases. *Food Chem.* 2016;213:98–107.
40. Memarpoor-Yazdia M, Asoodehb A, Chamania JK. A novel antioxidant and antimicrobial peptide from hen egg white lysozyme hydrolysates. *J Funct Foods.* 2012;4:278–86.
41. Power O, Jakeman P, FitzGerald RJ. Antioxidative peptides: enzymatic production, *in vitro* and *in vivo* antioxidant activity and potential applications of milk-derived antioxidative peptides. *Amino Acids.* 2013;44:797–820.
42. Lima CA, Campos JF, Filho JLM, Converti A, da Cunha MGC, Porto ALF. Antimicrobial and radical scavenging properties of bovine collagen hydrolysates produced by Penicillium aurantiogriseum URM 4622 collagenase. *J Food Sci Technol.* 2015;52:4459–66.
43. Osman A, Goda HA, Abdel-Hamid M, Badran SM, Otte J. Antibacterial peptides generated by alcalse hydrolysis of goat whey. *LWT-Food Sci Technol.* 2016;65:480–86.

44. Wald M, Schwarz K, Rehbein H, Bußmann B, Beermann C. Detection of antibacterial activity of an enzymatic hydrolysate generated by processing rainbow trout by-products with trout pepsin. *Food Chem*. 2016;205:221–28.
45. Froetschel MA. Bioactive peptides in digesta that regulate gastrointestinal function and intake. *J Anim Sci*. 1996;74:2500–8.
46. San Gabriel A, Uneyama H. Amino acid sensing in the gastrointestinal tract. *Amino Acids*. 2013;45:451–61.
47. Fernstrom JD. Large neutral amino acids: dietary effects on brain neurochemistry and function. *Amino Acids*. 2013;45:419–30.
48. Wu G, Fanzo J, Miller DD, Pingali P, Post M, Steiner JJ, et al. Production and supply of high-quality food protein for human consumption: sustainability, challenges and innovations. *Ann NY Acad Sci*. 2014;1321:1–19.
49. Wu G, Bazer FW, Cross HR. Land-based production of animal protein: impacts, efficiency, and sustainability. *Ann NY Acad Sci*. 2014;1328:18–28.
50. McCalla J, Waugh T, Lohry E. Protein hydrolysates/peptides in animal nutrition. In: Pasupuleki VK, Demain AL, editors. *Protein hydrolysates in biotechnology*. New York: Springer Science; 2010. p. 179–90.
51. Hou YQ, Yin YL, Wu G. Dietary essentiality of "nutritionally nonessential amino acids" for animals and humans. *Exp Biol Med*. 2015;240:997–1007.
52. Hou YQ, Yao K, Yin YL, Wu G. Endogenous synthesis of amino acids limits growth, lactation and reproduction of animals. *Adv Nutr*. 2016;7:331–42.
53. Lindemann MD, Cromwell GL, Monegue HJ, Cook H, Soltwedel KT, Thomas S, et al. Feeding value of an enzymatically digested protein for early-weaned pigs. *J Anim Sci*. 2000;78:318–27.
54. Kim SW, van Heugten E, Ji F, Lee CH, Mateo RD. Fermented soybean meal as a vegetable protein source for nursery pigs: I. Effects on growth performance of nursery pigs. *J Anim Sci*. 2010;88:214–24.
55. Zimmerman D. Interaction of intestinal hydrolysate and spray-dried plasma fed to weanling pigs, Iowa State University, Ames, IA, *Experiment* 9615, 1996.
56. Zimmerman D. The duration of carryover growth response to intestinal hydrolysate fed to weanling pigs, Iowa State University, Ames, IA, *Experiment* 9612, 1996.
57. Kim JH, Chae BJ, Kim YG. Effects of replacing spray dried plasma protein with spray dried porcine intestine hydrolysate on ileal digestibility of amino acids and growth performance in early-weaned pigs. *Asian-Aust J Anim Sci*. 2000;13:1738–42.
58. Stein H. *The effect of including DPS 50RD and DPS EX in the phase 2 diets for weanling pigs*. Brookings: South Dakota State University; 2002.
59. Lalles JP, Toullec R, Pardal PB, Sissons JW. Hydrolyzed soy protein isolate sustains high nutritional performance in veal calves. *J Dairy Sci*. 1995;78:194–204.
60. Opheim M, Sterten H, Øverland M, Kjos NP. Atlantic salmon (*Salmo salar*) protein hydrolysate – effect on growth performance and intestinal morphometry in broiler chickens. *Livest Sci*. 2016;187:138–45.
61. Frikha M, Mohiti-Asli M, Chetrit C, Mateos GG. Hydrolyzed porcine mucosa in broiler diets: effects on growth performance, nutrient retention, and histomorphology of the small intestine. *Poult Sci*. 2014;93:400–11.
62. Refstie S, Sahlström S, Bråthen E, Baeverfjord G, Krogedal P. Lactic acid fermentation eliminates indigestible carbohydrates and antinutritional factors in soybean meal for Atlantic salmon (*Salmo salar*). *Aquaculture*. 2005;246:331–45.

63. Khosravi S, Rahimnejad S, Herault M, Fournier V, Lee CR, Dio Bui HT, et al. Effects of protein hydrolysates supplementation in low fish meal diets on growth performance, innate immunity and disease resistance of red sea bream Pagrus major. *Fish Shellfish Immunol.* 2015;45:858–68.
64. Nagodawithana TW, Nelles L, Trivedi NB. Protein hydrolysates as hypoallergenic, flavors and palatants for companion animals. In: Pasupuleki VK, Demain AL, editors. *Protein hydrolysates in biotechnology.* New York: Springer Science; 2010. p. 191–207.
65. Kotzamanis YP, Gisbert E, Gatesoupe FJ, Zambonino Infante J, Cahu C. Effects of different dietary levels of fish protein hydrolysates on growth, digestive enzymes, gut microbiota, and resistance to Vibrio anguillarum in European sea bass (*Dicentrarchus labrax*) larvae. *Comp Biochem Physiol A.* 2007;147:205–14.
66. Martínez-Alvarez O, Chamorro S, Brenes A. Protein hydrolysates from animal processing by-products as a source of bioactive molecules with interest in animal feeding: a review. *Food Res Int.* 2015. doi: 10.1016/j.foodres.2015.04.005.
67. Ghosh PR, Fawcett D, Sharma SB, Poinern DEJ. Progress towards sustainable utilisation and management of food wastes in the global economy. *Int J Food Sci.* 2016, Article ID 3563478.
68. Irshad A, Sureshkumar S, Shalima Shukoor A, Sutha M. Slaughter house by-product utilization for sustainable meat industry-a review. *Int J Res Dev.* 2015;5:4725–734.
69. Food and Agriculture Organization [67]. http://www.fao.org/worldfoodsituation/csdb/en/. Accessed on 8 December 2016.
70. Animal feed prices. https://www.alibaba.com/product. Accessed on 8 December 2016.
71. Wang WW, Dai ZL, Wu ZL, Lin G, Jia SC, Hu SD, et al. Glycine is a nutritionally essential amino acid for maximal growth of milk-fed young pigs. *Amino Acids.* 2014;46:2037–45.

CHAPTER 12

Fermentation Process
The Factory of Bioactive Peptides

Belal J. Muhialdin, Mohammad Zarei, Kambiz Hassanzadeh, Chay Shyan Yea, and Raman Ahmadi

CONTENTS

Introduction	233
Fermentation and Fermented Foods	234
Peptides from Plant-Based Fermented Foods	236
Peptides from Animal-Based Fermented Foods	237
Starter Culture Role in the Peptides Production during Fermentation	240
Biological Activity of Peptides in Fermented Foods	240
Antimicrobial Activity	240
Antioxidant Activity	242
Antihypertensive Activity	243
Immunomodulation Activity	243
Applications of Peptides in Food Safety	244
Future Research and Conclusions	245
References	246

INTRODUCTION

Peptides with biological functions are known as "bioactive peptides" and defined as "peptide sequences within a protein that exert a beneficial effect on body functions and/or positively impact human health, beyond its known nutritional value" (Kitts and Weiler, 2003). Bioactive peptides consist of small numbers of amino acids (AAs) usually 2–20 AAs and are encoded in the primary structures of animal and plant proteins (Yao et al., 2014). They are released from the parent proteins during food processes such as acidification, fermentation, heating, and mechanical treating (Aluko, 2018). Several recent studies demonstrated the potential health benefits of bioactive peptides found in fermented foods due to their high acid stability that facilitates the passage through gastrointestinal digestion (Marco et al., 2017). On the other hand, bioactive peptides play an important role in extending the shelf life of fermented foods due to a broad range of antimicrobial activity (Muhialdin et al., 2020a).

Fermented foods are well known to contain great numbers of bioactive peptides especially those that have a high protein content in their raw materials such as milk, meat, fish, beans, and cereals (Xiang et al., 2019). The proteins are hydrolyzed by the indigenous enzymes and the enzymes of natural microflora of bacteria, yeast, and fungi (Matthews et al., 2004). With respect to the natural microflora, lactic acid bacteria (LAB) are the most dominant bacteria in the majority of fermented foods. Unlimited number of LABs have been historically used as starter cultures for the preparation of several fermented foods such as yogurt, cheese, kefir, salami, and kimchi (Tamang et al., 2016). They can improve the sensory properties of the fermented foods compared to their raw materials and delay the growth of spoilage and pathogenic microorganisms, thereby extending their shelf life (Karovičová and Kohajdova, 2007). The proteolytic enzyme system is very broad with several enzymes that can hydrolyze proteins. Proteins are polymers made of several AAs that have different molecular weights and functions (Thordarson et al., 2006). Therefore, the biological functions of the bioactive peptides highly depend on the AAs combination (Agyei et al., 2016).

This chapter shed light on the bioactive peptides present in animal-based and plant-based fermented foods. The process of producing fermented foods is described and the role of the starter cultures in the generating of peptides will be discussed. This chapter also addresses the currently known health benefits of the bioactive peptides in fermented foods based on *in vitro* studies and *in vivo* trails.

FERMENTATION AND FERMENTED FOODS

Fermentation is defined as "the process that applies starter cultures (microorganisms) for the transformation of raw materials into a palatable diet with stable shelf life" (Shiferaw Terefe and Augustin, 2020). Microbial fermentation is the most economical process for the generation of bioactive peptides. During the fermentation process, several chemical and physical changes take place due to the production of proteolytic enzymes and metabolites from the starter cultures (du Toit et al., 2011). The changes start with a significant drop of the pH value after a few hours as the organic acids are produced by the starter cultures to reach a pH between 3 and 4.5, depending on the fermented materials and the starter culture (Crafack et al., 2013). LABs are the most common microorganisms that are found in the majority of fermented foods (Tamang et al., 2016). Certain LAB strains belonging to the genera *Lactobacillus*, *Lactococcus*, and *Leuconostoc* have the status of Generally Recognized as Safe (GRAS). They are well known to produce organic acids including lactic acid, which is responsible for the pH drop and the pleasant taste of fermented foods. Moreover, they produce several antimicrobial compounds such as organic acids, alcohols, carbon dioxide, hydrogen peroxide, phenyllactic acid, bacteriocins, and peptides (Muhialdin et al., 2020b). Fermentation is classified by three major processes: spontaneous fermentation, which is the natural microflora of the raw materials; back-slopping, which uses a portion of a previous successful fermentation; and controlled fermentation, using a well-defined single or combination starter culture (Li and Gänzle, 2020). Back-slopping is the main method for producing traditional fermented foods to meet the consumer's acceptability because of the rich aromatic profile resulting from the dominant microflora from several previous successful fermentations. However, a controlled fermentation process is highly recommended to ensure the consistency of the product and it is biological function (Korhonen and Pihlanto, 2006). The fermentation technique plays a very important role in the biological functions of the generated peptides (Figure 12.1).

Great numbers of traditional fermented foods are consumed globally. The numerous varieties of fermented foods and beverages are related to several factors, such as the domestic

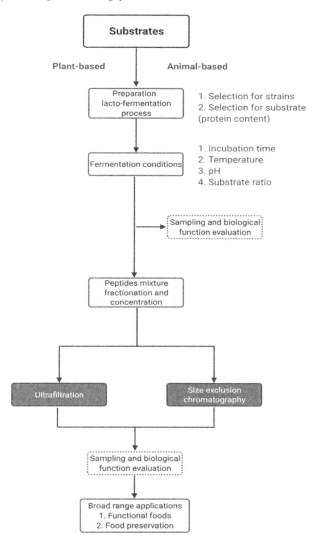

FIGURE 12.1 The steps to produce bioactive peptides via controlled lacto-fermentation.

acceptability, availability of raw materials, climate and cultivation, socioeconomic conditions, and techniques of fermentation (Shiferaw Terefe and Augustin, 2020). The fermented foods and beverages are divided into two major groups – animal-based (yogurt, cheese, kefir, sausage, salami) and plant-based (sauerkraut, kimchi, tempeh, natto, soy sauce, sourdough, tapai, kombucha, vinegar) (Tamang et al., 2020). Some of these fermented foods may contain probiotic microorganisms that have important health benefits for the consumer. The probiotic strains belong to yeasts or bacteria, but the most important groups belong to the LAB's. The species *Lactobacillus* is the most dominant in a broad range of fermented foods (Dewan and Tamang, 2007). Probiotics have been extensively studied for their health benefits through the modulation of gut microbiota (Chen et al., 2019; Arora et al., 2013). They contribute significantly to the health benefits of fermented foods. In addition, probiotics produce several bioactive compounds (postbiotic), such as free fatty acids, amino acids, and bioactive peptides (Chugh and Kamal-Eldin, 2020). Fermented foods have a high potential to provide health benefits to the consumer. For example, the antioxidant activity of fermented berries was higher by 30% compared to the non-fermented berries as a result of the release of the phenolic compounds from the berry tissue after fermentation (Curiel et al., 2015). Kuwaki et al. (2012) reported increased amounts of proteins, carbohydrates, and lipids, 18 types of AAs, and various vitamins including vitamin A, B1, B2, B6, B12, E, K, niacin, biotin, pantothenic acid, folic acid, five organic acids, and a large amount of dietary fiber and plant phytochemicals in fermented vegetable paste. In another study, lacto-fermented jackfruit juice demonstrated strong antimicrobial activity toward *Escherichia coli* 0157:H7, *Salmonella enterica* serovar Typhimurium, and *Staphylococcus aureus* due to the high content of bioactive compounds (Muhialdin et al., 2021a). Recently, it was proposed that fermented foods and their probiotics be included in the diet for their role in reducing the severity and mortality caused by Covid-19 infections by enhancing the host immune system function (Muhialdin et al., 2021b; Bousquet et al., 2020; Fonseca et al., 2020). A comprehensive description of traditional fermented foods has been profiled in previous published studies (Tamang et al., 2020; Soni and Dey, 2014).

PEPTIDES FROM PLANT-BASED FERMENTED FOODS

Several peptides have been isolated from plant-based foods including sourdough (Coda et al., 2012), fermented soy milk (Capriotti et al., 2015), chunghookjang (fermented soybean) (Yang et al., 2013), fermented buckwheat sprouts (Koyama et al., 2013), and fermented kenaf seeds (Arulrajah et al., 2020). These peptides were generated by digesting the parent proteins by the microorganisms applied for the fermentation. The microorganisms secrete their proteolytic enzymes and hydrolyze the proteins in the substrates (López-Otín and Bond, 2008). The bioactive peptides via the fermentation process had low molecular weight, cationic charge, and higher hydrophobicity ratio (Muhialdin et al., 2020a). The peptides can be isolated and purified using several methods, such as ultrafiltration, size exclusion chromatography, and preparative HPLC. The selection of the suitable plant-based substrate for peptides production will depend on its protein content. For example, beans such as soybeans, chickpeas, lentil, and cocoa beans are among the most studies peptides sources due to their high protein content (Domínguez-Pérez et al., 2020; Görgüç et al., 2020; Singh et al., 2014). On the other hand, agricultural by-products and non-edible seeds are good sources of proteins that can be converted into bioactive peptides with a broad range of biological functions via the fermentation process. Arulrajah et al. (2020) generated antioxidant peptides from the non-edible kenaf seed using a lacto-fermentation process. In another study, antioxidant peptides were generated using the by-products of palm oil (palm kernel cake) through solid-state fermentation. Plant-based substrates have two major kinds of proteins – storage proteins and structural proteins (Rahman and Lamsal, 2021). Therefore, the extraction of proteins from plants involves several stages and techniques. Moreover, plant-based proteins have a limitation in food applications because of their high content of allergen compounds (Breiteneder and Radauer, 2004). Verma et al. (2013) rated the risk of allergic reactions from bean consumption in order from the highest to lowest as the following: peanut > soybean > lentil > chickpea > pea > mung bean. Nevertheless, one advantage of the fermentation process is that it breaks down allergenic compounds in the plant thereby reducing the risk of inflammation (Yang et al., 2018). Therefore, the fermentation process is a great solution to generate low-molecular-weight peptides that have biological functions and are safe for food applications. A great number of studies have reported on the different biological functions associated with peptides generated in plant-based substrates via fermentation (Table 12.1).

PEPTIDES FROM ANIMAL-BASED FERMENTED FOODS

The bioactive peptides derived from animal-based substrates have been extensively studied for their structures and biological functions (Liu et al., 2020; Lee and Hur, 2017; Bhat et al., 2015) (Table 12.2). They are inactive in the parent proteins, thus they can be released and demonstrate biological functions after enzymatic or microbial digestion. The peptides derived from milk are the most abundant among other animal-based substrates due to the high global production of milk including cow (85%), buffalo (11%), goat (2.3%), sheep (1.4%), and camel milk (0.2%) (Vargas-Bello-Pérez et al., 2019). The majority of milk bioactive peptides are generated through fermentation to produce dairy products that are labeled as functional foods because of their potential health benefits. Several milk peptides show antihypertensive activity (Ahtesh et al., 2018), anti-inflammatory activity (Yvon et al., 2018), antioxidant activity (Ayyash et al., 2018), antimicrobial activity (Algboory and Muhialdin, 2021), obesity control (Bischoff et al., 2017), cardiovascular diseases control (Mendis, 2017), and immune system function (Aslam et al., 2020). Therefore, the consumption of fermented milk products is recommended to improve the metabolic health of the host (McGregor and Poppitt, 2013).

Eggs are also an important source of bioactive peptides with global production of 1,387 billion per year (Morris et al., 2018). Fan et al. (2019) studied peptides from egg white hydrolysate for its capacity to lower blood pressure in lab rats because of ACE inhibition activity. There are a few studies that demonstrate the improvement of biological activity of lacto-fermented compared to unfermented egg white (Nahariah et al., 2019; Matsuoka et al., 2017a). Fermented egg white protein that is rich in bioactive peptides showed a significant reduction in visceral fat and high potential to reduce the risk of obesity (Matsuoka et al., 2017a). In another study, lacto-fermented egg white significantly reduced serum cholesterol in adult subjects, suggesting a functional food that can reduce the risk of cardiovascular diseases (Matsuoka et al., 2017b). The peptides generated from egg white were recently reviewed for their anticancer activity and immunomodulation effects as well (Lee and Paik, 2019). Traditional fermented fish products including *Suanyu* contain peptide fractions with molecular weight of 0.18–0.5 kDa generated via the hydrolysis of fish proteins (Wang et al., 2017). Fermented fish peptides, including LDDPVFIH and VAAGRTDAGVH, demonstrated antioxidant activity and potential applications in functional foods (Najafian and Babji, 2019). The bioactive peptides isolated from different fermented sausages demonstrated a broad range of biological activities, such as antioxidant and ACE inhibitory activities (Gallego et al., 2018). Zhang et al. (2020) reported on the biological activities of peptides isolated from fermented meat and their role in preventing DNA damage in the host because of their antioxidant activity. Fermented meat products are traditional foods prepared via spontaneous fermentation, which leads to the hydrolysis of proteins by microbial enzymes and the generation of low-molecular-weight peptides. The generated peptides demonstrated biological functions beyond their nutritional role (Albenzio et al., 2017). There are many animal-based substrates that have been reported to contain bioactive peptides including fermented shrimp paste (Kleekayai et al., 2015), fermented blue mussel sauce (Rajapakse et al., 2005), and sea cucumber (Fuad et al., 2020). The starter cultures or the natural microflora are responsible for the

TABLE 12.1 Bioactive Peptides with Biological Functions Isolated from Plant-Based Fermented Substrates

Peptides	Source	Biological Activity	Testing Method	Reference
WLSYPMNPATGH, PRPPKPDAPR	Fermented rice cake	Antioxidant activity	In vitro, in silico	Wang et al. (2021)
AKVGLKPGGFFVLK, GSTIK, HGDRPR, TAHDDYK, LLLSK	Fermented kenaf seeds	Antimicrobial activity	In vitro	Arulrajah et al. (2020)
EAKPSFYLK, PVNNNAWAYATNFVPGK, AIGIFVKPDTAV	Fermented bitter beans	Antimicrobial activity	In vitro	Muhialdin et al. (2020c)
10 cationic peptides	Fermented palm kernel cake	Antifungal activity	In vitro	Asri et al. (2020)
INEGSLLLPH, SGGGGGVAGAATASR, GSGGGGGGFGGPRR, GGYQGGGYGGNSGGGY GNRG, GGSGGGGGSSSGRRP, and GDTVTVEFDTFLSR	Fermented beans	ACE inhibitor	In vitro	Jakubczyk et al. (2017)
HTSKALLDMLKRLGK	Fermented soybean meal	Antimicrobial activity	In vivo	Cheng et al. (2017)
SMATPHVAGAAALILS, KHPTWTNAQVRD, RLESTATYLGNSFYYGK	Fermented soybean (Natto)	Antimicrobial activity	In vitro	Kitagawa et al. (2017)
THPLPV, YVVFK	Fermented soy milk	ACE inhibitor	In silico	Capriotti et al. (2015)
LLPHHADADY	Fermented soy milk	Antioxidant activity	In silico	Capriotti et al. (2015)
KEDDEEEQGEEE	Fermented peas	ACE inhibitory	In vitro	Jakubczyk et al. (2013)
DVWY, FQ, VVG	Fermented buckwheat sprouts	Antihypertensive	In vitro	Koyama et al. (2013)
VFHAYSARGNYY, GNCPANWPSC, RNNYKSAGGK	Fermented vegetables	Antimicrobial activity	In vitro	Hu et al. (2013)
RGDDDDDDDD	Fermented sourdough	Immunomodulation activity	In vitro	Rizzello et al. (2012)
Unidentified peptides	Fermented mushroom	Antioxidant activity	In vitro	Sun et al. (2004)

TABLE 12.2 Bioactive Peptides with Biological Functions Isolated from Animal-Based Fermented Substrates

Substrate	Peptides	Source	Biological Activity	Reference
Milk	30 novel peptides	Fermented camel milk	Antimicrobial	Algboory & Muhialdin (2021)
	35 peptides	Fermented milk (kefir)	ACE inhibitory	Amorim et al. (2019)
	QEPVL, QEPV	Fermented milk	Immunomodulation activity	Jiehui et al. (2014)
	50 peptides	Fermented milk (yogurt)	ACE inhibitory	Kunda et al. (2012)
	RPKHPIKHQGLPQEV RPKHPIKHQGLPQEVLN ENLLR EVLNENLLRF FVAPFPEVFGK YQEPVLGPVRGPF YQEPVLGPVRGPFPI YQEPVLGPVRGPFIIV	Fermented milk (Fresco cheese)	ACE inhibitory	Torres-Llanez et al. (2011)
	LHLPLP	Fermented milk	Antihypertensive	Quirós et al. (2007)
	RPKHPIKHQ RPKHPIKHQGLPQ YPFPGPIPN MPFPKYPVQPF	Fermented milk (Gouda cheese)	ACE inhibitory	Saito et al. (2000)
Egg	Unidentified	Fermented egg white	Reduce visceral fat	Matsuoka et al. (2019)
	LAPYK, LKISQ, LKYAT, INKVVR, LFLIKH, and LGHWVY	Fermented egg white	ACE inhibitory	Fan et al. (2019)
	Unidentified	Fermented egg white	Reduce obesity	Matsuoka et al. (2017a)
	Unidentified	Fermented egg white	Reduce cholesterol	Matsuoka et al. (2017b)
Muscles	LDDPVFIH, VAAGRTDAGVH	Fermented fish	Antioxidant	Najafian & Babji (2019)
	AIPPHPYP, IAEVFLITDPK	Fermented fish	Antioxidant	Najafian & Babji (2018)
	SV, IF	Fermented shrimp paste	ACE inhibitory	Kleekayai et al. (2015)
	WP	Fermented shrimp paste	Antioxidant	Kleekayai et al. (2015)
	LKPNM	Fermented tuna	ACE inhibitory	Ryan et al. (2011)
	HFGBPFH	Fermented blue mussel sauce	Antioxidant	Rajapakse et al. (2005)
	EVMAGNLYPG	Fermented blue mussel sauce	ACE inhibitory	Je et al. (2005)

generation of these bioactive peptides due to the high acidity and the hydrolysis enzymes that help to digest the parent proteins into low-molecular-weight peptides.

STARTER CULTURE ROLE IN THE PEPTIDES PRODUCTION DURING FERMENTATION

Starter culture contains the microorganisms used to start the fermentation process (Rakhmanova et al., 2018). They play an important role in the development of texture and flavor of fermented foods due to chemical and physical changes. These microorganisms have a very complicated proteolytic system that contain several enzymes able to digest proteins among other macronutrients. With respect to all the starter cultures, LAB strains are the major species applied in more than 85% of the fermented foods and industrial fermentation (Brown et al., 2017). LAB strains have preferred characteristics including diversity, genetic stability, and production of aromatic compounds, and a great number of LAB strains have the status of Generally Regarded as Safe (GRAS) (Sacchini et al., 2017). There are approximately 150 LAB species known, but the main species applied in the food industry are *Lactobacillus*, *Lactococcus*, *Leuconostoc*, *Streptococcus*, and *Weissella* (Pasolli et al., 2020). LAB strains are found in most fermented foods and they can be easily isolated using De Man, Rogosa, and Sharpe (MRS) agar (Onggo and Fleet, 1993). The advantage of LAB strains is their ability to produce bioactive compounds such as organic acids, fatty acids, and bacteriocins and prevent the growth of unwanted microorganisms during fermentation (Muhialdin et al., 2020a). The American Type Culture Collection (ATCC) has hundreds of well profiled LAB strains that have been applied in industrial fermentation processes (van Beverwijk, 2019). However, the wild LAB strains offer a more complicated proteolytic system due to their environment and the competition with other microorganisms (Fadda et al., 2010). The proteolytic enzymes function to reduce the complexity of proteins in the substrates and generate simple low-molecular-weight peptides to support the growth of LAB cells (Daliri et al., 2017). The enzymes responsible for the hydrolysis of proteins belong to cell envelope-associated proteinases (CEPs) which are extracellular enzymes of large molecular weight (200 KDa) (Venegas-Ortega et al., 2019). The secretion of the CEPs is highly dependent on the fermentation conditions, such as pH, temperature, complexity of the substrates, and type and availability of the nitrogen (Vukotic et al., 2015). The diversity of the LAB proteolytic enzymes, complexity of the substrates, and the different fermentation conditions can explain the variety of bioactive peptides generated via lacto-fermentation. It is worth observing that enzyme production is influenced by LAB during the growth phase, the enzymatic activity is at its peak during the exponential phase and gradually decreases during the stationary phase (Williams et al., 2002). Therefore, the fermentation conditions and duration should be optimized for every strain to maximize the production of bioactive peptides for applications in the food industry.

BIOLOGICAL ACTIVITY OF PEPTIDES IN FERMENTED FOODS

Antimicrobial Activity

Antimicrobial activity is the most extensively studied biological function for bioactive peptides. A great number of studies have reported the significant effects of bioactive peptides on the growth inhibition of different pathogens, including bacteria (Asri et al., 2020; Aguilar-Toalá et al., 2017), yeasts (Jang et al., 2019; Bulgasem et al., 2016), fungi (Muhialdin 2020b; Slavokhotova et al., 2014), and viruses (Wang et al., 2010). The interest in bioactive peptides to control the growth of pathogens is due to their high potential as natural alternatives to antibiotics and their broad range of activity. The mechanisms of action of bioactive peptides significantly vary from one pathogen to another because of the differences in their cell wall components. The antibacterial activity has a well-established mechanism of action against Gram-positive and Gram-negative bacteria. Unlike antibiotics, antimicrobial peptides interact with the cell membrane of bacteria via electrostatic connection, and therefore it is very difficult for bacteria to develop resistance (Pfalzgraff et al., 2018). The majority of cationic peptides can cause pores in the targeted bacteria, which leads to leakage of the cytoplasmic content (Hale and Hancock, 2007). In addition, cationic peptides are attracted to the cell membrane (negative charge) and can disturb the cell functions (Huang et al., 2010). Fermented foods are rich sources for antimicrobial peptides. The antimicrobial peptides generated in dairy products, including yogurt and cheese, are very well known for their biological functions. Algboory and Muhialdin (2021) reported on the antibacterial activity of 30 novel peptides identified in fermented camel milk and their broad range of growth inhibition. The identified peptides were generated by the nonconventional starter culture and showed no match to the database. Moreno-Montoro et al. (2017) reported on the biological functions of fermented goat milk including the antimicrobial activity. In another study, a total of 46 antimicrobial peptides were identified in fermented sheep milk (de Lima et al., 2018). Cow milk, which is the most abundant, demonstrated the highest antimicrobial activity, and the fermented milk contained a high number of bioactive peptides (Aguilar-Toalá et al., 2017). In addition, bioactive peptides with strong antimicrobial activity were reported in several varieties of cheese (Atanasova et al., 2020; da Silva et al., 2019; Fialho et al., 2018). LABs are the natural microflora of milk from all animals and their proteolytic enzymes are well established to hydrolyze milk proteins and generate high numbers of bioactive peptides with different biological activities.

Plant-based fermented foods are the least studied for their antimicrobial peptides. Recently, the high demand for plant-based foods has led researchers to investigate the potential biological functions of the peptides isolated from fermented plant-based foods. The most studied plant-based antimicrobial peptides are those derived from sourdough. Several studies reported the presence of antimicrobial peptides in sourdough and their inhibition activity toward pathogenic bacteria and spoilage fungi (Luz et al., 2019; Gänzle, 2014; Rizzello et al., 2011). The peptides are mainly derived from the gluten which is the main protein of wheat (Ooms and Delcour, 2019). Fermented soybean products demonstrated a strong antimicrobial activity due to their bioactive peptides in foods such as miso (Shirako et al., 2020), natto (Taniguchi et al., 2019), tempeh (Ito et al., 2020), and fermented soy milk (Sanjukta, and Rai, 2016). Fermented vegetables are not well studied for the potential presence of antimicrobial peptides because of the assumption of low protein content. However, great numbers of LAB strains isolated from fermented vegetables such as kimchi, sauerkraut, and mixed pickles were associated with the production of antimicrobial peptides (Venegas-Ortega et al., 2019; Muhialdin et al., 2018; Hu et al., 2013). Agricultural industry by-products were used as substrates to generate antimicrobial peptides via fermentation especially PKC (protein kinase C) (Asri et al., 2020). Non-edible seeds are known for their high protein content such as foxtail millet, which showed strong antimicrobial activity and high peptide content after lacto-fermentation with *L. paracasei* Fn032 (Amadou et al., 2013). Emerging plant-based materials with high protein content have high potential as substrates for fermentation and production of antimicrobial peptides including mushrooms (Zhou et al., 2020), microalgae (Stack et al., 2020), and seaweeds (Chye et al., 2018). The preliminary studies show good antimicrobial activity for the peptides generated in these substrates with high potential in the food and pharmaceutical industries.

Antioxidant Activity

The natural metabolism activities and the external environmental stress can result in developing oxidants, such as ROS and free radicals, and they can react with other molecules including proteins and lipids, resulting in serious damage and inflammation (Khansari et al., 2009). The human body has a well-established defense mechanism against these oxidants – our immune system – that depends on the diet of the individual. Thus a diet that includes natural phytochemicals and peptides to boost cleansing performance is recommended. Natural antioxidants offer a safe alternative to synthetic antioxidants such as butylated hydroxyanisole (BHA), butylated hydroxytoluene (BHT) and propyl gallate. Several researchers have studied the potential antioxidant activity of bioactive peptides generated via fermentation (Muhialdin et al., 2020c; Coda et al., 2012). Milk peptides were among the most studied for their antioxidant activity as their parent proteins, including casein and whey proteins, were fully sequenced. The bioactive peptides are encrypted within the parent proteins sequences and released after enzymatic or microbial digestion (Power et al., 2013). Fermented dairy products, including yogurt, demonstrated strong antioxidant activity due to the presence of large numbers of peptides generated by the proteolytic activity of the starter cultures (Farvin et al., 2010). In another study, bovine and goat milk samples were subjected to proteolytic digestion using fungal enzymes, and the generated bioactive peptides showed significant antioxidant activity compared to the undigested milk samples (Zanutto-Elgui et al., 2019). These studies highlighted the role for proteolytic enzymes secreted by the starter cultures to generate low-molecular-weight peptides with high antioxidant activity. Another animal-based source for antioxidant peptides is egg white protein (Rao et al., 2012). Nimalaratne et al. (2015) reported on the enzymatic hydrolysis of egg white and the generation of 16 antioxidant peptides, including AEERYP and DEDTQAMP, which showed the highest activity. In another study, egg white was subjected to heat treatment and enzymatic digestion and resulted in the generating of low-molecular-weight peptides ($\leq$ 1KDa) with strong antioxidant activity (Wang et al., 2018). Fermented meat products were also reported for the presence of antioxidant peptides. Ohata et al. (2016) studied the bioactive peptides in fermented meat sauce and identified low molecular peptides (406.26 g/mol) that showed high antioxidant activity (90%). Fermented fish was also reported to contain antioxidant peptides, including IAEVFLITDPK and AIPPHPYP (Najafian, L. and Babji, 2018). The researchers suggested that the antioxidant activity was influenced by the low-molecular-weight peptides and the presence of hydrophilic AAs, including isoleucine, alanine, and proline. Antioxidant peptides were found in traditional fermented shrimp paste and the active peptides fraction had a molecular weight of 500 Da (Kleekayai et al., 2015).

On the other hand, several studies reported on the antioxidant activity of bioactive peptides generated in plant-based foods, such as soybeans (Singh and Vij, 2018), bitter beans (Muhialdin et al., 2020c), peanuts (Zhang et al., 2011), sourdough (Coda et al., 2012), and algae (Fan et al., 2014). The bioactive peptides generated using plant-based substrates have the advantage of low cost over animal-based proteins. Moreover, plant-based proteins have a broad range of varieties including beans, cereals, vegetables, mushrooms, and algae. The effects of fermentation on the antioxidant activity of plant-basted substrates and their bioactive compounds was highlighted in several studies (Hur et al., 2014). However, it is worth noting that the biological functions of antioxidant peptides may demonstrate other health benefits, as preventing oxidative stress can reduce inflammation. According to Sarmadi and Ismail (2010), antioxidant peptides may reduce the risk of type II diabetes via the scavenging of free radicals. The free radicals are related to cardiovascular diseases, and the antioxidant peptides may help to prevent these diseases by reducing their concentrations in the blood (Lakshmi et al., 2009). In addition, antioxidant peptides can reduce the risk of developing cancers caused by the free radicals that attack the body cells (Chi et al., 2015).

Antihypertensive Activity

Hypertension is an emerging disease that can increase the risk to develop cardiovascular and renal diseases (Garovic and Hayman 2007). The enzyme known as angiotensin I-converting enzyme (ACE), including ACE-I and ACE-II, plays a critical role in increasing blood pressure (Masuyer et al., 2012). Therefore, a great number of studies related to antihypertension activity have focused on inhibiting this enzyme to prevent high blood pressure. On the other hand, the endo- and exogenous reactive oxygen species (ROS) and free radicals have been associated with hypertension (Sies and Jones, 2020). Hence, the antioxidant activity of certain bioactive peptides is considered helpful in preventing high blood pressure (Singh et al., 2021). The two peptides, VPP and IPP, derived from fermented milk showed strong inhibitory effects toward ACE (Mizushima et al., 2004). In another study, the two peptides, IFL and WL, extracted from *tofuyo* (fermented tofu) demonstrated a high potential for their activity to reduce blood pressure (Kuba et al., 2003). Je et al. (2005) reported on the high ACE inhibitory activity of the novel peptide EVMAGNLYPG isolated from fermented blue mussel sauce. ACE inhibitory peptides were reported in several fermented foods. The ACE inhibitory peptides showed high stability in the digestive system and resistance to the proteolysis enzymes due to their low molecular weight and secondary structure (Tavares et al., 2011). On the other hand, they showed an important role in promoting cardiovascular health and preventing certain related diseases (Martínez-Sánchez et al., 2020). A lacto-fermentation process is suitable to produce ACE inhibitory peptides from low-cost plant-based substrates, such as seeds (Arulrajah et al., 2020), beans (Jakubczyk et al., 2017), and agricultural by-products (Wilson et al., 2011). Hernández-Ledesma et al. (2011) suggested the inclusion of these peptides in designer foods, also known as functional foods, to protect the host from the risk of high blood pressure. The ACE inhibitory mechanism of peptides from fermented foods is suggested to be due to its effect on the upregulating of ACE-II and improve the endothelial function, in addition to the reduction of oxidation and inflammation (Wu et al., 2017). Therefore, its biological function can be extended to prevent heart diseases.

Immunomodulation Activity

LABs and their antimicrobial compounds are reported to positively contribute to the immune system by forming barriers against foreign pathogens (Vieco-Saiz et al., 2019). Nevertheless, the bioactive peptides are the major metabolites in fermented foods that have direct and indirect effects on the immune system function of the host (Martinez-Villaluenga et al., 2017). There are two mechanisms to improve the immune system function via the bioactive peptides of fermented foods, including the modulation of gut microbiota in favor of friendly bacteria, and/or the antimicrobial activity toward the pathogens in the epithelial cells (Muhialdin et al., 2021b). Therefore, bioactive peptides from fermented foods are the key for balancing the gut microbiota and enhancing the immune system function. The mucosal immune system requires a balance between certain diets and immune tolerance to prevent pathogen colonization and initiation of inflammation (Fernández-Tomé et al., 2019). Bioactive peptides are a good source for AAs that provide the essential carbon and nitrogen for the immune cells (Fernández-Tomé et al., 2019). Low-molecular-weight peptides from fermented foods can be absorbed by the epithelial cells and help to reduce inflammation and prevent adhesion of pathogens. A great body of evidence demonstrated the potential role of dietary bioactive peptides to enhance the tolerance of the host immune system (Korhonen and Pihlanto, 2006). According to Blanco-Míguez et al. (2016), the advantages of natural bioactive peptides include their small size, good cell diffusion, low toxicity, low cost, and structure diversity. Therefore, research interest in determining the biological functions of bioactive peptides from fermented foods and their immunomodulation activity is on the rise. The two peptides QEPVL and QEPV, from fermented milk, were reported for their immunomodulation effects by increasing the lymphocyte proliferation rate (Jiehui et al., 2014). The bioactive peptides in fermented fish showed activity to reduce apoptosis and the damage in the gut caused by anti-inflammatory drugs (Marchbank et al., 2009). Plant-based fermented foods such as sourdough (Rizzello et al., 2012), soybean paste (Ashaolu, 2020), and oilseed by-products (Kadam and Lele, 2018) were also reported for the immunomodulation activity of their peptides. However, very limited studies have been carried out on human subjects and the current data are very limited. Therefore, future studies should consider clinical trials to determine the interaction between the bioactive peptides and the human immune system using defined doses of the identified peptides.

APPLICATIONS OF PEPTIDES IN FOOD SAFETY

Currently, the food industry depends heavily on the use of synthetic preservatives, such as propionic acid in bakeries, benzoic acid in jams, beverages and sauces; sorbic acid in cheese and bakeries; and nitrates and nitrites in meat products (Joute et al., 2016). According to the Food Drug Administration (FDA), synthetic preservatives are safe for human consumption (Rangan and Barceloux, 2009). However, recent studies demonstrated high risk of long-term consumption of chemically preserved foods and the potential to develop tumors due to its accumulation (Kushi et al., 2012). Venegas-Ortega et al. (2019) reviewed the role of bioactive peptides as natural preservatives to produce healthier processed foods. Their applications were considered by several researchers because of the broad range of antimicrobial activity toward pathogenic and spoilage microorganisms, including yeast, fungi, viruses, and bacteria (Muhialdin et al., 2021b; Asri et al., 2020, Arulrajah et al., 2020).

Bioactive peptides added to cheese, bread, and tomato paste extended the shelf-life 3–4 folds compared to the preservative-free foods and 2 folds compared to the chemically preserved foods (Muhialdin et al., 2011). In another study, LAB bioactive peptides inhibited the growth of several pathogens associated with the spoilage of seafood and extended the shelf life of treated samples (Rea et al., 2011). A great body of research has been carried out to determine the effects of bioactive peptides and their potential to replace synthetic preservatives or reduce their use in processed foods. The conventional method of bioactive peptides application is to be added with the food matrices at controlled concentrations during or after the process. The bioactive peptides have heat and acid tolerance and are stable within the food matrices. Thus, encapsulated peptides using starch-based coating materials showed extended stability and improved antimicrobial activity (de Vos et al., 2010). Recently, the addition of bioactive peptides to active packaging was proposed as a safe method to preserve solid foods, such as baked goods and meat products (Perez Espitia et al., 2012). The advantage of this packaging is the controlled release of the peptides to the surface of the preserved foods at a low rate, which helps to keep the food safe for a long time. The emerging technology is very promising and extensive research is currently ongoing to determine the suitable polymers to optimize the time and concentration of the bioactive peptides from the film to the surface of preserved foods.

FUTURE RESEARCH AND CONCLUSIONS

The application of peptides as a natural treatment for several diseases, such as cancer, infections, and cardiovascular diseases has attracted the interest of researchers from a broad range of fields. The addition of bioactive peptides to different foods can increase their biological functions including lowering blood pressure (Wu et al., 2017), lowering cholesterol (Nagaoka, 2019), reducing the risk of cardiovascular diseases (Cicero et al., 2017), and improving the immune system functions (da Cunha et al., 2017). Functional foods (medicinal foods) are an emerging class of food products that may play an important role in the future of the food industry (Goldberg, 2012). Among their advantages are low toxicity and the absence of side effects, especially for the natural peptides generated via fermentation (Agyei and Danquah, 2011). The bioactive peptides generated using certain LAB that have GRAS status have a high potential for applications in functional foods. The peptides are generated using low-cost substrates with high protein content and incorporated into the designed functional foods. The processing cost should be kept cost-effective and this can be achieved by minimizing the steps for the purification of the peptides. Therefore, it is recommended to use a mixture of bioactive peptides of the same molecular weight that is purified with simple methods, such as ultrafiltration and size exclusion chromatography (Muhialdin et al., 2020a).

The utilization of agricultural by-products and non-edible seeds as substrates of high protein content were proven to be efficient for lacto-fermentation and the generation of bioactive peptides with several biological functions (Arulrajah et al., 2020; Asri et al., 2020; Muhialdin et al., 2020c). On the other hand, the selection of LAB strains that have well known genetic information will ensure the consistency of the generated bioactive peptides and their biological functions. The development of powerful software facilitates the genomic analysis to understand the proteolytic system of the selected LAB strains (Hernandez-Valdes et al., 2020). As for the sequencing of the bioactive peptides, PEAKS STUDIO is applied to identify the sequences and with the advantage to identify novel peptides that have no match in the database via advanced algorithms and mathematical models (Tran et al., 2017). Future research will be focused on regulations to standardize the use of the bioactive peptides in functional foods and/or as natural food preservatives. According to the Food and Drug Administration (2020), natural peptides have a limited concentration that is approved to be used in functional foods. The development of functional foods will require deep studies to determine the concentration of the peptide that is needed to demonstrate the biological functions.

REFERENCES

Aguilar-Toalá, J. E., Santiago-López, L., Peres, C. M., Peres, C., Garcia, H. S., Vallejo-Cordoba, B., ... & Hernández-Mendoza, A. (2017). Assessment of multifunctional activity of bioactive peptides derived from fermented milk by specific *Lactobacillus plantarum* strains. *Journal of Dairy Science*, 100(1), 65–75.

Agyei, D., & Danquah, M. K. (2011). Industrial-scale manufacturing of pharmaceutical-grade bioactive peptides. *Biotechnology Advances*, 29(3), 272–277.

Agyei, D., Ongkudon, C. M., Wei, C. Y., Chan, A. S., & Danquah, M. K. (2016). Bioprocess challenges to the isolation and purification of bioactive peptides. *Food and Bioproducts Processing*, 98, 244–256.

Ahtesh, F. B., Stojanovska, L., & Apostolopoulos, V. (2018). Anti-hypertensive peptides released from milk proteins by probiotics. *Maturitas*, 115, 103–109.

Albenzio, M., Santillo, A., Caroprese, M., Della Malva, A., & Marino, R. (2017). Bioactive peptides in animal food products. *Foods*, 6(5), 35.

Algboory, H. L., & Muhialdin, B. J. (2021). Novel peptides contribute to the antimicrobial activity of camel milk fermented with *Lactobacillus plantarum* IS10. *Food Control*, 126, 108057. https://doi.org/10.1016/j.foodcont.2021.108057

Aluko, R. E. (2018). Food protein-derived peptides: Production, isolation, and purification. In Yada, R.Y. (ed.), *Proteins in food processing* (pp. 389–412). Woodhead Publishing.

Amadou, I., Le, G. W., Amza, T., Sun, J., & Shi, Y. H. (2013). Purification and characterization of foxtail millet-derived peptides with antioxidant and antimicrobial activities. *Food Research International*, 51(1), 422–428.

Amorim, F. G., Coitinho, L. B., Dias, A. T., Friques, A. G. F., Monteiro, B. L., de Rezende, L. C. D., ... & Quinton, L. (2019). Identification of new bioactive peptides from Kefir milk through proteopeptidomics: Bioprospection of antihypertensive molecules. *Food Chemistry*, 282, 109–119.

Arora, T., Singh, S., & Sharma, R. K. (2013). Probiotics: Interaction with gut microbiome and antiobesity potential. *Nutrition*, 29(4), 591–596.

Arulrajah, B., Muhialdin, B. J., Zarei, M., Hasan, H., & Saari, N. (2020). Lacto-fermented Kenaf (*Hibiscus cannabinus* L.) seed protein as a source of bioactive peptides and their applications as natural preservatives. *Food Control*, 110, 106969.

Ashaolu, T. J. (2020). Soy bioactive peptides and the gut microbiota modulation. *Applied Microbiology and Biotechnology*, 104, 9009–9017. https://doi.org/10.1007/s00253-020-10799-2

Aslam, H., Ruusunen, A., Berk, M., Loughman, A., Rivera, L., Pasco, J. A., & Jacka, F. N. (2020). Unravelled facets of milk derived opioid peptides: A focus on gut physiology, fractures and obesity. *International Journal of Food Sciences and Nutrition*, 71(1), 36–49.

Asri, N. M., Muhialdin, B. J., Zarei, M., & Saari, N. (2020). Low molecular weight peptides generated from palm kernel cake via solid state lacto-fermentation extend the shelf life of bread. *LWT-Food Science and Technology*, 134, 110206.

Atanasova, J., Dalgalarrondo, M., Iliev, I., Moncheva, P., Todorov, S. D., & Ivanova, I. V. (2020). Formation of free amino acids and bioactive peptides during the ripening of Bulgarian white brined cheeses. *Probiotics and Antimicrobial Proteins*, 13, 261–272. https://doi.org/10.1007/s12602-020-09669-0

Ayyash, M., Al-Nuaimi, A. K., Al-Mahadin, S., & Liu, S. Q. (2018). In vitro investigation of anticancer and ACE-inhibiting activity, α-amylase and α-glucosidase inhibition, and antioxidant activity of camel milk fermented with camel milk probiotic: A comparative study with fermented bovine milk. *Food Chemistry*, 239, 588–597.

Bhat, Z. F., Kumar, S., & Bhat, H. F. (2015). Bioactive peptides of animal origin: A review. *Journal of Food Science and Technology*, 52(9), 5377–5392.

Bischoff, S. C., Boirie, Y., Cederholm, T., Chourdakis, M., Cuerda, C., Delzenne, N. M., ... & Barazzoni, R. (2017). Toward a multidisciplinary approach to understand and manage obesity and related diseases. *Clinical Nutrition*, 36(4), 917–938.

Blanco-Míguez, A., Gutiérrez-Jácome, A., Pérez-Pérez, M., Pérez-Rodríguez, G., Catalán-García, S., Fdez-Riverola, F., ... & Sánchez, B. (2016). From amino acid sequence to bioactivity: The biomedical potential of antitumor peptides. *Protein Science*, 25(6), 1084–1095.

Bousquet, J., Anto, J., Czarlewski, W., Haahtela, T., Fonseca, S., Iaccarino, G., ... & Zuberbier, T. (2020). Loss of food fermentation in Westernized diet: A risk factor for severe COVID-19?. *Authorea Preprints*. doi: 10.22541/au.159526902.25301228.

Breiteneder, H., & Radauer, C. (2004). A classification of plant food allergens. *Journal of Allergy and Clinical Immunology*, 113(5), 821–830.

Brown, L., Pingitore, E. V., Mozzi, F., Saavedra, L., J. M. Villegas, & E. M. Hebert (2017). Lactic acid bacteria as cell factories for the generation of bioactive peptides. *Protein and Peptide Letters*, 24(2), 146–155.

Bulgasem, B. Y., Lani, M. N., Hassan, Z., Yusoff, W. M. W., & Fnaish, S. G. (2016). Antifungal activity of lactic acid bacteria strains isolated from natural honey against pathogenic Candida species. *Mycobiology*, 44(4), 302–309.

Capriotti, A. L., Caruso, G., Cavaliere, C., Samperi, R., Ventura, S., Chiozzi, R. Z., & Laganà, A. (2015). Identification of potential bioactive peptides generated by simulated gastrointestinal digestion of soybean seeds and soy milk proteins. *Journal of Food Composition and Analysis*, 44, 205–213.

Chen, J., Thomsen, M., & Vitetta, L. (2019). Interaction of gut microbiota with dysregulation of bile acids in the pathogenesis of nonalcoholic fatty liver disease and potential therapeutic implications of probiotics. *Journal of Cellular Biochemistry*, 120(3), 2713–2720.

Cheng, A. C., Lin, H. L., Shiu, Y. L., Tyan, Y. C., & Liu, C. H. (2017). Isolation and characterization of antimicrobial peptides derived from Bacillus subtilis E20-fermented soybean meal and its use for preventing Vibrio infection in shrimp aquaculture. *Fish & Shellfish Immunology*, 67, 270–279.

Chi, C. F., Hu, F. Y., Wang, B., Li, T., & Ding, G. F. (2015). Antioxidant and anticancer peptides from the protein hydrolysate of blood clam (*Tegillarca granosa*) muscle. *Journal of Functional Foods*, 15, 301–313.

Chugh, B., & Kamal-Eldin, A. (2020). Bioactive compounds produced by probiotics in food products. *Current Opinion in Food Science*, 32, 76–82.

Chye, F. Y., Ooi, P. W., Ng, S. Y., & Sulaiman, M. R. (2018). Fermentation-derived bioactive components from seaweeds: Functional properties and potential applications. *Journal of Aquatic Food Product Technology*, 27(2), 144–164.

Cicero, A. F., Fogacci, F., & Colletti, A. (2017). Potential role of bioactive peptides in prevention and treatment of chronic diseases: A narrative review. *British Journal of Pharmacology*, 174(11), 1378–1394.

Coda, R., Rizzello, C. G., Pinto, D., & Gobbetti, M. (2012). Selected lactic acid bacteria synthesize antioxidant peptides during sourdough fermentation of cereal flours. *Applied and Environmental Microbiology, 78*(4), 1087–1096.

Crafack, M., Mikkelsen, M. B., Saerens, S., Knudsen, M., Blennow, A., Lowor, S., ... & Nielsen, D. S. (2013). Influencing cocoa flavor using *Pichia kluyveri* and *Kluyveromyces marxianus* in a defined mixed starter culture for cocoa fermentation. *International Journal of Food Microbiology, 167*(1), 103–116.

Curiel, J. A., Pinto, D., Marzani, B., Filannino, P., Farris, G. A., Gobbetti, M., & Rizzello, C. G. (2015). Lactic acid fermentation as a tool to enhance the antioxidant properties of *Myrtus communis* berries. *Microbial Cell Factories, 14*(1), 1–10.

da Cunha, N. B., Cobacho, N. B., Viana, J. F., Lima, L. A., Sampaio, K. B., Dohms, S. S., ... & Dias, S. C. (2017). The next generation of antimicrobial peptides (AMPs) as molecular therapeutic tools for the treatment of diseases with social and economic impacts. *Drug Discovery Today, 22*(2), 234–248.

da Silva, D. D., de Lima, M. D. S. F., da Silva, M. F., da Silva, G. R., Campos, J. F., Albuquerque, W. W. C., ... & Porto, A. L. F. (2019). Bioactive water-soluble peptides from fresh buffalo cheese may be used as product markers. *LWT-Food Science and Technology, 108*, 97–105.

Daliri, E. B. M., Oh, D. H., & Lee, B. H. (2017). Bioactive peptides. *Foods, 6*(5), 32.

de Lima, M. D. S. F., da Silva, R. A., da Silva, M. F., da Silva, P. A. B., Costa, R. M. P. B., Teixeira, J. A. C., ... & Cavalcanti, M. T. H. (2018). Brazilian kefir-fermented sheep's milk, a source of antimicrobial and antioxidant peptides. *Probiotics and Antimicrobial Proteins, 10*(3), 446–455.

de Vos, P., Faas, M. M., Spasojevic, M., & Sikkema, J. (2010). Encapsulation for preservation of functionality and targeted delivery of bioactive food components. *International Dairy Journal, 20*(4), 292–302.

Dewan, S., & Tamang, J. P. (2007). Dominant lactic acid bacteria and their technological properties isolated from the Himalayan ethnic fermented milk products. *Antonie van Leeuwenhoek, 92*(3), 343–352.

Domínguez-Pérez, L. A., Beltrán-Barrientos, L. M., González-Córdova, A. F., Hernández-Mendoza, A., & Vallejo-Cordoba, B. (2020). Artisanal cocoa bean fermentation: From cocoa bean proteins to bioactive peptides with potential health benefits. *Journal of Functional Foods, 73*, 104134.

du Toit, M., Engelbrecht, L., Lerm, E., & Krieger-Weber, S. (2011). Lactobacillus: The next generation of malolactic fermentation starter cultures—An overview. *Food and Bioprocess Technology, 4*(6), 876–906.

Fadda, S., López, C., & Vignolo, G. (2010). Role of lactic acid bacteria during meat conditioning and fermentation: Peptides generated as sensorial and hygienic biomarkers. *Meat Science, 86*(1), 66–79.

Fan, H., Wang, J., Liao, W., Jiang, X., & Wu, J. (2019). Identification and characterization of gastrointestinal-resistant angiotensin-converting enzyme inhibitory peptides from egg white proteins. *Journal of Agricultural and Food Chemistry, 67*(25), 7147–7156.

Fan, X., Bai, L., Zhu, L., Yang, L., & Zhang, X. (2014). Marine algae-derived bioactive peptides for human nutrition and health. *Journal of Agricultural and Food Chemistry, 62*(38), 9211–9222.

Farvin, K. S., Baron, C. P., Nielsen, N. S., Otte, J., & Jacobsen, C. (2010). Antioxidant activity of yoghurt peptides: Part 2–characterisation of peptide fractions. *Food Chemistry, 123*(4), 1090–1097.

Fernández-Tomé, S., Hernández-Ledesma, B., Chaparro, M., Indiano-Romacho, P., Bernardo, D., & Gisbert, J. P. (2019). Role of food proteins and bioactive peptides in inflammatory bowel disease. *Trends in Food Science & Technology*, 88, 194–206.

Fialho, T. L., Carrijo, L. C., Júnior, M. J. M., Baracat-Pereira, M. C., Piccoli, R. H., & de Abreu, L. R. (2018). Extraction and identification of antimicrobial peptides from the Canastra artisanal minas cheese. *Food Research International*, 107, 406–413.

Fonseca, S., Rivas, I., Romaguera, D., Quijal, M., Czarlewski, W., Vidal, A., ... & Bousquet, J. (2020). Association between consumption of fermented vegetables and COVID-19 mortality at a country level in Europe. *MedRxiv*. doi: https://doi.org/10.1101/2020.07.06.20147025

Food & Drug Administration. (2020). Code of federal regulations, food for human consumption. https://www.accessdata.fda.gov/scripts/cdrh/cfdocs/cfcfr/cfrsearch.cfm?fr=172.320. Accessed on 3 May 2021.

Fuad, H., Hidayati, N., Darmawati, S., Munandar, H., Sulistyaningtyas, A. R., Nurrahman, N., ... & Stalis, N. E. (2020). Prospects of fibrinolytic proteases of bacteria from sea cucumber fermentation products as antithrombotic agent. In *BIO web of conferences* (Vol. 28). EDP Sciences.

Gallego, M., Mora, L., Escudero, E., & Toldrá, F. (2018). Bioactive peptides and free amino acids profiles in different types of European dry-fermented sausages. *International Journal of Food Microbiology*, 276, 71–78.

Gänzle, M. G. (2014). Enzymatic and bacterial conversions during sourdough fermentation. *Food Microbiology*, 37, 2–10.

Garovic, V. D., & Hayman, S. R. (2007). Hypertension in pregnancy: An emerging risk factor for cardiovascular disease. *Nature Clinical Practice Nephrology*, 3(11), 613–622.

Goldberg, I. (2012). *Functional foods: Designer foods, pharmafoods, nutraceuticals*. Springer Science & Business Media.

Görgüç, A., Gençdağ, E., & Yılmaz, F. M. (2020). Bioactive peptides derived from plant origin by-products: Biological activities and techno-functional utilizations in food developments–A review. *Food Research International*, 135, 109504. https://doi.org/10.1016/j.foodres.2020.109504

Hale, J. D., & Hancock, R. E. (2007). Alternative mechanisms of action of cationic antimicrobial peptides on bacteria. *Expert Review of Anti-infective Therapy*, 5(6), 951–959.

Hernández-Ledesma, B., del Mar Contreras, M., & Recio, I. (2011). Antihypertensive peptides: Production, bioavailability and incorporation into foods. *Advances in Colloid and Interface Science*, 165(1), 23–35.

Hernandez-Valdes, J. A., aan de Stegge, M., Hermans, J., Teunis, J., van Tatenhove-Pel, R. J., Teusink, B., ... & Kuipers, O. P. (2020). Enhancement of amino acid production and secretion by *Lactococcus lactis* using a droplet-based biosensing and selection system. *Metabolic Engineering Communications*, 11, e00133.

Hu, M., Zhao, H., Zhang, C., Yu, J., & Lu, Z. (2013). Purification and characterization of plantaricin 163, a novel bacteriocin produced by *Lactobacillus plantarum* 163 isolated from traditional Chinese fermented vegetables. *Journal of Agricultural and Food Chemistry*, 61(47), 11676–11682.

Huang, Y., Huang, J., & Chen, Y. (2010). Alpha-helical cationic antimicrobial peptides: Relationships of structure and function. *Protein & Cell*, 1(2), 143–152.

Hur, S. J., Lee, S. Y., Kim, Y. C., Choi, I., & Kim, G. B. (2014). Effect of fermentation on the antioxidant activity in plant-based foods. *Food Chemistry*, 160, 346–356.

Ito, M., Ito, T., Aoki, H., Nishioka, K., Shiokawa, T., Tada, H., ... & Takashiba, S. (2020). Isolation and identification of the antimicrobial substance included in tempeh using Rhizopus stolonifer NBRC 30816 for fermentation. *International Journal of Food Microbiology, 325*, 108645.

Jakubczyk, A., Karaś, M., Baraniak, B., & Pietrzak, M. (2013). The impact of fermentation and in vitro digestion on formation angiotensin converting enzyme (ACE) inhibitory peptides from pea proteins. *Food Chemistry, 141*(4), 3774–3780.

Jakubczyk, A., Karaś, M., Złotek, U., & Szymanowska, U. (2017). Identification of potential inhibitory peptides of enzymes involved in the metabolic syndrome obtained by simulated gastrointestinal digestion of fermented bean (*Phaseolus vulgaris* L.) seeds. *Food Research International, 100*, 489–496.

Jang, S. J., Lee, K., Kwon, B., You, H. J., & Ko, G. (2019). Vaginal lactobacilli inhibit growth and hyphae formation of *Candida albicans*. *Scientific Reports, 9*(1), 1–9.

Je, J. Y., Park, P. J., Byun, H. G., Jung, W. K., & Kim, S. K. (2005). Angiotensin I converting enzyme (ACE) inhibitory peptide derived from the sauce of fermented blue mussel, *Mytilus edulis*. *Bioresource Technology, 96*(14), 1624–1629.

Jiehui, Z., Liuliu, M., Haihong, X., Yang, G., Yingkai, J., Lun, Z., ... & Shaohui, Z. (2014). Immunomodulating effects of casein-derived peptides QEPVL and QEPV on lymphocytes in vitro and in vivo. *Food & Function, 5*(9), 2061–2069.

Joute, J. R., Chawhan, P., Rungsung, S., & Kirthika, P. (2016). Food additives and their associated health risks. *International Journal of Veterinary Sciences and Animal Husbandry, 1*(1), 1–5.

Kadam, D., & Lele, S. S. (2018). Value addition of oilseed meal: A focus on bioactive peptides. *Journal of Food Measurement and Characterization, 12*(1), 449–458.

Karovičová, Z. K. J., & Kohajdova, J. (2007). Fermentation of cereals for specific purpose. *Journal of Food and Nutrition Research, 46*(2), 51–57.

Khansari, N., Shakiba, Y., & Mahmoudi, M. (2009). Chronic inflammation and oxidative stress as a major cause of age-related diseases and cancer. *Recent Patents on Inflammation & Allergy Drug Discovery, 3*(1), 73–80.

Kitagawa, M., Shiraishi, T., Yamamoto, S., Kutomi, R., Ohkoshi, Y., Sato, T., ... & Yokota, S. I. (2017). Novel antimicrobial activities of a peptide derived from a Japanese soybean fermented food, Natto, against *Streptococcus pneumoniae* and *Bacillus subtilis* group strains. *AMB Express, 7*(1), 1–11.

Kitts, D. D., & Weiler, K. (2003). Bioactive proteins and peptides from food sources. Applications of bioprocesses used in isolation and recovery. *Current Pharmaceutical Design, 9*(16), 1309–1323.

Kleekayai, T., Harnedy, P. A., O'Keeffe, M. B., Poyarkov, A. A., CunhaNeves, A., Suntornsuk, W., & FitzGerald, R. J. (2015). Extraction of antioxidant and ACE inhibitory peptides from Thai traditional fermented shrimp pastes. *Food Chemistry, 176*, 441–447.

Korhonen, H., & Pihlanto, A. (2006). Bioactive peptides: Production and functionality. *International Dairy Journal, 16*(9), 945–960.

Koyama, M., Naramoto, K., Nakajima, T., Aoyama, T., Watanabe, M., & Nakamura, K. (2013). Purification and identification of antihypertensive peptides from fermented buckwheat sprouts. *Journal of Agricultural and Food Chemistry, 61*(12), 3013–3021.

Kuba, M., Tanaka, K., Tawata, S., Takeda, Y., & Yasuda, M. (2003). Angiotensin I-converting enzyme inhibitory peptides isolated from tofuyo fermented soybean food. *Bioscience, Biotechnology, and Biochemistry, 67*(6), 1278–1283.

Kunda, P. B., Benavente, F., Catalá-Clariana, S., Giménez, E., Barbosa, J., & Sanz-Nebot, V. (2012). Identification of bioactive peptides in a functional yogurt by micro liquid chromatography time-of-flight mass spectrometry assisted by retention time prediction. *Journal of Chromatography A*, *1229*, 121–128.

Kushi, L. H., Doyle, C., McCullough, M., Rock, C. L., Demark-Wahnefried, W., Bandera, E. V., ... & American Cancer Society 2010 Nutrition and Physical Activity Guidelines Advisory Committee. (2012). American Cancer Society Guidelines on nutrition and physical activity for cancer prevention: Reducing the risk of cancer with healthy food choices and physical activity. *CA: A Cancer Journal for Clinicians*, *62*(1), 30–67.

Kuwaki, S., Nakajima, N., Tanaka, H., & Ishihara, K. (2012). Plant-based paste fermented by lactic acid bacteria and yeast: Functional analysis and possibility of application to functional foods. *Biochemistry Insights*, *5*, BCI-S10529.

Lakshmi, S. V., Padmaja, G., Kuppusamy, P., & Kutala, V. K. (2009). Oxidative stress in cardiovascular disease. *NISCAIR Online Periodicals Repository*, *46*(6), 421–440.

Lee, J. H., & Paik, H. D. (2019). Anticancer and immunomodulatory activity of egg proteins and peptides: A review. *Poultry Science*, *98*(12), 6505–6516.

Lee, S. Y., & Hur, S. J. (2017). Antihypertensive peptides from animal products, marine organisms, and plants. *Food Chemistry*, *228*, 506–517.

Li, Q., & Gänzle, M. G. (2020). Host-adapted lactobacilli in food fermentations: Impact of metabolic traits of host adapted lactobacilli on food quality and human health. *Current Opinion in Food Science*, *31*, 71–80.

Liu, G., Li, S., Ren, J., Wang, C., Zhang, Y., Su, X., & Dai, Y. (2020). Effect of animal-sourced bioactive peptides on the in vitro development of mouse preantral follicles. *Journal of Ovarian Research*, *13*(1), 1–11.

López-Otín, C., & Bond, J. S. (2008). Proteases: Multifunctional enzymes in life and disease. *The Journal of Biological Chemistry*, *283*(45), 30433.

Luz, C., D'Opazo, V., Mañes, J., & Meca, G. (2019). Antifungal activity and shelf life extension of loaf bread produced with sourdough fermented by *Lactobacillus* strains. *Journal of Food Processing and Preservation*, *43*(10), e14126.

Marchbank, T., Elia, G., & Playford, R. J. (2009). Intestinal protective effect of a commercial fish protein hydrolysate preparation. *Regulatory Peptides*, *155*(1–3), 105–109.

Marco, M. L., Heeney, D., Binda, S., Cifelli, C. J., Cotter, P. D., Foligné, B., ... & Hutkins, R. (2017). Health benefits of fermented foods: Microbiota and beyond. *Current Opinion in Biotechnology*, *44*, 94–102.

Martínez-Sánchez, S. M., Gabaldón-Hernández, J. A., & Montoro-García, S. (2020). Unravelling the molecular mechanisms associated with the role of food-derived bioactive peptides in promoting cardiovascular health. *Journal of Functional Foods*, *64*, 103645.

Martinez-Villaluenga, C., Penas, E., & Frias, J. (2017). Bioactive peptides in fermented foods: Production and evidence for health effects. In Frias, J., Martinex-Villaluenga, C., & Penas, E. (eds), *Fermented foods in health and disease prevention* (pp. 23–47). Academic Press.

Masuyer, G., Schwager, S. L., Sturrock, E. D., Isaac, R. E., & Acharya, K. R. (2012). Molecular recognition and regulation of human angiotensin-I converting enzyme (ACE) activity by natural inhibitory peptides. *Scientific Reports*, *2*(1), 1–10.

Matsuoka, R., Kamachi, K., Usuda, M., Masuda, Y., Kunou, M., Tanaka, A., & Utsunomiya, K. (2019). Minimal effective dose of lactic-fermented egg white on visceral fat in Japanese men: A double-blind parallel-armed pilot study. *Lipids in Health and Disease*, *18*(1), 1–9.

Matsuoka, R., Kamachi, K., Usuda, M., Wang, W., Masuda, Y., Kunou, M., ... & Utsunomiya, K. (2017a). Lactic-fermented egg white improves visceral fat obesity in Japanese subjects—double-blind, placebo-controlled study. *Lipids in Health and Disease*, *16*(1), 1–9.

Matsuoka, R., Usuda, M., Masuda, Y., Kunou, M., & Utsunomiya, K. (2017b). Lactic-fermented egg white reduced serum cholesterol concentrations in mildly hypercholesterolemic Japanese men: A double-blind, parallel-arm design. *Lipids in Health and Disease*, *16*(1), 1–9.

Matthews, A., Grimaldi, A., Walker, M., Bartowsky, E., Grbin, P., & Jiranek, V. (2004). Lactic acid bacteria as a potential source of enzymes for use in vinification. *Applied and Environmental Microbiology*, *70*(10), 5715–5731.

McGregor, R. A., & Poppitt, S. D. (2013). Milk protein for improved metabolic health: A review of the evidence. *Nutrition & Metabolism*, *10*(1), 1–13.

Mendis, S. (2017). Global progress in prevention of cardiovascular disease. *Cardiovascular Diagnosis and Therapy*, *7*(Suppl 1), S32.

Mizushima, S., Ohshige, K., Watanabe, J., Kimura, M., Kadowaki, T., Nakamura, Y., ... & Ueshima, H. (2004). Randomized controlled trial of sour milk on blood pressure in borderline hypertensive men. *American Journal of Hypertension*, *17*(8), 701–706.

Moreno-Montoro, M., Olalla-Herrera, M., Rufián-Henares, J. Á., Martínez, R. G., Miralles, B., Bergillos, T., ... & Jauregi, P. (2017). Antioxidant, ACE-inhibitory and antimicrobial activity of fermented goat milk: Activity and physicochemical property relationship of the peptide components. *Food & Function*, *8*(8), 2783–2791.

Morris, S. S., Beesabathuni, K., & Headey, D. (2018). An egg for everyone: Pathways to universal access to one of nature's most nutritious foods. *Maternal & Child Nutrition*, *14* Supplement 3, e12679.

Muhialdin, B. J., Algboory, H. L., Kadum, H., Mohammed, N. K., Saari, N., Hassan, Z., & Hussin, A. S. M. (2020b). Antifungal activity determination for the peptides generated by *Lactobacillus plantarum* TE10 against *Aspergillus flavus* in maize seeds. *Food Control*, *109*, 106898.

Muhialdin, B. J., Algboory, H. L., Mohammed, N. K., Kadum, H., Hussin, A. S., Saari, N., & Hassan, Z. (2020a). Discovery and development of novel anti-fungal peptides against foodspoiling fungi. *Current Drug Discovery Technologies*, *17*(4), 553–561.

Muhialdin, B. J., Hassan, Z., & Saari, N. (2018). In vitro antifungal activity of lactic acid bacteria low molecular peptides against spoilage fungi of bakery products. *Annals of Microbiology*, *68*(9), 557–567.

Muhialdin, B. J., Hassan, Z., & Sadon, S. K. (2011). Antifungal activity of *Lactobacillus fermentum* Te007, *Pediococcus pentosaceus* Te010, *Lactobacillus pentosus* G004, and *L. paracasi* D5 on selected foods. *Journal of Food Science*, *76*(7), M493–M499.

Muhialdin, B. J., Hussin, A. S. M., Kadum, H., Hamid, A. A., & Jaafar, A. H. (2021a). Metabolomic changes and biological activities during the lacto-fermentation of jackfruit juice using *Lactobacillus casei* ATCC334. *LWT-Food Science and Technology*, *141*, 110940.

Muhialdin, B. J., Rani, N. F. A., & Hussin, A. S. M. (2020c). Identification of antioxidant and antibacterial activities for the bioactive peptides generated from bitter beans (*Parkia speciosa*) via boiling and fermentation processes. *LWT-Food Science and Technology*, *131*, 109776.

Muhialdin, B. J., Zawawi, N., Razis, A. F. A., Bakar, J., & Zarei, M. (2021b). Antiviral activity of fermented foods and their probiotics bacteria toward respiratory and alimentary tracts viruses. *Food Control*, *127*, 108140. https://doi.org/10.1016/j.foodcont.2021.108140

Nagaoka, S. (2019). Structure–function properties of hypolipidemic peptides. *Journal of Food Biochemistry*, *43*(1), e12539.

Nahariah, N., Hikmah, H., & Yuliati, F. N. (2019, March). Microbiological activities in fermented egg whites with different level of milk and fermentation times. In *IOP conference series: Earth and environmental science* (Vol. 247, No. 1, p. 012028). IOP Publishing.

Najafian, L., & Babji, A. S. (2018). Fractionation and identification of novel antioxidant peptides from fermented fish (pekasam). *Journal of Food Measurement and Characterization*, *12*(3), 2174–2183.

Najafian, L., & Babji, A. S. (2019). Purification and identification of antioxidant peptides from fermented fish sauce (Budu). *Journal of Aquatic Food Product Technology*, *28*(1), 14–24.

Nimalaratne, C., Bandara, N., & Wu, J. (2015). Purification and characterization of antioxidant peptides from enzymatically hydrolyzed chicken egg white. *Food Chemistry*, *188*, 467–472.

Ohata, M., Uchida, S., Zhou, L., & Arihara, K. (2016). Antioxidant activity of fermented meat sauce and isolation of an associated antioxidant peptide. *Food Chemistry*, *194*, 1034–1039.

Onggo, I., & Fleet, G. H. (1993). Media for the isolation and enumeration of lactic acid bacteria from yoghurts. *Australian Journal of Dairy Technology (Australia)*, *48*(2), 89–92.

Ooms, N., & Delcour, J. A. (2019). How to impact gluten protein network formation during wheat flour dough making. *Current Opinion in Food Science*, *25*, 88–97.

Pasolli, E., De Filippis, F., Mauriello, I. E., Cumbo, F., Walsh, A. M., Leech, J., ... & Ercolini, D. (2020). Large-scale genome-wide analysis links lactic acid bacteria from food with the gut microbiome. *Nature Communications*, *11*(1), 1–12.

Perez Espitia, P. J., de Fátima Ferreira Soares, N., dos Reis Coimbra, J. S., de Andrade, N. J., Souza Cruz, R., & Alves Medeiros, E. A. (2012). Bioactive peptides: Synthesis, properties, and applications in the packaging and preservation of food. *Comprehensive Reviews in Food Science and Food Safety*, *11*(2), 187–204.

Pfalzgraff, A., Brandenburg, K., & Weindl, G. (2018). Antimicrobial peptides and their therapeutic potential for bacterial skin infections and wounds. *Frontiers in Pharmacology*, *9*, 281.

Power, O., Jakeman, P., & FitzGerald, R. J. (2013). Antioxidative peptides: Enzymatic production, in vitro and in vivo antioxidant activity and potential applications of milk-derived antioxidative peptides. *Amino Acids*, *44*(3), 797–820.

Quirós, A., Ramos, M., Muguerza, B., Delgado, M. A., Miguel, M., Aleixandre, A., & Recio, I. (2007). Identification of novel antihypertensive peptides in milk fermented with *Enterococcus faecalis*. *International Dairy Journal*, *17*(1), 33–41.

Rahman, M. M., & Lamsal, B. P. (2021). Ultrasound-assisted extraction and modification of plant-based proteins: Impact on physicochemical, functional, and nutritional properties. *Comprehensive Reviews in Food Science and Food Safety*, *20*(2), 1457–1480.

Rajapakse, N., Mendis, E., Jung, W. K., Je, J. Y., & Kim, S. K. (2005). Purification of a radical scavenging peptide from fermented mussel sauce and its antioxidant properties. *Food Research International*, *38*(2), 175–182.

Rakhmanova, A., Khan, Z. A., & Shah, K. (2018). A mini review fermentation and preservation: Role of lactic acid bacteria. *MOJ Food Processing & Technology*, 6(5), 414–417.

Rangan, C., & Barceloux, D. G. (2009). Food additives and sensitivities. *Disease-A-Month*, 55(5), 292–311.

Rao, S., Sun, J., Liu, Y., Zeng, H., Su, Y., & Yang, Y. (2012). ACE inhibitory peptides and antioxidant peptides derived from in vitro digestion hydrolysate of hen egg white lysozyme. *Food Chemistry*, 135(3), 1245–1252.

Rea, M. C., Ross, R. P., Cotter, P. D., & Hill, C. (2011). Prokaryotic antimicrobial peptides: From genes to applications. In Drider, D., & Rebuffat, S. (eds), *Classification of bacteriocins from gram-positive bacteria* (pp. 29–55). Springer.

Rizzello, C. G., Cassone, A., Coda, R., & Gobbetti, M. (2011). Antifungal activity of sourdough fermented wheat germ used as an ingredient for bread making. *Food Chemistry*, 127(3), 952–959.

Rizzello, C. G., Nionelli, L., Coda, R., & Gobbetti, M. (2012). Synthesis of the cancer preventive peptide lunasin by lactic acid bacteria during sourdough fermentation. *Nutrition and Cancer*, 64(1), 111–120.

Ryan, J. T., Ross, R. P., Bolton, D., Fitzgerald, G. F., & Stanton, C. (2011). Bioactive peptides from muscle sources: Meat and fish. *Nutrients*, 3(9), 765–791.

Sacchini, L., Migliorati, G., Di Giannatale, E., Polmilo, F., & Rossi, F. (2017). LAB strains with bacteriocin synthesis genes and their applications. In Poltronieri, P. (ed.), *Microbiology in dairy processing challenges and opportunities* (pp. 161–163). John Wiley, UK.

Saito, T., Nakamura, T., Kitazawa, H., Kawai, Y., & Itoh, T. (2000). Isolation and structural analysis of antihypertensive peptides that exist naturally in Gouda cheese. *Journal of Dairy Science*, 83(7), 1434–1440.

Sanjukta, S., & Rai, A. K. (2016). Production of bioactive peptides during soybean fermentation and their potential health benefits. *Trends in Food Science & Technology*, 50, 1–10.

Sarmadi, B. H., & Ismail, A. (2010). Antioxidative peptides from food proteins: A review. *Peptides*, 31(10), 1949–1956.

Shiferaw Terefe, N., & Augustin, M. A. (2020). Fermentation for tailoring the technological and health related functionality of food products. *Critical Reviews in Food Science and Nutrition*, 60(17), 2887–2913.

Shirako, S., Kojima, Y., Hasegawa, T., Yoshikawa, T., Matsumura, Y., Ikeda, K., … & Sato, K. (2020). Identification of short-chain pyroglutamyl peptides in Japanese salted fermented soy paste (miso) and their anti-obesity effect. *Journal of Food Bioactives*, 12. https://doi.org/10.31665/JFB.2020.12251

Sies, H., & Jones, D. P. (2020). Reactive oxygen species (ROS) as pleiotropic physiological signalling agents. *Nature Reviews Molecular Cell Biology*, 21(7), 363–383.

Singh, B. P., Aluko, R. E., Hati, S., & Solanki, D. (2021). Bioactive peptides in the management of lifestyle-related diseases: Current trends and future perspectives. *Critical Reviews in Food Science and Nutrition*, 1–14. DOI: 10.1080/10408398.2021.1877109

Singh, B. P., & Vij, S. (2018). In vitro stability of bioactive peptides derived from fermented soy milk against heat treatment, pH and gastrointestinal enzymes. *LWT-Food Science and Technology*, 91, 303–307.

Singh, B. P., Vij, S., & Hati, S. (2014). Functional significance of bioactive peptides derived from soybean. *Peptides*, 54, 171–179.

Slavokhotova, A. A., Naumann, T. A., Price, N. P., Rogozhin, E. A., Andreev, Y. A., Vassilevski, A. A., & Odintsova, T. I. (2014). Novel mode of action of plant defense peptides–hevein-like antimicrobial peptides from wheat inhibit fungal metalloproteases. *The FEBS Journal*, *281*(20), 4754–4764.

Soni, S., & Dey, G. (2014). Perspectives on global fermented foods. *British Food Journal*, *116*(11), 1767–1787.

Stack, J., Le Gouic, A. V., & FitzGerald, R. J. (2020). Bioactive proteins and peptides from microalgae. *Encyclopedia of Marine Biotechnology*, *3*, 1443–1474.

Sun, J., He, H., & Xie, B. J. (2004). Novel antioxidant peptides from fermented mushroom Ganoderma lucidum. *Journal of Agricultural and Food Chemistry*, *52*(21), 6646–6652.

Tamang, J. P., Cotter, P. D., Endo, A., Han, N. S., Kort, R., Liu, S. Q., ... & Hutkins, R. (2020). Fermented foods in a global age: East meets West. *Comprehensive Reviews in Food Science and Food Safety*, *19*(1), 184–217.

Tamang, J. P., Watanabe, K., & Holzapfel, W. H. (2016). Diversity of microorganisms in global fermented foods and beverages. *Frontiers in Microbiology*, *7*, 377.

Taniguchi, M., Aida, R., Saito, K., Ochiai, A., Takesono, S., Saitoh, E., & Tanaka, T. (2019). Identification and characterization of multifunctional cationic peptides from traditional Japanese fermented soybean Natto extracts. *Journal of Bioscience and Bioengineering*, *127*(4), 472–478.

Tavares, T., del Mar Contreras, M., Amorim, M., Pintado, M., Recio, I., & Malcata, F. X. (2011). Novel whey-derived peptides with inhibitory effect against angiotensin-converting enzyme: In vitro effect and stability to gastrointestinal enzymes. *Peptides*, *32*(5), 1013–1019.

Thordarson, P., Le Droumaguet, B., & Velonia, K. (2006). Well-defined protein–polymer conjugates—synthesis and potential applications. *Applied Microbiology and Biotechnology*, *73*(2), 243–254.

Torres-Llanez, M. J., González-Córdova, A. F., Hernandez-Mendoza, A., Garcia, H. S., & Vallejo-Cordoba, B. (2011). Angiotensin-converting enzyme inhibitory activity in Mexican Fresco cheese. *Journal of Dairy Science*, *94*(8), 3794–3800.

Tran, N. H., Zhang, X., Xin, L., Shan, B., & Li, M. (2017). De novo peptide sequencing by deep learning. *Proceedings of the National Academy of Sciences*, *114*(31), 8247–8252.

van Beverwijk, A. L. (2019). Culture collection; why and wherefore. In Martin, S. M. (ed.), *Culture collections* (pp. 9–16). University of Toronto Press.

Vargas-Bello-Pérez, E., Márquez-Hernández, R. I., & Hernández-Castellano, L. E. (2019). Bioactive peptides from milk: Animal determinants and their implications in human health. *Journal of Dairy Research*, *86*(2), 136–144.

Venegas-Ortega, M. G., Flores-Gallegos, A. C., Martínez-Hernández, J. L., Aguilar, C. N., & Nevárez-Moorillón, G. V. (2019). Production of bioactive peptides from lactic acid bacteria: A sustainable approach for healthier foods. *Comprehensive reviews in Food Science and Food Safety*, *18*(4), 1039–1051.

Verma, A. K., Kumar, S., Das, M., & Dwivedi, P. D. (2013). A comprehensive review of legume allergy. *Clinical Reviews in Allergy & Immunology*, *45*(1), 30–46.

Vieco-Saiz, N., Belguesmia, Y., Raspoet, R., Auclair, E., Gancel, F., Kempf, I., & Drider, D. (2019). Benefits and inputs from lactic acid bacteria and their bacteriocins as alternatives to antibiotic growth promoters during food-animal production. *Frontiers in Microbiology*, *10*, 57.

Vukotic, G., Mirkovic, N., Jovcic, B., Miljkovic, M., Strahinic, I., Fira, D., ... & Kojic, M. (2015). Proteinase PrtP impairs lactococcin LcnB activity in *Lactococcus lactis* BGMN1-501: New insights into bacteriocin regulation. *Frontiers in Microbiology*, 6, 92.

Wang, J., Chen, C., Xu, Y., Jia, C., Zhang, B., Niu, M., ... & Xiong, S. (2021). Selection of antioxidant peptides from gastrointestinal hydrolysates of fermented rice cake by combining peptidomics and bioinformatics. *ACS Food Science & Technology*, 1(3), 443–452.

Wang, J., Chi, Y., Cheng, Y., & Zhao, Y. (2018). Physicochemical properties, in vitro digestibility and antioxidant activity of dry-heated egg white protein. *Food Chemistry*, 246, 18–25.

Wang, W., Xia, W., Gao, P., Xu, Y., & Jiang, Q. (2017). Proteolysis during fermentation of Suanyu as a traditional fermented fish product of China. *International Journal of Food Properties*, 20(sup1), S166–S176.

Wang, Y. D., Kung, C. W., & Chen, J. Y. (2010). Antiviral activity by fish antimicrobial peptides of epinecidin-1 and hepcidin 1–5 against nervous necrosis virus in medaka. *Peptides*, 31(6), 1026–1033.

Williams, A. G., Noble, J., Tammam, J., Lloyd, D., & Banks, J. M. (2002). Factors affecting the activity of enzymes involved in peptide and amino acid catabolism in non-starter lactic acid bacteria isolated from Cheddar cheese. *International Dairy Journal*, 12(10), 841–852.

Wilson, J., Hayes, M., & Carney, B. (2011). Angiotensin-I-converting enzyme and prolyl endopeptidase inhibitory peptides from natural sources with a focus on marine processing by-products. *Food Chemistry*, 129(2), 235–244.

Wu, J., Liao, W., & Udenigwe, C. C. (2017). Revisiting the mechanisms of ACE inhibitory peptides from food proteins. *Trends in Food Science & Technology*, 69, 214–219.

Xiang, H., Sun-Waterhouse, D., Waterhouse, G. I., Cui, C., & Ruan, Z. (2019). Fermentation-enabled wellness foods: A fresh perspective. *Food Science and Human Wellness*, 8(3), 203–243.

Yang, A., Zuo, L., Cheng, Y., Wu, Z., Li, X., Tong, P., & Chen, H. (2018). Degradation of major allergens and allergenicity reduction of soybean meal through solid-state fermentation with microorganisms. *Food & Function*, 9(3), 1899–1909.

Yang, H. J., Kwon, D. Y., Moon, N. R., Kim, M. J., Kang, H. J., & Park, S. (2013). Soybean fermentation with *Bacillus licheniformis* increases insulin sensitizing and insulinotropic activity. *Food & Function*, 4(11), 1675–1684.

Yao, Y., Yan, S., Han, J., Dai, Q., & He, P. A. (2014). A novel descriptor of protein sequences and its application. *Journal of Theoretical Biology*, 347, 109–117.

Yvon, S., Olier, M., Leveque, M., Jard, G., Tormo, H., Haimoud-Lekhal, D. A., ... & Eutamène, H. (2018). Donkey milk consumption exerts anti-inflammatory properties by normalizing antimicrobial peptides levels in Paneth's cells in a model of ileitis in mice. *European Journal of Nutrition*, 57(1), 155–166.

Zanutto-Elgui, M. R., Vieira, J. C. S., do Prado, D. Z., Buzalaf, M. A. R., de Magalhães Padilha, P., de Oliveira, D. E., & Fleuri, L. F. (2019). Production of milk peptides with antimicrobial and antioxidant properties through fungal proteases. *Food Chemistry*, 278, 823–831.

Zhang, Y., Hu, P., Xie, Y., Yang, P., Zheng, S., Tian, Y., ... & Feng, D. (2020). DNA damage protection and antioxidant activities of peptides isolated from sour meat co-fermented by *P. pentosaceus* SWU73571 and *L. curvatus* LAB26. *CyTA-Journal of Food*, 18(1), 375–382.

Zhang, Y., Zhang, H., Wang, L., Guo, X., Qi, X., & Qian, H. (2011). Influence of the degree of hydrolysis (DH) on antioxidant properties and radical-scavenging activities of peanut peptides prepared from fermented peanut meal. *European Food Research and Technology*, *232*(6), 941–950.

Zhou, J., Chen, M., Wu, S., Liao, X., Wang, J., Wu, Q., … & Ding, Y. (2020). A review on mushroom-derived bioactive peptides: Preparation and biological activities. *Food Research International*, *134*, 109230. https://doi.org/10.1016/j.foodres.2020.109230

CHAPTER 13

Subcritical Water Extraction and Microwave-Assisted Extraction

Leo M.L. Nollet

CONTENTS

Subcritical Water Extraction	259
Principles	259
Subcritical Water Hydrolysis and/or Extraction of Peptides	260
Microwave-Assisted Extraction	263
Principles	263
MAE Applications of Peptides	263
References	265

SUBCRITICAL WATER EXTRACTION

Principles

Subcritical water is an environmentally benign solvent which has the potential to provide an alternative to traditional methods of protein hydrolysis without the inclusion of expensive acids or enzymes.

Subcritical water (SCW) is defined as water maintained at a temperature of between 100 and 374 °C and at a pressure of less than 22.064 MPa, i.e., below its critical point (Figure 13.1). While in the subcritical state, water acquires unique properties, including a change in hydrogen bonding structure and an increased ionic product, K_w, which is 3 orders of magnitude greater than that of water at ambient conditions. The increase in ionic product drives the formation of hydronium (H_3O^+) and hydroxide (OH^-) ions. SCW therefore has the ability to act as either an acid or a base catalyst. Consequently, SCW is regarded as a "green" solvent and is attracting interest with regard to waste and biomass conversion, including hydrolysis of proteins.

Subcritical water extraction (SWE) technology is used for the extraction of active compounds from different biomass materials with low process cost, mild operating conditions, short process times, and environmental sustainability. SWE is also called superheated water extraction or pressurized hot water extraction (PHWE).

SWE uses water as an extraction solvent, which is safe, nontoxic, nonflammable and environmentally friendly. Water is easily available and cheap. The obtained extracts are safe, without a trace of any toxic solvents. SWE is characterized by higher diffusion into the plant matrix and increased mass-transfer properties in comparison to other extraction techniques. SWE can be applied for extraction of low-polar as well as nonpolar

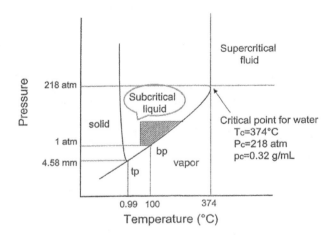

FIGURE 13.1 Phase diagram of water as a function of temperature and pressure (cross-hatched area indicates the preferred region (SWE)) (open access).

compounds. Its application has a low cost, the extraction solvent is easily available, and short extraction times minimize the cost of the extraction process.

Drawbacks of SWE may be:

- High investments costs;
- At elevated temperatures, the risk of unwanted reactions (caramelization, Maillard reactions) increases, and toxic compounds can be formed;
- At elevated temperatures, possible degradation of temperature-sensitive compounds can be expected.

Hydrolysis involves the reaction of an organic chemical with water to form two or more new substances and usually means the cleavage of chemical bonds by the addition of water. Subcritical water hydrolysis (SWH) is also termed hydrothermal liquefaction, hydrothermolysis, and aquathermolysis. It has the potential for breaking down cellulose and hemicellulose biopolymers, proteins, and different biomasses.

Subcritical Water Hydrolysis and/or Extraction of Peptides

To date, most studies on the subcritical water-mediated hydrolysis of proteins have focused on the production of amino acids, rather than the intermediate peptides.

Sereewatthanawut et al. [1] investigated the production of value-added protein and amino acids from deoiled rice bran by hydrolysis in subcritical water (SW) in the temperature range of 100–220 °C for 0–30 min. The results suggested that SW could effectively be used to hydrolyze deoiled rice bran to produce useful protein and amino acids. The amount of protein and amino acids produced are higher than those obtained by conventional alkali hydrolysis. The yields generally increased with increased temperature and hydrolysis time. However, thermal degradation of the product was observed when hydrolysis was carried out at higher temperature for extended periods of time. The highest yield of protein and amino acids were 219 ± 26 and 8.0 ± 1.6 mg/g of dry bran, and were obtained at 200 °C at a hydrolysis time of 30 min. Moreover, the product obtained at 200 °C after 30 min of hydrolysis exhibited high antioxidant activity and was shown to be suitable for use as culture medium for yeast growth.

Subcritical water hydrolysis was applied to obtain antioxidant and antimicrobial hydrolysates from tuna skin and isolated collagen [2]. Different temperatures (150–300 °C) with a pressure (50–100 bar) and reaction time (5 min) were employed to find the optimum conditions. Degree of hydrolysis was highest at 250 °C for both skin hydrolysate (SH) and collagen hydrolysate (CH). Antioxidant activities evaluated by four different assays increased with increasing temperature and were found to be highest at 280 °C. Both hydrolysates showed antimicrobial activity with the highest activity at 280 °C in CH. The structural and free amino acids in CH were found to be highest at 220 °C and 250 °C, respectively. Molecular weight profile of selected hydrolysate showed that low-molecular-weight peptides (<600 Da) and/or free amino acids are associated with functional activity.

The effect of subcritical water processing (SWP) temperature on the hydrolysis pattern and quality characteristics of soybean protein hydrolysates was investigated [3]. Two common cultivars of soybean were subjected to SWP at a temperature range of 150–250 °C and pressure of 22 MPa. The highest free amino group content and yield were obtained at a processing temperature of 190 °C for both the cultivars ($p < 0.05$). Gel electrophoresis analysis exhibited low molecular mass hydrolysates at this processing temperature. Increasing processing temperature generated lower molecular mass peptides, as revealed by gel permeable chromatography.

The optimization of processing temperature could produce peptide fractions for food applications.

The impact of subcritical water processing (SWP), in a temperature range of 240–300 °C, on the formation of hydrolysates from bovine serum albumin (BSA) was investigated [4]. SDS-PAGE analysis of the samples treated at all temperatures did not reveal any bands. Yield, as evaluated by the Kjeldahl and Biuret methods, varied and decreased beyond 280 °C ($P < 0.05$). The molecular mass of the hydrolysates was highly affected by temperature, with the formation of a new low-molecular-weight peak, as revealed by gel permeation chromatography. SWP at 280 °C caused the greatest release of free amino groups, as shown by the TNBS assay ($P < 0.05$). Regardless of the type of amino acids, the maximum amount was obtained at 280 °C. At this temperature, the generation of alanine and glycine was relatively high. The optimization of processing parameters may enhance the production of valuable peptides without the need for additional catalysts.

The use of enzymes to recover soluble peptides with functional properties from insoluble proteins could prove to be very expensive, implying high reaction times and low yields. In the study of Marcet et al. [5], the insoluble granular protein, previously delipidated, was hydrolyzed using enzymes (trypsin) as a comparison to the proposed alternative method: subcritical water hydrolysis using both nitrogen and oxygen streams. The result of the hydrolysis was characterized in terms of the yield and peptide size distribution as well as different functional properties. The SWH of the delipidated granules resulted in a higher recovery yield than that obtained by enzymatic hydrolysis in half the time. The foaming capacity of the peptides obtained by SWH was higher than that obtained by trypsin hydrolysis, although the foam stability was lower. Slight differences were detected between these peptides in terms of their emulsifying properties.

Powell et al. [6] were investigating the specificity of subcritical water with respect to the production of peptides from three model proteins, hemoglobin, bovine serum albumin, and β-casein. They compared the results with enzymatic digestion of proteins by trypsin. In addition, the effect of subcritical water treatment on two protein posttranslational modifications, disulfide bonds and phosphorylation, was investigated. The results show that high protein sequence coverages (>80%) can be obtained following SWH. These are comparable to those obtained following treatment with trypsin. Under mild subcritical water conditions (160 °C), all proteins showed favored cleavage of the Asp-X bond. The results for β-casein revealed favored cleavage of the Glu-X bond at subcritical water temperatures of 160 and 207 °C. That was similarly observed for bovine serum albumin at a subcritical water temperature of 207 °C. Subcritical water treatment results in very limited cleavage of disulfide bonds. Reduction and alkylation of proteins either prior to or post subcritical water treatment improve reported protein sequence coverages. The results for phosphoprotein β-casein show that, under mild subcritical water conditions, phosphorylation may be retained on the peptide hydrolysis products.

Chen et al. [7] showed the direct electrospray of subcritical water solutions of ubiquitin, hemoglobin, wheat germ agglutinin, and bradykinin by use of a high-pressure electrospray source; i.e., SCW conditions were created in the electrospray source. While the peptide bradykinin remained stable, the protein samples showed evidence of unfolding and degradation. At temperatures of 160 °C, a significant decrease in intact protein ions was observed, and at 180 °C, these were virtually eliminated. High resolution mass spectrometry of superheated ubiquitin confirmed the presence of peptide hydrolysis products.

Low-cost plant and algal biomass are increasingly demanded as a source of proteins. Subcritical water technology is an efficient green technique useful both for extraction and for hydrolysis of protein and other fractions (lipid, carbohydrates, phenolics) from these matrices. However, adequate selection of operational conditions is needed in order both to maximize their extraction yield and to avoid degradation into monomeric units and decomposition products. The review of Álvarez-Viñas et al. [8] summarizes the major features of subcritical water-based processes for the extraction/hydrolysis of protein. In order to valorize other valuable fractions from agro-food wastes and algal biomass, optimal conditions should be established as a compromise solution. Alternatively, stage-wise operation to sequentially obtain the target fractions could be desirable.

Food industry processing wastes are produced in enormous amounts every year. In particular, sub-critical water hydrolysis has been revealed as an interesting way for recovering high added-value molecules, and its applications have been broadly referred in the bibliography. Special interest has been focused on recovering protein hydrolysates in the form of peptides or amino acids, from both animal and vegetable wastes, by means of SWH.

The work of Marcet et al. [9] reviews the current state of art of using subcritical water hydrolysis for protein recovering from food industry wastes. Key parameters such as reaction time, temperature, amino acid degradation, and kinetic constants have been discussed. Besides, the characteristics of the raw material and the type of products that can be obtained depending on the substrate have been reviewed.

Wheat germ protein (WGP) was extracted with subcritical water and then hydrolyzed with Alcalase 2.4 L to obtain antioxidant hydrolysates [10]. Wheat germ peptides (WG-P, Mw < 1 kDa) were purified by using Sephadex G-15 column chromatography. The results showed that WG-P-4 possessed the strongest DPPH radical-scavenging activity in comparison with other peptides fractions. In addition, free amino acids and LC-MS/MS analysis showed that Gly-Pro-Phe, Gly-Pro-Glu, and Phe-Gly-Glu were the major peptides of WG-P-4.

MICROWAVE-ASSISTED EXTRACTION

Principles

Microwave-assisted extraction (MAE) is a process of using microwave energy to heat solvents in contact with a sample in order to partition analytes from the sample matrix into the solvent. Microwaves are non-ionizing electromagnetic waves on materials that have the ability to convert a part of the absorbed electromagnetic energy to heat energy.

Benefits of MAE are:

- Reduction in extraction time.
- Improved yield.
- Better accuracy.
- Suitable for thermolabile substances.

MAE Applications of Peptides

In the next paragraphs, a number of examples of MAE methods are given. This list is not exhaustive and many more examples can be found in the literature.

Hua et al. [11] developed a rapid microwave-assisted protein digestion technique based on classic acid hydrolysis reaction with 2% formic acid solution. In this mild chemical environment, proteins are hydrolyzed to peptides, which can be directly analyzed by MALDI-MS or ESI-MS without prior sample purification. Dilute formic acid cleaves proteins specifically at the C-terminal of aspartyl (Asp) residues within 10 min of exposure to microwave irradiation. By adjusting the irradiation time, the authors found that the extent of protein fragmentation can be controlled, as shown by the single fragmentation of myoglobin at the C-terminal of any of the Asp residues. The efficacy and simplicity of this technique for protein identification are demonstrated by the peptide mass maps of in-gel digested myoglobin and BSA, as well as proteins isolated from *Escherichia coli* K12 cells.

Accelerated proteolytic cleavage of proteins under controlled microwave irradiation has been achieved [12] Selective peptide fragmentation by endoproteases trypsin or lysine C led to smaller peptides that were analyzed by matrix-assisted laser desorption ionization (MALDI) or liquid chromatography-electrospray ionization (LC-ESI) techniques. The efficacy of this technique for protein mapping was demonstrated by the mass spectral analyses of the peptide fragmentation of several biologically active proteins, including cytochrome c, ubiquitin, lysozyme, myoglobin, and interferon α-2b. Most important, using this novel approach in the digestion of proteins occurs in minutes, in contrast to the hours required by conventional methods.

A protein can be readily acid-hydrolyzed within 1 min by exposure to microwave irradiation to form, predominantly, two series of polypeptide ladders containing either the N- or C-terminal amino acid of the protein, respectively [13]. Protein sequencing by mass analysis of polypeptide ladders subsequently controlled protein hydrolysis.

Lin et al. [14] describe the accelerated enzymatic digestion of several proteins in various solvent systems under microwave irradiation. The tryptic fragments of the proteins were analyzed by matrix-assisted laser desorption/ionization mass spectrometry. Under the influence of rapid microwave heating, these enzymatic reactions can proceed in a solvent such as chloroform, which, under traditional digestion conditions, renders the enzyme inactive. The digestion efficiencies and sequence coverages were increased when the trypsin digestions occurred in acetonitrile-, methanol-, and chloroform-containing solutions that were heated under microwave irradiation for 10 min using a commercial microwave applicator. The percentage of the protein digested under microwave irradiation increased with the relative acetonitrile content, but decreased as the methanol content was increased. These observations suggest that acetonitrile does not deactivate the enzyme during the irradiation period; in contrast, methanol does deactivate it. In all cases, the digestion efficiencies under microwave irradiation exceed those under conventional conditions.

Zhong et al. [15] report a microwave-assisted acid hydrolysis (MAAH) method for rapid protein degradation for peptide mass mapping and tandem mass spectrometric analysis of peptides for protein identification. It uses a 25% trifluoroacetic acid (TFA) aqueous solution to dissolve or suspend proteins, followed by microwave irradiation for 10 min. This detergent-free method generates peptide mixtures that can be directly analyzed by liquid chromatography (LC) matrix-assisted laser desorption ionization (MALDI) mass spectrometry (MS) – without the need of extensive sample cleanup.

Two-dimensional electrophoresis (2-DE) combined with mass spectrometry has significantly improved the possibilities of large-scale identification of proteins. However, 2-DE is limited by its inability to speed up the in-gel digestion process. A new approach to speed up the protein identification process utilizing microwave technology was developed [16]. Proteins excised from gels are subjected to in-gel digestion with endoprotease trypsin by microwave irradiation, which rapidly produces peptide fragments. The peptide fragments were further analyzed by matrix-assisted laser desorption/ionization technique for protein identification. The efficacy of this technique for protein mapping was demonstrated by the mass spectral analyses of the peptide fragmentation of several proteins, including lysozyme, albumin, conalbumin, and ribonuclease A. The method reduced the required time for in-gel digestion of proteins from 16 hours to as little as five minutes.

Jin et al. [17] focused on the preparation and characterization of the antioxidant peptides by microwave-assisted enzymatic hydrolysis of collagen from sea cucumber *Acaudina molpadioides* (ASC-Am) obtained from Zhejiang Province in China. The results exhibited the effects of microwave irradiation on hydrolysis of ASC-Am with different protease. Neutrase was selected from the four common proteases (papain, pepsin, trypsin, and neutrase) based on the highest content and DPPH-scavenging activity of hydrolysate Fa (Molecular weight < 1 kDa.

Antioxidant peptides have elicited interest for the versatility of their use in the food and pharmaceutical industries. Antioxidant peptides were prepared by microwave-assisted alkaline protease hydrolysis of collagen from sea cucumber (*Acaudina molpadioides*) [18] The results showed that microwave irradiation significantly improved the degree of hydrolysis of collagen and the hydroxyl radical (OH·)-scavenging activity of hydrolysate. The content and OH-scavenging activity of collagen peptides with molecular weight ≤ 1 kDa (CPS) in the hydrolysate obtained at 250 W increased significantly compared with the non-microwave-assisted control.

Wang et al. [19] reviewed the characterization, preparation, and purification of marine bioactive peptides and other molecules. In their review they give references of the microwave assisted extraction method applied on fish tissues, oysters, and shrimp for different molecules; and microwave-assisted acid hydrolysis of proteins for peptide mass mapping and tandem mass spectrometric analysis of peptides has been reported [5].

They state that in general, the compounds are extracted more selectively and quickly by this technique, with similar or better yields in comparison with conventional extraction processes. Meanwhile, this technique also uses less energy and solvent volume, has reduced costs, and is more environmentally friendly than traditional extraction processes.

REFERENCES

1. Sereewatthanawut I., Prapintip S., Watchiraruji K., Goto M., Sasaki M., Shotipruk A. Extraction of protein and amino acids from deoiled rice bran by subcritical water hydrolysis. *Bioresource Technology*, 2008, 99, 555–561. https://doi.org/10.1016/j.biortech.2006.12.030
2. Raju A., Chun B.-S. Subcritical water hydrolysis for the production of bioactive peptides from tuna skin collagen. *The Journal of Supercritical Fluids*, 2018, 141, 88–96. https://doi.org/10.1016/j.supflu.2018.03.006
3. Ramachandraiah K., Koh B.-B., Davaatseren M., Hong G.-P. Characterization of soy protein hydrolysates produced by varying subcritical water processing temperature. *Innovative Food Science & Emerging Technologies*, 2017, 43, 201–206. https://doi.org/10.1016/j.ifset.2017.08.011
4. Koh B.-B., Lee E.-J., Ramachandraiah, H.G.-P. Characterization of bovine serum albumin hydrolysates prepared by subcritical water processing. *Food Chemistry*, 2019, 278, 203–207. https://doi.org/10.1016/j.foodchem.2018.11.069
5. Marcet I., Álvarez C., Paredes B., Diaz M. Inert and oxidative subcritical water hydrolysis of insoluble egg yolk granular protein, functional properties, and comparison to enzymatic hydrolysis. *Journal of Agricultural and Food Chemistry*, 2014, 62(32), 8179–8186. https://doi.org/10.1021/jf405575c
6. Powell T., Bowra S., Cooper H.J. Subcritical water processing of proteins: An alternative to enzymatic digestion? *Analytical Chemistry*, 2016, 88(12), 6425–6432. https://doi.org/10.1021/acs.analchem.6b01013
7. Chen L.C., Rahman M.M., Hiraoka K. High pressure super-heated electrospray ionization mass spectrometry for sub-critical aqueous solution. *Journal of the American Society for Mass Spectrometry*, 2014, 25(11), 1862–1869. https://doi.org/10.1021/jasms.8b04657
8. Álvarez-Viñas M., Rodríguez-Seoane P., Flórez-Fernández N., et al. Subcritical water for the extraction and hydrolysis of protein and other fractions in biorefineries from agro-food wastes and algae: A review. *Food and Bioprocess Technology*, 2020. https://doi.org/10.1007/s11947-020-02536-4
9. Marcet I., Álvarez C., Paredes B., Díaz M. The use of sub-critical water hydrolysis for the recovery of peptides and free amino acids from food processing wastes. Review of sources and main parameters. *Waste Management*, 2016, 49, 364–371. https://doi.org/10.1016/j.wasman.2016.01.009
10. Zhang J., Wen C., Li C., Duan Y., Z. H., Ma H. Antioxidant peptide fractions isolated from wheat germ protein with subcritical water extraction and its transport across Caco-2 cells. *Journal of Food Science*, 2019, 84(8), 2139–2146. https://doi.org/10.1111/1750-3841.14720
11. Hua L., Low T.Y., Sze S.K. Microwave-assisted specific chemical digestion for rapid protein identification. *Proteomics*, 2006, 6, 586–591. https://doi.org/10.1002/pmic.200500304
12. Pramanik B.N., Mirza U.A., Ing Y.H., Liu Y.-H., Bartner P.L., Weber P.C., Bose A.K. Microwave-enhanced enzyme reaction or protein mapping by mass spectrometry: A new approach to protein digestion in minutes. *Protein Science*, 2002, 11(11), 2676. https://doi.org/10.1110/ps.0213702

13. Zhong H., Zhang Y., Wen Z., Li L. Protein sequencing by mass analysis of polypeptide ladders after controlled protein hydrolysis. *Nature Biotechnology*, 2004, 22, 1291–1296. https://doi.org/10.1038/nbt1011
14. Lin S.-S., Wu C.-H., Sun M.-C., Sun C.-M., Ho Y.-P. Microwave-assisted enzyme-catalyzed reactions in various solvent systems. *Journal of the American Society for Mass Spectrometry*, 2005, 16, 581–588. https://doi.org/10.1016/j.jasms.2005.01.012
15. Zhong H., Marcus S.L., Li L. Microwave-assisted acid hydrolysis of proteins combined with liquid chromatography MALDI MS/MS for protein identification. *Journal of the American Society for Mass Spectrometry*, 2005, 16, 471–481. https://doi.org/10.1016/j.jasms.2004.12.017
16. Juan H.-F., Chang S.-C., Huang H.-C., Chen S.-T. A new application of microwave technology to proteomics. *Proteomics*, 2005, 5(4), 840–842. https://doi.org/10.1002/pmic.200401056
17. Jin H.-X., Xu H.-P., Li Y., Zhang Q.-W., Xie H. Preparation and evaluation of peptides with potential antioxidant activity by microwave assisted enzymatic hydrolysis of collagen from sea cucumber acaudina molpadioides. *Marine Drugs*, 2019, 17(3), 169–183. https://doi.org/10.3390/md17030169
18. Li Y., Li J., Lin S.-J., Yang Z.-S., Jin H.-X. Preparation of antioxidant peptide by microwave-assisted hydrolysis of collagen and its protective effect against H2O2-induced damage of RAW264.7 cells. *Marine Drugs*, 2019, 17(11), 642–655. https://doi.org/10.3390/md17110642
19. Wang X., Yu, H., Xing, R., Li P. Characterization, preparation, and purification of marine bioactive peptides. *Hindawi BioMed Research International*, 2017, Article ID 9746720. https://doi.org/10.1155/2017/9746720

CHAPTER 14

Fractionation and Purification of Bioactive Peptides

Chay Shyan Yea, Raman Ahmadi, Mohammad Zarei, and Belal J. Muhialdin

CONTENTS

Introduction	267
Membrane Technology	268
Ultrafiltration and Nanofiltration	268
Electrodialysis	270
Electrophoresis	271
Gel-Based Electrophoresis	271
Capillary Electrophoresis	272
Chromatography	273
Reverse Phase High Performance Liquid Chromatography	273
Ion Exchange Chromatography	275
Size Exclusion Chromatography	276
Miscellaneous Chromatographic Techniques	277
Multidimensional Separation of Peptides	279
References	290

INTRODUCTION

Fractionation and purification represent the intermediate steps after the peptide has been released from the parent protein molecule as a result of food protein hydrolysis but prior to mass spectrometry and bioinformatics analysis (proteomic and peptidomic studies). Protein hydrolysate is a complex pool of peptide mixture containing fragments of different properties in terms of sequence, length, net charge, polarity, and molecular size. Among these peptides, some are essentially involved in the host's metabolism while others are of no biological significance. In order to understand the structure–activity relationship and the mechanism of action between peptide and other molecules, identifying the "right" peptide responsible for such interaction, from a vast pool of peptide fragments, is deemed crucial. Besides, the complexity of a peptide mixture obtained from hydrolysis or fermentation processes is beyond the capacity of mass spectrometry and bioinformatics tools for direct processing. These add-up factors drive the need for peptide separation prior to further analysis.

    This chapter discusses the peptide fractionation based on three major separation techniques: (1) membrane technology, (2) electrophoresis, and (3) chromatography, with different underlying principles. One important note is that, while authors have tried their best to differentiate the separation techniques, there is often no clear boundary between the compared techniques. With knowledge and technology that is advancing from time to time, setting up a machine/tool encompassing different separation mechanisms as one unit becomes feasible. For instance, gel electrophoresis and capillary electrophoresis is typically combined as capillary gel electrophoresis to exploit the advantages from both gel and capillary; electrodialysis coupled with ultrafiltration membrane applies both charge selectivity (from electrodialysis) and size exclusion ability (from porous membrane) to achieve separation; and hydrophilic interaction/cation exchange chromatography relies on a combination of hydrophilicity and net charge differences among molecules for separation. Each technique has its own strength and weakness, thus using another technique to complement the former technique would reduce the limitation/disadvantage associated with the former technique, providing higher resolving power to the system and ultimately improving the separation efficiency. In cases in which combining two separation mechanisms into a single machinery is not feasible, separation is typically performed in a two-step/multidimensional manner, i.e., separation using method A followed by method B. Upon successful separation, a bioinformatics study will follow, involving identification of the peptide structures/sequences responsible for the bioactivity of interest, such as, but not limited to, antihypertensive, antioxidative, antimicrobial, anti-stress, antitumor properties, and allergy response, along with elucidation of associated mechanisms/interactions.

MEMBRANE TECHNOLOGY

Membrane fractionation usually serves as, but is not limited to, the first step during a multi-step separation process. It helps to concentrate a desirable compound by reducing the complexity of a crude mixture via the removal of unwanted substances using a membranous barrier with specific molecular weight cutoff (MWCO). It offers high selectivity and low operating costs, requires only mild working conditions (low temperature, neutral pH), and is able to maintain the structural and physicochemical integrity of the peptides, thus allowing direct move-on to the next phase of separation without the hassle of pretreatment or washing. Unlike electrophoresis and chromatography that separate peptides in a sophisticated manner to yield multiple fractions, membrane fractionation produces only two fractions, namely permeate (containing compounds with M_w smaller than the cutoff, thus able to pass through) and retentate (containing compounds with M_w larger than the cutoff, thus retained).

Membrane fractionation is divided into (1) conventional filtration processing (microfiltration, ultrafiltration, nanofiltration) and (2) electrically assisted filtration (with electrodialysis being the main focus in current discussion). Among these, microfiltration is not suitable for peptide separation, for its pore size (100–10,000 nm) is meant for removal of large entities like yeast, bacteria cells, and macromolecules such as non-hydrolyzed protein, lipid, and fiber (Bazinet and Firdaous, 2009; Etzel and Arunkumar, 2016). The application of ultrafiltration, nanofiltration, and electrodialysis is detailed in the following section.

Ultrafiltration and Nanofiltration

Ultrafiltration (UF) and Nanofiltration (NF) are both pressure-driven processes, i.e., separation is performed based on membrane selectivity according to pressure difference. UF and NF differ in their pore size and separation mechanism. UF has a typical pore size of 1–100 nm, corresponding to the separation of molecules at MWCO 1–300 kDa, while NF has a smaller pore size of 1–10 nm to separate molecules at < 10 kDa, particularly in the range of 0.1–5 kDa (Bazinet and Firdaous, 2009; Oatley-Radcliffe et al., 2017). UF performs separation by means of molecular size, while NF separates by a combination of molecular size and charge. When a peptide mixture has similar molecular weight, for example, all peptides showing M_w < 10 kDa, separation by molecular size alone becomes difficult. NF, being able to offer separation by a combination of size and charge mechanism, is then introduced.

UF is commonly applied after protein hydrolysis to enrich bioactive peptides from a complex mixture of peptide pool containing desirable and undesirable compounds. Enrichment is crucial to increase the peptide purity, which is then important to improve the functional/biological activity. To achieve better separation, UF could be performed in a stepwise manner, using membranes with different MWCO, starting with high MWCO followed by low MWCO. A high MWCO membrane serves to eliminate non-hydrolyzed protein, fiber, polysaccharide, and other macromolecules while at the same time collecting high M_w peptides with emulsifying and stabilizing functional properties. Further fractionation using low MWCO membrane yields antioxidant, bioactive peptides, amino acids, and targeted compounds for nutraceutical/pharmaceutical application (Vollet Marson, Belleville, Lacour, and Dupas Hubinger, 2021).

On the other hand, NF is used to separate peptide fragments with M_w < 5 kDa, and NF with a < 1 kDa cutoff is particularly useful to separate short-chain oligopeptides collected after protein hydrolysis (Muro, Riera, and Fernández, 2013). In fact, it has been concluded that, irrespective of the complexity of an amino acid/peptide mixture, charge factor remains the most important criterion in order to have a successful separation of peptides with similar molecular weight (Muro et al., 2013). This finding signifies the essentiality of NF in peptide separation. Unlike UF, which sieves peptides in a straightforward manner that is based solely on molecular cutoff size, the selectivity of NF membrane is more complicated, since it relies not only on molecular size for exclusion but also on the electrostatic interaction between peptide–peptide and peptide–membrane as well. Briefly, at low feed concentration, charged peptides distribute selectively at the membrane interface, then are transported across the membrane via diffusion, convection, and electrophoretic mobility (Martin Orue, Bouhallab, and Garem, 1998; Muro et al., 2013). According to Donnan's theory, when a peptide possesses the same charge with the membrane (i.e., co-ion with the membrane), the peptide is electrically rejected for transportation across the membrane. In contrast, a peptide possessing opposite charge to the membrane (i.e., counter-ion), it is allowed to pass through. This then separates the peptide mixture. However, the exploitation of Donnan's theory to predict the behavior of an individual peptide in a peptide mixture consisting of charged peptides is very limited mainly because the physical property of the NF membrane is difficult to measure and the interaction between peptide and membrane remains complex and unclear, thus halting the mathematical modeling of such systems (Muro et al., 2013; Oatley-Radcliffe et al., 2017). In fact, a recent review by Vollet Marson et al. (2021) on the recovery of valuable compounds from yeast protein hydrolysate using membrane fractionation stated that, to date, there is no model that is universally satisfactory to describe the mass transfer and fouling mechanism during the fractionation process.

Two major limitations encountered during membrane separation of proteinaceous material are limited selectivity and fouling. Limited selectivity will result in low mass transfer, thus low separation efficiency. On the other hand, fouling is an undesirable yet spontaneous deposition/accumulation of substances including macromolecules, suspended particles, colloid, or bacteria on the membrane surface or entrapment in the membrane pore, causing blockage of the passageway and affecting the permeability of molecules, and ultimately reducing the membrane flux (amount of permeate allowed to pass through the membrane per unit of time) (Kumar and Ismail, 2015). Peptides are highly susceptible to fouling because they possess various charged groups that readily interact with charged membrane surfaces, ions, and water molecules in the solution (Pouliot, Gauthier, and Groleau, 2005; Vollet Marson et al., 2021). Additionally, it has been reported that the adsorption of peptides on membrane depends on the peptide characteristics, including charge, hydrophobicity, and molecular size (Gourley, Britten, Gauthier, and Pouliot, 1994), as well as membrane characteristics, including type of material (polyether sulfone, polyvinylidene fluoride, ethylene vinyl alcohol, cellulose acetate), zeta potential, roughness, thickness, and contact angle (Persico, Dhulster, and Bazinet, 2018). In particular, separation of peptides in the food area becomes tricky, as protein/peptide is a nutrient source that supports microbial growth, leading to the formation of biofilm and jeopardizes the microbiological safety of the collected permeate and retentate (Castro-Muñoz, Barragán-Huerta, Fíla, Denis, and Ruby-Figueroa, 2018; Vollet Marson et al., 2021). To tackle these issues, electrically assisted membrane fractionation has been introduced.

Electrodialysis

In electrically assisted membrane fractionation, an external electrical field is applied as an additional driving force to the pressure gradient to enhance separation efficiency. This is particularly useful in the separation of bioactive peptides, for these peptides generally have a low M_w of < 5 kDa, making separation by molecular size alone difficult, as a result of limited membrane selectivity. Two types of electrically-assisted fractionation are commonly used, electrodialysis (ED) and force-flow membrane electrophoresis. The former technique involves superimposing an electric field to a conventional membrane filtration unit, thus selectivity is based on the electrophoretic mobility of the molecule of interest as well as the molecular sieving effect by the membrane. The latter involves "forcing" a portion of the fluid containing the molecule of interest across a membrane to allow pass-through of the said molecule, and the selectivity is based on charge (Bazinet and Firdaous, 2009). Since force-flow technique is rarely applied in peptide separation, it is not discussed further.

In the context of peptide separation, ED remains one of the most promising techniques. It separates peptides by transporting charged species across at least one ion exchange membrane from a desalinated region (removal of ion) to a concentrated region (recovery of ion). The membrane is either cationic or anionic and is traditionally made of polyelectrolytes with fixed charges. Due to the stacking of ionic membrane and the fact that peptide is an organic substance, peptide fouling is prone to occur (Dlask and Václavíková, 2018). Thus, electrodialysis coupled with ultrafiltration membrane (EDUF) has come into play. EDUF has been widely used in the separation of bioactive peptides. This technique, comprises of both ultrafiltration and ion exchange membranes, being stacked into a conventional electrodialysis cell and was developed and patented by Bazinet, Amiot, Poulin, Labbé, and Tremblay (2004). Unlike other electrically-assisted membrane fractionation, EDUF applies no pressure gradient but depends on the electrophoretic mobility and molecular size difference among charged peptides to achieve separation (Doyen, Beaulieu, Saucier, Pouliot, and Bazinet, 2011). The mass transfer flow rate of peptide could be regulated by manipulating the supplied voltage, feed solution concentration, solution pH, and membrane property (compact vs. porous structure). In fact, the porous membrane has been found to be highly selective and is capable of separating targeted peptides from complex mixtures and residues from food and agro by-products (Vollet Marson et al., 2021). Przybylski et al. (2020) demonstrated the usefulness of EDUF to harness peptide-based food preservatives from slaughterhouse by-product (blood), while Durand et al. (2020) has shown the separation of anti-inflammatory peptide from herring fish, and Marie et al. (2019) has shown the separation of antihypertensive peptide from defatted flaxseed protein hydrolysate. These findings unanimously depict the feasibility of EDUF in the separation of bioactive peptides.

ELECTROPHORESIS

Electrophoresis, as its name suggests, applies electrical current (voltage) to induce movement among charged peptide molecules to achieve separation. Two electrophoretic techniques remain popular to date, namely, gel-based electrophoresis and capillary electrophoresis (CE). The former technique involves peptide separation on gel slabs (typically polyacrylamide gel added, with sodium dodecyl sulphate for protein/peptide works and agarose gel for DNA works), while the latter separates peptides in a fluid matrix using a fused silica capillary column (Burgi and Smith, 1995). Both techniques apply isoelectric focusing (IEF) as the core underlying principle to fractionate peptides based on isoelectric point (pI) over a continuous range of pH gradient. Having both a cationic and anionic terminal, a peptide is amphoteric in nature. When electrical current is passed through the gel/fluid matrix, a pH gradient is created. The charged peptide then migrates along the continuous pH gradient until reaching its pI and stops moving as a result of zero net charge, (i.e., reaching an equilibrium state with zero mobility). Peptides are now being separated based on their respective pIs.

In the 1980s, immobilized pH gradient (IPG) was introduced by Bjellqvist et al. (1982), successfully tackling protein smearing and uneven conductivity problems in conventional IEF. In IPG, a buffer matrix forming the pH gradient is covalently immobilized onto polyacrylamide gel, thus creating a more reproducible and stable pH gradient for improved separation (Kaplan, 2006; Righetti, Hamdan, Antonucci, Verzola, and Bossi, 2004). Formerly known as free-solution capillary zone electrophoresis (placement of protein solution within a narrow zone from an electrode), CE has advanced over time, replacing free solution with solution stabilized on supporting media such as silica gel, alumina, cellulose, agarose, and polyacrylamide gel. Switching from free solution to supporting media minimizes deleterious band-broadening from convection and diffusion associated with free solution and allows direct staining/fixing of the sample upon completion of electrophoresis (Kaplan, 2006). An interesting fact to note is that the two techniques overlap each other. There is no distinct boundary between the two techniques, due to the advancement in electrophoresis over the years (discovery of polyacrylamide gel and immobilized pH gradient) is applicable to both techniques and the fact that they are often coupled in a 2-dimensional electrophoresis to achieve better separation.

Gel-Based Electrophoresis

The most important and widely used gel electrophoresis in peptide work is sodium dodecyl sulphate polyacrylamide gel electrophoresis (SDS-PAGE). Acrylamide is polymerized and cross-linked (using bisacrylamide), forming a mechanically strong and chemically inert porous gel to achieve high resolution and efficient separation of peptides over a wide range of molecular mass (Reddy and Raju, 2012). The gel porosity decreases with increasing acrylamide content. However, since peptides are much smaller than proteins (< 10 kDa for peptide *vs.* 10–100 kDa for protein), increasing the acrylamide content alone does not clearly separate the peptides. Hence, tricine buffer was introduced to replace glycine in a conventional SDS-PAGE (Schägger and Von Jagow, 1987). Peptide separation further advanced when Sarfo, Moorhead, and Turner (2003) proposed to load peptide natively onto gel of low SDS concentration (below critical micelle concentration) using acrylamide at 30% T (total monomer concentration, i.e., gram of acrylamide and bisacrylamide per 100 mL) and 2.67% C (bisacrylamide concentration relative to total monomer concentration). The separation of small peptides (1–3 kDa) is made achievable via the enhancement of electrostatic interaction between SDS monomer and peptide under low voltage condition to result in a slow migration of peptide across acrylamide gel. Similarly, Zilberstein, Korol, Shlar, Righetti, and Bukshpan (2008) proposed to separate peptides (0.5–10 kDa) using a shallow positive charge gradient coupled with a large pore size polyacrylamide gel (4% T, 2.5% C, to minimize obstruction during peptide movement on the gel). The authors claimed this to be an "unorthodox" method, as larger peptides (carrying more negative charge on the surface upon denaturing by SDS) move further down the gel toward the positively charged electrode (anode) compared to smaller peptides (carrying lesser negative charge). This is in contrast to the conventional SDS-PAGE where smaller peptides would move faster, reaching the bottom of the gel quicker than larger ones. This approach exploits the "steady-state" condition of SDS-peptide micelles by creating a positive charge gradient within the gel matrix. As the micelle migrates along the charge gradient, it will move until it matches the surrounding charge density, eventually reaching a stop and remains stationary (i.e., a steady-state condition).

Unlike IEF that separates peptides based on charge, SDS-PAGE performs separation based solely on molecular size. This is made possible by the denaturing property of SDS. As a detergent with ionic head and alkyl tail, SDS binds to all nonpolar and charged groups on the peptides and makes them uniformly negatively charged. This causes all molecules to migrate in the same direction toward the anode (positive end), but at different speeds, driven solely by molecular size and is irrespective of their initial charges, whereby smaller peptides would migrate faster than larger, bulkier ones, reaching the bottom of the gel first. When the peptides move at different rates, they become separable. Finally, the gel is stained using Coomassie Blue G-250 or R-250 for band visualization. Silver staining is performed only when higher sensitivity is needed, for it is more expensive and laborious compared to Coomassie Blue staining.

Capillary Electrophoresis

Capillary Electrophoresis (CE) is conducted on a capillary made of high-purity, fused silica with an inner diameter of ≤ 100 μm. It was first introduced as a thermal controlling innovation to minimize convection and heat transfer problems in wide-tube electrophoresis, thus improving peak sharpness and resolution when detected using optical devices (Whatley, 2001). Due to the high surface-to-volume ratio inside the capillary, the occurrence of electroosmosis is significant. It is the spontaneous bulk movement of fluid near a charged surface in the presence of an electrical field (Whatley, 2001). When the electroosmotic force exceeds the electrophoretic force (the force that is supposed to guide the peptide movement direction), the molecule of interest would move in the opposite direction. This can be controlled by changing the buffer pH, whereby low pH leads to slow electroosmosis while high pH leads to fast electroosmosis. Thus, electroosmosis can be optimally manipulated to aid in difficult separation (Burgi and Smith, 1995).

In the context of food science, CE has been widely used in food analysis and foodomics study. Food analysis involves monitoring changes in food quality and safety aspects during processing and storage as well as detecting compounds such as carbohydrates, organic acids, flavonoids, pigments, vitamins, proteins/peptides, additives, and hazardous amines in food and beverages (Siren, 2015). In foodomics, an integrated area of food and nutrition, CE is mainly used to perform metabolite profiling that provides insights into the effects of functional food ingredients on health (Ibáñez C, 2016). It has been shown that, when coupled with mass spectrometry, CE could be used to study the antiproliferative effect of dietary polyphenol on colon cancer cell (Ibáñez et al., 2012). Besides, CE can be used to monitor the progress of protein hydrolysis and optimize the hydrolytic conditions to produce a representative peptide fingerprint that is essential for subsequent structural elucidation in peptidomic study (Burgi and Smith, 1995).

CHROMATOGRAPHY

Chromatography, indicative of "color writing", was first discovered by a Russian botanist, Mikhail Tsvett, who separated plant pigments using chalk ($CaCO_3$) packed in a glass column back in 1903. Liquid chromatography is a physical separation technique performed in a liquid phase. A mixture is separated by the adsorption/desorption efficiency between solute of interest (found in the mobile phase) and stationary phase. Stationary phase consists of adsorbents packed within a column while mobile phase is a carrier liquid, usually an organic solvent. High performance liquid chromatography (HPLC) is a modern form of liquid chromatography that uses a column that is neatly packed with small particles through which the mobile phase carrying the sample to be separated would flow at high pressure (Dong, 2006).

HPLC has become the central technique in peptide separation and has played a critical role in the rapid advancement of biological and biomedical sciences, since it started to gain the limelight in the 1970s. The enormous success of HPLC over the past 40 years can be attributed to several features, including high reproducibility, high recovery, and excellent resolution that can be achieved even for identically similar molecules. This is due to the fact that the interaction involved between solute (in the mobile phase) and stationary phase is based on forces that can be easily manipulated through changes in the elution conditions and pH. Manipulating the difference in structural properties, such as hydrophobicity, molecular size, and charge thus maximizes the separation potential, yielding a less complex mixture from a crude, highly complicated one, demonstrating the versatility of HPLC during peptide separation. The four major chromatographic techniques that are widely applied in peptide work are discussed as follows.

Reverse Phase High Performance Liquid Chromatography

To date, Reverse Phase High Performance Liquid Chromatography (RP-HPLC) remains the most popular chromatographic technique in peptide study, for it demonstrates the ability to work with mass spectrometry, enabling sequence identification that is crucial for peptidomic research, and offers superior resolution to separate polypeptide of nearly identical sequences. Rivier and McClintock (1983) reported that, RP-HPLC successfully resolved three insulin variants collected from bovine, porcine, and human, respectively, which differ only between 1 to 3 amino acids among each other. Besides, RP-HPLC has been demonstrated to separate two decapeptides that differ only by one single amino acid, whereby one peptide contains serine while the other contains threonine at the third position from the N-terminal (Carr, 2005). The high resolving power of RP-HPLC sets forth its fundamental role in peptide separation.

The term "reverse phase" is derived from the utilization of opposing phase polarity as compared to normal phase chromatography. In normal phase, the column is packed with a polar stationary phase (silica containing polar silanol groups) and coupled with a nonpolar mobile phase (organic solvent) to achieve separation. In contrast, the RP-HPLC column is packed with a nonpolar stationary phase (silica bonded with hydrocarbon alkyl chain) and pairs with polar mobile phase (water). The stationary phase is made up of spherical silica beads, usually 3 or 5 µm for an analytical column and 10 or 20 µm for a preparative column. The purpose of choosing silica is because it demonstrates physical robustness and remains stable under most conditions (except at pH > 6.5). The silica beads are then chemically attached to a hydrocarbon chain (usually a linear, aliphatic 18-carbon chain, thus the name C-18 is given) through covalent bonds to create a hydrophobic surface. It is on this thick layer of hydrocarbon surface that peptide adsorption/desorption occurs (Carr, 2005). Within the column, tiny beads are evenly distributed in a porous manner to increase surface area and facilitate binding of solutes from the mobile phase. For peptides separation, the pore size is generally 100–300 Å while protein separation requires a pore size >300 Å. As a rule of thumb, the solute diameter must at least be one-tenth the size of the pore diameter to avoid restricted diffusion of the solute and to allow the total surface area of adsorbents to be accessible (Aguilar, 2004).

RP-HPLC manipulates the difference in degree of hydrophobicity among peptide molecules to achieve separation. Bearing different sequences and numbers of amino acids, each peptide has a different hydrophobicity. The peptide sample to be separated is first dissolved in buffer A (usually water). The same buffer is used to condition the HPLC system prior to injection. All molecules that cannot bind will first be washed out. Within the column, the retained peptide interacts with the hydrophobic ligand through a chromatographic contact region at the surface of the stationary phase by partially adsorbing itself to the surface via strong hydrophobic interaction while exposing its remaining part to the surrounding mobile phase. When buffer B (usually organic solvent) is introduced into the system, a mobile phase mixture containing both buffer A and B is produced. Elution is either performed at a fixed ratio of buffer A:buffer B (known as isocratic elution) or at a gradually changing ratio of buffer A:buffer B (known as gradient elution) to alter the ionic strength. When the mobile phase reaches a specific concentration that matches the ionic strength of the peptide, the peptide then desorbs from the hydrophobic surface and elutes out from the column to be collected. The peptide with higher hydrophobicity will be eluted earlier than the peptide with lower hydrophobicity. Detection of peptides usually falls between the wavelength of 210–220 nm, which is specific for peptide bond, or at 280 nm, which corresponds to aromatic amino acids such as tryptophan and tyrosine, when the compound is known or suspected to contain a high amount of such amino acids.

During peptide separation, RP-HPLC usually requires an ion-pairing agent to improve the peak shape in the chromatogram. Ion-pairing agent is an acid added into the mobile phase at low concentration ranging from 0.005 to 0.01% (v/v). Trifluoroacetic acid (TFA), formic acid, and acetic acid are frequently used, with TFA showing the highest popularity because it produces better peak shape than the other two under most circumstances. Silica, used as the stationary phase, often contains metal impurities that will cause peak tailing and loss of resolution, jeopardizing the peak quality. Thus, TFA is added, for it could react with the basic moiety on peptides and keeps the pH below the pK_a of carboxyl side chain to maximize retention (Brandes, Claus, Bell, and Aurand, 2010). Also, the acidic nature of TFA will prevent the ionization of silanol groups on silica beads that is also acidic, suppressing ionic interactions between silica surfaces and peptides to facilitate sample separation and improve peak quality. However, in cases where mass spectrometry is needed, formic acid is preferred over TFA due to its higher compatibility. Alternatively, it is advisable to use a silica column of high purity when mass spectrometry follows, in order to keep the concentration of ion pairing agent the lowest possible. When a high concentration of ion pairing agent is used, it will cause signal reduction at the electrospray interface associated with the mass spectrometry unit, resulting in poor detection signal and sensitivity. Thus, high purity silica is preferred.

Ion Exchange Chromatography

Ion exchange chromatography (IEX) was introduced in the late 1940s to separate proteins (Velickovic, Ognjenovic, and Mihajlovic, 2012) and was then developed to separate other charged biomolecules, including peptides and nucleic acids, based on the net charge difference among the analyte of interest. Peptide, being built from different compositions of amino acids containing acidic and basic groups, behaves as a zwitterion and has its net charge easily altered when the surrounding pH changes. When the pH exceeds the (pI) of a peptide, the amino group will be deprotonated ($-NH_2$) while the carboxyl group will hold a negative charge (COO^-), resulting in net negative charge; when the pH falls below the pI, the amino group will be protonated and hold a positive charge ($-NH_3^+$) while the carboxyl group will become neutral ($-COOH$), resulting in net positive charge. IEX takes advantage of this net charge difference among peptide molecules to achieve separation.

Like RP-HPLC, IEX consists of two phases, i.e., a stationary phase containing charged functional groups that are covalently attached to a solid matrix and a mobile phase carrying the peptide mixture. There are two types of stationary phase, known as cation exchanger and anion exchanger, and both involve exchanging of counter ions with peptides of interest via ionic interaction at the stationary phase surface to achieve binding. Silica, synthetic resin, and polysaccharide (dextran, cellulose, agarose) are among the typical materials used to build the stationary phase. Among these materials, silica portrays a major drawback of instability at high and low pHs (pH < 2 and > 8) (Weston and Brown, 1997). In cation exchange chromatography, the solid matrix consists of negatively charged groups that are initially associated with cation (for example Na^+) from the mobile phase. These cations act as counter ions which are then "exchanged" with positively charged peptides during adsorption/binding. In anion exchange chromatography, the solid matrix is positively charged and reacts with negative counter ions (for example Cl^-) from the mobile phase. When the peptide sample is introduced, these negative counter ions will be displaced/exchanged with negatively charged peptides to allow binding to the column.

IEX separates peptides via two modes: (1) changing the pH of the mobile phase, i.e., by pH gradient and (2) changing the concentration of the displacing ions, i.e., by salt gradient. Both processes induce a gradually changing ionic strength in the mobile phase, allowing peptide molecules to desorb from the stationary phase at different times according to the changing pH/salt concentration. When the pH/salt matches the ionic interaction between peptide and stationary phase, the respective peptide will be eluted. The basic principle is explained as follows: when the peptide shows stronger affinity toward the stationary phase than toward the mobile phase, it remains adsorbed to the stationary phase and gets retained in the column. When the peptide-mobile phase interaction surpasses that of the peptide-stationary phase (via changing pH or salt concentration), the peptide gets eluted.

During peptide separation, a strong cationic exchanger (SCX) coupled with salt gradient is typically applied (Essader, Cargile, Bundy, and Stephenson Jr, 2005), whereby the peptide possessing the least positive net charge will be eluted first. One thing that allows the peptides to be resolved successfully under strong cationic conditions is its higher structural flexibility that enables spontaneous renaturation to withstand drastic and harsh conditions, in contrast to proteins, with limited flexibility to renature. Credited to this flexibility, peptide separation remains feasible even under pH as low as pH 2 (Velickovic et al., 2012). In recent years, the usage of pH gradient is gaining more interest over salt gradient, for it demonstrates higher buffer compatibility with mass spectrometry analysis that follows and requires no extensive washing steps (Zhou et al., 2007). Also, the feasibility of strong anion exchanger (SAX) has gained much popularity nowadays for peptide separation. While SCX has been widely used for quite some time, SAX has not been widely considered for peptidomic analysis until recent years (Giese, Ishihama, and Rappsilber, 2018; Ritorto, Cook, Tyagi, Pedrioli, and Trost, 2013). The majority of the peptides released from tryptic digestion (trypsin is the most common proteolytic enzyme, for it is an endopeptidase with high specificity and releases peptides with desirable size) have pIs with pH range of 3–5 as a result of posttranslational modification, such as phosphorylation and acetylation, which decreases the pIs of resulting peptides (Cargile, Sevinsky, Essader, Stephenson Jr, and Bundy, 2005), indicating that the peptide mixture obtained after tryptic digestion is likely acidic and negatively charged, thus made separable by SAX (Ritorto et al., 2013).

Size Exclusion Chromatography

The idea of size-based chromatographic separation first appeared in the 1950s when Synge and Tiselius (1950) observed the exclusion of small molecules from the pores of zeolite as a function of molecular size (Hong, Koza, and Bouvier, 2012). In modern separation, size exclusion chromatography (SEC) is often applied as the first stage of removing unwanted substances and cleaning up the sample, so that it contains the desirable compounds in a more concentrated form before the subsequent separation. The stationary phase in SEC consists of a porous gel matrix with well-defined pore size. As a rule of thumb, small pore size is required to achieve separation for small molecules and *vice versa*. In the case of large peptide separation (M_w of 44–100 kDa), a pore size of 200 Å diameter provides optimum resolution. On the other hand, separation of small peptides (M_w 1–44 kDa) is achieved at 125 Å (Hong, Koza, and Fountain, 2012). When a peptide mixture is loaded onto the column, extremely large particles that are unable to enter the pore quickly flow through the column and get eluted first as the void volume. In contrast, peptides that fall within the fractionation range get diffused into the pores, traverse a longer distance through the pores and remain momentarily within the column before elution. The smaller the peptide, the longer is the retention time, resulting in larger elution volume. Since the separation is based on the movement of particles according to size, isocratic elution at a constant gradient is usually performed in SEC. The mobile phase selection is important to prevent any unwanted interaction (either ionic or hydrophobic) between peptide and the solid matrix (the stationary phase) or between peptides themselves. For the separation of charged molecules, like protein and peptide, the buffer is usually denaturing and contains organic solvent, acid, or additives (Hong, Koza, and Fountain, 2012). The purpose of adding an organic solvent into the mobile phase is to reduce the hydrophobicity, which is particularly useful for peptides containing a high amount of hydrophobic amino acid residue (Irvine, 2003). The purpose of adding salt/pH adjustment is to minimize ionic interaction. For example, a buffer containing up to 100 mM NaCl could be used during protein/peptide separation. Higher salt concentration is not advisable as it would cause an adverse effect and jeopardize the separation efficiency (Aguilar, 2004).

The separation mechanism of SEC exploits the molecular size difference between molecules or, more specifically, the hydrodynamic radius (R_h) of the analyte. R_h is measured in angstrom units (Å) and gives an idea about the molecular size and shape, which could be further extrapolated to predict the molecular weight of the compound of interest. Therefore, SEC is sometimes used to estimate the molecular weight of an unknown compound, which is important during the characterization of small proteins and peptides. In order to provide an accurate estimation of the molecular weight, separation should be performed under ideal conditions. That it, secondary interactions, such as ionic interaction between analyte and free silanol from the solid matrix, as well as hydrophobic interaction between analyte and hydrophobic site on the solid matrix, should be kept as low as possible so as to minimize adsorption. The mobile phase is known to induce structural changes on peptides whereby, under native conditions, a peptide maintains a stable secondary structure and, in the presence of a denaturant, a peptide tends to form random coils. These structural changes then alter the R_h and elution volume, leading to an inaccurate molecular weight estimation. Therefore, the mobile phase should be carefully chosen to provide a minimally nonideal condition, ensuring a size-based peptide separation that gives a good prediction of the molecular weight (Hong, Koza, and Fountain, 2012).

Miscellaneous Chromatographic Techniques

Flash chromatography (FC) is an emerging technology that has started to gain attention only in the past 10 years. It is either performed under normal phase (polar stationary phase coupled with moderately nonpolar/organic mobile phase) or reverse phase (nonpolar stationary phase coupled with moderately polar mobile phase) conditions, depending on the sample compatibility (i.e., in which solvent does the sample dissolve best). FC is categorized as preparative chromatography due to its ability to isolate and purify a specific/desirable compound from a mixture in a large-scale manner. Unlike a conventional preparative column that has an average particle size of 5 μm as the solid matrix, FC uses much larger particles of 25–50 μm to construct the solid phase. This allows the accommodation of higher sample load in flash column, which then improves the purification efficiency, i.e., having the ability to separate more compounds per gram of solid matrix (Bickler, 2020). Besides, FC uses a column that is relatively shorter and wider, enabling a higher flow rate at lowered pressure compared to a conventional column. Also, a conventional column is made of stainless steel while a flash column is made of less expensive material, i.e., polypropylene and polyethylene, thus the cost to replace a flash column is lower than a conventional column. It has been estimated that the cost for a flash chromatography setup and a regular preparative HPLC setup is USD 207.00 and USD 2600.00, respectively (Bickler, 2020). This estimation strongly depicts the cost effectiveness offered by flash chromatography. Because of these benefits, FC is often used to purify compounds quickly and inexpensively. In cases where the collected fraction is not sufficiently pure, FC serves as an intermediate step ahead of RP-HPLC, purifying compounds up to 80% purity prior to further RP-HPLC to reach > 95% purity in the final compound. In short, depending on the purity needed, FC serves as either the sole purification step or as a pre-purification step prior to final HPLC purification (Sørensen, Mishra, Paprocki, Mehrotra, and Jensen, 2021). For peptides with high/similar hydrophobicity, RP-HPLC is sometimes not powerful enough to resolve the peptides and achieve good separation. Thus, the use of flash chromatography comes in handy. Its application is uncommon in peptide separation; however, it is not impossible. In fact, by using FC, Sørensen et al. (2021) has recently demonstrated the successful separation of short, medium, and long peptides that are highly hydrophobic and poorly soluble, setting ground for the potential of FC to be established as a rapid and inexpensive early-step peptide separation technique when performing peptidomic work.

Supercritical fluid chromatography (SFC), like the name suggests, involves the usage of a supercritical fluid, usually carbon dioxide (CO_2), as the mobile phase, which differs from the conventional chromatography that uses a liquid solvent (be it polar or nonpolar). CO_2 is compressed and heated above the critical temperature and pressure to reach a state known as "supercritical fluid", where increasing the temperature and/or pressure would not transform a liquid into a gas or a gas into a liquid. Here, CO_2 exists as an intermediate between liquid and gaseous phase with no distinctive boundary and thus SFC is recognised as a hybrid between gas chromatography and liquid chromatography. It offers several advantages, for instance, low viscosity and high diffusivity to allow a low back pressure and a high flow rate that would not be possible with a conventional liquid chromatography system. In fact, SFC has a proven shorter elution time (less than 12 min *vs.* 50 min, when using RP-HPLC) during the separation of five peptides having low M_w < 1 kDa in the work performed by Tognarelli, Tsukamoto, Caldwell, and Caldwell (2010). Also, CO_2 is relatively cheaper than the organic solvents used heavily in liquid chromatography; thus, it is more economical and cost-friendly. CO_2 demonstrates low toxicity and is thus safe for the application in the food industry. For instance, it is widely applied during the extraction of thermally labile unwanted components from black pepper (Sankar, 1989), essential oils (Tezel, Hortaçsu, and Hortaçsu, 2000), ginger, nutmegs and vanilla while retaining the characteristic flavors and aromas; decaffeination in coffee product (Saldaña, Zetzl, Mohamed, and Brunner, 2002); as well as enrichment of high-value nutraceuticals in food products (Datta, Auddy, and Amit, 2014). Besides being a well-known greenhouse gas, CO_2 is recycled in SFC. Instead of releasing the gas into the environment and causing adverse effects, SFC utilizes CO_2 to achieve separation, and the fact that it can be reused further strengthens its positive environmental impact, making it a "green and environmentally friendly" chromatographic technology.

Despite the fact that SFC has been commercially available since the early 1980s (Datta et al., 2014), its application was originally limited to nonpolar compounds, since supercritical CO_2 is nonpolar and has a very low elution strength. Then, the separation of hydrophilic and polar compounds by SFC became possible when multiple parameters, such as column type, presence of organic modifiers (methanol, acetonitrile), and acidic/basic additive (trifluoroacetic acid, acetic acid, ammonium acetate) in the mobile phase, were investigated in an attempt to elucidate conditions suitable to improve separation (Enmark et al., 2018; Tognarelli et al., 2010). It was reported that the selection and concentration of mobile phase additives is crucial to induce drastic improvements in the peak shape (J. Zheng, Pinkston, Zoutendam, and Taylor, 2006). With these peptides, chiral/achiral pharmaceutical and drug, and some other ionic analytes were successfully separated using SFC since the millennium years (Ashraf-Khorassani, Taylor, and Marr, 2000; Pinkston, Wen, Morand, Tirey, and Stanton, 2006; J. Zheng, Glass, Taylor, and Pinkston, 2005).

To date, the feasibility of flash chromatography and SFC to separate food protein hydrolysate (a complex peptide mixture after hydrolysis containing thousands or more copies of peptides at different length and sequence) remains unaddressed. Most works focus on the separation of simple peptide mixtures containing just a few sequences or the isolation of a compound with known structure. This leaves a knowledge gap on the large-scale separation of protein hydrolysates derived from food, particularly those showing biological activity significance. This is, thus, worth investigation by food scientists in the field.

MULTIDIMENSIONAL SEPARATION OF PEPTIDES

With advancing knowledge and technology in the field, many separation techniques based on different separation mechanisms, have been developed over the years to separate peptides efficiently in a sophisticated manner, from membrane technology to electrophoresis and liquid chromatography. While these techniques have been consistently used, each has its own limitations. For instance, high pH buffer (> pH 9) during RP-HPLC separation is not mass spectrometer-friendly and requires an extensive desalting step which likely results in severe sample loss (Mohammad, Jäderlund, and Lindblom, 1999). There is no single technique that has the full power to resolve peptides completely. For the separation of food-derived peptides, it becomes more complicated as the sample is hydrolyzed from a highly complex, native food protein sample. To address this limitation, separation is performed in an orthogonal, multidimensional manner that involves the coupling of several analytical separation techniques. Based on different mechanisms, these techniques resolve peptides in a way that each technique complements each other to improve the overall separation efficiency. While there is no limit to the number of possible dimensions, it is usually two-dimensional, with membrane separation as the first dimension and liquid chromatography (typically RP-HPLC) as the second dimension being the most popular combination (Acquah, Chan, Pan, Agyei, and Udenigwe, 2019). The selection of multidimensional approaches should be carefully manipulated to minimize issues, like sample loss and incompatibility with subsequent analysis.

The compilation of separation techniques performed either in single or multi-dimension, for obtaining bioactive peptides of different sequences from different sources of food protein, is tabulated in Table 14.1. Only edible proteins from food/animal feed sources are listed.

TABLE 14.1 Techniques Used to Achieve Separation of Peptide Sequences Responsible for Various Biological Activities from Different Food Proteins

No.	Sample Source	Technique Involved	Multi-dimensional or Single dimension	Peptide Sequence	Biological Activities	Ref.
	Buffalo skimmed milk	Size exclusion chromatography (SEC) using a FPLC	Single dimension	IPPK IVPN QPPQ DMPIQ LPVPQ APFPE FPGPIPK YPVEPFT YPFPGPIPK GPFPIIV	ACE inhibitory Antioxidant	(Abdel-Hamid, Otte, De Gobba, Osman, & Hamad, 2017)
	Camel milk	Gel permeation high performance liquid chromatography (GP-HPLC)	Single dimension	LPVPQ WK	Dipeptidyl peptidase-IV (DPP-IV) inhibitory	(Nongonierma, Paolella, Mudgil, Maqsood, & FitzGerald, 2017)
	Milk	UHPLC	Single dimension	IPP VPP	ACE inhibitory (+Antihypertensive in rat)	(Y. Chen et al., 2014)
	Whey protein concentrate	Polymeric MICROZA membrane Gel filtration chromatography	Multi-dimension	VAGTWY SAPLR VLDTDYK KIDAL ENSAEP IPAVFK VAGTWY IPAVF KIPAVF HTSGY	ACE inhibitory	(Alvarado, Muro, Illescas, Díaz, & Riera, 2019)

(Continued)

TABLE 14.1 (Continued)

No.	Sample Source	Technique Involved	Multi-dimensional or Single dimension	Peptide Sequence	Biological Activities	Ref.
	Parma dry cured ham	Semi-preparative RP-HPLC-UV	Single dimension	LGL ALM SFVTT GVVPL NSIM	ACE inhibitory	(Dellafiora et al., 2015)
	Dwarf gulper shark	Dialysis membrane HPLC RP-HPLC	Multi-dimension	VW	ACE inhibitory Antioxidant	(Ikeda et al., 2015)
	Cuttlefish (*Sepia officinalis*)	Sephadex G-25 gel filtration RP-HPLC	Multi-dimension	Numerous peptides have been identified without pinpointing which sequence was responsible for the activity, thus not listed.	ACE inhibitory (+Antihypertensive in rat)	(Balti et al., 2015)
	Loach (*Misgurnus anguillicaudatus*)	Ultrafiltration Gel filtration chromatography (GFC) RP-HPLC	Multi-dimension	AHLL	ACE inhibitory (+Antihypertensive in rat)	(Li, Zhou, Zeng, & Yu, 2016)
	Ostrich (*Struthio camelus*) egg white protein	RP-HPLC (semi-preparative)	Single dimension	WESLSRLLG	ACE inhibitory Antioxidant	(Asoodeh, Homayouni-Tabrizi, Shabestarian, Emtenani, & Emtenani, 2016)

(*Continued*)

TABLE 14.1 (Continued)

No.	Sample Source	Technique Involved	Multi-dimensional or Single dimension	Peptide Sequence	Biological Activities	Ref.
	Egg-yolk	Ultrafiltration Gel filtration chromatography (GFC) RP-HPLC	Multi-dimension	QSLVSVPGMS	Antioxidant ACE inhibitory	(Eckert et al., 2019)
	Soy protein	RP-HPLC	Single dimension	LIVTQ LIVT	Antihypertensive	(Vallabha & Tiku, 2014)
	Canary seeds milk	SDS-PAGE 2D-PAGE HPLC	Multi-dimension	LSLGT TDQPAG QQLQT FEPQLA KPQLYQPF	Antioxidant ACE inhibitory	(Valverde, Orona-Tamayo, Nieto-Rendón, & Paredes-López, 2017)
	Egg white proteins	Ultrafiltration Cation exchange chromatography (CEC) RP-HPLC HPLC	Multi-dimension	Numerous peptides have been identified without pinpointing which sequence was responsible for the activity, thus not listed.	ACE inhibitory	(H. Fan, Wang, Liao, Jiang, & Wu, 2019)
	Horse gram flour	Ultrafiltration Ion exchange chromatography Gel filtration chromatography RP-HPLC	Multi-dimension	Numerous peptides have been identified without pinpointing which sequence was responsible for the activity, thus not listed.	ACE inhibitory	(Bhaskar, Ananthanarayan, & Jamdar, 2019)

(Continued)

TABLE 14.1 (Continued)

No.	Sample Source	Technique Involved	Multi-dimensional or Single dimension	Peptide Sequence	Biological Activities	Ref.
	Rice bran	Ultrafiltration Gel filtration chromatography RP-HPLC	Multi-dimension	YSK	ACE inhibitory Antioxidant	(X. Wang, Chen, Fu, Li, & Wei, 2017)
	Chia seed	SDS-PAGE electrophoresis Size exclusion chromatography (SEC) by fast protein liquid chromatography (FPLC)	Multi-dimension	LIVSPLAGRL TAQEPTIRF PGLTIGDTIPNL LSLPNYHPNPRL IVSPLAGRL	ACE inhibitory	(San Pablo-Osorio, Mojica, & Urías-Silvas, 2019)
	Bovine casein	Ultrafiltration Size-exclusion chromatography RP-HPLC	Multi-dimension	MKP	Antihypertensive	(Yamada et al., 2013)
	Quinoa bran	Gel filtration chromatography RP-HPLC	Multi-dimension	RGQVIYVL ASPKPSSA QFLLAGR	ACE inhibitory Antioxidant	(Y. Zheng et al., 2019)
	α-lactalbumin from bovine milk	Affinity chromatography	Single dimension	TTFHTSGY GYDTQAIVQ	ACE inhibitory	(Villadóniga & Cantera, 2019)
	Salmon (*Salmo salar*) protein	RP-HPLC-MS/MS	Single dimension	FIKK HL IY LARL PHL PW VFPW VPW VY WNIP	Antioxidant	(Borawska, Darewicz, Pliszka, & Vegarud, 2016)

(*Continued*)

TABLE 14.1 (Continued)

No.	Sample Source	Technique Involved	Multi-dimensional or Single dimension	Peptide Sequence	Biological Activities	Ref.
	Edible microalgae (*Spirulina Maxima*)	Ultrafiltration Anion-exchange chromatography (DEAE) Gel-permeation chromatography RP-HPLC	Multi-dimension	LDAVNR MMLDF	Anti-inflammatory	(Vo, Ryu, & Kim, 2013)
	Tuna cooking juice	Ultrafiltration	Single dimension	KPEGMDPPLSE PEDRRDGAAGPK KLPPLLLAKLL MSGKLLAE PCTGR	Anticancer	(Hung, Yang, Kuo, & Hsu, 2014)
	Whole oyster tissues	Gel filtration chromatography (Sephadex G-25) UPLC-MS	Multi-dimension	LANAK PSLVGRPPVGKLTL VKVLLEHPVL	Antioxidant	(Umayaparvathi et al., 2014)
	Marine macroalga (*Ulva intestinalis*)	Ultrafiltration Gel exclusion chromatography RP-HPLC	Multi-dimension	FGMPLDR MELVLR	ACE inhibitory	(Sun et al., 2019)
	Scallop (*Patinopecten yessoensis*) female gonads	Sephadex G-25 gel filtration	Single dimension	HMSY PEASY	Antioxidant	(Wu et al., 2016)
	Shellfish (*Mytilus coruscus*)	Tangential flow filtration (TFF) Ion exchange chromatography (DEAE) RP-HPLC	Multi-dimension	GVSLLQQFFL	Anti-inflammatory	(E.-K. Kim, Kim, Hwang, Kang, et al., 2013)

(*Continued*)

TABLE 14.1 (Continued)

No.	Sample Source	Technique Involved	Multi-dimensional or Single dimension	Peptide Sequence	Biological Activities	Ref.
	Shellfish (*Ruditapes philippinarum*)	Tangential flow filtration (TFF) Ion exchange chromatography (DEAE) RP-HPLC	Multi-dimension	AVLVDKQCPD	Anticancer	(E.-K. Kim, Kim, Hwang, Lee, et al., 2013)
	Chickpea (*Cicer arietinum* L.)	Size exclusion chromatography (Sephadex G-25 gel filtration column) RP-HPLC	Multi-dimension	RQSHFANAQP	Antioxidant	(Kou et al., 2013)
	Tilapia (*Oreochromis niloticus*) frame protein	Ultrafiltration Ion exchange chromatography Gel filtration chromatography RP-HPLC	Multi-dimension	DCGY NYDEY	Antioxidant	(J. Fan, He, Zhuang, & Sun, 2012)
	Lentil	MPLC Semi-preparative RP-HPLC	Single dimension	LLSGTQNQPS-FLSGF NSLTLPILRYL TLEPNSVFLPVLLH	ACE inhibitory Antioxidant	(García-Mora et al., 2017)
	Common bean	UltrafleXtreme MALDI TOF/TOF-MS/MS	Single dimension	GLTSK LSGNK GEGSGA MPACGSS MTEEY	Anticancer	(Luna Vital, González de Mejía, Dia, & Loarca-Piña, 2014)
	Rapeseed (*Brassica campestris* L.)	Ultrafiltration Sephadex G-15 Gel filtration column RP-HPLC	Multi-dimension	WTP	Antitumor	(Lifeng Wang et al., 2016)

(*Continued*)

TABLE 14.1 (Continued)

No.	Sample Source	Technique Involved	Multi-dimensional or Single dimension	Peptide Sequence	Biological Activities	Ref.
	Soy protein hydrolysates	SDS-PAGE Co-immunoprecipitation	Multi-dimension	FEITPEKNPQ IETWNPNNKP VFDGEL	Topoisomerase II Inhibitory	(W. Wang, Rupasinghe, Schuler, & Gonzalez de Mejia, 2008)
	Soybean	XAD-2 adsorption chromatography Sephadex G-25 gel chromatography RP-HPLCs	Multi-dimension	XMLPSYSPY	Anticancer	(S. E. Kim et al., 2000)
	Black soybean meal byproduct	Ultrafiltration Gel filtration chromatography (Sephadex G-25) RP-HPLC	Multi-dimension	LVPK IVPK	Anticancer Antioxidant	(Z. Chen et al., 2019)
	Oat bran protein	Semi-preparative RP-HPLC	Single dimension	Numerous peptides have been identified without pinpointing which sequence was responsible for the activity, thus not listed.	Ion chelating Antioxidant	(Baakdah & Tsopmo, 2016)
	Oat bran protein	RP-HPLC	Single dimension	Numerous peptides have been identified without pinpointing which sequence was responsible for the activity, thus not listed.	Antioxidant	(Vanvi & Tsopmo, 2016)

(Continued)

TABLE 14.1 (Continued)

No.	Sample Source	Technique Involved	Multi-dimensional or Single dimension	Peptide Sequence	Biological Activities	Ref.
	Palm kernel cake protein	RP-HPLC Isoelectric focusing fractionation	Multi-dimension	WAFS WAF AWFS LPWRPATNVF	Antioxidant	(Zarei et al., 2014)
	Rapeseed protein	Ion-exchange chromatography Gel filtration chromatograph RP-HPLC	Multi-dimension	PAGPF	Antioxidant	(Zhang, Wang, Xu, & Gao, 2009)
	Rice bran protein	RP-HPLC	Single dimension	Numerous peptides have been identified without pinpointing which sequence was responsible for the activity, thus not listed.	Antioxidant	(Wattanasiritham, Theerakulkait, Wickramasekara, Maier, & Stevens, 2016)
	Soybean protein isolate	Ultrafiltration Size exclusion chromatography	Multi-dimension	Numerous peptides have been identified without pinpointing which sequence was responsible for the activity, thus not listed.	Antioxidant	(Beermann, Euler, Herzberg, & Stahl, 2009)
	Wheat gluten	Ion-exchange chromatography Gel filtration chromatography (Sephadex G-25 column followed by SP Sephadex C-25 column) RP-HPLC	Multi-dimension	LQPGQGQQG AQIPQQ	Antioxidant	(Suetsuna & Chen, 2002)

(Continued)

TABLE 14.1 (Continued)

No.	Sample Source	Technique Involved	Multi-dimensional or Single dimension	Peptide Sequence	Biological Activities	Ref.
	Amaranth proteins	Preparative RP-HPLC	Single dimension	AWEEREQGSR YLAGKPQQEH IYIEQGNGITGM TEVWDSNEQ	Antioxidant	(Orsini Delgado et al., 2016)
	Amaranth seed proteins	RP-HPLC	Single dimension	Numerous peptides have been identified without pinpointing which sequence was responsible for the activity, thus not listed.	ACE inhibitory Antithrombotic Antioxidant	(Ayala-Niño et al., 2019)
	Quinoa protein	SDS-PAGE RP-HPLC	Multi-dimension	LWREGM (F1) DKDYPK (F2) DVYSPEAG (F3) IFQEYI (F3) RELGEWGI (F3)	Antioxidant Anticancer	(Vilcacundo, Miralles, Carrillo, & Hernández-Ledesma, 2018)
	Rice bran proteins	RP-HPLC on a TSK gel ODS 120T column RP-HPLC using a C4- Wakosil column	Multi-dimension	AIRQGDVF VLEANPRSF YFPVGGDRPESF	Antioxidant	(Adebiyi, Adebiyi, Ogawa, & Muramoto, 2008)
	Mushroom (*Pleurotus cornucopiae*)	Ultrafiltration Sephadex G-25 column chromatography C_{18} SPE SCX SPE RP-HPLC	Multi-dimension	RLPSEFDLSAFLRA RLSGQTIEVTSEY-LFRH	Antihypertensive	(Jang et al., 2011)

(Continued)

TABLE 14.1 (Continued)

No.	Sample Source	Technique Involved	Multi-dimensional or Single dimension	Peptide Sequence	Biological Activities	Ref.
	Broccoli	Gel filtration chromatography (GFC) RP-HPLC	Multi-dimension	IPPAYTK LVLPGELAK TFQGPPHGIQVER LVLPGE LAK	ACE inhibitory (+Antihypertensive in rat)	(Dang et al., 2019)
	Bluefin leatherjacket (*Navodon septentrionalis*)	Ultrafiltration Anion-exchange chromatography Gel filtration chromatography RP-HPLC	Multi-dimension	GSGGL GPGGFI FIGP	Antioxidant	(Chi et al., 2015)
	Grass carp (*Ctenopharyngodon idella*)	Gel filtration chromatography RP-HPLC	Multi-dimension	PYSFK GFGPEL VGGRP	Antioxidant	(Cai et al., 2015)
	Common oat (*Avena sativa*)	RP-HPLC (Semi-preparative)	Single dimension	FFG IFFFL PFL WWK WCY FPIL CPA FLLA FEPL	ACE inhibitory Renin inhibitory DPP-IV inhibitor	(Bleakley, Hayes, O'Shea, Gallagher, & Lafarga, 2017)
	Umami peptide	UHPLC	Single dimension	CC CCNK HCHT AHSVRFY	ACE inhibitory Antioxidant	(Hao et al., 2020)
	Cottonseed proteins	Ultrafiltration Size exclusion chromatography (SEC)	Multi-dimension	Numerous peptides have been identified without pinpointing which sequence was responsible for the activity, thus not listed.	Antioxidant	(Liying Wang, Ma, Yu, & Du, 2021)

REFERENCES

Abdel-Hamid, M., Otte, J., De Gobba, C., Osman, A., & Hamad, E. (2017). Angiotensin I-converting enzyme inhibitory activity and antioxidant capacity of bioactive peptides derived from enzymatic hydrolysis of buffalo milk proteins. *International Dairy Journal*, 66, 91–98. https://doi.org/10.1016/j.idairyj.2016.11.006

Acquah, C., Chan, Y. W., Pan, S., Agyei, D., & Udenigwe, C. C. (2019). Structure-informed separation of bioactive peptides. *Journal of Food Biochemistry*, 43(1), e12765.

Adebiyi, A. P., Adebiyi, A. O., Ogawa, T., & Muramoto, K. (2008). Purification and characterisation of antioxidative peptides from unfractionated rice bran protein hydrolysates. *International Journal of Food Science & Technology*, 43(1), 35–43. https://doi.org/10.1111/j.1365-2621.2006.01379.x

Aguilar, M. I. (2004). Methods in molecular biology, HPLC of peptides and proteins: Basic theory and methodology. In M. I. Aguilar (Ed.), *HPLC of peptides and proteins: Methods and protocols* (pp. 1–22). Totowa, NJ: Humana Press Inc.

Alvarado, Y., Muro, C., Illescas, J., Díaz, M. D., & Riera, F. (2019). Encapsulation of antihypertensive peptides from whey proteins and their releasing in gastrointestinal conditions. *Biomolecules*, 9(5). https://doi.org/10.3390/biom9050164

Ashraf-Khorassani, M., Taylor, L., & Marr, J. (2000). Analysis of the sulfomycin component of alexomycin in animal feed by enhanced solvent extraction and supercritical fluid chromatography. *Journal of Biochemical and Biophysical Methods*, 43(1–3), 147–156.

Asoodeh, A., Homayouni-Tabrizi, M., Shabestarian, H., Emtenani, S., & Emtenani, S. (2016). Biochemical characterization of a novel antioxidant and angiotensin I-converting enzyme inhibitory peptide from *Struthio camelus* egg white protein hydrolysis. *Journal of Food and Drug Analysis*, 24(2), 332–342. https://doi.org/10.1016/j.jfda.2015.11.010

Ayala-Niño, A., Rodríguez-Serrano, G. M., González-Olivares, L. G., Contreras-López, E., Regal-López, P., & Cepeda-Saez, A. (2019). Sequence identification of bioactive peptides from amaranth seed proteins (*Amaranthus hypochondriacus* spp.). *Molecules*, 24(17). https://doi.org/10.3390/molecules24173033

Baakdah, M. M., & Tsopmo, A. (2016). Identification of peptides, metal binding and lipid peroxidation activities of HPLC fractions of hydrolyzed oat bran proteins. *Journal of Food Science and Technology*, 53(9), 3593–3601. https://doi.org/10.1007/s13197-016-2341-6

Balti, R., Bougatef, A., Sila, A., Guillochon, D., Dhulster, P., & Nedjar-Arroume, N. (2015). Nine novel angiotensin I-converting enzyme (ACE) inhibitory peptides from cuttlefish (Sepia officinalis) muscle protein hydrolysates and antihypertensive effect of the potent active peptide in spontaneously hypertensive rats. *Food Chemistry*, 170, 519–525. https://doi.org/10.1016/j.foodchem.2013.03.091

Bazinet, L., Amiot, J., Poulin, J. F., Labbé, D., & Tremblay, D. (2004). *Canada Patent No. WO/2005/082495*. Washington, DC: Patent Cooperation Treaty.

Bazinet, L., & Firdaous, L. (2009). Membrane processes and devices for separation of bioactive peptides. *Recent Patents on Biotechnology*, 3(1), 61–72.

Beermann, C., Euler, M., Herzberg, J., & Stahl, B. (2009). Anti-oxidative capacity of enzymatically released peptides from soybean protein isolate. *European Food Research and Technology*, 229(4), 637–644. https://doi.org/10.1007/s00217-009-1093-1

Bhaskar, B., Ananthanarayan, L., & Jamdar, S. (2019). Purification, identification, and characterization of novel angiotensin I-converting enzyme (ACE) inhibitory peptides from alcalase digested horse gram flour. *LWT*, *103*, 155–161. https://doi.org/10.1016/j.lwt.2018.12.059

Bickler, B. (2020). How does reversed-phase flash chromatography compare to prep HPLC? Retrieved from https://selekt.biotage.com/blog/how-does-reversed-phase-flash-chromatography-compare-to-prep-hplc

Bjellqvist, B., Ek, K., Righetti, P. G., Gianazza, E., Görg, A., Westermeier, R., & Postel, W. (1982). Isoelectric focusing in immobilized pH gradients: Principle, methodology and some applications. *Journal of Biochemical and Biophysical Methods*, *6*(4), 317–339.

Bleakley, S., Hayes, M., O' Shea, N., Gallagher, E., & Lafarga, T. (2017). Predicted release and analysis of novel ACE-I, renin, and DPP-IV inhibitory peptides from common oat (*Avena sativa*) protein hydrolysates using in silico analysis. *Foods*, *6*(12). https://doi.org/10.3390/foods6120108

Borawska, J., Darewicz, M., Pliszka, M., & Vegarud, G. E. (2016). Antioxidant properties of salmon (Salmo salar L.) protein fraction hydrolysates revealed following their ex vivo digestion and in vitro hydrolysis. *Journal of the Science of Food and Agriculture*, *96*(8), 2764–2772. https://doi.org/10.1002/jsfa.7441

Brandes, H. K., Claus, J. E., Bell, D. S., & Aurand, C. R. (2010). *Improved reversed-phase peptide separations on high-performance silica particles*. Pennslyvania, PA: Sigma-Aldrich Co.

Burgi, D., & Smith, A. J. (1995). Capillary electrophoresis of proteins and peptides. *Current Protocols in Protein Science*, *2*(1), 10.19.11–10.19.13.

Cai, L., Wu, X., Zhang, Y., Li, X., Ma, S., & Li, J. (2015). Purification and characterization of three antioxidant peptides from protein hydrolysate of grass carp (*Ctenopharyngodon idella*) skin. *Journal of Functional Foods*, *16*, 234–242. https://doi.org/10.1016/j.jff.2015.04.042

Cargile, B. J., Sevinsky, J. R., Essader, A. S., Stephenson Jr, J. L., & Bundy, J. L. (2005). Immobilized pH gradient isoelectric focusing as a first-dimension separation in shotgun proteomics. *Journal of Biomolecular Techniques*, *16*(3), 181.

Carr, D. (2005). *A guide to the analysis and purification of proteins and peptides by RP-HPLC*. Aberdeen, Scotland: Advanced Chromatography Technologies.

Castro-Muñoz, R., Barragán-Huerta, B. E., Fíla, V., Denis, P. C., & Ruby-Figueroa, R. (2018). Current role of membrane technology: From the treatment of agro-industrial by-products up to the valorization of valuable compounds. *Waste and Biomass Valorization*, *9*(4), 513–529.

Chen, Y., Liu, W., Xue, J., Yang, J., Chen, X., Shao, Y., ... Zhang, H. (2014). Angiotensin-converting enzyme inhibitory activity of *Lactobacillus helveticus* strains from traditional fermented dairy foods and antihypertensive effect of fermented milk of strain H9. *Journal of Dairy Science*, *97*(11), 6680–6692. https://doi.org/10.3168/jds.2014-7962

Chen, Z., Li, W., Santhanam, R. K., Wang, C., Gao, X., Chen, Y., ... Chen, H. (2019). Bioactive peptide with antioxidant and anticancer activities from black soybean [*Glycine max* (L.) Merr.] byproduct: Isolation, identification and molecular docking study. *European Food Research and Technology*, *245*(3), 677–689. https://doi.org/10.1007/s00217-018-3190-5

Chi, C.-F., Wang, B., Hu, F.-Y., Wang, Y.-M., Zhang, B., Deng, S.-G., & Wu, C.-W. (2015). Purification and identification of three novel antioxidant peptides from protein hydrolysate of bluefin leatherjacket (*Navodon septentrionalis*) skin. *Food Research International*, *73*, 124–129.

Dang, Y., Zhou, T., Hao, L., Cao, J., Sun, Y., & Pan, D. (2019). In vitro and in vivo studies on the angiotensin-converting enzyme inhibitory activity peptides isolated from broccoli protein hydrolysate. *Journal of Agricultural and Food Chemistry*, 67(24), 6757–6764. https://doi.org/10.1021/acs.jafc.9b01137

Datta, S., Auddy, R., & Amit, R. (2014). Supercritical fluid chromatography: A green approach for separation and purification of organic and inorganic analytes. In D. Inamuddin & A. Mohammad (Eds.), *Green chromatographic techniques* (pp. 55–80). Dordrecht, Netherlands: Springer Group.

Dellafiora, L., Paolella, S., Dall'Asta, C., Dossena, A., Cozzini, P., & Galaverna, G. (2015). Hybrid in silico/in vitro approach for the identification of angiotensin I converting enzyme inhibitory peptides from parma dry-cured ham. *Journal of Agricultural and Food Chemistry*, 63(28), 6366–6375. https://doi.org/10.1021/acs.jafc.5b02303

Dlask, O., & Václavíková, N. (2018). Electrodialysis with ultrafiltration membranes for peptide separation. *Chemical Papers*, 72(2), 261–271.

Dong, M. W. (2006). *Modern HPLC for practicing scientists*. Hoboken, NJ: John Wiley & Sons.

Doyen, A., Beaulieu, L., Saucier, L., Pouliot, Y., & Bazinet, L. (2011). Impact of ultrafiltration membrane material on peptide separation from a snow crab byproduct hydrolysate by electrodialysis with ultrafiltration membranes. *Journal of Agricultural and Food Chemistry*, 59(5), 1784–1792.

Durand, R., Pellerin, G., Thibodeau, J., Fraboulet, E., Marette, A., & Bazinet, L. (2020). Screening for metabolic syndrome application of a herring by-product hydrolysate after its separation by electrodialysis with ultrafiltration membrane and identification of novel anti-inflammatory peptides. *Separation and Purification Technology*, 235, 116205.

Eckert, E., Zambrowicz, A., Bobak, Ł., Zabłocka, A., Chrzanowska, J., & Trziszka, T. (2019). Production and identification of biologically active peptides derived from by-product of hen egg-yolk phospholipid extraction. *International Journal of Peptide Research and Therapeutics*, 25(2), 669–680. https://doi.org/10.1007/s10989-018-9713-x

Enmark, M., Glenne, E., Leśko, M., Langborg Weinmann, A., Leek, T., Kaczmarski, K., ... Fornstedt, T. (2018). Investigation of robustness for supercritical fluid chromatography separation of peptides: Isocratic vs gradient mode. *Journal of Chromatography A*, 1568, 177–187. https://doi.org/10.1016/j.chroma.2018.07.029

Essader, A. S., Cargile, B. J., Bundy, J. L., & Stephenson Jr, J. L. (2005). A comparison of immobilized pH gradient isoelectric focusing and strong-cation-exchange chromatography as a first dimension in shotgun proteomics. *Proteomics*, 5(1), 24–34.

Etzel, M., & Arunkumar, A. (2016). Novel membrane technologies for protein concentration and fractionation. In K. Knoerzer, P. Juliano, & G. Smithers (Eds.), *Innovative food processing technologies* (pp. 151–175). Cambridge, United Kingdom: Woodhead Publishing Ltd.

Fan, H., Wang, J., Liao, W., Jiang, X., & Wu, J. (2019). Identification and characterization of gastrointestinal-resistant angiotensin-converting enzyme inhibitory peptides from egg white proteins. *Journal of Agricultural and Food Chemistry*, 67(25), 7147–7156. https://doi.org/10.1021/acs.jafc.9b01071

Fan, J., He, J., Zhuang, Y., & Sun, L. (2012). Purification and identification of antioxidant peptides from enzymatic hydrolysates of tilapia (*Oreochromis niloticus*) frame protein. *Molecules*, 17(11). https://doi.org/10.3390/molecules171112836

García-Mora, P., Martín-Martínez, M., Angeles Bonache, M., González-Múniz, R., Peñas, E., Frias, J., & Martinez-Villaluenga, C. (2017). Identification, functional gastrointestinal stability and molecular docking studies of lentil peptides with dual antioxidant and angiotensin I converting enzyme inhibitory activities. *Food Chemistry*, *221*, 464–472. https://doi.org/10.1016/j.foodchem.2016.10.087

Giese, S. H., Ishihama, Y., & Rappsilber, J. (2018). Peptide retention in hydrophilic strong anion exchange chromatography is driven by charged and aromatic residues. *Analytical Chemistry*, *90*(7), 4635–4640.

Gourley, L., Britten, M., Gauthier, S., & Pouliot, Y. (1994). Characterization of adsorptive fouling on ultrafiltration membranes by peptides mixtures using contact angle measurements. *Journal of membrane Science*, *97*, 283–289.

Hao, L., Gao, X., Zhou, T., Cao, J., Sun, Y., Dang, Y., & Pan, D. (2020). Angiotensin I-converting enzyme (ACE) inhibitory and antioxidant activity of umami peptides after in vitro gastrointestinal digestion. *Journal of Agricultural and Food Chemistry*, *68*(31), 8232–8241. https://doi.org/10.1021/acs.jafc.0c02797

Hong, P., Koza, S., & Bouvier, E. S. (2012). A review size-exclusion chromatography for the analysis of protein biotherapeutics and their aggregates. *Journal of Liquid Chromatography & Related Technologies*, *35*(20), 2923–2950.

Hong, P., Koza, S., & Fountain, K. J. (2012). *Advances in size-exclusion chromatography for the analysis of small proteins and peptides: Evaluation of calibration curves for molecular weight estimation*. Milford, MA: Waters Corporation.

Hung, C.-C., Yang, Y.-H., Kuo, P.-F., & Hsu, K.-C. (2014). Protein hydrolysates from tuna cooking juice inhibit cell growth and induce apoptosis of human breast cancer cell line MCF-7. *Journal of Functional Foods*, *11*, 563–570. https://doi.org/10.1016/j.jff.2014.08.015

Ibáñez C, A. T., Valdés A, García-Cañas V, Cifuentes A, Simó C. (2016). Capillary electrophoresis in food and foodomics. In P. Schmitt-Kopplin (Ed.), *Capillary electrophoresis: Methods and protocols* (Vol. 1483, pp. 471–507). Berlin, Germany: Springer Science+Business Media.

Ibáñez, C., Simó, C., García-Cañas, V., Gómez-Martínez, Á., Ferragut, J. A., & Cifuentes, A. (2012). CE/LC-MS multiplatform for broad metabolomic analysis of dietary polyphenols effect on colon cancer cells proliferation. *Electrophoresis*, *33*(15), 2328–2336.

Ikeda, A., Ichino, H., Kiguchiya, S., Chigwechokha, P., Komatsu, M., & Shiozaki, K. (2015). Evaluation and identification of potent angiotensin-I converting enzyme inhibitory peptide derived from Dwarf Gulper Shark (*Centrophorus atromarginatus*). *Journal of Food Processing and Preservation*, *39*(2), 107–115. https://doi.org/10.1111/jfpp.12210

Irvine, G. B. (2003). High-performance size-exclusion chromatography of peptides. *Journal of Biochemical and Biophysical Methods*, *56*(1–3), 233–242.

Jang, J.-H., Jeong, S.-C., Kim, J.-H., Lee, Y.-H., Ju, Y.-C., & Lee, J.-S. (2011). Characterisation of a new antihypertensive angiotensin I-converting enzyme inhibitory peptide from *Pleurotus cornucopiae*. *Food Chemistry*, *127*(2), 412–418. https://doi.org/10.1016/j.foodchem.2011.01.010

Kaplan, B. (2006). Gel electrophoresis in protein and peptide analysis. In R.A. Meyers & C. Schoneich (Eds), *Encyclopedia of analytical chemistry: Applications, theory and instrumentation*. Hoboken USA: John Wiley & Sons, Ltd.

Kim, E.-K., Kim, Y.-S., Hwang, J.-W., Kang, S. H., Choi, D.-K., Lee, K.-H., … Park, P.-J. (2013). Purification of a novel nitric oxide inhibitory peptide d erived from enzymatic hydrolysates of *Mytilus coruscus*. *Fish & Shellfish Immunology*, 34(6), 1416–1420. https://doi.org/10.1016/j.fsi.2013.02.023

Kim, E.-K., Kim, Y.-S., Hwang, J.-W., Lee, J. S., Moon, S.-H., Jeon, B.-T., & Park, P.-J. (2013). Purification and characterization of a novel anticancer peptide derived from *Ruditapes philippinarum*. *Process Biochemistry*, 48(7), 1086–1090. https://doi.org/10.1016/j.procbio.2013.05.004

Kim, S. E., Kim, H. H., Kim, J. Y., Kang, Y. I., Woo, H. J., & Lee, H. J. (2000). Anticancer activity of hydrophobic peptides from soy proteins. *BioFactors*, 12(1–4), 151–155. https://doi.org/10.1002/biof.5520120124

Kou, X., Gao, J., Xue, Z., Zhang, Z., Wang, H., & Wang, X. (2013). Purification and identification of antioxidant peptides from chickpea (*Cicer arietinum* L.) albumin hydrolysates. *LWT - Food Science and Technology*, 50(2), 591–598. https://doi.org/10.1016/j.lwt.2012.08.002

Kumar, R., & Ismail, A. (2015). Fouling control on microfiltration/ultrafiltration membranes: Effects of morphology, hydrophilicity, and charge. *Journal of Applied Polymer Science*, 132(21), 42042.

Li, Y., Zhou, J., Zeng, X., & Yu, J. (2016). A novel ACE inhibitory peptide Ala-His-Leu-Leu lowering blood pressure sn Spontaneously hypertensive rats. *Journal of Medicinal Food*, 19(2), 181–186. http://doi.org/10.1089/jmf.2015.3483

Luna Vital, D. A., González de Mejía, E., Dia, V. P., & Loarca-Piña, G. (2014). Peptides in common bean fractions inhibit human colorectal cancer cells. *Food Chemistry*, 157, 347–355. https://doi.org/10.1016/j.foodchem.2014.02.050

Marie, G. C. U., Perreault, V., Henaux, L., Carnovale, V., Aluko, R. E., Marette, A., … Bazinet, L. (2019). Impact of a high hydrostatic pressure pretreatment on the separation of bioactive peptides from flaxseed protein hydrolysates by electrodialysis with ultrafiltration membranes. *Separation and Purification Technology*, 211, 242–251.

Martin Orue, C., Bouhallab, S., & Garem, A. (1998). Nanofiltration of amino acid and peptide solutions: Mechanisms of separation. *Journal of membrane Science*, 142(2), 225–233.

Mohammad, J., Jäderlund, B., & Lindblom, H. (1999). New polymer-based prepacked column for the reversed-phase liquid chromatographic separation of peptides over the pH range 2–12. *Journal of Chromatography A*, 852(1), 255–259.

Muro, C., Riera, F., & Fernández, A. (2013). Advancements in the fractionation of milk biopeptides by means of membrane processes. In B. Hernandez-Ledesma & C. C. Hisieh (Eds.), *Bioactive food peptides in health and disease* (pp. 241–266). London, United Kingdom: Intech Open Ltd.

Nongonierma, A. B., Paolella, S., Mudgil, P., Maqsood, S., & FitzGerald, R. J. (2017). Dipeptidyl peptidase IV (DPP-IV) inhibitory properties of camel milk protein hydrolysates generated with trypsin. *Journal of Functional Foods*, 34, 49–58. https://doi.org/10.1016/j.jff.2017.04.016

Oatley-Radcliffe, D. L., Walters, M., Ainscough, T. J., Williams, P. M., Mohammad, A. W., & Hilal, N. (2017). Nanofiltration membranes and processes: A review of research trends over the past decade. *Journal of Water Process Engineering*, 19, 164–171.

Orsini Delgado, M. C., Nardo, A., Pavlovic, M., Rogniaux, H., Añón, M. C., & Tironi, V. A. (2016). Identification and characterization of antioxidant peptides obtained by gastrointestinal digestion of amaranth proteins. *Food Chemistry*, 197, 1160–1167. https://doi.org/10.1016/j.foodchem.2015.11.092

Persico, M., Dhulster, P., & Bazinet, L. (2018). Redundancy analysis for determination of the main physicochemical characteristics of filtration membranes explaining their fouling by peptides. *Journal of Membrane Science, 563*, 708–717.

Pinkston, J. D., Wen, D., Morand, K. L., Tirey, D. A., & Stanton, D. T. (2006). Comparison of LC/MS and SFC/MS for screening of a large and diverse library of pharmaceutically relevant compounds. *Analytical Chemistry, 78*(21), 7467–7472.

Pouliot, Y., Gauthier, S. F., & Groleau, P. E. (2005). Membrane-based fractionation and purification strategies for bioactive peptides. In Y. Mine & F. Shahidi (Eds.), *Nutraceutical proteins and peptides in health and disease* (pp. 109–143). Boca Raton, FL: CRC Press.

Przybylski, R., Bazinet, L., Firdaous, L., Kouach, M., Goossens, J. F., Dhulster, P., & Nedjar, N. (2020). Harnessing slaughterhouse by-products: From wastes to high-added value natural food preservative. *Food Chemistry, 304*, 125448.

Reddy, P. R., & Raju, N. (2012). Gel-electrophoresis and its applications. In S. Magdeldin (Ed.), *Gel electrophoresis-Principles and basics* (pp. 15–32). London, United Kingdom: Intech Open Ltd.

Righetti, P. G., Hamdan, M., Antonucci, F., Verzola, B., & Bossi, A. (2004). Electrophoresis of proteins and peptides. *Journal of Chromatography Library, 69*, 633–668.

Ritorto, M. S., Cook, K., Tyagi, K., Pedrioli, P. G., & Trost, M. (2013). Hydrophilic strong anion exchange (hSAX) chromatography for highly orthogonal peptide separation of complex proteomes. *Journal of Proteome Research, 12*(6), 2449–2457.

Rivier, J., & McClintock, R. (1983). Reversed-phase high-performance liquid chromatography of insulins from different species. *Journal of Chromatography A, 268*, 112–119.

Saldaña, M. D., Zetzl, C., Mohamed, R. S., & Brunner, G. (2002). Decaffeination of guaraná seeds in a microextraction column using water-saturated CO_2. *The Journal of Supercritical Fluids, 22*(2), 119–127.

San Pablo-Osorio, B., Mojica, L., & Urías-Silvas, J. E. (2019). Chia seed (*Salvia hispanica* L.) pepsin hydrolysates inhibit angiotensin-converting enzyme by interacting with its catalytic site. *Journal of Food Science, 84*(5), 1170–1179. https://doi.org/10.1111/1750-3841.14503

Sankar, K. U. (1989). Studies on the physicochemical characteristics of volatile oil from pepper (*Piper nigrum*) extracted by supercritical carbon dioxide. *Journal of the Science of Food and Agriculture, 48*(4), 483–493.

Sarfo, K., Moorhead, G. B., & Turner, R. J. (2003). A novel procedure for separating small peptides on polyacrylamide gels. *Letters in Peptide Science, 10*(2), 127–133.

Schägger, H., & Von Jagow, G. (1987). Tricine-sodium dodecyl sulfate-polyacrylamide gel electrophoresis for the separation of proteins in the range from 1 to 100 kDa. *Analytical Biochemistry, 166*(2), 368–379.

Siren, H. (2015). Capillary electrophoresis in food analyses. In L. Nollet & F. Toldrá (Eds.), *Handbook of food analysis* (pp. 493–519). Boca Raton, FL: CRC Press.

Sørensen, K. K., Mishra, N. K., Paprocki, M. P., Mehrotra, A., & Jensen, K. J. (2021). High-performance reversed-phase flash chromatography purification of peptides and chemically modified insulins. *Chembiochem: A European Journal of Chemical Biology, 22*, 1818.

Suetsuna, K., & Chen, J. R. (2002). Isolation and characterization of peptides with antioxidant Activity derived from wheat gluten. *Food Science and Technology Research, 8*(3), 227–230. https://doi.org/10.3136/fstr.8.227

Sun, S., Xu, X., Sun, X., Zhang, X., Chen, X., & Xu, N. (2019). Preparation and identification of ACE inhibitory peptides from the marine macroalga Ulva intestinalis. *Marine Drugs*, *17*(3). https://doi.org/10.3390/md17030179

Synge, R. L. M., & Tiselius, A. (1950). Fractionation of hydrolysis products of amylose by electrokinetic ultrafiltration in an agar-agar jelly. *Biochemical Journal*, *2*, 41–42.

Tezel, A., Hortaçsu, A., & Hortaçsu, Ö. (2000). Multi-component models for seed and essential oil extractions. *The Journal of Supercritical Fluids*, *19*(1), 3–17.

Tognarelli, D., Tsukamoto, A., Caldwell, J., & Caldwell, W. (2010). Rapid peptide separation by supercritical fluid chromatography. *Bioanalysis*, *2*(1), 5–7.

Umayaparvathi, S., Arumugam, M., Meenakshi, S., Dräger, G., Kirschning, A., & Balasubramanian, T. (2014). Purification and characterization of antioxidant peptides from oyster (*Saccostrea cucullata*) hydrolysate and the anticancer activity of hydrolysate on human colon cancer cell lines. *International Journal of Peptide Research and Therapeutics*, *20*(2), 231–243. https://doi.org/10.1007/s10989-013-9385-5

Vallabha, V. S., & Tiku, P. K. (2014). Antihypertensive peptides derived from soy protein by fermentation. *International Journal of Peptide Research and Therapeutics*, *20*(2), 161–168.

Valverde, M. E., Orona-Tamayo, D., Nieto-Rendón, B., & Paredes-López, O. (2017). Antioxidant and antihypertensive potential of protein fractions from flour and milk substitutes from Canary seeds (*Phalaris canariensis* L.). *Plant Foods for Human Nutrition*, *72*(1), 20–25. https://doi.org/10.1007/s11130-016-0584-z

Vanvi, A., & Tsopmo, A. (2016). Pepsin digested oat bran proteins: Separation, antioxidant activity, and identification of new peptides. *Journal of Chemistry*, *2016*, 8216378. https://doi.org/10.1155/2016/8216378

Velickovic, T. C., Ognjenovic, J., & Mihajlovic, L. (2012). Separation of amino acids, peptides, and proteins by ion exchange chromatography. In D. Inamuddin & M. Luqman (Eds.), *Ion exchange technology II* (pp. 1–34). Dordrecht, Netherlands: Springer Group.

Vilcacundo, R., Miralles, B., Carrillo, W., & Hernández-Ledesma, B. (2018). In vitro chemopreventive properties of peptides released from quinoa (*Chenopodium quinoa* Willd.) protein under simulated gastrointestinal digestion. *Food Research International*, *105*, 403–411. https://doi.org/10.1016/j.foodres.2017.11.036

Villadóniga, C., & Cantera, A. M. B. (2019). New ACE-inhibitory peptides derived from α-lactalbumin produced by hydrolysis with Bromelia antiacantha peptidases. *Biocatalysis and Agricultural Biotechnology*, *20*, 101258. https://doi.org/10.1016/j.bcab.2019.101258

Vo, T.-S., Ryu, B., & Kim, S.-K. (2013). Purification of novel anti-inflammatory peptides from enzymatic hydrolysate of the edible microalgal Spirulina maxima. *Journal of Functional Foods*, *5*(3), 1336–1346. https://doi.org/10.1016/j.jff.2013.05.001

Vollet Marson, G., Belleville, M. P., Lacour, S., & Dupas Hubinger, M. (2021). Membrane fractionation of protein hydrolysates from by-products: Recovery of valuable compounds from spent yeasts. *Membranes*, *11*(1), 23.

Wang, L., Ma, M., Yu, Z., & Du, S.-k. (2021). Preparation and identification of antioxidant peptides from cottonseed proteins. *Food Chemistry*, *352*, 129399. https://doi.org/10.1016/j.foodchem.2021.129399

Wang, L., Zhang, J., Yuan, Q., Xie, H., Shi, J., & Ju, X. (2016). Separation and purification of an anti-tumor peptide from rapeseed (*Brassica campestris* L.) and the effect on cell apoptosis. *Food & Function*, *7*(5), 2239–2248. https://doi.org/10.1039/C6FO00042H

Wang, W., Rupasinghe, S. G., Schuler, M. A., & Gonzalez de Mejia, E. (2008). Identification and characterization of topoisomerase II inhibitory peptides from soy protein hydrolysates. *Journal of Agricultural and Food Chemistry*, 56(15), 6267–6277. https://doi.org/10.1021/jf8005195

Wang, X., Chen, H., Fu, X., Li, S., & Wei, J. (2017). A novel antioxidant and ACE inhibitory peptide from rice bran protein: Biochemical characterization and molecular docking study. *LWT*, 75, 93–99. https://doi.org/10.1016/j.lwt.2016.08.047

Wattanasiritham, L., Theerakulkait, C., Wickramasekara, S., Maier, C. S., & Stevens, J. F. (2016). Isolation and identification of antioxidant peptides from enzymatically hydrolyzed rice bran protein. *Food Chemistry*, 192, 156–162. https://doi.org/10.1016/j.foodchem.2015.06.057

Weston, A., & Brown, P. (1997). *HPLC and CE - Principles and practice*. Amsterdam, the Netherlands: Academic Press.

Whatley, H. (2001). Basic principles and modes of capillary electrophoresis. In J. R. Petersen & A. A. Mohammad (Eds.), *Clinical and forensic applications of capillary electrophoresis*. Totowa, NJ: Humana Press Inc.

Wu, H.-T., Jin, W.-G., Sun, S.-G., Li, X.-S., Duan, X.-H., Li, Y., ... Zhu, B.-W. (2016). Identification of antioxidant peptides from protein hydrolysates of scallop (*Patinopecten yessoensis*) female gonads. *European Food Research and Technology*, 242(5), 713–722. https://doi.org/10.1007/s00217-015-2579-7

Yamada, A., Sakurai, T., Ochi, D., Mitsuyama, E., Yamauchi, K., & Abe, F. (2013). Novel angiotensin I-converting enzyme inhibitory peptide derived from bovine casein. *Food Chemistry*, 141(4), 3781–3789. https://doi.org/10.1016/j.foodchem.2013.06.089

Zarei, M., Ebrahimpour, A., Abdul-Hamid, A., Anwar, F., Bakar, F. A., Philip, R., & Saari, N. (2014). Identification and characterization of papain-generated antioxidant peptides from palm kernel cake proteins. *Food Research International*, 62, 726–734. https://doi.org/10.1016/j.foodres.2014.04.041

Zhang, S. B., Wang, Z., Xu, S. Y., & Gao, X. F. (2009). Purification and characterization of a radical scavenging peptide from rapeseed protein hydrolysates. *Journal of the American Oil Chemists' Society*, 86(10), 959–966. https://doi.org/10.1007/s11746-009-1404-5

Zheng, J., Glass, T., Taylor, L., & Pinkston, J. D. (2005). Study of the elution mechanism of sodium aryl sulfonates on bare silica and a cyano bonded phase with methanol-modified carbon dioxide containing an ionic additive. *Journal of Chromatography A*, 1090(1–2), 155–164.

Zheng, J., Pinkston, J., Zoutendam, P., & Taylor, L. (2006). Feasibility of supercritical fluid chromatography/mass spectrometry of polypeptides with up to 40-mers. *Analytical Chemistry*, 78(5), 1535–1545.

Zheng, Y., Wang, X., Zhuang, Y., Li, Y., Tian, H., Shi, P., & Li, G. (2019). Isolation of novel ACE-inhibitory and antioxidant peptides from Quinoa bran albumin assisted with an in silico approach: Characterization, in vivo antihypertension, and molecular docking. *Molecules*, 24(24). https://doi.org/10.3390/molecules24244562

Zhou, H., Dai, J., Sheng, Q. H., Li, R. X., Shieh, C. H., Guttman, A., & Zeng, R. (2007). A fully automated 2-D LC-MS method utilizing online continuous pH and RP gradients for global proteome analysis. *Electrophoresis*, 28(23), 4311–4319.

Zilberstein, G., Korol, L., Shlar, I., Righetti, P. G., & Bukshpan, S. (2008). High-resolution separation of peptides by sodium dodecyl sulfate-polyacrylamide gel "focusing". *Electrophoresis*, 29(8), 1749–1752.

SECTION 4

Analysis of Bioactive Peptides

CHAPTER 15

Liquid Chromatography-Mass Spectrometry (LC-MS) Analysis of Bioactive Peptides

Evelien Wynendaele, Kevin Van der Borght, Nathan Debunne, and Bart De Spiegeleer

CONTENTS

Introduction	301
Sample Preparation	302
Liquid Chromatography	304
Mass Spectrometry	307
Data Analysis	312
References	312

INTRODUCTION

Peptidomics, i.e., the comprehensive, qualitative, and quantitative study of all peptides in a biological sample, has become an important part of food science and technology, as it can help in the characterization of wanted or unwanted food-derived peptides – not only in the original food but also after the consequential production, processing, storage, and digestion [1]. In past decades, different food-derived peptides have already been identified with multiple health beneficial properties, from antihypertensive and antithrombotic actions to antimicrobial, immunomodulatory, opioid, and antioxidant functions. Based on the information from Bioactive Peptide Database (BioPepDB), more than 4000 food-derived bioactive peptides have already been identified with various biological functions. Due to the low toxicity and high specificity, these bioactive peptides may be used in functional foods, nutra- and cosmeceuticals, food-grade bio-preservatives, and pharmaceuticals [2, 3].

Bioactive peptides, originating from food (e.g., animal or plant) proteins after digestion or hydrolysis, can be identified using different techniques, after which confirmatory *in vitro* and *in vivo* functionality studies may be carried out with the identified synthetic peptides [4]. Different separation methods are generally used during this peptide identification, including capillary electrophoresis (CE) and liquid chromatography (LC). CE is an attractive method due to its fast analysis time and minimum sample preparation steps, even in complex matrices; also, a low consumption of sample and reagents is inherent to this technique [5]. Moreover, CE has already proven its potential in a wide

DOI: 10.1201/9781003106524-19

range of food-related applications, as summarized by Papetti and Colombo in 2019 [6]. However, liquid chromatography is still the most widely used separation technique for peptide analysis, and is therefore further discussed in this chapter in detail. Different steps in the MS-based peptidomic analysis are discussed, including sample preparation, LC separation, detection, and data-acquisition using mass spectrometry (MS) as well as data analysis.

SAMPLE PREPARATION

In order to investigate the peptide composition in complex food or biological matrices, an adequate sample preparation is required as matrix components can interfere with the peptide LC-MS signal (matrix effect). The ideal sample preparation method should therefore eliminate unwanted analytes without losing a significant amount of the peptides of interest. Different possibilities exist; however, recovering the whole peptidome still remains a challenging task.

A first option for sample preparation is the use of solid-phase extraction (SPE) for peptides in solution. Different SPE-cartridges as well as different peptide separation protocols have been developed, thereby exploring different stationary-phase chemistries for peptide isolation (Table 15.1). SPE is found to be useful in targeted approaches after

TABLE 15.1 Examples of SPE Columns Used to Extract the Peptides from Their Matrix

Name of Column	Supplier	Chemistry	Sample	Ref
Sep-Pak C8	Waters	C_8	Plasma	[9]
Sep-Pak C18	Waters	C_{18}	Yoghurt/Seaweed/Crab/Olive water	[10-13]
Strata-X	Phenomenex	C_{18}	Urine/Milk/Yoghurt	[10, 14, 15]
Oasis HLB	Waters	N-vinylpyrrolidone and divinylbenzene	Urine/Milk/Dairy	[7, 15, 16]
Oasis MAX	Waters	Mixed-mode strong anion exchange (HLB-N$^+$R$_3$)	Plasma	[17]
Oasis WCX	Waters	Mixed-mode weak cation exchange (HLB-COO$^-$)	Plasma	[18]
Oasis MCX	Waters	Mixed-mode strong cation exchange (HLB-SO$_3^-$)	Cultures	[19]
Empore disks	3M	C_8/C_{18}	Serum	[20]
Bond Elut C18	Agilent	C_{18}	Saliva	[21]
Bond Elut SCX	Agilent	Mixed-mode strong cation exchange (-SO$_3^-$)	Pea hydrolysate	[22]
MonoSpin Amide	GL Sciences	HILIC-amide	Serum	[23]
HyperSep C18	Thermo Scientific	C_{18}	Serum	[24]
Acclaim 300 C18	Thermo Scientific	C_{18}	Wine	[25]

proper development and validation for the specific peptides but less so for untargeted peptidomic analyses due to the risk of peptide loss [7]. Reversed-phase (RP) SPE is the technique mostly used: a sample dissolved in a polar mobile phase is loaded onto the RP-SPE column, after which the non-retained impurities are eluted by washing with the same polar mobile phase; the peptides of interest are then eluted with a less polar mobile phase containing an organic modifier. Eluents can then be further concentrated using, e.g., freeze drying, vacuum/centrifugal evaporation, or nitrogen evaporation, again dissolved in a more polar solvent, compatible for the subsequent RP-LC-MS analysis [8]. Alternatively, hydrophilic interaction chromatography (HILIC) SPE is an interesting option as well, as solvent evaporation is then not required before the RP-LC-MS analysis, i.e., peptides are already present in a polar solvent after SPE elution, which perfectly fits a subsequent analytical RP-LC analysis.

Next to SPE, protein precipitation can also be used as a sample preparation step for peptide analysis in complex matrices. To remove the proteins, acid addition (e.g., trichloroacetic acid) as well as the addition of different organic solvents (e.g., methanol, acetone, or acetonitrile) can be used; a combination of these agents is also possible [23, 24, 26–28]. Although this is a very simple technique, peptide recovery can be low, due to the protein-adsorption phenomena and co-precipitation with the proteins [29]. For untargeted peptidomic analysis, the addition of acetonitrile (2:1 ratio) was found to be most efficient for protein precipitation, compared to methanol and ice-cold acetone [30, 31]. For targeted approaches, it is recommended to optimize the experimental conditions and systematically evaluate the different organic solvents (at different concentrations) to obtain the highest recovery for the investigated peptides [32]. As indicated by Taevernier et al., not only the amount of acetonitrile but also the percentage of formic acid can play a major role in protein precipitation during sample preparation: an acetonitrile range of $83 \pm 3\%$ and a formic acid content of $1 \pm 0.25\%$ were found to give an optimal loss of bovine serum albumin (BSA) in the solutions [33].

Ultrafiltration is another fast and easy-to-apply technique, to separate the low-molecular-weight (peptides) and high-molecular-weight (proteins) fractions. Different types of molecular weight cutoff membranes exist, which, however, do not permit complete separation of a specific mass range without partial loss or contamination with other undesired fractions [34, 28]. During ultrafiltration, several parameters such as centrifugation time, speed, pH, ionic strength, temperature, and organic solvent composition should be taken into account and optimized for this specific purpose. Moreover, the addition of acetonitrile is recommended to disrupt the protein–peptide interaction [35, 36].

Next to the protein–peptide interactions, peptides are known for their aspecific binding to surfaces as well, which becomes more prominent at low concentrations of purified peptides and/or when evaporating the solvent. The use of organic solvents, modified with acid or base, have already been proven beneficial in reducing aspecific adsorption. Moreover, the type of vial (i.e., glass vs. polypropylene vial) that is used and the storage duration can also be of influence: for some peptides, polypropylene appears to be most suitable, while for other peptides, less adsorption is observed when using glass vials for the expected duration of storage [37, 38]. Recently, a new LC-MS compatible anti-adsorption approach for peptides was developed based on a bovine serum albumin hydrolysate; this anti-adsorption diluent is easy to prepare and use and is compatible with acetonitrile and formic acid, which also increases the flexibility in its use [39].

LIQUID CHROMATOGRAPHY

When using liquid chromatography for peptide separation, different stationary phases can be applied. Reversed-phase liquid chromatography (RP-LC) remains the most widely used mode for peptide separation, thereby separating the molecules roughly based on differences in hydrophobicity. Favored RP-LC packings for the vast majority of peptide separations continue to be silica-based supports containing covalently bound octyl (C_8) or octadecyl (C_{18}) functionalities, with peptides being eluted from these stationary phases in order of increasing overall peptide hydrophobicity [40]. Mobile phases generally consist of aqueous trifluoroacetic acid (TFA) or formic acid (FA)/acetonitrile (ACN) or methanol systems. Low concentrations of the acid are added to decrease the pH and form ion pairs with the peptides, thereby improving the chromatographic peak shape (symmetry and peak width) [8]. For UV-detection, TFA is generally the strong acid of choice as it gives much sharper peaks and thus a better resolution compared to FA [41]. However, during mass spectrometry (MS) analysis, TFA causes ion suppression, due to the formed ion pairs with the peptides, and the use of FA becomes more favorable. Alternatively, the post-column addition of a mixture of propionic acid and isopropanol (TFA-Fix) can also be applied to overcome the problem of TFA ion suppression during MS analysis [42, 43].

The linear-solvent-strength (LSS) model for gradient elution, which is based on an isocratic retention behavior, describes the retention in function of the mobile phase composition and is therefore used for method development in RP-LC (equation 15.1).

$$\log k = \log k_w - S\varphi \qquad (15.1)$$

with φ the volume-fraction of organic solvent in a RP-LC mobile phase, k the retention factor, k_w the (extrapolated) value of k for $\varphi = 0$, and S a constant for a given compound and fixed experimental conditions (other than φ). This LSS model predicts a linear decrease in log k during the gradient with either time or the volume of mobile phase that has left/entered the column [44, 45]. Peptides possess S-values that range from 15 to 50, and therefore, both isocratic and gradient systems can, in principle, be used. However, for peptidomic analysis (with peptides with varying molecular weight, charge and length, and relatively high S values [46]), gradient elution is almost exclusively applied. When using isocratic conditions, a dramatic change in retention time is observed with small changes in organic modifier concentration (mobile phase); the operational range of organic modifier concentration, applicable for practical isocratic separation of different peptides to baseline resolution, is thus very narrow [47]. Moreover, the mobile phase must have a broad range of eluting power to ensure elution of all the bound peptides. In comparison, proteins have much higher S-values (ranging between 30 and 300) and therefore are also very sensitive to small changes in mobile phase composition. For small molecules (S-values ranging from 1 to 30), a robust system can be applied using an isocratic system (Figure 15.1).

Next to RP-LC, HILIC is applied in peptidomic analysis as well: it can be used for the separation of hydrophilic/polar peptides that demonstrate little or no retention on RP stationary phases [48]. Separating molecules on a HILIC column is primarily based on the partitioning of the polar molecule between a bulk eluent and a water-rich layer, immobilized at the polar stationary phase surface. Next, analyte adsorption to the stationary phase surface through various interactions, e.g., electrostatic interaction, hydrogen bonding, or hydrophobic interaction, also plays an important role in the retention

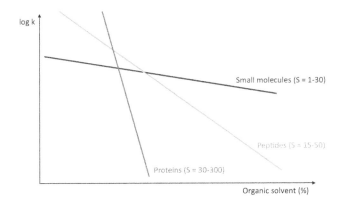

FIGURE 15.1 Use of gradient or isocratic elution in RP-LC based on the average change in log k per unit change in the volume fraction of organic solvent (S-value).

mechanism [49, 50]. The elution trend using gradient elution (low aqueous to high aqueous mobile-phase composition) is thus more related to a variety of physicochemical properties rather than hydrophobicity alone.

A wide range of HILIC stationary phases have been described, which can be divided into neutral, charged, or zwitterionic stationary phases [51]. Silica and silica modified with amino, diol, amide or zwitter-ionic groups are traditional polar stationary phases for HILIC; the formation of hydrogen bonds with unmodified silica or ion-exchange interactions with both unmodified and modified silanol groups are known interaction mechanisms for peptides [52, 53].

Peptide separation using a HILIC stationary phase is also dependent on the mobile phase composition: next to the choice of the organic solvent used, also the nature and the concentration of the buffer solution has an effect on the selectivity of the method. Buffers such as ammonium formate and ammonium acetate may therefore be recommended for peptide analysis to avoid peak tailing [53].

Janvier et al. showed that HILIC is a good alternative for the detection of peptides in comparison with RP-HPLC because of its ability to retain polar peptides and because of its compatibility with MS [48]. This was confirmed by Debunne, who investigated the retention times of 33 peptides using RP and HILIC LC-MS: a low correlation between both retention times was observed, illustrating the high degree of orthogonality between the two chromatographic systems (Figure 15.2) [54]. This orthogonality was also confirmed by Gilar et al. for the analysis of about 200 tryptic peptides: the HILIC-RP 2D-LC system, where peptide separation occurred on a HILIC stationary phase followed by a C_{18}-RP separation, demonstrated the highest degree of orthogonality compared to phenyl-C_{18}, pentafluorophenyl-C_{18}, C_{18}-C_{18}, size exclusion chromatography (SEC)-C_{18}, and strong cation exchange (SCX)-C_{18} combinations [55]. Tandem (2-dimensional) HILIC-RP liquid chromatography was also found to be a valuable and universal tool in the non-targeted screening of various types of complex food samples, including red wine, coffee, and meat extract, as it increases the covered polarity range compared to a classical one-dimensional chromatography [56].

A third mode of LC for peptide separation uses the net charge of the peptide, called ion-exchange chromatography (IEC). It separates the peptides based on charge differences, by using a gradient of increasing salt concentration (ionic strength); small, less-charged peptides thereby elute prior to the longer peptides. Both acidic (net negative

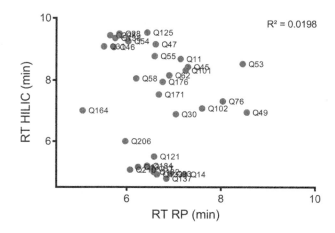

FIGURE 15.2 Retention time (RT) plot of peptides (numbers are from QuorumPeps database), comparing reversed phase (RP) chromatography to hydrophilic-interaction (HILIC) chromatography: a low correlation ($R^2 = 0.0198$) is observed, indicating a high degree of orthogonality between both chromatographic systems [54].

charge) and basic (net positive charge) peptides can be retained, using either primary, secondary, tertiary (weak anion-exchange), or quaternary amine (strong anion-exchange) groups adsorbed or covalently bound to a support, or carboxyl (weak cation-exchange) or sulfonate (strong cation-exchange) groups bound to a support matrix, respectively [40]. For cation-exchange chromatography, the applied pH should be below the isoelectric point (pI) of the peptides; for anion-exchange chromatography, it is recommended to use a pH above the peptide pI.

Next to the type of stationary phase, LC columns are characterized by different sizes (diameter and length) as well as dimensional particle characteristics such as particle size (distribution) and pore sizes. Wide pore silica (300 Å diameter) is mostly used for larger peptides and proteins, while small pore silica (100–130 Å diameter) can be used for separating smaller peptides (e.g., from protein digests). The latter are able to enter the small pores, thereby interacting with the surface, resulting in a good peptide separation; this is not the case for the larger peptides. However, wide-pore silicas also separate (smaller) peptides well and result in different selectivity and resolution [57]. Column length can also play a role in peptide separation, while this is less the case for proteins (high S-value): peptides interact less strongly with the hydrophobic reversed-phase surface than proteins, resulting in a better peak resolution with longer column (and gradient) length [58].

The design of analytical LC columns witnesses a continuous miniaturization, allowing not only smaller sample sizes to be analyzed but also an increased greenness (due to, e.g., lower solvent consumption), and improved performance in selectivity and sensitivity. As such, capillary- and nano-LC are using internal diameters of 100–500 μm and <100 μm, respectively. The capillary columns can either be filled (particle packed or monolithic) or open tubular (wall coated or porous layer). The potential of nano-LC-MS in peptide food analysis was demonstrated by the identification of the bioactive peptides produced during *in vitro* gastrointestinal digestion of soybean seeds (1173 peptides) and soy milk (1364–1422 peptides) [59]. Moreover, microfluidic LC-MS devices are being extensively developed, with several microchips already commercially available. The increase in sensitivity is exemplified by the identification of 65 glycosylated lactoferrins in goat milk using nano-LC-Chip-Q-TOF-MS [60].

MASS SPECTROMETRY

Most peptide detection and quantification occurs via ultraviolet (UV) detection and/or mass spectrometry (MS). While UV detection is an inexpensive analysis, the MS detector has become one of the most attractive techniques for peptide analysis, because of its high sensitivity and selectivity.

The first step in the mass spectrometric analysis is the production of gas phase ions of the molecules, generated by the ion source. These ions can whether or not be fragmented, followed by the separation of the ions according to their mass-to-charge (m/z) ratio by the mass analyzer and detected in proportion to their abundance, resulting in a mass spectrum. Different ionization sources exist, but electrospray Ionization (ESI) and (to a lesser extent) matrix-assisted laser desorption ionization (MALDI) are most frequently used for peptide characterization [61]; these soft ionization methods have made it possible to ionize large and thermally labile peptides and transfer them to the gas phase without dissociation [62]. For MALDI, the sample to be analyzed is dissolved in a matrix of small organic molecules (e.g., 2,5-dihydroxybenzoic acid), which show a strong absorption at the laser wavelength; this mixture is then dried before analysis, so the analyte molecules are embedded throughout the matrix, completely isolated from each other. Next, the sample is irradiated by a pulsed laser, leading to the ionization of the molecules. It shows a high efficiency, together with a high sensitivity, due to the presence of this matrix; a minimal amount of sample clusters as well as sample damage is observed, thereby making it an ideal way to characterize peptides. For ESI, the sample to be analyzed is dissolved in a solution (e.g., LC eluate), which is sprayed through a capillary tube with a high electric field (2–5 kV) at the end. This field induces a charge accumulation (redox reaction due to electron flows) at the liquid surface, located at the end of the capillary, which will break to form highly charged droplets. Nitrogen gas is then injected coaxially to evaporate the solvent molecules, thereby breaking down the droplets into smaller parts (Coulomb fission). In ESI-MS, high-molecular-weight molecules, such as larger peptides, typically carry multiple charges, thereby providing both an accurate molecular mass and structural information; small molecules will produce mainly monocharged ions [63].

After the ionization, the ions are further incorporated by the mass analyzer to be separated according to their m/z values. For peptide analysis, typically quadrupole, Time-of-flight (TOF), ion trap or Fourier transform ion cyclotron resonance (FTICR) mass analyzers are used [63]. Quadrupole mass analyzers resolve m/z by applying radio frequency (RF) and direct current (DC) voltages, allowing only a narrow mass/charge range to reach the detector. TOF analyzers accelerate the ions by using a short voltage gradient and measure the time ions take to traverse a field-free flight tube; the flight time is thereby proportional to the square root of the m/z [63]. An ion trap on the other hand is a device that uses an oscillating electric field to store ions. It works by using a RF quadrupolar field that traps the ions in two (linear ion trap) or three (the Paul ion trap) dimensions [63]. An electrostatic ion-trap (Orbitrap) stores ions in a stable flight path by balancing their electrostatic attraction by their inertia coming from an RF only trap [65]. With FTICR MS, high magnetic fields are used to trap the ions and cyclotron resonance to detect and excite them [63].

Generally, ion trap and quadrupole analyzers offer high sensitivity but limited resolution. Orbitrap and TOF, on the other hand, offer high resolving power [66]. A comparison of the different mass analyzers is given in Table 15.2. To increase the sensitivity and selectivity in targeted or untargeted peptidomic studies, it is common to combine mass analyzers in a hybrid way, such as in triple quadrupole, linear trap quadrupole–orbitrap

TABLE 15.2 The Resolution, Advantages and Disadvantages of the Mass Analysers Used for Peptide Analysis

Analyzer	Resolution	Advantages	Disadvantages
Quadrupole	2,000	Relatively cheap	Low resolution
		Sensitive SIR	Not suited for pulsed ionization
		Continuous	Small dynamic range
TOF	20-25,000	Fast analyser	Requires pulsed ionization
		Highest mass range	Intensive calibration
		Relatively high resolution	Large equipment
Ion-trap	2-15,000	Highly sensitive TICs	Lower resolution
		Relatively cheap	Small dynamic range
		Compact	Space charge effects
FT-ion cyclotron resonance	100-1,000,000	Ultra-high resolution	Very expensive
		Powerful capabilities	Small dynamic range
		High sensitivity	Hugh equipment
Orbitrap	50-250,000	High resolution	Expensive
			Less sensitive

(LTQ-orbitrap) [67], and quadrupole-TOF (Q-TOF) [68, 69]. An overview of studies using these combined MS techniques for the analysis of gluten peptides in foods was recently given by Alves et al. [70]. Also very recently, different (allergen) peptides were summarized by Pilolli et al., thereby indicating the different MS platforms that were used for peptide identification [71].

Ion mobility mass spectrometers (IMS) also require ionization of the analytes before they can be detected. In contrast to the other mass analyzers, where separation occurs according to the *m/z* value, in IMS, ions are separated based on their mobility in neutral buffer gas under the influence of an electric field [72]. Interfacing IMS with MS, i.e., ion mobility-mass spectrometry (IM-MS or IMS-MS), has gained much attention recently, as it is able to separate ions with different charges, structures and conformations, thereby increasing the selectivity [73]. Different instrumental platforms are currently available, depending on the electric field (E) applied (static or oscillating), absence or presence (parallel or perpendicular) of gas flow, ion-trap system (influencing the duty cycle and hence detection limits), and filter-possibilities (increasing selectivity). Currently, drift tube ion mobility spectrometry (DTIMS by, e.g., Agilent), traveling wave ion mobility spectrometry (TWIMS by, e.g., Waters), trapped ion mobility spectrometry (TIMS by, e.g., Bruker) and high-field asymmetric waveform ion mobility spectrometry (FAIMS by, e.g., Sciex) are commonly used. A detailed review of these different types of IMS can be found in [73, 74]. TIMS hybridized with MS, like timsTOF, is a relatively recent advance, which combines TIMS with ultra-high resolution TOF technology, thereby increasing the selectivity of the device and giving more confidence during compound identification [75–79]. The added value of ion mobility separation, including high resolution MS/MS, was clearly demonstrated by Gevaert et al.: using LC-MS, 638 unique peptides were identified in a pig cerebrolysin sample, with the number of identified peptides and unique sequences being higher when using ion mobility separation compared to the ion trap or TOF mass analyzers [80].

After passing through the mass analyzer, the ions are detected and transformed into a usable signal by a detector. Electron multiplier detectors, being the most widely used ion detectors in mass spectrometry, are based on the kinetic energy transfer of incident ions by collision with a surface that in turn generates secondary electrons, which are further amplified to give an electronic current. On the other hand, the detectors in FTICR or Orbitrap MS consist of a pair of metal plates within the mass analyzer region close to the ion trajectories; ions are then detected by the image current that they produce in a circuit connecting the plates [64].

An important aspect of mass spectrometry is the development of tandem MS (MS/MS) or MS^n instruments in which specific ions (protonated or other adducts) can be selected, fragmented, and the m/z of the fragment ions measured [81].

Gas-phase fragmentation of single or multiple charged peptide ions in tandem MS analysis is a powerful tool for the identification and structural elucidation. In the Roepstorff nomenclature, peptide fragment ions are indicated based on the location of dissociation and charge. Fragment ions comprising of the N-terminal amino acids are indicated by a, b, or c specifying the site of cleavage around the cleaved peptide bond, and numbered by the amino acids contained in the fragment (Figure 15.3). Analogously, fragment ions comprising of the C-terminal amino acids are indicated by x, y, and z and numbered accordingly [82]. Collision induced fragmentation (CID) is the most commonly used fragmentation method, but the use of electron-mediated fragmentation techniques is increasing. CID yields mostly b- and y-type ions, but additional a- and x-type ions are also formed [83]. Different modes, e.g., depending on the applied collision energy and activation time parameters, are possible, influencing the fragmentation pattern obtained [84–87]. Electron mediated fragmentation is achieved through electron capture dissociation (ECD) by the introduction of low energy electrons (<1 eV), or electron transfer dissociation (ETD) using radical anions (e.g., chemically ionized fluoranthene) as an electron donor. Electron mediated techniques yield c-type ions and z-type radical ions [88, 89]. Multi-fragmentation protocols combining ETD with CID in MS^n analysis can also be used to map posttranslational modifications such as disulfide bridging or glycosylation [85, 90, 91]. As indicated by Wen et al., with some soybean peptides, doubly charged parent ions can be dissociated to form singly charged fragments, which then can have a higher m/z value. This group identified 51 bioactive peptides by matching experimental product ions to those generated *in silico* (from protein hydrolysis), thereby focusing on a, b, and y fragments [92]; in contrast, Nimalaratne et al. identified 2 antioxidant peptides from chicken egg white, by only focusing on y fragments [93].

A frequently used instrument is a triple quadrupole MS (QqQ), where three quadrupoles are used: the first and third quadrupole are used as mass analyzers, while the second one is used as the collision cell (RF-only quadrupole), where the ions are activated by collision with neutral gas molecules (e.g., argon or helium), resulting in smaller fragments. In ion-traps, there is only one mass analyzer, which can be operated in different electronic modes, to (1) trap the ions in the sample, (2) electronically destabilize all ions, except the ion of interest, (3) activate this ion electronically to produce fragments, and (4) shift to analysis mode to read out the fragment ions by sequentially ejecting them from the trap for detection [63]. Alternatively, fragment ions can also be trapped and further fragmented, resulting in MS^n spectra [63, 64]. In tandem mass spectrometry, mainly four different scan modes are applied (Figure 15.4): product ion scan, precursor ion scan, neutral loss scan, and multiple (or selected) reaction monitoring. Product ion scanning ("daughter scan") consists of selecting a precursor ion (with chosen m/z value) and determining all product ions after fragmentation; this scan mode is frequently

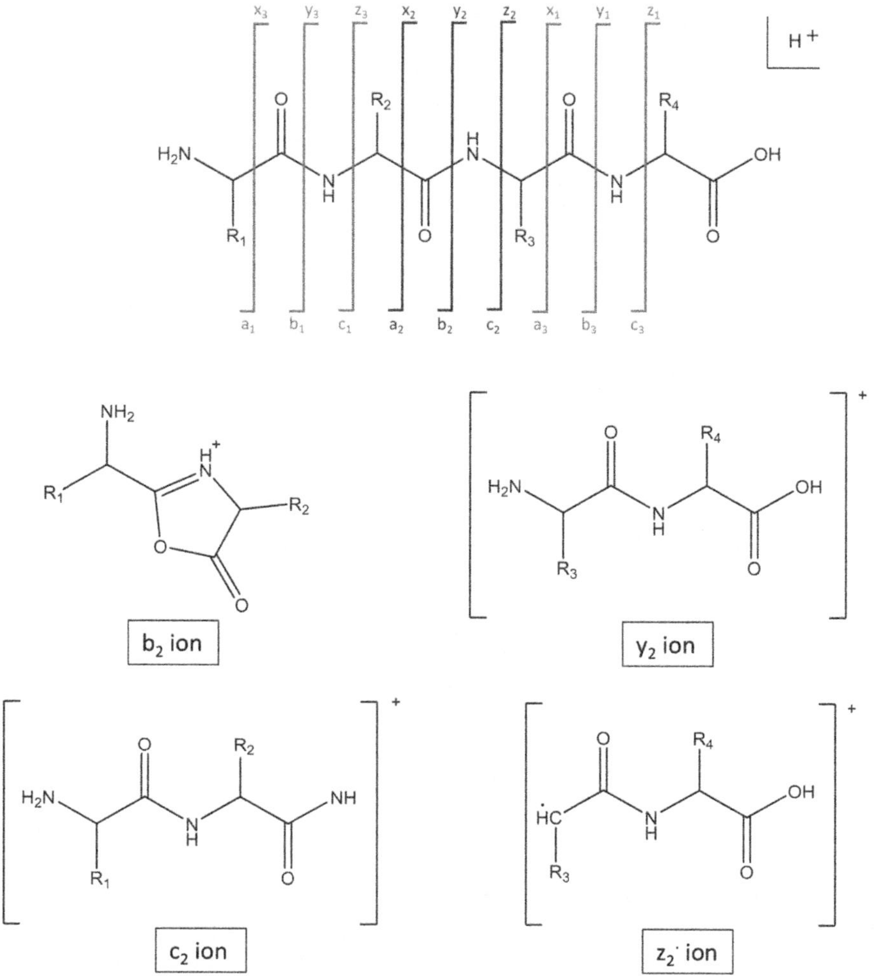

FIGURE 15.3 Roepstorff nomenclature of peptide fragmentation along the backbone (top) and an illustration of the most abundant ions resulting from CID and ETD/ECD fragmentation (bottom).

used for peptide sequencing analysis. Precursor ion scanning ("parent scan") consists of selecting a product ion with a specific m/z value and determining all precursor ions that produce this selected product ion. Neutral loss scanning consists of selecting a neutral fragment and detecting all the fragmentations leading to the loss of that fragment; a signal is thus only recorded if an ion undergoes fragmentation, producing a neutral ion that is equivalent to the mass difference of interest. However, the most commonly used scan mode in tandem mass spectrometry, using a triple quadrupole mass spectrometer, is multiple (or selected) reaction monitoring (SRM/MRM), given its high sensitivity and selectivity. MRM consists of selecting one or more fragmentation reactions (transitions): the ions (m/z value) selected in the first mass analyzer are only detected if they produce one of the selected product ions [72, 94]. For the allergen peptides present in milk, egg, peanut, soybean, hazelnut, and almond, different transitions were described and recently summarized by Pilolli et al. [71].

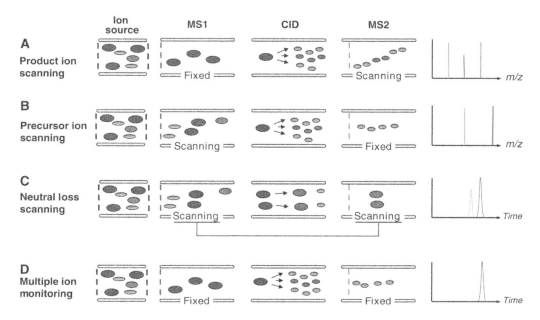

FIGURE 15.4 Representation of four different scan modes applied in tandem mass spectrometry: A, Product ion scanning; B, Precursor ion scanning; C, Neutral loss scanning; D, Multiple reaction monitoring (MRM). MS1: first mass analyzer, CID: collision induced dissociation, MS2: second mass analyzer. [95].

Next to these targeted acquisition modes (SRM/MRM), tandem mass spectrometry can be obtained in data-dependent acquisition (DDA) mode or data-independent acquisition (DIA) mode as well. In DDA, the mass spectrometer first selects the most intense peptide ions, after which these are fragmented and analyzed. This is the current standard technique for peptide identification [96]. In contrast, in DIA, the instrument focuses on a narrow mass window of precursor ions and acquires MS/MS data from all precursor ions detected within that window. In the next cycles, this mass window is stepped across the entire mass range, thereby systematically collecting MS/MS data from every mass and from all detected precursors; this allows for more sensitive and accurate quantification compared to DDA [97]. The most common method to generate DIA data is called SWATH (sequential window acquisition of all theoretical fragment ion spectra) and is performed on high-resolution LC-MS platforms (e.g., Q-TOF or Q-Orbitrap) for untargeted mass spectrometry analysis of complex samples. Untargeted peptide quantification using DIA is thereby comparable to targeted peptide quantification using MRM mass spectrometry. However, although in DIA the different mass windows collectively cover the entire *m/z* range, overall, only 1–3% of all incoming ions are isolated for mass analysis as a lot of ions are lost outside the selected mass windows. In order to increase this number, recently, a TIMS-quadrupole-TOF configuration was successfully developed, where the mobility separation of TIMS is synchronized with the quadrupole mass selection in a method called parallel accumulation-serial fragmentation (diaPASEF) [76, 98].

In order to increase tandem MS efficiency and sensitivity in LC-MS analysis, the LC mobile phase can be adjusted: supercharging agents like dimethyl sulfoxide (DMSO, 5%) can be added to the LC mobile phase solvents, thereby elevating the peptide charge states in ESI [99]. Moreover, in order to overcome TFA ion suppression, the supercharging agents m-NBA, 3-NPEA, TES or MOZ (0.1% concentrations) recently demonstrated to increase ESI-MS signal intensities and thus give lower detection limits, compared to only 0.1% TFA containing mobile phases [100].

DATA ANALYSIS

In peptidomics, where proteins and peptides are cleaved by an array of (unknown) proteolytic enzymes, peptide identification is not as "straightforward" as in proteomic analysis, where only highly specific proteolytic enzymes are used: searching the obtained tandem MS spectra against all possible peptide fragments in peptide spectral databases, instead of only against those matching the specificity pattern, hampers the search performance and quality of the results, making it often unfeasible. Moreover, also complex posttranslational modifications [101] are not taken into account, with the set of known or predicted proteins often being incomplete. Examples of such (proteomic) databases are X!Tandem, Mascot, SEQUEST, MS-GF+, MS-Fit, and OMSSA [28, 102, 103]. Lang et al. recently used the MaxQuant proteomic software against the UniProtKB Eucarida database, with no enzyme specificity, to identify peptides from defatted Antarctic krill powder, thereby setting the false discovery rate (FDR) cutoff to 0.01 (i.e., high confidence with 99% accuracy); as a result, two novel peptides were identified with specific enzyme-inhibiting properties [104].

Several databases exist which describe different properties of bioactive peptides, e.g., peptide sequence, physicochemical properties (length, isoelectric point), source and function [105]; some of them are even only focusing on food-derived peptides, like FeptideDB (food proteins) [106], FermFooDb (fermented food) [107], MBPDB (milk) [108] and BioPepDB [3]. These (combined) databases are mainly used to analyze the relationship between the peptide structure and their bioactivity (quantitative structure–activity relationship, QSAR), as well as to discover peptides with known bioactivity in (known or unknown) protein sources [105]. For all these bioactive peptide databases however, no mass spectrometry data are available, meaning that they cannot directly be used for the identification of bioactive peptides in LC-MS-analyzed samples [34]. Next to the peptide databases, small peptides can also be described (to a limited extent) in metabolite/metabolomics databases (e.g., METLIN) [109].

Next to MS spectral database searching for peptide identification, *de novo* identification can also be used, where peptide sequences are obtained by calculating the mass differences between the obtained MS fragments. Therefore, programs like Peaks [110] or PepNovo [111] can be used [28]. *De novo* sequencing of a heptapeptide from Lingzhi proteins, for example, was performed by Krobthong et al. by investigating the LC-MS/MS spectra using the series of a, b, x, and y fragment ions [112]. Algboory and Muhialdin, on the other hand, used PEAKS to identify the *de novo* peptide sequences in milk, after which they were subjected to the Uniprot database and the Antimicrobial Peptides Database (APD) to determine their similarity with previously identified peptides [113].

REFERENCES

1. Martini S, Solieri L, Tagliazucchi D. Peptidomics: New trends in food science. *Current Opinion in Food Science* 2021, 39, 51–59.
2. Carrasco-Castilla J, Hernández-Álvarez AJ, Jiménez-Martínez C, Gutiérrez-López GF, Dávila-Ortiz G. Use of proteomics and peptidomics methods in food bioactive peptide science and engineering. *Food Engineering Reviews* 2012, 4(4), 224–243.

3. Li Q, Zhang C, Chen H, Xue J, Guo X, Liang M, Chen M. BioPepDB: An integrated data platform for food-derived bioactive peptides. *International Journal of Food Sciences and Nutrition* 2018. DOI: 10.1080/09637486.2018.1446916.
4. Nongonierma AB, FitzGerald RJ. Strategies for the discovery and identification of food protein-derived biologically active peptides. *Trends in Food Science and Technology* 2017, 69, 289–305.
5. Ibanez C, Simo C, Garcia-Canas V, Cifuentes A, Castro-Puyana M. Metabolomics, peptidomics and proteomics applications of capillary electrophoresis-mass spectrometry in Foodomics: A review. *Analytica Chimica Acta* 2013, 802, 1–13.
6. Papetti A, Colombo R. High-performance capillary electrophoresis for food quality evaluation. *Evaluation Technologies for Food Quality* 2019, 301–377.
7. Sigdel TK, Nicora CD, Hsieh SC, Dai H, Qian WJ, Camp II DG, Sarwal MM. Optimization for peptide sample preparation for urine peptidomics. *Clinical Proteomics* 2014, 11(1), 7.
8. Debunne N, Verbeke F, Janssens Y, Wynendaele E, De Spiegeleer B. Chromatography of quorum sensing peptides: An important functional class of the bacterial peptidome. *Chromatographia* 2017. DOI: 10.1007/s10337-017-3411-2.
9. Aristoteli LP, Molloy MP, Baker MS. Evaluation of endogenous plasma peptide extraction methods for mass spectrometric biomarker discovery. *Journal of Proteome Research* 2007, 6(2), 571–581.
10. Kunda PB, Benavente F, Catalá-Clariana S, Giménez E, Barbosa J, Sanz-Nebot V. Identification of bioactive peptides in a functional yoghurt by micro liquid chromatography time-of-flight mass spectrometry assisted by retention time prediction. *Journal of Chromatography. Part A* 2012, 1229, 121–128.
11. Casado FJ, Montaño A, Carle R. Contribution of peptides and polyphenols from olive water to acrylamide formation in sterilized table olives. *LWT – Food Science and Technology* 2014, 59(1), 376–382.
12. André R, Guedes L, Melo R, Ascensão L, Pacheco R, Vaz PD, Serralheiro ML. Effect of food preparations on in vitro bioactivities and chemical components of Fucus vesiculosus. *Foods* 2020, 9(7), 955.
13. Menif EE, Offret C, Labrie S, Beaulieu L. Identification of peptides implicated in antibacterial activity of snow crab hepatopancreas hydrolysates by a bioassay-guided fractionation approach combined with mass spectrometry. *Probiotics and Antimicrobial Proteins* 2019, 11(3), 1023–1033.
14. Kononikhin AS, Starodubtseva NL, Bugrova AE, Shirokova VA, Chagovets VV, Indeykina MI, Popov IA, Kostyukevich YI, Vavina OV, Muminova KT, Khodzhaeva ZS, Kan NE, Frankevich VE, Nikolaev EN, Sukhikh GT. An untargeted approach for the analysis of the urine peptidome of women with preeclampsia. *Journal of Proteomics* 2016, 149, 38–43.
15. Català-Clariana S, Benavente F, Giménez E, Barbosa J, Sanz-Nebot V. Identification of bioactive peptides in hypoallergenic infant milk formulas by capillary electrophoresis-mass spectrometry. *Analytica Chimica Acta* 2010, 683(1), 119–125.
16. Moosmang S, Siltari A, Bolzer M, Kiechl S, Sturm S, Stuppner H. Development, validation, and application of a fast, simple, and robust SPE-based LC-MS/MS method for quantification of angiotensin I converting enzyme inhibiting tripeptides Val-Pro-Pro, Ile-Pro-Pro, and Leu-Pro-Pro in yoghurt and other fermented dairy products. *International Diary Journal* 2019, 97, 31–39.
17. Rein D, Ternes P, Demin R, Gierke J, Helgason T, Schön C. Artificial intelligence identified peptides modulate inflammation in healthy adults. *Food and Function* 2019, 10(9), 6030.

18. Gudlawar SK, Pilli NR, Siddiraju S, Dwivedi J. Highly sensitive assay for the determination of therapeutic peptide desmopressin in human plasma by UPLC-MS/MS. *Journal of Pharmaceutical Analysis* 2017, 7(3), 196–202.
19. Chen S, Huang G, Liao W, Gong S, Xiao J, Bai J, Hsiao WLW, Li N, Wu J. Discovery of the bioactive peptides secreted by Bifidobacterium using integrated MCX coupled with LC-MS and feature-based molecular networking. *Food Chemistry* 2021, 347, 129008.
20. González N, Iloro I, Durán JA, Elortza F, Suárez T. Evaluation of inter-day and inter-individual variability of tear peptide/protein profiles by MALDI-TOF MS analyses. *Molecular Vision* 2012, 18, 1572–1582.
21. La Barbera G, Capriotti AL, Cavaliere C, Ferraris F, Montone CM, Piovesana S, Chiozzi RZ, Laganà A. Saliva as a source of new phosphopeptide biomarkers: Development of a comprehensive analytical method based on shotgun peptidomics. *Talanta* 2018, 183, 245–249.
22. Li H, Aluko RE. Identification and inhibitory properties of multifunctional peptides from pea protein hydrolysate. *Journal of Agricultural and Food Chemistry* 2010, 58(21), 11471–11476.
23. Janssens Y, Debunne N, De Spiegeleer A, Wynendaele E, Planas M, Feliu L, Quarta A, Claes C, Van Dam D, De Deyn PP, Ponsaerts P, Blurton-Jones M, De Spiegeleer B. PapRIV, a BV-2 microglial cell activating quorum sensing peptide. *Scientific Reports* 2021, 11(1), 10723.
24. Debunne N, Wynendaele E, Janssens Y, De Spiegeleer A, Verbeke F, Tack L, Van Welden S, Goossens E, Knappe D, Hoffmann R, Van De Wiele C, Laukens D, Van Eenoo P, Van Immerseel F, De Wever O, De Spiegeleer B. The quorum sensing peptide EntF* promotes colorectal cancer metastasis in mice: A new factor in the microbiome-host interaction. *BioRXIV* 2020. DOI: 10.1101/2020.09.17.301044.
25. De Angelis E, Pilolli R, Monaci L. Coupling SPE on-line pre-enrichment with HPLC and MS/MS for the sensitive detection of multiple allergens in wine. *Food Control* 2017, 73, 814–820.
26. Dallas DC, Guerrero A, Khaldi N, Castillo PA, Martin WF, Smilowitz JT, Bevins CL, Barile D, German JB, Lebrilla CB. Extensive in vivo human milk peptidomics reveals specific proteolysis yielding protective antimicrobial peptides. *Journal of Proteome Research* 2013, 12(5), 2295–2304.
27. Manes NP, Gustin JK, Rue J, Mottaz HM, Purvine SO, Norbeck AD, Monroe ME, Zimmer JSD, Metz TO, Adkins JN, Smith RD, Heffron F. Targeted protein degradation by Salmonella under phagosome-mimicking culture conditions investigated using comparative peptidomics. *Molecular and Cellular Proteomics* 2007, 6(4), 717–727.
28. Dallas DC, Guerrero A, Parker EA, Robinson RC, Gan J, German JB, Barile D, Lebrilla CB. Current peptidomics: Applications, purification, identification, quantification, and functional analysis. *Proteomics* 2015, 15(5–6), 1026–1038.
29. Vitorino R, Barros AS, Caseiro A, Ferreira R, Amado F. Evaluation of different extraction procedures for salivary peptide analysis. *Talanta* 2012, 94, 209–215.
30. Tucholska M, Scozzaro S, Williams D, Ackloo S, Lock C, Siu KWM, Evans KR, Marshall JG. Endogenous peptides from biophysical and biochemical fractionation of serum analyzed by matrix-assisted laser desorption/ionization and electrospray ionization hybrid quadrupole time-of-flight. *Analytical Biochemistry* 2007, 370(2), 228–245.

31. Knudsen SB, Nielsen NJ, Qvist J, Andersen KB, Christensen JH. Generic multicriteria approach to determine the best precipitation agent for removal of biomacromolecules prior to non-targeted metabolic analysis. *Journal of Chromatography. Part B* 2021, 1167, 122567.
32. Esposito S, Mele R, Ingenito R, Bianchi E, Bonelli F, Monteagudo E, Orsatti L. An efficient liquid chromatography high resolution mass spectrometry approach for the optimization of the metabolic stability of therapeutic peptides. *Analytical and Bioanalytical Chemistry* 2017, 409(10), 2685–2696.
33. Taevernier L, Wynendaele E, D'Hondt M, De Spiegeleer B. Analytical quality-by-design approach for sample treatment of BSA-containing solutions. *Journal of Pharmaceutical Analysis* 2015, 5(1), 27–32.
34. Maes E, Oeyen E, Boonen K, Schildermans K, Mertens I, Pauwels P, Valkenborg D, Baggerman G. The challenges of peptidomics in complementing proteomics in a clinical context. *Mass Spectrometry Reviews* 2019, 38(3), 253–264.
35. Greening DW, Simpson RJ. A centrifugal ultrafiltration strategy for isolating the low-molecular weight (<or=25K) component of human plasma proteome. *Journal of Proteomics* 2010, 73(3), 637–648.
36. Finoulst I, Pinkse M, Van Dongen W, Verhaert P. Sample preparation techniques for the untargeted LC-MS-based discovery of peptides in complex biological matrices. *Journal of Biomedicine and Biotechnology* 2011. DOI: 10.1155/2011/245291.
37. Maes K, Smolders I, Michotte Y, Van Eeckhaut A. Strategies to reduce aspecific adsorption of peptides and proteins in liquid chromatography-mass spectrometry based bioanalyses: An overview. *Journal of Chromatography. Part A* 2014, 1358, 1–13.
38. Pezeshki A, Vergote V, Van Dorpe S, Baert B, Burvenich C, Popkov A, De Spiegeleer B. Adsorption of peptides at the sample drying step: Influence of solvent evaporation technique, vial material and solution additive. *Journal of Pharmaceutical and Biomedical Analysis* 2009, 49(3), 607–612.
39. Verbeke F, Bracke N, Debunne N, Wynendaele E, De Spiegeleer B. LC-MS compatible antiadsorption diluent for peptide analysis. *Analytical Chemistry* 2020, 92(2), 1712–1719.
40. Mant CT, Chen Y, Yan Z, Popa TV, Kovacs JM, Mills JB, Tripet BP, Hodges RS. HPLC analysis and purification of peptides. *Peptide Characterization and Application Protocols* 2007, 386, 3–55.
41. Jablonski JM, Wheat TE. *Practical Approaches to Peptide Isolation*. Waters Corporation 2017.
42. Shou WZ, Naidong W. Simple means to alleviate sensitivity loss by trifluoroacetic (TFA) mobile phases in the hydrophilic interaction chromatography-electrospray tandem mass spectrometric (HILIC-ESI/MS/MS) bioanalysis of basic compounds. *Journal of Chromatography. Part B* 2005, 825, 186–192.
43. D'Hondt M, Gevaert B, Wynendaele E, De Spiegeleer B. Implementation of a single quad MS detector in routine QC analysis of peptide drugs. *Journal of Pharmaceutical Analysis* 2016, 6(1), 24–31.
44. Snyder LR, Dolan JW. *High-Performance Gradient Elution: The Practical Application of the Linear-Solvent-Strength Model*. Wiley 2007.
45. den Uijl MJ, Schoenmakers PJ, Pirok BWJ, van Bommel MR. Recent applications of retention modelling in liquid chromatography. *Journal of Separation Science* 2020. DOI: 10.1002/jssc.202000905.

46. Vu H, Spicer V, Gotfrid A, Krokhin OV. A model for predicting slopes S in the basic equation for the linear-solvent-strength theory of peptide separation by reversed-phase high-performance liquid chromatography. *Journal of Chromatography. Part A* 2010, 1217(4), 489–497.
47. Mant CT, Lorne Burke TW, Hodges RS. Optimization of peptide separations in reversed-phase HPLC: Isocratic versus gradient elution. *Chromatographia* 1987, 24(1), 565–572.
48. Janvier S, De Sutter E, Wynendaele E, De Spiegeleer B, Vanhee C, Deconinck E. Analysis of illegal peptide drugs via HILIC-DAD-MS. *Talanta* 2017, 174, 562–571.
49. Kozlik P, Vaclova J, Kalikova K. Mixed-mode hydrophilic interaction/ion-exchange liquid chromatography – Separation potential in peptide analysis. *Microchemical Journal* 2021, 165, 106158.
50. Xu X, Gevaert B, Bracke N, Yao H, Wynendaele E, De Spiegeleer B. Hydrophilic interaction liquid chromatography method development and validation for the assay of HEPES zwitterionic buffer. *Journal of Pharmaceutical and Biomedical Analysis* 2017, 135, 227–233.
51. Van Dorpe S, Vergote V, Pezeshki A, Burvenich C, Peremans K, De Spiegeleer B. Hydrophilic interaction LC of peptides: Columns comparison and clustering. *Journal of Separation Science* 2010, 33(6–7), 728–739.
52. Kartsova LA, Bessonova EA, Somova VD. Hydrophilic interaction chromatography. *Journal of Analytical Chemistry* 2019, 74(5), 415–424.
53. Buszewski B, Noga S. Hydrophilic interaction liquid chromatography (HILIC) – A powerful separation technique. *Analytical and Bioanalytical Chemistry* 2012, 402(1), 231–247.
54. Debunne N. *Quorum Sensing Peptides as a Novel Microbiomal Causative Factor in Promoting Colorectal Cancer Metastasis: The EntF Metabolite Case.* Ghent University 2021.
55. Gilar M, Olivova P, Daly AA, Gebier JC. Orthogonality of separation in two-dimensional liquid chromatography. *Analytical Chemistry* 2005, 77(19), 6426–6434.
56. Hemmler D, Heinzmann SS, Wöhr K, Schmitt-Kopplin P, Witting M. Tandem HILIC-RP liquid chromatography for increased polarity coverage in food analysis. *Electrophoresis* 2018, 39(13), 1645–1653.
57. Carr DC. The role of chromatography in the characterisation and analysis of protein therapeutic drugs. *Special Issues* 2014, 32, 24–29.
58. Hsieh EJ, Bereman MS, Durand S, Valaskovic GA, MacCoss MJ. Effects of column and gradient lengths on peak capacity and peptide identification in nanoflow LC-MS/MS of complex proteomic samples. *Journal of the American Society for Mass Spectrometry* 2013, 24(1), 148–153.
59. Capriotti AL, Caruso G, Cavaliere C, Samperi R, Ventura S, Chiozzi RZ, Laganà A. Identification of potential bioactive peptides generated by simulated gastrointestinal digestion of soybean seeds and soy milk proteins. *Journal of Food Composition and Analysis* 2015, 44, 205–213.
60. Le Parc A, Dallas DC, Duaut S, Leonil J, Martin P, Barile D. Characterization of goat milk lactoferrin N-glycans and comparison with the N-glycomes of human and bovine milk. *Electrophoresis* 2014, 35(11), 1560–1570.
61. Agyei D, Tsopmo A, Udenigwe CC. Bioinformatics and peptidomics approaches to the discovery and analysis of food-derived bioactive peptides. *Analytical and Bioanalytical Chemistry* 2018, 410(15), 3463–3472.

62. Jonsson AP. Mass spectrometry for protein and peptide characterization. *Cellular and Molecular Life Sciences* 2001, 58(7), 868–884.
63. Wysocki VH, Resing KA, Zhang Q, Cheng G. Mass spectrometry of peptides and proteins. *Methods* 2005, 35(3), 211–222.
64. de Hoffmann E, Stroobant V. *Mass Spectrometry: Principles and Applications.* Wiley 2007.
65. Clarke W. *Mass Spectrometry in the Clinical Laboratory: Determining the Need and Avoiding Pitfalls.* Mass Spectrometry for the Clinical Laboratory 2017, 1–15.
66. Caron J, Chataigné G, Gimeno J, Duhal N, Goossens J, Dhulster P, Cudennec B, Ravallec R, Flahaut C. Food peptidomics of in vitro gastrointestinal digestions of partially purified bovine hemoglobin: Low-resolution versus high-resolution LC-MS/MS analyses. *Electrophoresis* 2016, 37(13), 1814–1822.
67. Olsen JV, Schwartz JC, Griep-Raming J, Nielsen ML, Damoc E, Denisov E, Lange O, Remes P, Taylor D, Splendore M, Wouters ER, Senko M, Makarov A, Mann M, Horning S. A dual pressure linear ion trap Orbitrap instrument with very high sequencing speed. *Molecular and Cellular Proteomics* 2009, 8(12), 2759–2769.
68. Serrano SMT, Zelanis A, Kitano ES, Tashima AK. Analysis of the snake venom peptidome. *Methods in Molecular Biology* 2018, 1719, 349–358.
69. Picariello G, Mamone G, Cutignano A, Fontana A, Zurlo L, Addeo F, Ferranti P. Proteomics, peptidomics, and immunogenic potential of wheat beer (weissbier). *Journal of Agricultural and Food Chemistry* 2015, 63(13), 3579–3586.
70. Alves TO, D'Almeida CTS, Scherf KA, Ferreira MSL. Modern approaches in the identification and quantification of immunogenic peptides in cereals by LC-MS/MS. *Frontiers in Plant Science* 2019, 10, 1470.
71. Pilolli R, Nitride C, Gillard N, Huet A, van Poucke C, de Loose M, Tranquet O, Larré C, Adel-Patient K, Bernard H, Mills CEN, Monaci L. Critical review on proteotypic peptide marker tracing for six allergenic ingredients in incurred foods by mass spectrometry. *Food Research International* 2020, 128, 108747.
72. Thomas SN. Mass spectrometry. In W Clarke & MA Marzinke (eds), *Contemporary Practice in Clinical Chemistry* 2019, 171–185. London: AACC Academic Press.
73. Luo MD, Zhou ZW, Zhu ZJ. The application of ion mobility-mass spectrometry in untargeted metabolomics: From separation to identification. *Journal of Analysis and Testing* 2020, 4(3), 163–174.
74. Ridgeway ME, Lubeck M, Jordens J, Mann M, Park MA. Trapped ion mobility spectrometry: A short review. *International Journal of Mass Spectrometry* 2018, 425, 22–35.
75. Yu F, Haynes SE, Teo GC, Avtonomov DM, Polasky DA, Nesvizhskii AI. Fast quantitative analysis of timsTOF PASEF data with MSFragger and IonQuant. *Molecular and Cellular Proteomics* 2020, 19(9), 1575–1585.
76. Meier F, Brunner AD, Frank M, Ha A, Bladau I, Voytik E, Kasper-Schoenefeld S, Lubeck M, Raether O, Bache N, Aebersold R, Collins BC, Röst HL, Mann M. diaPASEF: Parallel accumulation-serial fragmentation combined with data-independent acquisition. *Nature Methods* 2020, 17(12), 1229–1236.
77. Bush MF, Campuzano IDG, Robinson CV. Ion mobility mass spectrometry of peptide ions: Effects of drift gas and calibration strategies. *Analytical Chemistry* 2012, 84(16), 7124–7130.
78. Gelb AS, Jarratt RE, Huang Y, Dodds ED. A study of calibrant selection in measurement of carbohydrate and peptide ion-neutral collision cross sections by traveling wave ion mobility spectrometry. *Analytical Chemistry* 2014, 86(22), 11396–11402.

79. Li GY, Delafield DG, Li LJ. Improved structural elucidation of peptide isomers and their receptors using advanced ion mobility-mass spectrometry. *Trends in Analytical Chemistry* 2020, 124. DOI: 10.1016/j.trac.2019.05.048.
80. Gevaert B, D'Hondt M, Bracke N, Yao H, Wynendaele E, Vissers JP, De Cecco M, Claereboudt J, De Spiegeleer B. Peptide profiling of internet-obtained cerebrolysin using high performance liquid chromatography - Electrospray ionization ion trap and ultra high performance liquid chromatography - Ion mobility - Quadrupole time of flight mass spectrometry. *Drug Testing and Analysis* 2015, 7(9), 835–842.
81. Panchaud A, Affolter M, Kussmann M. Mass spectrometry for nutritional peptidomics: How to analyze food bioactives and their health effects. *Journal of Proteomics* 2012, 75(12), 3546–3559.
82. Roepstorff P, Fohlman J. Proposal for a common nomenclature for sequence ions in mass spectra of peptides. *Biomedical Mass Spectrometry* 1984, 11(11), 601.
83. Sleno L, Volmer DA. Ion activation methods for tandem mass spectrometry. *Journal of Mass Spectrometry* 2004, 39(10), 1091–1112.
84. Ichou F, Schwarzenberg A, Lesage D, Alves S, Junot C, Machuran-Mandard X, Tabet J. Comparison of the activation time effects and the internal energy distributions for the CID, PQD and HCD excitation modes. *Journal of Mass Spectrometry* 2014, 49(6), 498–508.
85. Penkert M, Hauser A, Harmel R, Fiedler D, Hackenberger CPR, Krause E. Electron transfer/higher energy collisional dissociation of doubly charged peptide ions: Identification of labile protein phosphorylations. *Journal of the American Society for Mass Spectrometry* 2019, 30(9), 1578–1585.
86. Diedrich JK, Pinto AFM, Yates III JR. Energy dependence of HCD on peptide fragmentation: Stepped collisional energy finds the sweet spot. *Journal of the American Society for Mass Spectrometry* 2013. DOI: 10.1007/s13361-013-0709-7.
87. Ramachandran S, Thomas T. A frequency-based approach to predict the low-energy collision-induced dissociation fragmentation spectra. *ACS Omega* 2020, 5(22), 12615–12622.
88. Zubarev RA. Electron capture dissociation tandem mass spectrometry. *Current Opinion in Biotechnology* 2004, 15(1), 12–16.
89. Zubarev RA, Kelleher NL, McLafferty FW. ECD of multiply charged protein cations. A non-ergodic process. *Journal of the American Chemical Society* 1998, 120(13), 3265–3266.
90. Huang L, Chiang C, Chen S, Wei S, Chen S. Complete mapping of disulfide linkages for etanercept products by multi-enzyme digestion coupled with LC-MS/MS using multi-fragmentations including CID and ETD. *Journal of Food and Drug Analysis* 2019, 27(2), 531–541.
91. Mechref Y. Use of CID/ETD mass spectrometry to analyze glycopeptides. *Current Protocols in Protein Science* 2012, Chapter 12: Unit-12.11.11.
92. Wen L, Jiang Y, Zhou X, Bi H, Yang B. Structure identification of soybean peptides and their immunomodulatory activity. *Food Chemistry* 2021, 359, 129970.
93. Nimalaratne C, Bandara N, Wu J. Purification and characterization of antioxidant peptides from enzymatically hydrolyzed chicken egg white. *Food Chemistry* 2015, 188, 467–472.
94. Giacometti J, Buretic-Tomljanovic A. Peptidomics as a tool for characterizing bioactive milk peptides. *Food Chemistry* 2017, 230, 91–98.
95. Domon B, Aebersold R. Mass spectrometry and protein analysis. *Science* 2006, 312(5771), 212–217.

96. Lahrichi SL, Affolter M, Zolezzi IS, Panchaud A. Food peptidomics: Large scale analysis of small bioactive peptides – A pilot study. *Journal of Proteomics* 2013, 88, 83–91.
97. Doerr A. DIA mass spectrometry. *Nature Methods* 2015, 12(1), 35.
98. Krasny L, Huang PH. Data-independent acquisition mass spectrometry (DIA-MS) for proteomic applications in oncology. *Molecular Omics* 2021, 17(1), 29.
99. Hahne H, Pachl F, Ruprecht B, Maier SK, Klaeger S, Helm D, Médard G, Wilm M, Lemeer S, Kuster B. DMSO enhances electrospray response, boosting sensitivity of proteomic experiments. *Nature Methods* 2013, 10(10), 989–992.
100. Nshanian M, Lakshmanan R, Chen H, Loo RRO, Loo JA. Enhancing sensitivity of liquid chromatography-mass spectrometry of peptides and proteins using supercharging agents. *International Journal of Mass Spectrometry* 2018, 427, 157–164.
101. Yao SX, Udenigwe CC. Peptidomics of potato protein hydrolysates: Implications of posttranslational modifications in food peptide structure and behavior. *Royal Society Open Science* 2018, 5(7). DOI: 10.1098/rsos.172425;172425.
102. Menschaert G, Vandekerckhove TTM, Baggerman G, Schoofs L, Luyten W, Van Criekinge W. Peptidomics coming of age: A review of contributions from a bioinformatics angle. *Journal of Proteome Research* 2010, 9(5), 2051–2061.
103. Agrawal H, Joshi R, Gupta M. Purification, identification and characterization of two novel antioxidant peptides from finger millet (*Eleusine coracana*) protein hydrolysate. *Food Research International* 2019, 120, 697–707.
104. Lang M, Song Y, Li Y, Xiang X, Ni L, Miao J. Purification, identification, and molecular mechanism of DPP-IV inhibitory peptides from defatted Antarctic krill powder. *Journal of Food Biochemistry* 2021. DOI: 10.1111/jfbc.13872.
105. Iwaniak A, Darewicz M, Minkiewicz P. Databases of bioactive peptides. In F Toldra & J Wu (eds), *Biologically Active Peptides* 2021, 309–330. Academic Press.
106. Panyayai T, Ngamphiw C, Tongsima S, Mhuantong W, Limsriprahan W, Choowongkomon K, Sawatdichaikul O. FeptideDB: A web application for new bioactive peptides from food protein. *Heliyon* 2019, 5(7), e02076.
107. Chaudhary A, Bhalla S, Patiyal S, Raghava GPS, Sahni G. FermFooDb: A database of bioactive peptides derived from fermented foods. *Heliyon* 2021, 7(4), e06668.
108. Nielsen SD, Beverly RL, Qu Y, Dallas DC. Milk bioactive peptide database: A comprehensive database of milk protein-derived bioactive peptides and novel visualization. *Food Chemistry* 2017, 232, 673–682.
109. Guijas C, Montenegro-Burke JR, Domingo-Almenara X, Palermo A, Warth B, Hermann G, Koellensperger G, Huan T, Uritboonthai W, Aisporna AE, Wolan DW, Spilker ME, Benton HP, Siuzdak G. Metlin: A technology platform for identifying knowns and unknowns. *Analytical Chemistry* 2018, 90(5), 3156–3164.
110. Ma B, Zhang K, Hendrie C, Liang C, Li M, Doherty-Kirby A, Lajoie G. PEAKS: Powerful software for peptide de novo sequencing by tandem mass spectrometry. *Rapid Communications in Mass Spectrometry* 2003, 17(20), 2337–2342.
111. Frank A, Pevzner P. PepNovo: De novo peptide sequencing via probabilistic network modeling. *Analytical Chemistry* 2005, 77(4), 964–973.
112. Krobthong S, Yingchutrakul Y, Samutrtai P, Choowongkomon K. The C-terminally shortened analogs of a hexapeptide derived from Lingzhi hydrolysate with enhanced tyrosinase-inhibitory activity. *Archiv der Pharmazie* 2021. DOI: 10.1002/ardp.202100204.
113. Algboory HL, Muhialdin BJ. Novel peptides contribute to the antimicrobial activity of camel milk fermented with *Lactobacillus plantarum* IS10. *Food Control* 2021, 126, 108057.

CHAPTER 16

Bioinformatic Analysis

Semih Ötleş, Bahar Bakar, and Burcu Kaplan Türköz

CONTENTS

Introduction	321
Sequence-Based Analysis in Bioactive Peptide Research	322
Structure-Based Analysis in Bioactive Peptide Research	328
Quantitative Structure–Activity Relationship	330
Molecular Docking	331
Molecular Dynamics	333
Conclusions and Future Prospects	333
References	334

INTRODUCTION

Bioinformatics is a collection of computer-based approaches designed to store, analyze, and compare biological data including genome, gene, transcript, protein–peptide sequence and structure, metabolic products, and taxonomy. Bioinformatics is a fundamental tool in several research areas focused on understanding molecular mechanisms in life sciences, including genomics, evolutionary genetics, diagnostic and therapeutic medical sciences, proteomics, and protein engineering (Can, 2014; Luscombe et al., 2011).

Recently, use of bioinformatics has been extended to food science and technology (Agyei, Bambarandage, and Udenigwe, 2018; Holton, Vijayakumar, and Khaldi, 2013). Current food science deals with highly complex interactions between food and the biological systems. Molecular nutrition, food immunology, and food metabolism are emerging fields which all focus on understanding the cellular effect of food molecules, including peptides and proteins. Proteins and peptides, among other biological molecules, are of prime importance in determining cellular fate, as they are considered the workers of the cell. Food proteins are not only important for their nutritional value and textural effects, but they exert many important cellular functions upon ingestion. Bioinformatics is very useful in analyzing the sequence–structure–function relationship of proteins and peptides, providing an estimate of their effect on human health.

Bioinformatics is used as a powerful prediction and analysis tool in bioactive peptide research. There are several algorithms and databases which can be used to predict and classify the possible bioactive peptides encrypted in protein sources, and development/optimization of experimental methods used for their production (Agyei, Bambarandage, and Udenigwe, 2018; Tu, Cheng, et al., 2018). Bioinformatic approaches are mainly used to select potential protein sources containing bioactive peptides and the suitable hydrolysis enzymes for peptide release. Furthermore, the bioactivity, allergenicity, toxicity, and sensory properties of a purified peptide can also be predicted based on its amino acid sequence. Integrating bioinformatic approaches with experimental validation offers great benefits: putative bioactivity of the protein source and the released peptides can be predicted, thereby minimizing the number of tests that must be performed on proteins and bioactive peptides. Moreover, bioinformatic analysis can help identify moonlighting peptides, which are peptides with more than one bioactivity. In this chapter, an overview of the use of bioinformatics in bioactive peptide research will be given with emphasis on sequence and structure homology, computational structure prediction, and structural analysis tools (Figure 16.1).

SEQUENCE-BASED ANALYSIS IN BIOACTIVE PEPTIDE RESEARCH

Bioinformatic analysis in bioactive peptide research is mainly focused on identification of promising protein sources based on their amino acid sequences. Protein sequences can be directly obtained from databases such as UniProt, and encoding gene sequences can be accessed from NCBI. These types of data are stored, categorized, and standardized in three sequence-based databases: NCBI (American National Biotechnology Information Center; https://www.ncbi.nlm.nih.gov/), EMBL-EBI (European Molecular Biology Laboratory; https://www.ebi.ac.uk/), and DDBJ (DNA Data Bank of Japan; https://www.ddbj.nig.ac.jp/index-e.html) (Cochrane, Karsch-Mizrachi, and Takagi, 2016). These databases can be searched using DNA/mRNA/protein sequence, molecule name, or function. Also, switching between the nucleic acid and the protein sequence is possible by various translation tools, like ExPASy (https://web.expasy.org/translate/), EMBL-EBI Sequence Translation (https://www.ebi.ac.uk/Tools/st/).

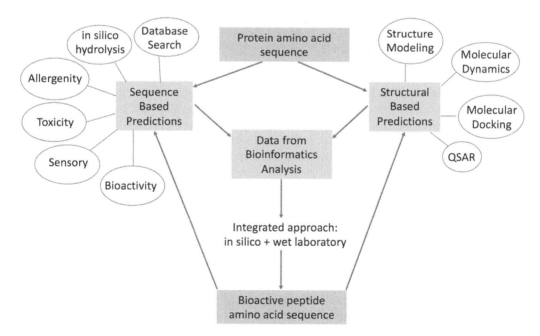

FIGURE 16.1 Overview of sequence and structure based bioinformatic approach in bioactive peptide research.

The functions of proteins and peptides are closely related to their amino acid sequences and sequence alignment is a very powerful tool in estimating their structural and/or functional similarities. The main principle here is the conservation of functionally and/or structurally important residues arising from evolutionary relationships or homology (Madden, 2002).

BLAST (Basic Local Alignment Search Tool; https://blast.ncbi.nlm.nih.gov/Blast.cgi) is a tool to search a sequence in a database in order to find similar sequences. The algorithm uses a local alignment strategy to find similarities between nucleic acid or protein sequences (Altschul et al., 1990). The sequence of interest (query) is entered into the provided space and users can select databases, organisms, program, and algorithm parameters. Output sequences are sorted according to their similarity to the query based on total score, E-value, and identity. E-value is an indicator of random similarity and lower E-values or the closer it is to zero, indicate that the match is beyond random. The alignments are scored based on matrices which contain scores for each pair of amino acids based on evolutionary conservation (Madden, 2002). Therefore, alignments are important in terms of highlighting evolutionarily conserved regions between two sequences. Conserved regions most of the time indicate structure and/or function similarity between proteins (Whisstock and Lesk, 2003).

In bioactive peptide research, the bioinformatic analysis starts by searching for the amino acid sequences of the proteins which might contain bioactive peptides. The amino acid sequences of the individual proteins present in the selected food source can be accessed from protein databases and these can be used to predict the presence of possible bioactive peptides. A list of commonly used bioactive peptide databases and tools are given in Table 16.1.

BIOPEP-UWM™ database of bioactive peptides (formerly BIOPEP) has been an important milestone for bioactive peptide research and is one of the most widely used tools in order to predict and analyze bioactive peptides. Sequence-based prediction by BIOPEP is useful in selecting protein source, bioactivity estimation, and bioactive peptide profile (Minkiewicz, Iwaniak, and Darewicz, 2019). The prediction takes into account the frequency of occurrence of previously identified bioactive peptides within the given sequence.

BIOPEP-Enzyme(s) Action, FeptideDB, and ExPASY-PeptideCutter tools can be used to perform *in silico* enzymatic hydrolysis (Gasteiger et al., 2005; Minkiewicz et al., 2008; Panyayai et al., 2019). The peptide release upon hydrolysis by the selected enzymes can be estimated using the knowledge of cleavage specificity of enzymes. The *in silico* hydrolysis approach is important as the occurrence does not necessarily mean that bioactive peptides will be released after enzymatic hydrolysis in the wet laboratory experiment. *In silico* hydrolysis can be used to simulate not only single enzyme hydrolysis but also multiple enzymes can be selected for the prediction of released bioactive peptides. Therefore, it is possible to estimate the release of bioactive peptides by the action of digestive enzymes, such as pepsin, chymotrypsin, and trypsin (Minkiewicz et al., 2008). The results of such analysis will include amounts and sequences of released bioactive peptides, functions of bioactive peptides, and degree of hydrolysis based on the protein primary structure. The selection of suitable enzyme(s) for the production of targeted bioactive peptides requires a high number of wet lab experiments, therefore, bioinformatics is used as a powerful tool to narrow down the experimental search space.

The bioactivity of peptides resulting from *in silico* hydrolysis can be evaluated using peptide databases including like BIOPEP, PepBank, PeptideDB, and BitterDB. The peptides and their bioactivities resulting from this *in silico* hydrolysis approach can then be

TABLE 16.1 Common Bioactive Peptide Databases and Tools

Databases/ Tools	Available Link/Descriptions	Reference
AgAbDb	http://196.1.114.46:8080/agabdb2/home.jsp Peptide and protein antigens and interactions database	(Kulkarni-Kale et al. 2014)
AHTPDB	http://crdd.osdd.net/raghava/ahtpdb/ Antihypertensive peptide databases	(Kumar et al. 2015)
AlgPred	http://crdd.osdd.net/raghava/algpred/ Allergenicity prediction tool	(Saha and Raghava 2006)
Allergen Nomenclature	http://allergen.org/ Allergen database	(King et al. 1994)
Allergen Online	http://www.allergenonline.org Allergen database and prediction tool	(Goodman et al. 2016)
Allergome	www.allergome.org Allergen database and prediction tool	(Mari et al. 2006)
AllerTOP v.2.0	https://www.ddg-pharmfac.net/AllerTOP/index.html Allergenicity prediction tool	(Dimitrov et al. 2014)
AntiBP2	http://crdd.osdd.net/raghava/antibp2/ Prediction of antimicrobial peptides	(Lata, Mishra, and Raghava 2010)
APD	https://wangapd3.com/main.php Antimicrobial peptides database	(Wang, Li, and Wang 2016)
AVPdb	http://crdd.osdd.net/servers/avpdb/index.php Antiviral peptide database	(Qureshi et al. 2014)
AVPpred	http://crdd.osdd.net/servers/avppred/submit.php Antiviral peptide prediction tool	(Thakur, Qureshi, and Kumar 2012)
BaAMPs	http://www.baamps.it/ Biofilm active antimicrobial peptides database	(Luca et al. 2015)
BACTIBASE	http://bactibase.hammamilab.org/about.php Bacteriocins database including alignment, physicochemical features, structure prediction tools	(Hammami, et al. 2010)
BactPepDB	http://bactpepdb.rpbs.univ-paris-diderot.fr/cgi-bin/home.pl Bacterial peptide database	(Rey, Deschavanne, and Tuffery 2014)
BAGEL4	http://bagel4.molgenrug.nl/ Bacteriocin peptide database	(van Heel et al. 2018)
BERT4Bitter	http://pmlab.pythonanywhere.com/BERT4Bitter	(Charoenkwan, Nantasenamat, et al. 2021)
BioPepDB	http://bis.zju.edu.cn/biopepdbr/index.php Food origin bioactive peptide database	(Li et al. 2018)
BIOPEP-UWM	http://www.uwm.edu.pl/biochemia/index.php/en/biopep Databases especially for food-based bioactive peptides and *in silico* hydrolysis epitopes	(Minkiewicz et al. 2019)
BitterDB	http://bitterdb.agri.huji.ac.il Bitter molecules and receptors database and prediction	(Dagan-Wiener et al. 2017)
BitterX	http://mdl.shsmu.edu.cn/BitterX/ Bitter molecules and receptors database and prediction	(Huang et al. 2016)

(*Continued*)

TABLE 16.1 (Continued)

Databases/ Tools	Available Link/Descriptions	Reference
CancerPPD	http://crdd.osdd.net/raghava/cancerppd/index.php Anticancer peptide database	(Tyagi et al. 2015)
DBAASP	https://dbaasp.org/ Antimicrobial peptide structure database and activity prediction tool	(Pirtskhalava et al. 2021)
EROP-Moscow	http://erop.inbi.ras.ru/ Bioactive peptide database	(Zamyatnin et al. 2006)
Expasy PeptideCutter	http://web.expasy.org/peptide_cutter/ *in silico* hydrolysis tool	(Gasteiger et al. 2005)
FeptideDB	http://www4g.biotec.or.th/FeptideDB/enzyme_digestion.php *in silico* hydrolysis and bioactivity prediction tool	(Panyayai et al. 2019)
IEDB	https://www.iedb.org/ Immune epitope database	(Sette et al. 2018)
IDPPIV-SCM	http://camt.pythonanywhere.com/iDPPIV-SCM DPP-IV inhibitory peptide prediction tool	(Charoenkwan, Kanthawong, et al. 2020)
MBPDB	http://mbpdb.nws.oregonstate.edu/ Milk bioactive peptide database	(Nielsen et al. 2017)
ParaPep	http://crdd.osdd.net/raghava/parapep/home.php Antiparasitic peptide database	(Mehta et al. 2014)
PepBank	http://pepbank.mgh.harvard.edu/ Bioactive peptide database	(Shtatland et al. 2007)
PepBDB	http://huanglab.phys.hust.edu.cn/pepbdb/ Biological peptide–protein complex structures database	(Wen et al. 2019).
PeptideDB	http://www.peptides.be/ Bioactive peptide database	(Liu et al. 2008)
PeptideRanker	http://distilldeep.ucd.ie/PeptideRanker/ Prediction of the probability of peptides bioactivity	(Mooney et al. 2012)
Pep-Calc.com	http://pepcalc.com/ Physicochemical features calculation tool	(Lear and Cobb 2016)
PlantPepDB	http://14.139.61.8/PlantPepDB/index.php Plant peptide database, physicochemical features calculation, and alignment tools	(Das et al. 2020)
PROSPER	https://prosper.erc.monash.edu.au/home.html Hydrolysis cleavage site estimation tool	(Song et al. 2012)
ProtParam	https://web.expasy.org/protparam/ Physicochemical features calculation tool	(Gasteiger et al. 2005)
QSPepPred	http://crdd.osdd.net/servers/qsppred/ Quorum sensing peptide prediction tool	(Rajput, Gupta, and Kumar 2015)
SPdb	http://proline.bic.nus.edu.sg/spdb/search.html Signal peptide database	(Choo, Tan, and Ranganathan 2005)
StraPep	http://isyslab.info/StraPep/ Bioactive peptide structure database	(Wang et al. 2018)

(*Continued*)

TABLE 16.1 (Continued)

Databases/Tools	Available Link/Descriptions	Reference
THPdb	https://webs.iiitd.edu.in/raghava/thpdb/index.html FDA-approved therapeutic peptide/protein database	(Usmani et al. 2017)
ToxinPred	http://crdd.osdd.net/raghava/toxinpred/ Toxicity prediction tool	(Gupta et al. 2013)
TumorHoPe	http://crdd.osdd.net/raghava/tumorhope/ Tumor recognize peptides database	(Kapoor et al. 2012)
YADAMP	http://yadamp.unisa.it/default.aspx Antimicrobial peptide database	(Piotto et al. 2012)

used as a reference point for the initial determination of protein source and hydrolysis enzyme to produce a specific bioactive peptide from a novel/convenient protein source. However, it is worth noting that *in silico* hydrolysis results must be experimentally validated due to several factors including the complexity of enzyme protein interactions and the tertiary structure of proteins. Since proteins have different three-dimensional shapes, it is not always straightforward to predict the pattern of enzymatic digestion (Tsai et al., 2002). The *in silico* hydrolysis predictions might provide realistic results only when protein tertiary structure is somehow denatured by pretreatments such as high temperature, ultrasound, ohmic heating, and microwave and pulsed electric field before enzymatic hydrolysis (Ulug, Jahandideh, and Wu, 2021).

A summary of sequence-based approaches in bioinformatic analysis is given in Figure 16.2

BIOPEP was used in several research studies to predict bioactive peptide potential of food sources and proteins. These include, prediction of DPP-IV inhibitory peptides (Liu, Cheng and Wu, 2019); ACE inhibitory peptides and antidiabetic peptides (Yap and Gan, 2020); ACE inhibitory, rennin inhibitory, and antioxidant peptides (Dave et al. 2016); antioxidant and antibacterial peptides (Sah et al. 2018); ACE inhibitory, antioxidant, and opioid peptides (Amigo, Martínez-Maqueda, and Hernández-Ledesma, 2020); and DPP-IV inhibitory, ACE inhibitory, and antioxidant peptides (de Fátima Garcia, de Barros, and de Souza Rocha, 2020). Other commonly used sequence-based bioinformatic tools are Peptide Ranker (Mooney et al. 2012) for the prediction of the likelihood of the bioactivity, and other tools for the prediction of toxicity, allergenicity, and bitterness of the bioactive peptides (Gupta et al., 2013; Saha and Raghava ,2006; Dagan-Wiener et

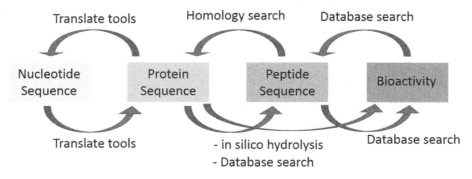

FIGURE 16.2 Sequence-based bioinformatics analysis in bioactive peptide research.

al., 2017; Huang et al., 2016). Bioinformatic analysis is therefore very valuable in selecting bioactive peptides which have the desired biological effect and do not have allergenic, toxic, or unwanted taste effects.

Determination of allergenicity risk is important for food researchers and must be carried out in the early phases of food product development studies as novel food products must not carry the risk of allergenicity (FAO/WHO 2001). Epitopes are the regions of allergens which interact with antibodies, and both the sequence and the structure of those epitopes are crucial for allergenicity. Therefore, allergenicity risk of a given bioactive peptide can be predicted from its amino acid sequence based on their degree of homology to known allergens. Allergen databases and tools can also be used to obtain information on biochemical properties, genetic connections, and literature studies about food allergens (Radauer, 2017). Widely known allergen databases and tools are presented in Table 16.1. For example, *in silico* analysis of casein-identified peptides were estimated as nonallergenic using AlgPred (Tu, Liu, et al. 2018). On the other hand, some antioxidant peptides from marine invertebrates were predicted as allergenic by AllerTOP and AlgPred tools (Chai et al., 2017). However, prediction of allergenicity by bioinformatic tools is not sufficient, and clinical trials must be conducted for allergen-free labeling of the final product containing the bioactive peptide.

Bioactive peptides may show toxic activity, and this risk should be taken into account in designing functional foods or nutraceuticals enriched with specific peptides. Therefore, *in silico* toxicity prediction of proteins and peptides is commonly used as a preliminary tool. ToxinPred (Gupta et al., 2013) is one of the prediction tools for toxic peptides and was used by several researchers (Yap and Gan, 2020). Some examples include antimicrobial peptides from rapeseed (Duan, Zhang, and Chen, 2021), bioactive peptides from tomato seed proteins (Kartal, Kaplan Türköz, and Otles, 2020) and ACE inhibitory peptides from yak casein (Lin et al., 2018).

Sensory properties of food are determined by appearance, texture, aroma, and taste, and are directly related to the chemical composition of the food. Human perceived tastes are salty, sour, sweet, bitter, and umami; the last three may be related to peptides and many peptides are reported to have strong bitter taste (Temussi, 2012; Iwaniak et al., 2016). BitterDB and BitterX are special databases where sequences of bitter peptides are stored and analyzed (Dagan-Wiener et al., 2017; Huang et al., 2016). In one study, chickpea DPP-IV inhibitory peptides were evaluated using BitterX and possible target bitter taste receptors were predicted (Chandrasekaran, Luna-Vital, and de Mejia, 2020). BitterX requires the SMILES (Simplified Molecular Input Line Entry System) code of the peptide as input; how to import SMILES strings are described elsewhere (Minkiewicz, Iwaniak, and Darewicz, 2017).

BIOPEP can also be used for prediction of bitterness of peptides as well as sour, sweet, umami, salty, enhancer (salt or umami enhancer), and suppressor (bitter suppressor) properties (Iwaniak et al., 2016). The sensory profile of ACE inhibitory peptides from goat milk caseins were analyzed using the BIOPEP-Profiles of Sensory Activity tool and bitter taste characteristics were predicted (Rani, Pooja, and Pal, 2017). Selamassakul et al. (2020) identified peptides from brown rice and predicted that some of them might have umami taste by the BIOPEP tool. Finally, a tool for the prediction of peptides which have umami taste was recently published (Charoenkwan, Yana, et al., 2020).

There are many other specialized databases such as PlantPepDB for peptides of plant origin, MBPDB for milk-derived peptides, AHTPDB for antihypertensive peptides and AntiBP2 for antibacterial peptides (Table 16.1). A list of other databases can be found in a recent review article (Minkiewicz, Iwaniak, and Darewicz, 2019). There are many studies using bioinformatics for the prediction and characterization of potential bioactive peptides from different food sources and are discussed in excellent review articles (Nongonierma and FitzGerald, 2017; Tu, Cheng, et al., 2018).

STRUCTURE-BASED ANALYSIS IN BIOACTIVE PEPTIDE RESEARCH

The homology at the primary structure (amino acid sequence) level gives valuable information about the biological function of proteins. But a complete understating of protein function is only possible when the tertiary structure (3D structure) of the protein is obtained. The amino acid sequence is the most important determinant of protein folding, therefore contains intrinsic information about the shape of a protein (Berg, Tymoczko, and Lubert, 2002). This sequence–structure–function relationship is very useful in bioinformatic analysis, as one can use an amino acid sequence to predict the tertiary structure, and the 3D structure can be used to predict the function of the peptide and protein of interest.

Tertiary structure of proteins can be experimentally calculated using macromolecular crystallography (MX), nuclear magnetic resonance (NMR) and cryo-electron microscopy (Cryo-EM) (Richardson and Richardson, 2014; Billeter, Wagner, and Wüthrich, 2008; Nwanochie and Uversky, 2019). Experimental protein structures are freely available in Protein Data Bank (PDB) (http://www.rcsb.org) and Protein Data Bank in Europe (PDBe) (http://www.ebi.ac.uk/pdbe/). Furthermore, solution scattering (SAS) is another important technique which gives information on the low-resolution structure of proteins in solution, and this information is very useful in hybrid modeling approaches (Gräwert and Svergun, 2020). SAS experimental data as well as protein models obtained can be accessed from SASBDB (www.sasbdb.org) (Valentini et al., 2015).

There have been tremendous developments in experimental protein structure determination in the past years, but there are still some challenges resulting from technical limitations depending on the target protein and also the time, money, and complex experimental setup requirements of these techniques.

Bioinformatic and computational biology approaches were developed in order to help overcome these issues, and today, computational protein structure modeling is a very powerful tool. The computational protein structure prediction uses mainly three different approaches – homology modeling, threading, and the ab initio method (Deng, Jia, and Zhang, 2018).

Protein structure contains valuable information to help accurately predict the presence of encrypted bioactive peptides. The location of bioactive peptide amino acids in the 3D shape of the protein will determine the accessibility of hydrolysis enzymes and will directly impact the number and type of bioactive peptides produced after enzymatic hydrolysis. There is only one bioinformatic tool, PROSPER, that takes into account secondary structure, solvent accessibility, and disorder predictions of the protein during *in silico* hydrolysis (Song et al., 2012). However, the tertiary structure of proteins is not taken into account in current sequence-based bioactive peptide predictions, which limits the accurate prediction of potential bioactive peptides. More research is needed to develop algorithms that can process atomic coordinates to predict enzymatic hydrolysis patterns.

Peptide chemical structures are of prime importance in determining which and how the peptide will interact with the target protein. The chemical structure or the primary structure can be obtained by mass spectrometry methods (Strack, 2015). Similar to proteins, some peptides can adopt well-defined tertiary structures, but most of the time the peptides will have more flexible structures which can adopt different conformations, depending on the environmental conditions and presence of an interacting partner. Experimental determination of peptide structures is different than that of proteins mainly due to the short length and high flexibility of peptides. Peptides can adopt different conformations in solution; therefore, it is not always possible to obtain an average structural model as is done for proteins (Thomas et al., 2006). Peptides with some level of structure can be crystallized, and their structure can be obtained using MX. The experimental details of peptide crystallography are described in a very detailed review article (Spencer and Nowick, 2015). NMR and SAS are also used in peptide structure determination, as these techniques are very suitable for flexible molecules and dynamic systems (Zhang et al., 2015; Gräwert and Svergun, 2020; Kozak et al., 2010; Hennig and Sattler, 2014). Other techniques, such as Circular Dichroism (CD), Electron Paramagnetic Resonance (EPR), and Fourier Transform Infrared Spectroscopy (FTIR), are also used to study peptide structure and biophysics (Thomas et al., 2006).

The available experimental peptide structures can be accessed from PDB, and also from a recent specific peptide structure database, StraPep (Wang et al., 2018). StraPep retrieves structure data from PDB and classifies these structures, depending on their bioactivities. SATPdb is another peptide structure database where structures of therapeutic peptides are stored and classified. Furthermore, it is possible to perform a structure-based similarity search against all the structures in the database if the structure of the bioactive peptide of interest is available (Singh et al., 2016). Therefore, the structural similarity of peptides with similar functions is also becoming available, which will provide important insight for future bioactive peptide selection and/or design studies.

Computational structure prediction is another approach to model peptide structures. Several tools for protein structure prediction can also be used for peptide structure

TABLE 16.2 Peptide and Protein Structure Prediction Tools

Tools	Available Link / Descriptions	Reference
I-TASSER	https://zhanglab.ccmb.med.umich.edu/I-TASSER/ (10–1500 amino acids)	(Roy, Kucukural, and Zhang 2010)
RaptorX	http://raptorx.uchicago.edu/ (26–1100 amino acids)	(Källberg et al. 2012)
QUARK	https://zhanglab.ccmb.med.umich.edu/QUARK (20–200 amino acids)	(Xu and Zhang 2012)
PEP-FOLD 3	https://bioserv.rpbs.univ-paris-diderot.fr/services/PEP-FOLD3/ (5–50 amino acids)	(Lamiable et al. 2016)
Robetta	https://robetta.bakerlab.org/queue.php (26–2001 amino acids)	(Kim, Chivian, and Baker 2004)
PEPstrMOD	https://webs.iiitd.edu.in/raghava/pepstrmod/ (7–25 amino acids)	(Singh et al. 2015)
PEP2D	http://crdd.osdd.net/raghava/pep2d/ Peptide secondary structure prediction	(Singh et al. 2019)
PSIPRED 4.0	http://bioinf.cs.ucl.ac.uk/psipred/ Secondary structure prediction	(Buchan and Jones 2019)
IUPred2A	https://iupred2a.elte.hu/ Prediction tool for disordered proteins	(Erdős and Dosztányi 2020)

modeling, and some online tools were developed specifically for peptides (Table 16.2). PEP-FOLD3 is a bioinformatics tool designed to predict peptide structure solely from sequence information, and results are generally available within hours (Lamiable et al., 2016). PEPstrMOD server is another tool that can be used to predict the tertiary structure of small peptides and also handles peptides having various modifications, like nonnatural residues and terminal modifications (Singh et al., 2015). There are also online bioinformatics tools that can be used for secondary structure prediction (Table 16.2).

The molecular details of protein–peptide interactions can be understood in detail by analyzing the structures of protein–peptide complexes. However, as already mentioned, due to the difficulties in experimental structure determination, the number of such complexes in databases is limited. One such database is PepBDB, which collects all available structures of peptide-mediated protein interactions with peptide lengths up to 50 residues in the Protein Data Bank (Wen et al., 2019).

Furthermore, several bioinformatic methods were developed for the prediction of the binding sites/residues on both partners. One of such an approach is PEP-Site Finder, which identifies the possible interacting regions on a given protein structure based on a peptide sequence (Saladin et al., 2014).

Despite all these efforts, the problem of bioactive peptide structure determination remains a challenge, due to the short lengths and unstable conformations of bioactive peptides in isolation. Therefore, use of bioinformatics for prediction of protein–peptide interactions, peptide binding sites on the protein, and estimation of peptide conformation upon interaction is gaining importance in bioactive peptide research.

The interactions between peptides and proteins with other molecules can be predicted and calculated using different structure-based computational approaches. QSAR, molecular docking, and molecular dynamics are the most widely used techniques. In the following sections, these computational-based approaches to screen, predict, and modify bioactivity of peptides will be described with examples.

Quantitative Structure–Activity Relationship

Quantitative structure–activity relationship (QSAR) is a method to obtain computational and mathematical models in order to find a correlation between structure/biophysical properties and functions/bioactivity of molecules (Verma, Coutinho, and Evans, 2010). QSAR is used to understand the relationship between structures and specific activity and to identify/design molecules with novel and/or optimum biological activity. Recently, QSAR was implemented in different food-derived bioactive peptide studies (Nongonierma and Fitzgerald, 2016a; Santos-Silva et al., 2020).

In general, making a QSAR model takes four steps (Nongonierma and Fitzgerald, 2016a). First, sequence, structural, and biological properties (such as half maximum inhibitory concentration, half maximum effective concentration) of peptides are collected from literature and databases. Second, is the selection of descriptors, which are basically the chemical characteristics of a molecule in numerical form. This step is quite well described for small molecules as size, charge, hydrophobicity, and other specific chemical information can be used to build numerical descriptors. On the other hand, for peptides, and especially proteins, selection and building descriptors is not straightforward; due to the complicated sequence–structure–function relationship, a small atomic change can result in a dramatic structural and/or functional change. Researchers developed different descriptor systems, where each amino acid is given a numerical value to describe relationship of amino acids and physicochemical properties (Collantes and Dunn, 1995; Hellberg et al., 1987; Sandberg et al., 1998; Tian, Lv, and Yang, 2012). These description dimensions can be grouped as 1D: molecular formula; 2D: topological properties; 3D: conformational features and 4D: orientation and time dependent. Depending on the chemistry and complexity of the peptide/protein of interest, more descriptors can be added for detailed features (Barley, Turner, and Goodacre, 2018). Third step is building a computer-aided mathematical model using these descriptors and coefficients of biological activity. The model is generally validated for statistical reliability at this stage, using available experimental information in peptide databases. Fourth step, after model-building and validation, the bioactivity potential of the peptide can be estimated (Nongonierma and Fitzgerald, 2016a; Wen et al., 2020).

An extensive list of available QSAR software can be accessed from the Swiss Institute of Bioinformatics – Clik2Drug web site (https://www.click2drug.org/index.php), and some are listed in Table 16.3.

Nongonierma and Fitzgerald (2016b) utilized QSAR models to obtain a correlation between the DPP-IV inhibitory activity of milk peptides with known IC50 values and the physicochemical properties of the peptides. They used two amino-acid description models, one used hydrophilicity, size, and charge properties as descriptors, and the other model used parameters including van der Waals volume, hydrophobicity of side chains, and net charge. Their results showed that different descriptors resulted in different model strengths, yet both models verified that the presence of a hydrophobic amino acid at the N terminus is related to DPP-IV activity.

As the number of studies using QSAR models increases, the relationship between bioactivity and structural/physicochemical properties of the peptides are better understood. Nongonierma and Fitzgerald (2016a) summarized structural requirements that are important over several bioactivities, such as DPP-IV inhibitory, ACE inhibitory, antioxidant, and antimicrobial properties. Recently, Manzanares et al. (2019) also summarized QSAR and structural analysis outputs including structural requirements of antihypertensive, antioxidant, and antidiabetic peptides. Liu, Cheng, and Wu (2019) indicated that DPP-IV inhibitory activity of peptides can be predicted using QSAR approaches and docking studies. Shao et al. (2021) used a successful 3D- QSAR model to screen ACE inhibitor peptides from silkworm cocoon waste in a recent study.

Molecular Docking

Molecular docking is a computer tool where one can simulate the interaction of two molecules by searching all possible atom interactions and narrows down the results based on energy minimization (Atilgan and Hu, 2010).

TABLE 16.3 QSAR Computer Program for Peptide

Tools	Available Link
CORAL	http://www.insilico.eu/CORAL/
Alvascience	https://www.alvascience.com/
OCHEM	https://ochem.eu/home/show.do
cQSAR	http://www.biobyte.com/bb/prod/cqsarad.html
SeeSAR	https://www.biosolveit.de/SeeSAR/

Accessed in March, 2021.

Once the structure of the peptide is obtained, molecular docking methods can be used to investigate the molecular mechanism of binding and atomic details of interaction of peptide with other molecules (Guedes, de Magalhães, and Dardenne, 2014). This approach will allow researchers to predict, design, and isolate peptides which might have higher affinities to the selected target molecules.

There are three different strategies for molecular docking used in bioactive peptide–protein interaction estimations. These are template-based, local, and global docking. Template-based is based on a template or existing information. Local docking uses existing information about details of molecular interaction and possible binding sites, whereas global docking is used when no information is available (Ciemny et al., 2018). Molecular docking is widely used in bioactive peptide research, and the available tools are listed in Table 16.4.

Several studies used docking to understand the binding mechanism of bioactive peptides to their target molecules; Jia et al. (2015) explained the ACE inhibitory mechanism of silkworm pupa protein; Zhu et al. (2021) predicted the binding mechanism of umami peptides from Atlantic cod to taste receptors, and Kong et al. (2021) investigated the molecular mechanism of DPP-IV inhibitory activity of peptides from walnut.

In peptide–protein docking studies, the target protein should be selected and the structure should be obtained from PDB or modeled and prepared for docking analysis (Tu, Cheng, et al., 2018). Sun et al. (2017) analyzed the ACE inhibitory mechanism of silkworm peptides by docking to a crystal structure of ACE-I-lisinopril complex (pdb id: 1O86) and they had to remove lisinopropyl and water molecules for preparing the structure for docking. Similarly, Nongonierma et al. (2018) removed the co-crystallized inhibitor from DPP-IV enzyme structure (pdb id: 1ORW) before the molecular docking analysis of a known DPP-IV inhibitory tripeptide (Ile-Pro-Ile) and its analogs.

TABLE 16.4 Molecular Docking Tools

Tools	Available Link	Reference
AutoDock	http://autodock.scripps.edu/	(Goodsell, Morris, and Olson 1996)
CABS-dock	http://biocomp.chem.uw.edu.pl/CABSdock/	(Kurcinski et al. 2015)
ClusPro	https://cluspro.bu.edu/peptide/index.php	(Porter et al. 2017)
HADDOCK	https://www.bonvinlab.org/software/bpg/peptides/	(Trellet, Melquiond, and Bonvin 2013)
FeptideDB B-AceP	http://www4g.biotec.or.th/FeptideDB/ligand_docking.php	(Panyayai et al. 2019)
GalaxyPepDock	https://seoklab.github.io/GalaxyPepDock/	(Lee et al. 2015)
HpepDock	http://huanglab.phys.hust.edu.cn/hpepdock/	(Zhou et al. 2018)
MDockPep	https://zougrouptoolkit.missouri.edu/mdockpep/	(Xu, Yan, and Zou 2018)
pepATTRACT	https://bioserv.rpbs.univ-paris-diderot.fr/services/pepATTRACT/	(de Vries et al. 2017)
PepSite2	http://pepsite2.russelllab.org/	(Petsalaki et al. 2009)
PEP-SiteFinder	https://bioserv.rpbs.univ-paris-diderot.fr/services/PEP-SiteFinder/#Overview	(Saladin et al. 2014)
PIPER-FlexPepDock	http://piperfpd.furmanlab.cs.huji.ac.il/	(Alam et al. 2017)

Molecular docking is a very helpful method for elucidating protein–peptide interactions, however, there are some limitations. The most important of these are: modeling difficulties arising from the flexibility of both protein and peptide structures, choosing the appropriate and correct model among the created docking models, and integrating the experimental data into the docking calculation tool (Ciemny et al., 2018).

Molecular Dynamics

Molecular Dynamics (MD) is a simulation technique that allows prediction of conformational dynamics of a molecule at the atomic level under certain conditions over a bounded time (Geng et al., 2019; Zhang, Aryee, and Simpson, 2020). The behavior of peptides and proteins in a given environment can be simulated using MD (Georgoulia and Glykos, 2019). In MD simulation, the force field of the system is calculated, and this calculation is critical for the reliability of the simulation. Force field equations consist of bond features (e.g., lengths, angles, dihedral angles) and nonbonded van der Waals and electrostatics interactions and some empirical constants (Geng et al., 2019).

MD is very useful in predicting the binding characteristics of two interacting molecules, such as ligand–protein interaction, and is heavily used in screening for active molecules in drug design (Zhao and Caflisch, 2015). The same principle applies to bioactive peptide research, where the details of binding of the peptide to its target cellular protein are important for the expected bioactivity. Therefore, MD simulations are also becoming very valuable tools in bioactive peptide research. Jiang et al. (2019) utilized docking (AutoDock) and MD simulation (GROMACS) methods to explain the molecular mechanisms of the interaction of two ACE-inhibitory peptides with the ACE protein, and the MD-predicted binding energies were found to be consistent with experimental data. Similarly, Kalyan et al. (2021) used docking (AutoDock) and MD simulation (GROMACS and AMBER force field) tools to explain the ACE inhibitory mechanism of various food-derived bioactive peptides. Commonly used molecular dynamics simulation tools are given in Table 16.5.

CONCLUSIONS AND FUTURE PROSPECTS

Bioinformatic analysis is very useful in bioactive peptide research to predict the bioactivity potential of proteins, the suitable enzymes for bioactive peptide release, the sequence of resulting peptides, and their potential bioactive properties. The use of sequence-based

TABLE 16.5 Molecular Dynamics Tools

Tools	Available link	Reference
Amber	https://ambermd.org/	(Case et al. 2005)
CHARMM	https://www.charmm.org//	(MacKerell et al. 1998)
Desmond	https://www.deshawresearch.com/index.html	(Bowers et al. 2006)
GROMACS	http://www.gromacs.org/	(Hess et al. 2008)
LAMMPS	https://lammps.sandia.gov/	(Plimpton 1995)
NAMD	www.ks.uiuc.edu/Research/namd	(Phillips et al. 2020)

bioinformatic analysis in bioactive peptide research became a routine approach and greatly reduces the time-consuming and expensive laboratory work. Bioinformatic tools can also be used to predict the allergenicity, toxicity, and sensory properties of bioactive peptides, which are especially important considering the potential use of purified bioactive peptides in functional and nutraceutical food products. One current limitation in predicting the presence of bioactive peptides in a given protein is the lack of tools that can take into account protein tertiary structure during *in silico* enzymatic hydrolysis. More research is needed to develop algorithms that can process atomic coordinates to predict *in silico* hydrolysis patterns. Structural bioinformatic analysis has recently gained more interest among bioactive peptide researchers. Structure prediction and analysis are important in understanding interactions of bioactive peptides with target molecules and improving their therapeutic/functional properties. The number of studies utilizing structural based analysis tools is increasing, and these studies highlight the high level of correlation between bioinformatic analysis and experimental results. Overall, integrating bioinformatic tools facilitates and accelerates bioactive peptide research. Furthermore, the increased use of those tools and feedbacks from users will allow developers to improve the available bioinformatic tools.

REFERENCES

Agyei, Dominic, Erandi Bambarandage, and Chibuike C. Udenigwe. 2018. "The Role of Bioinformatics in the Discovery of Bioactive Peptides." *Encyclopedia of Food Chemistry*, 337–44. https://doi.org/10.1016/b978-0-08-100596-5.21863-5.

Alam, Nawsad, Oriel Goldstein, Bing Xia, Kathryn Porter, Dima Kozakov, and Ora Schueler-Furman. 2017. "High-Resolution Global Peptide-Protein Docking Using Fragments-Based PIPER-FlexPepDock." *PLOS Computational Biology* 13(12): e1005905. https://doi.org/10.1371/journal.pcbi.1005905.

Altschul, Stephen, Warren Gish, Webb Miller, Eugene Myers, and David Lipman. 1990. "Basic Local Alignment Search Tool." *Journal of Molecular Biology* 215(3): 403–10. https://doi.org/10.1016/S0022-2836(05)80360-2.

Alvascience. n.d. "Software Solutions for Chemoinformatics and QSAR." Accessed March 16, 2021. https://www.alvascience.com/.

Amigo, Lourdes, Daniel Martínez-Maqueda, and Blanca Hernández-Ledesma. 2020. "In Silico and In Vitro Analysis of Multifunctionality of Animal Food-Derived Peptides." *Foods* . https://doi.org/10.3390/foods9080991.

Atilgan, Emrah, and Jianjun Hu. 2010. "Efficient Protein-Ligand Docking Using Sustainable Evolutionary Algorithms." In: *2010 10th International Conference on Hybrid Intelligent Systems, HIS 2010*, 113–18. IEEE. https://doi.org/10.1109/HIS.2010.5600082.

Barley, Mark H., Nicholas J. Turner, and Royston Goodacre. 2018. "Improved Descriptors for the Quantitative Structure-Activity Relationship Modeling of Peptides and Proteins." *Journal of Chemical Information and Modeling* 58(2): 234–43. https://doi.org/10.1021/acs.jcim.7b00488.

Berg, Jeremy M., John L. Tymoczko, and Stryer Lubert. 2002. "Protein Structure and Function." In: *Biochemistry*, 5th edition. New York: W. H. Freeman and Company. https://www.ncbi.nlm.nih.gov/books/NBK21177/.

Billeter, Martin, Gerhard Wagner, and Kurt Wüthrich. 2008. "Solution NMR Structure Determination of Proteins Revisited." *Journal of Biomolecular NMR* 42(3): 155–58. https://doi.org/10.1007/s10858-008-9277-8.

Bowers, Kevin J., David E. Chow, Huafeng Xu, Ron O. Dror, Michael P. Eastwood, Brent A. Gregersen, John Klepeis, et al. 11–17 Nov. 2006. "Scalable Algorithms for Molecular Dynamics Simulations on Commodity Clusters." In: *ACM/IEEE SC, 2006 Conference (SC'06)*, 43. Tampa, FL.

Buchan, Daniel W.A., and David T. Jones. 2019. "The PSIPRED Protein Analysis Workbench: 20 Years On." *Nucleic Acids Research* 47(W1): W402–7. https://doi.org/10.1093/nar/gkz297.

Can, Tolga. 2014. "Introduction to Bioinformatics." In: *MiRNomics: MicroRNA Biology and Computational Analysis*, edited by Malik Yousef and Jens Allmer, 1107: 51–71. New York: Springer Science+Business Media. https://doi.org/10.1007/978-1-62703-748-8.

Case, David A., Thomas E. Cheatham, Tom Darden, Holger Gohlke, Ray Luo, Kenneth M. Merz, Alexey Onufriev, Carlos Simmerling, Bing Wang, and Robert J. Woods. 2005. "The Amber Biomolecular Simulation Programs." *Journal of Computational Chemistry* 26(16): 1668–88. https://doi.org/10.1002/jcc.20290.

Chai, Tsun Thai, Yew Chye Law, Fai Chu Wong, and Se Kwon Kim. 2017. "Enzyme-Assisted Discovery of Antioxidant Peptides from Edible Marine Invertebrates: A Review." *Marine Drugs* 15(2): 1–26. https://doi.org/10.3390/md15020042.

Chandrasekaran, Subhiksha, Diego Luna-Vital, and Elvira Gonzalez de Mejia. 2020. "Identification and Comparison of Peptides from Chickpea Protein Hydrolysates Using Either Bromelain or Gastrointestinal Enzymes and Their Relationship with Markers of Type 2 Diabetes and Bitterness." *Nutrients* 12(12): 1–16. https://doi.org/10.3390/nu12123843.

Charoenkwan, Phasit, Sakawrat Kanthawong, Chanin Nantasenamat, Mehedi Hasan, and Watshara Shoombuatong. 2020. "IDPPIV-SCM: A Sequence-Based Predictor for Identifying and Analyzing Dipeptidyl Peptidase IV (DPP-IV) Inhibitory Peptides Using a Scoring Card Method." *Journal of Proteome Research* 19(10): 4125–36. https://doi.org/10.1021/acs.jproteome.0c00590.

Charoenkwan, Phasit, Chanin Nantasenamat, Md. Mehedi Hasan, Balachandran Manavalan, and Watshara Shoombuatong. 2021. "BERT4Bitter: A Bidirectional Encoder Representations from Transformers (BERT)-Based Model for Improving the Prediction of Bitter Peptides." *Bioinformatics, February*. https://doi.org/10.1093/bioinformatics/btab133.

Charoenkwan, Phasit, Janchai Yana, Chanin Nantasenamat, Md. Mehedi Hasan, and Watshara Shoombuatong. 2020. "IUmami-SCM: A Novel Sequence-Based Predictor for Prediction and Analysis of Umami Peptides Using a Scoring Card Method with Propensity Scores of Dipeptides." *Journal of Chemical Information and Modeling* 60(12): 6666–78. https://doi.org/10.1021/acs.jcim.0c00707.

Choo, Khar Heng, Tin Wee Tan, and Shoba Ranganathan. 2005. "SPdb – A Signal Peptide Database." *BMC Bioinformatics* 6(1): 249. https://doi.org/10.1186/1471-2105-6-249.

Ciemny, Maciej, Mateusz Kurcinski, Karol Kamel, Andrzej Kolinski, Nawsad Alam, Ora Schueler-Furman, and Sebastian Kmiecik. 2018. "Protein–Peptide Docking: Opportunities and Challenges." *Drug Discovery Today* 23(8): 1530–37. https://doi.org/10.1016/j.drudis.2018.05.006.

Clik2Drug. n.d. "Swiss Institute of Bioinformatics." Accessed March 16, 2021. https://www.click2drug.org/index.php.

Cochrane, Guy, Ilene Karsch-Mizrachi, and Toshihisa Takagi. 2016. "The International Nucleotide Sequence Database Collaboration." *Nucleic Acids Research* 44(D1): D48–50. https://doi.org/10.1093/nar/gkv1323.

Collantes, Elizabeth R., and William J. Dunn. 1995. "Amino Acid Side Chain Descriptors for Quantitative Structure-Activity Relationship Studies of Peptide Analogs." *Journal of Medicinal Chemistry* 38(14): 2705–13. https://doi.org/10.1021/jm00014a022.

CORAL. n.d. "CORAL-QSAR/QSPR." Accessed March 16, 2021 http://www.insilico.eu/CORAL/.

cQSAR. n.d. "BioByte." Accessed March 16, 2021. http://www.biobyte.com/bb/prod/cqsarad.html.

Dagan-Wiener, Ayana, Ido Nissim, Natalie Ben Abu, Gigliola Borgonovo, Angela Bassoli, and Masha Y. Niv. 2017. "Bitter or Not? BitterPredict, a Tool for Predicting Taste from Chemical Structure." *Scientific Reports* 7(1): 1–13. https://doi.org/10.1038/s41598-017-12359-7.

Das, Durdam, Mohini Jaiswal, Fatima Nazish Khan, Shahzaib Ahamad, and Shailesh Kumar. 2020. "PlantPepDB: A Manually Curated Plant Peptide Database." *Scientific Reports* 10(1): 1–8. https://doi.org/10.1038/s41598-020-59165-2.

Dave, Lakshmi A., Maria Hayes, Carlos A. Montoya, Shane M. Rutherfurd, and Paul J. Moughan. 2016. "Human Gut Endogenous Proteins as a Potential Source of Angiotensin-I-Converting Enzyme (ACE-I)-, Renin Inhibitory and Antioxidant Peptides." *Peptides* 76: 30–44. https://doi.org/10.1016/j.peptides.2015.11.003.

DDBJ. n.d. "DNA Data Bank of Japan." Accessed March 16, 2021 https://www.ddbj.nig.ac.jp/index-e.html.

Deng, Haiyou, Ya Jia, and Yang Zhang. 2018. Protein Structure Prediction. *International Journal of Modern Physics. Part B* 32(18). https://doi.org/10.1142/S021797921840009X.

Dimitrov, Ivan, Ivan Bangov, Darren R. Flower, and Irini Doytchinova. 2014. "AllerTOP v.2--a Server for *in silico* Prediction of Allergens." *Journal of Molecular Modeling* 20(6): 2278. https://doi.org/10.1007/s00894-014-2278-5.

Duan, Xiaojie, Min Zhang, and Fusheng Chen. 2021. "Prediction and Analysis of Antimicrobial Peptides from Rapeseed Protein Using In Silico Approach." *Journal of Food Biochemistry*, February, e13598. https://doi.org/10.1111/jfbc.13598.

EMBL-EBI. n.d. "The European Bioinformatics Institute." Accessed March 16, 2021 https://www.ebi.ac.uk/.

EMBL-EBI Sequence Translation. n.d. "European Bioinformatics Institute." Accessed March 16, 2021. https://www.ebi.ac.uk/Tools/st/.

Erdős, Gábor, and Zsuzsanna Dosztányi. 2020. "Analyzing Protein Disorder with IUPred2A." *Current Protocols in Bioinformatics* 70(1): e99. https://doi.org/10.1002/cpbi.99.

ExPASy. n.d. "Expasy: SIB Swiss Institute of Bioinformatics." Accessed March 16, 2021 https://web.expasy.org/translate/.

FAO/WHO. 2001. "Evaluation of Allergenicity of Genetically Modified Foods" *Report of a Joint FAO/WHO Expert Consultation on Allergenicity of Foods Derived from Biotechnology*. Rome, Italy.

Fátima Garcia, Bianca de, Márcio de Barros, and Thaís de Souza Rocha. 2020. "Bioactive Peptides from Beans with the Potential to Decrease the Risk of Developing Noncommunicable Chronic Diseases." *Critical Reviews in Food Science and Nutrition*: 1–19. https://doi.org/10.1080/10408398.2020.1768047.

Gasteiger, Elisabeth, Christine Hoogland, Alexandre Gattiker, S'everine Duvaud, Marc R. Wilkins, Ron D. Appel, and Amos Bairoch. 2005. "Protein Identification and Analysis Tools on the ExPASy Server." In *The Proteomics Protocols Handbook*, edited by John M Walker, 571–607. Totowa, NJ: Humana Press. https://doi.org/10.1385/1-59259-890-0:571.

Geng, Hao, Fangfang Chen, Jing Ye, and Fan Jiang. 2019. "Applications of Molecular Dynamics Simulation in Structure Prediction of Peptides and Proteins." *Computational and Structural Biotechnology Journal* 17: 1162–70. https://doi.org/10.1016/j.csbj.2019.07.010.

Georgoulia, Panagiota S., and Nicholas M. Glykos. 2019. "Molecular Simulation of Peptides Coming of Age: Accurate Prediction of Folding, Dynamics and Structures." *Archives of Biochemistry and Biophysics* 664(February): 76–88. https://doi.org/10.1016/j.abb.2019.01.033.

Goodman, Richard E., Motohiro Ebisawa, Fatima Ferreira, Hugh A. Sampson, Ronald van Ree, Stefan Vieths, Joseph L. Baumert, et al. 2016. "AllergenOnline: A Peer-Reviewed, Curated Allergen Database to Assess Novel Food Proteins for Potential Cross-Reactivity." *Molecular Nutrition and Food Research* 60(5): 1183–98. https://doi.org/10.1002/mnfr.201500769.

Goodsell, David S., Garrett Morris, and Arthur J. Olson. 1996. "Automated Docking of Flexible Ligands: Applications of AutoDock." *Journal of Molecular Recognition : JMR* 9(1): 1–5. https://doi.org/10.1002/(sici)1099-1352(199601)9:1<1::aid-jmr241>3.0.co;2-6.

Grawert, Tobias W., and Dmitri I. Svergun. 2020. "Structural Modeling Using Solution Small-Angle X-Ray Scattering (SAXS)." *Journal of Molecular Biology* 432(9): 3078–92. https://doi.org/10.1016/j.jmb.2020.01.030.

Guedes, Isabella A., Camila S. de Magalhães, and Laurent E. Dardenne. 2014. "Receptor-Ligand Molecular Docking." *Biophysical Reviews* 6(1): 75–87. https://doi.org/10.1007/s12551-013-0130-2.

Gupta, Sudheer, Pallavi Kapoor, Kumardeep Chaudhary, Ankur Gautam, Rahul Kumar, and Gajendra P.S. Raghava. 2013. "In Silico Approach for Predicting Toxicity of Peptides and Proteins." *PLOS ONE* 8(9): e73957. https://doi.org/10.1371/journal.pone.0073957.

Hammami, Riadh, Abdelmajid Zouhir, Christophe Le Lay, Jeannette Ben Hamida, and Ismail Fliss. 2010. "BACTIBASE Second Release: A Database and Tool Platform for Bacteriocin Characterization." *BMC Microbiology* 10(1): 22. https://doi.org/10.1186/1471-2180-10-22.

Hellberg, Sven, Michael Sjöström, Bert Skagerberg, and Svante Wold. 1987. "Peptide Quantitative Structure-Activity Relationships, a Multivariate Approach." *Journal of Medicinal Chemistry* 30(7): 1126–35. https://doi.org/10.1021/jm00390a003.

Hennig, Janosch, and Michael Sattler. 2014. "The Dynamic Duo: Combining NMR and Small Angle Scattering in Structural Biology." *Protein Science* 23(6): 669–82. https://doi.org/10.1002/pro.2467.

Hess, Berk, Carsten Kutzner, David van der Spoel, and Erik Lindahl. 2008. "GROMACS 4: Algorithms for Highly Efficient, Load-Balanced, and Scalable Molecular Simulation." *Journal of Chemical Theory and Computation* 4(3): 435–47. https://doi.org/10.1021/ct700301q.

Holton, Thérèse A., Vaishnavi Vijayakumar, and Nora Khaldi. 2013. "Bioinformatics: Current Perspectives and Future Directions for Food and Nutritional Research Facilitated by a Food-Wiki Database." *Trends in Food Science and Technology* 34(1): 5–17. https://doi.org/10.1016/j.tifs.2013.08.009.

Huang, Wenkang, Qiancheng Shen, Xubo Su, Mingfei Ji, Xinyi Liu, Yingyi Chen, Shaoyong Lu, Hanyi Zhuang, and Jian Zhang. 2016. "BitterX: A Tool for Understanding Bitter Taste in Humans." *Scientific Reports* 6: 23450. https://doi.org/10.0.4.14/srep23450.

Iwaniak, Anna, Piotr Minkiewicz, Małgorzata Darewicz, Krzysztof Sieniawski, and Piotr Starowicz. 2016. "BIOPEP Database of Sensory Peptides and Amino Acids." *Food Research International* 85: 155–61. https://doi.org/10.1016/j.foodres.2016.04.031.

Jia, Junqiang, Qiongying Wu, Hui Yan, and Zhongzheng Gui. 2015. "Purification and Molecular Docking Study of a Novel Angiotensin-I Converting Enzyme (ACE) Inhibitory Peptide from Alcalase Hydrolysate of Ultrasonic-Pretreated Silkworm Pupa (Bombyx mori) Protein." *Process Biochemistry* 50(5): 876–83. https://doi.org/10.1016/j.procbio.2014.12.030.

Jiang, Zhenyan, Hansi Zhang, Xuefeng Bian, Jingfeng Li, Jing Li, and Hui Zhang. 2019. "Insight into the Binding of ACE-Inhibitory Peptides to Angiotensin-Converting Enzyme: A Molecular Simulation." *Molecular Simulation* 45(3): 215–22. https://doi.org/10.1080/08927022.2018.1557327.

Källberg, Morten, Haipeng Wang, Sheng Wang, Jian Peng, Zhiyong Wang, Hui Lu, and Jinbo Xu. 2012. "Template-Based Protein Structure Modeling Using the RaptorX Web Server." *Nature Protocols* 7(8): 1511–22. https://doi.org/10.1038/nprot.2012.085.

Kalyan, Gazal, Vivek Junghare, Sourya Bhattacharya, and Saugata Hazra. 2021. "Understanding Structure-Based Dynamic Interactions of Antihypertensive Peptides Extracted from Food Sources." *Journal of Biomolecular Structure and Dynamics* 39(2): 635–49. https://doi.org/10.1080/07391102.2020.1715836.

Kapoor, Pallavi, Harinder Singh, Ankur Gautam, Kumardeep Chaudhary, Rahul Kumar, and Gajendra P.S. Raghava. 2012. "TumorHoPe: A Database of Tumor Homing Peptides." *PLOS ONE* 7(4): e35187. https://doi.org/10.1371/journal.pone.0035187.

Kartal, Canan, Burcu Kaplan Türköz, and Semih Otles. 2020. "Prediction, Identification and Evaluation of Bioactive Peptides from Tomato Seed Proteins Using In Silico Approach." *Journal of Food Measurement and Characterization* 14(4): 1865–83. https://doi.org/10.1007/s11694-020-00434-z.

Kim, David E., Dylan Chivian, and David Baker. 2004. "Protein Structure Prediction and Analysis Using the Robetta Server." *Nucleic Acids Research* 32: W526–31. https://doi.org/10.1093/nar/gkh468.

King, Te Piao, Donald Hoffman, Henning Lowenstein, David G. Marsh, Tomas A. E. Platts-Mills, and Wayne Thomas. 1994. "Allergen Nomenclature." *International Archives of Allergy and Immunology* 105(3): 224–33. https://doi.org/10.1159/000236761.

Kong, Xiangzhen, Lina Zhang, Weiguang Song, Caimeng Zhang, Yufei Hua, Yeming Chen, and Xingfei Li. 2021. "Separation, Identification and Molecular Binding Mechanism of Dipeptidyl Peptidase IV Inhibitory Peptides Derived from Walnut (*Juglans regia* L.) Protein." *Food Chemistry* 347: 129062.

Kozak, Maciej, Agnieszka Lewandowska, Stanisław Ołdziej, Sylwia Rodziewicz-Motowidło, and Adam Liwo. 2010. "Combination of SAXS and NMR Techniques as a Tool for the Determination of Peptide Structure in Solution." *The Journal of Physical Chemistry Letters* 1(20): 3128–31. https://doi.org/10.1021/jz101178t.

Kulkarni-Kale, Urmila, Snehal Raskar-Renuse, Girija Natekar-Kalantre, and Smita A. Saxena. 2014. "Antigen–Antibody Interaction Database (AgAbDb): A Compendium of Antigen–Antibody Interactions BT - Immunoinformatics." In: *Immunoinformatics*, edited by Rajat K De and Namrata Tomar, 149–64. New York: Springer. https://doi.org/10.1007/978-1-4939-1115-8_8.

Kumar, Ravi, Kumardeep Chaudhary, Minakshi Sharma, Gandharva Nagpal, Jagat Singh Chauhan, Sandeep Singh, Ankur Gautam, and Gajendra P.S. Raghava.

2015. "AHTPDB: A Comprehensive Platform for Analysis and Presentation of Antihypertensive Peptides." *Nucleic Acids Research* 43(D1): D956–62. https://doi.org/10.1093/nar/gku1141.

Kurcinski, Mateusz, Michal Jamroz, Maciej Blaszczyk, Andrzej Kolinski, and Sebastian Kmiecik. 2015. "CABS-Dock Web Server for the Flexible Docking of Peptides to Proteins without Prior Knowledge of the Binding Site." *Nucleic Acids Research* 43(W1): W419–24. https://doi.org/10.1093/nar/gkv456.

Lamiable, Alexis, Pierre Thévenet, Julien Rey, Marek Vavrusa, Philippe Derreumaux, and Pierre Tufféry. 2016. "PEP-FOLD3: Faster *de Novo* Structure Prediction for Linear Peptides in Solution and in Complex." *Nucleic Acids Research* 44(W1): W449–54. https://doi.org/10.1093/nar/gkw329.

Lata, Sneh, Nitish K. Mishra, and Gajendra P.S. Raghava. 2010. "AntiBP2: Improved Version of Antibacterial Peptide Prediction." *BMC Bioinformatics* 11 (1): S19. https://doi.org/10.1186/1471-2105-11-S1-S19.

Lear, Sam, and Steven L. Cobb. 2016. "Pep-Calc.Com: A Set of Web Utilities for the Calculation of Peptide and Peptoid Properties and Automatic Mass Spectral Peak Assignment." *Journal of Computer-Aided Molecular Design* 30(3): 271–77. https://doi.org/10.1007/s10822-016-9902-7.

Lee, Hasup, Lim Heo, Myeong Sup Lee, and Chaok Seok. 2015. "GalaxyPepDock: A Protein-Peptide Docking Tool Based on Interaction Similarity and Energy Optimization." *Nucleic Acids Research* 43(W1): W431–5. https://doi.org/10.1093/nar/gkv495.

Li, Qilin, Chao Zhang, Hongjun Chen, Jitong Xue, Xiaolei Guo, Ming Liang, and Ming Chen. 2018. "BioPepDB: An Integrated Data Platform for Food-Derived Bioactive Peptides." *International Journal of Food Sciences and Nutrition* 69(8): 963–68. https://doi.org/10.1080/09637486.2018.1446916.

Lin, Kai, Lan-wei Zhang, Xue Han, Liang Xin, Zhao-xu Meng, Pi-min Gong, and Da-you Cheng. 2018. "Yak Milk Casein as Potential Precursor of Angiotensin I-Converting Enzyme Inhibitory Peptides Based on In Silico Proteolysis." *Food Chemistry* 254: 340–47. https://doi.org/10.1016/j.foodchem.2018.02.051.

Liu, Feng, Geert Baggerman, Liliane Schoofs, and Geert Wets. 2008. "The Construction of a Bioactive Peptide Database in Metazoa." *Journal of Proteome Research* 7(9): 4119–31. https://doi.org/10.1021/pr800037n.

Liu, Rui, Jianming Cheng, and Hao Wu. 2019. "Discovery of Food-Derived Dipeptidyl Peptidase IV Inhibitory Peptides: A Review." *International Journal of Molecular Sciences* 20(3). https://doi.org/10.3390/ijms20030463.

Luca, Mariagrazia Di, Giuseppe Maccari, Giuseppantonio Maisetta, and Giovanna Batoni. 2015. "BaAMPs: The Database of Biofilm-Active Antimicrobial Peptides." *Biofouling* 31(2): 193–99. https://doi.org/10.1080/08927014.2015.1021340.

Luscombe, Nicholas M., Dov Greenbaum, Mark Gerstein, Frank K. Brown, Peter Willett, Jeremy G. Frey, Colin L. Bird, et al. 2011. "What Is Bioinformatics ? A Proposed Definition and Overview of the Field " 1(1): 1998. https://doi.org/10.1007/978-1-60761-839-3.

MacKerell, Alex D., Donald Bashford, Mathilde Bellott, Roland L. Dunbrack, Jeffrey D. Evanseck, Thomas M. Field, Saldanha Fischer, et al. 1998. "All-Atom Empirical Potential for Molecular Modeling and Dynamics Studies of Proteins." *The Journal of Physical Chemistry B* 102(18): 3586–616. https://doi.org/10.1021/jp973084f.

Madden, Tom. 2002. "The BLAST Sequence Analysis Tool." In: *The NCBI Handbook*, edited by Jo McEntyre and Jim Ostell, 281. Bethesda (MD): National Center for Biotechnology Information. (US). https://www.ncbi.nlm.nih.gov/books/NBK21101/.

Manzanares, Paloma, Mónica Gandía, Sandra Garrigues, and Jose F. Marcos. 2019. "Improving Health-Promoting Effects of Food-Derived Bioactive Peptides through Rational Design and Oral Delivery Strategies." *Nutrients* 11(10). https://doi.org/10.3390/nu11102545.

Mari, Adriano, Enrico Scala, Paola Palazzo, Stefano Ridolfi, Danila Zennaro, and Gabriele Carabella. 2006. "Bioinformatics Applied to Allergy: Allergen Databases, from Collecting Sequence Information to Data Integration. The Allergome Platform as a Model." *Cellular Immunology* 244(2): 97–100. https://doi.org/10.1016/j.cellimm.2007.02.012.

Mehta, Divya, Priya Anand, Vineet Kumar, Anshika Joshi, Deepika Mathur, Sandeep Singh, Abhishek Tuknait, et al. 2014. "ParaPep: A Web Resource for Experimentally Validated Antiparasitic Peptide Sequences and Their Structures." *Database : The Journal of Biological Databases and Curation* 2014. https://doi.org/10.1093/database/bau051.

Minkiewicz, Piotr, Jerzy Dziuba, Anna Iwaniak, Marta Dziuba, and Małgorzata Darewicz. 2008. "BIOPEP Database and Other Programs for Processing Bioactive Peptide Sequences." *Journal of AOAC International*: 965–80. https://doi.org/10.1093/jaoac/91.4.965.

Minkiewicz, Piotr, Anna Iwaniak, and Małgorzata Darewicz. 2017. "Annotation of Peptide Structures Using SMILES and Other Chemical Codes–Practical Solutions." *Molecules* . https://doi.org/10.3390/molecules22122075.

———. 2019. "BIOPEP-UWM Database of Bioactive Peptides: Current Opportunities." *International Journal of Molecular Sciences* 20(23): 5978. https://doi.org/10.3390/ijms20235978.

Mooney, Catherine, Niall J. Haslam, Gianluca Pollastri, and Denis C. Shields. 2012. "Towards the Improved Discovery and Design of Functional Peptides: Common Features of Diverse Classes Permit Generalized Prediction of Bioactivity." *PLOS ONE* 7(10). https://doi.org/10.1371/journal.pone.0045012.

NCBI. n.d. "The National Center for Biotechnology Information." Accessed March 16, 2021. https://www.ncbi.nlm.nih.gov/.

Nielsen, Søren Drud, Robert L. Beverly, Yunyao Qu, and David C. Dallas. 2017. "Milk Bioactive Peptide Database: A Comprehensive Database of Milk Protein-Derived Bioactive Peptides and Novel Visualization." *Food Chemistry* 232: 673–82. https://doi.org/10.1016/j.foodchem.2017.04.056.

Nongonierma, Alice B., Luca Dellafiora, Sara Paolella, Gianni Galaverna, Pietro Cozzini, and Richard J. FitzGerald. 2018. "In Silico Approaches Applied to the Study of Peptide Analogs of Ile-Pro-Ile in Relation to Their Dipeptidyl Peptidase IV Inhibitory Properties." *Frontiers in Endocrinology* 9(JUN): 1–15. https://doi.org/10.3389/fendo.2018.00329.

Nongonierma, Alice B., and Richard J. Fitzgerald. 2016a. "Learnings from Quantitative Structure-Activity Relationship (QSAR) Studies with Respect to Food Protein-Derived Bioactive Peptides: A Review." *RSC Advances* 6(79): 75400–413. https://doi.org/10.1039/c6ra12738j.

———. 2016b. "Structure Activity Relationship Modelling of Milk Protein-Derived Peptides with Dipeptidyl Peptidase IV (DPP-IV) Inhibitory Activity." *Peptides* 79: 1–7. https://doi.org/10.1016/j.peptides.2016.03.005.

Nongonierma, Alice B., and Richard J. FitzGerald. 2017. "Strategies for the Discovery and Identification of Food Protein-Derived Biologically Active Peptides." *Trends in Food Science and Technology* 69((November)): 289–305. https://doi.org/10.1016/J.TIFS.2017.03.003.

Nwanochie, Emeka, and Vladimir N. Uversky. 2019. "Structure Determination by Single-Particle Cryo-Electron Microscopy: Only the Sky (and Intrinsic Disorder) Is the Limit." *International Journal of Molecular Sciences*. https://doi.org/10.3390/ijms20174186.

OCHEM. n.d. "Online Chemical Database." Accessed March 16, 2021 https://ochem.eu/home/show.do.

Panyayai, Thitima, Chumpol Ngamphiw, Sissades Tongsima, Wuttichai Mhuantong, Wachira Limsripraphan, Kiattawee Choowongkomon, and Orathai Sawatdichaikul. 2019. "FeptideDB: A Web Application for New Bioactive Peptides from Food Protein." *Heliyon* 5(7): e02076. https://doi.org/10.1016/j.heliyon.2019.e02076.

Petsalaki, Evangelia, Alexander Stark, Eduardo García-Urdiales, and Robert B. Russell. 2009. "Accurate Prediction of Peptide Binding Sites on Protein Surfaces." *PLOS Computational Biology* 5(3): e1000335. https://doi.org/10.1371/journal.pcbi.1000335.

Phillips, James C., David J. Hardy, Julio D.C. Maia, John E. Stone, João V. Ribeiro, Rafael C. Bernardi, Ronak Buch, et al. 2020. "Scalable Molecular Dynamics on CPU and GPU Architectures with NAMD." *Journal of Chemical Physics* 153(4). https://doi.org/10.1063/5.0014475.

Piotto, Stefano P., Lucia Sessa, Simona Concilio, and Pio Iannelli. 2012. "YADAMP: Yet Another Database of Antimicrobial Peptides." *International Journal of Antimicrobial Agents* 39(4): 346–51. https://doi.org/10.1016/j.ijantimicag.2011.12.003.

Pirtskhalava, Malak, Anthony A. Amstrong, Maia Grigolava, Mindia Chubinidze, Evgenia Alimbarashvili, Boris Vishnepolsky, Andrei Gabrielian, Alex Rosenthal, Darrell E. Hurt, and Michael Tartakovsky. 2021. "DBAASP v3: Database of Antimicrobial/Cytotoxic Activity and Structure of Peptides as a Resource for Development of New Therapeutics." *Nucleic Acids Research* 49(D1): D288–97. https://doi.org/10.1093/nar/gkaa991.

Plimpton, Steve. 1995. "Fast Parallel Algorithms for Short-Range Molecular Dynamics." *Journal of Computational Physics* 117(6): 1–42. https://cs.sandia.gov/~sjplimp/papers/jcompphys95.pdf.

Porter, Kathryn A., Bing Xia, Dmitri Beglov, Tanggis Bohnuud, Nawsad Alam, Ora Schueler-Furman, and Dima Kozakov. 2017. "ClusPro PeptiDock: Efficient Global Docking of Peptide Recognition Motifs Using FFT." *Bioinformatics* 33(20): 3299–301. https://doi.org/10.1093/bioinformatics/btx216.

Qureshi, Abid, Nishant Thakur, Himani Tandon, and Manoj Kumar. 2014. "AVPdb: A Database of Experimentally Validated Antiviral Peptides Targeting Medically Important Viruses." *Nucleic Acids Research* 42(D1): D1147–53. https://doi.org/10.1093/nar/gkt1191.

Radauer, Christian. 2017. "Navigating through the Jungle of Allergens: Features and Applications of Allergen Databases." *International Archives of Allergy and Immunology* 173(1): 1–11. https://doi.org/10.1159/000471806.

Rajput, Akanksha, Amit Kumar Gupta, and Manoj Kumar. 2015. "Prediction and Analysis of Quorum Sensing Peptides Based on Sequence Features." *PLOS ONE* 10(3): e0120066. https://doi.org/10.1371/journal.pone.0120066.

Rani, Sapna, Km Pooja, and Gaurav Kumar Pal. 2017. "Exploration of Potential Angiotensin Converting Enzyme Inhibitory Peptides Generated from Enzymatic Hydrolysis of Goat Milk Proteins." *Biocatalysis and Agricultural Biotechnology* 11: 83–8. https://doi.org/10.1016/j.bcab.2017.06.008.

Rey, Julien, Patrick Deschavanne, and Pierre Tuffery. 2014. "BactPepDB: A Database of Predicted Peptides from a Exhaustive Survey of Complete Prokaryote Genomes." Database: 2014. https://doi.org/10.1093/database/bau106.

Richardson, Jane S., and David C. Richardson. 2014. "Biophysical Highlights from 54 Years of Macromolecular Crystallography." *Biophysical Journal* 106(3): 510–25. https://doi.org/10.1016/j.bpj.2014.01.001.

Roy, Ambrish, Alper Kucukural, and Yang Zhang. 2010. "I-TASSER: A Unified Platform for Automated Protein Structure and Function Prediction." *Nature Protocols* 5(4): 725–38. https://doi.org/10.1038/nprot.2010.5.

Sah, Baidya N.P., Todor Vasiljevic, Sandra McKechnie, and Osaana Donkor. 2018. "Antioxidative and Antibacterial Peptides Derived from Bovine Milk Proteins." *Critical Reviews in Food Science and Nutrition* 58(5): 726–40. https://doi.org/10.1080/10408398.2016.1217825.

Saha, Sudipto, and Gajendra Pal Singh Raghava. 2006. "AlgPred: Prediction of Allergenic Proteins and Mapping of IgE Epitopes." *Nucleic Acids Research* 34(suppl_2): W202–9. https://doi.org/10.1093/nar/gkl343.

Saladin, Adrien, Julien Rey, Pierre Thévenet, Martin Zacharias, Gautier Moroy, and Pierre Tufféry. 2014. "PEP-SiteFinder: A Tool for the Blind Identification of Peptide Binding Sites on Protein Surfaces." *Nucleic Acids Research* 42(W1): W221–6. https://doi.org/10.1093/nar/gku404.

Sandberg, Maria, Lennart Eriksson, Jörgen Jonsson, Michael Sjöström, and Svante Wold. 1998. "New Chemical Descriptors Relevant for the Design of Biologically Active Peptides. A Multivariate Characterization of 87 Amino Acids." *Journal of Medicinal Chemistry* 41(14): 2481–91. https://doi.org/10.1021/jm9700575.

Santos-Silva, Carlos André, Luisa Zupin, Marx Oliveira-Lima, Lívia Maria Batista Vilela, João Pacifico Bezerra-Neto, José Ribamar Ferreira-Neto, José Diogo Cavalcanti Ferreira, et al. 2020. "Plant Antimicrobial Peptides: State of the Art, In Silico Prediction and Perspectives in the Omics Era." *Bioinformatics and Biology Insights* 14. https://doi.org/10.1177/1177932220952739.

SeeSAR. n.d. "BioSolveIT." Accessed March 16, 2021. https://www.biosolveit.de/SeeSAR/.

Selamassakul, Orrapun, Natta Laohakunjit, Orapin Kerdchoechuen, Liping Yang, and Claudia S. Maier. 2020. "Bioactive Peptides from Brown Rice Protein Hydrolyzed by Bromelain: Relationship between Biofunctional Activities and Flavor Characteristics." *Journal of Food Science* 85(3): 707–17. https://doi.org/10.1111/1750-3841.15052.

Sette, Alessandro, Sheridan Martini, Daniel K. Wheeler, Randi Vita, Sandeep Kumar Dhanda, Swapnil Mahajan, James A. Overton, Jason R. Cantrell, and Bjoern Peters. 2018. "The Immune Epitope Database (IEDB): 2018 Update." *Nucleic Acids Research* 47(D1): D339–43. https://doi.org/10.1093/nar/gky1006.

Shao, Shan, Huaju Sun, Yaseen Muhammad, Hong Huang, Ruimeng Wang, Shuangxi Nie, Meiyun Huang, Ziyi Zhao, and Zhongxing Zhao. 2021. "Accurate Prediction for Adsorption Rate of Peptides with High ACE-Inhibitory Activity from Sericin Hydrolysate on Thiophene Hypercross-Linked Polymer Using CoMSIA in

3D-QSAR Model." *Food Research International* 141: 110144. https://doi.org/10.1016/j.foodres.2021.110 144.

Shtatland, Timur, Daniel Guettler, Misha Kossodo, Misha Pivovarov, and Ralph Weissleder. 2007. "PepBank - A Database of Peptides Based on Sequence Text Mining and Public Peptide Data Sources." *BMC Bioinformatics* 8(1): 280. https://doi.org/10.1186/1471-2105-8-280.

Singh, Harinder, Sandeep Singh, and Gajendra Pal Singh Raghava. 2019. "Peptide Secondary Structure Prediction Using Evolutionary Information." *BioRxiv*, no. 558791. https://doi.org/10.1101/558791.

Singh, Sandeep, Kumardeep Chaudhary, Sandeep Kumar Dhanda, Sherry Bhalla, Salman Sadullah Usmani, Ankur Gautam, Abhishek Tuknait, Piyush Agrawal, Deepika Mathur, and Gajendra P.S. Raghava. 2016. "SATPdb: A Database of Structurally Annotated Therapeutic Peptides." *Nucleic Acids Research* 44(D1): D1119–26. https://doi.org/10.1093/nar/gkv1114.

Singh, Sandeep, Gajendra P.S. Raghava, S. Kumaran, Abhishek Tuknait, Kumardeep Chaudhary, Balvinder Singh, and Harinder Singh. 2015. "PEPstrMOD: Structure Prediction of Peptides Containing Natural, Non-Natural and Modified Residues." *Biology Direct* 10(1): 1–19. https://doi.org/10.1186/s13062-015-0103-4.

Song, Jiangning, Hao Tan, Andrew J. Perry, Tatsuya Akutsu, Geoffrey I. Webb, James C. Whisstock, and Robert N. Pike. 2012. "PROSPER: An Integrated Feature-Based Tool for Predicting Protease Substrate Cleavage Sites." *PLOS ONE* 7(11): e50300. https://doi.org/10.1371/journal.pone.0050300.

Spencer, Ryan K., and James S. Nowick. 2015. "A Newcomer's Guide to Peptide Crystallography." *Israel Journal of Chemistry* 55(6-7): 698–710. https://doi.org/10.1002/ijch.201400179.

Strack, Rita. 2015. "Solving the Primary Structure of Peptides." *Nature Methods* 12(1): 11. https://doi.org/10.1038/nmeth.3524.

Sun, Huaju, Qing Chang, Long Liu, Kungang Chai, Guangyan Lin, Qingling Huo, Zhenxia Zhao, and Zhongxing Zhao. 2017. "High-Throughput and Rapid Screening of Novel ACE Inhibitory Peptides from Sericin Source and Inhibition Mechanism by Using in Silico and In Vitro Prescriptions." *Journal of Agricultural and Food Chemistry* 65(46): 10020–28. https://doi.org/10.1021/acs.jafc.7b04043.

Temussi, Piero A. 2012. "The Good Taste of Peptides." *Journal of Peptide Science* 18(2): 73–82. https://doi.org/10.1002/psc.1428.

Thakur, Nishant, Abid Qureshi, and Manoj Kumar. 2012. "AVPpred: Collection and Prediction of Highly Effective Antiviral Peptides." *Nucleic Acids Research* 40(W1): W199–204. https://doi.org/10.1093/nar/gks450.

Thomas, Annick, Sébastien Deshayes, Marc Decaffmeyer, Marie Hélène Van Eyck, Benoit Charloteaux, and Robert Brasseur. 2006. "Prediction of Peptide Structure: How Far Are We?. " *Proteins: Structure, Function, and Bioinformatics* 65(4): 889–97. https://doi.org/10.1002/prot.21151.

Tian, Feifei, Yonggang Lv, and Li Yang. 2012. "Structure-Based Prediction of Protein–Protein Binding Affinity with Consideration of Allosteric Effect." *Amino Acids* 43(2): 531–43. https://doi.org/10.1007/s00726-011-1101-1.

Trellet, Mikael, Adrien S.J. Melquiond, and Alexandre M.J.J. Bonvin. 2013. "A Unified Conformational Selection and Induced Fit Approach to Protein-Peptide Docking." *PLOS ONE* 8(3): e58769. https://doi.org/10.1371/journal.pone.0058769.

Tsai, Chun Jung, Patrizia Polverino de Laureto, Angelo Fontana, and Ruth Nussinov. 2002. "Comparison of Protein Fragments Identified by Limited Proteolysis and by Computational Cutting of Proteins." *Protein Science* 11(7): 1753–70. https://doi.org/10.1110/ps.4100102.

Tu, Maolin, Shuzhen Cheng, Weihong Lu, and Ming Du. 2018. "Advancement and Prospects of Bioinformatics Analysis for Studying Bioactive Peptides from Food-Derived Protein: Sequence, Structure, and Functions." *TrAC Trends in Analytical Chemistry* 105: 7–17. https://doi.org/10.1016/j.trac.2018.04.005.

Tu, Maolin, Hanxiong Liu, Ruyi Zhang, Hui Chen, Fengjiao Fan, Pujie Shi, Xianbing Xu, Weihong Lu, and Ming Du. 2018. "Bioactive Hydrolysates from Casein: Generation, Identification, and In Silico Toxicity and Allergenicity Prediction of Peptides." *Journal of the Science of Food and Agriculture* 98(9): 3416–26. https://doi.org/10.1002/jsfa.8854.

Tyagi, Atul, Abhishek Tuknait, Priya Anand, Sudheer Gupta, Minakshi Sharma, Deepika Mathur, Anshika Joshi, Sandeep Singh, Ankur Gautam, and Gajendra P.S. Raghava. 2015. "CancerPPD: A Database of Anticancer Peptides and Proteins." *Nucleic Acids Research* 43(D1): D837–43. https://doi.org/10.1093/nar/gku892.

Ulug, Sule Keskin, Forough Jahandideh, and Jianping Wu. 2021. "Novel Technologies for the Production of Bioactive Peptides." *Trends in Food Science and Technology* 108: 27–39. https://doi.org/10.1016/j.tifs.2020.12.002.

UniProt. n.d. "UniProt Knowledgebase." Accessed March 15, 2021. https://www.uniprot.org/.

Usmani, Salman Sadullah, Gursimran Bedi, Jesse S. Samuel, Sandeep Singh, Sourav Kalra, Pawan Kumar, Anjuman Arora Ahuja, Meenu Sharma, Ankur Gautam, and Gajendra P.S. Raghava. 2017. "THPdb: Database of FDA-Approved Peptide and Protein Therapeutics." *PLOS ONE* 12(7): e0181748. https://doi.org/10.1371/journal.pone.0181748.

Valentini, Erica, Alexey G. Kikhney, Gianpietro Previtali, Cy M. Jeffries, and Dmitri I. Svergun. 2015. "SASBDB, a Repository for Biological Small-Angle Scattering Data." *Nucleic Acids Research* 43(D1): D357–63. https://doi.org/10.1093/nar/gku1047.

van Heel, Auke J., Anne de Jong, Chunxu Song, Jakob H. Viel, Jan Kok, and Oscar P. Kuipers. 2018. "BAGEL4: A User-Friendly Web Server to Thoroughly Mine RiPPs and Bacteriocins." *Nucleic Acids Research* 46(W1): W278–81. https://doi.org/10.1093/nar/gky383.

Verma, Jitender, Vijay Khedkar Coutinho, and Coutinho Evans. 2010. "3D-QSAR in Drug Design - A Review." *Current Topics in Medicinal Chemistry*. https://doi.org/10.2174/156802610790232260.

Vries, Sjoerd de, Julien Rey, Christina Schindler, Martin Zacharias, and Pierre Tuffery. 2017. "The PepATTRACT Web Server for Blind, Large-Scale Peptide-Protein Docking." *Nucleic Acids Research* 45(W1): W361–64. https://doi.org/10.1093/nar/gkx335.

Waghu, Faiza Hanif, Ram Shankar Barai, Pratima Gurung, and Susan Idicula-Thomas. 2016. "CAMPR3: A Database on Sequences, Structures and Signatures of Antimicrobial Peptides." *Nucleic Acids Research* 44(D1): D1094–7. https://doi.org/10.1093/nar/gkv1051.

Wang, Guangshun, Xia Li, and Zhe Wang. 2016. "APD3: The Antimicrobial Peptide Database as a Tool for Research and Education." *Nucleic Acids Research* 44(D1): D1087–93. https://doi.org/10.1093/nar/gkv1278.

Wang, Jian, Tailang Yin, Xuwen Xiao, Dan He, Zhidong Xue, Xinnong Jiang, and Yan Wang. 2018. "StraPep: A Structure Database of Bioactive Peptides." Database. *The Journal of Biological Databases and Curation* 2018: 1–7. https://doi.org/10.1093/database/bay038.

Wen, Chaoting, Jixian Zhang, Haihui Zhang, Yuqing Duan, and Haile Ma. 2020. "Plant Protein-Derived Antioxidant Peptides: Isolation, Identification, Mechanism of Action and Application in Food Systems: A Review." *Trends in Food Science and Technology* 105: 308–22. https://doi.org/10.1016/j.tifs.2020.09.019.

Wen, Zeyu, Jiahua He, Huanyu Tao, and Sheng-You Huang. 2019. "PepBDB: A Comprehensive Structural Database of Biological Peptide-Protein Interactions." *Bioinformatics (Oxford, England)* 35(1): 175–77. https://doi.org/10.1093/bioinformatics/bty579.

Whisstock, James C., and Arthur M. Lesk. 2003. "Prediction of Protein Function from Protein Sequence and Structure." *Quarterly Reviews of Biophysics* 36(3): 307–40. https://doi.org/10.1017/S0033583503003901.

Xu, Dong, and Yang Zhang. 2012. "Ab Initio Protein Structure Assembly Using Continuous Structure Fragments and Optimized Knowledge-Based Force Field." *Proteins: Structure, Function and Bioinformatics* 80(7): 1715–35. https://doi.org/10.1002/prot.24065.

Xu, Xianjin, Chengfei Yan, and Xiaoqin Zou. 2018. "MDockPeP: An Ab-Initio Protein–Peptide Docking Server." *Journal of Computational Chemistry* 39(28): 2409–13. https://doi.org/10.1002/jcc.25555.

Yap, Pei Gee, and Chee Yuen Gan. 2020. "In Vivo Challenges of Anti-Diabetic Peptide Therapeutics: Gastrointestinal Stability, Toxicity and Allergenicity." *Trends in Food Science and Technology* 105: 161–75. https://doi.org/10.1016/j.tifs.2020.09.005.

Zamyatnin, Alexander A., Alexander S. Borchikov, Michail G. Vladimirov, and Olga L. Voronina. 2006. "The EROP-Moscow Oligopeptide Database." *Nucleic Acids Research* 34 (suppl_1): D261–66. https://doi.org/10.1093/nar/gkj008.

Zhang, Fan, Navid Adnani, Emmanuel Vazquez-Rivera, Doug R. Braun, Marco Tonelli, David R. Andes, and Tim S. Bugni. 2015. "Application of 3D NMR for Structure Determination of Peptide Natural Products." *The Journal of Organic Chemistry* 80(17): 8713–19. https://doi.org/10.1021/acs.joc.5b01486.

Zhang, Yi, Alberta N.A. Aryee, and Benjamin K. Simpson. 2020. "Current Role of in Silico Approaches for Food Enzymes." *Current Opinion in Food Science* 31: 63–70. https://doi.org/10.1016/j.cofs.2019.11.003.

Zhao, Hongtao, and Amedeo Caflisch. 2015. "Molecular Dynamics in Drug Design." *European Journal of Medicinal Chemistry* 91: 4–14. https://doi.org/10.1016/j.ejmech.2014.08.004.

Zhou, Pei, Bowen Jin, Hao Li, and Sheng You Huang. 2018. "HPEPDOCK: A Web Server for Blind Peptide-Protein Docking Based on a Hierarchical Algorithm." *Nucleic Acids Research* 46(W1): W443–50. https://doi.org/10.1093/nar/gky357.

Zhu, Wenhui, Wei He, Fei Wang, Ying Bu, Xuepeng Li, and Jianrong Li. 2021. "Prediction, Molecular Docking and Identification of Novel Umami Hexapeptides Derived from Atlantic Cod (*Gadus morhua*)." *International Journal of Food Science and Technology* 56(1): 402–12. https://doi.org/10.1111/ijfs.14655.

SECTION 5

Chemical Synthesis of Peptides

CHAPTER 17

Chemical Synthesis of Peptides

Javed Ahamad, Raja Kumar Parabathina, and Javed Ahmad

CONTENTS

Introduction	349
Beneficial Activities of Bioactive Peptides	350
Chemistry of Peptides	351
Chemical Synthesis of Peptides	351
Solid-Phase Peptide Synthesis	351
Linking of Amino Acids with Resins	352
Protection of Amino Acids	354
Coupling of Amino Acids	354
Deprotection	355
Removal of the Polymer	355
Advantages of SPPS	355
Chemical Synthesis of Difficult Peptides	356
Native Chemical Ligation Method	357
α-Ketoacid Hydroxylamine Ligation Method	357
Purification and Analysis of Peptides	357
Conclusion	358
References	360

INTRODUCTION

Proteins are macromolecules that are made up of amino acids joined by amide bonds. Proteins and peptides are the main components of foods, particularly those of animal origin. The two factors currently initiating the development of new and alternative protein research and product development are the rising global population and the need for the manufacture or alternative consumption of proteins (Stawikowski and Fields, 2002). Besides plant-based proteins from lentils, beans, peas, soy, etc. are reported to have health benefits including the cardiovascular system, glycemic regulation, etc. (Kay, 2012).

    Chemical synthesis, semi-synthesis, recombinant methods, and separation from natural sources are the major ways to get proteins and peptides industrially (Kimmerlin and Seebach, 2005; Hackenberger and Schwarzer, 2008). Bioactive peptides isolated from pulses, wheat, rice, soya, pumpkin, and even mushrooms have been identified in studies as potential sources. Although much of the research to date has focused on milk, cheese, and other dairy products as sources of bioactive proteins and peptides, synthetic bioactive peptides are gaining popularity (Korhonen, 2009).

Currently, bioactive peptides are produced from several protein sources using hydrolysis or fermentation with proteolytic lactic acid bacteria with commercially available enzymes. It is also often tedious, idiosyncratic, and inefficient to separate proteins from their natural source. In comparison, the production of proteins by recombinant DNA (rDNA), either in a heterologous host or *in vitro*, may provide access to large amounts of protein and allow 1 out of 20 typical amino-acid residues to be substituted for another one. However, the yield of properly folded proteins generated with rDNA is often restricted by aggregation. In addition, the constraints of the genetic code severely restrict the changes that are possible (Nilsson et al., 2005).

The new and advanced methods of chemical synthesis of proteins can solve many of the drawbacks of the existing methods of protein production (Borgia and Fields, 2000; Casi and Hilvert, 2003; Kent, 2003). In particular, chemical synthesis is easy to perform, easily automated, and facilitates purification. The use of current and emerging methods of protein synthesis could, therefore, promote research in all aspects of protein science (Wilken and Kent, 1998). The simple integration of non-natural functionality into proteins is made possible by chemical synthesis. The genetic code restricts natural protein components to approximately 20 α-amino acids. Similarly, methods that overcome this restriction but still rely on the ribosome are restricted to a subset of α-amino acids and α-hydroxy acids (Hendrickson et al., 2004; Mendel et al., 1995).

Due to the prodigious advancements in peptide synthesis that have happened over the past century, chemical synthesis of proteins is now possible. Fischer et al. (1901), first reported the synthesis of the dipeptide *glycyl glycine* and is also the first instance of the word "peptide" used to refer to an amino acid polymer (Fischer et al., 1901). Despite his inability to regulate its amino acid sequence, his 1907 synthesis of an octadecapeptide consisting of 15 glycine and 3 leucine residues was a remarkable achievement (Fischer, 1907). Several peptides, ranging from dipeptide sugar substitute aspartame to clinically used hormones, such as oxytocin, adrenocorticotropic, and calcitonin, are today produced by chemical synthesis (Pontiroli, 1998). So far, more than 400 peptides have entered into clinical trials. Therefore, a rapid, efficient, and reliable methodology for the chemical synthesis of these molecules is of utmost concern. This book chapter provides an aspect of chemical synthesis of bioactive peptides by solid phase peptide synthesis (SPPS) and also covers chemical synthesis of difficult peptides using new chemical ligation methods.

BENEFICIAL ACTIVITIES OF BIOACTIVE PEPTIDES

Food-derived bioactive peptides and proteins have a broad number of therapeutic effects, including antioxidant, immunomodulatory, antibacterial, anti-hypertensive, anti-carcinogenic, antidiabetic, anti-microbial, and anti-inflammatory effects (Pepe et al., 2016; Yang et al., 2014; Moller et al., 2008; Fujita et al., 2000; Cho et al., 2007; Schusdziarra et al., 1983; Sanchez de Medina F, Daddaoua et al., 2010). Consuming foods high in bioactive peptides, particularly those with multiple functions, could potentially have a huge impact on health and wellbeing, causing one to improve their health and thus avoid infections and diseases. The ability to prevent and treat chronic diseases through the consumption of functional foods could significantly reduce potential healthcare costs, which are expected to rise. Peptides with antihypertensive properties are among the most investigated in the field of functional foods. Many peptides have been studied *in vitro*, and a very few have also been tested in hypertensive animal models and human volunteers. In hypertensive rats, tripeptides from bonito muscle (IKP) and chicken muscle (IKW) have been shown to reduce blood pressure (Masuda et al., 1996). In an eight-week placebo-controlled study, regular intake of 100 ml of Calpise was shown to substantially lower blood pressure after four and eight weeks in hypertensive humans, relative to the control group (Satake et al., 2002).

Blood cholesterol levels have been recognized as a significant risk factor for cardiovascular disease, and comprehensive research is being conducted into strategies for preventing and lowering plasma cholesterol levels, including the use of bioactive peptides and proteins. Many food proteins, especially plant proteins like soybean, are proven to help lower serum cholesterol levels (Carroll, 1978).

There is considerable evidence from randomized clinical trials showing that substituting soy protein for animal protein reduces serum lipids, including total cholesterol, low-density lipoprotein (LDL) cholesterol, and triglycerides, while having no effect on high-density lipoprotein cholesterol levels (HDL) (Anderson et al., 1995). Osteoporosis is a chronic but relatively preventable disease, as adequate supplies of soluble or usable calcium can increase bone mineralization. Milk is a well-known calcium source and also includes casein, which enhances calcium absorption in the intestine and thereby serves as an osteoprotective (Kitts, 1994; Naito et al., 1972).

CHEMISTRY OF PEPTIDES

Twenty common α-amino acids are used in the biosynthesis of peptides and proteins. An amino group is bound to the α-carbon of a carboxyl group in a α-amino acid. Because of their distinct structure at the α-carbon, typical naturally occurring amino acids are referred to as *L*-amino acids. *D*-amino acids, one of its enantiomers, are highly unusual in nature. Except for cysteine, which has the *R* configuration, and glycine, which is achiral, *L*-amino acids have the *S* structure dependent on *R/S* designations. Amino acid bonds are referred to as amide bonds (Jaradat, 2017). The standard amino acids have two termini: an NH-group on the *N*-terminus and a COOH-group on the *C*-terminus (Figure 17.1).

CHEMICAL SYNTHESIS OF PEPTIDES

Bioactive peptides are currently synthesized by chemical means. Basically, two methods are mostly used to synthesize peptides: one is solid-phase peptide synthesis (SPPS), the other is solution peptide synthesis (SPS), although both are essentially based on the same principles. The protecting group (PG1) of the carboxy group of the first amino acid reacts in insoluble resin in SPPS, while the reaction occurs in a soluble medium in the SPS method of chemical synthesis of peptides and proteins (Borgia and Fields, 2000).

Solid-Phase Peptide Synthesis

Solid-phase peptide synthesis (SPPS) is a relatively new approach for synthesis of bioactive peptides and proteins (Canne et al., 1999). This technique of peptide synthesis has advantages over conventional methods as it gives higher yields, high product purity, faster, etc. Peptide synthesis is a complex technique in which desired amino acids are synthesized after multiple steps. There are various potential combinations produced in this method,

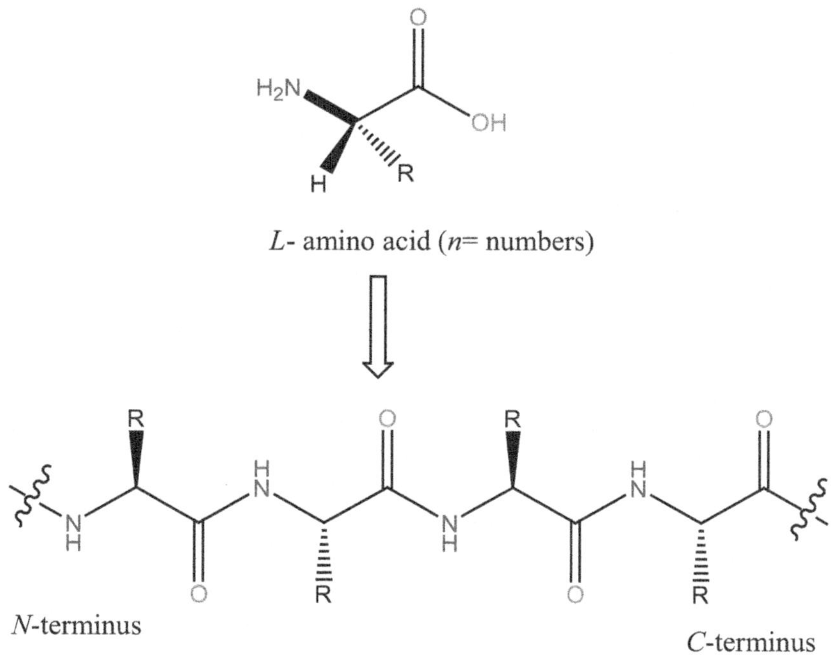

FIGURE 17.1 Chemical structure of *L*-amino acids and representation of *N*- and *C*-terminus.

from 20 common amino acids present in nature. Peptide synthesis is fascinating and challenging because of its complexity (Merrifield, 1986). When two amino acid solutions are combined, four different dipeptides can be formed (for example, the four dipeptides for a glycine and alanine mixture will be gly-gly, gly-ala, ala-gly, ala-ala. The free amino group or *N*-terminus is on the left-hand amino acid in this representation of peptides and the free carboxylic acid group is on the right-hand end of the *C*-terminus).

To ensure the synthesis of the desired dipeptide, the basic group of one amino acid and the acidic group of the other must both be made unable to react. This *deactivation* is known as the protection of reactive groups, and a group that is unable to react is called a *protected group* (Kemp and Carey, 1993). In classical peptide synthesis, the acids are protected, permitted to react, and then deprotected, after which one end of the dipeptide is protected and reacted with a new protected acid, and so on. In SPPS, the amino acid at one end of the peptide is bound to a water-insoluble polymer/resin and remains protected during the peptide's creation, suggesting that less protection steps are needed and that the reagents can be rinsed away without losing any of the peptide (Merrifield, 1996). The steps involved in the chemical synthesis of bioactive peptides in SPPS are discussed below and presented in Figure 17.2:

Linking of Amino Acids with Resins

The SPPS process involves binding the first amino acid to a resin, then elongating the peptide chain to generate the desired peptide. The first amino acid is temporarily anchored to the polymeric solid support by a linker (bifunctional spacer or handle) that is permanently attached to the polymeric solid support (Jaradat, 2017). The *N*-terminus and the *C*-terminus are the two ends of peptide chains, and which end is linked to the polymer depends on the polymer used. By reacting with a linkage agent on the amino acid and then reacting with

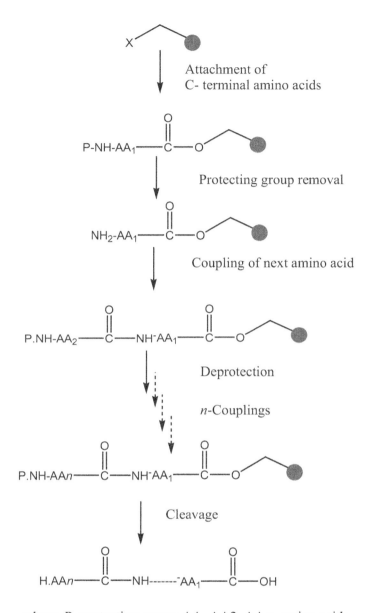

where, P- protecting group; AA, AA2, AAn- amino acids

FIGURE 17.2 Schematic representation of solid-phase peptide synthesis (SPPS).

the polymer on the other end of the linkage agent, the binding is achieved. This ensures that a peptide–polyamide bond can be formed that will not be hydrolyzed during subsequent peptide-formation reactions (Figure 17.3). The most common polymeric solid support used in resins is divinylbenzene-crosslinked polystyrene (1–2%) in the form of beads with diameters of about 50 microns. Commercially cross-linked polystyrene resins are available in 2 particle sizes: 35–75 microns and 75–150 microns (Vaino and Janda, 2000). Polymeric/resin supports used in the chemical synthesis of peptides and proteins include polyethylene glycol (PEG),

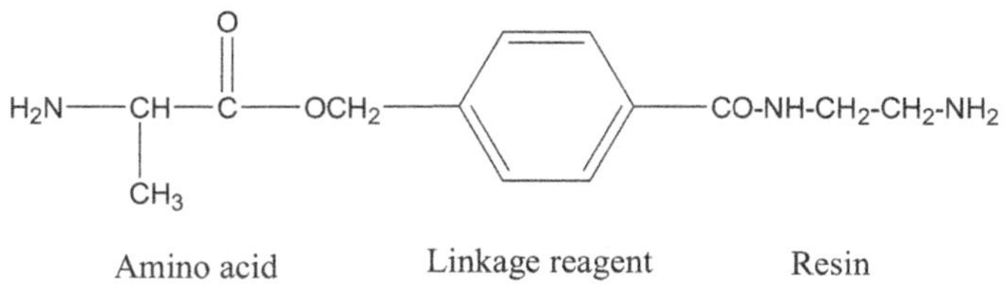

$$\text{H}_2\text{N}-\underset{\underset{\text{CH}_3}{|}}{\text{CH}}-\overset{\overset{\text{O}}{\|}}{\text{C}}-\text{OCH}_2-\text{C}_6\text{H}_4-\text{CO-NH-CH}_2\text{-CH}_2\text{-NH}_2$$

      Amino acid   Linkage reagent   Resin

FIGURE 17.3 Immobilization of the target peptide to a solid support *via* a linker.

polyethylene glycol-polystyrene (PEG-PS) graft resins, polyamides, such as cross-linked polystyrene or Kieselguhr, beaded copolymers of polyethylene glycol-acrylamide (PEGA), and polyethylene-polystyrene (PE-PS) films (García-Martín et al. 2006; Meldal, 1992; Atherton et al., 1981; Zalipsky et al., 1994; Berg et al., 1989).

Protection of Amino Acids

The second step in the chemical synthesis of peptides involves the protection of one end of amino acids to prevent reaction with acids. The most commonly used protecting reagents are *tert*-Butyloxycarbonyl (Boc), 9-Fluorenylmethoxycarbonyl (Fmoc), 2-(4-Biphenyl)-isopropoxycarbonyl (Bpoc), Dimethyl-3,5-dimethoxy benzyloxycarbonyl (Ddz), 4-Methoxybenzyloxycarbonyl, Benzyloxycarbonyl (Z), etc. Figure 17.4 shows the chemical structure of commonly used protecting agents in the synthesis of peptides.

Protecting reagents, such as *tert*-Butyloxycarbonyl (Boc) and 9-Fluorenylmeth oxycarbonyl (Fmoc) are the first choice in the synthesis of peptides. Merrifield (1964), utilized *tert*-Butyloxycarbonyl (Boc) as a protecting reagent instead of Benzyloxycarbonyl (Z) and chloromethylcopolystyrene-divinylbenzene as resin in the chemical synthesis of the peptide using SPPS method. The advent of Fmoc as a new protecting group significantly improved the chemical synthesis of peptides and proteins. Fmoc/t-Bu is now a preferred approach for the synthesis of peptides by the SPPS method (Carpino and Han, 1972; Camarero and Mitchell, 2005).

Coupling of Amino Acids

In the third step, the amide bonds are formed between the first protected amino acid with upcoming new amino acids. In this process, the first protected amino acid reacts with the next amino acid, that one attached with polyamide resins. The coupling steps occur many times and the upcoming amino acids are joined with the first protected amino acid. Since they undergo an acid–base reaction, forming an ammonium salt before any nucleophilic substitution takes place, it is difficult to attain an amide bond by treating a carboxylic acid with an amine. To solve this problem, a coupling reagent is used to activate the carboxy group by making the OH group a better leaving group, causing the carboxyl group to have a nucleophilic reaction by another residue's amino group. The rate of acylation, as well as the rate of side reactions, such as racemization, are determined by coupling reaction conditions (Coste et al., 1990; El-Faham and Albericio, 2011). A variety of coupling reagents, such as N,N'-Dicyclohexylcarbodiimide (DCC), N,N'-diisopropyl-carbodiimide (DIC), N-Ethyl-N'-(3-dimethyl amino propyl) carbodiimide-hydrochloride (EDC), N-hydroxybenzotriazole (HOBt), N-hydroxysuccinimide (HOSu), (benzotriazol-1-yloxy)tris(dimethylamino) phosphonium hexafluorophosphate (BOP), and 2-Succinimido,1,1,3,3)-tetramethyl uronium tetrafluoroborate (TSTU), are used in stepwise peptide chain elongation. The chemical structures of coupling reagents are given in Figure 17.5.

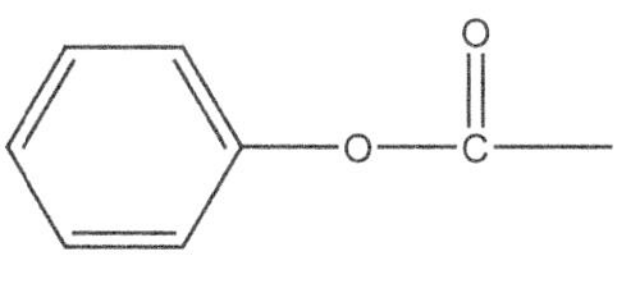

FIGURE 17.4 Protecting reagents commonly used in peptide synthesis.

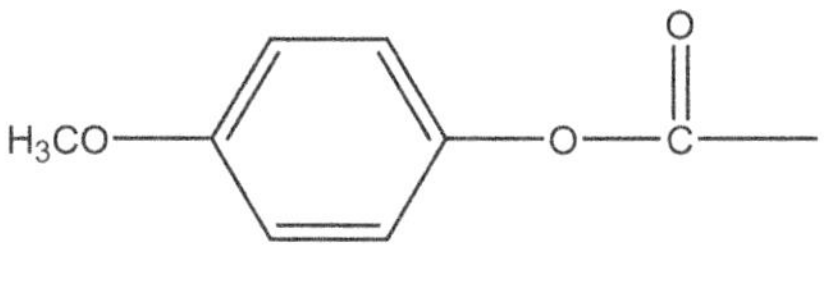

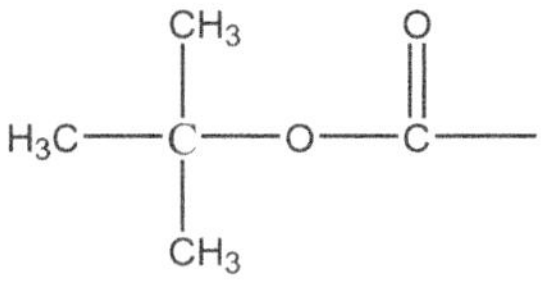

Deprotection

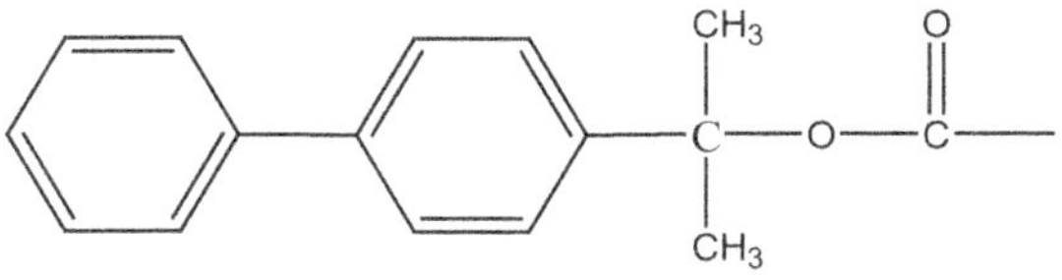

In the fourth step, the excess coupling reagents (such as DCC, DIC, EDC, HOBt, HOSu, BOP, etc.) are washed off with water or any suitable reagents such as DCM or DMF. Then the protecting reagents, such as Boc, Fmoc, Bpoc, Ddz, and benzyloxycarbonyl, are removed with piperidine. The above steps are repeated several times until the desired peptides have been formed.

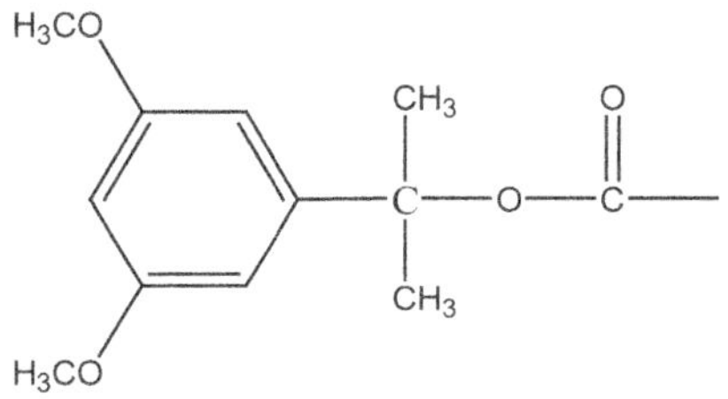

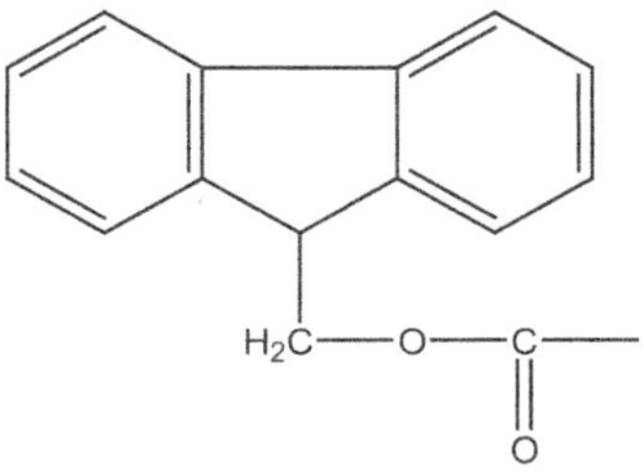

Removal of the Polymer

The polymer is removed in the final step of the chemical synthesis of peptides. After the peptide synthesis is completed, the polyamide must be separated from the peptides. This is accomplished by using a 95% trifluoroacetic acid (TFA) solution to cleave the polyamide–peptide bond. The side-chain protecting groups are also removed at this stage.

Advantages of SPPS

The main advantage of solid-phase peptide synthesis is that it gives the highest yield. Modern SPPS instrumentation improves coupling and deprotection and hence increases yields up to 99.99%. SPPS is also much faster than typical step-by-step solution synthesis. In less than a week, a 20 amino acid peptide can be synthesized utilizing SPPS in a 24-hour cycle. With the advent of automated synthesizers and sophisticated analytical and purification tools, peptide chemists can now produce peptides in the range of 20–50 amino acids in length and quantities of 20–100 milligrams. This is also more

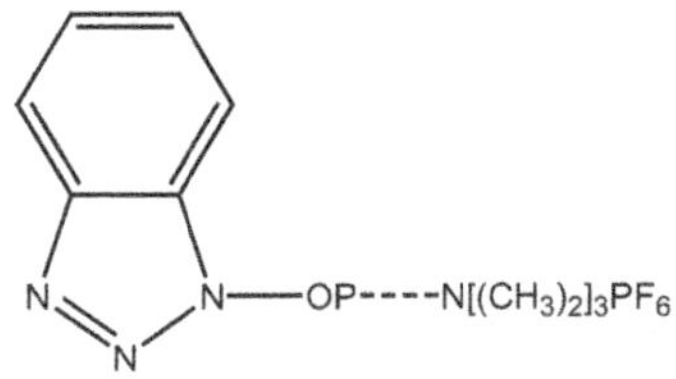

(1-H,1,2,3-Benzotriazol)-1-yloxytris (dimethylamino) phosphonium hexafluorophosphate) (BOP)

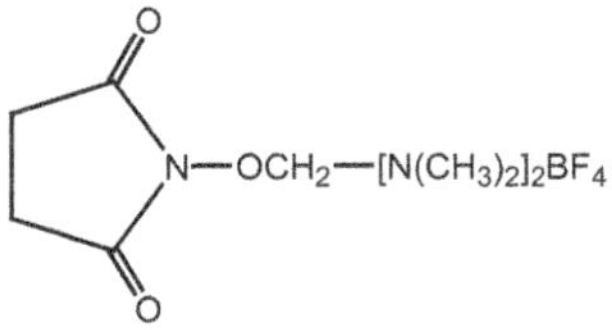

2-Succinimido,1,1,3,3)-tetramethyl uronium tetrafluoroborate (TSTU)

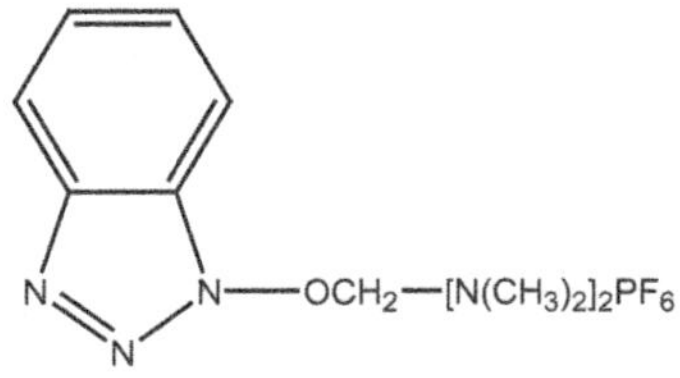

(2-1-H-Benzotriazol-1-yl) 1,1,3,3-tetraethyl uronium hexafluorophosphate) (HBTU)

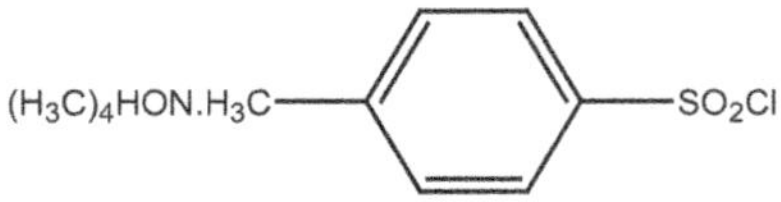

Tetrabutyl ammonium hydroxide, *p*-toulene sulphonyl chloride

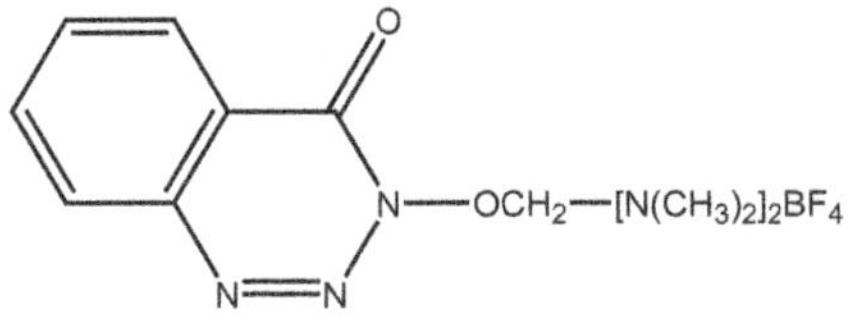

2-(3,4-Dihydro-4-oxo-1,2,3-benzotraiazin-3-yl)-1,1,3,3-tetra uronium tetrafluoroborate (TDBTU)

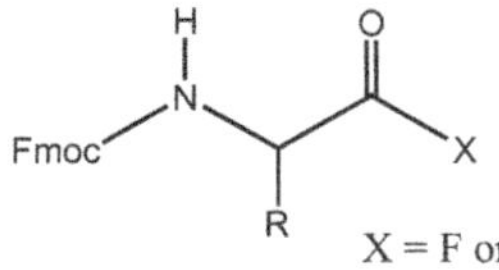

Fmoc amino acid halide

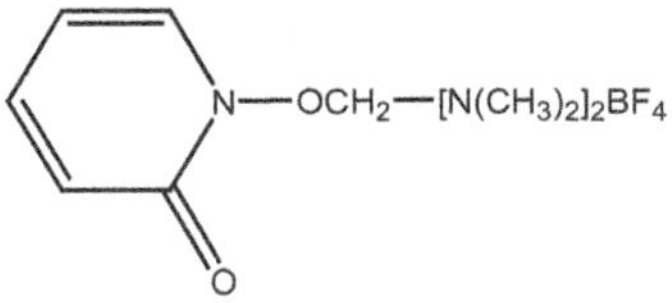

2-(2-Oxo-1-(2-H)-pyridyl-1,1,3,3)-tetramethyl uronium tetrafluoroborate

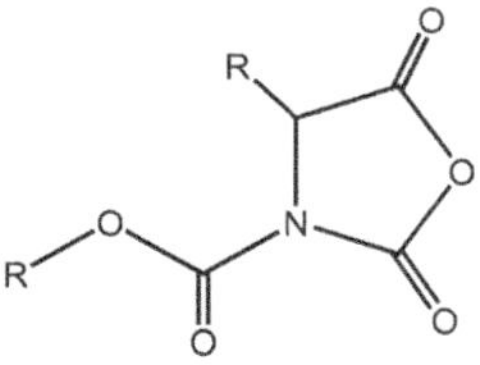

Urethane protected amino acid N-carboxy anhydride (UNCA)

FIGURE 17.5 Coupling reagents commonly used in peptide synthesis.

than appropriate for biochemists and biologists to undertake thorough pilot studies, and since they often only look at a specific peptide once, the velocity is especially useful. This velocity is sacrificed for purity when very large quantities of peptides are necessary (for example, in the industrial manufacture of peptide drugs). However, production rates are still high, with hundreds of grams of peptide produced per kilogram of polymer annually. Hundreds of thousands of doses are needed because only milligrams of a polymer are required per dose.

Chemical Synthesis of Difficult Peptides

Difficult peptides, a term coined in the 1980s to characterize the difficulty of the chemical synthesis of certain peptides. This refers to sequences with large enough inter- or intramolecular β-sheet interactions to form aggregates during peptide synthesis. Non-covalent hydrogen bonds on the peptide's backbone stabilize and mediate these structural interactions, which are favored according to the sequence (Kent, 1988). The ability of peptide chains to aggregate is translated from a list of common behavioral characteristics attributed to difficult sequences. The following are the main concerns with chemical synthesis using conventional methods: despite recouplings, incomplete amino-acylations (less than 15%) are common; sequence difficulties are aggravated when resin loading is high or when sterically hindered amino acids are present; and, more precisely, coupling reagents such as Fmoc, BOC, and others are extracted slowly or incompletely (Paradis-Bas et al., 2015). For the chemical synthesis of difficult peptides for large peptide sequences that cannot be achieved by conventional stepwise solid-phase synthesis, several strategies based on SPPS have been developed. Initially, fragment condensations, such as solid-phase or solution-phase protocols, were used. Both approaches require the conceptual and retro-synthetic separation of segments in the native series, which are then appropriately associated in a synthetic flow in the solid or solution phase. The synthesis of some peptides in which the binding of two fragments occurs by an unusual amide bond formation has been facilitated by recent native chemical ligation or chemo-selective ligation method of peptide synthesis (Dawson et al., 1994; Li et al., 2010).

Native Chemical Ligation Method

The native chemical ligation strategy (NCL) was introduced by Kent and coworkers in 1994. Peptides are synthesized in neutral buffered aqueous solutions in this method. The chemoselective ligation strategy is based on a reversible thioester exchange between a thioester intermediate C-terminal peptide and an N-terminal cysteine peptide to form another intermediate thioester that undergoes a rapid irreversible spontaneous intramolecular rearrangement to form a native amide bond and an intermediate peptide with a cysteine ligation residue (Dawson et al. 1994; Kent, 2009). The different steps of peptide synthesis by the native chemical ligation method were summarized in Figure 17.6.

α-Ketoacid Hydroxylamine Ligation Method

Bode et al. introduced the α-ketoacid hydroxylamine (KAHA) ligation method for the chemical synthesis of peptides (2006). This approach involves forming a native amide bond by decarboxylating hydroxylamines with α-ketoacids (Figure 17.7). This ligation uses an unprotected C-terminal peptide-ketoacid and an unprotected N-terminal peptide hydroxylamine (e.g., 5-oxaproline) to produce an intermediate depsipeptide at the ligation site. After treatment with a simple aqueous solution, the ester is rearranged to meet the native amide bond toward the intermediate peptide with a homoserine residue at the ligation site (Pusterla and Bode, 2012) (Figure 17.8).

Purification and Analysis of Peptides

High-performance liquid chromatography (HPLC) emerged as a powerful technique for identification, separation/isolation, and quantification of synthetic peptides because it offers high resolution and facilitates isolation. Reverse-phase HPLC (RP-HPLC) is commonly used for the purification of a peptide. Ion-exchange chromatography (IEC) and gel filtration liquid chromatography also have great utility in the purification of synthetic peptides (Sato et al., 2008). Matrix-assisted laser desorption/ionization time-of-flight mass spectrometry or ion-trap electrospray mass spectrometry will monitor the purification of peptides in real time. Synthetic peptides must be tested on a regular basis for

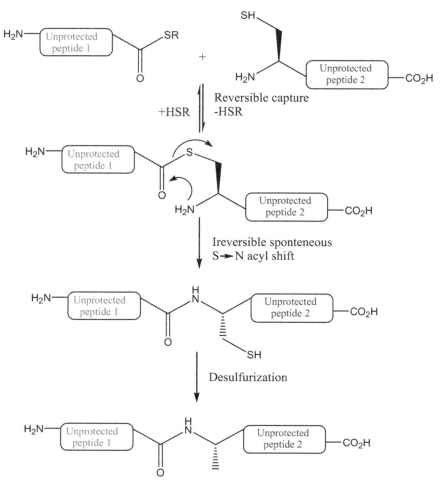

FIGURE 17.6 Chemical synthesis of peptide by the native chemical ligation (NCL) method.

the proper amino acid composition, and sequencing data can aid in certain situations. A combination of RP-HPLC and MS, with either Edman degradation sequence analysis or tandem-MS, provides the most effective synthetic peptide characterization (Del Mar Contreras et al., 2008).

CONCLUSION

Chemical synthesis of peptides and proteins has emerged as a valuable source of bioactive peptide. The chemically synthesized bioactive peptides currently used are aspartame, calcitonin, oxytocin, and adrenocorticotropic. Through various chemical synthesis methods, the availability of pure bioactive peptides and proteins is a reality and it also provides further materials for their investigation in *in-vitro* and *in-vivo* studies. As a

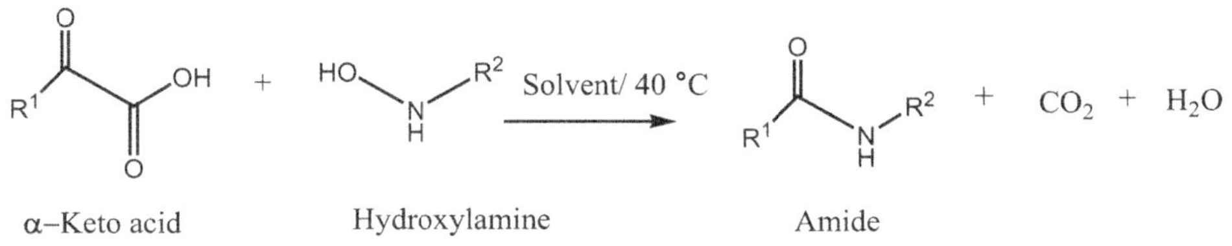

FIGURE 17.7 Formation of amide bond by the α-ketoacid hydroxylamine (KAHA) ligation method.

result, molecular biology and biochemistry research will benefit greatly from the use of chemically synthesized and carefully engineered proteins and peptides. The significance of the SPPS method in the chemical synthesis of peptides was demonstrated in this book chapter. Chemical ligation approaches have progressed to the point that it is now possible to synthesize difficult peptides. Despite advances in the overall chemical synthesis of proteins and peptides, synthetic chemists are still working on more cutting-edge approaches to greater protein access by simpler chemical methods.

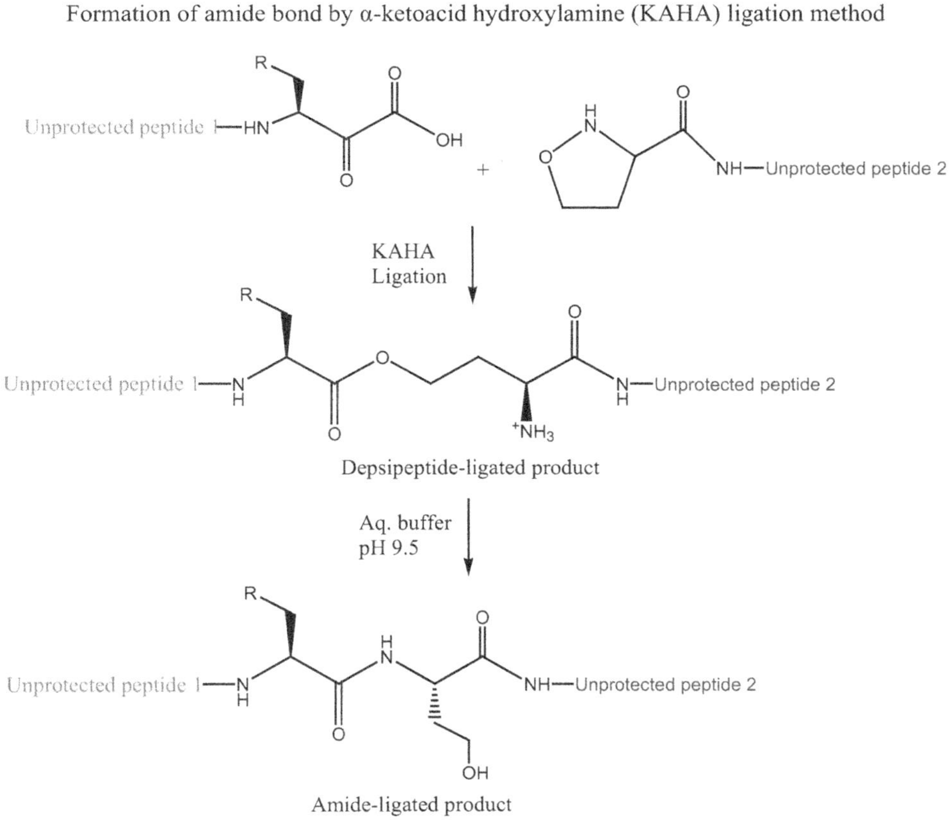

FIGURE 17.8 Chemical synthesis of peptide by α-ketoacid hydroxylamine (KAHA) ligation method.

REFERENCES

Anderson, J.W., Johnstone, B.M., Cook-Newell, M.E. (1995). Meta-analysis of the effects of soy protein intake on serum lipids. *N. Engl. J. Med.* 333(5): 276–282.

Atherton, E., Brown, E., Sheppard, R.C., Rosevear, A. (1981). A physically supported gel polymer for low pressure, continuous flow solid phase reactions. Application to solid phase peptide synthesis. *J. Chem. Soc. Chem. Commun.* 21(21): 1151–1152.

Berg, R.H., Almdal, K., Pedersen, W.B., Holm, A., Tam, J.P., Merrifield, R.B. (1989). Long-chain polystyrene-grafted polyethylene film matrix: A new support for solid-phase peptide synthesis. *J. Am. Chem. Soc.* 111(20): 8024–8026.

Bode, J.W., Fox, R.M., Baucom, K.D. (2006). Chemoselective amide ligations by decarboxylative condensations of N-alkylhydroxylamines and α-ketoacids. *Angew. Chem. Int. Ed.* 45(8): 1248–1252.

Borgia, J.A., Fields, G.B. (2000). Chemical synthesis of proteins. *Trends Biotechnol.* 18(6): 243–251.

Camarero, J.A., Mitchell, A.R. (2005). Synthesis of proteins by native chemical ligation using Fmoc-based chemistry. *Protein Pept. Lett.* 12(8): 723–728.

Canne, L.E., Botti, P., Simon, R.J., Chen, Y., Dennis, E.A., Kent, S.B.H. (1999). Chemical protein synthesis by solid phase ligation of unprotected peptide segments. *J. Am. Chem. Soc.* 121(38): 8720–8727.

Carpino, L.A., Han, G.Y. (1972). The 9-fluorenylmethoxycarbonyl amino protecting group. *J. Org. Chem.* 37(22): 3404–3409.

Carroll, K.K. (1978). The role of dietary protein in hypercholesterolemia and atherosclerosis. *Lipids* 13(5): 360–365.

Casi, G., Hilvert, D. (2003). Convergent protein synthesis. *Curr. Opin. Struct. Biol.* 13(5): 589–594.

Cho, S.J., Juillerat, M.A., Lee, C.H. (2007). Cholesterol lowering mechanism of soybean protein hydrolysate. *J. Agric. Food Chem.* 55(26): 10599–10604.

Coste, J., Le-Nguyen, D., Castro, B. (1990). PyBOP®: A new peptide coupling reagent devoid of toxic by product. *Tetrahedron Lett.* 31(2): 205–208.

Dawson, P.E., Muir, T.W., Clark-Lewis, I., Kent, S.B.H. (1994). Synthesis of proteins by native chemical ligation. *Science* 266(5186): 776–779.

Del Mar Contreras, M., Lopez-Exposito, I., Hernandez Ledesma, B., Ramos, M., Recio, I. (2008). Application of mass spectrometry to the characterization and quantification of food-derived bioactive peptides. *JOAC Int.* 91: 981–994.

El-Faham, A., Albericio, F. (2011). Peptide coupling reagents, more than a letter soup. *Chem. Rev.* 111(11): 6557–6602.

Fischer, E. (1907). Synthetical chemistry in its relation to biology (Faraday lecture). *J. Chem. Soc. Chem. Commun.* 91: 1749–1765.

Fischer, E., Fourneau, E. (1901). Ueber einige derivate des glykocolls. *Ber. Deutsch. Chem. Ges.* 34(2): 2868–2877.

Fujita, H., Yokoyama, K., Yoshikawa, M. (2000). Classification and antihypertensive activity of angiotensin I-converting enzyme inhibitory peptides derived from food proteins. *J. Food Sci.* 65(4): 564–569.

García-Martín, F., Quintanar-Audelo, M., García-Ramos, Y., Cruz, L.J., Gravel, C., Furic, R., et al. (2006). ChemMatrix, a poly(ethylene glycol)-based support for the solid phase synthesis of complex peptides. *J. Comb. Chem.* 8(2): 213–220.

Hackenberger, C.P.R., Schwarzer, D. (2008). Chemoselective ligation and modification strategies for peptides and proteins. *Angew. Chem. Int. Ed.* 47(52): 10030–10074.

Hendrickson, T.L., de Crecy-Lagard, V., Schimmel, P. (2004). Incorporation of nonnatural amino acids into proteins. *Annu. Rev. Biochem.* 73: 147–176.

Jaradat, D.M.M. (2017). Thirteen decades of peptide synthesis: Key developments in solid phase peptide synthesis and amide bond formation utilized in peptide ligation. *Amino Acids.* 50: 39–68. doi:10.1007/s00726-017-2516-0.

Kay, J.R-M. (2012). Food proteins as a source of bioactive peptides with diverse functions. *Br. J. Nutr.* 108: S149–S157.

Kemp, D.S., Carey, R.I. (1993). Synthesis of a 39-peptide and a 25-peptide by thiol capture ligations: Observation of a 40-fold rate acceleration of the intramolecular O,N-acyl-transfer reaction between peptide fragments bearing only cysteine protective groups. *J. Org. Chem.* 58(8): 2216–2222.

Kent, S. (2003). Total chemical synthesis of enzymes. *J. Pept. Sci.* 9(9): 574–593.

Kent, S.B. (1988). Chemical synthesis of peptides and proteins. *Annu. Rev. Biochem.* 57: 957–989.

Kent, S.B.H. (2009). Total chemical synthesis of proteins. *Chem. Soc. Rev.* 38(2): 338–351.

Kimmerlin, T., Seebach, D. (2005). '100 years of peptide synthesis': Ligation methods for peptide and protein synthesis with applications to β-peptide assemblies. *J. Pept. Res.* 65(2): 229–260.

Kitts, D.D. (1994). Bioactive substances in food: Identification and potential uses. *Can. J. Physiol. Pharmacol.* 72(4): 423–434.

Korhonen, H. (2009). Milk-derived bioactive peptides: From science to applications. *J. Funct. Foods* 1(2): 177–187.

Li, X., Lam, H.Y., Zhang, Y., Chan, C.K. (2010). Salicylaldehyde esterinduced chemoselective peptide ligations: Enabling generation of natural peptidic linkages at the serine/threonine sites. *Org. Lett.* 12(8): 1724–1727.

Masuda, O., Nakamura, Y., Takano, T. (1996). Antihypertensive peptides are present in aorta after oral administration of sour milk containing these peptides to spontaneously hypertensive rats. *J. Nutr.* 126(12): 3063–3068.

Meldal, M. (1992). Pega: A flow stable polyethylene glycol dimethyl acrylamide copolymer for solid phase synthesis. *Tetrahedron Lett.* 33(21): 3077–3080.

Mendel, D., Cornish, V.W., Schultz, P.G. (1995). Site-directed mutagenesis with an expanded genetic code. *Annu. Rev. Biophys. Biomol. Struct.* 24: 435–462.

Merrifield, B. (1986). Solid phase synthesis. *Science* 232(4748): 341–347.

Merrifield, B. (1996). The chemical synthesis of proteins. *Protein Sci.* 5(9): 1947–1951.

Merrifield, R.B. (1964). Solid phase peptide synthesis. III. An improved synthesis of bradykinin. *Biochemistry* 3: 1385–1390.

Moller, N.P., Scholz-Ahrens, K.E., Roos, N., Schrezenmeir, J. (2008). Bioactive peptides and proteins from foods: Indication for health effects. *Eur. J. Nutr.* 47(4): 171–182.

Naito, H., Kawakami, A., Inamura, T. (1972). *In vivo* formation of phosphopeptide with calcium-binding property in the small intestinal tract of the rat fed on casein. *Agric. Biol. Chem.* 36(3): 409–415.

Nilsson, B.L., Soellner, M.B., Raines, R.T. (2005). Chemical synthesis of proteins. *Annu. Rev. Biophys. Biomol. Struct.* 34: 91–118.

Paradis-Bas, M., Tulla-Puche, J., Albericio, F. (2015). The road to the synthesis of "difficult peptides". *Chem. Soc. Rev.* 45: 631–654. doi:10.1039/c5cs00680e.

Pepe, G., Sommella, E., Ventre, G., Scala, M.C., Adesso, S., Ostacolo, C., et al. (2016). Antioxidant peptides released from gastrointestinal digestion of "Stracchino" soft cheese: Characterization, invitro intestinal protection and bioavailability. *J. Funct. Foods* 26: 494–505.

Pontiroli, A.E. (1998). Peptide hormones: Review of current and emerging uses by nasal delivery. *Adv. Drug Deliv. Rev.* 29(1–2): 81–87. ISSN 0169-409X, https://doi.org/10.1016/S0169-409X(97)00062-8

Pusterla, I., Bode, J.W. (2012). The mechanism of the α-ketoacid-hydroxylamine amide-forming ligation. *Angew. Chem. Int. Ed.* 51(2): 513–516.

Sanchez de Medina, F., Daddaoua, A., Requena, P., Capitán-Cañadas, F., Zarzuelo, A., Dolores Suárez, M., Martínez-Augustin, O. (2010). New insights into the immunological effects of food bioactive peptides in animal models of intestinal inflammation. *Proc. Nutr. Soc.* 69(3): 454–462.

Satake, M., Enjoh, M., Nakamura, Y., Takano, T., Kawamura, Y., Arai, S., Shimizu, M. (2002). Transepithelial transport of the bioactive tripeptide, val-pro-pro, in human intestinal caco-2 cell monolayers. *Biosci. Biotechnol. Biochem.* 66(2): 378–384.

Sato, K., Iwai, K., Aito-Inoue, M. (2008). Identification of food derived bioactive peptides in blood and other biological samples. *JOAC Int.* 91(4): 995–1001.

Schusdziarra, V., Schick, A., De La Fuente, A., Specht, J., Klier, M., Brantl, V., Pfeiffer, E.F. (1983). Effect of *b*-casomorphins and analogs on insulin release in dogs. *Endocrinologist* 112: 885–889.

Stawikowski, M., Fields, G.B. (2002). Introduction to peptide synthesis. *Curr. Protoc. Protein Sci.* 26: 18.1.1–18.1.9. doi:10.1002/0471140864.ps1801s26.

Vaino, A.R., Janda, K.D. (2000). Solid-phase organic synthesis: A critical understanding of the resin. *J. Comb. Chem.* 2(6): 579–596.

Wilken, J., Kent, S.B. (1998). Chemical protein synthesis. *Curr. Opin. Biotechnol.* 9(4): 412–426.

Yang, S.C., Lin, C.H., Sung, C.T., Fang, J.Y. (2014). Antibacterial activities of bacteriocins: Application in foods and pharmaceuticals. *Front. Microbiol.* 5: 241.

Zalipsky, S., Chang, J.L., Albericio, F., Barany, G. (1994). Preparation and applications of polyethylene glycol-polystyrene graft resin supports for solid-phase peptide synthesis. *React. Polym.* 22(3): 243–258.

SECTION 6

Functions of Bioactive Peptides

CHAPTER 18

Antihypertensive Activity

Ritam Bandopadhyay, Pragya Shakti Mishra, and Awanish Mishra

CONTENTS

Introduction	365
Sources of Antihypertensive Peptides	366
Animal Sources	366
Dairy Products	366
Meat	367
Marine Sources	371
Egg	372
Plant Sources	372
Wheat	374
Rice	375
Amaranth and Quinoa	375
Soybean	375
Fruits and Vegetables	376
Legumes	377
Isolation of Antihypertensive Peptides	378
Enzymatic Hydrolysis	378
Fermentation	379
Genetic Engineering	379
Downstream Processing	379
High Pressure Processing	379
Ultrasound Processing	380
Microwave-Assisted Extraction	380
Supercritical Fluid Extraction	380
Role of Bioactive Peptide Antihypertensive Agent	381
Conclusion and Future Perspectives	383
References	383

INTRODUCTION

Bioactive peptides are specific 3–20 amino acid chains in a protein that serve various important physiological functions and have many health-promoting effects. The precursor protein does not carry significant biological effect, but after proteolytic activation, they exhibit their biological effect [1]. Bioactive peptides are released from a protein by various methods of hydrolysis including by enzymes, digestion with gastrointestinal enzymes, proteolysis in *in-vivo* experiments, or during preparation/processing of food (cooking, fermentation, ripening). Numerous bioactive peptides have been found in the last decade which show various beneficial physiological functions, like antidiabetic, antihypertensive, cholesterol-lowering, anticancer, antimicrobial activities.

Nature is the main source of such bioactive peptides. These are found in plants, animals, fungi, microbes, and various other protein sources. There are two methods by which bioactive peptides are discovered – the classical approach and the bioinformatics approach. In the classical approach, proteins are hydrolyzed by proteolytic enzymes. Small fragments of proteins are derived and these small fragments are then analyzed for their biological activity [2]. Alternatively, bacterial fermentation can also be used [3]. It relies on the data existing in a previously created database. The database is cross-checked to determine the frequency of existence of already identified bioactive peptides in a protein. Based on the results, specific enzymes are applied which can hydrolyze the parent protein in such a manner that the bioactive peptide sequence can be cleaved intact out of the parent protein. This method helps us identify known peptide sequences from unknown protein sequences. However, the main challenges in bioactive peptide development are to create a proper connection between ingestion of a bioactive peptide and observed positive health effects. In this chapter, emphasis has been given to antihypertensive bioactive peptides, their sources, isolation, analysis, and mechanisms of action.

SOURCES OF ANTIHYPERTENSIVE PEPTIDES

Food material of plant and animal origin which possess high protein content is the natural source of bioactive peptides. Isolation from this source has some additional benefits such as natural availability, price efficacy, efficient waste management; also, peptides are easily isolated natural sources, which is not the case in the development of synthetic drugs [4].

Animal Sources

Dairy Products
Milk is well-defined as a healthy drink, with a lot of health benefits. It offers various important substances which are needed for growth and development. Milk proteins are of high biological value and proteins derived from it provide various health benefits. Mellander in 1950 first reported the identification of milk-derived biopeptides [5]. Thereafter several researchers isolated various bioactive peptides from milk and examined their biological effects. [6]. Antihypertensive peptides were found in milk like VPP and IIP. These peptides were obtained after fermenting milk by *Lactococcus lactis* strains [7]. After single oral administration of these peptides to spontaneous hypertensive rats (SHR), reduction in systolic, diastolic, and total blood pressure was observed. A reduction in blood pressure by 15 to 18 mmHg compared to the control was observed in SHR after 6 to 12 h of treatment [8].

Another three antihypertensive peptides were found after fermentation of kombucha milk by *Lactobacillus casei* strain. The sequence of these three peptides was VAPFPEVFGK, LVYPFPGPLH, and FVAPEPFVFGKEK [9]. Goat milk was also reported to have three antihypertensive peptide sequences: PEQSLACQCL (from whey β-lactoglobulin), QSLVYPFTGPI (from β-casein) and ARHPHPHLSFM (from κ-casein).

These were isolated after the hydrolysis of milk obtained from goat by the gastric enzyme pepsin [10]. The presence of antihypertensive peptides in donkey milk was also detected. These were various genetic variants of β-lactoglobulins peptides. Three main kinds of genetic variations were detected: one genetic variant of β-LGI (β-LGIB), two genetic variants of β-LGII (β-LGIIB and β-LGIIC), and a third minor β-LGII variant (β-LGIID) [11]. Some other antihypertensive peptides have also been reported to be present in milk such as αS1-casein, αS2-casein, and α-lactalbumin.

Camel milk also possesses antihypertensive peptides. Camel whey protein (CWP), camel whey protein hydrolysates (CWPH), and its fractions SEC-F1, SCE-F2 were derived from camel milk. They were examined to have angiotensin converting enzyme (ACE)-inhibiting and free radical-scavenging activities. Although free radical-scavenging activity does not have any direct connection with the antihypertensive effect, it helps in the improvement of endothelial function. Camel whey protein is obtained from camel milk and CWPH was obtained after papain hydrolysis of CWP. CWP, CWPH, and fractions SEC-F1, SCE-F2 showed ACE inhibition with IC_{50} values of 576.7±6.5μg/mL, 410.8±2.7μg/mL, 469.3±4μg/mL, and 179.9±2.6 μg/mL, respectively [12]. Peptides from α/β-casein were obtained after fermentation of milk with *Lactobacillus helveticus* NK1, *Lactobacillus rhamnosus*, and *Lactobacillus reuteri* LR1 strains. Thus produced peptides possess antihypertensive activity and higher potency was recorded with *Lactobacillus rhamnosus* [13]. A list of the peptides obtained is given in Table 18.1. Lactic acid bacteria (LAB)-fermented dairy products contain two previously described casein-derived peptides: VPP and IPP [14]. These peptides show NO mediated vasodilation resulting in reduction of blood pressure [14]. Eight peptides – RHPEYAVSVLLR, GGAPPAGRL, GPPLPRL, ELKPTPEGDL, VLSELPEP, DAQSAPLRVY, RDMPIQAF, and LEQVLPRD – were isolated from whey protein. These peptides have shown potential ACE inhibitory effect, and DAQSAPLRVY and VLSELPEP were suggested as the most potent ACE inhibitors [15].

Meat

Meat is one of the most important animal-derived foods which contain proteins of the highest quality. Currently, most of the research work is focused on meat to identify various bioactive peptides. The interest of researches on bioactive peptide discovery from meat increases as it is an easily available product and a lot of by-products of meat are also available now in the market. This makes meat the most reliable source of proteins. Peptides are isolated from meat through the application of various proteolytic and digestive enzymes as trypsin, pepsin, and chymotrypsin. Twenty-two antihypertensive peptides were isolated from meat in an *in-vitro* study. Amongst them AKAPVA, PTPVP, and RPR showed better antihypertensive activity.

Meat derived from pork is also thought of as a reliable source of bioactive peptides [33]. Porcine blood protein hydrolysates were extracted by pepsin hydrolysis (8% pepsin at 37°C for 6 h) followed by ultrafiltration in Sephadex G-25 and purification by reversed phased HLPC. Thus derived peptide sequences (WVPSV, YTVF, and VVYPW) have antihypertensive effects. These peptides show antihypertensive effects by reducing the systolic blood pressure (SBP) [34]. Antihypertensive peptides were also found in collagen and troponin hydrolysates. Peptide sequences like NSIM, ALM, LGL, SFVTT, and GVVPL were identified in these tissues [35]. Three bioactive peptide sequences were found in beef (LSW, FGY, and YRQ) showing antihypertensive effects [36]. Peptide sequences KRQKYD, EKERERQ, KAPVA, PTPVT, RPR, GLSDGEWQ, GFHI, DFHING, and FHG were isolated from meat by enzymatic hydrolysis with the help of a protease enzyme

TABLE 18.1 Bioactive Peptides from Animal Sources

Source	Extraction Procedure	Sequence	Activity	IC$_{50}$	Ref
Seaweed	Hot water extraction, Enzymatic hydrolysis	IY, IW	ACEI	2.7 µM 1.5 µM	[16]
Seaweed	Chromatography	AKYSY	ACEI	1.52 µM	
Microalgae	Enzymatic hydrolysis (pepsin)	IAPG	ACEI	11.4 µM	
Yellow fin tuna	Chromatography	PTHIYGD	ACEI	2 µM	
Skipjack tuna	Chromatography	LRP	ACEI	1 µM	
Alaska Pollack skin	Enzymatic hydrolysis (serial protease)	GPL	ACEI	2.6 µM	
Chum salmon	Enzymatic hydrolysis (thermolysin)	VW	ACEI	2.5 µM	
Pink salmon	Enzymatic hydrolysis (papain)	IW	ACEI	1.2 µM	
Skate skin	Enzymatic hydrolysis (alcalase)	MVGSAPVL	ACEI	3.09 µM	
Small-spotted catshark	Enzymatic hydrolysis	VAMPF	ACEI	0.44 µM	
Pelagic thresher	Enzymatic hydrolysis (thermolysin)	IKW	ACEI	0.54 µM	
Marine shrimp	Enzymatic hydrolysis (protease)	IFVPAF	ACEI	3.4 µM	
Izumi shrimp	Enzymatic hydrolysis (protease)	ST	ACEI	4.03 µM	
Jellyfish	Enzymatic hydrolysis (pepsin, papain); Ultrafiltration	QPGPT	ACEI	80.67 µM	[17]
Sipuncula	Enzymatic hydrolysis (pepsin	AWLHPGAPKVF	ACEI	135 M	[16]
Pearl oyster	Enzymatic hydrolysis (pepsin)	ALAPQ	ACEI	167.5 µM	

(Continued)

TABLE 18.1 (Continued)

Source	Extraction Procedure	Sequence	Activity	IC$_{50}$	Ref
Pacific cod	Enzymatic hydrolysis (pepsin)	GASSGMPG	ACEI	6.9 μM	[18]
Chlorella vulgaris (microalgae)	Enzymatic hydrolysis (pepsin)	VVPPA IVVE FAL	ACEI	79.5 μM 315 μM 26.3 μM	[19]
Corbicula fluminea (molluscs)	Hydrolysis with Protamex and Flavourzyme	VKP VKK	ACEI	3.7 μM 1045 μM	[20]
milk	Hydrolysis by Endogenous enzymes	PYVRYL, LVYPFTGPIPN	ACEI		[21]
	Hydrolysis by Proteinase obtained from *Lactobacillus helveticus*	Fractions of Bovine αS1-casein, caprine β-CN, buffalo β-CN	ACEI		[22]
	Enzymatic hydrolysis by Protease of *Enterococcus faecalis*	LHLPLPL, AYFYPEL, RYLGY, YQKFPQY	ACEI	—	[23]
milk	Fermentation by *Lactobacillus helveticus* NK1 strain	α-casein-QHQKAMKPW, PWIQPKTKVIPYVRYL, IQPKTKVIPYVRYL, IQPKTKVIPY, KVIPYVRYL β-Casein-VVPPFLQPE, YPFPGPIPN, NIPPLTQTPV, DVENLHLPLLQSWM, LHLPLLQSW, EVLNENLLRF	ACEI	0.18 ± 0.02 mg/ml	[13]
	Fermentation by *Lactobacillus rhamnosus* F			0.21 ± 0.03 mg/ml	
	Fermentation by *Lactobacillus reuteri* LR1 strain			1.07 ± 0.13 mg/ml	
Manchego	Enzymatic hydrolysis	Ovine β-CN, KKYNVPQL, VRYL	ACEI	—	[24]

(*Continued*)

TABLE 18.1 (Continued)

Source	Extraction Procedure	Sequence	Activity	IC$_{50}$	Ref
Gouda	Enzymatic hydrolysis	αs1-CN, αs1-C f, β-CN f, β-CN f	ACEI	—	[25]
Spent hen muscle protein	hydrolysis with thermoase	VRP, LKY, VRY, KYKA, LKYKA	ACEI	0.034–5.77 μg/mL	[26]
		VKW, VHPKESF, VVHPKESF, VAQWRTKYETD AIQRTEELEEAKKK	ACE2 upregulation	—	
Meat	Proteolysis followed by oxidation	APPPAEVPEVHEEVH, PPPAEVPEVHEEVH, IPITAAKASRNIA, LPLGG, FAGGRGG, APPPPAEVP, SPLPPE, EGPQGPPGPVG, PGLIGARGPPGP	ACEI	—	[27]
cricket proteins	Hydrolysis with alcalase (3.0% E/S)	Cationic peptide Anionic peptide	ACEI	1.922 μg/mL 509.062 μg/mL	[28]
Ostrich egg	Anion exchange chromatography, alkaline hydrolysis	YV dipeptide	ACEI	63.97 μg/mL	[29]
Ostrich egg white protein	Enzymatic hydrolysis by pepsin and pancreatin	WESLSRLLG	ACEI	46.7 ± 1.4 mg/mL	[30]
Egg white hydrolysate	Gastrointestinal digestion by pepsin, trypsin	LAPYK, SVIRW, PKSVIRW, ADWAK	ACEI	Average- 5.29 ± 0.24 μM	[30]
monkfish waste	Enzymatic hydrolysis	Fish protein hydrolysates (FPH)	ACEI	931.3 ± 85.2 μg protein/mL	[31]
Tilapia gelatin protein	Enzymatic hydrolysis	LSGYGP	Non-competitive ACE inhibition	2.577 μM/L	[32]

Abbreviations (amino acids): A: alanine, R: arginine, N: asparagine, D: aspartic acid, C: cysteine, E: glutamic acid, Q: glutamine, G: glycine, H: histidine, I: isoleucine, L: leucine, K: lysine, M: methionine, F: phenylalanine, P: proline, S: serine, T: threonine, W: tryptophan, Y: tyrosine, V: valine, ACEI: angiotensin converting enzyme inhibitor.

exogenous enzyme, collagenase from *Aspergillus oryzae*. They were also reported to have antihypertensive activity [37]. Collagen proteins were also examined for antihypertensive activity. Peptide sequences PAGNPGADGQPGAKGANGAP, GAXGLXGP, GPRGF, VGPV, QGAR, LQGM, LQGMH, and LC were isolated from collagen by enzymatic hydrolysis. All of them have shown antioxidant activity [38]. Thermolysin-hydrolyzed collagen protein fractions from *Kacang* goat skin also show ACE enzyme inhibition activity [39]. Average IC_{50} values of the fractions obtained after hydrolysis were 0.83 mg/mL or 82.94 µg/mL [39].

ACE inhibitory peptides, VRP, LKY, VRY, KYKA, LKYKA, and ACE2 up-regulatory peptides VKW, VHPKESF, VVHPKESF and VAQWRTKYETDAIQRTEEL EEAKKK were also isolated from spent hen muscle protein. They were obtained after hydrolysis of spent hen muscle protein with thermoase [26]. The obtained ACE inhibitory peptides showed IC_{50} vale ranging 0.034–5.77 µg/mL and ACE2 up-regulatory actions which increases ACE2 levels by 0.52–0.84 folds [26]. Cricket protein hydrolysates produced by hydrolysis of cricket proteins with alcalase (3.0% E/S) also shows antihypertensive effects [28]. Two types of peptides were identified – cationic and anionic peptides. The cationic and anionic peptides showed ACE inhibitory activity of IC_{50} values of 1.922 µg/mL and 509.062 µg/mL, respectively [28]. Chicken foot protein-derived antihypertensive nonapeptide AVFQHNCQE was administered at a dose of 10 mg/kg BW and it was found to reduce blood pressure in SHR [40]. It reduces hypertension by improvement of endothelial function through increased NO release and decreased expression of ET-1(endothelin-1) and NADPH oxidase 4 [40].

Marine Sources

Marine creatures are a profuse source of distinct bioactive peptides of different sequences, with diverse physiological activities. The importance of marine organisms increased due to the presence of these biologically active peptides. The marine organisms represent one-half of the total global biodiversity. Because of this they are a huge source of novel compounds in the research field. Duck skin is a source of antihypertensive peptides. The by-product of the duck skin was subjected to hydrolysis by various proteolytic enzymes. WYPAAP peptide was obtained from it, which has been reported to have antihypertensive effects [41]. Several proteolytic enzyme preparations were used in the isolation such as collagenase, papain, protamex, and α-chymotrypsin [41]. Antihypertensive peptides, GMNNLTP and LEQ, isolated from marine microalgae (*Nannochloropsis oculata*), TFPHGP and HWTTQR isolated from muscle protein of seaweed pipefish, and 121 amino acids with a sequence of GDLGKTT TVSNWSPPKYKDTP was also isolated from marine sources [42]. Other cardio protective peptide sequences isolated from marine sequence are IY and IW from seaweed *Undaria pinnatifida* [43], and AKYSY from seaweed *Pyropia yezoensis* [44] etc. A complete list of cardio protective bioactive peptides is summarized in Table 18.1.

Marine collagen peptide (MCP) can serve the role of an antidiabetic and antihypertensive agent by targeting the pathogenesis of those diseases [45]. Specific antihypertensive tripeptide fragments IGP and VQP isolated from *Spirulina platensis* inhibit ACE enzyme, thus they can serve as antihypertensive peptides [46]. Also, the peptide GIVAGDVTPI, isolated from *Spirulina platensis*, showed reduction in blood pressure. Peptide GIVAGDVTPI shows direct endothelium-dependent vasodilation through PI3K/AKT/eNOS pathway resulting in increased NO release in the endothelium [47]. Thus, they can serve as bioactive peptides also.

Bioactive peptides derived from marine sources can serve as antidiabetic, antihypertensive, anticancer and cholesterol-lowering agents. Peptides VVQ, AFL, FAL, AQL, and VVPPA obtained from *C. vulgaris* show ACE inhibitory activity with IC_{50} values of 315.3 μM, 63.8 μM, 26.3 μM, and 57.1 μM, respectively [48]. Fish protein hydrolysates (FPH) from monkfish waste show ACE inhibitory activity, with IC_{50} values of 931.3 ± 85.2 μg protein/mL [31].These hydrolysates can be used as potent antihypertensive peptides. Various dipeptide and tripeptide sequences were also identified by Abachi et al. (2019) from marine sources and showed antihypertensive activities. Dipeptides DP, FL, GW, IF, LW, VF and tripeptides AFL, IVF, LVL show ACEI activity with IC_{50} values of 2.15 μg/ml, 1.33 μg/ml, 30.00 μg/ml, 1.67 μg/ml, 2.50 μg/ml, 2.70 μg/ml, and 63.80 μg/ml, 33.11 μg/ml, 12.30 μg/ml, respectively [49]. Some peptides isolated from tilapia skin were observed to have antihypertensive activities. Peptides of sequence LSGYGP obtained from tilapia skin gelatin hydrolysate protein show antihypertensive activities. They show noncompetitive inhibition of the ACE enzyme [32].

Egg

The egg is a common food with high protein content and consists of two parts – egg white and egg yolk. So, these two parts of eggs are investigated to find out various biological peptides in it. The bioactive peptides in eggs are mainly isolated by enzymatic digestion in the gastrointestinal tract. Antihypertensive peptides were isolated from eggs after hydrolysis by thermolysin, a digestive enzyme. These peptides showed the reduction of systolic and diastolic blood pressure in SHRs. A dipeptide, YV isolated from ostrich egg white (ovalbumin) was found to be a strong inhibitor of ACE with an IC_{50} value of 63.97μg/mL [29]. IRW was isolated from ovotransferrin, a protein obtained from the white portion of egg showing antihypertensive activities. When this was administered orally to SHRs, reduction of blood pressure was observed. When the mechanism of action was elucidated, upregulation of ACE2 and thus activation of ACE2/Ang (1-7)/MasR axis was identified as the precise mechanism [50].

Certain peptides derived from egg white, like Ovokinin (OA 358-365), decrease hypertension by exerting a vasorelaxant effect. This protein is obtained from ovalbumin after its digestion with peptic enzymes [51]. The purified fraction of hydrolysates obtained from egg white possesses peptides such as RVPSLM, TPSPR, DLQGK, AGLAPY, RVPSL, DHPFLF, HAEIN, QIGLF, HANENIF, VKELY, and TNGIIR, which were examined to have ACE-I inhibitory activity [52]. Hen's egg white lysozyme-derived peptides were also investigated to have calmodulin-dependent phosphodiesterase (CaMPDE) inhibitory activity and free radical-scavenging properties [53]. These help in vasodilation and reduce total peripheral resistance, leading to the antihypertensive effect. In an *in vivo* study, it was observed that cats fed the hydrolyzed poultry by-product meal diet tended to have lower serum ACE levels than those fed the commercial conventional poultry by-product meal diet [54]. Egg white hydrolysates can also be a potential source of bioactive peptides. Peptides obtained from egg white hydrolysate LAPYK, SVIRW, PKSVIRW, and ADWAK show ACEI activity with average IC_{50} values of 5.29 ± 0.24 μM and ORAC value of 1.50 ± 0.04 μM TE/μM [30].

Plant Sources

Plants (cereals, legumes, algae) are a major source of dietary proteins, especially in developing countries. Cereals consist of 6–15% protein. Pseudo-cereals are similar to cereals in terms of compositions and functionality. Cereals and pseudo-cereals consist of a wide range of biologically active peptides (see Table 18.2) which exert antihypertensive effects.

TABLE 18.2 Bioactive Antihypertensive Peptides of Plant Origin

Sources	Extraction	Peptide	Mechanism	IC$_{50}$ value	Ref
βCG of soy	Enzymatic hydrolysis	LAIPVNKP, LPHF	Inhibition of ACE enzyme	—	[55]
Glycinin of soy	Enzymatic hydrolysis	VLIVP, SPYP, WL	Inhibition of ACE enzyme	—	
Glycinin of soy (A4 and A5 fragments)	Enzymatic hydrolysis	NWGPLV	Inhibition of ACE enzyme	—	
Soy protein	Enzymatic hydrolysis	YVVFK; IPPGVPYWT, PNNKPFQ, NWGPLV, TPRVF	Hypotensive	—	
Soybean	Enzymatic hydrolysis	PGTAVFK	Hypotensive	26.5 μM	
		IVF, LLF, LNF, LSW	ACE inhibition	—	
soybean paste	Enzymatic hydrolysis by chymotrypsin	HHL	Hypotensive		[56]
soy milk	Enzymatic hydrolysis by protease	FFYY, WHP, FVP, LHPGDAQR	ACE inhibition	—	[105]
soybean	Fermentation by *Bacillus natto* or *subtili*	VAHINVGK, YVWK	ACE inhibition	—	[58]
Amaranth	—	GKP	Inhibition of ACE	0.352 mM	[59]
		LF		0.349 mM	
		FP		0.315 mM	
		YL		0.122 mM	
		RF		0.093 mM	
		HY		0.026 mM	
		VYVW		0.007 mM	
	Extracted from active fraction	LPP, LRP, VPP, TALEPT, HVIKPPS, SVFDEELS, ASANEPDEN, VEEEGNM	ACE inhibition	—	
		FNLPILR	Inhibition of renin enzyme	0.41 mM,	
		SNFNLPILR		2.50 mM	
		AFEDGFEWVSKF		1.47 mM	

(*Continued*)

TABLE 18.2 (Continued)

Sources	Extraction	Peptide	Mechanism	IC$_{50}$ value	Ref
	Glutelin hydrolysis	AY, FP, GY, MY, PR, VF, VW, VY	Increases endothelial NO production by inhibition of the degradation of bradikynin	—	[107]
Quinoa protein hydrolyzate (QPH)	Gastrointestinal digestion by pepsin, trypsin	FHPFPR NWFPLPR NIFRPF	ACEI	34.92 µM 16.77 µM 32.40 µM	[61]
Wheat gluten	Proteolysis by alcalase from *Pseudomonas aeruginosa*	SAGGYIW APATPSFW	ACEI	0.02 mg/mL 0.036 mg/mL	[62]
Rice bran	thermolysin digestion	LRA YY	ACEI	62.0±1.8µM 16.5±0.6µM	[63]
asparagus	Hydrolysis with alcalase	PDWFLLL ASQSIWLPGWL	ACEI	1.76 µmol/l 4.02 µmol/l	[64]
cauliflower	hydrolyzed with alcalase	APYDPDWYYIR SKGFTSPLF	ACEI	2.59 µmol/l 15.26 µmol/l	[65]
lemon basil seeds	Hydrolysis with alcalase	LGRNLPPI GPAGPAGL	ACEI	0.124 ± 0.02 mM 0.013 ± 0.001 mM	[66]
Corn Silk proteins	Enzymatic hydrolysis	SKFDNLYGCR (CSBps5)	Inhibits binding of AG-I to ACE	44.11 ± 1.04 µM (H-HL substrate) 81.71 ± 1.06 µM(Z-FHL substrate)	[67]
Cassia obtusifolia (sicklepod) seeds	Enzymatic hydrolysis by thermolysin	FHAPWK	ACE inhibition	16.83 ± 0.90 µM	[68]

Wheat

Wheat is a universal and essential food item. The inner layer of the wheat grain consists of a wide range of nutrients and essential amino acids. Triple peptide IVY isolated from wheat germ hydrolysate shows antihypertensive effects. Plasma amino-peptidase enzyme serves a central role in its activity. The peptidase enzyme acts on the IVY peptide and cleaves one amino acid resulting in the formation of VY peptide which is a potent ACE inhibitor [69]. Peptide sequences VALTGDNGHSDHVVHF, VDSLLTAAK, MDATALHYENQK, IGGIGTVPVGR, and SGGSYADELVSTAK isolated from crude wheat germ protein hydrolysate were found to have ACE inhibitory activity with IC$_{50}$ values of 189.3 µg/ml, 159.7µg/ml, 303.6µg/ml, 125.7µg/ml, and 128.2 µg/ml, respectively

[43]. SAGGYIW and APATPSFW are two peptides isolated from wheat gluten. They also show antihypertensive effects [62]. These peptides were isolated from wheat gluten with the help of a protease enzyme isolated from *Pseudomonas aeruginosa* [62]. Peptide sequences GEVPW, YMENF, and AFYRW, isolated from Tartary buckwheat albumin, and sequences TVGGAPAGRIVME and GNPIPREPGQVPAY, isolated from wheat germ protein, showed potent antioxidant activity [70]. This helps in the improvement of endothelial function, and that can be useful in treating hypertension.

Rice

Rice is the most common food in Asian countries. The proteins present in rice possess high biological value compared to the other cereal proteins. After hydrolysis of rice with pepsin, protein fractions F5-IV and F5-V (with peptide sequences FNVPSRYGIY, PWHNPRQGGF, and SPFWNINA) were obtained. They were examined to have antihypertensive activity [71]. Bran (a major by-product of rice milling) has also been reported to have antihypertensive effects. Peptide sequences LRA and YY, isolated from rice bran, have antihypertensive effects with IC_{50} value of 62.0±1.8μM and 16.5±0.6μM, respectively [63]. Trypsin-hydrolyzed rice bran also has antihypertensive and ACE inhibitory effects due to the presence of the bioactive peptide sequence YSK [72]. Trypsin-hydrolyzed rice bran has an antioxidant effect in various studies. After trypsin hydrolysis of rice residue proteins, peptide sequences RPNYTDA, TSQLLSDQ, TRTGDPFF, and NFHPQ were obtained. They possess antioxidant activity by DPPH radical scavenging [73]. Thus, all help in improvement of the endothelial function in hypertension. Peptides isolated from rice bran digests (mainly from albumin and glutelin) were found to have ACE inhibitory activity equivalent to 170.13±2.2 nmol captopril [74].

Amaranth and Quinoa

Amaranth is a pseudo-cereal that can be consumed as a green leafy vegetable. Quinoa is a flowering plant of the Amaranth family which is harvested because of its edible seeds. They have biologically active peptides which serve mainly as antihypertensive agents. Amaranth hydrolysate-enriched cookies have antihypertensive effects. Protein fraction F1-F6 extracted from defatted flour of *Amaranthus hypochondriacus* was also found to have antihypertensive effects. This shows blood pressure lowering effects due to renin inhibition [75].

Alcalase and trypsin-hydrolyzed quinoa bran has potent antihypertensive properties because of the presence of the peptide sequence RGQVIYVL (946.6 Da) [76]. *Lactobacillus casei* inoculated hydrolysate showed ACE inhibitory activity due to the presence of five peptide sequences LGGIWHL, VAHPVF, IRAMPVAV, ALFPTHR, and LAHMIVAGA [77]. Quinoa protein hydrolysate (QPH) produced by the gastrointestinal digestion of quinoa proteins, reduces SBP and diastolic blood pressure (DBP) in SHRs after 2–10 h of oral administration [61]. Three promising bioactive peptides, FHPFPR, NWFPLPR, and NIFRPF, were obtained from QPH and were further examined to have ACE-inhibition effects with IC_{50} values of 34.92, 16.77, and 32.40 μM, respectively [61].

Soybean

Soyabean is highly rich in protein content and relatively low in fats and cholesterol, compared to animal protein sources. Numerous bioactive peptides have been identified in soya bean having various physiologically beneficial properties, like lipid-lowering, antidiabetic, anticancer, antihypertensive, anti-inflammatory, and antioxidant activities.

A source of soybean-based antihypertensive peptides is soybean-based infant formula [78]. Three types of fragment proteins were isolated from this infant formulation having molecular weights of 3 kDa, 3–5 kDa, and 5–10 kDa. Peptide RPSYT, synthesized from the 5kDa fraction has the highest ACE inhibition activity [78]. Proteins obtained from soy are high in nitrogen content. Fermentation of soy protein with *Lactobacillus casei spp. pseudo plantarum*, has given two ACE inhibitory protein fractions F2 and F3. F2 fraction appeared to be a more potent ACE inhibitor than the F3 fraction. Two peptides, LIVTQ and LIVT, were synthesized from soy proteins by fluorenyl methyloxycarbonyl solid-phase peptide synthesis (SPPS). Peptides LIVTQ and LIVT showed ACE inhibition activity with IC_{50} values of 0.087 and 0.110 μM, respectively [79].

Fermentation of soybean milk produces protein fractions with biological activity. Two strains of *Lactobacillus casei* – CICC 20280 and CICC 23184 – were used in the fermentation of soybean milk. The ACE inhibition of these two fermented milks were tested and the results showed ACE inhibition with IC_{50} values of 1.13 and 0.89 mg/mL for CICC 20280 and CICC 23184, respectively [80]. The peptide and GABA content for CICC 20280 and CICC 23184 was found to be 3.97 ± 0.67 mg/m, 5.17 ± 0.22 mg/m and 21.71± 0.36mg/mL, 1.57 ± 0.21 mg/ml, respectively [80]. Peptide content in CICC 23184 was greater where GABA content in CICC 20280 was greater. CICC 23184 reduces systolic and diastolic blood pressure by 91.7 ± 5.9 and 74.2 ± 9.1 mmHg, respectively, whereas CICC 20280 reduces SBP and DBP by 122.4 ± 8.6 and 111.8 ± 6.4 mmHg, respectively [80]. It was also found that soy pulp (okara) extract showed ACE inhibitory activity in a concentration dependent manner and 7μl of the extract had efficacy as much as 5nM of captopril [81].

Two peptides, SY and GY, were identified by fragmented soybean screening. They decrease blood pressure in SHR by lowering the aldosterone level [82]. Peptides (VNP, LEPP, and WNPR, isolated after protease hydrolysis of soy milk) were reported to have ACE inhibitory potential [83]. βCG of soy, isolated after enzymatic hydrolysis have also been reported to exert antihypertensive effects in an *in-vitro* study.

Old-style Asian fermented soy foods like soybean paste, soy sauce, natto, and tempeh are rich in ACE inhibitory peptides [84]. Soy peptides like VAHINVGZK and YVWK (obtained from fermentation of soy proteins after fermenting with *Bacillus natto* or *Bacillus subtilis* and other peptides like PGTAVFK, IVF, LLF, LNF, LSW, LEF, YVVFK, IPPGVPYWT, PNNKPFQ, NWGPLV, and TPRVF were found to have antihypertensive properties [58]. LAIPVNKP and LPHF peptides in soy βCG and VLIVP, SPYP, and WL peptides in glycinin also have antihypertensive properties [55].

Peptide NWGPLV obtained from glycinin (A4 and A5) after enzymatic hydrolysis, shows antihypertensive activity in SHR models [85]. Two peptides, NWGPLV and TPRVF, were also examined to show antihypertensive effects by reducing the blood pressure in SHR [86].

Fruits and Vegetables

Fruits and vegetables are considered poor sources of proteins with content of (0.5–1.1)% and (0.2–3.9)%, respectively. Bioavailability of proteins obtained from fruits is quite low as they are mostly present in the seeds. Most research has been done to extract bioactive peptides from the peel of fruits. Peptides isolated from the pomegranate peel show antihypertensive activities of 75 ± 8 g/ml and 49 ± 3 g /ml, respectively [87]. After the enzymatic hydrolysis of asparagus with alkalis, two amino acid sequences PDWFLLL and ASQSIWLPGWL were found. They showed ACE inhibitory activity value of 1.76 μmol/L and 4.02 μmol/L, respectively, thus possessing antihypertensive activity [64]. Another study was done with virgin olive oil extracts using SHR models. When virgin olive oil (unfiltered) extracts were administered to SHR, a reduction in average blood pressure by 10 mmHg at 4 h (P < 0.01) and of 20 mmHg at 6 h was obtained for an initial blood pressure of 203.8 ± 1.8 mmHg [88]. Two antihypertensive peptides (APYDPDWYYIR and SKGFTSPLF) have been extracted from the stems and leaves of cauliflower – of which APYDPDWYYIR was a more potent ACE inhibitor [65]. Similarly, sweet potato contains various antihypertensive bioactive peptides which mainly include VSAIW, AIWGA, FVIKP, VVMPSTF, and FHDPMLR [89].

Legumes

Legumes (fabaceae) are high in protein, fiber, minerals and vitamins. Pepsin and pancreatin-treated pea protein hydrolysate exhibit 61.82% of ACE inhibition and pancreatin-treated pea protein hydrolysate exhibits 14.28% renin inhibition in SHRs, showing a reduction in systolic blood pressure values [90]. The peptides TVGMTAKF and QLLLQQ among the 12 peptides (TVGMTAKF, TVGMTAFK, FQQVPGPV, FQQPVVGPA, QLLLQQ, LLYQEPVLGPVR, VTSTGPVGH, LSAGGVGL, WGAFGK, QVAAAETR, GVGTGKPGER, and KENMLDK) isolated from flour produced from horse gram (lentils) show ACE inhibitory activity with IC_{50} values of 30.3±2.3 µM and 75.0±4.2 µM, respectively. They are synthesized by hydrolysis with alcalase [91]. Velvet bean extract was hydrolyzed differently with alcalase and flavoenzyme showing ACE inhibitory activity [92].

Peptide sequences GLTSK and GEGSGA, isolated from the non-digestible fractions of common bean (*Phaseolus vulgaris*), show antihypertensive activity. GLTSK and GEGSGA reduces the conversion of AGT to angiotensin I (AngI) by 38 and 28%, respectively, and molecular docking suggests that their mechanism of action is mainly through the interaction with the catalytic site of renin, the AngI-converting enzyme, and the AngII receptor, by hydrogen bonds, polar, hydrophobic, and cation-π interactions [93]. Corn silk bioactive peptides (CSBps) show antihypertensive effects. CSBps5 peptide with a sequence of SKFDNLYGCR (1258 Da), shows ACE inhibitory activity. The IC_{50} values measured were 44.11 ± 1.04 µM and 81.71 ± 1.06 µM for substrate proteins H-HL and Z-FHL, respectively. After docking analysis, it was seen that CSBps5 might inhibit AngI from joining to the catalytic pocket of ACE [94]. *Moringa oleifera* (Drumstick tree) originated bioactive peptides show antihypertensive effects by blocking the active sites of ACE, thus, blocking the action of ACE enzyme [95]. Peptide FHAPWK, isolated from thermolysin hydrolysate of *Cassia obtusifolia* (sicklepod) seeds, shows ACE inhibitory activity with IC50 values of 16.83 ± 0.90 µM [68]. Cowpea (*Vigna unguiculata*) extracts show antihypertensive activity. Alcalase hydrolysis of Mung bean (*Vigna radiata*) produces three fractions MBPHs-I (<3 kDa), MBPHs-II (3-10 kDa), and MBPHs-III (>10 kDa) of proteins. Amongst these three fractions, MBPHs-I was found to be the most potent ACE inhibitor with IC50 values of 4.66 µg/mL [96]. Peptides were derived from mung bean vicilin protein (MBVP). Alcalase-generated mung bean protein hydrolysate (AMBPH) and trypsin-generated mung bean protein hydrolysate (TMBPH) both have shown antihypertensive activity. MBVP, AMBPH, and TMBPH show ACE inhibitory activity with IC_{50} values of 0.66 mg/ml, 0.32 mg/ml, 0.54 mg/ml, respectively [97].

Sesame protein hydrolysate was obtained after hydrolysis of sesame seed protein isolate (SESPI) with pepsin and pancreatin. It shows ACE inhibition (<1 kDa peptide fraction) and renin inhibition (bigger peptides >3–5 and 5–10 kDa) activity [98]. Two peptides, LGRNLPPI (879.06 Dalton) and GPAGPAGL (639.347 Dalton), were obtained from defatted lemon basil seeds (DLBS) (hydrolysis of lemon basil seed oils with alcalase) having ACE inhibitory activity with IC_{50} values of 0.124 ± 0.02 mM and 0.013 ± 0.001 mM, respectively [66]. Non-germinated and germinated sorghum displayed ACE inhibition activity with an inhibitory percentage of 46.38 ± 2.21% and 15.91 ± 1.65%, respectively [99]. Crude hydrolysate of pea protein PPHT, obtained by hydrolysis of pea protein with thermoase, shows strong ACE inhibitory activity. Bowman-Birk family peptides (found in high quantities in leguminous seeds) decrease the BP in Wistar rats and SHR [100]. Pigeon pea (*Cajanus cajan*) (high in protein content of approximately 24%), is also a source of antihypertensive peptides. Pepsin-pancreatin-hydrolyzed pea protein (PPHPp) was obtained after pepsin and pancreatin hydrolysis of pigeon pea. PPHPp consists of antihypertensive peptides, which reduce systolic blood pressure (−34.6 mmHg) in SHR [101].

ISOLATION OF ANTIHYPERTENSIVE PEPTIDES

Bioactive peptides are peptide sequences of specific length having beneficiary effects on human health. They have various activities in human bodies like antihypertensive, immunomodulatory, antioxidative, anticancer, antidiabetic, antimicrobial and antithrombotic activities. There are numerous procedures for extracting peptides from proteins.

Enzymatic Hydrolysis

In this method, various bioactive hydrolysates are produced. This is done by mixing the protein source with various enzymes (alcalase, achymotrypsin, bromelain, cryotin F, flavourzyme, neutrase, orientase, papain, pepsin, pancreatin, pronase, protamex, protease N, protease A, thermolysin, trypsin, or validase). Many ACE inhibitory peptides have been produced by using gastrointestinal peptides like pepsin and trypsin.

Enzymes from plant sources (papain), animal sources (pepsin and trypsin) and microbial sources (thermolysin) have been used to isolate antihypertensive peptides. At the beginning of this method, proteins are hydrolyzed using one or more enzymes. During this step the temperature and pH is maintained at an optimal level. Temperature is maintained by thermally insulating the instruments and pH is maintained by the addition of various buffers. The obtained protein concentrates are then treated with four dissimilar proteolytic enzymes like alcalase, savinase, protamex, and corolase for various time durations. Then their ACE inhibitory and antioxidant properties are evaluated. Peptides with the highest ACE-inhibitory (IC_{50} = 0.18 mg/ml) and antioxidant activity (1.22 μmol of Trolox equivalent/mg of protein) were obtained after hydrolysis of hydrolysate by savinase for 2 h. Hydrolysis of Grass carp skin pieces with various enzymes, such as alcalase (60°C and pH 8.0), proteinase K (37°C and pH 8.0), collagenase 140 (37°C and pH 7.5), trypsin (after adjustment of the pH to 8.0), harvested bioactive peptides with antioxidant and antihypertensive activity [102].

Peptides LER and GAG were obtained from the muscle proteins of Black-bone silky fowl (*Gallus gallus domesticus* Brisson). Alcalase and papain were used to isolate those peptides from muscle proteins. ACE inhibitory peptides were also identified in beef *M. longissimus*. The hydrolysis is by thermolysin which, after fractionation, leads to the formation of 7 different fractions. Fraction V showed a maximum antihypertensive effect. Consequently, sub-fractionation of those fractions produces peptides like LSW, FGW, and WRQ. They show strong ACE inhibition activity with IC_{50} values of 0.89, 2.69, and 3.09 mM, respectively.

Fermentation

Fermentation is a process were the fermenting microbes (bacteria and yeast species) secret enzymes into their extracellular medium during their growth. Thus, when proteins are incubated with these microbes, they undergo proteolytic degradation and peptides are released into the medium. The bioactive peptides obtained from milk are mainly produced by this fermentation process (lactic acid bacteria). They can also be produced by hydrolysis with gastric enzymes. To augment the synthesis of these bioactive peptides, enzymatic digestion of them is done. Most of the commercially produced milk isolated antihypertensive agents are produced by fermenting them with *Lactobacillus helveticus*. Thus, fermentation is a very useful tool to produce functionally active milk products by fermentation with the help of lactic acid bacteria. These products can serve the role of antihypertensive agents [102].

Genetic Engineering

Recombinant DNA technique is also used extensively to produce antihypertensive amino acid sequences. Genetic engineering gives a better yield, and the manufacturing cost is also less compared to enzymatic hydrolysis. However, this method is not devoid of shortcomings. Undesirable expression of other peptides can be lethal for the host. To solve this problem separation of peptide of need is isolated or separated from other peptides by the application of enzymes which degrade or make other peptides non-functional. Antihypertensive peptides SLVYPFPGPI, NIPPLTQTPV, and DKIHPF were expressed by *Escherichia coli* successfully. Antihypertensive peptides expressed in the *Lactobacillus plantarum* NC8 181 strain by an inducible vector pSIP-409 showed ACE inhibitory activity. They were discovered from tuna frame protein and yellow fin sole frame protein. First *Lactobacillus plantarum* NC8 (RLP) was administered orally to SHR. Then, the expression of AngII was decreased, due to which the blood pressure in the examined animal was decreased. VLVPV peptide was discovered by the recombinant expression in modified transplastomic *Chlamydomonas reinhardtii*, were isolated by proteolytic cleavage, and purified by HPLC [102].

Downstream Processing

Downstream processing refers to the purification and recovery of peptide sequences. Usually, peptides with biological activity are extracted from the hydrolysates using various physical methods (like temperature assisted, solvent assisted and physical methods). The main drawback of these methods is that they are time- and energy-intensive, and most of the peptides can easily be destroyed by using elevated temperatures and solvents for extraction. Recent research is focused on development of some novel downstream techniques which are more time- and cost-effective and which can preserve the peptide of interest. In this section we are going to discuss various processes which are developed for this purpose. These novel techniques are employed in the case of cell structure digestion, as it helps in faster digestion. The enzymatic and novel techniques are used to speed up the digestion process for the extraction of biopeptides from a complex matrix.

High Pressure Processing

High pressure assisted extraction is a new technique using high pressure (generally 800 MPa but sometimes up to 1000 MPa) for improvement of extraction. After using high pressure, cells become more permeable due to large pressure difference between inner and outer parts of the cell membrane. This augmentation in the porosity of cells results in better penetration of the solvent, higher dissolution rates and increase in the rate of mass transfer. Thus, a high pressurized system enables isolation of bioactive peptides from numerous types of sources. An augmentation in proteolysis reaction and difference in the normal proteolysis pattern was observed in case of extractions, done under high pressure. This is because of reduction in binding properties of those intact proteins. This is observed in the case of β-lactoglobulins and ovalbumin. High pressure treatment of egg proteins derived from the white part at 800 MPa caused a superior susceptibility to pepsin digestion compared to temperature-assisted extraction at 95°C. A proteolytic treatment at 300 MPa results in deprivation of lentil proteins while increasing the number of peptides production [103].

Ultrasound Processing

The application of ultrasound (above 20 kHz) has been extensively tested to increase the enzymatic reaction or extraction yield. There are two main types of ultrasound devices. The first is an ultrasonic water bath and the second is an ultrasonic probe system with horn transducers. Ultrasonic water baths are inexpensive and are commonly used in laboratory settings. An ultrasonic probe system with horn transducers (used in batch or continuous mode) introduces vibrations directly into the sample. The main sound energy that is detected in ultrasound is the sound chamber. The phenomenon of the formation, diffusion, and induction of microbubbles in ultrasonically radiated fluids is called "sound cavitation". The formation and collapse of the chamber bubbles create macro-turbulence, high-velocity mid-cell collisions, and stress on the organism's microporous cells, causing the matrix to decompose. It breaks down the surface crust, while incisions and cells help to release bioactive compounds from the biological matrix, thereby increasing the extraction efficiency by enhancing mass transfer. Improving the enzymatic activity of amylose, glucose oxidase, cellulose and dextranase has been previously reported. Low-intensity ultrasound increases the reaction rate and alters the structure and modification of the α-helix and β-sheet fractions, increasing catalytic and specific stability [104].

Microwave-Assisted Extraction

Microwave-assisted extraction (MAE) involves the use of electromagnetic radiation at frequencies from 300 MHz to 300 GHz, which heats the solution in contact with a sample via dipolar rotation and an ionic conductor, thus, damaging the sample matrix. The use of microwave energy in the extraction process prevents weak hydrogen atoms, develops solvent infiltrations into the matrix, and facilitates solution. MAE has been reported to increase the yield of solid liquid extraction as well as the extraction of bioactive peptides from various matrices. However, the use of microwave aided extraction can improve the antioxidant capacity of bioactive polysaccharides from *Oricularia* anticoagulant (AAP). MAE has been used to extract bioactive sulfate polysaccharides from seaweed, including *Fucus vesiculosus* and *Ascophyllum nodosum* [105].

SUPERCRITICAL FLUID EXTRACTION

The efficiency of the traditional solid liquid extraction process can be improved by applying pressure and temperature to increase the extraction yield. The solubility properties, including density, variability and viscosity, are controlled by the application of pressure and temperature. This allows the use of environmentally friendly solvents such as water. Applying pressure and temperature causes the solvent to penetrate into the cell and to inhibit the cellular matrix. Supercritical CO_2 has a low viscosity. This improves diffusivity and extraction yield. Temperature and pressure to produce a soluble supercritical fluid and its use in the bioactive peptide extraction process are well documented. Some examples of supercritical fluid extraction processes are pressurized liquid extraction or rapid solvent extraction, high pressure solvent extraction, pressurized fluid extraction, and enhanced solvent extraction.

ROLE OF BIOACTIVE PEPTIDE ANTIHYPERTENSIVE AGENT

ACE enzyme inhibition is the main target of bioactive peptides to work as antihypertensive agents (Figure 18.1). The renin–angiotensin–aldosterone system (RAAS) plays the central role in maintenance of blood pressure in the human body. ACE, a heavily glycosylated membrane-bound zinc metalloprotease serves a central role in RAAS [106]. There are mainly two systems by which blood pressure is regulated: RAAS and the kinin-nitric oxide system [183]. RAAS starts with renin release from the JG (juxtaglomerular) cells of the kidney in response to various stimuli which command the body to increase the blood pressure. Renin then converts the angiotensinogen to Ang-I or AG-I. Angiotensinogen is

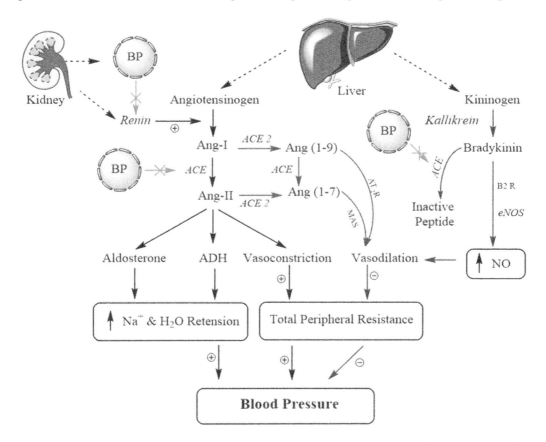

FIGURE 18.1   Schematic representation of antihypertensive targets of bioactive peptides.

an inactive dipeptide which does not exert any biological activity. Renin converts this to an active form, that is, AG-I. Then ACE, a membrane-bound enzyme, converts Ang-I to Ang-II. Then this AG-II acts on the AT-I receptor, leading to vasoconstriction, aldosterone release, and increased sodium absorption from the urine. The result is an increase in blood pressure i.e., hypertension. Thus, AG-II acts as a vasoactive peptide, which is responsible for vasoconstriction and rise of blood pressure [106]. There are also some hypotensive peptides that exert vasorelaxant action. Bradykinin is a hypotensive peptide which acts on BKB2 receptors, increasing NO release in the vascular endothelium. NO causes vasodilation and reduces the blood pressure. Ang-II also deactivates hypotensive peptides like bradykinin and kallidin, thus increasing blood pressure [106]. Many antihypertensive bioactive peptides decrease the blood pressure by increasing the production of vasodilatory factors, like NO and prostaglandins, and by decreasing the production of vasoconstrictor agents, like endothelin-1. Renin is an enzyme which cleaves dipeptide from angiotensinogen (inactive) to yield Ang-I (active); this in turn decreases the Ang-II level [107]. Thus, inhibition of renin decreases the production of AngII, which ultimately decreases the blood pressure.

ACE 2 is another peptidase enzyme present in the RAAS system, which hydrolyzes both Ang-I and Ang-II. But the affinity toward Ang-I is greater than Ang-II. Hydrolysis of Ang-I and Ang-II by ACE-2 produces Ang(1-7), which is a heptapeptide. Ang(1-7) binds to Mas-1 receptor (also known as MasR). This makes the ACE 2/Ang 1-7/MasR axis. Induction of this axis results in the decrease of blood pressure by increased production of vasodilatory factors [108]. Many bioactive peptides also directly increase the expression of ACE 2, thus inducing the ACE 2/Ang 1-7/MasR axis, which reduces blood pressure. Many antihypertensive peptides also directly increase endothelial NO release.

Bioactive peptides derived from milk after digestion with endogenous enzymes (PYVRYL, LVYPFTGPIPN), protease from *Enterococcus faecalis* (LHLPLPL, αs1-CN, αs1-CN, and αS2-CN) or proteinase of *Lactobacillus helveticus* PR4 (Bovine αS1-casein; αS1-CN, and β-CN; ovine αS1-CN and αS2-CN, caprine β-CN and αS2-CN; buffalo β-CN) show ACE inhibitory activity [21, 22, 23, 109]. Antihypertensive peptides isolated from meat after enzymatic digestion show ACE inhibitory activity [27]. Marine sources of antihypertensive peptides are from seaweed, microalgae, yellowfin tuna, skipjack tuna, Alaska pollack skin, chum salmon, pnk salmon, skate skin, small-spotted catshark, pelagic thresher, marine shrimp, izumi shrimp, jellyfish, sipuncula, pearl oyster, and Pacific cod [16–18]. Peptide sequences with ACE inhibitory activity isolated from crude wheat germ protein [62] are the peptides FNVPSRYGIY and PWHNPRQGGF, and SPFWNINA is the peptide from rice [71]. Peptides PGTAVFK, IVF, and LLF, isolated from soybean [55, 110], also serve as an ACE inhibitors. Protein fraction F1-F6, extracted from defatted flour of *Amaranthus hypochondriacus*, and peptides, such as IRLIIVLMPILMA, have shown antihypertensive effects by renin inhibition, thus reducing the blood pressure by increasing Ang-II [60, 75]. Many amaranth peptides also decrease the degradation of bradykinin. Thus, they induce NO release and cause vasodilation. Some bioactive peptides, like ADVFNPR, VVLYK, LPILR, and VIGPR, reduce blood pressure by decreasing vasoconstriction by reduction of endothelia-1 levels [111]. IRW isolated from egg white protein ovotransferrin decreases blood pressure by increasing ACE2 expression [50]. Glutelin-hydrolyzed amaranth peptides – AY, FP, GY, MY, PR, VF, VW, VY – increase endothelial NO production, thus acting as a blood pressure-lowering agent [60].

CONCLUSION AND FUTURE PERSPECTIVES

Hypertension is a serious health hazard and one of the most common disease conditions associated with the modern way of living with stress and unhealthy conditions. Awareness should be spread about hypertension, as it can lead to increased cardiovascular risk and other organ damage. Many synthetic drugs are now in use, but they are associated with acute or chronic side effects. Use of food and food-derived products in the treatment of hypertension is significant, as it can be a potential and side effect-free alternative to synthetically derived therapeutic agents. This report provides an insight about various food-derived bioactive peptides of animal and plant origin and various downstream processing techniques for isolation of those peptides.

Identification of bioactive peptides in food has given a new direction to functional food development processes and nutraceuticals. However more research is needed to understand their pharmacokinetics. Food-derived bioactive peptides have great potential for use as alternatives to chemical agents in the treatment of antihypertension. But more research is needed to identify various peptides, which act on various potential biological targets to reduce blood pressure, so that treatment can be diversified.

REFERENCES

1. Rizzello CG, Tagliazucchi D, Babini E, Rutella GS, Saa DL, Gianotti A. Bioactive peptides from vegetable food matrices: Research trends and novel biotechnologies for synthesis and recovery. *J Funct Foods* 2016 Dec 1;27:549–69.
2. Abdel-Hamid M, Otte J, De Gobba C, Osman A, Hamad E. Angiotensin I-converting enzyme inhibitory activity and antioxidant capacity of bioactive peptides derived from enzymatic hydrolysis of buffalo milk proteins. *Int Dairy J* 2017 Mar 1;66:91–8.
3. Koyama M, Hattori S, Amano Y, Watanabe M, Nakamura K. Blood pressure-lowering peptides from neo-fermented buckwheat sprouts: A new approach to estimating ACE-inhibitory activity. *PLOS ONE* 2014 Sep 15;9(9):e105802.
4. Baptiste DL, Hamilton JB, Foronda C, Sloand E, Fahlberg B, Pfaff T, Delva S, Davidson PM. Hypertension among adults living in Haiti: An integrative review. *J Clin Nurs* 2018 Jul;27(13–14):2536–45.
5. Berfenstam R, Jagenburg R, Mellander O. Protein hydrolysis in the stomachs of premature and full-term infants. *Acta Paediatr* 1955 Jul;44(4):348–54.
6. Baptiste DL, Hamilton JB, Foronda C, Sloand E, Fahlberg B, Pfaff T, Delva S, Davidson PM. Hypertension among adults living in Haiti: An integrative review. *J Clin Nurs* 2018 Jul;27(13–14):2536–45.
7. Park YW, Nam MS. Bioactive peptides in milk and dairy products: A review. *Korean J Food Sci Anim Resour* 2015;35(6):831.
8. Chen Y, Liu W, Xue J, Yang J, Chen X, Shao Y, Kwok LY, Bilige M, Mang L, Zhang H. Angiotensin-converting enzyme inhibitory activity of *Lactobacillus helveticus* strains from traditional fermented dairy foods and antihypertensive effect of fermented milk of strain H9. *J Dairy Sci* 2014 Nov;97(11):6680–92.
9. Elkhtab E, El-Alfy M, Shenana M, Mohamed A, Yousef AE. New potentially antihypertensive peptides liberated in milk during fermentation with selected lactic acid bacteria and kombucha cultures. *J Dairy Sci* 2017 Dec;100(12):9508–20.

10. Ibrahim HR, Ahmed AS, Miyata T. Novel angiotensin-converting enzyme inhibitory peptides from caseins and whey proteins of goat milk. *J Adv Res* 2017 Jan;8(1):63–71.
11. Vincenzetti S, Pucciarelli S, Polzonetti V, Polidori P. Role of proteins and of some bioactive peptides on the nutritional quality of donkey milk and their impact on human health. *Beverages* 2017 Sep;3(3):34.
12. Osman A, El-Hadary A, Korish AA, AlNafea HM, Alhakbany MA, Awad AA, Abdel- Hamid M. Angiotensin-I converting enzyme inhibition and antioxidant activity of papain-hydrolyzed camel whey protein and its hepato-renal protective effects in thioacetamide-induced toxicity. *Foods* 2021 Feb 20;10(2):468.
13. Begunova AV, Savinova OS, Glazunova OA, Moiseenko KV, Rozhkova IV, Fedorova TV. Development of antioxidant and antihypertensive properties during growth of *Lactobacillus helveticus*, *Lactobacillus rhamnosus* and *Lactobacillus reuteri* on cow's milk: Fermentation and peptidomics study. *Foods* 2020 Dec 23;10(1):17.
14. Adams C, Sawh F, Green-Johnson JM, Jones Taggart H, Strap JL. Characterization of casein-derived peptide bioactivity: Differential effects on angiotensin-converting enzyme inhibition and cytokine and nitric oxide production. *J Dairy Sci* 2020 Jul;103(7):5805–15.
15. Hussein FA, Chay SY, Ghanisma SBM, Zarei M, Auwal SM, Hamid AA, Ibadullah WZW, Saari N. Toxicity study and blood pressure-lowering efficacy of whey protein concentrate hydrolysate in rat models, plus peptide characterization. *J Dairy Sci* 2020 Mar;103(3):2053–64.
16. Pangestuti R, Kim SK. Bioactive peptide of marine origin for the prevention and treatment of non-communicable diseases. *Mar Drugs* 2017 Mar 9;15(3):67.
17. Liu X, Zhang M, Jia A, Zhang Y, Zhu H, Zhang C, Sun Z, Liu C. Purification and characterization of angiotensin I converting enzyme inhibitory peptides from jellyfish Rhopilema esculentum. *Food Res Int* 2013 Jan 1;50(1):339–43.
18. Je JY, Park PJ, Byun HG, Jung WK, Kim SK. Angiotensin I converting enzyme (ACE) inhibitory peptide derived from the sauce of fermented blue mussel, *Mytilus edulis*. *Bioresour Technol* 2005 Sep 1;96(14):1624–9.
19. Suetsuna K, Chen JR. Identification of antihypertensive peptides from peptic digest of two microalgae, *Chlorella vulgaris* and *Spirulina platensis*. *Mar Biotechnol (NY)* 2001 Jul;3(4):305–9.
20. Tsai JS, Lin TC, Chen JL, Pan BS. The inhibitory effects of freshwater clam (*Corbicula fluminea*, Muller) muscle protein hydrolysates on angiotensin I converting enzyme. *Process Biochem* 2006 Nov 1;41(11):2276–81.
21. Quirós A, Hernández-Ledesma B, Ramos M, Amigo L, Recio I. Angiotensin-converting enzyme inhibitory activity of peptides derived from caprine kefir. *J Dairy Sci* 2005 Oct;88(10):3480–7.
22. Minervini F, Algaron F, Rizzello CG, Fox PF, Monnet V, Gobbetti M. Angiotensin I-converting-enzyme-inhibitory and antibacterial peptides from *Lactobacillus helveticus* PR4 proteinase-hydrolyzed caseins of milk from six species. *Appl Environ Microbiol* 2003 Sep;69(9):5297–305.
23. del Mar Contreras M, Sanchez D, Sevilla MÁ, Recio I, Amigo L. Resistance of casein-derived bioactive peptides to simulated gastrointestinal digestion. *Int Dairy J* 2013 Oct 1;32(2):71–8.
24. Ruiz JÁ, Ramos M, Recio I. Angiotensin converting enzyme-inhibitory activity of peptides isolated from Manchego cheese. Stability under simulated gastrointestinal digestion. *Int Dairy J* 2004 Dec 1;14(12):1075–80.

25. Saito T, Nakamura T, Kitazawa H, Kawai Y, Itoh T. Isolation and structural analysis of antihypertensive peptides that exist naturally in Gouda cheese. *J Dairy Sci* 2000 Jul;83(7):1434–40.
26. Fan H, Wu J. Purification and identification of novel ACE inhibitory and ACE2 upregulating peptides from spent hen muscle proteins. *Food Chem* 2021 May 30;345:128867.
27. Bauchart C, Rémond D, Chambon C, Patureau Mirand P, Savary-Auzeloux I, Reynès C, Morzel M. Small peptides (<5kDa) found in ready-to-eat beef meat. *Meat Sci* 2006 Dec;74(4):658–66.
28. Hall F, Reddivari L, Liceaga AM. Identification and characterization of edible cricket peptides on hypertensive and glycemic in vitro inhibition and their anti-inflammatory activity on RAW 264.7 macrophage cells. *Nutrients* 2020 Nov 23;12(11):3588.
29. Khueychai S, Jangpromma N, Choowongkomon K, Joompang A, Daduang S, Vesaratchavest M, Payoungkiattikun W, Tachibana S, Klaynongsruang S. A novel ACE inhibitory peptide derived from alkaline hydrolysis of ostrich (*Struthio camelus*) egg white ovalbumin. *Process Biochem* 2018 Oct 1;73:235–45.
30. Zhang B, Liu J, Liu C, Liu B, Yu Y, Zhang T. Bifunctional peptides with antioxidant and angiotensin-converting enzyme inhibitory activity in vitro from egg white hydrolysates. *J Food Biochem* 2020 Sep;44(9):e13347.
31. Vázquez JA, Menduíña A, Nogueira M, Durán AI, Sanz N, Valcarcel J. Optimal production of protein hydrolysates from monkfish by-products: Chemical features and associated biological activities. *Molecules* 2020 Sep 6;25(18):4068.
32. Chen J, Ryu B, Zhang Y, Liang P, Li C, Zhou C, Yang P, Hong P, Qian ZJ. Comparison of an angiotensin-I-converting enzyme inhibitory peptide from tilapia (*Oreochromis niloticus*) with captopril: Inhibition kinetics, in vivo effect, simulated gastrointestinal digestion and a molecular docking study. *J Sci Food Agric* 2020 Jan 15;100(1):315–24.
33. Escudero E, Mora L, Fraser PD, Aristoy MC, Arihara K, Toldrá F. Purification and Identification of antihypertensive peptides in Spanish dry-cured ham. *J Proteomics* 2013 Jan 14;78:499–507.
34. Ren Y, Wan DG, Lu XM, Chen L, Zhang TE, Guo JL. Isolation and characterization of angiotensin I-converting enzyme inhibitor peptides derived from porcine hemoglobin. *Sci Res Essays* 2011 Dec 9;6(30):6262–9.
35. Saiga A, Tanabe S, Nishimura T. Antioxidant activity of peptides obtained from porcine myofibrillar proteins by protease treatment. *J Agric Food Chem* 2003 Jun 4;51(12):3661–7.
36. Choe J, Seol KH, Kim HJ, Hwang JT, Lee M, Jo C. Isolation and identification of angiotensin I-converting enzyme inhibitory peptides derived from thermolysin-injected beef M. longissimus. *Asian-Australas J Anim Sci* 2019 Mar;32(3):430–6.
37. Jang A, Jo C, Kang KS, Lee M. Antimicrobial and human cancer cell cytotoxic effect of synthetic angiotensin-converting enzyme (ACE) inhibitory peptides. *Food Chem* 2008 Mar 1;107(1):327–36.
38. Saiga AI, Iwai K, Hayakawa T, Takahata Y, Kitamura S, Nishimura T, Morimatsu F. Angiotensin I-converting enzyme-inhibitory peptides obtained from chicken collagen hydrolysate. *J Agric Food Chem* 2008 Oct 22;56(20):9586–91.
39. Pratiwi A, Hakim TR, Abidin MZ, Fitriyanto NA, Jamhari J, Rusman R, Erwanto Y. Angiotensin-converting enzyme inhibitor activity of peptides derived from Kacang goat skin collagen through thermolysin hydrolysis. *Vet World* 2021 Jan;14(1):161–7. doi: 10.14202/vetworld.2021.161-167.

40. Mas-Capdevila A, Iglesias-Carres L, Arola-Arnal A, Aragonès G, Aleixandre A, Bravo FI, Muguerza B. Evidence that nitric oxide is involved in the blood pressure lowering effect of the peptide AVFQHNCQE in spontaneously hypertensive rats. *Nutrients* 2019 Jan 22;11(2):225.
41. Lee S-j, Kim Y-s, Kim S-e, Kim E-k, Hwang J-w, Park T-k, Kim BK, et al. Purification and characterization of a novel angiotensin-I converting enzyme (ACE) inhibitory peptide derived from enzymatic hydrolysate of grass carp protein. *Peptides* 2012;33(1):52–58.
42. Wijesekara I, Qian ZJ, Ryu B, Ngo DH, Kim SK. Purification and identification of antihypertensive peptides from seaweed pipefish (*Syngnathus schlegeli*) muscle protein hydrolysate. *Food Res Int* 2011 Apr 1;44(3):703–7.
43. Karami Z, Peighambardoust SH, Hesari J, Akbari-Adergani B, Andreu D. Identification and synthesis of multifunctional peptides from wheat germ hydrolysate fractions obtained by proteinase K digestion. *J Food Biochem* 2019 Apr;43(4):e12800.
44. Suetsuna K. Purification and identification of angiotensin I-converting enzyme inhibitors from the red alga *Porphyra yezoensis*. *J Mar Biotechnol* 1998 Aug;6(3):163–7.
45. Zhu CF, Li GZ, Peng HB, Zhang F, Chen Y, Li Y. Effect of marine collagen peptides on markers of metabolic nuclear receptors in type 2 diabetic patients with/without hypertension. *Biomed Environ Sci* 2010 Apr;23(2):113–20.
46. He YY, Li TT, Chen JX, She XX, Ren DF, Lu J. Transport of ACE inhibitory peptides Ile-Gln-Pro and Val-Glu-Pro derived from *Spirulina platensis* across Caco-2 monolayers. *J Food Sci* 2018 Oct;83(10):2586–92.
47. Carrizzo A, Conte GM, Sommella E, Damato A, Ambrosio M, Sala M, Scala MC, Aquino RP, De Lucia M, Madonna M, Sansone F, Ostacolo C, Capunzo M, Migliarino S, Sciarretta S, Frati G, Campiglia P, Vecchione C. Novel potent decameric peptide of Spirulina platensis reduces blood pressure levels through a PI3K/AKT/eNOS-dependent mechanism. *Hypertension* 2019 Feb;73(2):449–57.
48. Ramos-Romero S, Torrella JR, Pagès T, Viscor G, Torres JL. Edible microalgae and their bioactive compounds in the prevention and treatment of metabolic alterations. *Nutrients* 2021 Feb 9;13(2):563.
49. Abachi S, Bazinet L, Beaulieu L. Antihypertensive and angiotensin-I-converting enzyme (ACE)-inhibitory peptides from fish as potential cardioprotective compounds. *Mar Drugs* 2019 Oct 29;17(11):613.
50. Wu J. A novel angiotensin converting enzyme 2 (ACE2) activating peptide: A reflection of 10 years of research on a small peptide Ile-Arg-Trp (IRW). *J Agric Food Chem* 2020 Dec 9;68(49):14402–8.
51. Dávalos A, Miguel M, Bartolomé B, López-Fandiño R. Antioxidant activity of peptides derived from egg white proteins by enzymatic hydrolysis. *J Food Prot* 2004 Sep;67(9):1939–44.
52. Yu Z, Liu B, Zhao W, Yin Y, Liu J, Chen F. Primary and secondary structure of novel ACE-inhibitory peptides from egg white protein. *Food Chem* 2012 Jul 15;133(2):315–22.
53. You SJ, Udenigwe CC, Aluko RE, Wu J. Multifunctional peptides from egg white lysozyme. *Food Res Int* 2010 Apr 1;43(3):848–55.
54. Zóia Miltenburg T, Uana da Silva M, Bosch G, Vasconcellos RS. Effects of enzymatically hydrolyzed poultry byproduct meal in extruded diets on serum angiotensin-converting enzyme activity and aldosterone in cats. *Arch Anim Nutr* 2021 Feb;75(1):64–77.

55. Chatterjee C, Gleddie S, Xiao CW. Soybean bioactive peptides and their functional properties. *Nutrients* 2018 Sep 1;10(9):1211.
56. Shin ZI, Yu R, Park SA, Chung DK, Ahn CW, Nam HS, Kim KS, Lee HJ. His-His-Leu, an angiotensin I converting enzyme inhibitory peptide derived from Korean soybean paste, exerts antihypertensive activity in vivo. *J Agric Food Chem* 2001 Jun;49(6):3004–9.
57. Tomatsu M, Shimakage A, Shinbo M, Yamada S, Takahashi S. Novel angiotensin I-converting enzyme inhibitory peptides derived from soya milk. *Food Chem* 2013 Jan 15;136(2):612–6.
58. Singh BP, Vij S, Hati S. Functional significance of bioactive peptides derived from soybean. *Peptides* 2014 Apr;54:171–9. doi: 10.1016/j.peptides.2014.01.022.
59. Nardo AE, Suárez S, Quiroga AV, Añón MC. Amaranth as a source of antihypertensive peptides. *Front Plant Sci* 2020 Sep 25;11:578631.
60. de la Rosa AP, Montoya AB, Martínez-Cuevas P, Hernández-Ledesma B, León-Galván MF, De León-Rodríguez A, González C. Tryptic amaranth glutelin digests induce endothelial nitric oxide production through inhibition of ACE: Antihypertensive role of amaranth peptides. *Nitric Oxide* 2010 Sep 15;23(2):106–11.
61. Guo H, Hao Y, Richel A, Everaert N, Chen Y, Liu M, Yang X, Ren G. Antihypertensive effect of quinoa protein under simulated gastrointestinal digestion and peptide characterization. *J Sci Food Agric* 2020 Dec;100(15):5569–76.
62. Zhang P, Chang C, Liu H, Li B, Yan Q, Jiang Z. Identification of novel angiotensin I-converting enzyme (ACE) inhibitory peptides from wheat gluten hydrolysate by the protease of *Pseudomonas aeruginosa*. *J Funct Foods* 2020;65:103751. doi: 10.1016/j.jff.2019.103751.
63. Shobako N, Ogawa Y, Ishikado A, Harada K, Kobayashi E, Suido H, Kusakari T, Maeda M, Suwa M, Matsumoto M, Kanamoto R, Ohinata K. A novel antihypertensive peptide identified in thermolysin-digested rice bran. *Mol Nutr Food Res* 2018 Feb;62(4):1700732. https://doi.org/10.1002/mnfr.201700732
64. Montone CM, Zenezini Chiozzi R, Marchetti N, Cerrato A, Antonelli M, Capriotti AL, Cavaliere C, Piovesana S, Laganà A. Peptidomic approach for the identification of peptides with potential antioxidant and anti-hyperthensive effects derived from asparagus by-products. *Molecules* 2019 Oct 8;24(19):3627.
65. Montone CM, Capriotti AL, Cavaliere C, et al. Peptidomic strategy for purification and identification of potential ACE-inhibitory and antioxidant peptides in *Tetradesmus obliquus* microalgae. *Anal Bioanal Chem* 2018;410:3573–86. doi: 10.1007/s00216-018-0925-x.
66. Kheeree N, Sangtanoo P, Srimongkol P, Saisavoey T, Reamtong O, Choowongkomon K, Karnchanatat A. ACE inhibitory peptides derived from de-fatted lemon basil seeds: Optimization, purification, identification, structure-activity relationship and molecular docking analysis. *Food Funct* 2020 Sep 23;11(9):8161–78.
67. Mas-Capdevila A, Iglesias-Carres L, Arola-Arnal A, Aragonès G, Aleixandre A, Bravo FI, Muguerza B. Evidence that nitric oxide is involved in the blood pressure lowering effect of the peptide AVFQHNCQE in spontaneously hypertensive rats. *Nutrients* 2019 Jan 22;11(2):225.
68. Shih YH, Chen FA, Wang LF, Hsu JL. Discovery and study of novel antihypertensive peptides derived from *Cassia obtusifolia* seeds. *J Agric Food Chem* 2019 Jul 17;67(28):7810–20.

69. Iwaniak A, Minkiewicz P, Darewicz M. Food-originating ACE inhibitors, including antihypertensive peptides, as preventive food components in blood pressure reduction. *Compr Rev Food Sci Food Saf* 2014 Mar;13(2):114–34.
70. Karami Z, Peighambardoust SH, Hesari J, Akbari-Adergani B, Andreu D. Identification and synthesis of multifunctional peptides from wheat germ hydrolysate fractions obtained by proteinase K digestion. *J Food Biochem* 2019 Apr;43(4):e12800.
71. Pinciroli M, Aphalo P, Nardo AE, Añón MC, Quiroga AV. Broken rice as a potential functional ingredient with inhibitory activity of renin and angiotensin-converting enzyme(ACE). *Plant Foods Hum Nutr* 2019 Sep;74(3):405–13.
72. Wang X, Chen H, Fu X, Li S, Wei J. A novel antioxidant and ACE inhibitory peptide from rice bran protein: Biochemical characterization and molecular docking study. *LWT* 2017 Jan 1;75:93–9.
73. Yan QJ, Huang LH, Sun Q, Jiang ZQ, Wu X. Isolation, identification and synthesis of four novel antioxidant peptides from rice residue protein hydrolyzed by multiple proteases. *Food Chem* 2015 Jul 15;179:290–5.
74. Uraipong C, Zhao J. In vitro digestion of rice bran proteins produces peptides with potent inhibitory effects on α-glucosidase and angiotensin I converting enzyme. *J Sci Food Agric* 2018 Jan;98(2):758–66.
75. Quiroga AV, Aphalo P, Nardo AE, Añón MC. In vitro modulation of renin-angiotensin system enzymes by amaranth (*Amaranthus hypochondriacus*) protein-derived peptides: Alternative mechanisms different from ACE inhibition. *J Agric Food Chem* 2017 Aug 30;65(34):7415–23.
76. Zheng Y, Wang X, Zhuang Y, Li Y, Tian H, Shi P, Li G. Isolation of novel ACE-inhibitory and antioxidant peptides from quinoa bran albumin assisted with an in silico approach: Characterization, in vivo antihypertension, and molecular docking. *Molecules* 2019 Dec 12;24(24):4562.
77. Obaroakpo JU, Liu L, Zhang S, Lu J, Pang X, Lv J. α-glucosidase and ACE dual inhibitory protein hydrolysates and peptide fractions of sprouted quinoa yoghurt beverages inoculated with Lactobacillus casei. *Food Chem* 2019 Nov 30;299:124985.
78. Puchalska P, Concepción García M, Luisa Marina M. Identification of native angiotensin-I converting enzyme inhibitory peptides in commercial soybean based infant formulas using HPLC-Q-ToF-MS. *Food Chem* 2014 Aug 15;157:62–9.
79. Vallabha S, Vishwanath, Tiku PK. Antihypertensive peptides derived 1123 from soy protein by fermentation. *Int J Pept Res Ther* 2014;20(2):161–8. doi: 10.1007/s10989-013-9377-5.
80. Bao Z, Chi Y. In vitro and in vivo assessment of angiotensin-converting enzyme (ACE) inhibitory activity of fermented soybean milk by *Lactobacillus casei* strains. *Curr Microbiol* 2016 Aug;73(2):214–9.
81. Nishibori N, Kishibuchi R, Morita K. Soy pulp extract inhibits angiotensin I-converting enzyme (ACE) activity in vitro: Evidence for its potential hypertension-improving action. *J Diet Suppl* 2017 May 4;14(3):241–51.
82. Nakahara T, Sano A, Yamaguchi H, Sugimoto K, Chikata H, Kinoshita E, Uchida R. Antihypertensive effect of peptide-enriched soy sauce-like seasoning and identification of its angiotensin I-converting enzyme inhibitory substances. *J Agric Food Chem* 2010 Jan 27;58(2):821–7.
83. Yoshikawa M. Bioactive peptides derived from natural proteins with respect to diversity of their receptors and physiological effects. *Peptides* 2015 Oct;72:208–25.

84. Hernández-Ledesma B, Amigo L, Ramos M, Recio I. Angiotensin converting enzyme inhibitory activity in commercial fermented products. Formation of peptides under simulated gastrointestinal digestion. *J Agric Food Chem* 2004 Mar 24;52(6):1504–10.
85. Kodera T, Nio N. Identification of an angiotensin I-converting enzyme inhibitory peptides from protein hydrolysates by a soybean protease and the antihypertensive effects of hydrolysates in 4 spontaneously hypertensive model rats. *J Food Sci* 2006;71(3):C164–C73.
86. Wang W, De Mejia EG. A new frontier in soy bioactive peptides that may prevent age-related chronic diseases. *Compr Rev Food Sci Food Saf* 2005 Oct;4(4):63–78.
87. Hernández-Corroto E, Marina ML, García MC. Extraction and identification by high resolution mass spectrometry of bioactive substances in different extracts obtained from pomegranate peel. *J Chromatogr A* 2019 Jun 7;1594:82–92.
88. Alcaide-Hidalgo JM, Margalef M, Bravo FI, Muguerza B, López-Huertas E. Virgin olive oil (unfiltered) extract contains peptides and possesses ACE inhibitory and antihypertensive activity. *Clin Nutr* 2020 Apr;39(4):1242–9.
89. Piovesana, Riccardo Zenezini Chiozzi, and Aldo Laganà. Characterization of antioxidant and angiotensin-converting enzyme inhibitory peptides derived from cauliflower by-products by multidimensional liquid chromatography and bioinformatics. *J Funct Foods* 2018;44(34):40–7.
90. Nazir MA, Mu T-H, Zhang M. Preparation and identification of angiotensin I-converting enzyme inhibitory peptides from sweet potato protein by enzymatic hydrolysis under high hydrostatic pressure. *Int J Food Sci Technol* 2020;55:482–89. doi: 10.1111/ijfs.14291.
91. Bhaskar B, Ananthanarayan L, Jamdar SN. Effect of enzymatic hydrolysis on the functional, antioxidant, and angiotensin I-converting enzyme (ACE) inhibitory properties of whole horse gram flour. *Food Sci Biotechnol* 2018 Jul 23;28(1):43–52.
92. Chel-Guerrero L, Galicia-Martínez S, Acevedo-Fernández JJ, Santaolalla-Tapia J, Betancur-Ancona D. Evaluation of hypotensive and antihypertensive effects of velvet bean (*Mucuna pruriens* L.) hydrolysates. *J Med Food* 2017 Jan;20(1):37–45.
93. Luna-Vital DA, Liang K, González de Mejía E, Loarca-Piña G. Dietary peptides from the non-digestible fraction of *Phaseolus vulgaris* L. decrease angiotensin II-dependent proliferation in HCT116 human colorectal cancer cells through the blockade of the renin-angiotensin system. *Food Funct* 2016 May 18;7(5):2409–19.
94. Li CC, Lee YC, Lo HY, Huang YW, Hsiang CY, Ho TY. Antihypertensive effects of corn silk extract and its novel bioactive constituent in spontaneously hypertensive rats: The involvement of angiotensin-converting enzyme inhibition. *Molecules* 2019 May 16;24(10):1886.
95. Khan H, Jaiswal V, Kulshreshtha S, Khan A. Potential angiotensin converting enzyme inhibitors from *Moringa oleifera*. *Recent Pat Biotechnol* 2019;13(3):239–48. doi: 10.2174/1872208313666190211114229.
96. Xie J, Du M, Shen M, Wu T, Lin L. Physico-chemical properties, antioxidant activities and angiotensin-I converting enzyme inhibitory of protein hydrolysates from mung bean (*Vigna radiate*). *Food Chem* 2019 Jan 1;270:243–50.
97. Gupta N, Srivastava N, Bhagyawant SS. Vicilin-A major storage protein of mungbean exhibits antioxidative potential, antiproliferative effects and ACE inhibitory activity. *PLOS ONE* 2018 Feb 6;13(2):e0191265.

98. Aondona MM, Ikya JK, Ukeyima MT, Gborigo TJA, Aluko RE, Girgih AT. In vitro antioxidant and antihypertensive properties of sesame seed enzymatic protein hydrolysate and ultrafiltration peptide fractions. *J Food Biochem* 2021 Jan;45(1):e13587.
99. Arouna N, Gabriele M, Pucci L. The impact of germination on sorghum nutraceutical properties. *Foods* 2020 Sep 2;9(9):1218.
100. de Freitas MAG, Amaral NO, Álvares ADCM, de Oliveira SA, Mehdad A, Honda DE, Bessa ASM, Ramada MHS, Naves LM, Pontes CNR, Castro CH, Pedrino GR, de Freitas SM. Blood pressure-lowering effects of a Bowman-Birk inhibitor and its derived peptides in normotensive and hypertensive rats. *Sci Rep* 2020 Jul 15;10(1):11680.
101. Olagunju AI, Omoba OS, Enujiugha VN, Alashi AM, Aluko RE. Antioxidant properties, ACE/renin inhibitory activities of pigeon pea hydrolysates and effects on systolic blood pressure of spontaneously hypertensive rats. *Food Sci Nutr* 2018 Aug 22;6(7):1879–89.
102. Kaur A, Kehinde BA, Sharma P, Sharma D, Kaur S. Recently isolated food-derived antihypertensive hydrolysates and peptides: A review. *Food Chem* 2021 Jun 1;346:128719.
103. Hayes M, Tiwari BK. Bioactive carbohydrates and peptides in foods: An overview of sources, downstream processing steps and associated bioactivities. *Int J Mol Sci* 2015 Sep 17;16(9):22485–508.
104. Wang J, Sun B, Liu Y, Zhang H. Optimisation of ultrasound-assisted enzymatic extraction of arabinoxylan from wheat bran. *Food Chem* 2014 May 1;150:482–8.
105. Rodríguez-Jasso RM, Mussatto SI, Pastrana L, Aguilar CN, Teixeira JA. Fucoidan-degrading fungal strains: Screening, morphometric evaluation, and influence of medium composition. *Appl Biochem Biotechnol* 2010 Dec;162(8):2177–88.
106. Khan MY, Kumar V. Mechanism & inhibition kinetics of bioassay-guided fractions of Indian medicinal plants and foods as ACE inhibitors. *J Tradit Complement Med* 2018 Apr 30;9(1):73–84.
107. Malomo SA, Onuh JO, Girgih AT, Aluko RE. Structural and antihypertensive properties of enzymatic hemp seed protein hydrolysates. *Nutrients* 2015 Sep 10;7(9):7616–32.
108. Santos RA, Ferreira AJ, Verano-Braga T, Bader M. Angiotensin-converting enzyme 2, angiotensin-(1–7) and Mas: New players of the renin-angiotensin system. *J Endocrinol* 2013 Jan 18;216(2):R1–RR17.
109. Pujiastuti DY, Ghoyatul Amin MN, Alamsjah MA, Hsu J-L. Marine organisms as potential sources of bioactive peptides that inhibit the activity of asngiotensin I-converting enzyme: A review. *Molecules* 2019;24:2541. doi: 10.3390/molecules24142541.
110. Kuba M, Tana C, Tawata S, Yasuda M. Production of angiotensin I-converting enzyme inhibitory peptides from soybean protein with Monascus purpureus acid proteinase. *Process Biochem* 2005 May 1;40(6):2191–6.
111. Zheng Y, Li Y, Zhang Y, Ruan X, Zhang R. Purification, characterization, synthesis, in vitro ACE inhibition and in vivo antihypertensive activity of bioactive peptides derived from oil palm kernel glutelin-2 hydrolysates. *J Funct Foods* 2017 Jan 1;28:48–58.

CHAPTER 19

Bioactive Peptides in Neurodegenerative Diseases

Kambiz Hassanzadeh, Marco Feligioni, Mohammad Zarei, Belal J. Muhialdin, Rita Maccarone, Massimo Corbo, and Lucia Buccarello

CONTENTS

Introduction	391
Neurodegenerative Diseases: Definition and Classification	392
Alzheimer's Disease	392
Parkinson's Disease	395
Huntington's Disease	396
Amyotrophic Lateral Sclerosis	397
Bioactive Peptides	398
Bioactive Peptides in Neurodegenerative Disease	398
Role of Bioactive Peptides in Regulation of Reactive Oxygen Species Levels	398
Bioactive Peptides Against Glutamate Excitotoxicity and Cell Death-Signaling	400
Bioactive Peptides and Modulation of Gut Microbiota	401
Bioactive Peptides Against Inflammation	402
Conclusion	403
References	403

INTRODUCTION

Bioactive peptides, isolated small fragments of proteins, are involved in several biological activities and significantly contribute to human physiological responses. Recent years have witnessed an increasing number of studies focused on the application of these peptides as therapeutic agents in disease management. A wide range of therapeutic effects including antimicrobial, anticancer, antidiabetic, antioxidant and inhibition of neurodegenerative diseases (NDDs) have been proposed for bioactive peptides effects (Baig et al., 2018).

The existing agents for prevention or treatment of NDDs are insufficient, and the number of drugs approved is limited by the high failure rates in clinical trials (Morris et al., 2014). Several lines of evidence indicate that peptides are pivotal tools for NDDs research, including basic and clinical studies, and can be used to study the properties of misfolded proteins and/or peptides.

This chapter offers a brief introduction to bioactive peptide application in neurodegenerative diseases. We discuss the successful application of synthetic peptides and their natural counterparts in drug discovery along with their drawbacks and limitations. In addition, we discuss the therapeutic utilities and versatilities of peptide inhibitors in various neurodegenerative disorders as therapeutic drug candidates.

NEURODEGENERATIVE DISEASES: DEFINITION AND CLASSIFICATION

Neurodegenerative diseases (NDDs) are progressive degenerative conditions characterized by loss of neurons within the brain associated with protein deposition and changes in neurochemical properties in the central nervous system (CNS) and peripheral organs (Kovacs, 2019). Considering the increase in the elderly population in recent years, age-dependent disorders such as NDDs are becoming increasingly prevalent (Heemels, 2016). Major NDDs include Parkinson's disease (PD), Alzheimer's disease (AD), Huntington's disease (HD), amyotrophic lateral sclerosis (ALS), frontotemporal dementia, and spinocerebellar ataxias. These diseases are diverse in their clinical manifestations: some affect a person's movement ability, others affect cognitive ability, memory, speech, and even breathing (Gitler et al., 2017). However, they share some common neuropathological mechanisms. A comprehensive understanding of the causes and mechanisms of these diseases, is crucial for enabling scientists to develop suitable treatments.

Misfolded proteins, aggregation, and accumulation of proteins are the main pathological events in neurodegenerative disease therefore uncovering the role of the unfolded protein response seems to be one important strategy in the treatment of these diseases (Cornejo & Hetz, 2013). Furthermore, changes in protein elimination mechanisms, such as the ubiquitin-proteasome system and the autophagy-lysosome pathway, have high impact on the pathogenesis (Nijholt et al., 2011).

Several molecular pathways contribute to the chronic damage of neurons, for instance, chronic excitotoxicity induced by glutamate has been reported for progressive neurodegeneration in AD, ALS, and HD (Lewerenz & Maher, 2015). Other important pathways are energetic dysregulation, metabolic changes, dysregulation of ion homeostasis, and adaptations that have shown to play roles in neurodegenerative disorders (Von Bernhardi and Eugenín, 2012).

The role of misfolded proteins and their deposition in neurodegenerative disease have been established over past decades (Table 19.1).

Alzheimer's Disease

Dementia is identified as acquired loss of cognition in multiple cognitive domains severe enough to interfere with social or occupational activities of daily living. It has been reported that two-thirds of cases of dementia in people age 65 and older are caused by Alzheimer's disease (AD). Onset of disease before 65 years old is unusual and seen in less than 10% of AD patients. This type of neurodegenerative disease progressively impairs the behavioral and cognitive functions such as memory, attention, comprehension, judgment, etc. (Arvanitakis et al., 2019).

The early presenting sign is episodic short-term memory loss and relative sparing of long-term memory. In the early stages of disease, impairment in executive activity ranges from subtle to significant. Then language disorder and impairment of visuospatial skills might be represented. Also, neuropsychiatric symptoms, such as agitation, apathy,

TABLE 19.1 Classification of Neurodegenerative Disease

Category	Disease Type		Protein(s) Involved	Brain Affected Area*
α-Synucleinopathy (McCann et al., 2014)	Parkinson's disease (PD)		α-Synuclein	Basal ganglia including: Caudate nucleus Putamen Globus pallidus Subthalamic nucleus Substantia nigra
	Dementia with Lewy bodies (DLB)		α-Synuclein	Amygdala Cerebral cortex Dorsal motor nucleus Hippocampus (CA2) Locus coeruleus Olfactory bulb Substantia nigra
	Multiple system atrophy (MSA)		α-Synuclein	Putamen Substantia nigra Pontine nuclei Medulla (inferior olivary nucleus) Cerebellum
Tauopathy (Ganguly & Jog, 2020b)	Primary	Progressive supranuclear palsy (PSP)	Tau	Subthalamic nucleus Substantia nigra Superior colliculus Cerebellar dentate
		Corticobasal degeneration (CBD)	Tau	Frontoparietal association cortices Neostriatum Substantia nigra
		Argyrophilic grain disease (AGD)	Tau	Limbic structures
		globular glial tauopathy (GGT)	Tau	Frontotemporal cortex
		Pick disease (PiD)	Tau	Basal forebrain Frontal and temporal lobes Limbic structures Striatum
		Neurofibrillary tangle (NFT)-dementia or primary age-related tauopathy, PART	Tau	medial temporal lobe, particularly the hippocampal formation and adjacent regions

(Continued)

TABLE 19.1 (Continued)

Category	Disease Type		Protein(s) Involved	Brain Affected Area*
	Secondary	Alzheimer's disease (AD): Secondary to Amyloid deposition	Tau, Aβ	Basal forebrain Frontal and temporal lobes Limbic structures Locus coeruleus Olfactory bulb
		Chronic traumatic encephalopathy (CTE): Secondary to repeated trauma	Tau	Cortical sulci
TDP-43 Proteinopathy (Brettschneider et al., 2015)	Frontotemporal lobar degeneration (FTLD)-TDP		TDP-43	Frontal and temporal cortices Basal ganglia Substantia nigra
	Amyotrophic lateral sclerosis (ALS)		TDP-43	Motor cortex Brainstem motor neurons Spinal cord motorneurons
FUS Proteinopathy (Brettschneider et al., 2015)	Frontotemporal lobar degeneration (FTLD)-FUS		FUS	Motor cortex Brainstem motor neurons Spinal cord motor neurons
	Amyotrophic lateral sclerosis (ALS)			
Huntington's disease	Huntington's disease		Huntingtin	neocortex, entorhinal cortex, subiculum, hippocampal pydamidal neurons, and striatum

* Dugger & Dickson, 2017; Ganguly & Jog, 2020a; Reiner et al., 2011

psychosis, disinhibition, and social withdrawal are also common in the mid to late stages. In the late stages of disease, motor tasks and olfactory dysfunction, parkinsonian, dystonia, and akathisia, and also sleep disturbances occur (Kumar et al., 2021).

There are macroscopic and microscopic features for Alzheimer's disease. In fact, there is no single feature or combination of features as a specific marker, but some features are highly suggestive in the diagnosis of AD. It has been documented that a moderate cortical atrophy is present in the brain of patients. The frontal and temporal cortices show enlarged sulcai spaces with atrophy of the gyri, whereas the primary motor and somatosensory cortices appear unaffected (Perl, 2010). A growing atrophy in the posterior cortical areas has also been observed in the AD brain, most notable the precuneus and posterior cingulate gyrus, driven partly by functional imaging studies (Rami et al., 2012).

There are some other microscopic features like amyloid plaques, neuritic plaques and neurofibrillary tangles (NFTs). Senile amyloid plaques are formed through the extracellular nonvascular accumulation of Aβ42 and Aβ40 peptides as a result of imbalance in the production and clearance pathways and abnormal processing of APP (amyloid precursor protein) by the β- and γ-secretases (DeTure and Dickson, 2019). These small peptides (4 kDa) fold into a highly fibrillogenic beta-pleated sheet structure.

Cored neuritic plaques are another microscopic feature containing tau protein (Tau-positive neurites) usually have a central zone of dense amyloid, sometimes forming a compact core (Dickson, 1997), while some neuritic plaques contain activated microglia and reactive astrocytes. In addition, some of the dystrophic neurites associated with neuritic plaques contain tau filaments, which can have a paired helical filament morphology and are liable to be observed with electron microscopy (Serrano-Pozo et al., 2011).

Formation of NTFs is another hallmark of AD. It was first described as "neurofibrils", which forms thick bundles near the surface of affected neurons (Ryan et al., 2015). Amyloid plaques, especially cored neuritic plaques, and NTFs containing filamentous tau proteins are required for neuropathologic diagnosis of AD (DeTure and Dickson, 2019).

Like other neurodegenerative disease, so far, there is no specific cure for Alzheimer's disease, although there are some available drugs that may improve the symptoms. Only very recently can we state that the FDA has approved, after almost 30 years, a novel and very promising drug against AD which is a monoclonal antibody, called Aducanumab, able to contrast the b-amyloid fibrils and inducing a significant reduction of amyloid plaques (https://doi.org/10.1002/alz.047259). This is the first approved drug directly contrasting amyloid plaques formation. Aducanumab seems to provide clinically meaningful benefit in association with amyloid lowering (Cummings et al., 2021). Aducanumab has move into clinical phase 4 trials, so now we will wait for observations from larger cohorts.

Classically, AD is clinically treated with cholinergic agents which improve and maintain cognitive abilities in patients with mild-to-moderate AD. Also, some neuroprotective supplements are used as complementary medicine. Therefore, novel treatments that preserve cognitive ability and prevent the progression of AD are needed.

Parkinson's Disease

As the second most common neurodegenerative disease, Parkinson's disease is characterized by a progressive loss of dopaminergic neurons in the substantia nigra pars compacta. Rest tremor, rigidity, bradykinesia and postural instability, as well as a variety of other motor and non-motor symptoms, including anxiety and depression, are clinical symptoms of PD (Obeso et al., 2017). Both genes and environmental/lifestyle factors contribute to the pathogenesis of PD. Age is known as the most important risk factor, since the disease onset is mostly around 60 years (Simon et al., 2020); therefore, age-dependent biological failures, like telomere dysfunction, epigenetic changes, mitochondrial defects, ubiquitin-proteasome system (UPS), and autophagy-lysosomal system play a role in the pathophysiology of Parkinson's disease (González-Casacuberta et al., 2019; Pohl and Dikic, 2019).

It is now well established that PD is a systemic disorder and alpha-synuclein is a key player in disease progression. Alpha-synuclein, a neuronal protein of 140 amino acids long was first isolated and sequenced in 1988. It is mainly located in the presynaptic compartment, but it has been identified also in the nucleus, therefore, accounting for the name "synuclein" (SYNapse + NUCLEus, synuclein) (Maroteaux et al., 1988). The exact physiological activity of alpha-synuclein is still unclear, but its connection to the pool of synaptic vesicles and synaptic transmissions suggests its important role in the neurotransmitter release process, synaptic function, and plasticity (Lashuel et al., 2013). Accumulation of aggregated alpha-synuclein within neurons results in the reduction of synaptic proteins, progressive decrease in neuronal excitability, and, finally, cell death (Volpicelli-Daley et al., 2011).

The treatment strategy for Parkinson's patients is mainly symptomatic, focused on addressing motor issues, such as tremor, rigidity, bradykinesia, and non-motor symptoms, like constipation, cognition, mood, and sleep disturbances. So far, no drug capable of modifying the disease exists in the market. To ameliorate the motor symptoms, dopamine-based therapies are used. While non-motor symptoms require other approaches, like cholinesterase inhibitors for cognition or use of selective serotonin reuptake inhibitors for psychiatric symptoms (Armstrong and Okun, 2020).

Studies to treat the non-motor symptoms of PD have shown that rivastigmine, donepezil, galantamine, and memantine, which are known as cholinesterase inhibitors and N-methyl D-aspartate receptor antagonists, can improve the cognitive functions (Feldman et al., 2001; Szeto and Lewis, 2016). However, additional pharmacological studies should be carried out before these drugs are considered safe and tolerable, as there are potential adverse effects, such as diarrhea, headache, and dizziness (Feldman et al., 2001; Szeto and Lewis, 2016).

Huntington's Disease

As the most common inherited neurodegenerative disease, Huntington's disease (HD) is identified by uncontrolled excessive motor movements and cognitive and emotional deficits (Paulson and Albin, 2011; Roos, 2010). HD is a predominantly inherited, neuropsychiatric disorder which affects generations of afflicted families. It always starts in adulthood and progresses slowly over the years with common and typical symptoms. The discovery of the HD mutation has also shed light on the possible mechanisms involved in disease. The significant variability in clinical symptoms and the different range in age of onset, even among affected members in the same family, are now recognized to stem directly from the type of mutation in HD (a dynamic repeat in a polyglutamine-encoding CAG tract) (Paulson and Albin, 2011). Onset of the disease is defined by the presence of a motor disorder, usually the involuntary movements known as chorea. In some patients, other motor abnormalities may lead to a diagnosis of HD (Louis et al., 2000). In some patients, symptoms begin even before the age of 20 years with early behavioral disorders and learning problems at school (Juvenile Huntington's disease, JHD) (Roos, 2010).

Clinical symptoms of this "neuropsychiatric" disorder, including progressive movement disorder, progressive cognitive disturbance culminating in dementia, and various behavioral disturbances that often precede diagnosis, can vary depending on the state of disease (Paulson and Albin, 2011).

HD is characterized by the overexpression of a misfolded protein called huntingtin and caused by CAG codon repeat expansion in its own gene. CAG encodes the amino acid glutamine, so the expansion results in an abnormally long glutamine tract within the N-terminus of the huntingtin protein (Htt). Abnormal Htt has shown to be insoluble and can accumulate, causing cell death. HD is part of a large family of other pathology collectively termed as "polyglutamine" diseases because they are caused by the overexpression of repeated CAG sequences (Zheng and Diamond, 2012). Although Huntingtin is expressed in many tissues, the clinical symptoms of HD reflect a CNS disorder, and the histopathologic abnormalities are almost limited to the brain. However, there are investigations to evaluate the effects of expanded huntingtin in other organs (Borovecki et al., 2005; Chiang et al., 2007).

Like other NDDs, there is not a cure for Huntington's. There are some potential compounds or strategies considered for prevention or symptomatic treatments, including neuroprotective and antiapoptotic compounds (Wei et al., 2001), transglutaminase inhibitors (Dubinsky and Gray, 2006), histone deacetylase inhibitors (Gardian et al., 2005), antioxidants, mitochondrial enhancers, etc.

Amyotrophic Lateral Sclerosis

Amyotrophic lateral sclerosis (ALS), a fatal motor neuron disorder, is characterized by progressive loss of the upper and lower motor neurons (UMNs and LMNs) at the spinal or bulbar level (Rowland and Shneider, 2001). Finding the molecular mechanisms of neurodegeneration in ALS will help us to understand the disease's progress. Moreover, clarification of molecular mechanisms yields insight into discovering newer therapeutic approaches.

Mutation of the gene encoding the superoxide dismutase 1 (SOD1), an antioxidant enzyme, has been reported as one of the most common causes of ALS (De vos et al., 2007; Ivanova et al., 2014). Misfolding in the mutated SOD1 enzyme can lead to aggregation in the motor neurons within the central nervous system (Forsberg et al., 2011). There are also other genes which are known to cause familial ALS, including TARDBP (encodes TAR DNA-binding protein 43 (TDP-43)); FUS (encodes for fusion in sarcoma), ANG (codes for angiogenin, ribonuclease, and the RNAase A family 5), OPTN (codes for optineurin) and C9orf72 (Chiò et al., 2011; Corrado et al., 2010; Gijselinck et al., 2012; Kabashi et al., 2011; Majounie et al., 2012).

There are four main types of ALS phenotypical expressions (Kiernan et al., 2011; Vucic and Kiernan, 2007; Wijesekera and Leigh, 2009):

1. Limb-onset ALS: both upper motor neuron (UMN) and lower motor neuron (LMN) signs in the limbs are presenting.
2. Bulbar onset ALS: speech and swallowing difficulties followed by limb weakening in the later stages of the disease.
3. Primary lateral sclerosis (PLS) with pure UMN involvement.
4. Progressive muscular atrophy (PMA) with pure LMN involvement.

As this classification illustrates, the main clinical feature in ALS is based on a combination of UMN and LMN damage. It has been demonstrated that almost 70% of the cases among patients are affected with the limb-onset ALS. Bulbar onset accounts for 25% of the cases, and 5% of the cases have initial trunk or respiratory involvement (Kiernan et al., 2011). During the ALS progression, a distinctive feature of a combination of upper motor and LMN degeneration signs within the same CNS region appears in patients (Gordon et al., 2006), and eventually the main cause of death in patients is respiratory failure as the result of pulmonary complications (Corcia et al., 2008).

ALS, such as Parkinson's disease, is suggested to be due to both environmental and genetic factors, and clinical trials have been conducted giving the same treatment to patients with ALS or PD (Lomen-Hoerth et al., 2002; Wijesekera and Leigh, 2009).

The past three decades of ALS research has improved our understanding of the pathophysiology of disease. However, the translation of these efforts into effective treatments has been disappointing so far, and most cases with ALS didn't have the chance to participate in clinical trials (Kiernan et al., 2021). Although numerous novel potential treatments are being tested in Phase 1 to Phase 3 clinical trials (www.clinicaltrials.gov), no effective treatment currently exists that is able to stop the progression of ALS. The only disease-modifying treatment approved for clinical use by FDA, Riluzole, demonstrated to extend the life span in ALS patients (Miller et al., 1999). In primary investigations, a 38.6% reduction in mortality was reported (Bensimon et al., 1994), and it has been associated with 35% improvement in survival with the 100 mg dose (Lacomblez et al., 1996).

BIOACTIVE PEPTIDES

Bioactive peptides are identified as chains of amino acid residues, derived from hydrolyzed food protein present in nature. These peptides can be extracted from different food-derived animal, plants, and marine protein sources. They are health-protective molecules that, similar to hormones, can control many important body functions (Korhonen and Pihlanto, 2006; Lee and Hur, 2019; Udenigwe and Aluko, 2012). They are characterized by many properties with, for example, antioxidant, antihypertensive, anti-inflammation, antitumor, antimicrobial, and neuroprotective effects (Assadollahi et al., 2019; Blondelle and Lohner, 2000; Kodera and Nio, 2006; Kudo et al., 2009; Lee and Hur, 2019; Majumder et al., 2016; McCann et al., 2014).

Bioactive Peptides in Neurodegenerative Disease

Like other biogenic compounds, the potency and efficacy of these peptides completely depends on the structure and amino acid composition. In this section, we provide a review on understanding the role of bioactive peptides and their applications in the management of neurodegenerative diseases. To this aim, the effects of bioactive peptides are described based on their mechanism of action.

Role of Bioactive Peptides in Regulation of Reactive Oxygen Species Levels

Reactive oxygen species (ROS) are highly reactive chemical molecules formed by the partial reduction of oxygen. Examples of these compounds include superoxide anion (O_2^-), hydroxyl radical ($^\bullet OH$), and hydrogen peroxide (H_2O_2) (Ray et al., 2012).

ROS and free radicals can attack and interact with the membrane lipids, protein, and DNA in the cell. They can be of endogenous or exogenous origin, but in both cases, oxidation by free radicals in the body have influence on human health and may cause several chronic diseases like cancer, cardiovascular diseases, diabetes, and neurodegenerative disorders (Dong et al., 2008).

On the other hand, we know that the body is supported by several defensive antioxidant systems that can scavenge and transform ROS into harmless species (Yeung et al., 2002). The antioxidant enzymatic system includes catalase glutathione peroxidase (GSH-Px) (CAT), superoxide dismutase (SOD), and glutathione reductase (GR). There are also nonenzymatic antioxidants which are able to scavenge and eradicate oxidative stress (McCord, 1993). Keeping the balance between ROS and body antioxidant is critical for avoiding oxidative stress and preventing disease (Ghazizadeh et al., 2020).

The antioxidant activity of peptides isolated from natural compounds and food sources was reported by Marcuse in 1960 for the first time (Gomez-Guillen et al., 2011).

Later, several studies were done to confirm the antioxidant properties of these compounds extracted from plant and animal sources (Baltic et al., 2014). Despite numerous research in this area, the exact antioxidant mechanism of these peptides still has not been completely elucidated. Compared with natural proteins, small peptides display more antioxidant activity and could be absorbed in the gastrointestinal system without further digestion (Liu et al., 2016).

Basically, antioxidant peptides can be classified into two main categories – endogenous and exogenous. Endogenous peptides are produced naturally inside the cells such as glutathione, carnosine, anserine, GHL; whereas, the exogenous types are derived from food protein sources during gastrointestinal digestion (Gallego et al., 2020; Zhang et al., 2020).

It is believed that antioxidant peptides are free radical scavengers and they are able to prevent lipid peroxidation and chelate transition metal ions (Jiménez-Colmenero et al., 2010; Rajapakse et al., 2005; Suetsuna et al., 2000).

Oxidative stress, as the main cause of ROS generation, has an impact on neuronal cell death. There are some hydrolysates/peptides isolated from animal and egg proteins which are able to provide neuroprotection through inhibiting oxidative stress factors (Gu et al., 2018; Zhang et al., 2015). It has been reported that a peptide (Met-Glu-Ile-Phe-Val-Lys-Thr-Lys-Thr-Gly) could protect PC-12 cells through prevention of nitric oxide and ROS generation and lipid peroxidation. Also, whey protein hydrolysates prevented DNA fragmentation by ROS. Stimulation of cellular antioxidant enzymes, like superoxide dismutase (SOD) and catalase, was reported as the main mechanism of action. Moreover, a small peptide, Pro-Ala-Tyr-Ser-Cys (PAYSC), obtained from anchovy hydrolysate, inhibited ROS and malondialdehyde (MDA) production (Lee and Hur, 2019; Zhao et al., 2017). There are other reports indicating that hydrolysates isolated from plants showed significant antioxidant activities by inhibiting ROS production and enhanced catalase and SOD enzymes activities in H_2O_2-treated neuronal cells (Chen et al., 2015; Zhu et al., 2013).

Several factors, like method of protein isolation, peptide concentration, degree of hydrolysis, and type of protease used in the process, affect the antioxidant properties of peptides (Erdmann et al., 2008; Liu et al., 2010). Other properties such as type and number of amino acids, as well as the order of amino acid sequence play a key role in their antioxidant activity (Erdmann et al., 2008; Rajapakse et al., 2005).

It has been reported that Tyr, Trp, Met, Lys, Cys, and His are the amino acids that contribute to antioxidant activity (Baltic et al., 2014). For instance, the imidazole group of histidine amino acid has been found to be related to the hydrogen donating, lipid peroxyl radical-trapping and the metal-chelating properties, and the cysteine SH group has been reported to have the main role in interaction with free radicals (Baltic et al., 2014; Sarmadi and Ismail, 2010). Recently, it has been demonstrated that the presence of hydrophobic amino acids, such as proline, tryptophan, valine, and phenylalanine in the small peptides (2–20 amino acids), provide potent antioxidant activity (Ketnawa et al., 2018; Tadesse and Emire, 2020; Yang et al., 2019). Histidine-containing dipeptides, carnosine (ß-alanyl-L-histidine), and anserine (N-ß-alanyl-1-methyl-L-histidine) are the most studied hydrophilic antioxidants existing in meat and in some fish. Carnosine and anserine protect DNA against L-3, 4-dihydroxyphenylalanine Fe (III)-induced damage. Also, oral administration of L-carnosine was able to increase the total antioxidant capacity of human serum (Di Bernardini et al., 2011).

In addition, the configuration of amino acids has been shown to play a role for example, substitution of L-His by D-His in a peptide, resulting in diminishing antioxidative activity (Baltic et al., 2014).

Type of meat, as the source of peptide, has been considered as a main factor in antioxidant activity, for example the concentration of carnosine in chicken is 500 mg/kg, while in pork, it is 2700 mg/kg in, but anserine is present in higher concentration in chicken muscle (Purchas and Busboom, 2005; Young et al., 2013).

There are reports on peptides extracted from collagen. In an experimental study, porcine collagen was treated with pepsin, and then the derived hydrolysate was treated with papain, protease from bovine pancreas (PP), and a cocktail of three enzymes (PP and bacterial proteases from *Streptomyces* and *Bacillus polymyxa*). Results indicated that the hydrolysate which was treated with the cocktail of three enzymes showed the highest level of antioxidant activity, and four antioxidant peptides were extracted from this hydrolysate (Li et al., 2007). In another study, a 36-amino acid residue peptide was isolated from bovine tendon collagen α1, that was able to scavenge free radical and chelate metals (Banerjee et al., 2012).

It is noteworthy that regarding the antioxidant peptides, protein quality and its amount, as well as cost of the raw materials are important for finding the cost benefit peptide. Thus, economical and available materials containing a high amount of protein may feasibly be used to prepare the antioxidant peptides (Tadesse and Emire, 2020).

Another parameter that should be taken into account in antioxidant activity is the ability of the peptide in the chelating of pro-oxidative metals, such as zinc, copper, and cobalt. The type of complex, bound peptide, and metal ion is important and eventually induce different biological functions (Baltic et al., 2014; Young et al., 2013).

Bioactive Peptides Against Glutamate Excitotoxicity and Cell Death-Signaling

Glutamate excitotoxicity is another mechanism of neuronal cell death induced by oxidative stress in neurodegenerative disease, such as AD and PD (Swerdlow, 2012). In physiological conditions, transmission of glutamate is associated with energy stability, thus impairment of energy metabolism leads to neuronal cell damage. Moreover, accumulation of glutamate and unnecessary stimulation of its receptors leads to cell death through different mechanisms including ROS generation, mitochondrial dysfunction, and overload of calcium. In pathological condition, excessive release of glutamate into the synaptic space stimulates glutamate receptors of N methyl-D-asparate (NMDA), which in turn leads to an influx of calcium and sodium and eventually depolarization of the postsynaptic neuron (Wu and Tymianski, 2018).

It has been demonstrated that a peptide blocking the interaction of death-associated protein kinase 1 (DAPK1) with glutamate receptor, called NR2B, protects against neuronal death in mice (Tu et al., 2010). In addition, it has been reported that cell permeable peptides including the major calpain cleavage site, are able to prevent the excitotoxicity through attenuating Kidins 220 calpain processing. The mechanism is related to the NMDA receptor toxicity pathway (Gamir-Morralla et al., 2015).

It is well known that overstimulation of presynaptic N-methyl-D-aspartate (NMDA) receptors triggers and supports glutamate release, while postsynaptic NMDA receptors are responsible for the subsequent apoptotic cascade (Centeno et al., 2007). Previously, we found that presynaptic c-Jun N-terminal kinase 2 (JNK2) specifically controls NMDA-evoked glutamate release (Nisticò et al., 2015), and recently, we demonstrated that an interaction between Syntaxin-1a and JNK2 is essential to this mechanism. Therefore, we designed and produced a new cell-permeable peptide, "JGRi1", a 26 amino acid cell-permeable peptide. The 12 residues (GRKKRRQRRRPP) of the HIV-1 Tat protein that confers cell permeability have been linked to the effector portion (IEQSIEQEEGLNRS) as a part of the N-terminal amino acid sequence of STX1a which corresponds to the part of the minimal contact area with JNK2. This peptide could block the JNK2/STX1a interaction specifically and was able to prevent the presynaptic NMDA receptor signaling. JGRi1 diminishes the NMDA-evoked glutamate release both in *in-vitro* and *ex-vivo* experiments while also being able to widely diffuse throughout brain tissue via intraperitoneal administration. In this study, we concluded that JGRi1 acts as a pharmacological tool that promotes neuroprotection (Marcelli et al., 2019).

There are several reports indicating that peptides from natural sources have been reported to interact with key factors in apoptosis (programmed cell death), like Bax and Bcl-2 protein, caspases, as well as factors involved in DNA fragmentation (Chen et al., 2015; Jin et al., 2013). A small 24-amino acid peptide, Humanin (HN), was able to inhibit the binding of amyloid beta (Aβ), a known AD inducer, to the death receptors and eventually inhibited the neuronal cell death induced by proteins related to AD genes (Harada et al., 2004; Hashimoto et al., 2001). In addition, peptides containing Gly-Pro-Arg and Arg-Glu-Arg can inhibit the neuronal cell death induced by Aβ through counteracting Aβ-induced activation of the caspase 3 and p53 pathways (Mileusnic et al., 2004). The peptide Met-Gln-Ile-Phe-Val-Lys-Thr-Leu-Thr-Gly blocked the cytochrome C release and controlled expression of the genes involved in apoptosis, such as those encoding cleaved caspase-3, Bax, and iNOS (Kim et al., 2010). In addition, hydrophobic whey protein hydrolysates not only increased Bcl-2 (an anti-apoptotic protein) but also were able to decrease the intracellular calcium levels, caspase-3 expression, and PARP cleavage (Jin et al., 2013). The MQIPVLTLTG peptides isolated from venison protein and hydrolysates extracted from lantern fish, plant, and egg proteins showed to prevent apoptosis in *in vitro* studies (Kim et al., 2010; W. Li et al., 2016).

Bioactive Peptides and Modulation of Gut Microbiota

Several lines of evidence indicate that altering the regular diet by influencing the microbiota-gut-brain axis affects neurodegenerative diseases (Gubert et al., 2020). It is well known that the composition of diet plays a key role in the health of the gut as well as the gut microbiota composition and balance, which affect the mutual signaling between the gut and brain (Wang et al., 2018). In this area, it has been recently reported that cognitive dysfunction and inflammation in older people could be improved using Mediterranean diet as a gut microbiota modulator (Ghosh et al., 2020). Nagpal and colleagues, using a modified Mediterranean-ketogenic diet, modulated the composition of gut microbiota and their metabolites which were associated with improved Alzheimer's disease biomarkers in patient cerebrospinal fluid (Nagpal et al., 2019). In addition, there are some peptides able to relieve the symptoms of NDDs. For example, peptides isolated from sesame cake (PSC) enhanced antioxidative stress by reducing the ROS levels and Aβ deposition in a transgenic *Caenorhabditis elegans* model of AD. This mechanism might contribute to the delay of AD onset in this model (Ma et al., 2017). Also, PSC significantly decreased the α-synuclein aggregation in a PD *C. elegans* model, indicating that PSC, as a beneficial tool, ameliorated PD-related pathologies (Ma et al., 2020). Therefore, peptides obtained from protein digestion in the diet or consumed as supplements can contribute to the gut homeostasis by regulating the gut microbiota and ROS homeostasis.

Peptides may improve the dysbiosis of gut homeostasis through modifying the gut ROS level. In fact, the imbalance between ROS amount and antioxidant systems leads to apoptosis and also affects gut microbiota, which have been shown to be associated with neurodegenerative disease (Cryan et al., 2019; Dumitrescu et al., 2018).

Moreover, peptide treatment increased some gut probiotics (Zhang et al., 2020). These probiotics can produce various metabolites with antioxidant activity, like glutathione, folate, and butyrate and eventually are able to protect the host against oxidative stress injury through eradicating ROS accumulation (Wu et al., 2021).

Taken together, several lines of evidence indicating that alteration in gut microbiota can change the enteric and peripheral inflammatory pathways, which in turn induce neuroinflammation and neurodegeneration (Pellegrini et al., 2018, 2020; Wu et al., 2021), therefore targeting this pathway using some peptides in the diet could be a helpful approach to preventing neurodegenerative disease.

Bioactive Peptides Against Inflammation

Inflammation is the physiologic response to the damage which is characterized by augmented endothelial permeability and infiltration of leukocytes into extravascular tissue. Neuroinflammation is a defensive process that initially can protect the CNS by eliminating or preventing various pathogens (Wyss-Coray and Mucke, 2002). Therefore, it has beneficial effects via inducing tissue repair and removing cellular debris (Russo and McGavern, 2016). However, an excessive and uncontrolled inflammatory process often leads to chronic diseases. Endogenous factors, such as genetic mutation and protein aggregation, or environmental stimulus like trauma, infection, and drugs, might be the reasons for persistence of inflammatory responses (Glass et al., 2010) which may lead to NDDs (Kempuraj et al., 2016).

Neuroinflammation is a major mechanism involved in the onset and progression of several neurodegenerative disorders like AD and PD. In fact, it is responsible for an abnormal secretion of proinflammatory cytokines to induce signaling pathways contributing to neurodegeneration. Once the inflammatory cascade is activated by damage signals, the synaptic dysfunction occurs through several molecular mechanisms. This process generates a positive feedback mechanism promoting even more cellular damage and death (Guzman-Martinez et al., 2019). (Figure 19.1)

Gut inflammation is another mechanism to induce an immune response and inflammatory processes which could contribute to neurodegenerative disease like PD pathology (Guzman-Martinez et al., 2019).

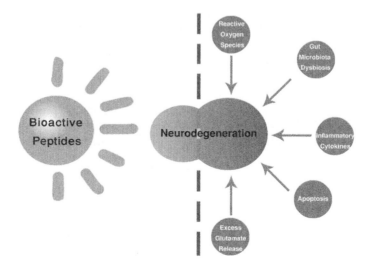

FIGURE 19.1 Bioactive peptides counteract different pathways involved in neurodegeneration.

In recent decades peptides and protein hydrolysates derived from food sources like meat, egg, milk, fish, and soybeans (to name a few) have all been evaluated for their potential beneficial effects on immune system and inflammation. For instance, bioactive peptides from milk, as a rich source of caseins and whey proteins, have been tested for this purpose. The tripeptide, Val-Pro-Pro (VPP), have been reported to have the ability of reducing leukocyte-endothelial interactions, mainly through inhibition of proinflammatory c-Jun N-terminal kinase (JNK), a member of the MAP kinase superfamily pathway (Aihara et al., 2009). Casein hydrolysates obtained through enzymatic digestion have also been evaluated for anti-inflammatory properties. For example, digestion of casein with corolase produces compounds exerting anti-inflammatory effects on activated macrophages (Nielsen et al., 2012). It has been reported that milk contains transforming growth factor-beta (TGF-beta), IL-10, and immunoglobulins, a number of anti-inflammatory and immunomodulatory agents (Rutherfurd-Markwick, 2012). Moreover, tripeptides IRW and IQW, obtained from ovotransferrin (an egg white component), were able to down-regulate the cytokine-induced inflammatory protein expression in vascular endothelium, by modulation of the NF-κB pathway (Majumder et al., 2013). In an *in vitro* study, Fitzgerald and colleagues showed that a fish hydrolysate preparation could induce proliferation and migration in intestinal epithelial cells, which may contribute to anti-inflammatory and healing properties (Fitzgerald et al., 2005).

CONCLUSION

Currently, people and governments all over the world are concerned about the rapid progress of noncommunicable diseases, such as cancers, diabetes, and cardiovascular and neurodegenerative diseases. Therefore, the interest in health-promoting foods has dramatically increased. Based on the evidence reviewed here, we think that bioactive peptides isolated from food sources are reasonable tools to counteract oxidative stress, apoptosis and cell death pathway, inflammation and eventually neurotoxicity. Although these compounds are promising as potential anti-neurodegenerative agents, further investigations are needed to understand their pharmacokinetic dynamics and to clarify these beneficial effects in order to effectively translate the research from the bench to the bedside.

REFERENCES

Aihara, K., Ishii, H., & Yoshida, M. (2009). Casein-derived tripeptide, Val-Pro-Pro (VPP), modulates monocyte adhesion to vascular endothelium. *Journal of Atherosclerosis and Thrombosis*, 16(5), 594–603. https://doi.org/10.5551/jat.729

Armstrong, M. J., & Okun, M. S. (2020). Diagnosis and treatment of Parkinson disease: A review. *JAMA - Journal of the American Medical Association*, 323(6), 548–560. https://doi.org/10.1001/jama.2019.22360

Arvanitakis, Z., Shah, R. C., & Bennett, D. A. (2019). Diagnosis and management of dementia: Review. *JAMA - Journal of the American Medical Association*, 322(16), 1589–1599. https://doi.org/10.1001/jama.2019.4782

Assadollahi, V., Hassanzadeh, K., Abdi, M., Alasvand, M., Nasseri, S., & Fathi, F. (2019). Effect of embryo cryopreservation on derivation efficiency, pluripotency, and differentiation capacity of mouse embryonic stem cells. *Journal of Cellular Physiology*, 234(12). https://doi.org/10.1002/jcp.28759

Baig, M. H., Ahmad, K., Saeed, M., Alharbi, A. M., Barreto, G. E., Ashraf, G. M., & Choi, I. (2018). Peptide based therapeutics and their use for the treatment of neurodegenerative and other diseases. *Biomedicine and Pharmacotherapy*, *103*, 574–581. https://doi.org/10.1016/j.biopha.2018.04.025

Baltic, M., Boskovic, M., Ivanovic, J., Janjic, J., Dokmanovic, M., Markovic, R., & Baltic, T. (2014). Bioactive peptides from meat and their influence on human health. *Tehnologija Mesa*, *55*(1), 8–21. https://doi.org/10.5937/tehmesa1401008b

Banerjee, P., Suseela, G., & Shanthi, C. (2012). Isolation and identification of cryptic bioactive regions in bovine achilles tendon collagen. *Protein Journal*, *31*(5), 374–386. https://doi.org/10.1007/s10930-012-9415-8

Bensimon, G., Lacomblez, L., & Meininger, V. (1994). A controlled trial of riluzole in amyotrophic lateral sclerosis. *New England Journal of Medicine*, *330*(9), 585–591. https://doi.org/10.1056/nejm199403033300901

Blondelle, S. E., & Lohner, K. (2000). Combinatorial libraries: A tool to design antimicrobial and antifungal peptide analogues having lytic specificities for structure-activity relationship studies. *Biopolymers*, *55*(1). https://doi.org/10.1002/1097-0282(2000)55:1

Borovecki, F., Lovrecic, L., Zhou, J., Jeong, H., Then, F., Rosas, H. D., Hersch, S. M., Hogarth, P., Bouzou, B., Jensen, R. V., & Krainc, D. (2005). Genome-wide expression profiling of human blood reveals biomarkers for Huntington's disease. *Proceedings of the National Academy of Sciences of the United States of America*, *102*(31), 11023–11028. https://doi.org/10.1073/pnas.0504921102

Brettschneider, J., Del Tredici, K., Lee, V. M. Y., & Trojanowski, J. Q. (2015). Spreading of pathology in neurodegenerative diseases: A focus on human studies. *Nature Reviews Neuroscience*, *16*(2), 109–120. https://doi.org/10.1038/nrn3887

Centeno, C., Repici, M., Chatton, J. Y., Riederer, B. M., Bonny, C., Nicod, P., Price, M., Clarke, P. G. H., Papa, S., Franzoso, G., & Borsello, T. (2007). Role of the JNK pathway in NMDA-mediated excitotoxicity of cortical neurons. *Cell Death and Differentiation*, *14*(2), 240–253. https://doi.org/10.1038/sj.cdd.4401988

Chen, H., Zhao, M., Lin, L., Wang, J., Sun-Waterhouse, D., Dong, Y., Zhuang, M., & Su, G. (2015). Identification of antioxidative peptides from defatted walnut meal hydrolysate with potential for improving learning and memory. *Food Research International*, *78*, 216–223. https://doi.org/10.1016/j.foodres.2015.10.008

Chiang, M. C., Chen, H. M., Lee, Y. H., Chang, H. H., Wu, Y. C., Soong, B. W., Chen, C. M., Wu, Y. R., Liu, C. S., Niu, D. M., Wu, J. Y., Chen, Y. T., & Chern, Y. (2007). Dysregulation of C/EBPα by mutant Huntingtin causes the urea cycle deficiency in Huntington's disease. *Human Molecular Genetics*, *16*(5), 483–498. https://doi.org/10.1093/hmg/ddl481

Chiò, A., Borghero, G., Pugliatti, M., Ticca, A., Calvo, A., Moglia, C., Mutani, R., Brunetti, M., Ossola, I., Marrosu, M. G., Murru, M. R., Floris, G., Cannas, A., Parish, L. D., Cossu, P., Abramzon, Y., Johnson, J. O., Nalls, M. A., Arepalli, S., ... Volanti, P. (2011). Large proportion of amyotrophic lateral sclerosis cases in Sardinia due to a single founder mutation of the TARDBP gene. *Archives of Neurology*, *68*(5), 594–598. https://doi.org/10.1001/archneurol.2010.352

Corcia, P., Pradat, P. F., Salachas, F., Bruneteau, G., le Forestier, N., Seilhean, D., Hauw, J. J., & Meininger, V. (2008). Causes of death in a post-mortem series of ALS patients. *Amyotrophic Lateral Sclerosis*, *9*(1), 59–62. https://doi.org/10.1080/17482960701656940

Cornejo, V. H., & Hetz, C. (2013). The unfolded protein response in Alzheimer's disease. *Seminars in Immunopathology*, 35(3), 277–292. https://doi.org/10.1007/s00281-013-0373-9

Corrado, L., Del Bo, R., Castellotti, B., Ratti, A., Cereda, C., Penco, S., Sorarù, G., Carlomagno, Y., Ghezzi, S., Pensato, V., Colombrita, C., Gagliardi, S., Cozzi, L., Orsetti, V., Mancuso, M., Siciliano, G., Mazzini, L., Comi, G. P., Gellera, C., … Silani, V. (2010). Mutations of FUS gene in sporadic amyotrophic lateral sclerosis. *Journal of Medical Genetics*, 47(3), 190–194. https://doi.org/10.1136/jmg.2009.071027

Cryan, J. F., O'riordan, K. J., Cowan, C. S. M., Sandhu, K. V., Bastiaanssen, T. F. S., Boehme, M., Codagnone, M. G., Cussotto, S., Fulling, C., Golubeva, A. V., Guzzetta, K. E., Jaggar, M., Long-Smith, C. M., Lyte, J. M., Martin, J. A., Molinero-Perez, A., Moloney, G., Morelli, E., Morillas, E., … Dinan, T. G. (2019). The microbiota-gut-brain axis. *Physiological Reviews*, 99(4), 1877–2013. https://doi.org/10.1152/physrev.00018.2018

Cummings, J., Aisen, P., Lemere, C., Atri, A., Sabbagh, M., & Salloway, S. (2021). Aducanumab produced a clinically meaningful benefit in association with amyloid lowering. *Alzheimer's Research and Therapy*, 13(1). https://doi.org/10.1186/s13195-021-00838-z

De vos, K. J., Chapman, A. L., Tennant, M. E., Manser, C., Tudor, E. L., Lau, K. F., Brownlees, J., Ackerley, S., Shaw, P. J., Mcloughlin, D. M., Shaw, C. E., Leigh, P. N., Miller, C. C. J., & Grierson, A. J. (2007). Familial amyotrophic lateral sclerosis-linked SOD1 mutants perturb fast axonal transport to reduce axonal mitochondria content. *Human Molecular Genetics*, 16(22), 2720–2728. https://doi.org/10.1093/hmg/ddm226

DeTure, M. A., & Dickson, D. W. (2019). The neuropathological diagnosis of Alzheimer's disease. *Molecular Neurodegeneration*, 14(1), 32. https://doi.org/10.1186/s13024-019-0333-5

Di Bernardini, R., Rai, D. K., Bolton, D., Kerry, J., O'Neill, E., Mullen, A. M., Harnedy, P., & Hayes, M. (2011). Isolation, purification and characterization of antioxidant peptidic fractions from a bovine liver sarcoplasmic protein thermolysin hydrolyzate. *Peptides*, 32(2), 388–400. https://doi.org/10.1016/j.peptides.2010.11.024

Dickson, D. W. (1997). Neuropathological diagnosis of Alzheimer's disease: A perspective from longitudinal clinicopathological studies. *Neurobiology of Aging*, 18(4) (suppl.). https://doi.org/10.1016/S0197-4580(97)00065-1

Dong, S., Zeng, M., Wang, D., Liu, Z., Zhao, Y., & Yang, H. (2008). Antioxidant and biochemical properties of protein hydrolysates prepared from Silver carp (Hypophthalmichthys molitrix). *Food Chemistry*, 107(4), 1485–1493. https://doi.org/10.1016/j.foodchem.2007.10.011

Dubinsky, R., & Gray, C. (2006). CYTE-I-HD: Phase I dose finding and tolerability study of cysteamine (cystagon in Huntington's disease). *Movement Disorders*, 21(4), 530–533. https://doi.org/10.1002/mds.20756

Dugger, B. N., & Dickson, D. W. (2017). Pathology of neurodegenerative diseases. *Cold Spring Harbor Perspectives in Biology*, 9(7). https://doi.org/10.1101/cshperspect.a028035

Dumitrescu, L., Popescu-Olaru, I., Cozma, L., Tulbă, D., Hinescu, M. E., Ceafalan, L. C., Gherghiceanu, M., & Popescu, B. O. (2018). Oxidative stress and the microbiota-gut-brain axis. *Oxidative Medicine and Cellular Longevity*, 2018. https://doi.org/10.1155/2018/2406594

Erdmann, K., Cheung, B. W. Y., & Schröder, H. (2008). The possible roles of food-derived bioactive peptides in reducing the risk of cardiovascular disease. *Journal of Nutritional Biochemistry*, *19*(10), 643–654. https://doi.org/10.1016/j.jnutbio.2007.11.010

Feldman, H., Gauthier, S., Hecker, J., Vellas, B., Subbiah, P., Whalen, E., & The Donepezil MSAD Study Investigators Group. (2001). A 24-week, randomized, double-blind study of donepezil in moderate to severe Alzheimer's disease. *Neurology*, *57*(4), 613–620. https://doi.org/10.1212/WNL.57.4.613

Fitzgerald, A. J., Rai, P. S., Marchbank, T., Taylor, G. W., Ghosh, S., Ritz, B. W., & Playford, R. J. (2005). Reparative properties of a commercial fish protein hydrolysate preparation. *Gut*, *54*(6), 775–781. https://doi.org/10.1136/gut.2004.060608

Forsberg, K., Andersen, P. M., Marklund, S. L., & Brännström, T. (2011). Glial nuclear aggregates of superoxide dismutase-1 are regularly present in patients with amyotrophic lateral sclerosis. *Acta Neuropathologica*, *121*(5), 623–634. https://doi.org/10.1007/s00401-011-0805-3

Gallego, M., Mauri, L., Aristoy, M. C., Toldrá, F., & Mora, L. (2020). Antioxidant peptides profile in dry-cured ham as affected by gastrointestinal digestion. *Journal of Functional Foods*, *69*, 103956. https://doi.org/10.1016/j.jff.2020.103956

Gamir-Morralla, A., López-Menéndez, C., Ayuso-Dolado, S., Tejeda, G. S., Montaner, J., Rosell, A., Iglesias, T., & Díaz-Guerra, M. (2015). Development of a neuroprotective peptide that preserves survival pathways by preventing Kidins220/ ARMS calpain processing induced by excitotoxicity. *Cell Death and Disease*, *6*(10), e1939–e1939. https://doi.org/10.1038/cddis.2015.307

Ganguly, J., & Jog, M. (2020a). Tauopathy and movement disorders—Unveiling the chameleons and mimics. *Frontiers in Neurology*, *11*, 599384. https://doi.org/10.3389/fneur.2020.599384

Ganguly, J., & Jog, M. (2020b). Tauopathy and movement disorders—Unveiling the chameleons and mimics. *Frontiers in Neurology*, *11*, 599384. https://doi.org/10.3389/fneur.2020.599384

Gardian, G., Browne, S. E., Choi, D. K., Klivenyi, P., Gregorio, J., Kubilus, J. K., Ryu, H., Langley, B., Ratan, R. R., Ferrante, R. J., & Beal, M. F. (2005). Neuroprotective effects of phenylbutyrate in the N171-82Q transgenic mouse model of Huntington's disease. *Journal of Biological Chemistry*, *280*(1), 556–563. https://doi.org/10.1074/jbc.M410210200

Ghazizadeh, H., Saberi-Karimian, M., Aghasizadeh, M., Sahebi, R., Ghazavi, H., Khedmatgozar, H., Timar, A., Rohban, M., Javandoost, A., & Ghayour-Mobarhan, M. (2020). Pro-oxidant–antioxidant balance (PAB) as a prognostic index in assessing the cardiovascular risk factors: A narrative review. *Obesity Medicine*, *19*, 100272. https://doi.org/10.1016/j.obmed.2020.100272

Ghosh, T. S., Rampelli, S., Jeffery, I. B., Santoro, A., Neto, M., Capri, M., Giampieri, E., Jennings, A., Candela, M., Turroni, S., Zoetendal, E. G., Hermes, G. D. A., Elodie, C., Meunier, N., Brugere, C. M., Pujos-Guillot, E., Berendsen, A. M., De Groot, L. C. P. G. M., Feskins, E. J. M., … O'Toole, P. W. (2020). Mediterranean diet intervention alters the gut microbiome in older people reducing frailty and improving health status: The NU-AGE 1-year dietary intervention across five European countries. *Gut*, *69*(7), 1218–1228. https://doi.org/10.1136/gutjnl-2019-319654

Gijselinck, I., Van Langenhove, T., van der Zee, J., Sleegers, K., Philtjens, S., Kleinberger, G., Janssens, J., Bettens, K., Van Cauwenberghe, C., Pereson, S., Engelborghs, S., Sieben, A., De Jonghe, P., Vandenberghe, R., Santens, P., De Bleecker, J., Maes, G.,

Bäumer, V., Dillen, L., ... Van Broeckhoven, C. (2012). A C9orf72 promoter repeat expansion in a Flanders-Belgian cohort with disorders of the frontotemporal lobar degeneration-amyotrophic lateral sclerosis spectrum: A gene identification study. *The Lancet Neurology*, *11*(1), 54–65. https://doi.org/10.1016/S1474-4422(11)70261-7

Gitler, A. D., Dhillon, P., & Shorter, J. (2017). Neurodegenerative disease: Models, mechanisms, and a new hope. *DMM Disease Models and Mechanisms*, *10*(5), 499–502. https://doi.org/10.1242/dmm.030205

Glass, C. K., Saijo, K., Winner, B., Marchetto, M. C., & Gage, F. H. (2010). Mechanisms underlying inflammation in neurodegeneration. *Cell*, *140*(6), 918–934. https://doi.org/10.1016/j.cell.2010.02.016

Gomez-Guillen, M. C., Gimenez, B., Lopez-Caballero, M. E., & Montero, M. P. (2011). Functional and bioactive properties of collagen and gelatin from alternative sources: A review. *Food Hydrocolloids*, *25*(8), 1813–1827. https://doi.org/10.1016/j.foodhyd.2011.02.007

González-Casacuberta, I., Juárez-Flores, D. L., Morén, C., & Garrabou, G. (2019). Bioenergetics and autophagic imbalance in patients-derived cell models of Parkinson disease supports systemic dysfunction in neurodegeneration. *Frontiers in Neuroscience*, *13*, 894. https://doi.org/10.3389/fnins.2019.00894

Gordon, P. H., Cheng, B., Katz, I. B., Pinto, M., Hays, A. P., Mitsumoto, H., & Rowland, L. P. (2006). The natural history of primary lateral sclerosis. *Neurology*, *66*(5), 647–653. https://doi.org/10.1212/01.wnl.0000200962.94777.71

Gu, H., Song, I. B., Han, H. J., Lee, N. Y., Cha, J. Y., Son, Y. K., & Kwon, J. (2018). Antioxidant activity of royal jelly hydrolysates obtained by enzymatic treatment. *Korean Journal for Food Science of Animal Resources*, *38*(1), 135–142. https://doi.org/10.5851/kosfa.2018.38.1.135

Gubert, C., Kong, G., Renoir, T., & Hannan, A. J. (2020). Exercise, diet and stress as modulators of gut microbiota: Implications for neurodegenerative diseases. *Neurobiology of Disease*, *134*. https://doi.org/10.1016/j.nbd.2019.104621

Guzman-Martinez, L., Maccioni, R. B., Andrade, V., Navarrete, L. P., Pastor, M. G., & Ramos-Escobar, N. (2019). Neuroinflammation as a common feature of neurodegenerative disorders. *Frontiers in Pharmacology*, *10*, 1008. https://doi.org/10.3389/fphar.2019.01008

Harada, M., Habata, Y., Hosoya, M., Nishi, K., Fujii, R., Kobayashi, M., & Hinuma, S. (2004). N-Formylated humanin activates both formyl peptide receptor-like 1 and 2. *Biochemical and Biophysical Research Communications*, *324*(1), 255–261. https://doi.org/10.1016/j.bbrc.2004.09.046

Hashimoto, Y., Niikura, T., Tajima, H., Yasukawa, T., Sudo, H., Ito, Y., Kita, Y., Kawasumi, M., Kouyama, K., Doyu, M., Sobue, G., Koide, T., Tsuji, S., Lang, J., Kurokawa, K., & Nishimoto, I. (2001). A rescue factor abolishing neuronal cell death by a wide spectrum of familial Alzheimer's disease genes and Aβ. *Proceedings of the National Academy of Sciences of the United States of America*, *98*(11), 6336–6341. https://doi.org/10.1073/pnas.101133498

Heemels, M. T. (2016). Neurodegenerative diseases. *Nature*, *539*(7628), 179. https://doi.org/10.1038/539179a

Ivanova, M. I., Sievers, S. A., Guenther, E. L., Johnson, L. M., Winkler, D. D., Galaleldeen, A., Sawaya, M. R., Hart, P. J., & Eisenberg, D. S. (2014). Aggregation-triggering segments of SOD1 fibril formation support a common pathway for familial and sporadic ALS. *Proceedings of the National Academy of Sciences of the United States of America*, *111*(1), 197–201. https://doi.org/10.1073/pnas.1320786110

Jiménez-Colmenero, F., Sánchez-Muniz, F. J., & Olmedilla-Alonso, B. (2010). Design and development of meat-based functional foods with walnut: Technological, nutritional and health impact. *Food Chemistry, 123*(4), 959–967. https://doi.org/10.1016/j.foodchem.2010.05.104

Jin, M. M., Zhang, L., Yu, H. X., Meng, J., Sun, Z., & Lu, R. R. (2013). Protective effect of whey protein hydrolysates on H2O2-induced PC12 cells oxidative stress via a mitochondria-mediated pathway. *Food Chemistry, 141*(2), 847–852. https://doi.org/10.1016/j.foodchem.2013.03.076

Kabashi, E., Bercier, V., Lissouba, A., Liao, M., Brustein, E., Rouleau, G. A., & Drapeau, P. (2011). Fus and tardbp but not sod1 interact in genetic models of amyotrophic lateral sclerosis. *PLOS Genetics, 7*(8). https://doi.org/10.1371/journal.pgen.1002214

Kempuraj, D., Thangavel, R., Natteru, P. A., Selvakumar, G. P., Saeed, D., Zahoor, H., Zaheer, S., Iyer, S. S., & Zaheer, A. (2016). Neuroinflammation induces neurodegeneration. *Journal of Neurology, Neurosurgery and Spine, 1*(1). http://www.ncbi.nlm.nih.gov/pubmed/28127589

Ketnawa, S., Wickramathilaka, M., & Liceaga, A. M. (2018). Changes on antioxidant activity of microwave-treated protein hydrolysates after simulated gastrointestinal digestion: Purification and identification. *Food Chemistry, 254*, 36–46. https://doi.org/10.1016/j.foodchem.2018.01.133

Kiernan, M. C., Vucic, S., Cheah, B. C., Turner, M. R., Eisen, A., Hardiman, O., Burrell, J. R., & Zoing, M. C. (2011). Amyotrophic lateral sclerosis. *The Lancet, 377*(9769), 942–955. https://doi.org/10.1016/S0140-6736(10)61156-7

Kiernan, M. C., Vucic, S., Talbot, K., McDermott, C. J., Hardiman, O., Shefner, J. M., Al-Chalabi, A., Huynh, W., Cudkowicz, M., Talman, P., Van den Berg, L. H., Dharmadasa, T., Wicks, P., Reilly, C., & Turner, M. R. (2021). Improving clinical trial outcomes in amyotrophic lateral sclerosis. *Nature Reviews Neurology, 17*(2), 104–118. https://doi.org/10.1038/s41582-020-00434-z

Kim, E. K., Lee, S. J., Moon, S. H., Jeon, B. T., Kim, B., Park, T. K., Han, J. S., & Park, P. J. (2010). Neuroprotective effects of a novel peptide purified from venison protein. *Journal of Microbiology and Biotechnology, 20*(4), 700–707. https://doi.org/10.4014/jmb.0909.09033

Kodera, T., & Nio, N. (2006). Identification of an anqiotensin I-converting enzyme inhibitory peptides from protein hydrolysates by a soybean protease and the antihypertensive effects of hydrolysates in spontaneously hypertensive model rats. *Journal of Food Science, 71*(3), C164–C173. https://doi.org/10.1111/j.1365-2621.2006.tb15612.x

Korhonen, H., & Pihlanto, A. (2006). Bioactive peptides: Production and functionality. *International Dairy Journal, 16*(9), 945–960. https://doi.org/10.1016/j.idairyj.2005.10.012

Kovacs, G. G. (2019). Molecular pathology of neurodegenerative diseases: Principles and practice. *Journal of Clinical Pathology, 72*(11), 725–735. https://doi.org/10.1136/jclinpath-2019-205952

Kudo, K., Onodera, S., Takeda, Y., Benkeblia, N., & Shiomi, N. (2009). Antioxidative activities of some peptides isolated from hydrolyzed potato protein extract. *Journal of Functional Foods, 1*(2), 170–176. https://doi.org/10.1016/j.jff.2009.01.006

Kumar, A., Sidhu, J., Goyal, A., Tsao, J. W., & Svercauski, J. (2021). Alzheimer disease (nursing). *StatPearls.* http://www.ncbi.nlm.nih.gov/pubmed/33760564

Lacomblez, L., Bensimon, G., Leigh, P. N., Guillet, P., & Meininger, V. (1996). Dose-ranging study of riluzole in amyotrophic lateral sclerosis. *Lancet*, *347*(9013), 1425–1431. https://doi.org/10.1016/S0140-6736(96)91680-3

Lashuel, H. A., Overk, C. R., Oueslati, A., & Masliah, E. (2013). The many faces of α-synuclein: From structure and toxicity to therapeutic target. *Nature Reviews Neuroscience*, *14*(1), 38–48. https://doi.org/10.1038/nrn3406

Lee, S. Y., & Hur, S. J. (2019). Mechanisms of neuroprotective effects of peptides derived from natural materials and their production and assessment. *Comprehensive Reviews in Food Science and Food Safety*, *18*(4), 923–935. https://doi.org/10.1111/1541-4337.12451

Lewerenz, J., & Maher, P. (2015). Chronic glutamate toxicity in neurodegenerative diseases-What is the evidence? *Frontiers in Neuroscience*, *9*. https://doi.org/10.3389/fnins.2015.00469

Li, B., Chen, F., Wang, X., Ji, B., & Wu, Y. (2007). Isolation and identification of antioxidative peptides from porcine collagen hydrolysate by consecutive chromatography and electrospray ionization-mass spectrometry. *Food Chemistry*, *102*(4), 1135–1143. https://doi.org/10.1016/j.foodchem.2006.07.002

Li, W., Zhao, T., Zhang, J., Wu, C., Zhao, M., & Su, G. (2016). Comparison of neuroprotective and cognition-enhancing properties of hydrolysates from soybean, walnut, and peanut protein. *Journal of Chemistry*, *2016*. https://doi.org/10.1155/2016/9358285

Liu, M., Wang, Y., Liu, Y., & Ruan, R. (2016). Bioactive peptides derived from traditional Chinese medicine and traditional Chinese food: A review. *Food Research International*, *89*, 63–73. https://doi.org/10.1016/j.foodres.2016.08.009

Liu, Q., Kong, B., Xiong, Y. L., & Xia, X. (2010). Antioxidant activity and functional properties of porcine plasma protein hydrolysate as influenced by the degree of hydrolysis. *Food Chemistry*, *118*(2), 403–410. https://doi.org/10.1016/j.foodchem.2009.05.013

Lomen-Hoerth, C., Anderson, T., & Miller, B. (2002). The overlap of amyotrophic lateral sclerosis and frontotemporal dementia. *Neurology*, *59*(7), 1077–1079. https://doi.org/10.1212/WNL.59.7.1077

Louis, E. D., Anderson, K. E., Moskowitz, C., Thorne, D. Z., & Marder, K. (2000). Dystonia-predominant adult-onset Huntington disease: Association between motor phenotype and age of onset in adults. *Archives of Neurology*, *57*(9), 1326–1330. https://doi.org/10.1001/archneur.57.9.1326

Ma, X., Cui, X., Li, J., Li, C., & Wang, Z. (2017). Peptides from sesame cake reduce oxidative stress and amyloid-β-induced toxicity by upregulation of SKN-1 in a transgenic Caenorhabditis elegans model of Alzheimer's disease. *Journal of Functional Foods*, *39*, 287–298. https://doi.org/10.1016/j.jff.2017.10.032

Ma, X., Li, J., Cui, X., Li, C., & Wang, Z. (2020). Dietary supplementation with peptides from sesame cake alleviates Parkinson's associated pathologies in Caenorhabditis elegans. *Journal of Functional Foods*, *65*, 103737. https://doi.org/10.1016/j.jff.2019.103737

Majounie, E., Renton, A. E., Mok, K., Dopper, E. G. P., Waite, A., Rollinson, S., Chiò, A., Restagno, G., Nicolaou, N., Simon-Sanchez, J., van Swieten, J. C., Abramzon, Y., Johnson, J. O., Sendtner, M., Pamphlett, R., Orrell, R. W., Mead, S., Sidle, K. C., Houlden, H., … Logroscino, G. (2012). Frequency of the C9orf72 hexanucleotide repeat expansion in patients with amyotrophic lateral sclerosis and frontotemporal dementia: A cross-sectional study. *The Lancet Neurology, 11*(4), 323–330. https://doi.org/10.1016/S1474-4422(12)70043-1

Majumder, K., Chakrabarti, S., Davidge, S. T., & Wu, J. (2013). Structure and activity study of egg protein ovotransferrin derived peptides (IRW and IQW) on endothelial inflammatory response and oxidative stress. *Journal of Agricultural and Food Chemistry, 61*(9), 2120–2129. https://doi.org/10.1021/jf3046076

Majumder, K., Mine, Y., & Wu, J. (2016). The potential of food protein-derived anti-inflammatory peptides against various chronic inflammatory diseases. *Journal of the Science of Food and Agriculture, 96*(7), 2303–2311. https://doi.org/10.1002/jsfa.7600

Marcelli, S., Iannuzzi, F., Ficulle, E., Mango, D., Pieraccini, S., Pellegrino, S., Corbo, M., Sironi, M., Pittaluga, A., Nisticò, R., & Feligioni, M. (2019). The selective disruption of presynaptic JNK2/STX1a interaction reduces NMDA receptor-dependent glutamate release. *Scientific Reports, 9*(1), 1–12. https://doi.org/10.1038/s41598-019-43709-2

Maroteaux, L., Campanelli, J. T., & Scheller, R. H. (1988). Synuclein: A neuron-specific protein localized to the nucleus and presynaptic nerve terminal. *Journal of Neuroscience, 8*(8), 2804–2815. https://doi.org/10.1523/jneurosci.08-08-02804.1988

McCann, H., Stevens, C. H., Cartwright, H., & Halliday, G. M. (2014). α-synucleinopathy phenotypes. *Parkinsonism and Related Disorders, 20*(suppl.1). https://doi.org/10.1016/S1353-8020(13)70017-8

McCord, J. M. (1993). Human disease, free radicals, and the oxidant/antioxidant balance. *Clinical Biochemistry, 26*(5), 351–357. https://doi.org/10.1016/0009-9120(93)90111-I

Mileusnic, R., Lancashire, C. L., & Rose, S. R. R. (2004). The peptide sequence Arg-Glu-Arg, present in the amyloid precursor protein, protects against memory loss caused by Aβ and acts as a cognitive enhancer. *European Journal of Neuroscience, 19*(7), 1933–1938. https://doi.org/10.1111/j.1460-9568.2004.03276.x

Miller, R. G., Rosenberg, J. A., Gelinas, D. F., Mitsumoto, H., Newman, D., Sufit, R., Borasio, G. D., Bradley, W. G., Bromberg, M. B., Brooks, B. R., Kasarskis, E. J., Munsat, T. L., & Oppenheimer, E. A. (1999). Practice parameter: The care of the patient with amyotrophic lateral sclerosis (An evidence-based review): Report of the Quality Standards Subcommittee of the American Academy of Neurology. *Neurology, 52*(7), 1311–1323. https://doi.org/10.1212/wnl.52.7.1311

Morris, G. P., Clark, I. A., & Vissel, B. (2014). Inconsistencies and controversies surrounding the amyloid hypothesis of Alzheimer's disease. *Acta Neuropathologica Communications, 2*(1), 135. https://doi.org/10.1186/s40478-014-0135-5

Nagpal, R., Neth, B. J., Wang, S., Craft, S., & Yadav, H. (2019). Modified Mediterranean-ketogenic diet modulates gut microbiome and short-chain fatty acids in association with Alzheimer's disease markers in subjects with mild cognitive impairment. *EBioMedicine, 47*, 529–542. https://doi.org/10.1016/j.ebiom.2019.08.032

Nielsen, D. S. G., Theil, P. K., Larsen, L. B., & Purup, S. (2012). Effect of milk hydrolysates on inflammation markers and drug-induced transcriptional alterations in cell-based models. *Journal of Animal Science, 90*(suppl.4), 403–405. https://doi.org/10.2527/jas.53953

Nijholt, D. A. T., De Kimpe, L., Elfrink, H. L., Hoozemans, J. J. M., & Scheper, W. (2011). Removing protein aggregates: The role of proteolysis in neurodegeneration. *Current Medicinal Chemistry*, *18*(16), 2459–2476. https://doi.org/10.2174/092986711795843236

Nisticò, R., Florenzano, F., Mango, D., Ferraina, C., Grilli, M., Di Prisco, S., Nobili, A., Saccucci, S., D'Amelio, M., Morbin, M., Marchi, M., Mercuri, N. B., Davis, R. J., Pittaluga, A., & Feligioni, M. (2015). Presynaptic c-Jun N-terminal kinase 2 regulates NMDA receptor-dependent glutamate release. *Scientific Reports*, *5*. https://doi.org/10.1038/srep09035

Obeso, J. A., Stamelou, M., Goetz, C. G., Poewe, W., Lang, A. E., Weintraub, D., Burn, D., Halliday, G. M., Bezard, E., Przedborski, S., Lehericy, S., Brooks, D. J., Rothwell, J. C., Hallett, M., DeLong, M. R., Marras, C., Tanner, C. M., Ross, G. W., Langston, J. W., … Stoessl, A. J. (2017). Past, present, and future of Parkinson's disease: A special essay on the 200th Anniversary of the Shaking Palsy. *Movement Disorders*, *32*(9), 1264–1310. https://doi.org/10.1002/mds.27115

Paulson, H. L., & Albin, R. L. (2011). Huntington's disease: Clinical features and routes to therapy. In: *Neurobiology of Huntington's Disease: Applications to Drug Discovery*. CRC Press/Taylor & Francis. http://www.ncbi.nlm.nih.gov/pubmed/21882418

Pellegrini, C., Antonioli, L., Calderone, V., Colucci, R., Fornai, M., & Blandizzi, C. (2020). Microbiota-gut-brain axis in health and disease: Is NLRP3 inflammasome at the crossroads of microbiota-gut-brain communications? *Progress in Neurobiology*, *191*. https://doi.org/10.1016/j.pneurobio.2020.101806

Pellegrini, C., Antonioli, L., Colucci, R., Blandizzi, C., & Fornai, M. (2018). Interplay among gut microbiota, intestinal mucosal barrier and enteric neuro-immune system: A common path to neurodegenerative diseases? *Acta Neuropathologica*, *136*(3), 345–361. https://doi.org/10.1007/s00401-018-1856-5

Perl, D. P. (2010). Neuropathology of Alzheimer's disease. *Mount Sinai Journal of Medicine*, *77*(1), 32–42. https://doi.org/10.1002/msj.20157

Pohl, C., & Dikic, I. (2019). Cellular quality control by the ubiquitin-proteasome system and autophagy. *Science*, *366*(6467), 818–822. https://doi.org/10.1126/science.aax3769

Purchas, R. W., & Busboom, J. R. (2005). The effect of production system and age on levels of iron, taurine, carnosine, coenzyme Q10, and creatine in beef muscles and liver. *Meat Science*, *70*(4), 589–596. https://doi.org/10.1016/j.meatsci.2005.02.008

Rajapakse, N., Mendis, E., Jung, W. K., Je, J. Y., & Kim, S. K. (2005). Purification of a radical scavenging peptide from fermented mussel sauce and its antioxidant properties. *Food Research International*, *38*(2), 175–182. https://doi.org/10.1016/j.foodres.2004.10.002

Rami, L., Sala-Llonch, R., Solé-Padullés, C., Fortea, J., Olives, J., Lladó, A., Pea-Gómez, C., Balasa, M., Bosch, B., Antonell, A., Sanchez-Valle, R., Bartrés-Faz, D., & Molinuevo, J. L. (2012). Distinct functional activity of the precuneus and posterior cingulate cortex during encoding in the preclinical stage of Alzheimer's disease. *Journal of Alzheimer's Disease*, *31*(3), 517–526. https://doi.org/10.3233/JAD-2012-120223

Ray, P. D., Huang, B. W., & Tsuji, Y. (2012). Reactive oxygen species (ROS) homeostasis and redox regulation in cellular signaling. *Cellular Signalling*, *24*(5), 981–990. https://doi.org/10.1016/j.cellsig.2012.01.008

Reiner, A., Dragatsis, I., & Dietrich, P. (2011). Genetics and neuropathology of Huntington's disease. *International Review of Neurobiology, 98*, 325–372. https://doi.org/10.1016/B978-0-12-381328-2.00014-6

Roos, R. A. C. (2010). Huntington's disease: A clinical review. *Orphanet Journal of Rare Diseases, 5*(1). https://doi.org/10.1186/1750-1172-5-40

Rowland, L. P., & Shneider, N. A. (2001). Amyotrophic lateral sclerosis. *New England Journal of Medicine, 344*(22), 1688–1700. https://doi.org/10.1056/NEJM200105313442207

Russo, M. V., & McGavern, D. B. (2016). Inflammatory neuroprotection following traumatic brain injury. *Science, 353*(6301), 783–785. https://doi.org/10.1126/science.aaf6260

Rutherfurd-Markwick, K. J. (2012). Food proteins as a source of bioactive peptides with diverse functions. *British Journal of Nutrition, 108*, S149–S157. https://doi.org/10.1017/S000711451200253X

Ryan, N. S., Rossor, M. N., & Fox, N. C. (2015). Alzheimer's disease in the 100 years since Alzheimer's death. *Brain, 138*(12), 3816–3821. https://doi.org/10.1093/brain/awv316

Sarmadi, B. H., & Ismail, A. (2010). Antioxidative peptides from food proteins: A review. *Peptides, 31*(10), 1949–1956. https://doi.org/10.1016/j.peptides.2010.06.020

Serrano-Pozo, A., Frosch, M. P., Masliah, E., & Hyman, B. T. (2011). Neuropathological alterations in Alzheimer disease. *Cold Spring Harbor Perspectives in Medicine, 1*(1). https://doi.org/10.1101/cshperspect.a006189

Simon, D. K., Tanner, C. M., & Brundin, P. (2020). Parkinson disease epidemiology, pathology, genetics, and pathophysiology. *Clinics in Geriatric Medicine, 36*(1), 1–12. https://doi.org/10.1016/j.cger.2019.08.002

Suetsuna, K., Ukeda, H., & Ochi, H. (2000). Isolation and characterization of free radical scavenging activities peptides derived from casein. *Journal of Nutritional Biochemistry, 11*(3), 128–131. https://doi.org/10.1016/S0955-2863(99)00083-2

Swerdlow, R. H. (2012). Mitochondria and cell bioenergetics: Increasingly recognized components and a possible etiologic cause of Alzheimer's disease. *Antioxidants and Redox Signaling, 16*(12), 1434–1455. https://doi.org/10.1089/ars.2011.4149

Szeto, J. Y. Y., & Lewis, S. J. G.. (2016). Current treatment options for Alzheimer's disease and Parkinson's disease dementia. *Current Neuropharmacology, 14*(4), 326–338.

Tadesse, S. A., & Emire, S. A. (2020). Production and processing of antioxidant bioactive peptides: A driving force for the functional food market. *Heliyon, 6*(8), e04765. https://doi.org/10.1016/j.heliyon.2020.e04765

Tu, W., Xu, X., Peng, L., Zhong, X., Zhang, W., Soundarapandian, M. M., Balel, C., Wang, M., Jia, N., Zhang, W., Lew, F., Chan, S. L., Chen, Y., & Lu, Y. (2010). DAPK1 interaction with NMDA receptor NR2B subunits mediates brain damage in stroke. *Cell, 140*(2), 222–234. https://doi.org/10.1016/j.cell.2009.12.055

Udenigwe, C. C., & Aluko, R. E. (2012). Food protein-derived bioactive peptides: Production, processing, and potential health benefits. *Journal of Food Science, 77*(1). https://doi.org/10.1111/j.1750-3841.2011.02455.x

Volpicelli-Daley, L. A., Luk, K. C., Patel, T. P., Tanik, S. A., Riddle, D. M., Stieber, A., Meaney, D. F., Trojanowski, J. Q., & Lee, V. M. Y. (2011). Exogenous α-synuclein fibrils induce Lewy body pathology leading to synaptic dysfunction and neuron death. *Neuron, 72*(1), 57–71. https://doi.org/10.1016/j.neuron.2011.08.033

Von Bernhardi, R., & Eugenín, J. (2012). Alzheimer's disease: Redox dysregulation as a common denominator for diverse pathogenic mechanisms. *Antioxidants and Redox Signaling*, *16*(9), 974–1031. https://doi.org/10.1089/ars.2011.4082

Vucic, S., & Kiernan, M. C. (2007). Abnormalities in cortical and peripheral excitability in flail arm variant amyotrophic lateral sclerosis. *Journal of Neurology, Neurosurgery, and Psychiatry*, *78*(8), 849–852. https://doi.org/10.1136/jnnp.2006.105056

Wang, S., Harvey, L., Martin, R., van der Beek, E. M., Knol, J., Cryan, J. F., & Renes, I. B. (2018). Targeting the gut microbiota to influence brain development and function in early life. *Neuroscience and Biobehavioral Reviews*, *95*, 191–201. https://doi.org/10.1016/j.neubiorev.2018.09.002

Wei, H., Qin, Z. H., Senatorov, V. V., Wei, W., Wang, Y., Qian, Y., & Chuang, D. M. (2001). Lithium suppresses excitotoxicity-induced striatal lesions in a rat model of Huntington's disease. *Neuroscience*, *106*(3), 603–612. https://doi.org/10.1016/S0306-4522(01)00311-6

Wijesekera, L. C., & Leigh, P. N. (2009). Amyotrophic lateral sclerosis. *Orphanet Journal of Rare Diseases*, *4*(1). https://doi.org/10.1186/1750-1172-4-3

Wu, Q. J., & Tymianski, M. (2018). Targeting nmda receptors in stroke: New hope in neuroprotection Tim Bliss. *Molecular Brain*, *11*(1). https://doi.org/10.1186/s13041-018-0357-8

Wu, S., Bekhit, A. E. D. A., Wu, Q., Chen, M., Liao, X., Wang, J., & Ding, Y. (2021). Bioactive peptides and gut microbiota: Candidates for a novel strategy for reduction and control of neurodegenerative diseases. *Trends in Food Science and Technology*, *108*, 164–176. https://doi.org/10.1016/j.tifs.2020.12.019

Wyss-Coray, T., & Mucke, L. (2002). Inflammation in neurodegenerative disease - A double-edged sword. *Neuron*, *35*(3), 419–432. https://doi.org/10.1016/S0896-6273(02)00794-8

Yang, X. R., Zhang, L., Ding, D. G., Chi, C. F., Wang, B., & Huo, J. C. (2019). Preparation, identification, and activity evaluation of eight antioxidant peptides from protein hydrolysate of hairtail (*Trichiurus japonicas*) muscle. *Marine Drugs*, *17*(1), 23. https://doi.org/10.3390/md17010023

Yeung, S. Y., Lan, W. H., Huang, C. S., Lin, C. P., Chan, C. P., Chang, M. C., & Jeng, J. H. (2002). Scavenging property of three cresol isomers against H2O2, hypochlorite, superoxide and hydroxyl radicals. *Food and Chemical Toxicology*, *40*(10), 1403–1413. https://doi.org/10.1016/S0278-6915(02)00102-3

Young, J. F., Therkildsen, M., Ekstrand, B., Che, B. N., Larsen, M. K., Oksbjerg, N., & Stagsted, J. (2013). Novel aspects of health promoting compounds in meat. *Meat Science*, *95*(4), 904–911. https://doi.org/10.1016/j.meatsci.2013.04.036

Zhang, J., Du, H., Zhang, G., Kong, F., Hu, Y., Xiong, S., & Zhao, S. (2020). Identification and characterization of novel antioxidant peptides from crucian carp (*Carassius auratus*) cooking juice released in simulated gastrointestinal digestion by UPLC-MS/MS and in silico analysis. *Journal of Chromatography B: Analytical Technologies in the Biomedical and Life Sciences*, *1136*. https://doi.org/10.1016/j.jchromb.2019.121893

Zhang, Q. X., Jin, M. M., Zhang, L., Yu, H. X., Sun, Z., & Lu, R. R. (2015). Hydrophobicity of whey protein hydrolysates enhances the protective effect against oxidative damage on PC 12 cells. *Journal of Dairy Research*, *82*(1), 1–7. https://doi.org/10.1017/S0022029914000405

Zhang, Z., He, S., Cao, X., Ye, Y., Yang, L., Wang, J., Liu, H., & Sun, H. (2020). Potential prebiotic activities of soybean peptides Maillard reaction products on modulating gut microbiota to alleviate aging-related disorders in D-galactose-induced ICR mice. *Journal of Functional Foods*, *65*, 103729. https://doi.org/10.1016/j.jff.2019.103729

Zhao, T., Su, G., Wang, S., Zhang, Q., Zhang, J., Zheng, L., Sun, B., & Zhao, M. (2017). Neuroprotective effects of acetylcholinesterase inhibitory peptides from anchovy (*Coilia mystus*) against glutamate-induced toxicity in PC12 cells. *Journal of Agricultural and Food Chemistry*, *65*(51), 11192–11201. https://doi.org/10.1021/acs.jafc.7b03945

Zheng, Z., & Diamond, M. I. (2012). Huntington disease and the huntingtin protein. *Progress in Molecular Biology and Translational Science*, *107*, 189–214. https://doi.org/10.1016/B978-0-12-385883-2.00010-2

Zhu, K. X., Guo, X., Guo, X. N., Peng, W., & Zhou, H. M. (2013). Protective effects of wheat germ protein isolate hydrolysates (WGPIH) against hydrogen peroxide-induced oxidative stress in PC12 cells. *Food Research International*, *53*(1), 297–303. https://doi.org/10.1016/j.foodres.2013.05.007

CHAPTER 20

Antimicrobial Activity of Bioactive Peptides and Their Applications in Food Safety
A Review

Mohammad Hossein Maleki, Hooman Jalilvand Nezhad, Nima Keshavarz Bahadori, Milad Daneshniya, and Zahra Latifi

CONTENTS

Introduction	415
Extraction and Obtaining the Bioactive Peptides	416
Diversity of Antimicrobial Bioactive Peptides	416
The Mechanism of Action of Antimicrobial Peptides	417
Barrel Stave Model	418
Toroidal Model	419
Carpet Model	419
Application of Antimicrobial Peptides	420
Application of Bioactive Peptides with Antimicrobial Activity in Food Safety	420
Conclusion	422
References	422

INTRODUCTION

There are many natural ingredients in foods that assist in preventing diseases, slowing down the progression of diseases, and even treating diseases. These properties of natural ingredients originate from existing proteins and peptides, causing these foods to be known as medicinal foods. In recent years, the attention of researchers has been paid to bioactive peptides and their properties due to the high cost of production and, ultimately, the high cost of synthetic and chemical drugs for consumers. In addition to the high cost of production, the side effects of synthetic drugs are considerably high and may deteriorate the patient's condition due to their physical weakness. Bioactive peptides are defined as peptide sequences within proteins that can have positive effects on body performance and/or human health beyond their known nutritional value [1,2]. Bioactive peptides can be prepared by proteolytic hydrolysis or during food processing such as baking, ripening, or fermenting processes. Bioactive peptides typically contain 3–20 amino acids, and

DOI: 10.1201/9781003106524-26

the sequence strength and composition of these amino acids determine their biological activity [3].

Extraction and Obtaining the Bioactive Peptides

Microbial fermentation and enzymatic hydrolysis or a combination of both are commonly employed to obtain bioactive peptides. The biological function of bioactive peptides is more appropriate compared to their parent protein; hence, these peptides are released and extracted using some laboratory methods. After identifying the protein source, specific and nonspecific proteases can be used to release the desired peptides. Many factors, such as the conditions of the hydrolysis process, the degree of protein hydrolysis, the time of hydrolysis, the type of enzyme, and the size of the desired peptide, affect this process. There are two significant obstacles to the industrial production of peptides by enzymatic hydrolysis method. The first obstacle is the presence of these peptides in an intricate complex with amino acids, oligopeptides, fibers, and other compounds. The dependence of the bioactivity of these peptides on some of their physicochemical properties, such as electrical charge, is another obstacle to the production of peptides by the enzymatic hydrolysis method; therefore, the development of plant peptides requires methods such as ultra-filtration or chromatography for purification and filtration [4]. During the extraction by fermentation, the lactic acid bacteria are used to release bioactive peptides. Also, proteolytic enzymes isolated from these lactic acid bacteria can be used in the enzymatic hydrolysis method. Furthermore, fungi such as *Aspergillus* can also be used along with bacteria to produce these peptides during fermentation. Three basic mechanisms create the function of the proteolytic system of these bacteria. First, the activity of several proteolytic enzymes in the cell wall causes the decomposition of proteins into the peptides with between 4 and 30 amino acids. Second, there is a transition system that includes binding proteins and two permeases to form a transport channel and two ATPases to provide system energy. Third, a group of intracellular peptidases works together for the decomposition of peptides transmitted to amino acids [5].

Diversity of Antimicrobial Bioactive Peptides

The diversity of antimicrobial peptides is considerably high, which has made it challenging to categorize them, so a general classification is applied based on their substructure. The linear peptides are those, such as the cecropin from the silkworm [6] and manganin from the African clawed frog (*Xenopus laevis*) [7]. This arrangement is made only after entering the membrane, and then the secondary structure of the alpha helical amphipathic is accepted [8]. The peptides such as Bactenecin [9] and Defensins [10] are relatively stiff non-parallel β-sheets that are encapsulated by disulfide bonds and surrounded by cationic and hydrophobic fragments. A large family of linear peptides characterized by the dominance of one or two amino acids (e.g., tryptophan-rich indolicidin in bovine neutrophils and proline-arginine-rich PR39 in porcine neutrophils) [11,12], detaches the hydrophobic and hydrophilic side chains in the surrounding area of peptide scaffolds in the membrane. A combination of multiple peptides containing several structural groups is often expressed in defense tissues of multicellular organisms. The changes that occur after translation include proteolytic processing and, in some cases, glycosylation (non-enzymatic addition of sugar to amino groups of the protein) [13], the amidation of terminal carboxyl, and isomerization of amino acid [14], and halogenation (the incorporation of X-ions into organic compounds) [15]. Some peptides are derived from larger proteins such as Buforin II from Histone 2A [16] and Lactoferricin from Lactoferrin by the proteolysis process [17]. Due to the high diversity of sequences, it is rarely possible to find a similar peptide sequence from two different species of animals, even the species of frogs, insects, or mammals (there are exceptions such as peptides isolated from highly protected proteins, e.g., Buforin II). However, significant protection of the amino acid sequence in precursor molecules can be identified between specific classes of different peptides of different species as well as in antimicrobial peptides of a particular species [18]. This feature indicates the limitations in the sequences that exist in translation, secretion, or intracellular transition, and membrane-degradation peptide groups. The Cathelicidins have significantly shown this feature [19]. What is the reason for this diversity? The single mutations can dramatically cause diversity by altering the bioactivity of each peptide. The adaptation of species to particular microbial environments existing in their habitat (such as microbes related to food sources) is probably the reason for this diversity [18, 20]. It is normal that a particular species of a living organism would be exposed to the microbes with ineffective peptides over time. They can cause such casualties, but it can cope with microbes and survive microbial agents through the emergence of people who had useful mutations. Since the acquisitive immune system is flexible, this system can discover new environments by a species and use new and more food resources. However, acquisitive immune factors spend more time on maintaining and reducing the time of response to an attack compared to the equipment of the central immune systems (such as antimicrobial peptides). The most important features of the acquisitive immune system include specific responses to the pathogen and antigen, delayed contact and maximum response, post-exposure immunological memory (thus activated in vaccination of this system), and it can be found only in vertebrates. However, in the central immune system, the response is nonspecific, and without immunological memory, the contact is immediately led to a maximum response, and it is found in all animals [20]. Due to the variety developed in the synthetic laboratory, it can be said that approximately all active molecules consist of hydrophilic, hydrophobic and cationic amino acids arranged in a single molecule that can be organized into an amphipathic molecule [21]. The natural peptides are composed of D-amino acids instead of L- amino acids because the isomers of D-amino acids resist the protease enzymes while fully maintaining the antibiotic properties of the peptides [21]. The rapid digestion of antimicrobial peptides by proteases in the blood flow and cells is one of the major challenges in all of these peptides, which can also be a significant obstacle to using them as effective medicines. The proteases simply decompose the peptide bonds. Reshaping and substituting D-amino acids with L-amino acids is a straightforward way to increase the stability of peptides. Short linear or amphiphilic annular peptides, which contain both L- and D-amino acids, have the potential to be produced with different degrees of selectivity and antimicrobial activity [22,23]. Recently, an antioxidant-resistant antibacterial peptide has been produced, which is composed of β-amino acids [24,25].

THE MECHANISM OF ACTION OF ANTIMICROBIAL PEPTIDES

An extensive range of methods and tools have been applied to study the mechanism of action of antimicrobial peptides. There is not only one technique to determine and formulate the mechanism of action of antimicrobial peptides. The placement of the polar heads of phospholipids on the membrane of the cell and the load distribution on the peptides are important factors in the reaction of the peptide with the membrane. In prokaryotic cells (bacterial cells), hydrophilic antimicrobial peptides recognize the anionic lipids on the outer surface of the bacterial membrane. In eukaryotic cells, these anionic lipids are placed on the cytoplasmic side of the membrane; this structural feature is a reason for the relatively higher cell-killing activity of antimicrobial peptides against bacterial cells than against eukaryotic cells [26]. The death of a bacterium is caused by the formation of pores in the bacterial membrane by three processes: 1) binding the antimicrobial peptides to the bacterial membrane, 2) accumulating antimicrobial peptides within the membrane, 3) forming the pores for perforation and killing of the cell. Several models explain the increase in membrane permeability through the action of antimicrobial peptides. Three well-known models are considered as potential mechanisms of antimicrobial action of bioactive peptides: barrel stave, toroidal model, and carpet model [26,27,28] (Figure 20.1).

Barrel Stave Model

According to this model, antimicrobial peptides are accumulated after binding with bacterial membranes and form dimers and multimers. A barrier-like structure is formed on

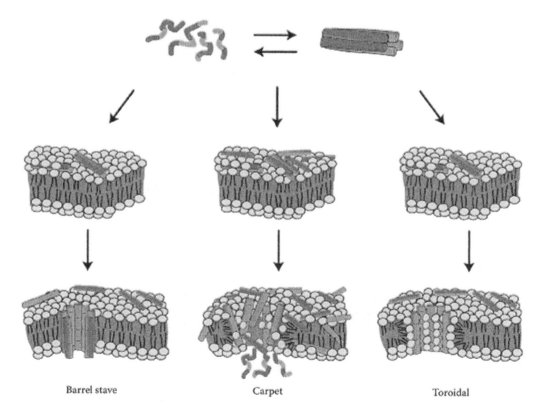

FIGURE 20.1 The potential mechanisms of actions for the antimicrobial activity of bioactive peptides [31].

the bacterial membrane by the accumulation of these peptides. These multimers form some pores in two layers of the bacterial membrane and ultimately lead to cell death [29].

Toroidal Model

The path of pore formation in this model is similar to the barrel stave model. The connection of the outer and inner layers of lipids with the peptide toward the two layers of the peptide is a distinct feature of this model. The model is used for most antimicrobial peptides (e.g., melittin) [29].

Carpet Model

In this model, the peptides first cover the outer surface of the membrane in a carpet-like pattern and then act as a detergent, where lipid bilayers are degraded after the concentration of these peptides reaches the threshold. The pores are filled with micelle-like units [30].

Other models related to the performance of antimicrobial peptide include antibacterial properties of many antimicrobial peptides due to pore formation in lipid bilayers. Barrel-stave (A), wedge (B), pore toroidal (C), carpet (D), and aggregate channel (E) models are also referred to as pore-forming mechanisms by antimicrobial peptides (Figure 20.2) [32]. However, other mechanisms of action of antimicrobial peptides have been described. In the molecular electroporation model, peptides can produce an excellent electrostatic potential for pore formation. In the Sinking Float model, the equilibrium in lipid bilayers is impaired after the peptide infiltrates. Such peptides can form temporary pores that are pernicious to bacteria. Defensins and Catalysidines can inactivate bacterial lipopolysaccharides by binding to a specific part of the molecule. Many peptides act by inhibiting the processes inside the cell of the microorganism. In the joining Canal model, the transfer of peptides between lipid bilayers is performed without pore formation. DNA synthesis, protein biosynthesis, or both processes are inhibited by some peptides [27].

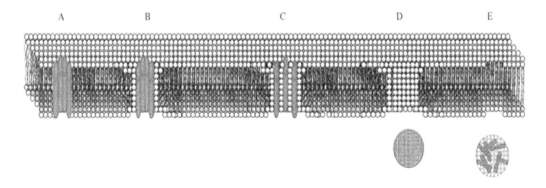

FIGURE 20.2 The models for pore formation by antimicrobial peptides [32]. See text for A, B, C, D, and E.

APPLICATION OF ANTIMICROBIAL PEPTIDES

Due to the growing problem of resistance against conventional antibiotics and the need for new antibiotics, antimicrobial peptides can be used as new approaches for the treatment of human disease [33]. Most of the medicinal uses of antimicrobial peptides were local, and antimicrobial peptides as a new systemic drug (prescription drugs that act throughout the body) have not been prescribed because of uncertainty about the long-term toxicology and relative safety of local treatment [34]. The factors that prevent the growth of antimicrobial peptides as systemic therapy are that they are active in laboratory conditions, where they only act at considerably high doses in animal infection models, which are usually close to toxic doses of the peptide [35]. Antimicrobial peptides have different applications as antiseptic agents. An extensive range of antibiotics originating from antimicrobial peptides is used as "chemical condoms" to prevent the transmission of sexually transmitted diseases such as *Neisseria*, *Chlamydia*, HIV, and *Herpes simplex* virus [36]. Antimicrobial peptides are used in radiotherapy to detect bacterial and fungal infections from sterile inflammation, given the specific binding of antimicrobial peptides to the membrane of pathogens [37]. Antimicrobial peptides are probably capable of enhancing the ability of *in vitro* antibiotics by facilitating the access of antibiotics into the bacterial cell, a phenomenon for which the polymyxin cationic peptide molecule is known [38]. One of the major concerns in the use of medical devices such as intravenous catheters is the microbial contamination of the surfaces of synthetic polymeric materials. Many pieces of research have been carried out to design novel approaches to prevent microbial contamination and understand the mechanism of absorption and microbial proliferation on the surfaces of materials. The antibiotic binding is an effective, safe, and cost-effective method to reduce the contamination of intravascular catheters in ICU. The use of antimicrobial peptides such as meganine in polymeric materials of these devices through covalent bonding is one of the successful approaches in this regard [39]. One of the critical capabilities of some insect-derived antimicrobial peptides is the control of several plant pathogens; it has been proved that they can be effective through transgenic plants. The expression of transgenic genes from antimicrobial peptides, especially insect-derived antimicrobial peptides, may lead to effective strategies in plants to cope with pathogens for protecting current plants [41,40].

Application of Bioactive Peptides with Antimicrobial Activity in Food Safety

The ways to prevent or slow down the microbial growth in foods and ultimately to prevent food spoilage are adding synthetic preservatives, such as salts of benzoic acid, sorbates, and nitrites. The limitations of the impact on all pathogens, the lack of effective and safe preservatives, and the tendency of consumers toward purchasing products with minimal processing are among the challenges of using synthetic preservatives [42]. For these reasons, carrying out research to find alternative synthetic preservatives seems to be essential [43]. In the meantime, antibacterial peptides, especially the bacteriocin group, are excellent options [44]. Bacteriocins are ribosomal synthesized bioactive peptide compounds in the form of peptide complexes or released on the extracellular surface that have a bacteriostatic effect on other species. The use of bacteriocins as biological preservatives began about two decades ago [45, 46]. These protein metabolites typically have molecular weights below 10 kDa [47]. For the use of bacteriocins as QPS (Qualified Presumption of Safety), they

TABLE 20.1 Some Bioactive Peptides with Antimicrobial Activity [51]

Peptide Sequence		Function
LRLKKYKVPQL	Lysine-Arginine-Lysine-Lysine-Lysine-Tyrosine-Lysine-Valine-Proline-Glutamine-Leucine	Interacts with bacteria to cause inhibition
PGTAVFK	Proline-Glycine-Threonine-Alanine-Valine-henylalanine-Lysine	Causes bacteria and yeast membrane destruction
KVGIN KVAGT VRT PGDL LPMH EKF IRL	Lysine-Valine-Glycine-soleucine-Asparagine Lysine-Valine-Alanine-Glycine-Threonine Valine-Arginine-Threonine-Proline-Glycine-Aspartic acid-Leucine Leucine-Proline-Methionine-Histidine Glutamic acid-Lysine-Phenylalanine Isoleucine-Arginine-Leucine	Inhibits *Listeria ivanovii* and *Escherichia coli* growth
Lp-Def1		Interacts with and impairs mitochondrial functions in *Candida albicans*
Maize α-hairpinins		Binds to microbial DNA to cause cell death

should be heat-resistant and effective against pathogenic and food-spoiling bacteria and not endanger the health of consumers [48,49]. Pediocin and nisin are the only bacteriocins that are commercially available nowadays. Nisin is used in the dairy industry to increase milk storage time in tropical countries as well as in canned products to eliminate pathogenic bacteria. Nisin is the first bacteriocin that has obtained approval from the U.S. Food and Drug Administration (FDA) and is employed as a food preservative in many countries [49,50]. Pediocin is also used to maintain safety and increase the storage time of cheese, salads, and meats [52]. Table 20.1 lists the peptide sequences of some bioactive peptides with antimicrobial activity and their functions.

In 1993, bacteriocins were divided into four groups by Klaenhamer [53]. The peptides of group I, which contain 19 to 38 amino acids and are also called as lantibiotics, are heat-resistant and have molecular weights of less than 5 Da. The members of this group include lanthionine and its derivatives. Nisin is the chief antibiotic of this group of bacteriocins.

Group II of bacteriocins contains heat-resistant peptides with a molecular weight of less than 10 kDa and a spiral and amphiphilic structure. According to the study by Drider et al. [54], this group is divided into three classes: IIA, IIB, and IIC. The class IIA, which consists of 37 to 48 amino acids, includes peptides such as Pediocin, Enterocin, and Sakacins [55]. Class IIB contains heterodimer bacteriocins, which means that they consist of two combined peptides. These peptides have low activities in separate modes [56]. The bacteriocins of class IIC, such as Entrocine AS-48 and Circularin A, have an annular structure [57]. The bacteriocins of class IIC are heat-sensitive peptides with a molecular weight of more than 30 kDa and also have a protein structure, complex activity, and a different mechanism of action compared to the bacteriocins of other classes.

CONCLUSION

The bioactive peptides can be identified as specific amino acid sequences that have beneficial physiological effects. They are actually protein components that are inactive within the protein structure and exhibit different physiological functions when released by hydrolysis. The effects of these peptides include antihypertensive, antimicrobial, anticoagulant, antioxidant, bone protection, and growth enhancement effects, which are different depending on the type and amino acid sequences of the peptides. The side effects of synthetic antimicrobial preservatives, as well as antibiotics, have attracted researchers' attention to bioactive peptides with antimicrobial activity, and many studies have been carried out in this regard. However, lack of stable and scalable methods for producing bioactive peptides from different food or non-food sources, inadequate knowledge of gastrointestinal stability, or peptide absorption by them, and lack of applicable clinical trials to provide basic evidence for potential health claims are among the reasons for the delay in commercialization and the practical applications of these peptides, despite conducting many pieces of research in this field. In general, it can be hoped that the use of active peptides of various properties, especially peptides with antimicrobial activities as substitutes for chemical preservatives in foods to be achievable in the near future with the help of research being carried out worldwide.

REFERENCES

1. D.D. Kitts, K. Weiler, "Bioactive proteins and peptides from food sources. Applications of bioprocesses used in isolation and recovery," *Current Pharmaceutical Design*, 9(16), pp.1309–1323, 2003. https://doi.org/10.2174/1381612033454883
2. M.H. Maleki, M. Daneshniya, N. Keshavarz bahadori, M.R. Hassanjani, Z. Latifi, "A review of biological properties of bioactive peptides: Antioxidant activity," in *3rd International Congress of Science Engineering and Technology*, Hamburg, 2020.
3. Z.F. Bhat, S. Kumar, H.F. Bhat, "Bioactive peptides of animal origin: A review," *Journal of Food Science and Technology*, 52(9), pp.5377–5392, 2015. https://doi.org/10.1007/s13197-015-1731-5
4. Z. Hafeez, C. Cakir-Kiefer, E. Roux, C. Perrin, L. Miclo, A. Dary-Mourot, "Strategies of producing bioactive peptides from milk proteins to functionalize fermented milk products," *Food Research International*, 63, pp.71–80, 2014. https://doi.org/10.1016/j.foodres.2014.06.002
5. M. Ortiz-Martinez, R. Winkler, S. García-Lara, "Preventive and therapeutic potential of peptides from cereals against cancer," *Journal of Proteomics*, 111, pp.165–183, 2014. https://doi.org/10.1016/j.jprot.2014.03.044
6. H. Steiner, D. Hultmark, Å. Engström, H. Bennich, H.G. Boman, "Sequence and specificity of two antibacterial proteins involved in insect immunity," *Nature*, 292(5820), pp.246–248, 1981.
7. M. Zasloff, "Magainins, a class of antimicrobial peptides from Xenopus skin: Isolation, characterization of two active forms, and partial cDNA sequence of a precursor," *Proceedings of the National Academy of Sciences of the United States of America*, 84(15), pp.5449–5453, 1987. https://doi.org/10.1073/pnas.84.15.5449
8. B. Bechinger, M. Zasloff, S.J. Opella, "Structure and orientation of the antibiotic peptide magainin in membranes by sol id-state nuclear magnetic resonance

spectroscopy," *Protein Science*, 2(12), pp.2077–2084, 1993. https://doi.org/10.1002/pro.5560021208
9. D. Romeo, B. Skerlavaj, M. Bolognesi, R. Gennaro, "Structure and bactericidal activity of an antibiotic dodecapeptide purified from bovine neutrophils," *Journal of Biological Chemistry*, 263(20), pp.9573–9575, 1988.
10. M.E. Selsted, S.S. Harwig, T. Ganz, J.W. Schilling, R.I. Lehrer, "Primary structures of three human neutrophil defensins," *The Journal of Clinical Investigation*, 76(4), pp.1436–1439, 1985. https://doi.org/10.1172/JCI112121
11. M.E. Selsted, M.J. Novotny, W.L. Morris, Y.Q. Tang, W. Smith, J.S. Cullor, "Indolicidin, a novel bactericidal tridecapeptide amide from neutrophils," *Journal of Biological Chemistry*, 267(7), pp.4292–4295, 1992.
12. B. Agerberth, J.Y. Lee, T. Bergman, M. Carlquist, H.G. Boman, V. Mutt, H. Jörnvall, "Amino acid sequence of PR-39: Isolation from pig intestine of a new member of the family of proline-arginine-rich antibacterial peptides," *European Journal of Biochemistry*, 202(3), pp.849–854, 1991. https://doi.org/10.1111/j.1432-1033.1991.tb16442.x
13. P. Bulet, J.L. Dimarcq, C. Hetru, M. Lagueux, M. Charlet, G. Hegy, J.A. Hoffmann, J.A. Hoffmann, "A novel inducible antibacterial peptide of Drosophila carries an O-glycosylated substitution," *Journal of Biological Chemistry*, 268(20), pp.14893–14897, 1993.
14. J.J. Smith, S.M. Travis, E.P. Greenberg, M.J. Welsh, "Cystic fibrosis airway epithelia fail to kill bacteria because of abnormal airway surface fluid," *Cell*, 85(2), pp.229–236, 1996. https://doi.org/10.1016/S0092-8674(00)81099-5
15. A. Shinnar, T. Urell, M. Rao, E. Sooner, W. Lane, M. Zasloff, "Peptide chemistry, structure and biology," in *Proceedings of the 14th American Peptide Symposium* (Kaumaya, P., and Hodges R., eds), Mayflower Scientific Ltd., Kingswinford, UK, pp.189–191, 1996.
16. H.S. Kim, H. Yoon, I. Minn, C.B. Park, W.T. Lee, M. Zasloff, S.C. Kim, "Pepsin-mediated processing of the cytoplasmic histone H2A to strong antimicrobial peptide buforin I," *Journal of Immunology*, 165(6), pp.3268–3274, 2000. https://doi.org/10.4049/jimmunol.165.6.3268
17. H. Ulvatne, L.H. Vorland, "Bactericidal kinetics of 3 lactoferricins against *Staphylococcus aureus* and *Escherichia coli*," *Scandinavian Journal of Infectious Diseases*, 33(7), pp.507–511, 2001. https://doi.org/10.1080/00365540110026692
18. M. Simmaco, G. Mignogna, D. Barra, "Antimicrobial peptides from amphibian skin: What do they tell us?," *Peptide Science*, 47(6), pp.435–450, 1998. https://doi.org/10.1002/(SICI)1097-0282(1998)47:6<435::AID-BIP3>3.0.CO;2-8
19. M. Zanetti, R. Gennaro, M. Scocchi, B. Skerlavaj, "Structure and biology of cathelicidins," in *The Biology and Pathology of Innate Immunity Mechanisms*, Springer, Boston, MA, pp.203–218, 2002. https://doi.org/10.1007/0-306-46831-X_17
20. H.G. Boman, "Innate immunity and the normal microflora," *Immunological Reviews*, 173(1), pp.5–16, 2000. https://doi.org/10.1034/j.1600-065X.2000.917301.x
21. W.L. Maloy, U.P. Kari, "Structure–activity studies on magainins and other host defense peptides," *Biopolymers: Original Research on Biomolecules*, 37(2), pp.105–122, 1995. https://doi.org/10.1002/bip.360370206
22. S. Fernandez-Lopez, H.S. Kim, E.C. Choi, M. Delgado, J.R. Granja, A. Khasanov, K. Kraehenbuehl, G. Long, D.A. Weinberger, K.M. Wilcoxen, M.R. Ghadiri, "Antibacterial agents based on the cyclic D, L-α-peptide architecture," *Nature*, 412(6845), pp.452–455, 2001. https://doi.org/10.1038/35086601

23. Z. Oren, Y. Shai, "Cyclization of a cytolytic amphipathic α-helical peptide and its diastereomer: Effect on structure, interaction with model membranes, and biological function," *Biochemistry*, 39(20), pp.6103–6114, 2000. https://doi.org/10.1021/bi992408i
24. Y. Hamuro, J.P. Schneider, W.F. DeGrado, "De novo design of antibacterial β-peptides," *Journal of the American Chemical Society*, 121(51), pp.12200–12201, 1999. https://doi.org/10.1021/ja992728p
25. E.A. Porter, X. Wang, H.S. Lee, B. Weisblum, S.H. Gellman, "Non-haemolytic β-amino-acid oligomers," *Nature*, 404(6778), pp.565–565, 2000. https://doi.org/10.1038/35007145
26. K. Matsuzaki, "Why and how are peptide–lipid interactions utilized for self-defense? Magainins and tachyplesins as archetypes," *Biochimica et Biophysica Acta (BBA)- Biomembranes*, 1462(1–2), pp.1–10, 1999. https://doi.org/10.1016/S0005-2736(99)00197-2
27. Y. Shai, "Mechanism of the binding, insertion and destabilization of phospholipid bilayer membranes by α-helical antimicrobial and cell non-selective membrane-lytic peptides," *Biochimica et Biophysica Acta (BBA)-Biomembranes*, 1462(1–2), pp.55–70, 1999. https://doi.org/10.1016/S0005-2736(99)00200-X
28. L. Yang, T.M. Weiss, R.I. Lehrer, H.W. Huang, "Crystallization of antimicrobial pores in membranes: Magainin and protegrin," *Biophysical Journal*, 79(4), pp.2002–2009, 2000. https://doi.org/10.1016/S0006-3495(00)76448-4
29. H.V. Westerhoff, D. Juretić, R.W. Hendler, M. Zasloff, "Magainins and the disruption of membrane-linked free-energy transduction," *Proceedings of the National Academy of Sciences of the United States of America*, 86(17), pp.6597–6601, 1989. https://doi.org/10.1073/pnas.86.17.6597
30. G. Bierbaum, H.G. Sahl, "Induction of autolysis of staphylococci by the basic peptide antibiotics Pep 5 and nisin and their influence on the activity of autolytic enzymes," *Archives of Microbiology*, 141(3), pp.249–254, 1985. https://doi.org/10.1007/BF00408067
31. M.C. Chappell, "Angiotensins: From endocrine to intracrine functions," *Bioactive Peptides*, pp. 3–20, 2009.
32. A.B. Snyder, R.W. Worobo, "Chemical and genetic characterization of bacteriocins: Antimicrobial peptides for food safety," *Journal of the Science of Food and Agriculture*, 94(1), pp.28–44, 2014. https://doi.org/10.1002/jsfa.6293
33. A. Asoodeh, M. Homayouni-Tabrizi, H. Shabestarian, S. Emtenani, S. Emtenani, "Biochemical characterization of a novel antioxidant and angiotensin I-converting enzyme inhibitory peptide from Struthio camelus egg white protein hydrolysis, " *Journal of Food and Drug Analysis*, 24(2), pp.332–342, 2016. https://doi.org/10.1016/j.jfda.2015.11.010
34. M. Zasloff, *In from Development of Novel Antimicrobial Agents: Emerging Strategies* (Ed. Lohner, K.), Horizon Scientific, Wymondham, UK, pp.261–270, 2001.
35. R.P. Darveau, M.D. Cunningham, C.L. Seachord, L. Cassiano-Clough, W.L. Cosand, J. Blake, C.S. Watkins, "Beta-lactam antibiotics potentiate magainin 2 antimicrobial activity in vitro and in vivo. Antimicrobial agents and chemotherapy," 35(6), pp.1153–1159, 1991. https://doi.org/10.1128/AAC.35.6.1153
36. B. Yasin, M. Pang, J.S. Turner, Y. Cho, N.-N. Dinh, A.J. Waring, R.I. Lehrer, E.A. Wagar, "Evaluation of the inactivation of infectious herpes simplex virus by

host-defense peptides," *European Journal of Clinical Microbiology and Infectious Diseases*, 19(3), pp.187–194, 2000. https://doi.org/10.1007/s100960050457
37. M.M. Welling, A. Paulusma-Annema, H.S. Balter, E.K. Pauwels, P.H. Nibbering, "Technetium-99m labelled antimicrobial peptides discriminate between bacterial infections and sterile inflammations," *European Journal of Nuclear Medicine*, 27(3), pp.292–301, 2000. https://doi.org/10.1007/s002590050036
38. A. Giacometti, O. Cirioni, F. Barchiesi, G. Scalise, "In-vitro activity and killing effect of polycationic peptides on methicillin-resistant Staphylococcus aureus and interactions with clinically used antibiotics," *Diagnostic Microbiology and Infectious Disease*, 38(2), pp.115–118, 2000. https://doi.org/10.1016/S0732-8893(00)00175-9
39. S.L. Haynie, G.A. Crum, B.A. Doele, "Antimicrobial activities of amphiphilic peptides covalently bonded to a water-insoluble resin," *Antimicrobial Agents and Chemotherapy*, 39(2), pp.301–307, 1995. https://doi.org/10.1128/AAC.39.2.301
40. G. DeGray, K. Rajasekaran, F. Smith, J. Sanford, H. Daniell, "Expression of an antimicrobial peptide via the chloroplast genome to control phytopathogenic bacteria and fungi," *Plant Physiology*, 127(3), pp.852–862, 2001. https://doi.org/10.1104/pp.010233
41. M. Osusky, G. Zhou, L. Osuska, R.E. Hancock, W.W. Kay, S. Misra, "Transgenic plants expressing cationic peptide chimeras exhibit broad-spectrum resistance to phytopathogens," *Nature Biotechnology*, 18(11), pp.1162–1166, 2000. https://doi.org/10.1038/81145
42. K. Keymanesh, S. Soltani, S. Sardari, "Application of antimicrobial peptides in agriculture and food industry," *World Journal of Microbiology and Biotechnology*, 25(6), pp.933–944, 2009. https://doi.org/10.1007/s11274-009-9984-7
43. J.L. Parada, C.R. Caron, A.B.P. Medeiros, C.R. Soccol, "Bacteriocins from lactic acid bacteria: Purification, properties and use as biopreservatives," *Brazilian Archives of Biology and Technology*, 50(3), pp.512–542, 2007. https://doi.org/10.1590/S1516-89132007000300018
44. T. Rydlo, J. Miltz, A. Mor, "Eukaryotic antimicrobial peptides: Promises and premises in food safety," *Journal of Food Science*, 71(9), pp.R125–R135, 2006. https://doi.org/10.1111/j.1750-3841.2006.00175.x
45. I.F. Nes, S.S. Yoon, D.B. Diep, "Ribosomally synthesiszed antimicrobial peptides (bacteriocins) in lactic acid bacteria: A review," *Food Science and Biotechnology*, 16(5), pp.675–690, 2007.
46. M.A. Riley, J.E. Wertz, "Bacteriocins: Evolution, ecology, and application," *Annual Reviews in Microbiology*, 56(1), pp.117–137, 2002. https://doi.org/10.1146/annurev.micro.56.012302.161024
47. M. Mirzaei, S. Mirdamadi, M.R. Ehsani, M. Aminlari, E. Hosseini, "Purification and identification of antioxidant and ACE-inhibitory peptide from *Saccharomyces cerevisiae* protein hydrolysate," *Journal of Functional Foods*, 19, pp.259–268, 2015. https://doi.org/10.1016/j.jff.2015.09.031
48. S. Leroy, I. Lebert, R. Talon, "Microorganisms in traditional fermented meats," in *Handbook of Fermented Meat and Poultry*, pp.97–105, 2014. https://doi.org/10.1002/9781118522653.ch12
49. R. Hammami, A. Zouhir, J.B. Hamida, I. Fliss, "BACTIBASE: A new web-accessible database for bacteriocin characterization," *BMC Microbiology*, 7(1), p.89, 2007. https://doi.org/10.1186/1471-2180-7-89

50. S.Y.H. Tafreshi, S. Mirdamadi, D. Norouzian, S. Khatami, S. Sardari, "Effect of non-nutritional factors on nisin production," *African Journal of Biotechnology*, 9(9), 2010. https://doi.org/10.5897/AJB10.1409
51. E.B.M. Daliri, D.H. Oh, B.H. Lee, "Bioactive peptides," *Foods*, 6(5), p.32, 2017. https://doi.org/10.3390/foods6050032
52. J. Zhang, G. Liu, P. Li, Y. Qu, "Pentocin 31-1, a novel meat-borne bacteriocin and its application as biopreservative in chill-stored tray-packaged pork meat," *Food Control*, 21(2), pp.198–202, 2010. https://doi.org/10.1016/j.foodcont.2009.05.010
53. T.R. Klaenhammer, "Genetics of bacteriocins produced by lactic acid bacteria," *FEMS Microbiology Reviews*, 12(1–3), Issue(1–3), pp.39–85, 1993. https://doi.org/10.1111/j.1574-6976.1993.tb00012.x
54. D. Drider, G. Fimland, Y. Héchard, L.M. McMullen, H. Prévost, "The continuing story of class IIa bacteriocins," *Microbiology and Molecular Biology Reviews*, 70(2), pp.564–582, 2006. https://doi.org/10.1128/MMBR.00016-05
55. A.K. Bhunia, M.C. Johnson, B. Ray, "Direct detection of an antimicrobial peptide ofPediococcus acidilactici in sodium dodecyl sulfate-polyacrylamide gel electrophoresis," *Journal of Industrial Microbiology*, 2(5), pp.319–322, 1987. https://doi.org/10.1007/BF01569434Y
56. Y. Kawai, K. Ishii, K.U. Arakawa, B. Saitoh, J. Nishimura, J. Nishimura, H. Kitazawa, Y. Yamazaki, Y. Tateno, T. Itoh, T. Saito, "Structural and functional differences in two cyclic bacteriocins with the same sequences produced by lactobacilli," *Applied and Environment Microbiology*, 70(5), pp.2906–2911, 2004. https://doi.org/10.1128/AEM.70.5.2906-2911.2004
57. J.W. Mulders, I.J. Boerrigter, H.S. Rollema, R.J. Siezen, W.M. de Vos, "Identification and characterization of the lantibiotic nisin Z, a natural nisin variant," *European Journal of Biochemistry*, 201(3), pp.581–584, 1991. https://doi.org/10.1111/j.1432-1033.1991.tb16317.x

5th International Conference on Researchers in Science & Engineering & 2nd International Congress on Civil, Architecture and Urbanism in Asia.06 August 2020, Kasem Bundit University, Bangkok, Thailand.

CHAPTER 21

Opioid Activity

Sureal Ahmad Sheikh, Pragya Shakti Mishra,
Ritam Bandopadhyay, and Awanish Mishra

CONTENTS

Introduction	427
Sources of Opioid Peptides	428
Food Sources	428
Animal Sources	428
Plant Sources	429
Production of Bioactive Peptides	431
Enzymatic Hydrolysis	432
Microbial Fermentation	432
Chemical Synthesis	432
Integrated Approach toward Discovery of Opioid Peptides	433
Mechanisms of Opioid Peptides	433
Functions of Opioid Peptides in Human Health	434
Antihypertensive Peptides	434
Antioxidative Peptides	435
Antimicrobial Peptides	435
Immunomodulatory Peptides	436
Other Functions	436
Conclusion and Future Perspectives	436
References	437

INTRODUCTION

The human body is prone to various types of infections, which lead to abnormal health and eventually weaken the body's defensive system. The use of synthetic drugs or conventional medications for curing a specific type of disease is not always attractive because there are chances of toxicity or resistance. There have been many developments of functional foods for the replacement of conventional medications as they are extremely safe and have fewer side effects [1]. So, there is a continuous effort to improve the food industry and food-derived products for the benefit of human health.

Bioactive peptides are those substances derived from various food sources and are known to improve health-related issues in humans. Due to their small size, bioactive peptides inhibit protein–protein interactions and can be obtained by proteolytic cleavage of a larger protein molecule. They consist of 2–20 amino-acid residues and have different targets to act upon, such as digestive, cardiovascular, or endocrine systems and also have antidiabetic, antimicrobial, antihypertensive, and immunomodulatory functions [2]. Opioid peptides are substances consisting of 5–80 amino acids [3]. They diffuse locally and with a concentration much lower than neurotransmitters. There are different types of opioid receptors that bind with opioid peptides such as μ, κ, and δ for their activation, and these are responsible for various functions, including the sleep cycle, for the transmission of messages through neurons (neurotransmission) and pain perception [4].

There are various sources, such as plants, animals, and microbes through which bioactive peptides can be obtained. The major source of bioactive peptides is through food proteins, and those peptides find their use in functional foods, nutraceuticals, and natural health products. In this chapter, we highlight bioactive peptides which are obtained from food sources, their analysis, and functions in the human body.

In 1902, a substance was discovered that stimulated the secretion of pancreatic enzymes and was found in the intestinal lining known as secretin. After this discovery, many other peptides were reported using the same mechanism of sequencing which led to the concept of peptides [5]. In 1950, the first discovery of bioactive peptides from a food source was found when phosphorylated casein peptides were discovered in rachitic infants with vitamin D deficiency [6]. More research has been done on the isolation of bioactive peptides from various sources.

Sources of Opioid Peptides

Food Sources

Food is considered a source of not only dietary compounds but also of biologically active compounds that may exert a beneficial effect on the overall health of an individual. Bioactive peptides can be seen highly present in complex food proteins [7]. The chief sources of bioactive peptides include bovine milk, cheese, and many other dairy products [8]. Examples of animal sources from which bioactive peptides are obtained include eggs, meat, bovine blood, and fish sources, such as herring, tuna, sardines and salmon. There are also plant sources through which these peptides could be obtained, such as wheat, maize, soya, mushroom, pumpkin, and sorghum. Further BPs can be found exclusively in fermented foods [9]. As per the literature, the opioid peptides obtained from animal sources show their activity via the μ receptor, and the plant-derived peptides show their activity by delta opioid receptors [10]. Soymorphins are the exception.

Animal Sources

A large number of bioactive peptides are obtained through animal sources and have been reported to produce a significant health benefit to humans. Serum albumin, which is present in the blood, acts as a main reservoir of bioactive peptides and other proteins. Meat is another important source of nutrients, such as antioxidants, fiber, mostly protein, as well as bioactive peptides. Creatine, carnosine, and glutathione are some of examples of bioactive peptides obtained from meat [11].

> **Milk:** Milk is considered a great source of bioactive peptides, as it includes various nutrients such as lactose, essential vitamins, and minerals [12]. Milk proteins play an essential part in various biological activities. For example, antibacterial activity is shown by lactoferrin. Colostrum, which is present in mother's milk, helps in the postnatal development of a baby. Lactoferrin also has its role in showing immunomodulating effects and has some antimicrobial properties. Cytokine production is also influenced by lactoferrin-derived peptides.

Fermentation by lactic acid bacteria is used to obtain functional foods containing many bioactive peptides [13]. Fermented milk, comprising bioactive peptides, lowers blood pressure in hypertensive patients. Whey protein also contains a number of bioactive peptides, such as α-lactalbumin and β-lactoglobulin, which have some antihypertensive properties. Cow milk contains plenty of bioactive peptides. Casein phosphopeptides are generated by the hydrolytic action of pepsin on milk and provide calcium and other minerals such as zinc, iron, and copper.

Casein from human milk and β-casein from buffalo milk are reliable sources of exogenous opioid peptides. Peptide Arg-Tyr-Leu-Gly-Tyr-Leu-Glu, obtained from α-casein, shows opioid activity [14]. Fermentation of milk with lactic acid bacteria is the most preferred source of opioid peptides. Opioid peptide sequences Arg-Tyr-Leu-Gly-Tyr-Leu-Glu and Arg-Tyr-Leu-Gly-Tyr-Leu are obtained from β-casein of buffalo milk [15]. Various peptide sequences, like Tyr-Leu-Gly-Tyr-Leu-Glu, Tyr-Leu-Gly-Tyr-Leu, and Arg-Tyr-Leu-Gly-Tyr-Leu-Glu, are obtained from α-casein, showing potent opioid activity [16]. Peptide sequences Tyr-Gly-Leu-Phe-NH$_2$ and Arg-Tyr-Tyr-Gly-Tyr-OCH$_3$, are derived from human milk, and lactalbumin and lactoferrin also show opioid activity [17, 18].

Dairy products as a source of bioactive peptides: Dairy products are another source of bioactive peptides. These products possess a considerable number of bioactive peptides encrypted within their proteins. Though they are derived from milk, proteins from other dairy sources exhibit characteristic physiological functions. Colostrum is an example that contains a high number of immunoglobulins which play an important role in the growth and development of an organism. Antimicrobial proteins are present in high concentration. Bioactive peptides can be obtained using digestive enzymes during the fermentation of milk. Dairy products, like milk protein hydrolysates, fermented milk, and cheese varieties, are chief sources of bioactive peptides. Fermented sour-milk products are reported to possess antihypertensive activities. [19].

Peptides obtained from parmesan cheese and cheddar cheese have opioid activity. The peptides obtained from these sources fall under the category of beta casomorphins. Peptide sequence Tyr-Pro-Phe-Pro-Gly-Pro-Ile obtained from beta casomorphine 7 is a potent opioid peptide [20].

Plant Sources

Plenty of peptides are derived predominantly from plant sources, such as gluten exorphins. Some grains, such as wheat, barley, and oats contain such exorphins. Gluteomorphins act as opioid antagonists on δ-type opioid receptor. There are various gluteomorphins, such as A4, A5, B4, B5, and 7. They are believed to have a strong effect on the central nervous system as they cross the blood stream [21]. The peripheral nervous system as well as memory has been associated with gluteomorphin A5. They also have an anticonceptive effect. Learning abilities, concentration, as well as problems related with digestion in autistic patients – especially children – can be improved by removing casein and gluten proteins. They are associated with casomorphins and gluteomorphins, which are responsible for autism spectrum disorder [22]. Peptide sequence Tyr-Pro-Gln-Pro-Gln-Pro-Phe, isolated from α-gliadin, shows opioid activity [23].

Soy proteins also contain bioactive peptides and possess antimicrobial activity. They have many other physiological roles. Soybean β-conglycinin β-subunit, from which soymorphins are derived, act as opioid agonists and have anxiolytic activity. Oral administration in mice of different types of soymorphins, such as soymorphin-5, -6, and -7 produce an anxiolytic effect [24]. These may supress food intake, especially soymorphin-7, whereas soymorphin-5 is associated with decreasing tryglycerides both in plasma and liver in mice with diabetes. Glucose and lipid metabolism can also be controlled by soymorphins. Other important bioactive peptides are rubiscolins (rubiscolin 5 and rubiscolin 6), which are majorly found in spinach protein. Oral administration of rubiscolin 6 has been reported to produce anxiolytic and analgesic effects [25].

Hordein protein obtained from barley is another a reliable source of opioid peptides. Recently, wheat-derived peptides like gluten exorphins, gliadorphin from gliadin, gluten morphine from glutenin were also discovered to have good opioid activity.

Soybean-derived opioid peptides are also a reliable plant-derived source. They are known as soymorphins, and they act on μ-opioid receptors. Various types of soymorphin have been identified, like soymorphin-5, soymorphin-6, soymorphin-7, etc. Among them, soymorphin-5 was found to have the highest opioid agonistic activity. The peptide sequence of soymorphin-5 was found to be Tyr-Pro-Phe-Val-Val [26].

Opioid peptides are divided into two categories: endogenous and exogenous peptides.

Endogenous peptides are produced naturally in the human body (mammals) and could be secreted as hormones or neurotransmitters (enumerated in Table 21.1), which majorly include endorphins, endomorphins, and dynorphins [27]. Enkephalins were the first discovered in 1975 as endogenous ligand for opioid receptors. Typical opioid peptides are those having the Try-Gly-Gly-Phe sequence at their N-terminal. Different opioid receptor types have different physiological and pharmacological functions to perform. Endorphins are further subdivided into 4 groups (α, β, γ, and σ) and can be found in different parts of the body, such as the hypothalamus, the

TABLE 21.1 Endogenous Opioid Peptides

Opioid Receptor	Endogenous Opioid Peptides	Sequence	Opioid Drugs	References
κ receptor	Dynorphin A	Tyr-Gly-Gly-Phe-Leu-Arg-Arg-Ile-Arg-ProLys-LeuLys-Trp-Asp-Asn-Gln	Ethylketocyclazocine (EKC)	[29]
	Dynorphin B	Phe-Gly-Gly-Phe-Thr-Gly-Ala-Arg-Lys-SerAla-ArgLys-Leu-Ala-Asn-Gln		[30]
μ receptor	Endorphin-1	Tyr-Pro-Trp-Phe-NH$_2$	Morphine, Fentanyl, Sufentanil	[21]
	Endorphin -2	Tyr-Pro-Phe-Phe-NH$_2$		
	β endorphin	Tyr-Gly-Gly-Phe-Met-Thr-Ser-Glu-Lys-SerGln-ThrPro-Leu-Val-Thr-Leu-Phe-Lys-AsnAla-Ile-Ile-LysAsn-Ala-Tyr-Lys-Lys-Gly-Glu		[31]
δ receptor	Met-enkephalin	Tyr-Gly-Gly-Phe-Met	Deltorphin [D-Pen2, D-Pen5]	[3]
	Leu-enkephalin	Tyr-Gly-Gly-Phe-Leu		

TABLE 21.2 Exogenous Opioid Peptides from Foods

source	Name of peptides	Sequences	References
Bovine milk β-casein	β casomorphine 4	Tyr-Pro-Phe-Pro	[32]
	β casomorphine 5	Tyr-Pro-Phe-Pro-Gly	
	β casomorphine 6	Tyr-Pro-Phe-Pro-Gly-Pro	
	β casomorphine 7	Tyr-Pro-Phe-Pro-Gly-Pro-Ile	
Bovine milk β-casein	Neocasomorphine 6	Tyr-Pro-Val-Glu-Pro-Phe	[33]
Bovine milk α-lactalbumin	$α_b$-lactorphin	Tyr-Gly-Leu-Phe-NH2	[14]
Beta casein of human milk	$β_h$-casomorphine-4	Tyr-Pro-Phe-Val	[14]
	$β_h$-casomorphine-5	Tyr-Pro-Phe-Val-Glu	
	$β_h$-casomorphine-7	Tyr-Pro-Phe-Val-Glu-Pro-Ile	
	$β_h$-casomorphine-8	Tyr-Pro-Phe-Val-Glu-Pro-Ile-Pro	
Lactalbumin	α lactorphin	Tyr-Gly-Leu-Phe-NH$_2$	[34]
Bovine lactoferrin	Lactoferrsoxin A	Tyr-Leu-Gly-Ser-Gly-Tyr-OCH	[35]
	Lactoferrsoxin B	Arg-Tyr-Tyr-Gly-Tyr-OCH$_3$	
	Lactoferrsoxin C	Lys-Tyr-Leu-Gly-Pro-Gln-Tyr-OCH$_3$	
Soy proteins	Soymorphin 5	Tyr-Pro-Phe-Val-Val	[24]
	Soymorphin 6	Tyr-Pro-Phe-Val-Val-Asn	
	Soymorphin 7	Tyr-Pro-Phe-Val-Val-Asn-Ala	
Weight HMW gluten	Gluten exorphin A4	Gly-Tyr-Tyr-Pro	[10]
	Gluten exorphin A5	Gly-Tyr-Tyr-Pro-Thr	
	Gluten exorphin B4	Tyr-Gly-Gly-Trp	
	Gluten exorphin B5	Tyr-Gly-Gly-Trp-Leu	
Spinach RuBisCO	Rubiscolin-5	Tyr-Pro-Leu-Asp-Leu	[36]
	Rubiscolin-6	Tyr-Pro-Leu-Asp-Leu-Phe	

nervous system, as well as the brain. β-endomorphin helps in neurotransmission and is also considered the most powerful endogenous peptide [28]. Its functions include alleviation of stress, pain in the body, and anxiety behavior.

Exogenous peptides are the peptides derived outside the body and are mainly known as exorphins because they have an activity like morphine (listed in Table 21.2). They can be obtained through different food sources, such as plants and animals. The peptides obtained from animal proteins have the affinity to bind to μ receptors whereas those obtained from plant proteins bind to σ receptor with the exception of soymorphins. There are various sources of food derived exogenous peptides; for instance, casein which is derived from human milk, supplies β-casein and β-casomorphin.

PRODUCTION OF BIOACTIVE PEPTIDES

Various methods exist through which bioactive peptides from food can be obtained. Some of the methods are given in following paragraphs.

Enzymatic Hydrolysis

This is an important method for the production of bioactive peptides. In this method, we have hydrolysis of proteins by enzymes. Depending upon the optimal temperature and pH of different enzymes, they can be added either simultaneously or sequentially [37]. Bioactive peptides with a specific molecular weight are released by these proteolytic enzymes.

Phenolic compounds are known for their antimicrobial, antioxidant, antihypertensive, and antidiabetic activities. These compounds interfere with biological functions, also during the production of bioactive peptides. It is necessary to remove these nonprotein bioactive compounds. Various methods used for their extraction are, e.g., ethanol extraction or supercritical extraction.

Proteolysis is a process which may affect hydrolysis depending on the type of enzyme. This leads to proton generation which causes fluctuation of the pH of the medium. A buffer can be used to perform the proteolysis. Addition of acid or alkali can adjust the pH, but this may lead to a high salt concentration in the hydrolysates.

There are other factors that may affect the production of peptides such as the enzyme type used, temperature, and duration of hydrolysis. Centrifugation of the mixture is done to remove low-molecular-weight peptides in the supernatant – after the enzymatic digestion, that is. Freeze-drying, cross-flow membrane filtration, or column chromatography are used for obtaining the peptides. Gel-filtration can be used to separate low-molecular peptides, depending on size.

Microbial Fermentation

Bacteria or yeasts secrete proteolytic enzymes that lead to the breakdown of bioactive peptides. Factors on which hydrolysis depend are concentration of substrate, strains of the microbes used, and time of fermentation.

Different proteolytic systems in microorganisms may lead to differences in the functionality of protein hydrolysates. Bacteria may differ in their mechanisms of proteolysis thus generating peptides with different bioactivities. Various fungi can be used in the generation of supernatant comprising peptides. These can be hydrolyzed further to get shorter peptide sequences [38].

Chemical Synthesis

There are two key chemical approaches for peptide synthesis: solution phase synthesis (SPS) and solid phase peptide synthesis (SPPS). In SPS, coupling of single amino acids is performed in a solution. Compressing the shorter fragments of the target peptide produces the long peptides [23]. The SPS approach is also called "fragments condensation process". High purity of the target peptide can be achieved through the SPS method. This method is very economical, but its only drawback is the long reaction time. It also requires a resin as a support for growing a peptide chain.

At present, SPPS is widely used for obtaining therapeutic peptides. It has advantages over SPS because has lower manufacturing costs and is less time-consuming [39]. Protein chains can be synthesized by a chemical ligation process.

The best and most efficient method for ligating the peptides is native chemical ligation (NCL). It allows two unprotected peptide fragments in which one contains N-terminal cysteine to be joined by a covalent peptide bond at the site of ligation and to form a thioester linked intermediate. NCL also helps to overcome the drawbacks of SPPS by allowing researchers to synthesize peptides that are greater than 50 amino acids in length.

Integrated Approach toward Discovery of Opioid Peptides

Integrated approach or integrated bioinformatics approach to identify and evaluate activity of a candidate peptide. This method is an integration of classical chemical synthesis and bioinformatics.

In the chemical synthesis approach, a protein which is likely to contain a bioactive peptide is synthesized. Then that protein is subjected to enzymatic degradation resulting in the production of fractions of proteins. From these various fractions, the active fractions which show desirable activity are identified. Then the various peptides of that active fraction are analyzed with, e.g., mass spectroscopy or HPLC. After identifying the active peptides, an *in vivo* study is performed.

In the bioinformatics approach, various databases are searched for active peptide identification. Such databases are BIOPEP (predicts the precursors of bioactive peptides), NeuroPIpred and NeuroPP (neuropeptide databases), ToxinPred (predicts toxicity of peptides), I-TASSER (structure and function prediction), and NCBI (peptide sequence information). Next *in-silico* QSAR (quantitative structure–activity relationship) studies of those peptides are executed to confirm their structure and function. The following step is molecular docking or virtual analysis of these peptides. In this step, various parameters, like interaction residues, interaction type, and energy produced after docking are evaluated. Finally, a statistical improvement of those peptides is done according to the results of molecular docking.

In the integrated approach, after the statistical improvement step, the protein substrate and enzyme are identified and then subjected to chemical analysis. So, by this integrated approach, we are identifying the active peptides with the help of bioinformatics. The *in vivo* analysis of those peptides is done by the chemical approach. This eliminates various disadvantages associated with each method [40].

MECHANISMS OF OPIOID PEPTIDES

A receptor is a protein that binds to a chemical messenger and triggers a final cell reaction. In the case of opioid peptides, they bind to receptors or are converted into small peptides or amino acids. Existing receptors are G-protein pair receptors (GPCR) [41]. When the opioid binds to the peptide receptor molecule, changes inside the cell occur, such as enzyme activation, ion channel openings, and gene expression. Receptors associated with opioids are GPCRs. They are also known as heptahelical receptors, as they have seven transmembrane spanning extracellular loops across the cell membrane. When opioid peptides attach to the outer part of the receptor, it acts by altering the structure of the receptor protein [26, 42].

Opioid peptide signaling systems are more complex than other common neurotransmitters where many receptors do not bind and only one ligand works with different receptors, such as acetylcholine. Although most of them are connected to multiple receptors, they are also able to attach to receptor subtypes, making them more complicated in signaling processes. However, despite the complications of the receptors and the mechanism of action of these bioactive peptides, they still play a promising and effective role in various clinical trials [41].

During digestion, digestive enzymes convert proteins into peptides and amino acids in the stomach. Many factors contribute to the transport and absorption of peptides in GIT, including PK_A, peptide size, and pH microclimate. Absorption and intestinal changes affect the area where the peptide encounters the GIT and therefore affect absorption. Peptides larger than di-tripeptides have not been found to penetrate easily into healthy individuals except in conditions such as gastrointestinal disorders. It has been found that there is no complete inhibition in the intestinal absorption and the gluten present in cereal flour can cross the intestinal epithelium, including exorphins A5 and A4, while the migration process is unclear. In mammals, four systems of transport of peptide exist: PTS-1, PTS-2, PTS-3, and PTS-4. All four peptide transport systems transport opioid peptides, including Try-MFT-1, met-enkephalin, and leu-enkephalin [43]. Opioids found in the digestive tract in the gut initially interact with receptors in the enteric nervous system and thus affect GIT functions. The enteric nervous system has a network of nerve cells located in the wall of the tract, which controls movement and secretion and regulates breakdown, assimilation, and immunomodulation.

Due to peptidase activity, the half-life of opioid peptides in the blood is shorter. The opioid peptides are also available endogenously. Leu-enkephalin and dynorphin-A has half-life of 6.7 minutes and 1 minute, respectively [44]. Dermorphine shows a longer half-life than enkephalins, and the half-life can be extended by binding these peptides to protein-carrying proteins, such as transferrine or albumin [45]. Therefore, the study of the opioid peptide profile in the bloodstream needs to be investigated for future research benefits. Further, glycosylation of opioid peptides provides an important measure for modulation of biological membranes interaction that alters their pharmacokinetic and pharmacodynamic properties. Glycosylation process of opioid peptides makes them a substrate of Glut-1 transporters, which further enhance CNS bioavailability and potentiate anticonceptive activity [46].

FUNCTIONS OF OPIOID PEPTIDES IN HUMAN HEALTH

Many functions are performed by the opioid peptides; they intervene in hypertension, diabetes, inflammation, cancer, hyperlipidemia, immune disorders, and oxidative stress. Various functions of opioid peptides are listed in Table 21.3.

Antihypertensive Peptides

Various side effects are associated with the synthetic antihypertensive drugs, such as dizziness, cough, and headache. So, a lot of research is going on to find peptides with antihypertensive properties. These peptides are safer to use as they are derived from food directly and bind to tissues with higher affinity than synthetic drugs.

Lactobacillus helveticus fermentates contain angiotensin-I converting enzymes (ACE), such as IPAVF, AHKAL, and APLRV [55]. Valyl-prolyl-proline (Val-Pro-Pro) and isoleucyl-prolyl-proline (Ile-Pro-Pro) are the two biologically active peptides obtained from fermented milk. These may lower the blood pressure in spontaneously hypertensive rats. Some of the dietary proteins obtained from milk contain many peptides which are released during gastrointestinal digestion. They have antihypertensive, immunomodulatory, and antimicrobial properties.

TABLE 21.3 Functions of Various Opioid Peptides

Peptides	Animal Model	Dose	Effects	References
Rubiscolin 5, Rubiscolin 6	Mice	3 nM/animal	Anticonception	[36]
Rubiscolin-6	Mice	100mg/kg	Enhances memory consolidation	[25]
Beta casomorphine 4, Beta casomorphine 5, Beta casomorphine 6, Beta casomorphine 7	Rat	60-2000 nM	Analgesic and naloxone Reversible activity	[47]
Beta casomorphine 5	Mice	1 mg/kg	Improvement of learning and memory, analgesic	[48]
Beta casomorphine 7	Rat	0.1-020 nM	Increased food intake	[49]
Gluten exorphin C	mice	5 mg/kg	Increased learning behavior and reduce anxiety	[50]
Gluten exorphin B5	Rat	3 mg/kg	Increased prolactin secretion	[51]
soymorphin- 5,6, and 7	Mice	10–30 mg/kg	Anxiolytic activity	[24]
soymorphin- 5 and 7	Mice	30 mg/kg	Appetite suppression and anorexigenic activity	[52]
soymorphin- 5 amide	Rat	5 mg/kg	Reduced anxiety	[53]
Beta casomorphine 7	Diabetic rats	-----	Reduced hyperglycemia and free radical mediated oxidative stress	[54]

Antioxidative Peptides

A lot of research is going on to study the effect of antioxidative peptides and their impact on the overall health of individuals. About 5–16 amino-acid residues are present in antioxidant peptides. These are obtained from foods and are safe with low cost, high activity, and are easily absorbed. Some examples of amino acids that are known to produce antioxidant activity are Tyr, His, Trp, Met, Cys, and Lys.

According to an experimental study performed on male Wistar rats, administration of soy protein isolate (SPI), showed paraquat (PQ)-induced oxidative stress reduction and showed to prevent the rise of serum TBARS concentration [56]. Two peptides, Leu/Ile-Lys and Phe-lys, have strong oxygen radical capacities and are the yielded protein hydrolysates of *Hylarana guentheri* [57].

Antimicrobial Peptides

Antimicrobial peptides consist of 12–80 amino acids with a molecular weight of 1–5 kDa. They are widely distributed in nature and provide innate immunity by protecting humans from different pathogens, such as bacteria, parasites, fungi, and viruses. They bind to the target cell membrane, causing its disintegration and ultimately leading to apoptosis (cell death). The antimicrobial properties of different peptides depend upon various physicochemical properties, such as their size, solubility, charge, and hydrophobicity. They also possess a rich content of cysteine or glycine, which is the key feature of antibacterial peptides [58].

A sequence of peptide ASHLGHHALDHLLK (H2), which is sourced from *Holothuria tubulosa*, inhibits the growth of *L. monocytogenes*. Another peptide sequence KTC ENLADTYKGPPPFFTTG (phaseococcin), obtained from *Phaseolus coccineus*, inhibits HIV reverse transcriptase activity [59].

Immunomodulatory Peptides

These peptides are found in different sources, such as soybeans, milk, and honey. β-casein and α lactalbumin are the two peptides that provide resistance to bacterial infections in mice. Immunomodulatory peptides increase the function of immune cells *viz.*, proliferation of lymphocytes, synthesis of antibodies, and enhancement in the activity of natural killer cells [60]. Some of the allergic reactions are also decreased by immunomodulatory peptides and the mucosal immunity in GIT can also be increased by the use of these peptides [61].

Other Functions

Other activities of opioid peptides include anticonception, enhancement of memory, analgesia, and learning improvement. Rubiscolin 5,6 obtained from the Spanish Rubisco plant showed anticonception effect in mice [36]. Rubiscolin-6 also shows to enhance memory consolidation in mice [25]. β-casomorphine 4,5,6,7 shows analgesic and naloxone reversible activity in rat models [47]. Improvement of learning and memory is also shown by β-casomorphine 5 in mice [48].

Appetite stimulation is an important activity of opioid peptides. β-casomorphine 7 stimulates food intake after 6 hours of intravenous injection in rats [49]. Gluten exorphin C and B5 shows to have effects on sleep, learning, and memory. Gluten exorphin C shows improved learning behavior and reduces anxiety in mice [50]. Whereas gluten B5 shows increased prolactin secretion in mice, thus helping in breastfeeding [51]. Soymorphin shows various physiological effects. Soymorphin 5,6,7 shows anxiolytic effects in mice when given at doses of 10–30 mg/kg through oral or intra-peritoneal route [24]. Soymorphin 5,7 shows reduction in food intake, i.e., appetite suppression and anorexigenic activity in mice when given at a dose of 30 mg/kg through oral route [52]. Amide derivative of soymorphin 5 decreases anxiety in rats after 30 min of intra-peritoneal injection [53]. Apart from these CNS effects β-casomorphine 7 exerts hyperglycemic activity in diabetic rats by improving the insulin levels and also shows free radical-mediated oxidative stress by increasing the level of superoxide dismutase [54].

CONCLUSION AND FUTURE PERSPECTIVES

Opioid peptides help in many physiological processes and function as neurotransmitters or neurohormones that help in stress-related conditions, sedations, nociception, appetite, and gastrointestinal digestion. There are plenty of bioactive peptides present in proteins of various food matrixes and different other natural sources. With the help of enzymatic hydrolysis, a large number of bioactive peptides are produced having antimicrobial, antihypertensive, immunomodulatory, antioxidant, antithrombic, and many other activities. As a result of this, bioactive peptides find their use in different medical conditions, such as hypertension, type II diabetes, and obesity.

Despite their massive use, there are many obstacles to overcome especially from the technological viewpoint. They must be produced at a large scale without losing activity. The identification, isolation, and purification of new bioactive peptides by scientists and using them commercially will improve human health significantly.

REFERENCES

1. Patil S.M., Sujay S., Chandana Kumari V.B., Tejaswini M., Sushma P., Prithvi S., Ramith R. Bioactive peptides: Its production and potential role on health. *Int J Innov. Sci., Eng. Technol.* 2020;7(1):2348–7968.
2. Park Y.W., Nam M.S. Bioactive peptides in milk and dairy products: A review. *Korean J. Food Sci. Anim. Resour.* 2015;35(6):831.
3. Hughes J., Smith T., Kosterlitz H., Fothergill L.A., Morgan B., Morris H. Identification of two related pentapeptides from the brain with potent opiate agonist activity. *Nature* 1975;258(5536):577–579.
4. Wang Y., Van Bockstaele E.J., Liu-Chen L.-Y. In vivo tracking of endogenous opioid receptors. *Life Sci.* 2008;83(21–22):693–699.
5. Bayliss W.M., Starling E.H. The mechanism of pancreaticsecretion. *J. Physiol.* 1902;28(5):325–353.
6. Mellander O. The physiological importance of the casein phosphopeptidecalcium salts. II. Peroral calcium dosage of infants. Some aspects of the pathogenesis ofrickets. *Acta Soc. Bot. Pol.* 1950;55:247–255.
7. Meisel H., Bockelmann W. Bioactive peptides encrypted in milk proteins:proteolytic activation and thropho-functional properties. *Antonie Leeuwenhoek* 1999;76(1–4):207–215.11.
8. Choi J., Sabikhi L., Hassan A., Anand S. Bioactive peptides in dairy products. *Int. J. Dairy Technol.* 2012;65(1):1–12.
9. Sánchez A., Vázquez A. Bioactive peptides: A review. *Food Qualityand Saf.* 2017;1(1):29–46.
10. Yoshikawa M., Takahashi M., Yang S. Delta opioid peptides derived from plant proteins. *Curr. Pharm. Des.* 2003;9(16):1325–1330.
11. Arihara K., Ohata M. Bioactive compounds in meat. In: *Meat Biotechnology.* Springer, New York, 2008, pp. 231–249.
12. Pereira P.C. Milk nutritional composition and its role in human health. *Nutrition* 2014;30(6):619–627.
13. Abd El-Salam M.H., El-Shibiny S. Bioactive peptides of buffalo, camel, goat, sheep,mare, and yak milks and milk products. *Food Rev. Int.* 2013;29(1):1–23.
14. Teschemacher H., Koch G., Brantl V. Milk protein-derived opioid receptor ligands. *Biopolymers* 1997;43(2):99–117. doi: 10.1002/(SICI)1097-0282(1997)43:2<99::AID-BIP3>3.0.CO;2-V.
15. Henschen A., Brantl V., Teschemacher H., Lottspeich F. β-casomorphins–Novel ppioid peptides derived from bovine casein–isolatio n and structure. In: *Endogenous*

and Exogenous Opiate Agonists and Antagonists. Elsevier, Amsterdam, The Netherlands, 1980, pp. 233–236.
16. Loukas S., Varoucha D., Zioudrou C., Streaty R.A., Klee W.A. Opioid activities and structures of alphacasein-derived exorphins. *Biochemistry* 1983;22(19):4567–4573.
17. Yoshikawa M., Tani F., Yoshimura T., Chiba H. Opioid peptides from milk proteins. *Agric. Biol. Chem.* 1986;50:2419–2421.
18. Tani F., Iio K., Chiba H., Yoshikawa M. Isolation and characterization of opioid antagonist peptides derived from human lactoferrin. *Agric. Biol. Chem.* 1990;54(7):1803–1810.
19. Tidona F., Criscione A., Guastella A.M., Zuccaro A., Bordonaro S., Marletta D. Bioactive peptides in dairy products. *Ital. J. Anim. Sci.* 2009;8(3):315–340.
20. Bell S.J., Grochoski G.T., Clarke A.J. Health implications of milk containing β-casein with the A2 genetic variant. *Crit. Rev. Food Sci. Nutr.* 2006;46(1):93–100.
21. Garg S., Nurgali K., Mishra V.K. Food proteins as source of opioid peptides—A review. *Curr. Med. Chem.* 2016;23(9):893–910.
22. Zaky E.A. Autism spectrumdisorder (ASD): The past, the present,and the future. *J. Child andAdolescent Behav.* 2017;5:1–4.
23. Pruimboom L., De Punder K. The opioid effects of gluten exorphins: Asymptomatic celiac disease. *J. Health Popul. Nutr.* 2015;33:24.
24. Ohinata K., Agui S., Yoshikawa M. Soymorphins, novel μ opioid peptides derived from soy β-conglycinin β-subunit, have anxiolytic activities. *Biosci. Biotechnol. Biochem.* 2007;71(10):2618–2262.
25. Yang S., Kawamura Y., Yoshikawa M. Effect of rubiscolin, a δ opioid peptide derived from RuBisCO, on memory consolidation. *Peptides* 2003;24(2):325–328.
26. Liu Z., Udenigwe C.C. Role of food-derived opioid peptides in the central nervous and gastrointestinal systems. *J. Food Biochem.* 2019;43(1):e12629.
27. Janecka A., Fichna J., Janecki T. Opioid receptors and their ligands. *Curr. Top. Med. Chem.* 2004;4(1):1–17.
28. Kaur J., Kumar V., Sharma K., Kaur S., Gat Y., Goyal A., Tanwar B. Opioid peptides:an overview of functional significance. *Int. J. Pept. Res. Ther.* 2020;26(1):33–41.
29. Goldstein A., Fischli W., Lowney L.I., Hunkapiller M., Hood L. Porcine pituitary dynorphin: Complete amino acid sequence of the biologically active heptadecapeptide. *Proc. Natl. Acad. Sci. U. S. A.* 1981;78(11):7219–7223. doi: 10.1073/pnas.78.11.7219.
30. Chavkin C., James I.F., Goldstein A. Dynorphin is a specific endogenous ligand of the kappa opioid receptor. *Science* 1982;215(4531):413–415. doi: 10.1126/science.6120570.
31. Li C.H., Chung D. Isolation and structure of an untriakontapeptide with opiate activity from camel pituitary glands. *Proc. Natl. Acad. Sci. U. S. A.* 1976;73(4):1145–1148. doi: 10.1073/pnas.73.4.1145.
32. Brantl V., Teschemacher H., Bläsig J., Henschen A., Lottspeich F. Opioid activities of beta-casomorphins. *Life Sci.* 1981;28(17):1903–1909. doi: 10.1016/0024-3205(81)90297-6.
33. Jinsmaa Y., Yoshikawa M. Enzymatic release of neocasomorphin and β-casomorphin from bovine β-casein. *Peptides* 1999;20(8):957–962. doi: 10.1016/S0196-9781(99)00088-1.
34. Yoshikawa M., Tani F., Yoshimura T., Chiba H. Opioid peptides from milk proteins. *Agric. Biol. Chem.* 1986;50:2419–2421. doi: 10.1080/00021369.1986.10867763.

35. Tani F., Iio K., Chiba H., Yoshikawa M. Isolation and characterization of opioid antagonist peptides derived from human lactoferrin. *Agric. Biol. Chem.* 1990;54(7):1803–1810.
36. Yang S., Yunden J., Sonoda S., Doyama N., Lipkowski A.W., Kawamura Y., Yoshikawa M. Rubiscolin, a δ selective opioid peptide derived from plant RuBisCO. *FEBS Lett.* 2001;509(2):213–217. doi: 10.1016/S0014-5793(01)03042-3.
37. Norris R., FitzGerald R.J. Antihypertensive peptides from foodproteins. In: *Bioactive Food Peptides in Health and Disease.* InTech Publishers, Rijeka, Croatia, 2013, pp. 45–72.
38. Daliri B.-M.E., Oh D.H., Lee B.H. Bioactive peptides. *Foods* 2017;6(5):32.
39. Chandrudu S., Simerska P., Toth I. Chemical methods for peptide and protein production. *Molecules* 2013;18(4):4373–4388.
40. Tyagi A., Daliri E.B., Kwami Ofosu F., Yeon S.J., Oh D.H. Food-derived opioid peptides in human health: A review. *Int. J. Mol. Sci.* 2020;21(22):8825.
41. Fricker L.D. *Colloquium Series on Neuropeptides.* Morgan & Claypool Publishers, Williston, VT, 2012.
42. Law P.Y., Wong Y.H., Loh H.H. Molecular mechanisms and regulation of opioid receptor signaling. *Annu. Rev. Pharmacol. Toxicol.* 2000;40:389–430. doi: 10.1146/annurev.pharmtox.40.1.38.
43. Ganapathy V., Miyauchi S. Transport systems for opioid peptides in mammalian tissues. *AAPS J.* 2005;7(4):E852–E856. doi: 10.1208/aapsj070482.
44. Wang J., Hogenkamp D.J., Tran M., Li W.-Y., Yoshimura R.F., Johnstone T.B., Shen W.-C., Gee K.W. Reversible lipidization for the oral delivery of leu-enkephalin. *J. Drug Target.* 2006;14(3):127–136. doi: 10.1080/10611860600648221.
45. Dennis M.S., Zhang M., Meng Y.G., Kadkhodayan M., Kirchhofer D., Combs D., Damico L.A. Albumin binding as a general strategy for improving the pharmacokinetics of proteins. *J. Biol. Chem.* 2002;277(38):35035–35043. doi: 10.1074/jbc.M205854200.
46. Rodriguez R.E., Rodriguez F.D., Sacristán M.P., Torres J.L., Valencia G., Garcia Antón J.M. New glycosylpeptides with high antinociceptive activity. *Neurosci. Lett.* 1989;101(1):89–94. doi: 10.1016/0304-3940(89)90446-1. PMID: 2549457.
47. Matthies H., Stark H., Hartrodt B., Ruethrich H.-L., Spieler H.-T., Barth A., Neubert K. Derivatives of β-casomorphins with high analgesic potency. *Peptides* 1984;5(3):463–470. doi: 10.1016/0196-9781(84)90070-6.
48. Sakaguchi M., Koseki M., Wakamatsu M., Matsumura E. Effects of systemic administration of β-casomorphin-5 on learning and memory in mice. *Eur. J. Pharmacol.* 2006;530(1–2):81–87. doi: 10.1016/j.ejphar.2005.11.014.
49. Lin L., Umahara M., York D., Bray G. β-casomorphins stimulate and enterostatin inhibits the intake of dietary fat in rats. *Peptides* 1998;19(2):325–331. doi: 10.1016/S0196-9781(97)00307-0.
50. Belyaeva Y.A., Dubynin V., Stovolosov I., Dobryakova Y.V., Bespalova Z.D., Kamenskii A. Effects of acute and chronic administration of exorphin C on behavior and learning in white rat pups. *Mosc. Univ. Biol. Sci. Bull.* 2009;64(2):66–70. doi: 10.3103/S0096392509020035.
51. Fanciulli G., Dettori A., Demontis M.P., Tomasi P.A., Anania V., Delitala G. Gluten exorphin B5 stimulates prolactin secretion through opioid receptors located outside the blood-brain barrier. *Life Sci.* 2005;76(15):1713–1719. doi: 10.1016/j.lfs.2004.09.023.

52. Kaneko K., Iwasaki M., Yoshikawa M., Ohinata K. Orally administered soymorphins, soy-derived opioid peptides, suppress feeding and intestinal transit via gut μ1-receptor coupled to 5-HT1A, D2, and GABAB systems. *Am. J. Physiol. Gastrointest. Liver Physiol.* 2010;299(3):G799–G805. doi: 10.1152/ajpgi.00081.2010.
53. Chesnokova E., Saricheva N., Dubynin V., Kamenskij A., Kalikhevich V., Adermasova Z. Behavioral effect of soymorphin-5-amide in rats. *Mosc. Univ. Biol. Sci. Bull.* 2014;69(3):103–107. doi: 10.3103/S0096392514030055.
54. Yin H., Miao J., Zhang Y. Protective effect of beta-casomorphin-7 on type 1 diabetes rats induced with streptozotocin. *Peptides* 2010;31(9):1725–1729. doi: 10.1016/j.peptides.2010.05.016.
55. Ahn J., Park S., Atwal A., Gibbs B., Lee B. Angiotensin II converting enzyme (ACE) inhibitory peptides from whey fermented by Lactobacillus species. *J. Food Biochem.* 2009;33(4):587–602.
56. Takenaka A., Annaka H., Kimura Y., Aoki H., Igarashi K. Reduction of paraquat-induced oxidative stress in rats by dietary soy peptide. *Biosci. Biotechnol. Biochem.* 2003;67(2):278–283.
57. Agrawal H., Joshi R., Gupta M. Isolation, purification and characterization of antioxidative peptide of pearl millet (*Pennisetum glaucum*) protein hydrolysate. *Food Chem.* 2016;204:365–372.
58. Pelegrini P.B., Del Sarto R.P., Silva O.N., Franco O.L., Grossi-De-Sá M.F. Antibacterial peptides from plants: What they are and how they probably work. *Biochem. Res. Int.* 2011;2011:1–9.
59. Bintsis T. Foodborne pathogens. *AIMS Microbiol.* 2017;3(3):529–563.
60. Mesaik M.A., Dastagir N., Uddin N., Rehman K., Azim M.K. Characterization of immunomodulatory activities of honey glycoproteins and glycopeptides. *J. Agric. Food Chem.* 2015;63(1):177–184.
61. Korhonen H., Pihlanto A. Food-derived bioactive peptides--Opportunities for designing future foods. *Curr. Pharm. Des.* 2003;9(16):1297–1308.

CHAPTER 22

Immunomodulating Activity

Lourdes Santiago-López, A. Alejandra López-Pérez, Lilia M. Beltrán-Barrientos, Adrián Hernández-Mendoza, Belinda Vallejo-Cordoba, and Aarón F. González-Córdova

CONTENTS

Introduction	441
Overview of the Immune System and Protein Hydrolysates	442
Immunomodulatory Effect of Food-Derived Peptides	443
In vitro Studies	443
In vivo Assays	447
Clinical Trials	451
Mechanism of Action	452
Conclusions	453
Acknowledgment	453
Funding	454
References	454

INTRODUCTION

Over the recent decades, interest has risen in food not only for being a source of nutrients and having a role in disease prevention, but also for having the ability to help in the treatment of various diseases. In this regard, several foods with high protein content have been studied for being a rich source of potentially bioactive peptides (2–20 amino acids), which may be released by different pathways (e.g., during fermentation by specific lactic acid bacteria, through enzymatic hydrolysis, and during gastrointestinal digestion) (Daliri et al., 2017). In this last aspect, several scientific articles evidence that peptides may be obtained from different sources (e.g., dairy products, meat, vegetables, seafood, eggs, and cereals) (Jakubczyk et al., 2020). Moreover, their biological effects are determined by the amino acid composition and sequence. Although the correlation between structure and functional properties yet is not well established, their functional properties have been observed in several *in vitro* and *in vivo* assays and clinical trials (Daliri et al., 2017; Kang et al., 2019).

Different peptides have been shown to possess antihypertensive, antimicrobial, antioxidative, antithrombotic, and immunomodulatory effects (Sánchez et al., 2017). The immunomodulatory properties such as proliferation, antibody and cytokine production, and regulation and phagocytic activity of peptides have been mainly described in *in vitro* studies (Rémond et al., 2016). Furthermore, several immunomodulatory peptides have

DOI: 10.1201/9781003106524-28

also shown anti-inflammatory and antioxidant properties. In this regard, oxidative stress has been related to immune activities, and specific peptides may ameliorate the inflammation by decreasing the inflammatory cytokines and by suppressing the inflammatory molecules (Chakrabarti et al., 2014; Liu et al., 2020). In this sense, several *in vivo* studies have reported the immunomodulatory properties of peptides derived from α-, β-casein, and α-lactalbumin from milk, which showed the capacity to stimulate the proliferation of lymphocytes, and increased the hemolytic antibodies and the phagocytic capacity in the splenocytes and macrophages of mice (Kiewiet et al., 2018). Other peptides derived from rice, soybean, and eggs have also presented immunomodulatory properties, since they enhanced IL-10 cytokine, cell proliferation, up-regulated the production of IL-2 cytokine and TNF, and decreased the reactive oxygen species (Zhu et al., 2020). Although the mechanism of action of food-derived immunomodulatory peptides are still under study, there is evidence that supports peptides may modulate or regulate the immune system, being this is an attractive opportunity to manage immune-related diseases. This chapter reviews several studies that evaluate food-derived peptides and their effect on the immune system and their potential mechanism of action.

OVERVIEW OF THE IMMUNE SYSTEM AND PROTEIN HYDROLYSATES

The immune system is the host defense mechanism that serves to remove potentially harmful substances known as pathogens; which is composed of the innate and adaptive system. The innate system includes the activation of complements, phagocytosis, natural killer (NK) activity, mast cells, dendritic cells, and neutrophils; meanwhile, the lymphocyte proliferation of T and B cells belong to the adaptive system (Beutler, 2004; Steinman and Hemmi, 2006). In this sense, the communication between these two defense systems may be linked by the signaling of cytokines or by the interactions between different cell types (e.g., dendritic cells, NK cells, neutrophils) (Rivera et al., 2016). In this regard, the immunomodulatory effect may be defined as any substance that may modify the response of the immune system by increasing (immunostimulators) or by decreasing (immunosuppressives) some inflammatory mediators of the innate and the adaptive system (Bascones-Martinez et al., 2014). Some results evaluated in several studies are the capacity of peptides to decrease the pro-inflammatory (IL-6, IL-1B, IFN, TFN), or to enhance the anti-inflammatory (IL-10) cytokines (Muñoz-Carrillo et al., 2018). In addition, the high or low phagocytic activity response provides the body with the ability to protect environmental pathogenic bacteria, fungi, and malignant cells and contribute to adaptive immunity by presenting antigens to lymphocytes and to eliminate unwanted host cells in order to maintain tissue homeostasis (Lim et al., 2017).

Overall, the immunomodulatory properties of food-derived peptides or hydrolysates have been evaluated in *in vitro* experiments, principally, casein-derived peptides; however, other protein hydrolysates from whey, soy, egg, wheat, rice, fish, sheep, oyster, and others have been explored (Gauthier et al., 2006; Santiago-López et al., 2016; Chalamaiah et al., 2018). The effects that have been reported for various peptide sequences are dependent on the food matrix, as well as the type of hydrolysis used (enzymatic or fermentation by microorganisms) (Figure 22.1).

Milk proteins are the most studied source with potential immunomodulatory effect. Hydrolysates from casein, α-lactalbumin, β-lactoglobulin, and serum albumin obtained with tripsin, α-chimotripsin, pepsin, and pancreatin enhanced the phagocytic activity of peritoneal macrophages of mice (Kazlauskaite et al., 2005; Biziulevicius et al., 2006) and

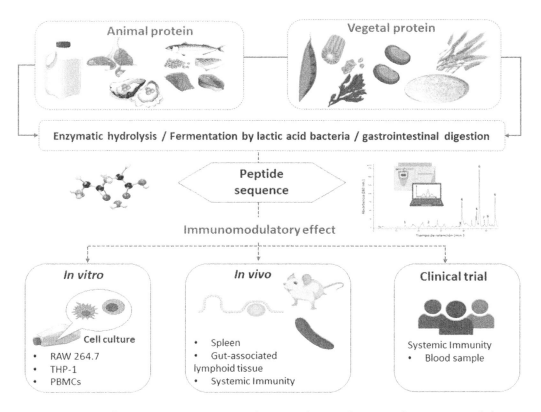

FIGURE 22.1 Schematic representation showing the production of immunomodulatory protein hydrolysates or peptides from food proteins that have been evaluated *in vitro* and *in vivo* (mice and clinical trials).

enhanced the weight of the thymus and the spleen (Pan et al., 2013). Additionally, whey hydrolysates (< 10 kDa) stimulated the production of serum IgA and IFN-γ in healthy mice (Saint-Sauveur et al., 2008). Finally, the immunomodulatory effect has been evaluated in several *in vitro* and *in vivo* studies (mice and clinical trials), which are described in detail in the following section.

IMMUNOMODULATORY EFFECT OF FOOD-DERIVED PEPTIDES

In vitro Studies

Cell line studies allow us to determine the response of several main immune markers (e.g., cytokines, nitric oxide, adhesion molecules, phagocytic activity) after treatment with food-derived hydrolysates or peptides (Ndiaye et al., 2012; Pan et al., 2013; Chalamaiah et al., 2018; Xu et al., 2019). The immunomodulatory effect of peptides has been evaluated in *in vitro* testing using cell lines, such as RAW 264.7 (macrophage cells), U937 and THP-1 (human monocyte cell lines), and human peripheral blood mononuclear cells (PBMCs). Moreover, the immunomodulatory properties of food-derived peptides have been shown to inhibit the phagocytic activity of mice-derived macrophages or in RAW264.7 cell line, suppressed or enhanced the proliferation of lymphocytes, induced

TNF-α, NF-κB levels, down-regulated NO/PG-E$_2$ and iNOS/COX-2 in macrophages, and showed an anti-inflammatory effect (Kiewiet et al., 2018). Table 22.1 shows *in vitro* and *ex vivo* assays of food-derived peptides and their potential immunomodulatory effect.

As previously mentioned, peptides with immunomodulatory effects derived from milk proteins, have been among the first and widely studied in different *in vitro* models. In this regard, casein hydrolysates by specific enzymes and hydrolysates derived from fermentation with lactic acid bacteria, increased the production of hemolytic antibodies in mice splenocytes or rat and increased the phagocytic capacity of murine macrophages (Korhonen and Pihlanto 2003; Gauthier et al., 2006; FitzGerald and Murray 2006; Möller et al., 2008).

Studies with whey protein isolates hydrolyzed with trypsin-chymotrypsin, enhanced the proliferation of murine splenocytes in cultures stimulated with concanavalin-A, and inhibited ConA-induced cytokines secretion. These results showed that acid or neutral fractions were able to stimulate splenocyte proliferation (Saint-Sauveur et al., 2008). Similarly, whey protein lactoglobulin hydrolyzed with alcalase and trypsin increased splenocytes proliferation of mice (Ma et al, 2014), activated the monocytes by increasing TNF-α, and induced the TGF-β and Treg differentiation of PBMC of humans (Rodríguez-Carrio et al., 2014), which shows the ability of these peptides to stimulate the innate response.

For instance, the peptide GYPMYPLPR derived from rice albumin using trypsin showed the promotion of phagocytosis activity for human polymorphonuclears leukocytes and enhanced the superoxide anion (Takahashi et al., 1994). Moreover, the peptide YGIYPR from rice using trypsin enzyme enhanced the proliferation on macrophage RAW 264.7 cells, where the hydrophobic amino acids and Arg were correlated with immunomodulatory activity, as it increased the interaction of the peptides and cytomembrane of cells (Xu et al., 2019). These results showed that the effect obtained *in vitro* was related to the structure of peptides, and the immune response was dependent of the type of line cells used in the assays. Both studies showed the immunomodulatory activity for two peptides derived from rice but in different type cells.

Likewise, the immunomodulatory activity of peptides of α-zein hydrolysates from maize with thermolysin and pepsin were evaluated in human monocyte cell line U937 cells. In this regard, the specific peptide FLPFNQL showed the highest activity reducing IL-6 production. The authors suggested that the presence of FL amino acids at the N-terminal contributes to the potential immunomodulatory activity *in vitro* (Liu et al., 2020). The IL-6 is a cytokine involved in the regulation of acute-phase response to injury and infection. Then, their dysregulation could result in several immune diseases (Tanaka et al., 2014).

On the other hand, extracts from *Porphyra tenera* (edible seaweeds), obtained by action of proteases and carbohydrase showed no cytotoxicity in RAW264.7 macrophages, and inhibited lipopolysaccharides (LPS)-induced nitric oxide (NO) production in RAW264.7 macrophages. These results suggest that the natural compounds from herbs may be an alternative in the search of immunomodulatory peptides (Senevirathne et al., 2010). Similarly, hydrolysates from *Phorphyra columbina* protein using flavourzyme and fungal protease concentrate increased IL-10 secretion in splenocytes, macrophages, and in lymphocytes. The effect of the hydrolysate on IL-10 production was mediated by activation of JNK (C-jun N-terminal kinase), mitogen-activated protein kinase (MAPK) and nuclear factor-kB (NF-κB) dependent pathways in T-lymphocytes (Cian et al., 2012). Furthermore, peptides from *Spirulina maxima* (LDAVNR and MMLDF) obtained by

TABLE 22.1 *In vitro* Studies Using Peptides of Food-Derived Protein Hydrolysates with Immunomodulatory Effect

Source	Enzymes/ Microorganisms	Fraction/Peptide Sequence	Model of Study	Key Findings	Reference
Whey protein	trypsin/chymotrypsin	Short-chain peptides	Murine spleenocytes	Stimulate the proliferation of murine spleen lymphocytes at lower concentrations (0.5–500 µg mL−1).	Mercier et al. 2004
	Bifidobacterium lactis NCC362 enzymes	Peptide fragments	Mice spleenocytes	Up-regulated IFN-γ and IL-10 production and down-regulated IL-4 secretion.	Prioult et al. 2005
Soybean protein	Trypsin	MITLAIPVNK-PGR	Human blood	Stimulated phagocytosis of human neutrophils	Tsuruki et al. 2003
	Alcalase	Peptides < 1.2 kDa	RAW264.7 macrophage cells	Inhibited the NO, iNOS, PGE-2, COX-2, and TNF-α	Vernaza et al. 2012
Salmon	Alcalase, Flavourzyme, Neutrase, Protamex, pepsin and trypsin	Peptides of 1-2 kDa PAY	RAW264.7 macrophage cells	Inhibiting NO and TNF-α, IL-6, and IL-1β	Ahn et al. 2012, 2015
	Alcalase 2.4 L	Hydrolysates lower molecular weight	Mice spleenocytes	↑The proliferation of lymphocytes, and phagocytosis of the peritoneal macrophages of mice	Kong et al. 2008
Alaska pollock	Trypsin	NGMTY, NGLAP, WT	Mice spleenocytes	↑ Proliferation of lymphocytes	Hou et al. 2016
Yellow field pea proteins	Thermolysin	Peptides < 3 kDa	LPS/IFN-γ-activated RAW 264.7 NO− macrophages	Inhibition of NO production and TNF-α, and IL-6. ↑ Phagocytic activity of their peritoneal macrophages and stimulated the gut mucosa, IgA+ cells, IL-4+, IL-10+, IFN-γ+ cells in the small intestine lamina propria. Regulation of IL-6 secretion by IEC via TLR-2 and TLR-4.	Ndiaye et al. 2012

(*Continued*)

TABLE 22.1 (Continued)

Source	Enzymes/ Microorganisms	Fraction/Peptide Sequence	Model of Study	Key Findings	Reference
Amaranth protein	Papain, trypsin, pepsin, alkaline protease, neutral protesase	Peptides < 2 kDa	THP-1 and RAW264.7 macrophage cells	Production of NO, TNF-α, PGE-2 and COX-2	Montoya-Rodriguez et al. 2012
Shellfish protein	Flavourzyme, neutrase, alcalase, papain, pepsin, chymotrypsin, trypsin	GVSLLQQFFL	RAW264.7 macrophage cells	Inhibited the NO production	Kim et al. 2013
Wheat germ globulin	Alcalase, neutrase, papain, pepsin and trypsin	Peptides 0.3-1.450 kDa	RAW264.7 macrophage cells	↑ Lymphocyte proliferation, phagocytosis and secretion of TNF-α, IL-6 and NO production.	Wu et al. 2016
Egg yolk livetins	Pepsin	Peptides < 10 kDa	RAW264.7 macrophage cells	Inhibited the TNF-α, IL-6, IL-1β and NO	Chalamaiah and Wu 2017
Tilapia, casein and pea	Proteinase (*Virgibacillus halodenitrificans* SK1-3-7)	Hydrolysates	THP-1 cells	↑ Innate immunity ↓ IL-1β, IL-6, IL-8, TNF-α, COX-2 production	Toopchman et al. 2017

NO = nitric oxide, iNOS = inducible nitric oxide synthase, PGE-2 = prostaglandin-E2, COX-2 = cyclooxygenase-2, IEC = intestinal epithelial cells hydrolysis with trypsin, pepsin, and α-chymotrypsin, inhibited the histamine production from RBL-2H3 mast cells and IL-8 from EA.hy926 endothelial cells (Vo et al., 2013).

Eggs have also shown to be a potential source of bioactive peptides, which may be derived from proteins as ovalbumin, ovotransferrin, ovomucin, lyzozyme, and avidin (Anderson 2015). In this sense, a study reported by Chalamaiah and Wu (2017) showed the anti-inflammatory effect of egg yolk livetin fractions in RAW264.7 macrophages cells. The results showed that hydrolysates obtained with pepsin inhibited the NO, IL-1β, IL-6, TNF-α, and the expression of inducible nitric oxide iNOS. Meanwhile, alcalase hydrolysates were more effective in inhibiting the prostaglandin-E2 (PGE-2) and the expression of COX-2. Furthermore, livetin and its enzymatic hydrolysates significantly enhanced ($p < 0.05$) the phagocytic activity in RAW264.7 cells. In general, it may be concluded that the immunomodulatory effect is related to the specificity of the enzyme cleavage sites, which led to the generation of specific peptides with particular amino acid sequences.

In vivo Assays

In order to demonstrate the immunomodulatory effect of peptides or hydrolysates of food-derived proteins, *in vivo* studies with different animal models have been employed, mainly mice (Egusa and Otani, 2009; Cai et al., 2013; Mallet et al., 2014; Chalamaiah et al., 2015; Kim et al., 2017). Most studies have evaluated the dose- and time-dependent immunomodulatory effects of different peptide sequences (Table 22.2). In general, the main markers evaluated were lymphocyte proliferation ($CD4^+$, and $CD8^+$), peritoneal macrophage phagocytosis, NK cell activity, serum antibodies (IgA, IgM and IgG), mucosal immunity (S-IgA), and cytokines associated to response Th1, Th2, Treg, and Th17 (Chalamaiah et al., 2018).

In this sense, tryptic casein hydrolysates (1.0 mg/g per body weight (BW)) were administered to BALB/c mice, and after five days, the phagocytic activity regarding peritoneal macrophages and blood phagocytic cells were enhanced in the groups treated with hydrolysates (Kazlauskaitė et al., 2005). Similar results were reported for casein, α-lactalbumin, β-lactoglobulin, ovalbumin, and serum albumin hydrolysates with a mix of enzymes (Biziulevičius et al., 2006).

In another study, healthy female BALB/C mice were fed with egg yolk peptide digest (EYLLPD) at a dose of 121.5 µg of protein/day for seven days. The results showed an increase in IgA^+ cells, IL-6 secretion by the small intestinal epithelial cells, and the modulation of the Th1/Th2 response at the mucosal level (Nelson et al., 2007). The IL-6 cytokine is important for the innate and the adaptive response due to the pleiotropic nature, which promotes the development of pathogenic Th17 cells and the maturation of B-lymphocytes (Choy and Rose-John 2017). The results of this study showed that EYLLPD might have played an important role in preventing bacterial infection. Additionally, hydrolysates from Rohu egg (0.25, 0.5, and 1 g/kg BW) were administered to female BALB/c mice for 45 days. The immunological markers, such as splenic NK cell cytotoxicity, macrophage phagocytosis, and level of IgA, were increased. In particular, hydrolysates with pepsin and alcalase showed effect on IgA, whereas trypsin hydrolysates showed effect in $CD4^+$ and $CD8^+$ cells of spleen (Chalamaiah et la., 2014). On the other hand, hydrolysates of ovalbumin, lysozyme, ovomucoid, and whole egg protein were obtained by hydrolysis of pepsin, neutrase, and alcalase. A reduction of lymphocyte proliferation and IL-13, IL-10 and TNF-α in murine spleen and mesenteric lymph node cells stimulated with concanavalin A were observed (Lozano-Ojalvo et al., 2016).

TABLE 22.2 *In vivo* Studies Using Peptides of Food-Derived Protein Hydrolysates with Immunomodulatory Effects

Source	Type of Hidrolisis	Enzymes/ Microorganisms	Peptide Sequence	Model of Study	Key Findings	Reference
Milk / Dairy products	Enzymatic	Trypsin	Hydrolysate	BALB/c mice	↑Phagocytosing capacity	Kazlauskaitė et al. 2005
		Papain and Trypsin	Peptides < 10 kDa	ICR mice	Improve the level of hemolysin in serum and phagocytosis of macrophages. ↓Type I hypersensitivity by decreasing IgE levels, IL-4 in serum and histamine and bicarbonate in peritoneal mast cells. ↑TGF-β in the serum.	Pan et al. 2013
		No reported	Peptides 5 kDa	C3H/HeOuJ mice	↓ Acute allergic skin response and mast cell degranulation. ↑ Increased Foxp3+ cell numbers in the MLN	Kiewiet et al. 2017a
	Microbial Fermentation	*Lactobacillus helveticus*	Peptidic fraction	BALB/c mice (infection with *Escherichia coli* H10407)	↑Intestinal and serum IgA levels, number of IgA-secreting B lymphocytes in the intestinal lamina propria, stimulation of Th2 response (IL-4 vs. IFNγ)	LeBlanc et al. 2004
		Milk fermentation products of *L. helveticus* R389	No reported	BALB/c mice	↑Expression of TRPV6 channels in the duodenum. Expression calcineurin in the small intestine. Upregulate of IL-2 and TNF production. ↑Number of mucosal mast cells and goblet cells.	Vinderola et al. 2007
		Enterococcus faecalis hydrolysis	Casein hydrolysate	Human aortic endothelial cells	Inhibition the production of MCP-1, IL-8 and expression of VCAM-1, ICAM-1 and E-selectin. Attenuated the adhesion of human monocytes to activated endothelial cells. Inhibiting the NF-κB pathway through a PPAR-γ dependent mechanism.	Marcone et al. 2015

(*Continued*)

TABLE 22.2 (Continued)

Source	Type of Hidrolisis	Enzymes/ Microorganisms	Peptide Sequence	Model of Study	Key Findings	Reference
Fermented pacific (Seacure®)	Fermentation	Yeast	Protein concentrate	BALB/c mice	↑ Phagocytic activity of peritoneal macrophages, the number of IgA+ cells, IL-4, IL-6, IL-10, IFN-γ, TNF-α, in the lamina propria of the small intestine	Duarte et al. 2006
Egg yolk digests	Enzymatic	Pepsin	EYLLPD	BALB/c mice	↑IL-6 secretion by small intestinal epithelial cells, IgA+ cells, orchestrating the Th1/Th2 response.	Nelson et al. 2007
Soybean	Enzymatic	*Rhizopus oryzae* neutral protease preparation	Hydrolysate	C3H/HeN mice	Mitogenic activity toward C3H/HeN mouse spleen cells. ↑ The number of spleen CD11b+, CD49b+, interleukin IL-12+CD11b+, and interferon IFN-γ+CD49b+, IFN-gamma+CD4+, IL-12+CD11b+ cells, while that of spleen IL-4+CD4+ cells was largely unchanged and cytotoxic activity.	Egusa and Otani 2009.
Egg white	Enzymatic	Aminopeptidase	Glu-Trp-Pro	Porcine model (DSS induced colitis)	Attenuated DSS-induced clinical symptoms, including weight loss, mucosal and submucosal inflammation, crypt distortion, and colon muscle thickening. ↓ Intestinal permeability, Intestinal expression of TNF, IL-6, IL-1β, IFN-γ, IL-8, and IL-17. ↑Mucin gene expression.	Lee et al. 2009

(*Continued*)

TABLE 22.2 (Continued)

Source	Type of Hidrolisis	Enzymes/ Microorganisms	Peptide Sequence	Model of Study	Key Findings	Reference
Oyster	Enzymatic	Bromelian	hydrolysate of oyster protein	BALB/c mice	↑Spleen lymphocyte proliferation and the activity of NK cells. Improve intestinal absorption, increase food utilization ratio, and maintain the normal physiological function of mice.	Cai et al. 2013
Shark protein	Enzymatic	Trypsin and chymotrypsin	Hydrolysate	BALB/c mice (infection with *Escherichia coli* H10407)	↑IgA$^+$ cells and IL-6, TNF, TGF, and IL-10.	Mallet et al. 2014
Common carp egg	Enzymatic	Pepsin, alcalase	Hydrolysate	BALB/c mice	↑Secretory immunoglobulin A in the gut, the splenic NK cell cytotoxicity, macrophage phagocytosis and IgA, percentages of CD4$^+$ and CD8$^+$ cells in the spleen.	Chalamaiah et al. 2015
Coix glutelin protein	Enzymatic	Trypsin	Peptides < 3 kDa	ICR mice	Stimulated mice splenocytes and peritoneal macrophages. ↑The spleen index	Ling-Ling et al. 2017
Oyster	Enzymatic hydrolysis	Bromelain, pepsin, trypsin	Hydrolysate	C3H/HeOuJ mice	↑Regulatory B and T cells in the spleen, Th1 and Th17 in the Peyer's patches	Kiewiet et al. 2017
Tuna cooking	Enzymatic hydrolysis	Enzyme A and B	Hydrolysate	BALB/c mice	↑Weight of the spleen and thymus and the proliferation of splenocytes. ↑IL-10, IL-2, serum IgG1 and IgG2a levels in serum.	Kim et al. 2017b

MLN = mesenteric lymph nodes, DSS = dextran sodium sulfate, NK = natural killer, VCAM-1 = vascular cell adhesion molecule-1, MCP-1 = monocyte chemoattractant protein-1, ICAM-1 = intercellular adhesion molecule-1

On the other hand, Wang et al. (2010) evaluated the effect of oyster hydrolysates on the antitumor activity and the immunostimulating effect in female BALB/c mice. The oyster hydrolysates were administered by oral gavage once a day at 1 mg/g for 14 days. The supplementation of hydrolysates inhibited the transplantable sarcoma-S180, enhanced the weight coefficients of the thymus and the spleen, and increased the NK activity, the proliferation of lymphocytes, and the phagocytic rate of macrophages. Results showed the potential of hydrolysates of oyster for tumor therapy and a potential immunomodulatory effect. Alternatively, marine oligopeptide from chum salmon were evaluated in female ICR mice at different doses. Interestingly, the group supplemented with oligopeptides, proliferation of the lymphocytes was increased, as well as the percentage of $CD4^+$ cells and cytokines IL-2, IFN-γ, and IL-5, IL-6 of Th1 and Th2 response (Yang et al., 2009). Alaska pollock hydrolysates obtained by glutamic acid biosensor, enhanced humoral, cellular and nonspecific immunity, IL-2, IL-4, and IL-6 lymphocyte proliferation rates in immunosuppressed mice. The sequence identified was PTGADY, which may be responsible for the immunomodulatory effect (Hou et al., 2016).

Clinical Trials

In the search for establishing the immunomodulatory effects of bioactive peptides, few clinical studies have been documented. The review published by Kiewiet et al. (2018) documented three clinical trials, which evaluated the effect of wheat gluten, salmon fish protein, and soybean protein.

Horiguchi et al. (2005) reported the first study on the effect of wheat gluten hydrolysates on the immune system in healthy human subjects. Although the type of study was not specified, it was performed with only nine volunteers, which were grouped into two (n = 5) and (n = 4) groups, for the treated and control groups, respectively. The test group consumed 3 g per day of gluten hydrolysates (Glutamine Peptide GP-1) during six days, whereas the control group did not receive any treatment. The immunological parameters were evaluated in blood samples. In this sense, the results showed that the intake of gluten hydrolysates increased ($p < 0.018$) the NK cell activity, and no adverse effects were reported.

A randomized, double-blind, parallel, 3-arm, multicentric study was reported by Nesse et al. (2011) to evaluate the clinical effect of fish protein hydrolysates (Amizate®) on the immunoglobulin concentrations (G, A, and M levels), CD4 / CD8 ratios, and hemoglobin in serum. Malnourished Indian schoolchildren aged 6–8 years with mild and moderate malnutrition (Gomez´s classification) were randomly assigned into 3-arm (n = 146 subjects) groups. The study was performed during 4 months, where the Arm-A consumed 120 mL of a chocolate drink fortified with 3 g of Amizate®, while the Arm-B consumed chocolate drink fortified with 6 g of Amizate® and the Arm-C consumed only the chocolate drink as placebo. Results showed that there was no significant difference between groups on the immunoglobulins concentration ($p > 0.118$), the $CD4^+$ and the $CD8^+$ and the hemoglobin levels. Interestingly, the Amizate® improved body mass index values, thus improving the health of malnourished children, without any adverse alteration of immune response.

In the same way, in a cross-over double-blind study (Yimit et al., 2012) examined the effect of soybean peptides on the immune and brain function, as well as the neurochemistry in 10 healthy volunteers 20–25 years old. For this purpose, a dosage of 8 g of dried powder of hydrolyzed soy protein were suspended in 200 mL of water for each volunteer and 24 h later, blood samples were collected. An effect on $CD3^+$, $CD4^+$, $CD8^+$, $CD11b^+$, $CD14^+$, $CD19^+$, and $CD56^+$ lymphocytes subsets cell counts ($p < 0.05$) were observed. In particular, CD11b+ and $CD14^+$ were associated with macrophage activity, $CD3^+$ and $CD4^+$ with immature T cells, and $CD19^+$ with B-cell activity. Furthermore, a decrease of adrenalin ($p < 0.01$) and an enhanced level of dopamine ($p < 0.05$), and θ, α-2, and β-L frequency bands, were related to functional brain activation patterns. Authors concluded that soybean peptides may modulate cellular immune systems, regulate neurotransmitters, and boost brain function.

MECHANISM OF ACTION

The molecular mechanisms underlying the immunomodulatory properties of food-derived peptides are not fully understood and only few studies have focused on the elucidation. Nevertheless, it has been suggested that their beneficial effects seem to be mediated by binding to specific immune-receptors. In this sense, the immunomodulatory effect is visualized to promote the macrophage activation, the phagocytosis stimulation, and the production of cytokines and antibodies. Additionally, the enhancement of the NK cells $CD4^+$, $CD8^+$, $CD11b^+$, and $CD56^+$ and activation of transcription factors, such as NF-kB and MAPK, and the inhibition of pro-inflammatory mediators have also been suggested (Kiewit et al., 2018; Yang et al., 2020).

In particular, peptides with immunomodulatory effects, the amino acid sequence, the length, the charge, and the hydrophobicity are important for the binding to specific receptors on the immune cells. Some peptides containing hydrophobic (Gly, Val, Leu, Pro, Phe), negatively charged amino acid (Glu), and aromatic amino acid (Tyr) in their sequence, could determine the immunological effect (Chalamaiah et al., 2018). Furthermore, hydrophobic peptides from β-lactoglobulin and α-lactalbumin bearing 2–3 positive charges promote the proliferation of mouse splenocyte cells, and short peptides enhance TNF-α, whereas soy peptides of low molecular and more positive charge are most effective in the proliferation (Kong et al., 2008; Saint-Sauveur et al., 2008).

Kiewit et al. (2018) suggested three mechanisms associated to peptides. The peptides can directly stimulate specific receptors, peptide transporter Pep-T1, via endocytosis. Furthermore, Santiago-Lopez et al. (2016) summarized that δ- or κ-type opioid receptors may be a target for immunomodulatory peptides, specifically the presence of Arg in the N- or C-terminal and regulate the peripheral immune system to promote the cell differentiation, antibody production, and phagocytosis processes (Figure 22.2).

Overall, peptides can be recognized by specific receptors known as toll-like receptors (TLR), which are found on immune cells and in the intestinal epithelium. Peptides derived from milk proteins activated the receptors TLR2, 3, 4, 5, 7, 8, and 9, leading to the production of some cytokines TNF-α, IL-10, and IL-8 in PBMC cells (Kiewiet et al., 2017b), whereas the casein phosphopeptides induced the proliferation and IL-6 production in $CD19^+$ cells after the administration of an anti-TLR4 antibody (Tobita et al., 2006). A second mechanism is mediated by activation of the peptide transporter PepT1 expressed in the small intestine. The activity has been evaluated for soy peptides (KVP and VPY), and whey peptide (IPAV) to inhibit the pro-inflammatory cytokines, NF-kB and MAPK, extracellular signal-regulated kinase, and c-Jun N-terminal kinase (Adibi, 2003). A third mechanism has been associated with the endocytosis process (nonspecific form of vesicle mediated internalization) for larger food-derived peptides. In this process, hydrophobic peptides with β-turns structure interacts with the cell membrane involved

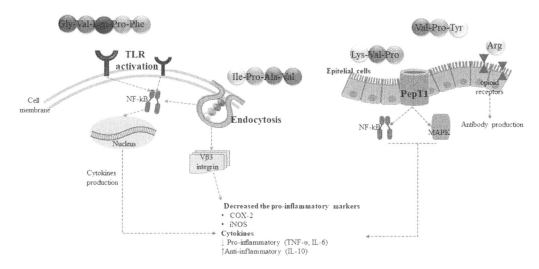

FIGURE 22.2 Reported mechanisms of action for Food-derived immunomodulatory peptides. (Adapted from Santiago-López et al., 2016); Kiewiet et al., 2018.)

for the internalization of the peptide. This effect is influenced by the size, hydrophobicity, and charge of the peptides. This mechanism was reported for lunasin peptide from soy, which inhibited αVβ3 integrin-mediated pro-inflammatory markers and downregulation of the NF-κB pathway (Cam et al., 2013).

CONCLUSIONS

The immune system plays an important role as the body's protective system, and food-derived peptides have demonstrated the ability to stimulate this system. There are a variety of *in vitro* and *in vivo* studies that show the effect of peptides either at a single dose, or through frequent consumption. However, only three documented clinical studies have been made. In fact, the reported effects on different immunological markers showed no significant difference. These results could be attributed to the design of the studies, as well as the number of subjects; therefore, the implementation of randomized controlled clinical studies remains an area of research opportunity. Additionally, the mechanisms of action of these peptides remain hypothesized, and only a few studies have reported the specific sequence to which the immunomodulatory effect has been attributed. Thus, analytical techniques such as sodium, dodecyl sulfate polyacrylamide gel electrophoresis (SDS-PAGE), size exclusion chromatography, and high-performance liquid chromatography-mass spectrometer (HPLC-MS) may complement the molecular weight distribution and determine the amino acid sequence.

ACKNOWLEDGMENT

The authors gratefully acknowledge CONACyT for the postdoctoral scholarship (research project 10980) granted to L. Santiago-López (Estancias Posdoctorales por México en Atención a la Contingencia del COVID-19).

FUNDING

This study was supported by the Mexican Council of Science and Technology (CONACyT, Mexico City, Mexico) research project CB2017-2018, A1-S-53161 & 10980.

REFERENCES

Adibi, S. A. 2003. Regulation of expression of the intestinal oligopeptide transporter (Pept-1) in health and disease. *American Journal of Physiology. Gastrointestinal and Liver Physiology* 285(5): G779–G788. https://doi.org/10.1152/ajpgi.00056.2003.

Ahn, C. B., Je, J. Y., and Cho, Y. S. 2012. Antioxidant and anti-inflammatory peptide fraction from salmon byproduct protein hydrolysates by peptic hydrolysis. *Food Research International* 49(1): 92–98. https://doi.org/10.1016/j.foodres.2012.08.002.

Ahn, C. B., Cho, Y. S., and Je, J. Y. 2015. Purification and anti-inflammatory action of tripeptide from salmon pectoral fin byproduct protein hydrolysate. *Food Chemistry* 168: 151–156. https://doi.org/10.1016/j.foodchem.2014.05.112.

Anderson, C. J. 2015. Bioactive egg components and inflammation. *Nutrients* 7(9): 7889–7913. https://doi.org/10.3390/nu7095372.

Bascones-Martinez, A., Mattila, R., Gomez-Font, R., and Meurman, J. H. 2014. Immunomodulatory drugs: Oral and systemic adverse effects. *Medicina Oral, Patologia Oral y Cirugia Bucal* 19(1): e24–e31. https://doi.org/10.4317/medoral.19087.

Beutler, B. 2004. Innate immunity: An overview. *Molecular Immunology* 40(12): 845–859. https://doi.org/10.1016/j.molimm.2003.10.005.

Biziulevičius, G. A., Kislukhina, O. V., Kazlauskaitė, J., and Zukaite, V. 2006. Food-protein enzymatic hydrolysates possess both antimicrobial and immunostimulatory activities: A 'cause and effect'theory of bifunctionality. *FEMS Immunology and Medical Microbiology* 46(1): 131–138. https://doi.org/10.1111/j.1574-695X.2005.00019.x.

Cai, B., Pan, J., Wu, Y. et al. 2013. Immune functional impacts of oyster peptide-based enteral nutrition formula (OPENF) on mice: A pilot study. *Chinese Journal of Oceanology and Limnology* 31(4): 813–820. https://doi.org/10.1007/s00343-013-2311-z.

Cam, A., Sivaguru, M., and Gonzalez de Mejia, E. 2013. Endocytic mechanism of internalization of dietary peptide lunasin into macrophages in inflammatory condition associated with cardiovascular disease. *PLOS ONE* 8(9): e72115. https://doi.org/10.1371/journal.pone.0072115.

Chakrabarti, S., Jahandideh, F., and Wu, J. 2014. Food-derived bioactive peptides on inflammation and oxidative stress. *BioMed Research International* 2014: 608979. https://doi.org/10.1155/2014/608979.

Chalamaiah, M., Hemalatha, R., Jyothirmayi, T. et al. 2014. Immunomodulatory effects of protein hydrolysates from rohu (*Labeo rohita*) egg (roe) in BALB/c mice. *Food Research International* 62: 1054–1061. https://doi.org/10.1016/j.foodres.2014.05.050.

Chalamaiah, M., Hemalatha, R., Jyothirmayi, T. et al. 2015. Chemical composition and immunomodulatory effects of enzymatic protein hydrolysates from common carp (*Cyprinus carpio*) egg. *Nutrition* 31(2): 388–398. https://doi.org/10.1016/j.nut.2014.08.006.

Chalamaiah, M., and Wu, J. 2017. Anti-inflammatory effects of egg yolk livetins (α, β, and γ-livetin) fraction and its enzymatic hydrolysates in lipopolysaccharide-induced RAW 264.7 macrophages. *Food Research International* 100(1): 449–459. https://doi.org/10.1016/j.foodres.2017.07.032.

Chalamaiah, M., Yu, W., and Wu, J. 2018. Immunomodulatory and anticancer protein hydrolysates (peptides) from food proteins: A review. *Food Chemistry* 245: 205–222. https://doi.org/10.1016/j.foodchem.2017.10.087.

Choy, E., and Rose-John, S. 2017. Interleukin-6 as a multifunctional regulator: Inflammation, immune response, and fibrosis. *Journal of Scleroderma and Related Disorders* 2(2_suppl): S1–S5. https://doi.org/10.5301/jsrd.5000265.

Cian, R. E., López-Posadas, R., Drago, S. R. et al. 2012. A *Porphyra columbina* hydrolysate upregulates IL-10 production in rat macrophages and lymphocytes through an NF-κB, and p38 and JNK dependent mechanism. *Food Chemistry* 134(4): 1982–1990. https://doi.org/10.1016/j.foodchem.2012.03.134.

Daliri, E. B., Oh, D. H., and Lee, B. H. 2017. Bioactive peptides. *Foods* 6(5): 32. https://doi.org/10.3390/foods6050032.

Duarte, J., Vinderola, G., Ritz, B. et al. 2006. Immunomodulating capacity of commercial fish protein hydrolysate for diet supplementation. *Immunobiology* 211(5): 341–350. https://doi.org/10.1016/j.imbio.2005.12.002.

Egusa, S., and Otani, H. 2009. Soybean protein fraction digested with neutral protease preparation, "Peptidase R", produced by *Rhizopus oryzae*, stimulates innate cellular immune system in mouse. *International Immunopharmacology* 9(7–8): 931–936. https://doi.org/10.1016/j.intimp.2009.03.020.

Fitzgerald, R. J., and Murray, B. A. 2006. Bioactive peptides and lactic fermentations. *International Journal of Dairy Technology* 59(2): 118–125. https://doi.org/10.1111/j.1471-0307.2006.00250.x.

Gauthier, S. F., Pouliot, Y., and Saint-Sauveur, D. 2006. Immunomodulatory peptides obtained by the enzymatic hydrolysis of whey proteins. *International Dairy Journal* 16(11): 1315–1323. https://doi.org/10.1016/j.idairyj.2006.06.014.

Horiguchi, N., Horiguchi, H., and Suzuki, Y. 2005. Effect of wheat gluten hydrolysate on the immune system in healthy human subjects. *Bioscience, Biotechnology, and Biochemistry* 69(12): 2445–2449. https://doi.org/10.1271/bbb.69.2445.

Hou, H., Fan, Y., Wang, S. et al. 2016. Immunomodulatory activity of Alaska pollock hydrolysates obtained by glutamic acid biosensor–Artificial neural network and the identification of its active central fragment. *Journal of Functional Foods* 24: 37–47. https://doi.org/10.1016/j.jff.2016.03.033.

Jakubczyk, A., Karaś, M., Rybczyńska-Tkaczyk, K. et al. 2020. Current trends of bioactive peptides-new sources and therapeutic effect. *Foods* 9(7): 846. https://doi.org/10.3390/foods9070846.

Kang, H. K., Lee, H. H., Seo, C. H., and Park, Y. 2019. Antimicrobial and immunomodulatory properties and applications of marine-derived proteins and peptides. *Marine Drugs* 17(6): 350. https://doi.org/10.3390/md17060350.

Kazlauskaitė, J., Biziulevičius, G. A., Žukaitė, V. et al. 2005. Oral tryptic casein hydrolysate enhances phagocytosis by mouse peritoneal and blood phagocytic cells but fails to prevent induced inflammation. *International Immunopharmacology* 5(13–14): 1936–1944. https://doi.org/10.1016/j.intimp.2005.06.015.

Kiewiet, M. B. G., van Esch, B. C., Garssen, J. et al. 2017. Partially hydrolyzed whey proteins prevent clinical symptoms in a cow's milk allergy mouse model and enhance regulatory T and B cell frequencies. *Molecular Nutrition and Food Research* 61(11): 1700340. https://doi.org/10.1002/mnfr.201700340.

Kiewiet, M. B., Faas, M. M., and De Vos, P. 2018. Immunomodulatory protein hydrolysates and their application. *Nutrients* 10(7): 904. https://doi.org/10.3390/nu10070904.

Kim, E. K., Kim, Y. S., Hwang, J. W. et al. 2013. Purification of a novel nitric oxide inhibitory peptide derived from enzymatic hydrolysates of *Mytilus coruscus*. *Fish and Shellfish Immunology* 34(6): 1416–1420. https://doi.org/10.1016/j.fsi.2013.02.023.

Kim, M. J., Kim, K. B., Sung, N. Y. et al. 2017. Immune-enhancement effects of tuna cooking drip and its enzymatic hydrolysate in Balb/c mice. *Food Science and Biotechnology* 27(1): 131–137. https://doi.org/10.1007/s10068-017-0278-9.

Kong, X., Guo, M., Hua, Y. et al. 2008. Enzymatic preparation of immunomodulating hydrolysates from soy proteins. *Bioresource Technology* 99(18): 8873–8879. https://doi.org/10.1016/j.biortech.2008.04.056.

Korhonen, H., and Pihlanto, A. 2003. Food-derived bioactive peptides-opportunities for designing future foods. *Current Pharmaceutical Design* 9(16): 1297–1308. https://doi.org/10.2174/1381612033454892.

Lee, M., Kovacs-Nolan, J., Archbold, T. et al. 2009. Therapeutic potential of hen egg white peptides for the treatment of intestinal inflammation. *Journal of Functional Foods* 1(2): 161–169. https://doi.org/10.1016/j.jff.2009.01.005.

Lim, J. J., Grinstein, S., and Roth, Z. 2017. Diversity and versatility of phagocytosis: Roles in innate immunity, tissue remodeling, and homeostasis. *Frontiers in Cellular and Infection Microbiology* 7: 191. https://doi.org/10.3389/fcimb.2017.00191.

Liu, P., Liao, W., Qi, X. et al. 2020. Identification of immunomodulatory peptides from zein hydrolysates. *European Food Research and Technology* 246(5): 931–937. https://doi.org/10.1007/s00217-020-03450-x.

Ling-Ling, L., Li, B., Hui-Fang, J. et al. 2017. Immunomodulatory activity of small molecular ($\leq$ 3 kDa) Coix glutelin enzymatic hydrolysate. *CYTA-Journal of Food* 15(1): 41–48. https://doi.org/10.1080/19476337.2016.1201147.

Lozano-Ojalvo, D., Molina, E., and López-Fandiño, R. 2016. Hydrolysates of egg white proteins modulate T-and B-cell responses in mitogen-stimulated murine cells. *Food and Function* 7(2): 1048–1056. https://doi.org/10.1039/c5fo00614g.

Ma, J. J., Mao, X. Y., Wang, Q., et al. 2014. Effect of spray drying and freeze drying on the immunomodulatory activity, bitter taste and hygroscopicity of hydrolysate derived from whey protein concentrate. *LWT - Food Science and Technology* 56(2): 296–302. https://doi.org/10.1016/j.lwt.2013.12.019.

Mallet, J. F., Duarte, J., Vinderola, G. et al. 2014. The immunopotentiating effects of shark-derived protein hydrolysate. *Nutrition* 30(6): 706–712. https://doi.org/10.1016/j.nut.2013.10.025.

Marcone, S., Haughton, K., Simpson, P. J. et al. 2015. Milk-derived bioactive peptides inhibit human endothelial-monocyte interactions via PPAR-γ dependent regulation of NF-κB. *Journal of Inflammation* 12(1): 1–13. https://doi.org/10.1186/s12950-014-0044-1.

Mercier, A., Gauthier, S. F., and Fliss, I. 2004. Immunomodulating effects of whey proteins and their enzymatic digests. *International Dairy Journal* 14(3): 175–183. https://doi.org/10.1016/j.idairyj.2003.08.003.

Möller, N. P., Scholz-Ahrens, K. E., Roos, N., and Schrezenmeir, J. 2008. Bioactive peptides and proteins from foods: Indication for health effects. *European Journal of Nutrition* 47(4): 171–182. https://doi.org/10.1007/s00394-008-0710-2.

Montoya-Rodríguez, A., de Mejía, E. G., Dia, V. P. et al. 2014. Extrusion improved the anti-inflammatory effect of amaranth (*Amaranthus hypochondriacus*) hydrolysates in LPS-induced human THP-1 macrophage-like and mouse RAW 264.7 macrophages by preventing activation of NF-κ B signaling. *Molecular Nutrition and Food Research* 58(5): 1028–1041. https://doi.org/10.1002/mnfr.201300764.

Muñoz-Carrillo, J. L., Contreras-Cordero, J. F., Gutiérrez-Coronado, O. et al. 2018. Cytokine profiling plays a crucial role in activating immune system to clear infectious pathogens. In: *Immune Response Activation and Immunomodulation*, eds. R. Tyagi and P. S. Bisen, 1–30. London UK: IntechOpen. https://doi.org/10.5772/intechopen.80843.

Ndiaye, F., Vuong, T., Duarte, J., Aluko, R.T., and Matar, C. 2012. Anti-oxidant, anti-inflammatory and immunomodulating properties of an enzymatic protein hydrolysate from yellow field pea seeds. *European Journal of Nutrition* 51(1): 29–37. https://doi.org/10.1007/s00394-011-0186-3.

Nelson, R., Katayama, S., Mine, Y. et al. 2007. Immunomodulating effects of egg yolk low lipid peptic digests in a murine model. *Food and Agricultural Immunology* 18(1): 1–15. https://doi.org/10.1080/09540100601178623.

Nesse, K. O., Nagalakshmi, A. P., Marimuthu, P., and Singh, M. 2011. Efficacy of a fish protein hydrolysate in malnourished children. *Indian Journal of Clinical Biochemistry : IJCB* 26(4): 360–365. https://doi.org/10.1007/s12291-011-0145-z.

Pan, D. D., Wu, Z., Liu, J. et al. 2013. Immunomodulatory and hypoallergenic properties of milk protein hydrolysates in ICR mice. *Journal of Dairy Science* 96(8): 4958–4964. https://doi.org/10.3168/jds.2013-6758.

Prioult, G., Pecquet, S., and Fliss, I. 2005. Allergenicity of acidic peptides from bovine β-lactoglobulin is reduced by hydrolysis with *Bifidobacterium lactis* NCC362 enzymes. *International Dairy Journal* 15(5): 439–448. https://doi.org/10.1016/j.idairyj.2004.09.001.

Rémond, D., Savary-Auzeloux, I., and Boutrou, R. 2016. Bioactive peptides derived from food proteins. In: *The Molecular Nutrition of Amino Acids and Proteins*, ed. D. Dardevet, 3–11. Ceyrat, France: Academic Press. https://doi.org/10.1016/C2014-0-02227-7.

Rivera, A., Siracusa, M. C., Yap, G. S., and Gause, W. C. 2016. Innate cell communication kick-starts pathogen-specific immunity. *Nature Immunology* 17(4): 356–363. https://doi.org/10.1038/ni.3375.

Rodríguez-Carrio, J., Fernández, A., Riera, F. A., and Suárez, A. 2014. Immunomodulatory activities of whey β-lactoglobulin tryptic-digested fractions. *International Dairy Journal* 34(1): 65–73. https://doi.org/10.1016/j.idairyj.2013.07.004.

Saint-Sauveur, D., Gauthier, S. F., Boutin, Y., and Montoni, A. 2008. Immunomodulating properties of a whey protein isolate, its enzymatic digest and peptide fractions. *International Dairy Journal* 18(3): 260–270. https://doi.org/10.1016/j.idairyj.2007.07.008.

Sánchez, A., and Vázquez, A. 2017. Bioactive peptides: A review. *Food Quality and Safety* 1(1): 29–46. https://doi.org/10.1093/fqsafe/fyx006.

Santiago-López, L., Hernández-Mendoza, A., Vallejo-Cordoba, B., et al. 2016. Food-derived immunomodulatory peptides. *Journal of the Science of Food and Agriculture* 96(11): 3631–3641. https://doi.org/10.1002/jsfa.7697.

Senevirathne, M., Ahn, C. B., and Je, J. Y. 2010. Enzymatic extracts from edible red algae, *Porphyra tenera*, and their antioxidant, anti-acetylcholinesterase, and anti-inflammatory activities. *Food Science and Biotechnology* 19(6): 1551–1557. https://doi.org/10.1007/s10068-010-0220-x.

Steinman, R. M., and Hemmi, H. 2006. Dendritic cells: Translating innate to adaptive immunity. In: *From Innate Immunity to Immunological Memory*, ed. B. Pulendran and R. Ahmed, 17–58. Berlin, Heidelberg. https://doi.org/10.1007/3-540-32636-7.

Takahashi, M., Moriguchi, S., Yoshikawa, M., and Sasaki, R. 1994. Isolation and characterization of oryzatensin: A novel bioactive peptide with ileum-contracting and immunomodulating activities derived from rice albumin. *Biochemistry and Molecular Biology International* 33(6): 1151–1158.

Tanaka, T., Narazaki, M., and Kishimoto, T. 2014. IL-6 in inflammation, immunity, and disease. *Cold Spring Harbor Perspectives in Biology* 6(10): a016295. https://doi.org/10.1101/cshperspect.a016295.

Tobita, K., Kawahara, T., and Otani, H. 2006. Bovine beta-casein (1–28), a casein phosphopeptide, enhances proliferation and IL-6 expression of mouse CD19+ cells via toll-like receptor 4. *Journal of Agricultural and Food Chemistry* 54(21): 8013–8017. https://doi.org/10.1021/jf0610864.

Toopcham, T., Mes, J. J., Wichers, H. J., and Yongsawatdigul, J. 2017. Immunomodulatory activity of protein hydrolysates derived from *Virgibacillus halodenitrificans* SK1-3-7 proteinase. *Food Chemistry* 224: 320–328. https://doi.org/10.1016/j.foodchem.2016.12.041.

Tsuruki, T., Kishi, K., Takahashi, M., et al. 2003. Soymetide, an immunostimulating peptide derived from soybean β-conglycinin, is an fMLP agonist. *FEBS Letters* 540(1–3): 206–210. https://doi.org/10.1016/s0014-5793(03)00265-5.

Vernaza, M. G., Dia, V. P., De Mejia, E. G., and Chang, Y. K. 2012. Antioxidant and antiinflammatory properties of germinated and hydrolysed Brazilian soybean flours. *Food Chemistry* 134(4): 2217–2225. https://doi.org/10.1016/j.foodchem.2012.04.037.

Vinderola, G., Matar, C., and Perdigón, G. 2007. Milk fermentation products of *L. helveticus* R389 activate calcineurin as a signal to promote gut mucosal immunity. *BMC Immunology* 8(1): 1–10. https://doi.org/10.1186/1471-2172-8-19.

Vo, T. S., Ryu, B., and Kim, S. K. 2013. Purification of novel anti-inflammatory peptides from enzymatic hydrolysate of the edible microalgal *Spirulina maxima*. *Journal of Functional Foods* 5(3): 1336–1346. https://doi.org/10.1016/j.jff.2013.05.001.

Wang, Y. K., He, H. L., Wang, G. F., et al. 2010. Oyster (*Crassostrea gigas*) hydrolysates produced on a plant scale have antitumor activity and immunostimulating effects in BALB/c mice. *Marine Drugs* 8(2): 255–268. https://doi.org/10.3390/md8020255.

Wu, W., Zhang, M., Sun, C., et al. 2016. Enzymatic preparation of immunomodulatory hydrolysates from defatted wheat germ (*Triticum vulgare*) globulin. *International Journal of Food Science and Technology* 51(12): 2556–2566. https://doi.org/10.1111/ijfs.13238.

Xu, Z., Mao, T. M., Huang, L., et al. 2019. Purification and identification immunomodulatory peptide from rice protein hydrolysates. *Food and Agricultural Immunology* 30(1): 150–162. https://doi.org/10.1080/09540105.2018.1553938.

Yang, Q., Cai, X., Huang, M., and Wang, S. 2020. A specific peptide with immunomodulatory activity from *Pseudostellaria heterophylla* and the action mechanism. *Journal of Functional Foods* 68: 103887. https://doi.org/10.1016/j.jff.2020.103887.

Yang, R., Zhang, Z., Pei, X., et al. 2009. Immunomodulatory effects of marine oligopeptide preparation from Chum Salmon (*Oncorhynchus keta*) in mice. *Food Chemistry* 113(2): 464–470. https://doi.org/10.1016/j.foodchem.2008.07.086.

Yimit, D., Hoxur, P., Amat, N., et al. 2012. Effects of soybean peptide on immune function, brain function, and neurochemistry in healthy volunteers. *Nutrition* 28(2): 154–159.

Zhu, W., Ren, L., Zhang, L., et al. 2020. The potential of food protein-derived bioactive peptides against chronic intestinal inflammation. *Mediators of Inflammation* 2020: 6817156. https://doi.org/10.1155/2020/6817156.

CHAPTER 23

Bioactive Peptides
Cytomodulatory Activity

Carlotta Giromini and Mariagrazia Cavalleri

CONTENTS

Introduction	461
Digestibility, Bioaccessibility and Bioavailability of Food Peptides	462
Cytomodulatory Activity of Food Protein and Peptides	464
Conclusion	476
References	477

INTRODUCTION

In recent years, a strong sense of awareness has spread toward the role of diet, intended not only as a source of molecules to satisfy nutritional needs but also aimed at providing substances able of enhancing the health status of animals and humans (Contor et al., 2001). Accordingly, the concept of functional foods was clearly established as a key nutritional approach. A food can be regarded as functional when it demonstrates to beneficially affect one or more target functions of the body in a way which is relevant to either the state of well-being and health or the reduction of the risk of a disease (Roberfroid, 2002).

Peptides encrypted into food proteins were considered to be of the most interesting bioactive molecules (Bhat et al., 2015). Dietary proteins may carry out a wide range of nutritional, functional, and biological activities by means of their biologically active peptides which can be released during food processing or through the digestive enzymes of the host (Korhonen et al., 2003).

A bioactive molecule is defined as "a food component that can affect biological processes or substrates and, hence, have an impact on body function or condition and ultimately health" (Schrezenmeir et al., 2000). Such peptides may be considered bioactive if they are assumed within the ingested food and cause a positive measurable impact on the organism (Moller et al. 2008).

The need for health-promoting foods which can potentially carry in the body bioactive compounds has increased over the years, largely due to the sharp rise of mortality rate in developed countries due to hypertension and dyslipidemia (Cicero et al., 2017), cancer, and diabetes (D'Souza et al., 2019), as well as obesity (Pérez-Gregorio et al., 2020).

Bioactive peptides usually contain 2–20 amino-acid residues that are generally rich in hydrophobic amino acids (Moller et al., 2008), with variable chain lengths and low molecular weight in comparison to protein mass (Saadi et al., 2015). They are inactive in the parent protein substrates, but bioactive peptides are released during gastrointestinal digestion or food processing (Bath et al., 2015), through different proteases like pepsin, trypsin, chymotrypsin, and pancreatin (Saadi et al., 2014). Thus, these active compounds act as independent entities and potential metabolism modulators (Kohoronen et al., 2003) interacting with transporter agents to reach different targets (Saadi et al., 2014).

Among the effects of bioactive peptides, their immunomodulatory, anti-inflammatory, antioxidant, and antimicrobial properties are the most studied (Baldi et al., 2005). In recent years, particular interest has been directed toward their potential role in blood pressure reduction (Fekete et al., 2016; 2018; Giromini et al., 2017) and lipid metabolism disorders (Cicero et al., 2017). In addition, they found large application in the regulation of immunity and the prevention of infections (Yang et al., 2009; Cicero et al., 2017) and in cancer preventive therapies (Xiao et al., 2015).

Peptides oral administration involves the complex process of digestion, so that they must be resistant to the action of digestive enzymes and be able to cross the intestinal epithelial barrier to reach their main preferred locations in order to exert their beneficial effects (Amigo et al., 2020). Much of the research in this regard has been focused on studying the bioavailability of these active systems and their stability under upper gastrointestinal digestion conditions (Saadi et al., 2014). The bioavailability of peptides is firstly limited by their release and solubility from the food matrix, becoming available for intestinal absorption (Anson et al. 2010).

Once they reach the target tissue or organ, bioactive peptides have been involved in cell differentiation, proliferation, or apoptosis (Hartmann et al., 2007). In fact, the primary bioactive effect of food peptides occurs at the cellular level by modulating the different function (cyto-modulation). The cytomodulatory effect comprises all effects mediated by a compound on a cell (e.g., modulation of viability, growth and proliferation, DNA stability, apoptosis, and necrosis).

This chapter describes the cytomodulatory effects of the major peptides deriving from both animal and plant sources on different target cells. The concepts of digestibility, bioavailability, and bioaccessibility as prerequisites to determine the peptide bioactivity (and cytomodulatory activity) will also be discussed.

DIGESTIBILITY, BIOACCESSIBILITY AND BIOAVAILABILITY OF FOOD PEPTIDES

To study peptide bioactivity, the concepts of digestibility, bioavailability, and bioaccessibility need to be carefully considered, as they represent prerequisites of bioactivity at a target organ or tissue.

First of all, digestion is a complex process that involves the simultaneous action of digestive enzymes, pH, and temperature on ingested food ingredients (Amigo et al., 2020). *In vivo* protein digestion begins when food proteins reach the stomach, where the pH ranges from 1 to 5, and their transit time is of 30 min to 3 h. The gastric juice primarily promotes protein digestion (Giromini et al., 2019a) that is influenced by several factors, such as the type of proteins, which may determine different behavior under gastrointestinal digestion, the different enzyme specificity determining the type of peptides released (Amigo et al., 2020), physicochemical characteristics (i.e., low pH and temperature, high osmolality, viscosity, fiber content, and energy density), food composition, physiological factors (Singh et al., 2015), endogenous secretions, and motility (Bouzerzour et al.,

2012). In this regard protein digestibility is of great importance in estimating the protein availability for intestinal absorption after digestion, reflecting on the efficiency of protein utilization on diet (Almeida et al., 2015). The degradation pattern depends on the peptide's length. Indeed, the smaller size of peptides has been related to a higher absorption rate, resulting in higher bioactivity when compared with bigger peptides (Perez-Gregorio et al., 2020).

The bioaccessibility of a protein is a key parameter for the determination of its bioavailability. The fraction of ingested protein that is effectively released from the food matrix and available for intestinal absorption in the gut after digestion is defined bioaccessible, whereas bioavailability is defined as the quantity of ingested protein that becomes available at circulation level for utilization in normal physiologic functions (Shi et al., 2000; Guerra et al., 2012; Giromini et al., 2019a). The simulated digestion process for the bioaccessibility evaluation is typically based on *in vitro* procedures (Parada et al., 2007), while bioavailabiliy is usually determined by *in vivo* assays, which consider individual variability, physiological state, dose, and the co-presence of other food components. Moreover, bioactivity represents all events linked to how the bioactive compound reaches the target tissue, interacts with biomolecules, and how it is metabolized and transformed to induce a physiologic response (Giromini et al., 2019b).

All these factors are strictly interconnected. Bioavailable peptides need to be stable under the dependence of gastrointestinal parameters, food processing, and matrix complexity, which, otherwise, may affect the efficiency of the digestion and absorption processes. Bioavailability is the result of 3 main steps: protein digestibility or solubility in the gastrointestinal tract, absorption by the intestinal epithelial cells, and tissue distribution (Giromini et al., 2019a). These peptides may become bioavailable and exert their functions only after their release from the entire protein substrate and their activation by enzymatic hydrolysis (i.e., during gastrointestinal digestion by digestive enzymes) (Meisel et al., 2003) or by microbial fermentation (Korhonen et al., 2006) or *in vitro* during food processing or ripening by isolated or microbial enzymes such as *Lactobacillus helveticus* (Moller et al., 2008). A bioactive peptide with a great natural bioresistance to deal with the digestion process (i.e., resistance to degradation by proteases and serum peptidases) (Sun et al., 2020), can boast a good rate of bioavailability (Saadi et al., 2015) and, subsequently, of bioactivity. Most of the food products' absorption occurs in the intestine (He et al., 2018), mostly in the jejunum (Amigo et al., 2020), thus, it is known to be the largest interface between the organism and the external environment (Mayer, 2000). Di- and tripeptides and amino acids from ingested proteins are easily absorbed by intestinal epithelial cells; however, orally administered long-chain bioactive peptides may have difficulties in crossing the intestinal epithelial cell barrier and lose the ability to act *in vivo*. Thus, if these peptides pass through the gastrointestinal tract remaining intact and accessible for their absorption at the intestinal level, they may be able to affect the major body systems.

The main factors limiting the bioaccessibility of peptides are hydrolytic effects of digestive proteases, interaction with other nutritional components of the food matrix, reduction in permeation across mucin because of charge effects, gastrointestinal environment (Udenigwe et al., 2021) and the different behavior of proteins under gastrointestinal conditions (Amigo et al., 2020). The bioaccessibility is also strictly connected with the digestibility of the protein/peptide, indeed their stability to digestion has to be assessed even before their absorption rate (Sanchez-Rivera et al., 2014). All these parameters together with the physicochemical properties and primary structure of peptides determine the subsequent level of their absorption and reception of the systemic circulation (Amigo et al., 2020).

Therefore, the studying of strategic ways to improve the bioaccessibility rate of biopeptides holds a great importance to enhance the bioavailability itself. The bioactivity of food peptides could be tested through *in vitro* biochemical assays, cell culture, *in vivo* studies in animal models, and/or with clinical trials (Chakrabarti et al., 2018).

Several *in vitro* methods/models have been developed to simulate the digestion processes and to assess the behavior of food components (Minekus et al., 2014), to investigate the intestinal transport and bioavailability of biopeptides (Tretola et al., 2020). *In vitro* digestion, coupled with cell culture model, represents a widely recognized system to preliminarily check food peptide bioactivities with a high predictive power.

CYTOMODULATORY ACTIVITY OF FOOD PROTEIN AND PEPTIDES

Bioactive peptides/proteins with potential cytomodulatory activity can be derived from many animal sources, such as milk (casein and whey), eggs, fish, and meat (Moller et al., 2008) (Table 23.1).

To date, milk proteins are considered the most important source of animal-based bioactive peptides (Perez-Gregorio et al., 2020), which may act as specific signals triggering viability of cancer cells and may possess cytomodulatory activities (Gobbetti et al., 2007). Most of the studies are focused on the apoptotic or cell-growth inhibitory effects of milk-derived peptides against different tumor cell lines, naming different β casein fragments (Meisel and FitzGerald, 2003; Perego et al., 2012; Meisel H. 2001; Zhao et al., 2014), α casomorphin (Kampa et al., 1997) and k-casecidin (Matin et al., 2002). On the other hand, some cell growth-stimulating casein fragments have also been reported by Azuma et al. (1989) and Naguane et al. (1989), which could enhance DNA synthesis in BALB/c3T3 cells. Plaisancié et al. (2013-2015) reported that β-CN (94-123) and derived fragments (94–108,117–123) increased the proliferation of goblet and paneth cells. Purup et al. (2018) and Giromini et al. (2019b) demonstrated that both casein and whey-based isolates exerted proliferative effects on intestinal cells and stimulatory effects on cell metabolic activity in undifferentiated HT29-MTX-E12 cells, respectively. On the contrary, Phelan et al. (2009) showed that the exposure to increasing concentrations of casein hydrolysates resulted in a reduced Jurkat T cell viability and growth. Whey protein-derived peptides, as lactoferricins, arrest the cell cycle and induce apoptosis as well in different cancer cell types (Deng et al., 2013; Mader et al., 2007; De Mejía et al., 2010). Bovine colostrum and milk whey-derived growth factors have a stimulatory effect on the proliferation of epidermal, epithelial, embryonic, and connective tissue cells, while TGF-β1 and TGF-β2 decrease the proliferation of lymphocytes (Korhonen, H.J., 2010). Moreover, colostrum whey proteins decrease the adhesion of Caco-2/15 cells but increase the migration of IPEC-J2 cells, while Mozzarella cheese whey proteins induce morphological changes in IPEC-J2 cells (Blais et al., 2014). A yogurt α-lactalbumin obtained by membrane dialysis demonstrated an antiproliferative effect on intestinal Caco-2 and IEC-6 cells at a dose of 3.4 µg mL^{-1} protein (Ganjam et al., 1997). Interestingly, a complex of human α-lactalbumin and oleic acid showed chemo-preventive properties *in vitro* through the activation of apoptosis over 40 different lymphoma and carcinoma cell lines (Fast et al., 2005). Finally, a partially purified peptide from the acid whey portion of Buffalo cheese was found to affect the proliferation of Caco-2 cells by De Simone et al. (2009).

A number of studies have demonstrated the antiproliferative and cytotoxic activity of egg proteins and associated peptides against cancer cells (Lee et al., 2019). Lysozyme

TABLE 23.1 Examples of Studies Investigating the Cytomodulatory Activity of Protein/Peptides from Animal Sources

Source	Protein	Peptide	Cyto-modulatory Effect	Animals/Cell Lines	Reference
MILK and dairy food					
MILK	β Casein	β casokinin (f177-183)	↑ DNA-synthesis	mouse fibroblast cells BALB/c3T3	Nagaune et al., 1989
		β-casomorphin-7	↑ Apoptosis	human leukemia cells (HL-60)	H., M., & J., F. (2003)
		β-casomorphin-5	↑ Apoptosis	intestinal tumor HT-29 and AZ-97 cells	Perego et al., 2012
		CPP β-CN (f1-25/28)4P fragments 1–18 and 105–117	↑ Apoptosis ↑ DNA synthesis	HL-60 cells BALB/c3T3 cells	Meisel H. 2001 Azuma et al., 1989
		β-CN (94-123), derived fragments (94–108) and (117–123)	Changes in epithelial differentiation: ↑ cell proliferation	Goblet cells and Paneth cells	Plaisancié et al., 2013 Plaisancié et al., 2015
		β-Casobapt-ZW EPVLGPVRGP	↑ Apoptosis	Caco-2 cells	Zhao et al., 2014
	Casein	Hydrolisates	↓ viability and growth	Jurkat T cells	Phelan et al., 2009
	α casein	α-Casomorphin HIQKED(V)	↓ Cell proliferation	human prostate cancer cell lines (LNCaP, DU145, and PC3)	Kampa et al., 1997
	K casein	k-casecidin	Cytotoxic activity	human leukemia cells lines	Matin, et al., 2002
	Whey proteins	Lactoferrin	↓ cell growth	breast cancer (MDA-MB-231) and nasopharyngeal carcinoma cells	Deng et al., 2013
		Lactoferricin LfcinB Lactoferricin	↑ Apoptosis ↑ Apoptosis, modulation of gene expression, ↓ angiogenesis, arrest cell cycle	Jurkat T-leukemia cells different cancer cell types	Mader et al., 2007 De Mejía et al., 2010

(Continued)

TABLE 23.1 (Continued)

Source	Protein	Peptide	Cyto-modulatory Effect	Animals/Cell Lines	Reference
Milk	Whey protein/casein enzymatic hydrolysates		↑ cell proliferation/migration	Normal rat small intestinal epithelial (IEC-6) cells	Purup et al., 2018
	Whey protein/casein proteins		↑ metabolic activity	Intestinal goblet HT29-MTX-E12 cells	Giromini et al., 2019b
Bovine colostrum, milk and whey		Growth factors: BTC (beta cellulin), EGF (epidermal growth factor).	↑ cell proliferation, ↑ wound healing and bone resorption	epidermal, epithelial and embryonic cells	Korhonen, H.J. (2010)
		TGF-β1 and TGF-β2 (transforming growth factor)	↑ proliferation ↓ proliferation	connective tissue cells lymphocytes and epithelial cells	
Colostrum	Whey proteins	-	↓ Adhesion ↑ Migration	Caco-2/15 cells IPEC-J2 cell	Blais et al., 2014
Mozzarella cheese		-	Induce morphological changes	IPEC-J2 cells	
Yogurt		α-lactalbumin	↓ cell division	IEC-6 cells and Caco-2 cells lines	Ganjam et al., 1997
Buffalo cheese	Acid whey	Partially purified peptide	↓ cell proliferation	Caco-2 cells	De Simone et al., 2009
EGGS					
Egg white	Ovalbumin self-assembled nanostructured lysozyme-based particle	Gln-Ile-Gly-Leu-Phe (QIGLF)	↑ cell viability ↑ Cell death	Caco-2 cells MCF-7 cells	Ding et al., 2014 Mahanta et al., 2015

(*Continued*)

TABLE 23.1 (Continued)

Source	Protein	Peptide	Cyto-modulatory Effect	Animals/Cell Lines	Reference
	Ovomucin	β-subunit	Cytotoxic effects	SEKI cell (human melanoma cell) and 3LL (Lewis lung cancer cell) and carcoma-180 (SR-180) cells	Yokota et al., 1999
		α-subunit (70 kDa) fragment (OVMa70F)	Antitumor effect. ↓ Angiogenesis. Accumulation of neutrophils, macrophages and lymphocytes.	tumor tissues	Oguro et al., 2000
	Ovotransferrin	two-step enzyme hydrolysates	Cytotoxic activity	AGS, LoVo, HT-29, HeLa and other cancer cell lines	Moon et al., 2013 Lee et al., 2017
	Phosvitin		↓ cell growth	cervix (HeLa), breast (MCF-7), stomach (AGS), lung (A549 and SK-MES-1), liver (HepG2), and larynx (Hep-2) cancer cell lines	Moon et al., 2014
Egg yolk	Yolk proteins		↓ cell proliferation	colorectal cancer in rats	Azuma et al., 2000
			Suppressed aberrant crypt foci formation. ↓ cell proliferation through SCFA production	rat model of azoxymethane (AOM)-induced colon cancer	Ishikawa et al., 2009

(Continued)

TABLE 23.1 (Continued)

Source	Protein	Peptide	Cyto-modulatory Effect	Animals/Cell Lines	Reference
FISH					
Giant squid		water-soluble peptide (YPEP)	↑ proliferation, alkaline phosphatase activity and collagen content	preosteoblastic MC3T3-E1 cell	Kim et al., 2011
		Esperase hydrolysate	Cytotoxic	MCF-7 and glioma cell lines	Alemán et al., 2011
Red Sea Moses sole *Pardachirus marmoratus*		Pardaxin	Mitochondrial dysfunction and apoptosis	breast cancer (MDA-MB-231 and MDA-MB-453), colon cancer (SW1116 and SW620), gastric cancer (NCI-N87 and AGS), hepatic cancer (HepG2 and HA22T/VGH), lung cancer (A549), renal cancer (BFTC909), and glioma (U-87MG).	Ting et al., 2014
Marine sponge *Reniochalina stalagmitis*		Reniochalistatins E	cytotoxic	RPMI-8226, MGC-803, HL-60, HepG2, and HeLa cell lines	Zhan et al., 2014
Mediterranean tunicate *Aplidium albicans*		Aplidine	↓ cell viability	HeLa tumor cells	García-Fernández et al., 2002
Marine mollusk *Megathura crenulata*		Keyhole limpet hemocyanin (KLH)	↑ Apoptosis	treated SEG-1 cells (Human oesophageal adenocarcinoma cells)	McFadden et al., 2003

(*Continued*)

TABLE 23.1 (Continued)

Source	Protein	Peptide	Cyto-modulatory Effect	Animals/Cell Lines	Reference
Arca subcrenata		polypeptide P2	↓ Cell proliferation	HeLa and HT-29 cells in S-180 tumor-bearing mice	Hu et al., 2012
Marine ascidians		Vitilevuamide	Cytotoxic	P388 lymphocytic leukaemia	Edler et al., 2002
Anchovy		440.9 Da hydrophobic peptide	↑ apoptosis	human U937 lymphoma cells	Lee et al., 2003
Marine sponge *Jaspis johnstoni*		Jaspamide (Jasplakinolide)	↑ apoptosis ↓ Bcl-2 protein expression, ↑ Bax levels	Jurkat T cells	Zheng et al., 2011
Mollusk *Dolabella auricularia*		Dolastin	Anticancer activity. ↓ tubulin polymerization and tubulin dependent GTP hydrolysis	P388 cell line	Aneiros et al., 2004
Cod		Glycopeptide designated Thomsen-Friedenreich disaccharide100	Blocks gal3-mediated angiogenesis, tumor-endothelial cell interactions and metastasis	prostate cancer cells in mice	Guha et al., 2013
Sardine	hydrolysate	Val - Tyr (angiotensin I-converting enzyme inhibitory peptide)	Antiproliferation action: ↓ cell growth	serum-stimulated vascular smooth muscle cell (VSMC)	Matsui et al., 2005
Bullacta exarata	Hydrolysate with trypsin	BEPT II and BEPT II-1	↓ Cell proliferation in a time- and dose-dependent manner	Prostate cancer cell (PC-3)	Ma et al., 2013

(*Continued*)

TABLE 23.1 (Continued)

Source	Protein	Peptide	Cyto-modulatory Effect	Animals/Cell Lines	Reference
Chum salmon (*Oncorhynchus keta*)	skin gelatin	Hydrolysates by Alcalase, papain or Neutrase	↑ cell growth, ↑ cell cycle progression, ↓ cell apoptosis	hFOB1.19 = human fetal osteoblast cell	Fu et al., 2013
MEAT					
Bovine meat	beef sarcoplasmic protein hydrolysates	GFHI	cytotoxic effect ↓ cell viability	MCF-7 cells AGS cells	Jang et al, 2008
		GLSDGEWQ	↓ proliferation	AGS cells	
Pork plasma	Hydrolysates	approximately 2–15 amino acids in length	↑ metabolic activity ↑ cell proliferation	Skeletal muscle cell extracted from beef longissimus thoracis	Andreassen et al., 2020
Animal Muscle tissues		Carnosine	↑ Healing of wounds and ulcers, ↑ expression of transcription factors ↑ expression of vimentin	cultured rat fibroblasts	Kovacs-nolan et al., 2010 Ikeda et al. 1999
Bovine meat myofibrillar and connective tissue		ACE-inhibitory peptide hydrolysates	Non-cytotoxic	Vero cell lines	Ryder et al., 2016

has been shown to have tumor growth and metastasis inhibition capacity (Sava et al., 1991) and a self-assembled nanostructured lysozyme (snLYZ), synthesized using a simple desolvation technique, displayed 95% MCF-7 breast cancer cell lines death through a reactive oxygen species-based mechanism (Mahanta et al., 2015). Both α and β ovomucin subunit fragments (Oguro et al., 2000; Yokota et al., 1999) have been reported to elicit cytotoxic and antitumor effects by inhibiting cancer cell growth, as well as two-step enzyme hydrolysates of ovotransferrin (Moon et al., 2013; Lee et al., 2017), phosvitin (Moon et al., 2014) and antiproliferative yolk proteins (Azuma et al., 2000; Ishikawa et al., 2009). Moreover, the QIGLF ovoalbumin peptide at high concentration of 5 mM offered a strong protection by enhancing H_2O_2-decreased Caco-2 cells viability (Ding et al., 2014). Finally, a water-soluble peptide YPEP from egg yolk showed an increasing proliferation of preosteoblastic MC3T3-E1 cells (Kim et al., 2011).

Many bioactive peptides are derived from different marine species. Among the bioactive peptides identified in this group, the great part has a strong cytotoxic activity by triggering the canonical apoptotic pathways and resulting in alterations in cell morphology, mitochondrial dysfunctions, and blocking tubulin polymerization. For example, peptides from squid gelatine have cytotoxic activity on human breast carcinoma (MCF-7) and glioma cell lines (Alemàn et al., 2011), while the peptide reniochalistatin E from the sponge *Reniochalina stalagmites* showed a cytotoxic activity toward several tumor cell lines (Zhan et al., 2014). BEPT II e BEPT II-1 from *Bullacta exarata* have also demonstrated apoptotic activity toward prostate cancer cells (Ma et al., 2013). The polypeptide P2 of *Arca subcrenata* has been demonstrated *in vivo* to have an antiproliferative action on HeLa and HT-29 cells in S-180 tumor-bearing mice (Hu et al., 2012). On the contrary, Fu et al. (2013) have found that hydrolysates of skin gelatine of chum salmon are able to enhance human fetal osteoblast cell growth and cycle progression, inhibiting apoptosis.

At present, there is limited research available demonstrating the cytomodulatory effects of bioactive peptides derived from meat. Jang et al. (2008) have showed that peptides derived from beef sarcoplasmic protein hydrolysates had a cytotoxic activity against diverse human cancer cell lines. GFHI had the greatest cytotoxic effect on MCF-7 cells and decreased the cell viability of AGS cells, while GLSDGEWQ strongly inhibited the proliferation of AGS cells, but none of the found peptides had any cytotoxic effect on A549 lung carcinoma cell lines. Andreassen et al. (2020) demonstrated that some hydrolysates of pork plasma enhanced skeletal muscle cells growth, with more than 150% increase in metabolic activity and 50% increase in cell proliferation when cultured in serum-free conditions for 3 days compared with control cells cultured with full serum conditions. In addition, carnosine, enhanced the expression of transcription factors involved in cell differentiation and formation of periodontal tissues and stimulated the epithelialization of gastric and duodenal ulcers as well (Kovacs-nolan et at., 2010). Furthermore, carnosine has been shown to stimulate the expression of vimentin in cultured rat fibroblasts, which is thought to play a fundamental role in maintaining cell structure and integrity (Ikeda et al., 1999). Finally, all peptide-containing hydrolysates from both meat myofibrillar and connective tissue extracts showed no significant reduction in Vero cell viability; consequently, they may have a potential application in the production of health-promoting products (Ryder et al., 2016) (Table 23.2).

Bioactive peptides from plant origin also received marked attention (Pérez-Gregorio et al., 2020). Diverse walnut-derived peptides showed significant inhibitory activity on MCF-7, Caco-2 and HeLa cancer cell lines, without affecting non-cancerous IEC-6 cells (Ma et al., 2015). A cytotoxic activity against KB cells mediated by cherimoya fruit has been found also by Wélé et al. (2005). Many peptides derived from plant sources showed

TABLE 23.2 Examples of Studies Investigating Cytomodulatory Activity of Protein/Peptides from Plant Sources

Plant Source	Protein	Peptide	Cyto-modulatory Effects	Tissue/Cell Lines	Reference
Walnut (*Juglans regia* L.)	Protein hydrolysated by papain enzyme	CTLEW (Cys–Thr–Leu–Glu–Trp)	↑ apoptosis and autophagy ↓ cancer cells growth	MCF-7 cells	Ma et al., 2015
	Glutelin and prolamin	-	Antiproliferative activity	Caco-2 and HeLa cancer cells	Carrillo et al., 2017
cherimoya fruit (*Annona cherimola, Annona glauca*)		Hydrolysate obtained with pepsin and corolase enzymes	Strong anticancer activity	PC-3 (prostate) and K-562 (leukemia) cancer cell UACC62 (melanoma)	Wélé et al., 2005
		cherimolacyclopeptides glaucacylopeptide	cytotoxic activity	KB (human nasopharyngeal carcinoma) cell culture system	
Cabbage (*Brassica alboglabra, B. parachinensis*)		napin-like peptides ~14.5 kDa	antiproliferative activity	leukemia L1210 cells	Ngai et al., 2004
Common Beans (*Phaseolus vulgaris*)	Protein hydrolysates	vulgarinin ~7 kDa peptide, defensin-type	↓ proliferation	leukemia cell lines L1210 and M1 and breast cancer cell line MCF-7	Mejia et al., 2005
	non-digestible fractions (NDF)	GLTSK, LSGNK, GEGSGA, MPACGSS and MTEEY	antiproliferative effect by modifying molecules involved in either cell cycle arrest or apoptosis.	human colorectal cancer cells (HCT116, RKO and KM12L4)	Vital et al., 2014

(*Continued*)

TABLE 23.2 (Continued)

Plant Source	Protein	Peptide	Cyto-modulatory Effects	Tissue/Cell Lines	Reference
Vigna sesquipedalis		Sesquin, defensin-like	Antiproliferative activity	breast cancer (MCF-7) cells and leukemia M1 cells. Human immunodeficiency virus-type 1 reverse transcriptase	Wong et al., 2005
Tepary bean		Lectins	↓ Colony formation	C33-A and Sw480 cell lines	Valadez-Vega et al., 2011
Phaseolus coccineus seeds		Lectin, called PCL	Antiproliferative effect ↑ apoptosis and necrosis.	L929 cells	Chen et al., 2009
Buckwheat (*Fagopyrum esculentum*)		buckwheat PI 5 to 7 kDa, Bowman-Birk class	↓ cell growth	leukemia cells	Park et al., 2004
Rice bran		Glu-Gln-Arg-Pro-Arg (ENPRP)	↓ cell growth	Colon cancer cells (Caco-2, HCT-116), breast cancer cells (MCF-7, MDA-MB-231), liver cancer cells (HepG-2)	Kannan et al., 2010
			↑ Apoptosis. Morphological changes, DNA fragmentation, caspases activation.	human breast cancer cells (MCF-7)	Li, R. (2014)

(*Continued*)

TABLE 23.2 (Continued)

Plant Source	Protein	Peptide	Cyto-modulatory Effects	Tissue/Cell Lines	Reference
Amaranth (*Amaranthus hypochondriacus*) Seed	Glutelin extracts digested with trypsin	Lunasin-like peptide	↑ Apoptosis	HeLa cells	Silva-Sánchez et al., 2008
Amaranthus mantegazzianus seed	Protein isolate		Antiproliferative effects ↓ cell adhesion ↑ apoptosis	osteosarcoma UMR106 cell line	Barrio et al., 2010
Soybean		Lunasin	Cytotoxic effect	L1210 Leukaemia cells	de Mejia et al., 2010
Soybean		Lunasin	↓ tumor volume	Lymphoma mice model	Chang et al., 2014
		Lunasin	Antiproliferative effects	colorectal cancer HT-29 cells	Fernández-Tomé et al., 2018
Abrus precatorius	agglutinin protein (10kDAGP)	Lunasin fragment SKWQHQQDSC anionic peptide fraction	Antiproliferative activities	Ehrlich's ascites carcinoma (EAC) and B16 melanoma (B16M)	Behera et al., 2014
Soybean	soy protein isolate		↑↓ metabolic activity in a dose-dependent manner	HT29-MTX-E12 cells	Giromini et al., 2019b
Hazelnut *Corylus avellana* L.	Meal proteins		Antiproliferative activity	CCC221/DLD-1 colon cancer cells	Aydemir et al., 2014

(*Continued*)

TABLE 23.2 (Continued)

Plant Source	Protein	Peptide	Cyto-modulatory Effects	Tissue/Cell Lines	Reference
Rapeseed	defatted rapeseed meal	peptide fraction (RSP2	↓ viability, morphological changes, ↑ apoptosis	HeLa cells	Xue et al., 2011
Spirulina platensis	whole proteins hydrolyzed using three gastrointestinal endopeptidases (pepsin, trypsin and chymotrypsin)	Fraction Tr2 Fraction Tr1 → new peptide HVLSRAPR	anti-proliferation activities ↓ cell growth	cancer cells (MCF-7, HepG-2 and SGC-7901) HT-29 cancer cells	Wang et al., 2017
Cycas revoluta		Ala–Trp–Lys–Leu–Phe–Asp–Asp–Gly–Val	↓ proliferation, disruption of nucleosome structures, ↑ apoptosis by direct DNA binding.	human epidermoid cancer (Hep2) and colon carcinoma cells (HCT15)	Mandal et al., 2012
Algae Chlorella vulgaris	microalgae protein waste hydrolysated by pepsin	VECYGPNRPQF	↓ proliferation, post-G1 cell cycle arrest.	AGS (human gastric cancer) cells	Sheih et al., 2010

an antiproliferative activity on different cancer cell line types, i.e., cabbage (Ngai et al., 2004), common beans (Mejia et al., 2005; Vital et al., 2014), *Vigna sesquipedalis* (Wong et al., 2005), tepary bean (Valadez-Vega et al., 2011), and *Phaseolus coccineus* (Chen et al., 2009). In addition, a pentapeptide Glu-Gln-Arg-Pro-Arg, extracted from rice bran, demonstrated to cause the 84% inhibition of the growth of colon cancer cells, 80% of the growth of breast cancer cells, and 84% of liver cancer cells at a dose of 600–700 µg·mL^{-1} (Kannan et al., 2010), while a buckwheat peptide inhibited the growth of leukemia cells (Park et al., 2004). The induction of apoptosis by the caspase activation pathway against MCF-7 cells was another important effect of rice bran peptide studied by Li, (2014) and also the effect of peptides from amaranth seeds against HeLa cells (Silva-Sànchez et al., 2008) and UMR106 cell lines (Barrio et al., 2010). Among the peptides of plant origin, particular interest has been directed toward lunasin, a 43-mer peptide present in the seeds of different plants, but most notably it can be isolated from soybean. This peptide acts preventively at different cancer stages, mainly by internalizing inside cells and inhibiting histone acetylation in the nucleus (Hernandez-Ledesma et al., 2013). In L1210 leukemic cells, it has shown a cytotoxic effect, inducing cell cycle arrest in G2/M phase and apoptosis through the activation of caspase-3 (de Mejia et al., 2010; de Mejia & Dia, 2010), and in colorectal cancer HT-29 cells, it showed an antiproliferative effect (Fernàndez-Tomé et al., 2018). Lunasin's anticarcinogenic activity has been studied also *in vivo* in mice models showing a reduced lymphoma volume (Chang et al., 2014). Moreover, the algae-derived peptide VECYGPNRPQF had strong dose-dependent antiproliferative effect and induced a post-G1 cell cycle arrest in AGS cells. However, no cytotoxicity was observed in WI-38 lung fibroblasts cells *in vitro* (Sheih et al., 2010). Interestingly, a soy protein had an hormetic response in HT29-MTX-E12 intestinal cells because at the lowest concentration it stimulated metabolic activity of the cells, which is inhibited at the highest concentration (Giromini et al., 2019b). Other plant-derived peptides demonstrated to inhibit proliferation (Behera et al., 2014; Aydemir et al., 2014; Wang et al., 2017) and to induce apoptosis (Xue et al., 2011; Mandal et al., 2012) on different cancer cell lines.

Finally, based on the aforementioned studies performed in animal and plant foods, the research interest on the development of peptide-enriched products is increasing, however, the translation of the new findings into practical and commercial application remains immature (Amigo et al., 2020). Some of the reasons behind this delay are: the lack of suitable technologies for industrial-scale production of such peptides (Korhonen et al., 2003) from different food sources; the lack of studies for their optimum utilization during passage through the gastrointestinal tract (Korhonen et al., 2003), screening methods to measure their long term effects on health (Bath et al., 2015) and relevant indicators or biomarkers which are useful to predict potential benefits relating to a target function in the host's organism (Diplock et al., 2000); the lack of knowledge of their specific mechanisms of action and gastrointestinal stability (Chakrabarti et al., 2018). Among the above reasons, one of the most important is the weak correlation of the *in vitro* bioactivities of peptides with the *in vivo* functions (Giromini et al., 2019a, Cheli et al., 2015).

CONCLUSION

It has been recognized worldwide that food-derived bioactive peptides are valuable ingredients for functional foods and/or nutraceuticals production. The study of food peptides effect at cellular level revealed important information for their application as functional ingredients. In fact, the understanding of the peptide mode of action at the cellular and molecular level may provide valuable inputs for the correct formulation. The cell-based system allowed the preliminary study of food peptide bioactivities and investigation of their mode of action with high predictive power. Even though it is generally recognized that the translation of *in vitro* activity to *in vivo* effects is not always realistic. This discrepancy is due to the molecular characteristics of peptides, gastrointestinal conditions as well as other dietary and non-dietary factors. This lack of correlation represents a key issue related to efficacy of bioactive peptides. The chapter revealed that both animal and plant proteins have encrypted bioactive peptides with different functions at the body level. This suggests the key role of a varied diet to maximize the beneficial effect of each bioactive peptide in the organism.

REFERENCES

Alemán, A., Pérez-Santín, E., Bordenave-Juchereau, S., Arnaudin, I., Gómez-Guillén, M. C., & Montero, P. (2011). Squid gelatin hydrolysates with antihypertensive, anticancer and antioxidant activity. *Food Research International*, 44(4), 1044–1051.

Almeida, C. C., Guerra Monteiro, M. L., da Costa-Lima, B. R. C., Alvares, T. S., & Conte-Junior, C. A. (2015). In vitro digestibility of commercial whey protein supplements. *LWT - Food Science and Technology*, 61(1), 7–11, ISSN 0023-6438.

Amigo, L., & Hernández-Ledesma, B. (2020). Current evidence on the bioavailability of food bioactive peptides. *Molecules (Basel, Switzerland)*, 25(19), 4479.

Andreassen, R. C., Pedersen, M. E., Kristoffersen, K. A., & Rønning, S. B. (2020). Screening of by-products from the food industry as growth promoting agents in serum-free media for skeletal muscle cell culture. *Food and Function*, 11(3), 2477–2488.

Aneiros, A., & Garateix, A. (2004). Bioactive peptides from marine sources: Pharmacological properties and isolation procedures. *Journal of Chromatography B*, 803(1), 41–53.

Anson, N. M., Havenaar, R., Bast, A., & Haenen, G. R. (2010). Antioxidant and anti-inflammatory capacity of bioaccessible compounds from wheat fractions after gastrointestinal digestion. *Journal of Cereal Science*, 51(1), 110–114.

Aydemir, L. Y., Gökbulut, A. A., Baran, Y., & Yemenicioğlu, A. (2014). Bioactive, functional and edible film-forming properties of isolated hazelnut (*Corylus avellana* L.) meal proteins. *Food Hydrocolloids*, 36, 130–142.

Azuma, N., Nagaune, S., Ishino, Y., Mori, H., Kaminogawa, S., & Yamauchi, K. (1989). DNA-synthesis stimulating peptides from human β-casein. *Agricultural and Biological Chemistry*, 53(10), 2631–2634.

Azuma, N., Suda, H., Iwasaki, H., Yamagata, N., Saeki, T., Kanamoto, R., & Iwami, K. (2000). Antitumorigenic effects of several food proteins in a rat model with colon cancer and their reverse correlation with plasma bile acid concentration. *Journal of Nutritional Science and Vitaminology*, 46(2), 91–96.

Baldi, A., Ioannis, P., Chiara, P., Eleonora, F., Roubini, C., & Vittorio, D. (2005). Biological effects of milk proteins and their peptides with emphasis on those related to the gastrointestinal ecosystem. *The Journal of Dairy Research*, 72(S1), 66.

Barrio, D. A., & Añón, M. C. (2010). Potential antitumor properties of a protein isolate obtained from the seeds of Amaranthus mantegazzianus. *European Journal of Nutrition*, 49(2), 73–82.

Behera, B., Devi, K. S. P., Mishra, D., Maiti, S., & Maiti, T. K. (2014). Biochemical analysis and antitumor effect of Abrus precatorius agglutinin derived peptides in Ehrlich's ascites and B16 melanoma mice tumor model. *Environmental Toxicology and Pharmacology*, 38(1), 288–296.

Bhat, Z. F., Kumar, S., & Bhat, H. F. (2015). Bioactive peptides of animal origin: A review. *Journal of Food Science and Technology*, 52(9), 5377–5392.

Blais, M., Pouliot, Y., Gauthier, S., Boutin, Y., & Lessard, M. (2014). A gene expression programme induced by bovine colostrum whey promotes growth and wound-healing processes in intestinal epithelial cells. *Journal of Nutritional Science*, 3, e57. doi: 10.1017/jns.2014.56

Bouzerzour, K., Morgan, F., Cuinet, I., Bonhomme, C., Jardin, J., Le Huërou-Luron, I., & Dupont, D. (2012). *In vivo* digestion of infant formula in piglets: Protein digestion kinetics and release of bioactive peptides. *British Journal of Nutrition*, 108(12), 2105–2114.

Carrillo, W., Gómez-Ruiz, J. A., Ruiz, A. L., & Carvalho, J. E. (2017). Antiproliferative activity of walnut (*Juglans regia* L.) proteins and walnut protein hydrolysates. *Journal of Medicinal Food*, 20(11), 1063–1067.

Chakrabarti, S., Guha, S., & Majumder, K. (2018). Food-derived bioactive peptides in human health: Challenges and opportunities. *Nutrients*, 10(11), 1738.

Chang, H. C., Lewis, D., Tung, C. Y., Han, L., Henriquez, S. M. P., Voiles, L., et al. (2014). Soypeptide lunasin in cytokine immunotherapy for lymphoma. *Cancer Immunology Immunotherapy*, 63(3), 283–295.

Cheli, F., Giromini, C., & Baldi, A. (2015). Mycotoxin mechanisms of action and health impact: 'in vitro' or 'in vivo' tests, that is the question. *World Mycotoxin Journal*, 8(5), 573–589.

Chen, J., Liu, B., Ji, N., Zhou, J., Bian, H., Li, C., et al. (2009). A novel sialic acid-specific lectin from *Phaseolus coccineus* seeds with potent antineoplastic and antifungal activities. *Phytomedicine*, 16(4), 352–360.

Cicero, A., Fogacci, F., & Colletti, A. (2017). Potential role of bioactive peptides in prevention and treatment of chronic diseases: A narrative review. *British Journal of Pharmacology*, 174(11), 1378–1394.

Contor, L. (2001). Functional food science in Europe. *Nutrition, Metabolism, and Cardiovascular Diseases*, 4(Suppl), 20–23.

De Mejía, E. G., & Dia, V. P. (2010). The role of nutraceutical proteins and peptides in apoptosis, angiogenesis, and metastasis of cancer cells. *Cancer Metastasis Reviews*, 29(3), 511–528.

De Mejia, E. G., Wang, W., & Dia, V. P. (2010). Lunasin, with an arginine-glycine-aspartic acid motif, causes apoptosis to L1210 leukemia cells by activation of caspase-3. *Molecular Nutrition and Food Research*, 54(3), 406–414.

De Simone, C., Picariello, G., Mamone, G., Stiuso, P., Dicitore, A., Vanacore, D., Chianese, L., Addeo, F., & Ferranti, P. (2009). Characterisation and cytomodulatory properties of peptides from Mozzarella di Bufala Campana cheese whey. *Journal of Peptide Science*, 15(3), 251–258.

Deng, M., Zhang, W., Tang, H., Ye, Q., Liao, Q., Zhou, Y., et al. (2013). Lactotransferrin acts as a tumor suppressor in nasopharyngeal carcinoma by repressing AKT through multiple mechanisms. *Oncogene*, 32(36), 4273–4283.

Ding, L., Zhang, Y., Jiang, Y., Wang, L., Liu, B., & Liu, J. (2014). Transport of egg white ACE-inhibitory peptide, Gln-Ile-Gly-Leu-Phe, in human intestinal Caco 2 cell monolayers with cytoprotective effect. *Journal of Agricultural and Food Chemistry*, 62(14), 3177–3182.

Diplock, A. T., Aggett, P. J., Ashwell, M., Bornet, F., Fern, E. B., & Roberfroid, M. B. (2000). Scientific concepts of functional foods in Europe: Consensus document. In: *Functional Foods II-Claims and Evidence* (eds Buttriss, J., Saltmarsh, M.). The Royal Society of Chemistry, Cambridge, 8–59.

D'Souza, M. J., Li, R. C., Gannon, M. L., & Wentzien, D. E. (2019). 1997–2017 leading causes of death information due to diabetes, neoplasms, and diseases of the circulatory system, issues cautionary weight-related lesson to the US population at large. *IEEE Network*, 1–6. doi: 10.1109/ICESI.2019.8863033

Edler, M. C., Fernandez, A. M., Lassota, P., Ireland, C. M., & Barrows, L. R. (2002). Inhibition of tubulin polymerization by vitilevuamide, a bicyclic marine peptide, at a site distinct from colchicine, the vinca alkaloids, and dolastatin 10. *Biochemical Pharmacology*, 63(4), 707–715.

Fast, J., Mossberg, A. K., Nilsson, H., Svanborg, C., Akke, M., & Linse, S. (2005). Compact oleic acid in HAMLET. *FEBS Letters*, 579(27), 6095–6100.

Fekete, A. A., Giromini, C., Chatzidiakou, Y., Givens, D. I., & Lovegrove, J. A. (2016). Whey protein lowers blood pressure and improves endothelial function and lipid biomarkers in adults with prehypertension and mild hypertension: Results from the chronic Whey2Go randomized controlled trial. *The American Journal of Clinical Nutrition*, 104(6), 1534–1544.

Fekete, Á. A., Giromini, C., Chatzidiakou, Y., Givens, D. I., & Lovegrove, J. A. (2018). Whey protein lowers systolic blood pressure and Ca-caseinate reduces serum TAG after a high-fat meal in mildly hypertensive adults. *Scientific Reports*, 8(1), 1–9.

Fernández-Tomé, S., Sanchón, J., Recio, I., & Hernández-Ledesma, B. (2018). Transepithelial transport of lunasin and derived peptides: Inhibitory effects on the gastrointestinal cancer cells viability. *Journal of Food Composition and Analysis*, 68, 101–110.

Fu, Y., & Zhao, X. H. (2013). *In vitro* responses of hFOB1.19 cells toward chum salmon (Oncorhynchus keta) skin gelatin hydrolysates in cell proliferation, cycle progression and apoptosis. *Journal of Functional Foods*, 5(1), 279–288.

Ganjam, L. S., Thornton Jr, W. H., Marshall, R. T., & MacDonald, R. S. (1997). Antiproliferative effects of yogurt fractions obtained by membrane dialysis on cultured mammalian intestinal cells. *Journal of Dairy Science*, 80(10), 2325–2329.

García-Fernández, L. F., Losada, A., Alcaide, V., Alvarez, A. M., Cuadrado, A., González, L., et al. (2002). Aplidine induces the mitochondrial apoptotic pathway via oxidative stress-mediated JNK and p38 activation and protein kinase C delta. *Oncogene*, 21(49), 7533–7544.

Giromini, C., Cheli, F., Rebucci, R., & Baldi, A. (2019a). Invited review: Dairy proteins and bioactive peptides: Modeling digestion and the intestinal barrier. *Journal of Dairy Science*, 102(2), 929–942.

Giromini, C., Fekete, Á. A., Givens, D. I., Baldi, A., & Lovegrove, J. A. (2017). Short-communication: A comparison of the *in vitro* angiotensin-1-converting enzyme inhibitory capacity of dairy and plant protein supplements. *Nutrients*, 9(12), 1352.

Giromini, C., Lovegrove, J. A., Givens, D. I., Rebucci, R., Pinotti, L., Maffioli, E., Tedeschi, G., Sundaram, T. S., & Baldi, A. (2019b). In vitro-digested milk proteins: Evaluation of angiotensin-1-converting enzyme inhibitory and antioxidant activities, peptidomic profile, and mucin gene expression in HT29-MTX cells. *Journal of Dairy Science*, 102(12), 10760–10771.

Gobbetti, M., Minervini, F., & Rizzello, C. G. (2007). Bioactive peptides in dairy products. In: *Handbook of Food Products Manufacturing: Health, Meat, Milk, Poultry, Seafood, and Vegetables* (ed. Y. H. Hui). John Wiley & Sons, Hoboken, NJ, 489–517.

Guerra, A., Etienne-Mesmin, L., Livrelli, V., Denis, S., Blanquet-Diot, S., & Alric, M. (2012). Relevance and challenges in modeling human gastric and small intestinal digestion. *Trends in Biotechnology*, 30(11), 591–600.

Guha, P., Kaptan, E., Bandyopadhyaya, G., Kaczanowska, S., Davila, E., Thompson, K., et al. (2013). Cod glycopeptide with picomolar affinity to galectin-3 suppresses T-cell apoptosis and prostate cancer metastasis. *Proceedings of the National Academy of Sciences of the United States of America*, 110(13), 5052–5057.

Hartmann, R., & Meisel, H. (2007). Food-derived peptides with biological activity: From research to food applications. *Current Opinion in Biotechnology*, 18(2), 163–169.

He, Y., Shen, L., Ma, C., Chen, M., Pan, Y., Yin, L., Zhou, J., Lei, X., Ren, Q., Duan, Y., Zhang, H., & Ma, H. (2018). Protein hydrolysates' absorption characteristics in the dynamic small intestine in vivo. *Molecules (Basel, Switzerland)*, 23(7), 1591.

Hernandez-Ledesma, B., Hsieh, C. C., & de Lumen, B. O. (2013). Chemopreventive properties of Peptide Lunasin: A review. *Protein and Peptide Letters*, 20(4), 424–432.

Hu, X., Song, L., Huang, L., Zheng, Q., & Yu, R. (2012). Antitumor effect of a polypeptide fraction from Arca subcrenata *in vitro* and in vivo. *Marine Drugs*, 10(12), 2782–2794.

Ikeda, D., Wada, S., Yoneda, C., Abe, H., & Watabe, S. (1999). Carnosine stimulates vimentin expression in cultured rat fibroblasts. *Cell Structure and Function*, 24(2), 79–87.

Ishikawa, S. I., Asano, T., Takenoshita, S., Nozawa, Y., Arihara, K., & Itoh, M. (2009). Egg yolk proteins suppress azoxymethane-induced aberrant crypt foci formation and cell proliferation in the colon of rats. *Nutrition Research*, 29(1), 64–69.

Jang, A., Jo, C., Kang, K. S., & Lee, M. (2008). Antimicrobial and human cancer cell cytotoxic effect of synthetic angiotensin-converting enzyme (ACE) inhibitory peptides. *Food Chemistry*, 107(1), 327–336.

Kampa, M., Bakogeorgou, E., Hatzoglou, A., Damianaki, A., Martin, P. M., & Castanas, E. (1997). Opioid alkaloids and casomorphin peptides decrease the proliferation of prostatic cancer cell lines (LNCaP, PC3 and DU145) through a partial interaction with opioid receptors. *European Journal of Pharmacology*, 335(2–3), 255–265.

Kannan, A., Hettiarachchy, N. S., Lay, J. O., & Liyanage, R. (2010). Human cancer cell proliferation inhibition by a pentapeptide isolated and characterized from ricebran. *Peptides*, 31(9), 1629–1634.

Kim, H. K., Lee, S., & Leem, K. H. (2011). Protective effect of egg yolk peptide on bone metabolism. *Menopause*, 18(3), 307–313.

Korhonen, H., & Pihlanto, A. (2003). Food-derived bioactive peptides--Opportunities for designing future foods. *Current Pharmaceutical Design*, 9(16), 1297–1308.

Korhonen, H., & Pihlanto, A. (2006). Bioactive peptides: Production and functionality. *International Dairy Journal*, 16(9), 945–960.

Korhonen, H. J. (2010). Health-promoting proteins and peptides in colostrum and whey. In: *Bioactive Proteins and Peptides as Functional Foods and Nutraceuticals* (eds Y. Mine, E. Li-Chan and B. Jiang), pp. 151–168. Ames, IA: Wiley-Blackwell.

Kovacs-nolan, J., & Mine, Y. (2010). Animal muscle-based bioactive peptides. In: *Bioactive Proteins and Peptides as Functional Foods and Nutraceuticals* (eds Y. Mine, E. Li-Chan and B. Jiang), pp. 225–232. Ames, IA: Wiley-Blackwell.

Lee, J. H., Moon, S. H., Kim, H. S., Park, E., Ahn, D. U., & Paik, H. D. (2017). Antioxidant and anticancer effects of functional peptides from ovotransferrin hydrolysates. *Journal of the Science of Food and Agriculture*, 97(14), 4857–4864.

Lee, J. H., & Paik, H. D. (2019). Anticancer and immunomodulatory activity of egg proteins and peptides: A review. *Poultry Science*, 98(12), 6505–6516.

Lee, Y. G., Kim, J. Y., Lee, K. W., Kim, K. H., Lee, H. J., & Ann, N. Y. (2003). Peptides from anchovy sauce induce apoptosis in a human lymphoma cell (U937) through the increase of caspase-3 and -8 activities. *Academic Science*, 1010, 399–404.

Li, R. (2014). Investigation of rice bran derived anti-cancer pentapeptide for mechanistic potency in breast cancer cell models. *Theses and Dissertations*, 2249.

Ma, J., Huang, F., Lin, H., & Wang, X. (2013). Isolation and purification of a peptide from *Bullacta exarata* and its impaction of apoptosis on prostate cancer cell. *Marine Drugs*, 11(1), 266–273.

Ma, S., Huang, D., Zhai, M., Yang, L., Peng, S., Chen, C., et al. (2015). Isolation of a novel bio-peptide from walnut residual protein inducing apoptosis and autophagy on cancer cells. *BMC Complementary and Alternative Medicine*, 15, 1–14.

Mader, J. S., Richardson, A., Salsman, J., Top, D., de Antueno, R., Duncan, R., & Hoskin, D. W. (2007). Bovine lactoferricin causes apoptosis in Jurkat T-leukemia cells by sequential permeabilization of the cell membrane and targeting of mitochondria. *Experimental Cell Research*, 313(12), 2634–2650.

Mahanta, S., Paul, S., Srivastava, A., Pastor, A., Kundu, B., & Chaudhuri, T. K. (2015). Stable self-assembled nanostructured hen egg white lysozyme exhibits strong antiproliferative activity against breast cancer cells. *Colloids and Surfaces, Part B: Biointerfaces*, 130, 237–245.

Mandal, S. M., Migliolo, L., Das, S., Mandal, M., Franco, O. L., & Hazra, T. K. (2012). Identification and characterization of a bactericidal and proapoptotic peptide from *Cycas revoluta* seeds with DNA binding properties. *Journal of Cellular Biochemistry*, 113(1), 184–193.

Matin, M. A., & Otani, H. (2002). Cytotoxic and antibacterial activities of chemically synthesized κ-casecidin and its partial peptide fragments. *Journal of Dairy Research*, 69(2), 329–334.

Matsui, T., Ueno, T., Tanaka, M., Oka, H., Miyamoto, T., Osajima, K., & Matsumoto, K. (2005). Antiproliferation action of an angiotensin I - Converting enzyme inhibitory peptides, Val-Tyr, via an L-type Ca2+ channel inhibition in cultured vascular smooth muscle cells. *Hypertension Research*, 28, 545–552.

Mayer, E. A. (2000). The neurobiology of stress and gastrointestinal disease. *Gut*, 47(6), 861–869.

McFadden, D. W., Riggs, D. R., Jackson, B. J., & Vona-Davis, L. (2003). Keyhole limpet hemocyanin, a novel immune stimulant with promising anticancer activity in Barrett's esophageal adenocarcinoma. *American Journal of Surgery*, 186(5), 552–555.

Meisel, H. (2001). Bioactive peptides from milk proteins: A perspective for consumers and producers. *Australian Journal of Dairy Technology*, 56, 83–91.

Meisel, H., & FitzGerald, R. J. (2003). Biofunctional peptides from milk proteins: Mineral binding and cytomodulatory effects. *Current Pharmaceutical Design*, 9(16), 1289–1295.

Mejia, E. G., Valadez-Vega, M. D. C., Reynoso-Camacho, R., & Loarca-Pina, G. (2005). Tannins, trypsin inhibitors and lectin cytotoxicity in tepary (*Phaseolus acutifolius*) and common (*Phaseolus vulgaris*) beans. *Plant Foods for Human Nutrition*, 60(3), 137–145.

Minekus, M., Alminger, M., Alvito, P., Ballance, S., Bohn, T., Bourlieu, C., et al. (2014). A standardised static *in vitro* digestion method suitable for food - An international consensus. *Food and Function*, 5(6), 1113–1124.

Möller, N. P., Scholz-Ahrens, K. E., Roos, N., & Schrezenmeir, J. (2008). Bioactive peptides and proteins from foods: Indication for health effects. *European Journal of Nutrition*, 47(4), 171–182.

Moon, S. H., Lee, J. H., Lee, M., Park, E., Ahn, D. U., & Paik, H. D. (2014). Cytotoxic and antigenotoxic activities of phosvitin from egg yolk. *Poultry Science*, 93(8), 2103–2107.

Moon, S. H., Lee, J. H., Lee, Y. J., Chang, K. H., Paik, J. Y., Ahn, D. U., & Paik, H. D. (2013). Screening for cytotoxic activity of ovotransferrin and its enzyme hydrolysates. *Poultry Science*, 92(2), 424–434.

Naguane, S., Azuma, N., Ishino, Y., Mori, H., Kaminogawa, S., & Yamauchi, K. (1989). DNA-synthesis peptide from bovine β-casein. *Agricultural and Biological Chemistry*, 53, 3275–3278.

Ngai, P. H. K., & Ng, T. B. (2004). A napin-like polypeptide with translation inhibitory, trypsin-inhibitory, antiproliferative and antibacterial activities from kale seeds. *Journal of Peptide Research*, 64(5), 20.

Oguro, T., Watanabe, K., Tani, H., Ohishi, H., & Ebina, T. (2000). Morphological observations on antitumor activities of 70 kDa fragment in α-subunit from pronase-treated ovomucin in a double grafted tumor system. *Food Science and Technology Research*, 6(3), 179–185.

Parada, J., & Aguilera, J. M. (2007). Food microstructure affects the bioavailability of several nutrients. *Journal of Food Science*, 72(2), R21–R32.

Park, S., & Ohba, H. (2004). Suppressive activity of protease inhibitors from buckwheat seeds against human T-acute lymphoblastic leukemia cell lines. *Applied Biochemistry and Biotechnology*, 117(2), 65–74.

Perego, S., Cosentino, S., Fiorilli, A., Tettamanti, G., & Ferraretto, A. (2012). Casein phosphopeptides modulate proliferation and apoptosis in HT-29 cell line through their interaction with voltage-operated L-type calcium channels. *Journal of Nutritional Biochemistry*, 23(7), 808–816.

Pérez-Gregorio, R., Soares, S., Mateus, N., & de Freitas, V. (2020). Bioactive peptides and dietary polyphenols: Two sides of the same coin. *Molecules (Basel, Switzerland)*, 25(15), 3443.

Phelan, M., Aherne-Bruce, S. A., O'Sullivan, D., FitzGerald, R. J., & O'Brien, N. M. (2009). Potential bioactive effects of casein hydrolysates on human cultured cells. *International Dairy Journal*, 19(5), 279–285.

Plaisancié, P., Boutrou, R., Estienne, M., Henry, G., Jardin, J., Paquet, A., & Léonil, J. (2015). β-casein (94–123)-derived peptides differently modulate production of mucins in intestinal goblet cells. *Journal of Dairy Research*, 82(1), 36–46.

Plaisancié, P., Claustre, J., Estienne, M., Henry, G., Boutrou, R., Paquet, A., & Léonil, J. (2013). A novel bioactive peptide from yogurts modulates expression of the gel-forming MUC2 mucin as well as population of goblet cells and Paneth cells along the small intestine. *Journal of Nutritional Biochemistry*, 24(1), 213–221.

Purup, S., Nielsen, S. D., Le, T. T., Bertelsen, H., Sorensen, J., & Larsen, L. B. (2018). Wound healing properties of commercial milk hydrolysates in intestinal cells. *International Journal of Peptide Research and Therapeutics*, 25(2), 483–491.

Roberfroid, M. (2002). Global view on functional foods: European perspectives. *British Journal of Nutrition*, 88(S2), S133–S138.

Ryder, K., Bekhit, A. E.-D., McConnell, M., & Carne, A. (2016). Toward generation of bioactive peptides from meat industry waste proteins: Generation of peptides using commercial microbial proteases. *Food Chemistry*, 208, 42–50.

Saadi, S., Saari, N., Anwar, F., Abdul Hamid, A., & Ghazali, H. M. (2015). Recent advances in food biopeptides: Production, biological functionalities and therapeutic applications. *Biotechnology Advances*, 33(1), 80–116.

Sanchez-Rivera, L., Martinez-Maqueda, D., Cruz-Huerta, E., Miralles, B., & Recio, I. (2014). Peptidomics for discovery, bioavailability and monitoring of dairy bioactive peptides. *Food Research International*, 63, 170–181.

Sava, G., Ceschia, V., Pacor, S., & Zabucchi, G. (1991). Observations on the antimetastatic action of lysozyme in mice bearing Lewis lung carcinoma. *Anticancer Research*, 11(3), 1109–1113.

Schrezenmeir, J., Korhonen, H., Williams, C., Gill, H. S., & Shah, N. (2000). Foreword. *British Journal of Nutrition*, 84(S1), S1–S166.

Sheih, I-C., Fang, T. J., Wu, T.-K., Lin, P.-H., & Lin, P. H. (2010). Anticancer and antioxidant activities of the peptide fraction from algae protein waste. *Journal of Agricultural and Food Chemistry*, 58(2), 1202–1207.

Shi, J., & Le Maguer, M. (2000). Lycopene in tomatoes: Chemical and physical properties affected by food processing. *Critical Reviews in Biotechnology*, 20(4), 293–334.

Silva-Sánchez, C., De La Rosa, A. B., León-Galván, M. F., De Lumen, B. O., de León-Rodríguez, A., & De Mejía, E. G. (2008). Bioactive peptides in amaranth (*Amaranthus hypochondriacus*) seed. *Journal of Agricultural and Food Chemistry*, 56(4), 1233–1240.

Singh, H., Ye, A., & Ferrua, M. J. (2015). Aspects of food structures in the digestive tract. *Current Opinion in Food Science*, 3, 85–93.

Sun, X., Acquah, C., Aluko, R. E., & Udenigwe, C. C. (2020). Considering food matrix and gastrointestinal effects in enhancing bioactive peptide absorption and bioavailability. *Journal of Functional Foods*, 64, 103680.

Ting, C. H., Huang, H. N., Huang, T. C., Wu, C. J., & Chen, J. Y. (2014). The mechanisms by which pardaxin, a natural cationic antimicrobial peptide, targets the endoplasmic reticulum and induces c-FOS. *Biomaterials*, 35(11), 3627–3640.

Tretola, M., Bee, G., & Silacci, P. (2020). Gallic acid affects intestinal-epithelial-cell integrity and selected amino-acid uptake in porcine *in vitro* and *ex vivo* permeability models. *British Journal of Nutrition*, 126(4), 492–500. doi:10.1017/S0007114520004328

Udenigwe, C. C., Raliat, O., Okagu, I. U., & Obeme-Nmom, J. I. (2021). Bioaccessibility of bioactive peptides: Recent advances and perspectives. *Current Opinion in Food Science*, 39, 182–189.

Valadez-Vega, C., Alvarez-Manilla, G., Riverón-Negrete, L., García-Carrancá, A., Morales-González, J. A., Zuñiga-Pérez, C., et al. (2011). Detection of cytotoxic activity of lectin on human colon adenocarcinoma (Sw480) and epithelial cervical carcinoma (C33-A). *Molecules*, 16(3), 2107–2118.

Vital, D. A. L., De Mejía, E. G., Dia, V. P., & Loarca-Piña, G. (2014). Peptides in common bean fractions inhibit human colorectal cancer cells. *Food Chemistry*, 157, 347–355.

Wang, Z., & Zhang, X. (2017). Isolation and identification of antiproliferative peptides from Spirulina platensis using three-step hydrolysis. *Journal of the Science of Food and Agriculture*, 97(3), 918–922.

Wélé, A., Zhang, Y. J., Brouard, J. P., Pousset, J. L., & Bodo, B. (2005). Two cyclopeptides from the seeds of *Annona cherimola*. *Phytochemistry*, 66(19), 2376–2380.

Wong, J. H., & Ng, T. B. (2005). Sesquin, a potent defensin-like antimicrobial peptide from ground beans with inhibitory activities toward tumor cells and HIV-1 reverse transcriptase. *Peptides*, 26(7), 1120–1126.

Xiao, Y. F., Jie, M. M., Li, B. S., Hu, C. J., Xie, R., Tang, B., & Yang, S. M. (2015). Peptide-based treatment: A promising cancer therapy. *Journal of Immunology Research*, 761820, 1–13. https://doi.org/10.1155/2015/761820

Xue, Z., Yu, W., Liu, Z., Wu, M., & Wang, J. (2011). Induction of apoptosis in cervix neoplasms hela cells by a rapeseed peptide hydrolysate fraction. *Journal of Food Biochemistry*, 35(4), 1283–1297.

Yang, Y., Zhang, Z., Pei, X., Han, H., Wang, J., Wang, L., et al. (2009). Immunomodulatory effects of marine oligopeptide preparation from chum salmon (*Oncorhynchus keta*) in mice. *Food Chemistry*, 113(2), 464–470.

Yokota, T., Ohishi, H., & Watanabe, K. (1999). Antitumor effects of beta subunit from egg white ovomucin on xenografted sarcoma-180 cells in mice. *Food Science and Technology Research*, 5(3), 279–283.

Zhan, K. X., Jiao, W. H., Yang, F., Li, J., Wang, S. P., Li, Y. S., et al. (2014). Reniochalistatins A–E, cyclic peptides from the marine sponge *Reniochalina stalagmitis*. *Journal of Natural Products*, 77(12), 2678–2684.

Zhao, H., Zhou, F., Wang, L., Fengling, B., Dziugan, P., Walczak, P., & Zhang, B. (2014). Characterization of a bioactive peptide with cytomodulatory effect released from casein. *European Food Research and Technology*, 238(2), 315–322.

Zheng, L.-H., Wang, Y.-J., Sheng, J., Wang, F., Zheng, Y., Lin, X.-K., & Sun, M. (2011). Antitumor peptides from marine organisms. *Marine Drugs*, 9(10), 1840–1859.

CHAPTER 24

Other Biological Functions

Leo M.L. Nollet

CONTENTS

Peptides: Antidiabetic, Anticoagulation, and Mineral Binding Properties 485
    Antidiabetic Properties 485
    Anticoagulation Properties 489
    Mineral Binding Properties 499
References 501

PEPTIDES: ANTIDIABETIC, ANTICOAGULATION, AND MINERAL BINDING PROPERTIES

Peptides from food have diverse functions. As seen in Chapters 17 to 22, a wide area of functions is largely discussed.

In this chapter, a number of lesser-known functions of bioactive peptides are detailed. The first part gives a summary of research on antidiabetic and anti-obesity peptides from diverse sources.

Coagulation and anticoagulation are complex processes. In the second part, the roles of these peptide activities are discussed.

In the last part, peptides with mineral properties are reviewed.

Antidiabetic Properties

Cholecystokinin (CCK or CCK-PZ) is a peptide hormone of the gastrointestinal system responsible for stimulating the digestion of fat and protein. Cholecystokinin, officially called pancreozymin, is synthesized and secreted by enteroendocrine cells in the duodenum, the first segment of the small intestine. Its presence causes the release of digestive enzymes and bile from the pancreas and gallbladder, respectively, and also acts as a hunger suppressant.

Currently, there is great interest in the nutraceutical and pharmaceutical applications of bioactive peptides obtained from marine species and marine by-products due to their wide range of biological activities. Cudennec and Ravallec [1] wrote a mini-review focusing on the potential of marine peptides to be used as appetite suppressive molecules in the prevention and/or in the treatment of obesity syndrome. An important and promising aspect in the fight against obesity is the study of anorexigenic gut hormones, like cholecystokinin (CCK) and glucagon-like peptide 1 (GLP-1), that are synthetized by enteroendocrine cells in the presence of nutrients in the gastrointestinal tract. Recent works which have focused on the interaction between marine peptides and CCK and GLP-1 are here reviewed.

Today, the increasing prevalence of diabetes is considered a main risk for human health worldwide. Except for milk- and bean-derived peptides, there is limited evidence on the potential management of type 2 diabetes mellitus using bioactive peptides – even less for marine-derived organisms. Recent advances in the understanding of the regulation of glucose metabolism using bioactive peptides from natural proteins, such as protection and reparation of pancreatic β-cells, enhancing glucose-stimulated insulin secretion and influencing the sensitivity of insulin and the signaling pathways, and inhibition of bioactive peptides to dipeptidyl peptidase IV, α-amylase, and α-glucosidase activities were summarized [2]. This paper tried to understand the underlying mechanism involved and the structure characteristics of bioactive peptides responsible for antidiabetic activities to prospect the utilization of rich marine organism proteins.

Cudennec et al. [3] investigated the action of fish protein hydrolysates (FPH) from blue whiting (*Micromesistius poutassou*) and brown shrimp (*Penaeus aztecus*) on cholecystokinin release from intestinal endocrine cells (STC-1). They demonstrated for the first time that FPH were able to highly stimulate CCK-releasing activity from STC-1 cells and that this stimulation was mainly due to peptide molecules. The partial purification of CCK-stimulating peptides showed that their molecular weight ranged between 1000 and 1500 Da for fish and crustacean FPH. Finally, in an aim to industrially produce hydrolysates enriched in CCK-stimulating molecules, the authors tested the effects of membrane processes (ultrafiltration and nanofiltration) on active peptide enrichments.

To find appetite suppressive molecules derived from fish protein hydrolysates, both *in vitro* and *in vivo* experiments were performed by Cudennec at al. [4] in order to demonstrate that hydrolysates produced from blue whiting muscle (BWMH) possess satiating properties. For the first time, it was demonstrated that a protein hydrolysate obtained from marine source was able to enhance CCK and glucagon-like peptide-1 (GLP-1) secretion in an STC-1 cell line [4]. To demonstrate that these *in vitro* activities also exist *in vivo*, the effect of BWMH preload administration in rats and its repercussion on food intake and metabolic plasma marker levels was investigated. Results showed that BWMH reduced short-term food intake which was correlated to an increase in the CCK and GLP-1 plasma levels. Moreover, it was demonstrated that the chronic administration of BWMH led to a decrease in body weight gain.

Bioactive peptides show significant potential for use in health management strategies, particularly as components of drugs and functional foods for diabetes treatment. Many antidiabetic bioactive peptides have been isolated and validated. The aim of the review of Yan et al. [5] was to update the state of knowledge of the origin, structural characteristics, and action. Additionally, the potential mechanisms of bioactive peptides on key enzymes and proteins, such as α-amylase, α-glucosidase, glucagon-like peptides and dipeptidyl peptidase-IV, that participate in glycemic level control from the intake of carbohydrates to blood glucose regulation were overviewed. This knowledge should facilitate research and industrial efforts to better understand and evaluate the potential of bioactive peptides with antidiabetic properties for blood glucose level management.

A study aimed at identifying new peptide sequences involved in gut hormone secretion released by protein *in vitro* gastrointestinal digestion was undertaken [6]. Targeted gut hormones were cholecystokinin and glucagon-Like Peptide 1. The activity of DPP-IV (dipeptidyl peptidase-IV) was also considered, as it strongly modulates GLP-1 action. Successive purification steps were performed to isolate peptide fractions involved in these bioactivities whose sequence was determined by LC-MS-MS. Three peptide sequences,

ANVST, TKAVEH, and KAAVT, were pointed out for their stimulating effects on GLP-1 secretion. The sequence VAAA was isolated for its DPP-IV inhibitory properties. Two peptide groups were strongly involved in CCK release and shared a certain occurrence of aromatic amino-acid residues.

Two cuttlefish (*Sepia officinalis*) viscera protein hydrolysates were obtained with different enzymes extracted from cuttlefish and smooth hound (*Mustellus mustellus*) [7]. Their ability to stimulate the secretion of cholecystokinin and glucagon-like peptide 1, using the enteroendocrine STC-1 cell line and inhibit the DPP-IV activity during a simulated gastrointestinal digestion was assayed. The physicochemical parameters of hydrolysates and their effects on intestinal cell viability were also determined. The hydrolysate obtained with cuttlefish enzymes (CVPH1) appeared to be the most promising for all assessed bioactivities. Thus, CVPH1 was able to stimulate CCK and active GLP-1-releasing activities of enteroendocrine cells without any cytotoxicity and to inhibit DPP-IV activity. Moreover, these actions were enhanced after gastrointestinal digestion, and CVPH1 was also able to inhibit the intestinal DPP-IV activity of Caco-2 cells.

An *in vitro* GI digestion of bovine hemoglobin was carried out, and the bioactivity of the digests on CCK and GLP-1 secretion and DPP-IV activity was measured [8]. Intestinal digests exhibited the most potent action on gut hormone release and DPP-IV activity inhibition. They also had the ability to promote hormone gene expression. As a conclusion, two fractions from the final intestinal digest led to the greatest GLP-1 secretion increase and DPP-IV activity inhibition.

A human *in vitro* static simulated gastrointestinal digestion model was adapted to a dog, which demonstrated the promising effects of a tilapia by-product hydrolysate on the regulation of food intake and glucose metabolism [9]. Promising effects on intestinal hormone secretion and DPP-IV inhibitory activity were evidenced. New bioactive peptides were identified, able to stimulate CCK and GLP-1 secretions and inhibit the DPP-IV activity after a transport study through a Caco-2 cell monolayer.

Recent research trends and scientific knowledge in seaweed protein-derived peptides with particular emphasis on production, isolation, and potential health impacts in prevention of hypertension, diabetes, and oxidative stress are summarized in reference [10].

The major source of blood glucose is the hydrolysis of dietary carbohydrates by carbohydrate-hydrolyzing enzymes and subsequent absorption by the small intestine. Therefore, inhibition of these enzymes, such as α-amylase and α-glucosidase, is one of the promising therapies, especially in the treatment of type II diabetes. The 2 insulinotropic incretin hormones, GLP-1 and glucose independent insulinotropic polypeptide (GIP), are used to enhance glucose-dependent insulin secretion and to regulate postprandial blood glucose level. The cleavage of the N-terminal X-Proline and X-Alanine sequence from these peptide hormones by DPP-IV inactivates these incretin hormones. Therefore, another alternative approach to develop antidiabetic therapies is based on the development of incretin analogues and DPP-IV inhibitors. Although it is not well evidenced for the inhibition mechanism of DPP-IV by seaweed derived peptides, many are reported that the analog has been designed based on the knowledge of the enzyme's substrate specificity. Many are dipeptide derivatives, and are small, low-molecular-weight compounds with good oral availability, and they function as competitive inhibitors of the enzyme [11]. Numerous bioactive peptides with various biofunctions were discovered and isolated from marine organisms. NutripeptinR and Hydro MN PeptideR from marine protein hydrolysate are among these bioactive peptides. These peptides are capable of lowering the postprandial blood glucose level and alleviating type-II diabetes.

Recently, there are some studies focused on antidiabetic enzymatic protein hydrolysates and peptides. For example, Suetsuna and Saito [12] reported pepsin decomposed boiled laver mixture (*Pyropia yezoensis*) hydrolysate fractions that exhibited blood sugar-lowering activity. Similarly, Harnedy and FitzGerald [13] have identified potential precursors for generation of peptides with DPP IV inhibitory activity from 3 hydrolyzed fractions of *Palmaria palmata* protein (aqueous, alkaline, and the mixture of aqueous and alkaline protein fraction hydrolysates by alcalase 2.4 l, flavourzyme 500 l, and corolase PP enzymes. These hydrolysates have shown dipeptidyl peptidase (DPPIV inhibitory activity with IC50 values of 2.52 ± 0.05, >5.00, and 1.65 ± 0.12 mg/mL for aqueous protein fraction hydrolysate; >5.00, 4.60 ± 0.09, and 3.16 ± 0.07 mg/mL for alkaline protein fraction hydrolysate; 4.24 ± 0.02, >5.00, and 2.26 ± 0.09 mg/mL for aqueous and alkaline combined protein fraction hydrolysate of alcalase, flavourzyme, and corolase PP enzymes, respectively. These results suggest the potential of seaweed peptides for their antidiabetic properties.

The peptides of soy protein obtained by enzymatic digestion with proteases were analyzed for their antidiabetic, antihypertensive, and antioxidant activities [14]. Peptides prepared with alkaline proteinase (AP) exhibited the highest α-glucosidase inhibitory activity compared with those from papain and trypsin digestion. AP hydrolysates also exhibited DPP-IV inhibitory, angiotensin-converting enzyme (ACE) inhibitory, and antioxidant activities. Gastrointestinal digestion of peptides enhanced α-glucosidase, DPP-IV, and ACE inhibitory activities compared with AP hydrolysates. AP peptides showing the highest α-glucosidase inhibitory activity were purified by anion-exchange and size-exclusion chromatography and were identified using tandem MS. The authors found three novel α-glucosidase inhibitory peptides with sequences LLPLPVLK, SWLRL, and WLRL with IC50 of 237.43 ± 0.52, 182.05 ± 0.74, and 162.29 ± 0.74 μmol/L, respectively.

Functional foods containing peptides offer the possibility to modulate the absorption of sugars and insulin levels to prevent diabetes. The potential of germinated soybean peptides to modulate postprandial glycemic response through inhibition of DPP-IV, salivary α-amylase, and intestinal α-glucosidases was investigated [15]. A protein isolate from soybean sprouts was digested by pepsin and pancreatin. Protein digest and peptide fractions obtained by ultrafiltration (<5, 5–10, and >10 kDa) and subsequent semi-preparative reverse phase liquid chromatography (F1, F2, F3, and F4) were screened for *in vitro* inhibition of DPP-IV, α-amylase, maltase, and sucrase activities. Protein digest inhibited DPP-IV (IC_{50} = 1.49 mg/mL), α-amylase (IC_{50} = 1.70 mg/mL), maltase, and sucrase activities of α-glucosidases (IC_{50} = 3.73 and 2.90 mg/mL, respectively). Peptides of 5–10 and >10 kDa were more effective at inhibiting DPP-IV (IC_{50} = 0.91 and 1.18 mg/mL, respectively), while peptides of 5–10 and <5 kDa showed a higher potency to inhibit α-amylase and α-glucosidases. Peptides in F1, F2, and F3 were mainly fragments from β-conglycinin, glycinin, and P34 thiol protease. The analysis of structural features of peptides in F1–F3 allowed the tentative identification of potential antidiabetic peptides.

Milk protein hydrolysate possessing free-radical-scavenging and anti-inflammatory activities have many beneficial effects on the increase of the glucose-induced insulin secretion and reduction in postprandial glycemia [16].

Egg and soy proteins generate bioactive peptides with multiple biological effects, exerting nutritional and physiological benefits. The review of de Campos Zani et al. [17] focuses on the antidiabetic and anti-obesity effects of egg- and soy-derived peptides and hydrolysates, *in vivo* and *in vitro*, relevant to these conditions. Studies using the intact protein were considered only when comparing the results with the hydrolysate or peptides. *In vivo* evidence suggests that bioactive peptides from egg and soy can potentially be used to manage elements of glucose homeostasis in metabolic syndrome; however, the mechanisms of action on glucose and insulin metabolism, and the interaction between peptides and their molecular targets, remain unclear. Optimizing the production of egg- and soy-derived peptides and standardizing the physiological models to study their effects on diabetes and obesity could help to clarify the effects of these bioactive peptides in metabolic syndrome-related conditions.

Novel peptides from albumin were identified, evaluated, and validated for their antidiabetic activity against α-glucosidase and α-amylase [18]. Albumin hydrolysate was purified and identified, and tandem MS was adapted to characterize the amino acid sequences of peptides from the hydrolysate. In addition, the antidiabetic effects of the peptides with α-glucosidase and α-amylase inhibitory activity have been performed. Results also suggested that peptide KLPGF had α-glucosidase inhibitory activity with an IC50 of 59.5 ± 5.7 µmol l–1 and α-amylase inhibitory activity with an IC50 of 120.0 ± 4.0 µmol l–1.

Wang et al. [19] evaluated the antidiabetic activity of hydrolyzed peptides derived from *Juglans mandshurica* Maxim. fruits in insulin-resistant HepG2 cells and type 2 diabetic mice.

Proteins from watermelon (*Citrullus lanatus* L.) seed were isolated using an acid-induced precipitation method and then hydrolyzed using pepsin, trypsin, and alcalase [20]. The hydrolysates were investigated for *in vitro* antioxidant and α-amylase inhibitory properties. The results indicate that these multidirectional bioactivities of watermelon seed protein hydrolysates may serve as useful tools in the formulation of antidiabetic agents.

Hydrolysates with high α-glucosidase inhibitory activity (α-GIA) were obtained from hemp seed protein (HSP) by alcalase treatment at the degree of hydrolysis (DH) of 27.24±0.88% [21]. Interestingly, the hydrolysates at DH≤9.68±0.45% could be an α-glucosidase activator instead. Two novel α-glucosidase inhibitory peptides with sequences of Leu-Arg (287.2 Da) and Pro-Leu-Met-Leu-Pro (568.4 Da) were identified.

In 2020, Kehinde and Sharma published an excellent review on recently isolated antidiabetic isolates and multiple peptides from multiple food sources [22]. Tables 24.1 to 24.5 [22] detail dairy sources of antidiabetic peptides (Table 24.1), egg-related sources (Table 24.2), marine sources (Table 24.3), cereal and pseudocereal sources (Table 24.4), and leguminous sources (Table 24.5).

Anticoagulation Properties

Casein (CN) has been regarded as an excellent protein source for preparing bioactive peptides. The casein peptides released in the mouse gastrointestinal tract were evaluated. The 10-week-old mouse was orally administered with 5 mg casein [60]. After 0.5 h, the peptides in the stomach and small intestine of the mouse were extracted and analyzed by ultra-performance liquid chromatography-quadrupole time-of-flight mass spectrometry (UPLC-Q-TOF-MS/MS). A total of 343 peptides were identified, and 98, 36, 181, and 28 peptides were derived from α_{s1}-, α_{s2}-, β-, and κ-CN, respectively. Then, *in silico* methods were adopted to predict the potential anticoagulant peptide, including PeptideRanker, Innovagen. A novel anticoagulant peptide, AVPYPQR (β-CN, fragment 177–183), was screened, and its anticoagulant activity was verified. *In vitro* anticoagulant assay showed that the peptide AVPYPQR can observably prolong activated partial thromboplastin time (APTT), prothrombin time (PT), and thrombin time (TT), which indicated that the peptide AVPYPQR exerts its anticoagulant activity in the intrinsic, extrinsic, and

TABLE 24.1 Dairy Sources of Antidiabetic Protein Isolates, Hydrolysates, and Peptides [22]

Source	Extraction Method(S)	Extraction Tool	Separation/ Purification/ Fractionation Technique(s)	Hydrolysate Name/ Peptide Sequence	Molecular Weight	Antidiabetic Mechanism	Inhibition Value	Inhibition Value	Reference
Milk protein and protein hydrolysate	Enzymatic	Trypsin	HPLC			Reduction in blood plasma glucose level *in vivo*			[23]
Whey proteins α-lactalbumin β-lactoglobulin Lactoferrin Bovine serum albumin Whey protein isolate	Enzymatic	Pepsin	Gel electrophoresis		1.0–26.6kDA	DPP-IV and α-glucosidase inhibition	DPP-IV (IC$_{50}$) (mg/mL) 0.036±0.002 1.279±0.100 0.379±0.035 0.513±0.056 0.075±0.006	α-glucosidase (ic$_{50}$) (mg/ml) - 3.5±0.04 - - 4.5±0.6.	[24]
Dairy protein (Whey Protein Isolate WPI)	Enzymatic	Pepsin	Ultrafiltration			DPP-IV inhibition	IC$_{50}$ (mg/mL) 0.075		[25]
Whey proteins (β-lactoglobulin hydrolysates)	*In silico* analysis	BIOPEP database		IPA		DPP-IV inhibition *in vitro*	Conc.(μm) 80-100	% inhibition 20-4010-100	[26]
Milk casein from sheep	Enzymatic Hydrolysis	Trypsin, Pepsin, Chymotrypsin		Chymotrypsin hydrolysates (CH) Trypsin hydrolysates (PH)		α-amylase inhibition *in vitro*	%inhibition for raw (H) after 240mins incubation 38.69±0.44 35.16±0.51 44.38±0.59	% inhibition for boiled (H) after 240mins incubation 34.84±0.44 32.89±0.65 39.62±0.3.	[27]
Milk (β-lactoglobulin)	Enzymatic	Trypsin	RP-HPLC	VAGTWY	696.6Da	DPP-IV inhibition and glucose reduction *in vivo*	174μM		[28]

(Continued)

TABLE 24.1 (Continued)

Source	Extraction Method(S)	Extraction Tool	Separation/Purification/Fractionation Technique(s)	Hydrolysate Name/Peptide Sequence	Molecular Weight	Antidiabetic Mechanism	Inhibition Value	Inhibition Value	Reference
Gouda cheese			RP-HPLC	LPQNIPPL LPQ VPITPTL VPITPT		Reduction in glucose level in vivo and DPP-IV inhibition in vivo and in vitro	DPP-IV IC_{50} (μM) 4.6E+01 8.2E+01 1.1E+02 1.3E+02		[29]
Milk proteins	Chemical synthesis		Reverse Phase-Ultra Performance Liquid Chromatography	EK GL AL VA WV FLHL		DPP-IV inhibition in vitro	DPP-IV IC_{50} (mg/mL) 0.654±0.001 0.492±0.115 0.178±0.014 0.032±0.001 0.020±0.001 0.111±0.003 0.038±0.001		[30]
Milk proteins	Enzymatic hydrolysis	Pepsin and pancreatin	RP-HPLC	LKTPEGDL LPYPY IPIQY IPI WR WPI digests		DPP-IV inhibition	Conc. (mg/mL) 12.5 75 50 12.5 75 62.5μg/mL	IC_{50} (Basolateral solution after 2hrs) 18±2 12±4 15±3 14±5 46±8 12±3	[31]
Whey Protein Concentrate rich in Bovine β-Lactoglobulin	Enzymatic hydrolysis	Trypsin	Semi preparative RP-HPLC	IPVAF TPEVDDEALEK IPAVFK VLVLDTDYK		DPP-IV inhibition in vitro	IC_{50} (μM) 44.7±3.6 319.5±4.0 143.0±1.3 424.4±31.5		[32]
Bovine Whey Protein Isolate and α-Lactoglobulin	Enzymatic hydrolysis	Pepsin		WLAHKALCSEKLDQ LAHKALCSEKL TKCEVFRE LCSEKLDQ LKTPEGDL LKPTEGDLEIL	838.02 1212.47 1011.16 935.06 969.1 1324.54	DPP-IV inhibition in vitro	IC_{50} (μM) 141 165 166 186 45 57		[33]

TABLE 24.2 Egg-Related Sources of Antidiabetic Peptides [22]

Source	Extraction Method(s)	Extraction Tool	Separation/ Purification/ Fractionation Techniques	Hydrolysate Name/Peptide Sequence	Molecular Weight	Antidiabetic Mechanism	Inhibition Value	Inhibition Value	Reference
Egg white protein	Enzymatic	Alcalase	HPLC	RVPSLM TPSPR DLQGK AGLAPY RVPSL DHPLFLF HAEIN QIGLF		α-amylase and α-glucosidase inhibition *in vitro*	α-amylase (IC_{50}) >150μmol/L	α-glucosidase (IC_{50}) 23.07 μmol/L 40.02 μmol/L >150 μmol/L	[34]
Egg yolk protein by-product	Enzymatic	Proteinase from *C. ficifolia* Asian pumpkin pulp	Chemical synthesis	RASDPLLSV RNDDLNYIQ LAPSLPGKPKPD AGTTCLFTPLALPYDYSH		α-glucosidase and DPP-IV inhibition *in vitro*	α-glucosidase - 1065.6μmol/L -	DPP-IV(μmol/L) 426.25 350-400 361.5 -	[35]
Egg yolk protein	Enzymatic hydrolysis	Pepsin	Exchange chromatography and RP-HPLC	YINQMPQKSRE YINQMPQKSREA VTGRFAGHPAAQ		DPP-IV and α-glucosidase Inhibition *in vitro*	DPP-IV IC_{50} (μg/mL) 222.8 355.8 1402.2	α-glucosidase IC_{50} (μg/mL) 1694.3 454.6 365.4	[36]

TABLE 24.3 Marine Sources of Antidiabetic Hydrolysates and Peptides [22]

Source	Extraction Method(s)	Extraction Tool	Separation/ Purification/ Fractionation Techniques	Hydrolysate Name/Peptide Sequence	Molecular Weight	Antidiabetic Mechanism	Inhibition Value	Reference
Wild marine fish	Enzymatic	Mixture of 25% pepsin, 35% trypsin, 35% chymotrypsin, 5% pancreatic lipase	Filtration through ceramic membranes (200μm)	Marine collagen peptides	130–3000Da	Decrease in blood fasting glucose level and blood fasting insulin level of diabetic patients		[37]
Fish skin gelatin (Halibut, Tilapia, Hake, and Milkfish)	Enzymatic	Flavourzyme from *Aspergillus oryzae*	Ultrafiltration	Halibut SPGSSGPQGFTG GPVGPAGNPGANGLN PPGTGPRGQPNIGF Tilapia IPGDPGPPGPPGP LPGERGRPGAPGP GPKGDRGLPGPPGRDGM	862.32 1021.42 1261.44 919.53 1026.58 1358.76	DDP-IV inhibition (*in vivo* and *in vitro*), enhancement of GLP-1 and insulin secretion *in vivo*	DPP-IV(μM) 101.6 81.3 146.7 65.4 76.8 89.6	[38]
Goby fish protein	Enzymatic	*B. mojavensis* (A21) and alkaline protease extraction from trigger fish intestine		Hydrolysates from treatments with B. mojavensis A21 proteases and triggerfish crude alkaline proteases		Decrease in serum glucose level, α-amylase activities and hepatic glycogenesis *in vivo*		[39]
Salmon skin gelatin	Enzymatic Hydrolysis	Alcalase Bromelain Flavourzyme	Fractionation by ultrafiltration and purification by HPLC	GPAE GPGA	300.4Da 372.4Da	DPP-IV inhibition *in vitro*	IC$_{50}$(μM) 49.6 41.9	[40]
Tuna cooking juice	Enzymatic	Orientase and Protease XXIII	Gel filtration and HPLC	PACGGFYISGRPG CAYQWGRPVNRIR PGVGGPMGPIGPCYQ	1304.6Da 1690.8Da 1412.7Da	DPP-IV inhibition *in vitro*	IC$_{50}$(μM) 96.4 78.0 116.1	[41]

TABLE 24.4 Cereal and Pseudocereal Sources of Antidiabetic Hydrolysates and Peptides

Source	Extraction Methods	Extraction Tool	Separation/ Purification/ Fractionation Techniques	Hydrolysate Name/ Peptide Sequence	Molecular Weight	Antidiabetic Mechanism	Antidiabetic Mechanism	Inhibition Value	Inhibition Value	Reference
Defatted rice bran	Enzymatic	Umamienzyme G (UG) and Bioprase SP (BSP)	Gel filtration chromatography	UG peptides BSP peptides		DPP-IV inhibition *in vitro*		IC_{50} (mg/mL) 2.3 ± 0.1 26.4 ± 2.3		[42]
Rice bran protein and Sake lees	Enzymatic	Denazyme AP protease from *Aspergillus oryzae*	Gel filtration chromatography	Rice protein hydrolysate Sakelees hydrolysate		α-glucosidase and DPP-IV inhibition *in vitro*		DPP-IV IC_{50} (mg/mL) 1.45 ± 0.13 1.28 ± 0.18		[43]
Brewers' spent grain	Enzymatic	Alcalase, SGID	Membrane fractionation and RP-HPLC	BSG hydrolysate ILDL ILLPGAQDGL		DPP-IV inhibition *in vitro*		IC_{50} 3.57 ± 0.19 mg/mL $112.1\mu M$ $145.5\mu M$		[44]
Brewers' spent grain	Enzymatic	Alcalase	Ultrafiltration	BSG hydrolysate		α-glucosidase inhibition *in vitro*		At concentration of 4.0mg/mL inhibited 21.42% α-glucosidase		[45]
Amaranthus grain	Enzymatic	Alcalase	Gel filtration	Albumin hydrolysate fraction after 48hrs (AHF48) Globulin hydrolysate fraction after 48hrs (GBHF48) Glutelin hydrolysate fraction after 48hrs (GLHF48)		DPP-IV inhibition *in vivo*	Fraction 2nd AHF48 3rd AHF48 3rd GBH48 4th GBH48 2nd GLH48 3rd GLH48	IC_{50} (mg/mL) 8.34 ± 0.09 1.98 ± 0.01 5.6 ± 0.1 0.25 ± 0.04 1.95 ± 0.08 0.12 ± 0.006		[46]
Amaranthus seed	Enzymatic, SGID	Trypsin, trypsin-pancreatin mixture	Ultrafiltration and ultracentrifugation	STHASGFFFHPT STNYFLJSCLLFVLFNGGCMGEG GLTEVWDSNEQEF	1482.6Da 2428.7Da 1553.6Da	DPP-IV inhibition *in vitro*				[47]
Quinoa protein	SGID	Pepsin and pancreatin	Ultrafiltration and RP-HPLC	IQAEGGLT DKDYPK GEHGSDGNV	787.4 764.4 870.4	DPP-IV α-amylase α-glucosidase inhibition *in vitro*	%DPP-IV inhibition at 250 μM 17.05 ± 0.06 - -	%α-amylase inhibition at 250μM - 6.86 ± 0.16 -	%α-glucosidase inhibition at 250μM 55.85 ± 0.26 22.16 ± 0.6 30.84 ± 0.69	[48]

(Continued)

TABLE 24.4 (Continued)

Source	Extraction Methods	Extraction Tool	Separation/ Purification/ Fractionation Techniques	Hydrolysate Name/ Peptide Sequence	Molecular Weight	Antidiabetic Mechanism	Antidiabetic Mechanism	Inhibition Value	Inhibition Value	Reference
Quinoa	Enzymatic	Papain and papainlike enzyme	Gel Permeation High Performance Liquid Chromatography (GP-HPLC)	Quinoa protein hydrolysates from Papain enzyme (QPH-P) Quinoa protein hydrolysates from Papain-like enzyme (QPH-PL)		DPP-IV inhibition *in vitro*	IC_{50} 0.88±0.05 0.98±0.04			[49]
Oat protein	Enzymatic hydrolysis	Alcalase	Ultrafiltration	FLQPNLDEH DLELQNNVHPH TPNAGVSGAAAGAGGKH	556.757	Increased insulin secretion and sensitivity, lowering of blood glucose *in vivo*				[50]
Cumin seeds (*Cuminum cyminum*)	Enzymatic hydrolysis	Commercial enzyme Protamex	Gel Elution Fraction Entrapment Electrophoresis SGELFREE°	FFRSKLLSDGAAAAKGALLPQYW RCMAFLLSDGAAAAQQLLPQYW DPAQPNYPWTAVLVFRH		α-amylase inhibition *in vitro*	IC_{50} (μM) 0.04 0.002 0.05			[51]

TABLE 24.5 Leguminous Sources of Antidiabetic Peptides [22]

Source	Extraction Method(s)	Extraction Tool	Separation/ Purification/ Fractionation Techniques	Hydrolysate Name/ Peptide Sequence	Molecular Weight (Da)	Antidiabetic Mechanism	Reference
Black beans	Enzymatic	Alcalase	Ultrafiltration	AKSPLF ATNNPLF FEELN LSVSVL	661.37 661.34 650.29 616.37	Reduced glucose uptake, blockage of glucose transport and DPP-IV inhibition *in vitro* and *in silico*	[52]
Common beans	Enzymatic, germination and SGID	Alcalase from *Bacillus lichenifor- mis*, pepsin and pancreatin Alcalase and bromelain	Ultrafiltration	RGPLVNPDPKPFL	1448.81	DPP-IV inhibition *in vitro*	[53]
Hard-to-cook beans	Enzymatic	Alcalase and bromelain	Ultrafiltration	LLSL WVVL AIVLL	445 515 527	Glucose uptake and DPP-IV inhibition *in vitro*	[54]
Common beans (Pinto Durgo and Black 8025)	Enzymatic, SGID	Alcalase, bromelain, pepsin, pancreatin	Ultrafiltration	FFL QLGGH LLSL QQEG WGVFN EPHGK HVQNQ NDEPASG	452.23 510.25 444.29 460.19 621.29 566.28 624.29 688.26	Insulin secretion, glucose uptake and DPP-IV inhibition *in vitro*	[55]

(*Continued*)

TABLE 24.5 (Continued)

Source	Extraction Method(s)	Extraction Tool	Separation/ Purification/ Fractionation Techniques	Hydrolysate Name/ Peptide Sequence	Molecular Weight (Da)	Antidiabetic Mechanism	Reference
Black beans (*Phaseolus vulgaris* L.)	Enzymatic	Trypsin flavourzyme, proteinase k, thermolysin, alcalase, pepsin, papain, chymotrypsin	Dialysis	FEELN AKSPLF EGLELLLLLAG	627.3 661.4 1252.8	α-amylase, α-glucosidase and DPP-IV inhibition *in vitro* and *in silico*	[56]
7S74oybeans (Aglycin peptide)	Chemical synthesis		HPLC	ASCNGVCPFEM PPCGSSACRCIP VGLVVGYCRHSG	3742.3	Blood glucose level reduction and enhancing insulin signal at gene levels *in vivo*	[57]
Meju, unsalted soybeans)	Fermentation	*Bacillus subtilis* and *Aspergillus oryzae*	Peptides profile determined by Ultra performance liquid chromatography (UPLC)			Increased glucose tolerance and enhanced insulin sensitivity *in vivo*	[58]
Soybean condiment	Fermentation					α-amylase and α-glucosidase inhibition *in vivo* and *in vitro*	[59]

common pathways. Meanwhile, the cell viability of this peptide was estimated on the human umbilical vein endothelial cells (HUVECs). The physicochemical characteristics of this peptide have been assayed by PepDraw and ExPASy-ProtParam. The study indicated that casein could be a valuable source for preparing bioactive peptides by gastrointestinal (GI) tract digestion.

Various bioactive peptides are identified from casein hydrolysates. YQEPVLGPVR (PICA), a novel antithrombotic peptide derived from beta-casein (fragment 193–202), was identified by high-performance liquid chromatography – that is, liquid chromatography-mass spectrometry/mass spectrometry [61]. The anticoagulation activity assay showed that this peptide has strong anticoagulant activity. It was proved that the peptide did not interact with the active site of thrombin to inhibit thrombin, and that it inhibited thrombin activity by binding the exosite-1 of thrombin, which was also confirmed by the fibrinogen clotting time assay. It was shown that PICA prolonged fibrinogen clotting time in a dose-dependent manner. Secondary structures of the thrombin–PICA complex were also measured by circular dichroism to prove that PICA can combine with thrombin. Moreover, Discovery Studio 2017 R2 software was used for molecular docking to provide the potential mechanism for the antithrombotic activity of the peptide.

Many kinds of bioactive components with nutritional and pharmaceutical activities in *Mytilus edulis* were reported [62]. Eight different parts of *Mytilus edulis* tissues, i.e., the foot, byssus, pedal retractor muscle, mantle, gill, adductor muscle, viscera, and other parts, were separated and the proteins from these tissues were prepared. A total of 277 unique peptides from the hydrolysates of different proteins were identified by UPLC-Q-TOF-MS/MS, and the molecular weight distribution of the peptides in different tissues was investigated by sodium dodecyl sulfate-polyacrylamide gel electrophoresis (SDS-PAGE). The bioactivity of the peptides was predicted through the Peptide Ranker database and molecular docking. Moreover, the peptides from the adductor muscle were chosen to do the active validation of anticoagulant activity. The active mechanism of three peptides from the adductor muscle, VQQELEDAEERADSAEGSLQK, RMEADIAAMQSDLDDALNGQR, and AAFLLGVNSNDLLK, were analyzed by Discovery Studio 2017, which also explained the anticoagulant activity of the hydrolysates of proteins from adductor muscle.

Optimization of the thrombin inhibitory activities of different enzymatic hydrolysates derived from *Mytilus edulis* was conducted, and then an optimal hydrolysis condition by trypsin (5000 u/g) was determined as follows, digested at 45 °C and pH 8.5 for 2 h with a protein concentration of 25 mg/mL [63]. Thrombin inhibitory activity was proved to be 76.92 ± 4.66% under this condition. A total of 39 peptides were identified in the hydrolysate by UPLC-Q-TOF–MS/MS, and all the peptides were predicted to be nontoxic by *in silico* predictive approaches. Twenty-six peptides were predicted to be anticoagulant peptides by the molecular docking method, and the peptide 26 (Lys-Asn-Ala-Glu-Asn-Glu-Leu-Gly-Glu-Val-Thr-Val-Arg) was predicted to be a better anticoagulant peptide through both structure–activity relationship and affinity activity to thrombin. The interactional positions between peptide and thrombin were also involved in the interaction site on the S1 pocket of thrombin and strongly promoted its thrombin inhibitory activity. The firmly non-bonded interactions made the bound of peptide and thrombin firmly. Eventually, the chemical identification and activity verification of synthetic peptide 26 were conducted, and the thrombin inhibitory activity was 89.96 ± 5.30% at the concentration of 9 mg/mL.

A newly discovered anticoagulant peptide was isolated, purified, and identified from the pepsin hydrolysate of oyster (*Crassostrea gigas*) which could potentially prolong the activated partial thromboplastin time and thrombin time [64]. The anticoagulant peptide with a 1264.36 Da molecular mass was similar to the amino-acid sequence of the C-terminal segment (DFEEIPEEYLQ) of hirudin (a potent thrombin inhibitor). The peptide specifically inhibited a vital blood coagulation factor: thrombin. The molecular docking energy scores of the anticoagulant peptide with the active site, exosite-I and exosite-II of thrombin were 132.355 kcal mol^{-1}, 151.266 kcal mol^{-1} and 147.317 kcal mol^{-1}, respectively. The anticoagulant peptide interacted with thrombin by competing with fibrinogen for an anion-binding exosite I. In the anticoagulant peptide–thrombin complex, there are seven hydrogen bonds, and reciprocity exists between hydrogen atoms and oxygen atoms, and electrostatic and hydrophobic interactions are also involved. Such abundant interactions may be accountable for the high affinity and specificity of the anticoagulant peptide.

A novel food-derived anticoagulant, heptapeptides (P-3-CG), was isolated and characterized from oyster (*Crassostrea gigas*) pepsin hydrolysate [65]. P-3-CG competed with fibrinogen against thrombin active domain by a spontaneous and exothermic reaction which was entropically driven. The residue Lys7 of P-3-CG anchored thrombin S_1 pocket strongly, which inhibited fibrinogen binding to the thrombin, then blocked the conversion of fibrinogen to fibrin. The fibrinogen clotting time was prolonged to 27.55 s, and the reciprocally authenticated results of dynamic light scattering and scanning electron microscope further explained the fibrinogen clotting time extension. Inhibition of amidolytic activity of thrombin was affected significantly by reaction time and P-3-CG concentration. Furthermore, P-3-CG prolonged activated partial thromboplastin time significantly *in vitro/vivo* and decreased the mortality which was confirmed by pulmonary pathological slide results.

Peptides from protein hydrolysate of a mixture of chicken combs and wattles (CCWs) were obtained through enzymatic hydrolysis, and their anticoagulant and inhibitory effects on angiotensin I-converting enzyme (ACE) were investigated [66]. The protein hydrolysate exhibited anticoagulant capacity by the intrinsic pathway (activated partial thromboplastin time) and potent ACE-inhibitory activity. The peptides were sequenced by LC-MS to identify those with higher inhibitory potential. From the pool of sequenced peptides, the following three peptides were selected and synthesized based on their low molecular weight and the presence of amino acids with ACE-inhibitory potential at the C-terminus: peptide I (APGLPGPR), peptide II (Piro-GPPGPT), and peptide III (FPGPPGP). Peptide III (FPGPPGP) showed the highest ACE-inhibitory capacity among the peptides selected. In conclusion, a peptide (FPGPPGP) of unknown sequence was identified as having potent ACE-inhibitory capacity.

Peptides from hematophagous (blood-feeding) and venomous organisms have been recognized as potential anticoagulant agents. Of late, peptides derived from the hydrolysis of food proteins, including edible seaweed, milk, and seed proteins, have also shown to possess promising *in vitro* anticoagulant activity [67]. To overcome the problems associated with regular anticoagulants, peptides targeting vital steps in the clotting cascade have been studied. The authors focused on anticoagulant peptides with known targets, inhibiting crucial factors in the coagulation cascade such as FXa, FXIa, FXIIa, and FVIIa/TF complex, as well as peptides with unknown targets.

Mineral Binding Properties

The ability of casein phosphopeptides (CPPs) to bind and transport minerals has been previously studied. However, the single bioactive peptide responsible for the effects of

CPPs has not been identified. This study was to purify calcium-binding peptides from CPPs and to determine their effects on calcium and magnesium uptake by Caco-2 cell monolayers. Five monomer peptides designated P1–P5 were isolated, and the amino-acid sequences were determined using LC-MS/MS [68]. Compared with the CPP-free control, all five monomeric peptides exhibited significant enhancing effects on the uptake of calcium and magnesium ($P < 0.05$). Interestingly, when calcium and magnesium were presented simultaneously with P5, magnesium was taken up with priority over calcium in the Caco-2 cell monolayers. For example, at 180 min, the amount of transferred magnesium and calcium was 78.4 ± 0.95 μg/well and 2.56 ± 0.64 μg/well, respectively, showing a more than 30-fold difference in the amount of transport caused by P5.

A novel calcium-binding peptide from casein hydrolysate was purified using reversed-phase high performance liquid chromatography and sequenced by high-performance liquid chromatography-mass spectrometry (MS)/MS [69]. The amino acid sequence of the calcium-binding peptide was identified as VLPVPQK (N- to C-terminal, MW = 779.4960 Da). The calcium-binding characteristics of VLPVPQK were further investigated using UV absorption spectroscopy, zeta potential, and isothermal titration calorimetry (ITC). The results showed that VLPVPQK has a strong calcium binding activity (129.46 mg g^{-1}), 312% higher than that of 3-hour enzymatic hydrolysates. VLPVPQK could chelate calcium with a 1:3 stoichiometry, causing a decrease in the positive charge of the peptide–Ca^{2+} complex. Furthermore, VLPVPQK could effectively enhance calcium transport and absorption in a concentration-dependent manner in Caco-2 cell monolayers, suggesting that VLPVPQK has the potential to be developed as a nutraceutical additive.

There are several peptides with specific sequences which form complexes by binding with minerals like Ca, P, and any others in a solution, at intestinal pH [70]. The greater anionic character of these peptides results in the formation of soluble complexes as they become resistant to further proteolytic attack, preventing the formation of insoluble complexes with minerals. Peptides having the sequence Ser-P:Ser-P:Ser-P:Glu-IleVal-Pro-Asn are isolated from αS1-casein. Milk caseins are known to stabilize calcium and phosphate ions, a reason for which several phosphopeptides have been identified from enzymatic digest of milk proteins, including αS1-casein comprising fragments (f43–f58, f59–f79, and f43–f49), αS2- casein possessing two fragments (f1–f24 and f46–f70) and β-casein with four fragments (f1–f28, f2–f28, f1–f25, and f33–f48) [71]. When casein proteins are subjected to tryptic digestion, casein phosphopeptides (CPPs) are released from the N-terminus polar domain containing a cluster of phosphorylated seryl residues and are responsible for interaction between calcium phosphate and casein, resulting in the formation of casein micelle. Also, these phosphorylated peptides bind to calcium and phosphate ions increasing their bioavailability [72]. However, *in vivo* studies showed that CPPs were unable to enhance the absorption of calcium ion in the intestine.

Bioactive peptides derived from egg proteins are reported to display various biological activities, ACE inhibitory (antihypertensive), antioxidant, antimicrobial, anti-inflammatory, antidiabetic, and iron-/calcium-binding activities [73]. More importantly, simulated *in vitro* gastrointestinal digestion has indicated that consumption of egg proteins has physiological benefits due to the release of such multifunctional peptides.

The enzymatic hydrolysis of tomato seed protein hydrolysates (TSPH) was performed using alcalase and a two-factor response surface methodology [74]. The best conditions were 131.4 min and 3% enzyme/substrate (E/S) for antioxidant activity; 174.5 min and 2.93% E/S for ACE inhibition; and 66.79 min and 2.27% E/S for the calcium binding. Antioxidant and ACE hydrolysates were characterized by higher solubility, zeta potential, and thermal stability, while properties of the calcium binding hydrolysate were only minimally affected by the enzymatic hydrolysis. Gel electrophoresis showed that molecular weights of polypeptides in the calcium binding TSPH were higher compared to those in ACE and antioxidant TSPHs. This was due to the low degree of hydrolysis of the calcium binding hydrolysate.

REFERENCES

1. Cudennec B., Ravallec R. Biological active peptides from marine sources related to gut hormones. *Current Protein and Peptide Science*, 2013, 14(3), 231–234.
2. Xia E.-Q., Zhu S.-S., He M.-J., Luo F., Fu C.Z., Zou T.B. Marine peptides as potential agents for the management of type 2 diabetes mellitus—A prospect. *Marine Drugs*, 2017, 15(4), 88. https://doi.org/10.3390/md15040088
3. Cudennec B., Ravallec-Plé R., Courois E., Fouchereau-Peron M. Peptides from fish and crustacean by-products hydrolysates stimulate cholecystokinin release in STC-1 cells. *Food Chemistry*, 2008, 111(4), 970–975. https://doi.org/10.1016/j.foodchem.2008.05.016
4. Cudennec B., Fouchereau-Peron M., Ferry F., Duclos E., Ravallec R. *In vitro* and *in vivo* evidence for a satiating effect of fish protein hydrolysate obtained from blue whiting (Micromesistius poutassou) muscle. *Journal of Functional Foods*, 2012, 4(1), 271–277. https://doi.org/10.1016/j.jff.2011.12.003
5. Yan J., Zhao J., Yang R., Zhao W. Bioactive peptides with antidiabetic properties: A review. Food Science+. *International Journal of Food Science and Technology*, 2019, 54(6), 1909–1919. https://doi.org/10.1111/ijfs.14090
6. Caron J., Cudennec B., Domenger D., Belguesmia Y., Flahaut C., Kouach M., Lesage J., Goossens J.-F., Dhulster P., Ravallec R. Simulated GI digestion of dietary protein: Release of new bioactive peptides involved in gut hormone secretion. *Food Research International*, 2016, 89(1), 382–390. https://doi.org/10.1016/j.foodres.2016.08.033
7. Cudennec B., Balti R., Ravallec R., Caron J., Bougatef A., Dhulster P., Nedjar N. *In vitro* evidence for gut hormone stimulation release and dipeptidyl-peptidase IV inhibitory activity of protein hydrolysate obtained from cuttlefish (*Sepia officinalis*) viscera. *Food Research International*, 2015, 78, 238–245. https://doi.org/10.1016/j.foodres.2015.10.003
8. Caron J., Domenger D., Belguesmia Y., Kouach M., Lesage J., Goossens J.-F., Dhulster P., Ravallec R., Cudennec B. Protein digestion and energy homeostasis: How generated peptides may impact intestinal hormones? *Food Research International*, 2016, 88(B), 310–318. https://doi.org/10.1016/j.foodres.2015.12.018
9. Theysgeur S., Cudennec B., Deracinois B., Perrin C., Guiller I., Lepoudère A., Flahaut C., Ravallec R. New bioactive peptides identified from a tilapia by-product hydrolysate exerting effects on DPP-IV activity and intestinal hormones regulation after canine gastrointestinal simulated digestion. *Molecules*, 2021, 26(1), 136. https://doi.org/10.3390/molecules26010136
10. Admassu H., Abdalbasit M., Gasmalla A., Yang R., Zhao W. Bioactive peptides derived from seaweed protein and their health benefits: Antihypertensive, antioxidant, and antidiabetic properties. *Journal of Food Science*, 2018, 83(1), 6–16. https://doi.org/10.1111/1750-3841.14011
11. Stöckel-Maschek A., Stiebitz B., Faust J., Born I., Kähne T., Gorrell M.D., Neubert K. Different inhibition mechanisms of dipeptidyl peptidase IV by tryptophan containing peptides and amides. In: Hildebrandt et al. (editors). *Dipeptidyl Aminopeptidases*

in Health and Disease. New York: Kluwer Academic/Plenum Publishers, 2003, pp. 69–72.

12. Suetsuna K., Saito M. Enzyme-decomposed materials of laver and uses thereof. US Patent No. 6217879B1, 2001.
13. Harnedy P., FitzGerald R. *In vitro* assessment of the cardioprotective, antidiabetic and antioxidant potential of Palmaria palmata protein hydrolysates. *Journal of Applied Phycology*, 2013, 25(6), 1793–1803. https://doi.org/10.1007/s10811-013-0017-4
14. Wang R., Zhao H., Pan X., Orfila C., Lu W., Ma Y. Preparation of bioactive peptides with antidiabetic, antihypertensive, and antioxidant activities and identification of α-glucosidase inhibitory peptides from soy protein. *Food Science and Nutrition*, 2019, 7(5), 1848–1856. https://doi.org/10.1002/fsn3.1038
15. Gonzalez-Montoya M., Hernández-Ledesma B., Mora-Escobedo R., Martínez-Villaluenga C. Bioactive peptides from germinated soybean with antidiabetic potential by inhibition of dipeptidyl peptidase-IV, α-amylase, and α-glucosidase enzymes. *International Journal of Molecular Sciences*, 2018, 19(10), 2883. https://doi.org/10.3390/ijms19102883
16. El-Sayed M., Awad S. Milk bioactive peptides: Antioxidant, antimicrobial and antidiabetic activities. *Advances in Biologicalchemistry*, 2019, 7(1), 22–23. https://doi.org/10.11648/j.ab.20190701.15
17. de Campos Zani S.C., Wu J., Chan C.B. Egg and soy-derived peptides and hydrolysates: A review of their physiological actions against diabetes and obesity. *Nutrients*, 2018, 10(5), 549. https://doi.org/10.3390/nu10050549
18. Yu Z., Yin Y., Zhao W., Liu J., Chen F. Antidiabetic activity peptides from albumin against α-glucosidase and α-amylase. *Food Chemistry*, 2012, 135(3), 2078–2085. https://doi.org/10.1016/j.foodchem.2012.06.088
19. Wang J., Du K., Fang L., Liu C., Min W., Liu J. Evaluation of the antidiabetic activity of hydrolyzed peptides derived from *Juglans mandshurica* Maxim. fruits in insulin-resistant HepG2 cells and type 2 diabetic mice. *Journal of Food Biochemistry*, 2018, 42(3), e12518. https://doi.org/10.1111/jfbc.12518
20. Arise R., Yekeen A., Ekun O. *In vitro* antioxidant and *a*-amylase inhibitory properties of watermelon seed protein hydrolysates. *Environmental and Experimental Biology*, 2016, 14(4), 163–172. https://doi.org/10.22364/eeb.14.23
21. Ren Y., Liang K., Jin Y., Zhang M., Chen Y., Wu H., Lai F. Identification and characterization of two novel α-glucosidase inhibitory oligopeptides from hemp (*Cannabis sativa* L.) seed protein. *Journal of Functional Foods*, 2016, 26, 439–450. https://doi.org/10.1016/j.jff.2016.07.024
22. Kehinde B.A., Sharam P. Recently isolated antidiabetic hydrolysates and peptides from multiple food sources: A review. *Critical Reviews in Food Science and Nutrition*, 2020, 60(2), 1–18. https://doi.org/10.1080/10408398.2018.1528206
23. El-Sayed M.I., Awad S., Wahba A., El Attar A., Yousef M.I., Zedan M. *In vivo* antidiabetic and biological activities of milk protein and milk protein hydrolysate. *Advanced Research*, 2016, 4(2), 154. https://doi.org/10.4172/2329-888X.1000154
24. Lacroix I.M.E., Li-Chan E.C.Y. Inhibition of dipeptidyl peptidase (DPP)-IV and a-glucosidase activities by pepsin-treated whey proteins. *Journal of Agricultural and Food Chemistry*, 2013, 61(31), 7500–7506. https://doi.org/10.1021/jf401000s
25. Lacroix I., Li-Chan E. Dipeptidyl peptidase-IV inhibitory activity of dairy protein hydrolysates. *International Dairy Journal*, 2012, 25(2), 97–102. https://doi.org/10.1016/j.idairyj.2012.01.003

26. Tulipano G., Sibilia V., Caroli A.M., Cocchi D. Whey proteins as source of dipeptidyl dipeptidase IV (dipeptidyl peptidase-4) inhibitors. *Peptides*, 2011, 32(4), 835–838. https://doi.org/10.1016/j.peptides.2011.01.002
27. Jan F., Kumar S., Jha R. Effect of boiling on the antidiabetic property of enzyme treated sheep milk casein. *Veterinary World*, 2016, 9(10), 1152–1156. https://doi.org/10.14202/vetworld.2016.1152-1156
28. Uchida M., Ohshiba Y., Mogami O. Novel dipeptidyl peptidase-4-inhibiting peptide derived from b-lactoglobulin. *Journal of Pharmacological Sciences*, 2011, 117(1), 63–66. https://doi.org/10.1254/jphs.11089SC
29. Uenishi H., Kabuki T., Seto Y., Serizawa A., Nakajima H. Isolation and identification of casein-derived dipeptidyl-peptidase 4 (DPP-4)-inhibitory peptide LPQNIPPL from gouda-type cheese and its effect on plasma glucose in rats. *International Dairy Journal*, 2012, 22(1), 24–30. https://doi.org/10.1016/j.idairyj.2011.08.002
30. Nongonierma A.B., FitzGerald R.J. Dipeptidyl peptidase IV inhibitory and antioxidative properties of milk protein derived dipeptides and hydrolysates. *Peptides*, 2013, 39, 157–163. https://doi.org/10.1016/j.peptides.2012.11.016
31. Lacroix I.M., Chen X.M., Kitts D., Li-Chan E.C.Y. Investigation into the bioavailability of milk protein-derived peptides with dipeptidyl-peptidase IV inhibitory activity using caco-2 cell monolayers. *Food and Function*, 2017, 8(2), 701–709. https://doi.org/10.1039/C6FO01411A
32. Silveira S.T., Martínez-Maqueda D., Recio I., Hernández-Ledesma B. Dipeptidyl peptidase-IV inhibitory peptides gener ated by tryptic hydrolysis of a whey protein concentrate rich in b-lactoglobulin. *Food Chemistry*, 2013, 141(2), 1072–1077. https://doi.org/10.1016/j.foodchem.2013.03.056
33. Lacroix I.M.E., Li-Chan E.C.I. Isolation and characterization of peptides with dipeptidyl peptidase-IV inhibitory activity from pepsin-treated bovine whey proteins. *Peptides*, 2014, 54, 39–48. https://doi.org/10.1016/j.peptides.2014.01.002
34. Yu Z., Zhao W., Liu J., Lu J., Chen F. QIGLF, a novel angiotensin I-converting enzyme-inhibitory peptide from egg white protein. *Journal of the Science of Food and Agriculture*, 2011, 91(5), 921–926. https://doi.org/10.1002/jsfa.4266
35. Zambrowicz A., Eckert E., Pokora M., Bobak L., Dąbrowska A., Szołtysik M., Trziszka T. and Chrzanowska J. Antioxidant and antidiabetic activities of peptides isolated from a hydrolysate of an egg-yolk protein by-product prepared with a proteinase from Asian pumpkin (*Cucurbita ficifolia*). *RSC Advances*, 2015, 5(14), 10460–10467. https://doi.org/10.1039/C4RA12943A
36. Zambrowicz A., Pokora M., Setner B., Dąbrowska A., Szołtysik M., Babij K., Szewczuk Z., Trziszka T., Lubec G., Chrzanowska J. Multifunctional peptides derived from an egg yolk protein hydrolysate: Isolation and characterization. *Amino Acids*, 2015, 47(2), 369–380. https://doi.org/10.1007/s00726-014-1869-x
37. Zhu C.F., Li G.Z., Peng H.B., Zhang F., Chen Y., Li Y. Treatment with marine collagen peptides modulates glucose and lipid metabolism in Chinese patients with type 2 diabetes mellitus. *Applied Physiology, Nutrition, and Metabolism*, 2010, 35(6), 797–804. https://doi.org/10.1139/H10-075
38. Wang T.Y., Hsieh C.H., Hung C.C., Jao C.L., Chen M.C., Hsu K.C. Fish skin gelatin hydrolysates as dipeptidyl peptidase IV inhibitors and glucagon-like peptide-1 stimulators improve glycemic control in diabetic rats: A comparison between warm- and cold-water fish. *Journal of Functional Foods*, 2015, 19, 330–340. https://doi. https://doi.org/10.1016/j.jff.2015.09.037

39. Nasri R., Abdelhedi O., Jemil I., Daoued I., Hamden K., Kallel C., Elfeki A., Lamri-Senhadji M., Boualga A., Nasri N. Ameliorating effects of goby fish protein hydrolysates on high-fat-high-fructose diet-induced hyperglycemia; oxidative stress and deterioration of kidney function in rats. *Chemico-Biological Interactions*, 2012, 242, 71–80. https://doi.org/10.1016/j.cbi.2015.08.003
40. Li-Chan E.C.Y., Hunag S.L., Jao C.L., Ho K.P., Hsu K.C. Peptides derived from Atlantic salmon skin gelatin as dipeptidyl-peptidase IV inhibitors. *Journal of Agricultural and Food Chemistry*, 2012, 60(4), 973–978. https://doi.org/10.1021/jf204720q
41. Huang S.L., Jao C.L., Ho K.P., Hsu K.C. Dipeptidyl-peptidase IV inhibitory activity of peptides derived from tuna cooking juice hydrolysates. *Peptides*, 2012, 35(1), 114–121. https://doi.org/10.1016/j.peptides.2012.03.006
42. Hatanaka T., Inoue Y., Arima J., Kumagai Y., Usuki H., Kawakami K., Kimura M., Mukaihara T. Production of dipeptidyl peptidase IV inhibitory peptides from defatted rice bran. *Food Chemistry*, 2012, 134(2), 797–802. https://doi.org/10.1016/j.foodchem.2012.02.183
43. Hatanaka T., Uraji M., Fujita A., Kawakami K. Antioxidation activities of rice-derived peptides and their inhibitory effects on dipeptidylpeptidase-IV. *International Journal of Peptide Research and Therapeutics*, 2015, 21(4), 479–485. https://doi.org/10.1007/s10989-015-9478-4
44. Connolly A., O'Keeffe M., Nongonierma A., Piggott C., FitzGerald R. Isolation of peptides from a novel brewers spent grain protein isolate with potential to modulate glycemic response. *International Journal of Food Science and Technology*, 2017, 52(1), 146–153. https://doi.org/10.1111/ijfs.13260
45. Lin H., Li L., Tian Y., Zhang X., Li B. Protein hydrolysate from brewer's spent grain and its inhibitory ability of a-glucosidase. *Advanced in Materials Research*, 2012, 581–582, 138–141. https://doi.org/10.4028/www.scientific.net/AMR.581-582.138
46. Jorge S.-S., Rau l.R.-B., Isabel G., Edith P., Bernardo E., César A., Gerardo D., Rubén R. Dipeptidyl peptidase IV inhibitory activity of protein hydrolyzates from (*Amaranthus hypochondriacus* L.) grain and their influence on post- prandial glycemia in streptozotocin-induced diabetic mice. *African Journal of Traditional, Complementary and Alternative Medicines*, 2015, 12(1), 90–98. https://doi.org/10.4314/ajtcam.v12i1.13
47. Velarde-Salcedo A.J., Barrera-Pacheco A., Lara-González S., Montero-Morán G.M., Díaz-Gois A., González de Mejia E., Barba de la Rosa A.P. In vitro inhibition of dipeptidyl peptidase IV by peptides derived from the hydrolysis of amaranth (*Amaranthus hypochondriacus* L.) proteins. *Food Chemistry*, 2013, 136(2), 758–764. https://doi.org/10.1016/j.foodchem.2012.08.032
48. Vilcacundo R., Villaluenga M.C., Ledesma H.B. Release of dipeptidyl peptidase IV, a-amylase and a-glucosidase inhibitory peptides from quinoa (*Chenopodium quinoa* willd.) during *in vitro* simulated gastrointestinal digestion. *Journal of Functional Foods*, 2017, 35, 531–539. https://doi.org/10.1016/j.jff.2017.06.024
49. Nongonierma A.B., Le Maux S., Dubrulle C., Barre C., Fitzgerald R.J. Quinoa (*Chenopodium quinoa* willd.) protein hydrolysates with *in vitro* dipeptidyl peptidase IV (DPP-IV) inhibitory and antioxidant properties. *Journal of Cereal Science*, 2015, 65, 11112–11118. https://doi.org/10.1016/j.jcs.2015.07.004
50. Zhang H., Wang J., Liu Y., Sun B. Peptides derived from oats improve insulin sensitivity and lower blood glucose in streptozotocin-induced diabetic mice. *Journal of Biomedical Science*, 2015, 4(1). http://doi.org/10.4172/2254-609x.100007

51. Siow H.L., Gan C.Y. Extraction, identification, and structure–activity relationship of antioxidative and a-amylase inhibitory peptides from cumin seeds (*Cuminum cyminum*). *Journal of Functional Foods*, 2016, 22, 1–12. https://doi.org/10.1016/j.jff.2016.01.011
52. Mojica L., Luna-Vital D.A., de Mejia Gonzalez E. Black bean peptides inhibit glucose uptake in caco-2 adenocarcinoma cells by blocking the expression and translocation pathway of glucose transporters. *Toxicology Reports*, 2018, 5, 552–560. https://doi.org/10.1016/j. toxrep.2018.04.007
53. Rocha T., Hernandez L.M.R., Mojica L., Johnson H.M., Chang Y., Mejía E. Germination of Phaseolus vulgaris and alcalase hydrolysis of its proteins produced bioactive peptides capable of improving markers related to type-2 diabetes *in vitro*. *Food Research International*, 2015, 76, 150–159. http://doi.org/10.1016/j.foodres.2015.04.041
54. Oseguera-Toledo M.E., de Mejia E.G., Amaya S.L. Hard-to-cook bean (*Phaseolus vulgaris* L.) proteins hydrolyzed by alcalase and bromelain produced bioactive peptide fractions that inhibit tar- gets of type-2 diabetes and oxidative stress. *Food Research International*, 2015, 76(3), 839–851. https://doi.org/10.1016/j.foodres.2015.07.046
55. Oseguera-Toledo M.E., de Mejia G.E., Sivaguru M., Llano A.S. Common bean (*Phaseolus vulgaris* L.) protein-derived peptides increased insulin secretion, inhibited lipid accumulation, increased glucose uptake and reduced the phosphatase and tensin homologue activation *in vitro*. *Journal of Functional Foods*, 2016, 27, 160–177. https:. https://doi.org/10.1016/j.jff.2016.09.001
56. Mojica L., de Mej 1a E.G. Optimization of enzymatic production of antidiabetic peptides from black bean (*Phaseolus vulgaris* L.) proteins, their characterization and biological potential. *Food and Function*, 2016, 7(2), 713–727. https://doi.org/10.1039/c5fo01204j
57. Lu J., Zeng Y., Hou W., Zhang S., Li L., Luo X., Xi W., Chen Z., Xiang M. The soybean peptide aglycin regulates glucose homeostasis in type 2 diabetic mice via IR/IRS1 pathway. *The Journal of Nutritional Biochemistry*, 2012, 23(11), 1449–1457. https://doi.org/ 10.1016/j.jnutbio.2011.09.007
58. Yang H., Kwon D., Kim M., Kang S., Park S. Meju, unsalted soybeans fermented with *Bacillus subtilis* and *aspergilus oryzae*, potentiates insulinotropic actions and improves hepatic insulin sensitivity in diabetic rats. *Nutrition and Metabolism*, 2012, 9(1), 37. https://doi.org/10.1186/1743-7075-9-37
59. Ademiluyi A.O., Oboh G., Boligon A.A., Athayde M.L. Effect of fermented soybean condiment supplemented diet on a-amylase and a-glucosidase activities in streptozotocin-induced diabetic rats. *Journal of Functional Foods*, 2014, 9, 1–9. https://doi.org/10.1016/j.jff.2014.04.003
60. Tu M., Liu H., Cheng S., Mao F., Chen H., Fan F., Lu W., Du M. Identification and characterization of a novel casein anticoagulant peptide derived from *in vivo* digestion. *Food and Function*, 2019, 10(5), 2552–2559. https://doi.org/10.1039/C8FO02546K
61. Liu H., Tu M., Cheng S., Chen H., Wang Z., Du M. An anticoagulant peptide from beta-casein: Identification, structure and molecular mechanism. *Food and Function*, 2019, 10(2), 886–892. https://doi.org/10.1039/C8FO02235F
62. Qiao M., Tu M., Chen H., Mao F., Yu C., Du M. Identification and *in silico* prediction of anticoagulant peptides from the enzymatic hydrolysates of Mytilus edulis proteins. *International Journal of Molecular Sciences*, 2018, 19(7), 2100. https://doi.org/10.3390/ijms19072100

63. Feng L., Tu M., Qiao M., Fan F., Chen H., Song W., Du M. Thrombin inhibitory peptides derived from *Mytilus edulis* proteins: Identification, molecular docking and *in silico* prediction of toxicity. *European Food Research and Technology*, 2018, 244(2), 207–217. https://doi.org/10.1007/s00217-017-2946-7
64. Cheng S., Tu M., Chen H., Xu Z., Wang Z., Liu H., Zhao G., Zhu B., Du M. Identification and inhibitory activity against α-thrombin of a novel anticoagulant peptide derived from oyster (*Crassostrea gigas*) protein. *Food and Function*, 2018, 9(12), 6391–6400. https://doi.org/10.1039/C8FO01635F
65. Cheng S., Tu M., Liu H., An Y., Du M., Zhu B. A novel heptapeptide derived from *Crassostrea gigas* shows anticoagulant activity by targeting for thrombin active domain. *Food Chemistry*, 2021, 334, 127507. https://doi.org/10.1016/j.foodchem.2020.127507
66. Kênia Alencar Bezerra T., Thalles J., Gomes de Lacerda J., Ramos Salu B., Vilela Oliva M.L., Juliano M.A., Bertoldo Pacheco M.T., Sueky Madruga M. Identification of angiotensin I-converting enzyme-inhibitory and anticoagulant peptides from enzymatic hydrolysates of chicken combs and wattles. *Journal of Medicinal Food*, 2019, 22(12), 1294–1300. https://doi.org/10.1089/jmf.2019.0066
67. Syed A.A., Mehta A. Target specific anticoagulant peptides: A review. *International Journal of Peptide Research and Therapeutics*, 2018, 24(1), 1–12. https://doi.org/10.1007/s10989-018-9682-0
68. Cao Y., Miao J.-Y., Liu G., Luo Z., Xia Z., Liu F., Yao M., Cao X., Sun S., Lin Y., Lan Y., Xiao H. Bioactive peptides isolated from casein phosphopeptides enhance calcium and magnesium uptake in Caco-2 cell monolayers. *Journal of Agriculture and Food Chemistry*, 2017, 65(11), 2307–2314. https://doi.org/10.1021/acs.jafc.6b05711
69. Liao W., Liu S., Liu X., Duan S., Xiao S., Yang Z., Cao Y., Miao J. The purification, identification and bioactivity study of a novel calcium-binding peptide from casein hydrolysate. *Food and Function*, 2019, 10(12), 7724–7732. https://doi.org/10.1039/C9FO01383K
70. Bhandar D., Rafiq S., Gat Y., Gat P., Waghmare R., Kumar V.A. Review on bioactive peptides: Physiological functions, bioavailability and safety. *International Journal of Peptide Research and Therapeutics*, 2020, 26(1), 139–150. https://doi.org/10.1007/s10989-019-09823-5
71. Reynolds E.C. U.S. Patent No. 6,780,844. U.S. Patent and Trademark Office, Washington, DC, 2004.
72. Meisel H., FitzGerald R.J. Biofunctional peptides from milk proteins: Mineral binding and Cytomodulatory effects. *Current Pharmaceutical Design*, 2003, 9(16), 1289–1295. https://doi.org/10.2174/1381612033454847
73. Liu Y.-F., Oey I., Bremer P., Carne A., Silcock P. Bioactive peptides derived from egg proteins: A review. *Critical Reviews in Food Science and Nutrition*, 2018, 58(15), 2508–2530. https://doi.org/10.1080/10408398.2017.1329704
74. Meshginfar N., Mahoonak A.S., Hosseinian F., Tsopmo A. Physicochemical, antioxidant, calcium binding, and angiotensin converting enzyme inhibitory properties of hydrolyzed tomato seed proteins. *Journal of Food Biochemistry*, 2019, 43(2), e12721. https://doi.org/10.1111/jfbc.12721

SECTION 7

Regulatory Status of Bioactive Peptides

CHAPTER 25

Regulatory Status of Bioactive Peptides

Faraat Ali and Javed Ahmad

CONTENTS

Introduction	509
Approved Bioactive Peptides in the Market	510
Significance of Bioactive Peptides in the Treatment of Diseases	511
Global Market of Bioactive Peptides	511
Regulatory Perspectives of Bioactive Peptides	513
Challenges to Market Authorization and Commercialization	515
Concluding Remarks	516
References	516

INTRODUCTION

Bioactive peptides (BP) are recognized as an important class of molecules, occupying a space between traditional organic compounds and high-molecular-weight biopharmaceuticals. Peptides are polymers of amino acids monomers with fewer than 50 residues that are connected by a peptide (amide, -CONH-) link and have no three-dimensional structure [1–3]. Since proteins are made up of one or more polypeptide chains linked to coenzymes and cofactors or other macromolecules, peptides are differentiated from proteins by their scale. Over 7000 naturally occurring peptides have been discovered, all of which play a role in several physiological processes as hormones, neurotransmitters, growth factors, ion channel ligands, and antimicrobials. BP has shown promise in the treatment of a variety of diseases, namely autoimmune diseases, hormone disorders, infections, and metabolic diseases, such as diabetes of different origins [4–9].

The BP sector has evolved at an unprecedented rate over the last decade, with the therapeutic peptide demand expected to hit $25 billion by 2018. Synthetic drug administration has become less appealing due to significant problems with toxicity, low response rates, and the rise of resistance among many patients. More specifically, an unhealthy diet is one of the four major lifestyle risk factors for noncommunicable diseases (NCDs), and a global policy that promotes a balanced diet has become a key component of the World Health Organization's action plan for NCD prevention and control [10]. As a result, the production of functional foods with unique health benefits is increasing. Because of their small scale, ease of processing, exquisite efficacy, and specificity, BP have proved to be especially useful tools for inhibiting protein–NC interactions. Essential BP have been discovered in both dietary and nondietary sources over the years. The quest for BP in foods has largely relied on empirical and bioinformatic approaches, or a combined approach to obtain *in vitro* results. The peptides are inactive as long as they are in their parent protein structure, but they become active when enzymes or microbial fermentation cleave them intact [11]. Screening wide phase validated libraries for novel BP that bind to targets of interest and have desirable biological properties is another method of bioactive peptide discovery. These methods produce peptides that are used to grow therapeutic molecules. The schematic workflow from peptide identification to market authorization has been represented in Figure 25.1 [12].

An overview of the BP landscape is given here, including historical perspectives, regulatory status, challenges to market authorization and commercialization, and benchmarks are discussed.

APPROVED BIOACTIVE PEPTIDES IN THE MARKET

Nature remains the most important source of BP since several biologically active peptides can be found in almost all living organisms. Recombinant libraries, as well as physicochemical libraries, are other sources of BP [13]. The existence of a robust product pipeline has the ability for substantial prospect development [14–16]. BP granted approval by the

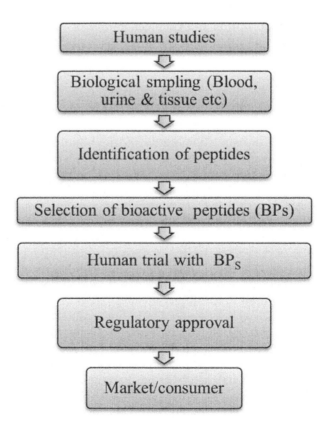

FIGURE 25.1 Workflow for dietary peptide identification and validation to market authorization.

U.S. Food and Drug Administration (FDA) from 2018 to March 2021 is represented in Table 25.1 [17–19].

SIGNIFICANCE OF BIOACTIVE PEPTIDES IN THE TREATMENT OF DISEASES

The latest progress in bioactive peptides is a promising and increasing research area. This isn't too shocking, since amino-acid residues are responsible for the control and direction of all aspects of cell processes and organize most intercellular communications [20]. After all, the data on bioactive peptides currently available have many limitations.

To begin with, further research into the conceptual pathways that control the presumed activities of individual peptides is required. Most peptides tend to have several modes of action, suggesting that they have cytoprotective properties. Additionally, it is often difficult to relate a particular peptide's mechanism of action to a complex of polypeptides. Individual parts of diverse peptide-rich protein hydrolysate must be categorized to learn something about their actions, specific receptors, and activation signaling pathways in facilitating some of their potential benefits. After that, extraction techniques should be standardized. *In vitro* fractionation, which incorporates analytical separation of protein-digested fractions with *in vivo* identification of various biological processes to classify the receptor responsible for the effect, may be used to find a solution [21].

While several *in vivo* studies have demonstrated the efficacy of bioactive peptides, most of them are still in the early stages, especially in terms of clinical evidence. Antihypertensive peptides were shown to amplify the trigger wave velocity in individuals, which is a significant predictive factor for cardiovascular events. [22, 23]. Even if the protein hydrolysates appear to be healthy, the presence of immunogenic proteins and peptides within them can cause or intensify allergic reactions [24].

Anti-inflammatory peptides have also been shown to minimize infection variables in humans, but the evidence is limited and their safety profile remains unclear. Many of the peptides have been examined for their immunomodulatory, antioxidant, analgesic, and probably microbial activity. Furthermore, the effects of a number of these peptides have been observed in animal studies and on a variety of cancer cell lines (but not non-tumoral ones). Nevertheless, only phase I trials have been conducted, and further investigation is necessary to validate the effectiveness, safety, and immunogenicity in humans.

GLOBAL MARKET OF BIOACTIVE PEPTIDES

Over the last few years, the global demand for BP has grown in both size and economic value. The broad applications of BP in metabolic and autoimmune disorders are major factors driving their development [25–27]. The global BP market was estimated at US$21.5 billion in 2016 and is projected to expand at a CAGR (compound annual growth rate) of 9.4 % by 2025 [28]. The existence of a large product pipeline opens the door to major future expansion [29]. Over the last three years, the FDA has approved nearly 117 new medicines and therapeutic biological products. As a result of the significant paradigm change toward peptide production and their extension into a range of therapeutic categories, quality evaluation of the new BP entering the market is critical to ensure both safety and effectiveness [30–32].

TABLE 25.1 Bioactive Peptides Granted Approval by the US-FDA from 2018 to March 2021 [17–19]

S. No.	BPs Name (Brand name)	Active Molecule Name	Indication	Characteristics	Year	Route of Administration
1.	Lutathera	Lutetium Lu 177-dotatate	Theranostics, Neuroendocrine tumors/Pancreatic cancer	Lu 177 chelation by Dotatate and bind to Tyrosine 3-octreotate	2018	Intravenous
2.	68 Gallium DOTA-TOC	68Ga DOTA-TOC	Neuroendocrine tumors, diagnostic	68Ga chelated by DOTA bound to Tyr3-octreotide	2019	Intravenous
3.	Scenesse®	Afamelanotide	Skin damage and pain	13 aa lineal peptide analogs of α-MSH	2019	Subcutaneous Implant
4.	Vyleesi	Bremelanotide	Women hypoactive sexual desire	7 aa cyclic peptide analogs of α-MSH	2019	Subcutaneous
5.	PADCEV	Enfortumab Vedotin-Ejfv	Cancers Nectin-4	ADC with a synthetic analog of marine natural peptide dolastatin 10	2019	Intravenous infusion
6.	Polivy	Polatuzumab Vedotin-Piiq	B-cell lymphoma diffused	ADC with a synthetic analog of dolastatin 10	2019	Intravenous infusion
7.	Tepezza	Teprotumumab-trbw	To treat thyroid eye disease	type I insulin-like growth factor receptor inhibitor	2020	Intravenous infusion
8.	Vyepti	Eptinezumab jjmr	Treatment of migraine in seniors	CGRP Antibody	2020	Intravenous
9.	Evkeeza	evinacumab-dgnb	For the treatment of homozygous familial hypercholesterolemia	blockade of ANGPTL3 lowers TG and HDL-C by rescuing LPL and EL activities, respectively.	2021	Intravenous infusion

REGULATORY PERSPECTIVES OF BIOACTIVE PEPTIDES

FDA is a regulatory scientific body that considers polymers as peptides consisting of 40 or fewer amino acids. Peptides come in the category of an important class of drugs. So, the FDA eases the evaluations of these medicinal products and their therapeutic equivalents. A peptide manufacturer faces lots of challenges during chemical synthesis or recombinant DNA technology. It is also difficult to produce a generic peptide drug that is the equivalent to a brand name. Nowadays, 100 peptide drugs are marketed in Europe, the US, and Japan, with $15 billion to $20 billion annual global sales of the same [33].

BP have specific amino-acid sequences and as functional foods and nutraceuticals have a vital role as pharmaceuticals [34]. These products are intended to supplement the basic diet and contain the ingredient in the form of amino acids. In India, the Food Safety and Standard Act 2005 (FSSAI) drafts the rules and regulations concerning the quality, efficacy, and safety of these products. In association with the FDA, the PFA (Prevention of Food Adulteration Act, also India) controls food and drug adulteration. Other regulatory authorities of India that provide guidelines regarding the requirements of safety and labeling of BP are ICMR (Indian Council of Medical Research) and DBT (Department of Biotechnology) [35]. They are known as generally recognized as safe (GRAS) because of their pharmaceutical importance and widespread human consumption as a dietary supplement [36].

In India, traditional Indian Ayurvedic Medicines (IAM) are the common forms of functional foods and nutraceuticals. Most of these products are directly available to the consumer without any prescription. India is the major contributor to the international functional food and nutraceutical market [37, 38].

In Australia, National Food Authority (NFA) reported that 32% of the people purchase products based on nutritive value rather than their safety assurance. Food Standards Code (FSC) drafts general regulatory guidelines for bioactive compounds and other commodity-specific labeling, e.g., dietary products. Food Standards Australia and New Zealand (FSANZ) regulate standards for food production, labeling, and advertising in Australia and New Zealand [39, 40]. Apart from FSC, there is a Code of Practice on Nutrient Claims (COPONC) developed by the NFA to ensure the right availability of the product to the public and accuracy about the labeling of the product [41].

In Japan, the International Union of Food Science and Technology holds the scientific symposium with scientific regulatory bodies, i.e., the Ministry of Health and Welfare, the Ministry of Agriculture, Forestry and Fisheries, and the Japan section of the international life science institute (ILSI) [42]. More than 60 BP have been approved as pharmaceuticals in Japan. However, over 260 have been tested in human clinical trials and approximately 150 are in active clinical trials [43]. Food for Specified Health Uses (FOSHU) is the authorized policy established in 1993 by the Ministry of Health and Welfare (MHW) to permit bioactive products (functional food) legally. In Japan, all bioactive compounds are known as functional foods instead of nutraceuticals or food supplements. In 2001, other regulatory systems controlled by the Japanese government are "foods with health claims" (FHC) and "foods with nutrient function claims" (FNFC) to avoid false or misleading label claims [44].

FOSHU marketed products led to saturation after 2007, as the approval is not meant for product sale. However, another regulatory system was established in 2015 as "Foods with function claims" (here called New Functional Foods) [45]. As of now, more than 1000 FOSHU products are available in the market, and their health claims are approved by the Japanese Ministry of Health and Welfare. In the near future, FOSHU categories containing anti-fatigue or skin health-promoting foods are expected to be in high demand by Japanese consumers [46, 47].

As of 2017, approximately 68 BP have been approved in the United States and Europe. In total 484 peptides are reported in the dataset. More than 155 peptides are under active clinical studies [43]. BP are considered novel foods (EU reg. 2015/2283). European Food Safety Authority (EFSA) set up the guidance manual on the preparation and use of novel food under the EFSA NDA Panel, 2016. Moreover, EU reg. 2017/2469 is the regulation directive implemented by the Commission. [48].

European regulations for novel foods and bioactive compounds are summarized here [49, 50]:

- EC Regulation on Food Safety (EC 178/2002)
- EC Regulation on Genetically Modified Food and Feed (EC 1829/2003)
- EC Regulation on Novel Foods and Novel Food Ingredients (EC 258/97)
- EC Regulation on Nutrition and Health Claims made on foods (EC 1924/2006)
- EC Regulation on traceability and labeling of food produced from genetically modified organisms (EC 1830/2003)

A wide variety of sausages are produced in the different regions of Europe. These sausages have importance as they produce free amino acids and BP on proteolysis. On 19th January 2007, the regulation on nutrition and health claims was made under EFSA. However, any claim relating to the effect of BP in foods has not yet been authorized by the commission [51, 52].

The Canadian Food and Drugs Act and Regulations come into force in 1953. The Food Directorate of the Health Protection Branch of Health Canada (1996) defined nutraceuticals as medicinal products, as they are not associated with food. Health Canada released a proposed framework in October 2001 to allow product-specific health claims for foods. An authorized claim identification number (CIN) is mentioned on the product label [53].

BP can be known as functional food or nutraceuticals, as they elicit health benefits and act as additives in the development of functional products. In Canada, the Food and Drug Regulations (FDR) Act authorized the nutrient content and health claims of various active compounds. The Canadian Food Inspection Agency and Health Canada (2009) control the disease risk reduction claims relative to the consumption of food. Natural Health Product Directorate (NHPD) was established in 2004 by Health Canada. NHPs are natural health products (like vitamins, minerals, and peptides) that are regulated by Health Canada, which ensures their safety, effectiveness and quality. Separately, NPHD is associated with the evaluation of the health claim of NHP and their related products [54, 55].

As per a Statistics Canada survey conducted in 2002, the major portion of Canadian functional food and nutraceuticals are exported to Japan, the U.S., and the EU. Canadian companies produce and distribute a wide range of nutraceutical and functional food products globally. According to Agriculture and Agri-Food Canada (AAFC, 2006), nutraceutical and functional foods have great potential for a good health care system [37].

In the United States, health-related claims of a bioactive product are regulated under the Federal Drug and Cosmetic Act (FDCA) 1938, Nutrition Labeling and Education Act (NLEA) 1990, the Dietary Supplement Health and Education Act (DSHEA) 1994, and the Food and Drug Administration Modernization Act (FDAMA) 1997. NLEA controls the health claims of food and dietary supplements to reduce the risk of various unwanted diseases or other harms. DSHEA is the scientific authority that has a vital role in the regulation of claims of all dietary supplements required for the specific effect on individual well-being. However, it does not control therapeutic or specific disease prevention claims. According to DSHEA a dietary supplement is "a product (other than tobacco) that contains dietary ingredients such as amino acids, vitamins, minerals or herbs intended to supplement the diet" [56]. The Center for Food Safety and Applied Nutrition (CFSAN) is an agency (under the FDA) that is responsible for the regulation of dietary supplements. Manufacturers must notify CFSAN not later than 30 days after a product is first marketed about any health claims [57].

The FDA recommendeds a daily intake of dietary supplements to overcome risks associated with cardiovascular diseases [58]. THPDB (https://webs.iiitd.edu.in/raghava/thpdb/keyword.php) is a database of the FDA-approved therapeutic peptides. Two other significant databases, i.e., ChEMBL and Drug Bank also provide information on FDA-approved protein and peptide therapeutics. [59, 60].

BP are known as health foods in China. The China Food and Drug Administration (CFDA) is the regulatory body that governs the registration and assessment of healthy foods. As per the system, health food is defined as any food product claiming specific health functions, but not intended as drugs for disease treatment. Only CFDA-approved institutions are permitted for the testing of BP. [61].

The CFDA recognized Health Food Expert Committee examines the safety evaluation and approves the applications based on toxicology studies, stability studies, functionality, and detailed inspection of the manufacturing process. Hence, after the approval of the Committee, these BP are permitted to the market by CFDA [62]. The bioactivity assessment of peptides from traditional Chinese medicine and traditional Chinese food benefits the pharmaceutical and food industry to a great extent [63].

The guideline for test procedures and acceptance criteria for the biological and biotechnological product is described under ICH (International Council for Harmonisation) Q6B. This is a set of specifications for proteins and polypeptides. Various other regulatory guidelines for the manufacture of drug products and drug substance of therapeutics have been explained thereunder. Moreover, ICH also guides regulatory agencies and industries regarding regulatory expectations on the development, implementation, and assessment of various biological products [64]. ICH also authorized other technical requirements for pharmaceuticals for human use [65]. As per the ICH Q3A guideline, related substances present in the drug product are to be reported as impurities while the ICH guideline Q6A recommends a specific stability-indicating procedure to determine the content of the drug substance [66].

CHALLENGES TO MARKET AUTHORIZATION AND COMMERCIALIZATION

Therapeutic peptides are processed peptide-based medicines used in pharmacological treatments, while hydrolyzed food proteins (fermented and enzyme-treated) may be used as functional foods in nonpharmacological treatments. Before the goods can be commercialized, they must overcome several hurdles in both cases.

Although certain functional commercial food risk factors may contribute to the denaturation of protein, most research may not evaluate the implications of storage on predicted peptide profiling or the probability of a change in the bioactive components of the peptides after hydrogenation. Non-reproducible peptide patterns can be produced because of processing conditions, such as temperature, fermentation time, or substrate hydrolysis, particularly when the substrate contains a mixture of proteins [67]. After being decrypted by heat, the antibacterial activity of α-lactalbumin [68] and lysozyme [69] increased. As a result, to produce foods with unique health effects, the optimal conditions under which functional foods could be produced to have full health benefits must be researched and implemented.

The amount of protein hydrolysis that occurs during the processing of BP in foods is crucial in the production of functional products. However, since peptide materials serve as substances for further hydrolysis, kinetic modeling of protein hydrolysis using enzymes or microbial fermentation is difficult. As some studies have shown, multiple sequential hydrolyses can lead to peptides with reduced or missing activities because of degradation [70].

Food-derived BP are usually not independent peptides of high purity, which is partially due to the associated low cost and limited yield [71]. Unique peptide substances, on the other hand, can lose their possible greater therapeutic effects with other food components such as polyphenols or other peptides after purification, lowering their potency.

BP led to an increased risk of becoming toxic, allergenic, or carcinogenic compounds in the food mix as they combine with other essential nutrients, such as carbohydrates or fats. When ingested, the peptides can bring about undesirable side reactions that cause unexpected physiological effects. Since *in vitro* findings do not always match *in vivo* results, further *in vivo* studies on the potency and safety of bioactive peptides are needed before they are commercialized. Consumer acceptance of BP in foods is influenced by their sensory properties. The bitter taste of some BP, for example, makes some food products less appealing to consumers. To address this problem, it is critical that genomics modeling be used in BP research.

CONCLUDING REMARKS

BP have a lot of potential as useful functional ingredients in balanced diets to combat the global epidemic of non-communicable diseases. The performance is attributed to the peptides' specificity, potency, and safety. Microbial fermentation will continue to be a useful approach for producing a diverse range of BP in foods, as microbial proteolytic systems (particularly lactic acid bacteria) produce many peptides with various potentials throughout fermentation. As a result, new strains must be characterized genomically and proteomically to predict their protease profiles and to promote functional foods advancement. Computational biology, in combination with appropriate high-throughput peptide screen technology, will continue to be important tools in the quest for BP. In the near future, modular therapeutic peptide biology combined with cost-effective scalable innovations will enable the development of potentially effective multipurpose therapeutic peptides with improved pharmacokinetics, enhanced specific targeting, and lower costs.

REFERENCES

1. Vlieghe, P., Lisowski, V., Martinez, J. and Khrestchatisky, M. 2010. Synthetic therapeutic peptides: Science and market. *Drug Discov Today* 15(1–2): 40–56.

2. Ladner, R.C., Sato, A.K., Gorzelany, J. and de Souza, M. 2004. Phage display-derived peptides as therapeutic alternatives to antibodies. *Drug Discov Today* 9(12): 525–529.
3. McGregor, D.P. 2008. Discovering and improving novel peptide therapeutics. *Curr Opin Pharmacol* 8(5): 616–619.
4. Fosgerau, K. and Hoffmann, T. 2015. Peptide therapeutics: Current status and future directions. *Drug Discov Today* 20(1): 122–128.
5. Padhi, A., Sengupta, M., Sengupta, S., et al. 2014. Antimicrobial peptides and proteins in mycobacterial therapy: Current status and future prospects. *Tuberculosis (Edinb)* 94(4): 363–373.
6. Buchwald, H., Dorman, R.B., Rasmus, N.F., et al. 2014. Effects on GLP-1, PYY, and leptin by direct stimulation of terminal ileum and cecum in humans: Implications for ileal transposition. *Surge Obes Relat Dis* 10(5): 780–786.
7. Giordano, C., Marchio, M., Timofeeva, E. and Biagini, G. 2014. Neuroactive peptides as putative mediators of antiepileptic ketogenic diets. *Front Neurol* 5: 63.
8. Souery, W.N. and Bishop, C.J. 2018. Clinically advancing and promising polymer-based therapeutics. *Acta Biomater* 67: 1–20.
9. Marx, V. 2005. Watching peptide drugs grow up. *Chem Eng News* 83: 17–24.
10. Mendis, S., Davis, S. and Norrving, B. 2015. Organizational update: The world health organization global status report on noncommunicable diseases 2014; one more landmark step in the combat against stroke and vascular disease. *Stroke* 46(5): e121–122.
11. Manzanares, P., Salom, J.B., Garcia-Tejedor, A., et al. 2015. Unraveling the mechanisms of action of lactoferrin-derived antihypertensive peptides: ACE inhibition and beyond. *Food Funct* 6(8): 2440–2452.
12. Mesaik, M.A., Dastagir, N., Uddin, N., et al. 2015. Characterization of immunomodulatory activities of honey glycoproteins and glycopeptides. *J Agric Food Chem* 63(1): 177–184.
13. Uhlig, T., Kyprianou, T., Martinelli, F.G., et al. 2014. The emergence of peptides in the pharmaceutical business: From exploration to exploitation. *EuPA Open Proteom* 4: 58–69.
14. Albericio, F. and Kruger, H.G. 2012. Therapeutic peptides. *Future Med Chem* 4(12): 1527–1531.
15. Dorpe, S.V., Verbeken, M., Wynendaele, E., et al. 2011. Purity profiling of peptide drugs. *J Bioanal Biomed* 6(003). doi:10.4172/1948-593X.S6-003
16. Wu, L.C., Chen, F., Lee, S.L., et al. 2017. Building parity between brand and generic peptide products: Regulatory and scientific considerations for quality of synthetic peptides. *Int J Pharm* 518(1–2): 320–334.
17. Torre, D.L., Beatriz, G., Albericio, F., et al. 2020. Peptide therapeutics 2.0. *Molecules* 25(10): 2293. https://doi.org/10.3390/molecules25102293.
18. https://www.fda.gov/drugs/new-drugs-fda-cders-new-molecular-entities-and-new-therapeutic-biological-products/novel-drug-approvals-2020.
19. https://www.fda.gov/drugs/new-drugs-fda-cders-new-molecular-entities-and-new-therapeutic-biological-products/novel-drug-approvals-2021.
20. Craik, D.J., Fairlie, D.P., Liras, S. and Price, D. 2013. The future of peptide-based drugs. *Chem Biol Drug Des* 81(1): 136–147.
21. Sato, K., Egashira, Y., Ono, S., Mochizuki, S., Shimmura, Y., Suzuki, Y., et al. 2013. Identification of a hepatoprotective peptide in wheat gluten hydrolysate against D-galactosamine-induced acute hepatitis in rats. *J Agric Food Chem* 61(26): 6304–6310.

22. Cicero, A.F., Rosticci, M., Gerocarni, B., Bacchelli, S., Veronesi, M., Strocchi, E., et al. 2011. Lactotripeptides effect on office and 24-h ambulatory blood pressure, blood pressure stress response, pulse wave velocity and cardiac output in patients with high-normal blood pressure or first-degree hypertension: A randomized double-blind clinical trial. *Hypertens Res* 34(9): 1035–1040.
23. Cicero, A.F., Colletti, A., Rosticci, M., Cagnati, M., Urso, R., Giovannini, M., et al. 2016. Effect of lactotripeptides (isoleucine-proline-proline/valine-proline-proline) on blood pressure and arterial stiffness changes in subjects with suboptimal blood pressure control and metabolic syndrome: A double-blind, randomized, crossover clinical trial. *Metab Syndr Relat Disord* 14(3): 161–166.
24. Franck, P., Moneret Vautrin, D.A., Dousset, B., Kanny, G., Nabet, P., Guénard-Bilbaut, L., et al. 2002. The allergenicity of soybean-based products is modified by food technologies. *Int Arch Allergy Immunol* 128(3): 212–219.
25. Lau, J.L. and Dunn, M.K. 2017. Therapeutic peptides: Historical perspectives, current development trends, and future directions. *Bioorg Med Chem* 26(10): 2700–2707.
26. Craik, D.J., Fairlie, D.P., Liras, S. and Price, D. 2013. The future of peptide-based drugs. *Chem Biol Drug Des* 81(1): 136–147.
27. Muheem, A., Shakeel, F., Jahangir, M.A., et al. 2016. A review on the strategies for oral delivery of proteins and peptides and their clinical perspectives. *Saudi Pharm J* 24(4): 413–428.
28. Semalty, A., Semalty, M., Singh, R., et al. 2007. Properties and formulation of oral drug delivery systems of protein and peptides. *Indian J Pharm Sci* 69(6): 741–747.
29. Currier, J.R., Galley, L.M., Wenschuh, H., et al. 2008. Peptide impurities in commercial synthetic peptides and their implications for vaccine trial assessment. *Clin Vaccin Immunol* 15(2): 267–276.
30. Anon. 2017. Peptide therapeutics market by application (cancer, cardiovascular disorder, metabolic disorder, respiratory disorder, pain, dermatology), by type (generic, innovative) by type of manufacturers (in-house, outsourced), and segment forecasts, 2018–2025. *Grand View Research*. https://www.grandviewresearch.com/industry-analysis/peptide-therapeutics-market#.
31. Mullard, A. 2018. 2017 FDA drug approvals. *Nat Rev Drug Discov* 17(2): 81–85.
32. FDA. 2020. *FDA Approved Drug Products*. FDA.
33. https://www.fda.gov/drugs/regulatory-science-action/impact-story-developing-tools evaluate-complex-drug-products-peptides.
34. Patil, P., Wadehra, A., Garg, V., et al. 2015. Biofunctional properties of milk protein derived bioactive peptides-A review. *Asian J Dairy Food Res* 34(4): 253–258.
35. Baldi, A., Sharma, S. and Arora, M. 2013. Comparative insight of regulatory guidelines for probiotics in USA, India and Malaysia: A critical review. *Int J Biotechnol Wellness Ind* 2: 51–64.
36. Shah, N., Patel, A. and Shah, D. 2017. *Bioactive Components Produced by Food Grade Bacteria-An Overview*. In the Souvenir of National Seminar Dynamism in Dairy Industry and Consumer demands organized by SMC College of Dairy Science in association with Indian dairy association at AAU, Anand, Gujarat, India on Feb 4–5, 2017.
37. Acharya, S.N., Basu, S.K. and Thomas, J.E. 2007. Prospects for growth in global nutraceutical and functional food markets: A Canadian perspective. *Aust J Basic Appl Sci* 1: 637–649.
38. Patil, S.B., Jayaprakasha, G.K., Chidambara, K.N. and Vikram, A. 2009. Bioactive compounds: Historical perspectives, opportunities, and challenges. *J Agric Food Chem* 57(18): 8142–8160.

39. Williams, P.G., Yeatman, H., Zakrzewski, S., et al. 2003. Nutrition and related claims used on packaged Australian foods implications for regulation. *Asia Pac J Clin Nutr* 12(2): 138–150.
40. Korhonen, H. 2009. Bioactive milk proteins and peptides: From science to functional applications. *J Funct Foods* 1(2): 177–187.
41. Gianfranceschi, G.L., Gianfranceschi, G., Quassinti, L. and Bramucci, M. 2018. Biochemical requirements of bioactive peptides for nutraceutical efficacy. *J Funct Foods* 47: 252–263.
42. Arai, S., Osawa, T., Ohigashi, H., et al. 2001. A mainstay of functional food science in Japan-history, present status and future outlook. *Biosci Biotehnol Biochem* 65(1): 1–13.
43. Dunn, M.K. and Lau, J.L. 2018. Therapeutic peptides: Historical perspectives, current development trends, and future directions. *Bioorg Med Chem* 26(10): 2700–2707.
44. Roupas, P. and Williams, P. 2007. Regulatory aspects of bioactive dairy ingredients. *Bull Int Dairy Fed* 413: 16–26.
45. Yamamoto, N. and Iwatani, S. 2019. Functional food products in Japan: A review. *Food Sci Hum Well* 8: 96–101.
46. Gevaert, B., Veryser, L., Verbeke, F., et al. 2016. Fish hydrolysates: A regulatory perspective of bioactive peptides. *Protein Pept Lett* 23(12): 1052–1060.
47. Lafarga, T. and Hayes, M. 2016. Bioactive protein hydrolysates in the functional food ingredient industry: Overcoming current challenges. *Food Rev Int* 33: 559–583.
48. Turck, D., Bresson, J.-L., Burlingame, B., et al. 2018. Safety of shrimp peptide concentrates as a novel food pursuant to regulation (EU) 2015/2283. *EFSA J* 16: e05267.
49. Enzing, C., Ploeg, M., Barbosa, M., et al. 2014. *Microalgae-Based Products for the Food and Feed Sector: An Outlook for Europe*. JRC Scientific and Policy Report.
50. Caradonia, F., Battaglia, V., Righi, L., et al. 2019. Plant biostimulant regulatory framework: Prospects in Europe and current situation at international level. *J Plant Growth Regul* 38(2): 438–448.
51. Gallego, M., Mora, L., Escudero, E. and Toldrá, F. 2018. Bioactive peptides and free amino acids profiles in different types of European dry-fermented sausages. *Int J Food Microbiol* 276: 71–78.
52. Sieber, R. and Walther, B. 2011. Bioactive proteins and peptides in foods. *Int J Vitam Nutr Res* 81(2–3): 181–191.
53. Fitzpatrick, K.C. 2004. Regulatory issues related to functional foods and natural healthproducts in Canada: Possible implications for manufacturers of conjugated linoleic acid[1-3]. *Am J Clin Nutr* 79(6) supplement: 1217S–1220S.
54. Malla, S., Hobbs, J.E. and Kofi Sogah, E. 2013. *Functional Foods and Natural Health Products Regulations in Canada and around the World: A Summary.* CAIRN.
55. Hewitt, L.M., Parrott, J.L. and McMaster, M.E. 2006. A decade of research on the environmental impacts of pulp and paper mill effluents in Canada: Sources and characteristics of bioactive substances. *J Toxicol Environ Health B Crit Rev* 9(4): 341–356.
56. Gilani, G.S., Xiao, C. and Lee, N. 2008. Need for accurate and standardized determination of amino acids and bioactive peptides for evaluating protein quality and potential health effects of foods and dietary supplements. *J AOAC Int* 91(4): 894–900.

57. Chatterjee, C., Gleddie, S. and Xiao, C.-W. 2018. Soybean bioactive peptides and their functional properties. *Nutrients* 10(9): 1211.
58. Ganaie, T.A., Mukhtar, H., Allaic, F.M., et al. 2017. Fermentable bioactive peptides as a functional molecule. *Asian J Adv Basic Sci* 5: 154–167.
59. Usmani, S.S., Bedi, G., Samuel, J.S., et al. 2017. THPdb: Database of FDA-approved peptide and protein therapeutics. *PLOS ONE* 12(7): e0181748.
60. Cicero, A.F.G., Fogacci, F. and Colletti, A. 2017. Potential role of bioactive peptides in prevention and treatment of chronic diseases: A narrative review. *Br J Pharmacol* 174(11): 1378–1394.
61. Chalamaiah, M., Ulug, S.K., Hong, H. and Wu, J. 2019. Regulatory requirements of bioactive peptides (protein hydrolysates) from food proteins. *J Funct Foods* 58: 123–129.
62. Wong, K.L., Wong, R.N.S., Zhang, L., et al. 2014. Bioactive proteins and peptides isolated from Chinese medicines with pharmaceutical potential. *Chin Med* 9: 19.
63. Liu, M., Wang, Y., Liu, Y. and Ruan, R. 2016. Bioactive peptides derived from traditional Chinese medicine and traditional Chinese food: A review. *Food Res Int* 89(1): 63–73.
64. ICH guidelines. https://www.ich.org/page/efficacy-guidelines.
65. Bersi, G., Barberis, S.E., Origone, A.L., et al. 2018. Bioactive peptides as functional food ingredients. In A.M. Grumezescu and A.M. Holban (eds), *Role of Materials Science in Food Bioengineering*, 147–186. Academic Press.
66. Vergote, V., Burvenich, C., Van de Wiele, C. and De Spiegeleer, B. 2009. Quality specifications for peptide drugs: A regulatory pharmaceutical approach. *J Pept Sci* 15(11): 697–710.
67. Lacroix, I.M.E. and Li-Chan, E.C.Y. 2012. Dipeptidyl peptidase-IV inhibitory activity of dairy protein hydrolysates. *Int Dairy J* 25(2): 97–102.
68. Agyei, D., Ongkudon, C.M., Wei, C.Y., et al. 2016. Bioprocess challenges to the isolation and purification of bioactive peptides. *Food Bioprod Process* 98: 244–256.
69. Takahashi, H., Tsuchiya, T., Takahashi, M., et al. 2016. Viability of murine Norovirus in salads and dressings and its inactivation using heat denatured lysozyme. *Int J Food Microbiol* 233: 29–33.
70. Naqash, S.Y. and Nazeer, R.A. 2013. Antioxidant and functional properties of protein hydrolysates from pink perch (*Nemipterus japonicus*) muscle. *J Food Sci Technol* 50(5): 972–978.
71. Lemes, A.C., Sala, L., Ores, J.D.C., et al. 2016. A review of the latest advances in encrypted bioactive peptides from protein-rich waste. *Int J Mol Sci* 17(6): 950–194.

Index

A

AAs, *see* Amino acids
ACE, *see* Angiotensin-converting enzyme
ACE-inhibitory peptides, 86, 218
Acid hydrolysis, 214, 264
Acquisitive immune system, 417
AD, *see* Alzheimer's disease
AGS cells, 471
AHA, *see* American Heart Association
Albumin, 59
Algal bioactive peptide generation, 39
Algal proteins
   macroalgal proteins, 32–34
   microalgal proteins, 34–36
Alkaline hydrolysis, 214
Alkaline proteinase (AP), 488
α-Amino acids, 351
α-Amylase inhibitors, 130
α-Ketoacid hydroxylamine (KAHA) ligation, 357, 359
Alpha-synuclein, 395
α-Zein hydrolysate, 133
ALS, *see* Amyotrophic lateral sclerosis
Alzheimer's disease (AD)
   cored neuritic plaques, 395
   dementia, 392
   macroscopic and microscopic features, 394–395
   memory loss, 392
   NTFs, 395
American Heart Association (AHA), 15
American Type Culture Collection (ATCC), 240
Amino acids (AAs), 210, 213, 218, 227
AMPs, *see* Antimicrobial peptides
Amyotrophic lateral sclerosis (ALS)
   gene mutation, 397
   neuron disorder, 397
   types, 397
ANCs, *see* Antinutritional compounds
Angiotensin-converting enzyme (ACE), 132, 488, 499, 500
Angiotensin I-converting enzyme, 4, 61, 103, 182, 183, 243
Angiotensin II (Ang II), 61

Animal-based by-products, 183
Animal peptides, 224–225
Animal protein sources, 196, 199
Animal sources, 5–6, 428
   antihypertensive activity
      dairy products, 366–367
      egg, 372
      marine, 371–372
      meat, 367–371
   opioid activity
      dairy products, 429
      milk, 428–429
Anticancer activity, 16–17
Anticancer peptides, 46–47
Anticoagulation properties, 489, 498–499
Antidiabetic peptides, 46, 64–65
Antidiabetic properties, 485–489
   cereal and pseudocereal sources of, 489, 494–495
   dairy sources of, 489–491
   egg-related sources of, 489, 492
   leguminous sources of, 489, 496–497
   marine sources of, 489, 493
Antihypertensive activity, 14–15, 243
   animal sources
      dairy products, 366–367
      egg, 372
      marine, 371–372
      meat, 367–371
   future perspectives, 383
   isolation of, 378
      downstream processing, 379–380
      enzymatic hydrolysis, 378
      fermentation, 379
      genetic engineering, 379
   myosin sources, 62
   plant sources, 372–374
      amaranth and quinoa, 375
      fruits and vegetables, 376–377
      legumes, 377–378
      rice, 375
      soyabean, 375–376
      wheat, 374–375
   schematic representation, 381
   sources, 366

supercritical fluid extraction, 380–381
troponin sources, 62–63
Antihypertensive agent role, 381–382
Antihypertensive downstream processing, 379
  high pressure, 379–380
  microwave-assisted extraction (MAE), 380
  ultrasound, 380
Antihypertensive peptides, 40, 61–62, 218–220, 434
Anti-inflammatory activity, 9, 14
Anti-inflammatory peptides, 511
Antimicrobial activity, 240–242
  action mechanism of, 417–418
    barrel stave model, 418–419
    carpet model, 419
    toroidal model, 419
  application of, 420
    food safety, 420–421
  bioactive peptides, 421
  diversity of, 416–417
  extraction and obtaining, 416
  pore formation models, 419
Antimicrobial peptides (AMPs), 64, 100, 219, 220, 223, 435–436
Antinutritional compounds (ANCs), 154, 155
Antioxidant activity, 242–243
Antioxidant peptides, 39–40, 63–64
Antioxidative peptides, 219–221, 435
Antithrombotic peptides, 63
AP, see Alkaline proteinase
Aquaculture, 47
Arca subcrenata, 471
Arg-Pro-Arg peptide, 6
Arg-Val-Pro-Ser-Leu peptide, 5
Ark shell *(Scapharca subcrenata)*, 115
Asn-Arg-Tyr-His-Glu (NRYHE), 168
Aspergillus, 416
ATCC, see American Type Culture Collection
Autolysis, 100–101
Autolytic technologies, 101
Automatic Edman degradation, 38
AVPYPQR peptide, 489
2,2'-Azino-bis(3-ethylbenzothiazolin-6-sulfonic acid) (ABTS) method, 134

B

Back-slopping, 234
Bacteriocins, 420–421
Bacterium death, 418
Barrel stave model, 418–419
Beneficial activities, 350–351
Bioactive molecule, 461
Bioactive Peptide Database (BioPepDB), 8, 157, 301

and tools, 324–326
Bioactive peptides (BP), 3, 301, 398, 427–428
  applications, 9
  challenges and future applications, 47–48
  and clinical applications, 4
  defined, 3, 56
  history, 56–57
  neurodegenerative diseases (NDDs)
    glutamate excitotoxicity and cell death-signaling, 400–401
    gut microbiota modulation, 401–402
    inflammation, 402–403
    neurodegeneration pathways, 402
    neurodegenerative disease, 398
    reactive oxygen species (ROS), 398–400
  production, 60–61
    to commercialization, 7–8
  research (see Bioactive peptides (BP) research)
  sources, 5, 57
    blood, 59–60
    meat and meat products, 57–59
  stability and bioavailability, 8–9
  types
    antidiabetic peptides, 64–65
    antihypertensive peptides, 61–63
    antimicrobial peptides, 64
    antioxidant peptides, 63–64
    antithrombotic peptides, 63
Bioactive peptides (BP) research
  sequence-based analysis, 322–328
  structure-based analysis, 328–330
    molecular docking, 331–333
    molecular dynamics (MD), 333
    quantitative structure–activity relationship (QSAR), 330–331
Bioinformatic analysis
  description, 321–322
  future prospects, 333–334
  sequence-based analysis
    allergenicity risk, 327
    Basic Local Alignment Search Tool (BLAST), 323
    BIOPEP, 326–327
    BIOPEP-UWM™ database, 323
    in silico hydrolysis approach, 323, 326
    NCBI, 322
    sensory properties, 327
  structure-based analysis
    molecular docking, 331–333
    molecular dynamics (MD), 333
    peptide chemical structures, 328–329
    protein–peptide interactions, 330
    protein structure, 328

quantitative structure–activity
relationship (QSAR), 330–331
tertiary structure, 328
Biological activities, 17–18, 47
Biological functions
    anticoagulation properties, 489, 498–499
    antidiabetic properties, 485–489
        cereal and pseudocereal sources of, 489, 494–495
        dairy sources of, 489–491
        egg-related sources of, 489, 492
        leguminous sources of, 489, 496–497
        marine sources of, 489, 493
    mineral binding properties, 499–501
BioPepDB, *see* Bioactive Peptide Database
Blood, 59–60
Blood coagulation, 47, 63
Blue mussel *(Mytilus edulis)*, 104
Blue whiting muscle (BWMH), 486
Body's defensive system, 427
Bones, 58–59
Bovine cruor, 64
Bowman-Birk inhibitor, 16
BP, *see* Bioactive peptides
Bradykinin, 61, 132
Bromelain, 196
Bullacta exarata, 471
BWMH, *see* Blue whiting muscle
By-products, 4, 6, 57–59
    animal-based, 183
    plant-based, 181–182

C

Canadian Food and Drugs Act and Regulations, 514
Cantaloupe crude enzyme extracts, 102
Capillary electrophoresis (CE), 272–273
Carpet model, 419
Casein (CN), 76, 429, 489
Casein-derived peptides, 77
Casein phosphopeptides (CPPs), 86, 499, 500
Cavitation process, 60
CE, *see* Capillary electrophoresis
Cell envelope-associated proteinases (CEPs), 240
Center for Food Safety and Applied Nutrition (CFSAN), 515
CEPs, *see* Cell envelope-associated proteinases
Cereals
    antidiabetic activity, 130–132
    antihypertensive activity, 132–133
    antioxidant activity, 134
    immunomodulatory activity, 133
    opioid activity, 133–134
    proteins, 129–130
CFDA, *see* China Food and Drug Administration
CFSAN, *see* Center for Food Safety and Applied Nutrition
CH, *see* Collagen hydrolysate
Charcoal powder, 212
Cheese, 85–86
Chemical condoms, 420
Chemical hydrolysis, 7, 60, 100, 189, 190
Chemical synthesis
    difficult peptides, 356–357
        α-ketoacid hydroxylamine (KAHA) ligation, 357, 359
        native chemical ligation (NCL) strategy, 357, 358
    solid-phase peptide synthesis (SPPS), 351–352
        advantages, 355–356
        amino acids coupling, 354
        amino acids linking, resins, 352–354
        amino acids protection, 354
        deprotection, 355
        polymer removal, 355
        schematic representation, 353
Chemistry of peptides, 351
China Food and Drug Administration (CFDA), 515
Cholecystokinin (CCK), 144, 485, 486
Chromatography, 273
    ion exchange chromatography (IEX), 275–276
    miscellaneous chromatographic techniques, 277–279
    reverse phase high performance liquid chromatography (RP-HPLC), 273–275
    size exclusion chromatography (SEC), 276–277
Chronic hyperglycemia, 46
Chymotrypsin, 196
Code of Practice on Nutrient Claims (COPONC), 513
Collagen, 57, 59
Collagen hydrolysate (CH), 261
Computational modeling, 164
Conus, 100
COPONC, *see* Code of Practice on Nutrient Claims
Cored neuritic plaques, 395
Cow milk, 241
CPAP peptide, 46
CPP, *see* Critical process parameters
CPPs, *see* Casein phosphopeptides

Critical process parameters (CPP), 102
Crude/peptide-enriched protein hydrolysates, 38
Cryptic peptides, 76
CVPH1, 487
Cysteine proteases, 102
Cytomodulatory activity
  protein/peptides from animal sources, 464–470
  protein/peptides from plant sources, 471–476

D
Dairy probiotics, 80
Dairy products, 5, 366–367, 429
  cheese, 85–86
  kefir, 86–87
  milk, 75–79
  yogurt, 79–85
Degree of hydrolysis (DH), 102, 192, 194–195, 212–214
Delivery system strategies, 10–13
Dementia, 392
Denaturation, 81
DH, *see* Degree of hydrolysis
Diabetes mellitus, 46, 64, 130
Dietary Supplement Health and Education Act (DSHEA), 515
Difficult peptides, 356–357
  α-ketoacid hydroxylamine (KAHA) ligation, 357, 359
  native chemical ligation (NCL) strategy, 357, 358
Dipeptidyl peptidase-4 (DPP-4), 130, 170, 486–488
  blood glucose regulation, 131
  inhibitory activity, 46, 83
Dipeptidyl peptidase-4 inhibitor (DPP-4i), 64
DPP-4, *see* Dipeptidyl peptidase 4
DPPH-scavenging peptides, 40

E
ED, *see* Electrodialysis
EDUF, *see* Electrodialysis coupled with ultrafiltration membrane
EFSA, *see* European Food Safety Authority
Eggs, 5
Egg yolk peptide digest (EYLLPD), 447
Electrodialysis (ED), 270–271
Electrodialysis coupled with ultrafiltration membrane (EDUF), 270–271
Electrophoresis (CE), 271
  capillary electrophoresis (CE), 272–273
  gel-based electrophoresis, 271–272
Electrospray ionization (ESI) source, 38
Endogenous opioid peptides, 429–431
Endogenous peptides, 5
Enzymatic digestion, 7
Enzymatic hydrolysis (EH), 35, 102, 154, 416
Enzymatic protein hydrolysis, 100, 190–194
  animal protein sources, 196, 199
  degree of hydrolysis (DH), 194–195
  enzymatic protein hydrolysis, 190–194
  marine protein sources, 198, 199
  plant protein sources, 197, 199
  proteases used in bioactive peptides, 195–198
  protein hydrolysis, 189–190
Enzyme-catalyzed hydrolysis, 190
Epinecidin-1, 100
ESI, *see* Electrospray ionization source
Ethanol, 212
European Commission (EC), 57
European Food Safety Authority (EFSA), 514
Exogenous enzymatic hydrolysis, 101–103
Exogenous enzymes, 195
Exogenous opioid peptides, 431
EYLLPD, *see* Egg yolk peptide digest

F
Fast atom bombardment (FAB)-MS techniques, 38
FC, *see* Flash chromatography
FDA, *see* Food and Drug Administration
FDR Act, *see* Food and Drug Regulations Act
Fermentation, 233–234, 429
  biological activity of peptides
    antihypertensive activity, 243
    antimicrobial activity, 240–242
    antioxidant activity, 242–243
    immunomodulation activity, 243–244
  definition, 234
  fermented foods, 234–236
  future research, 245
  peptides applications, food safety, 244–245
  peptides from animal-based fermented foods, 237, 239, 240
  peptides from plant-based fermented foods, 236–238
  starter culture role, 240
Fermentation parameters, 166
Fermentative processes, 103
Fermented dairy foods, 79
Fermented food products, 103
Field-grown meat, 141
Fish protein hydrolysates (FPH), 486
Flash chromatography (FC), 277–278
FLPFNQL peptide, 444

Food and Drug Administration (FDA), 16, 142, 513
Food and Drug Regulations (FDR) Act, 514
Food by-products, 181
Food for Specified Health Uses (FOSHU), 513, 514
Food matrix, 8, 82
Food peptides
　bioaccessibility and bioavailability, 463–464
　cytomodulatory activity
　　protein/peptides from animal sources, 464–470
　　protein/peptides from plant sources, 471–476
　digestibility, 462–463
Food proteins, 55–56, 98, 102
Food safety, 244–245, 420–421
Food Safety and Standard Act 2005 (FSSAI), 513
Foods for specified health uses (FOSHU), 142
Food Standards Australia and New Zealand (FSANZ), 513
Food Standards Code (FSC), 513
Food waste, 181
FOSHU, see Foods for specified health uses
FPH, see Fish protein hydrolysates
Fractionation and purification
　chromatography, 273
　　ion exchange chromatography (IEX), 275–276
　　miscellaneous chromatographic techniques, 277–279
　　reverse phase high performance liquid chromatography (RP-HPLC), 273–275
　　size exclusion chromatography (SEC), 276–277
　electrophoresis (CE), 271
　　capillary electrophoresis (CE), 272–273
　　gel-based electrophoresis, 271–272
　membrane technology
　　electrodialysis (ED), 270–271
　　ultrafiltration and nanofiltration, 268–270
　multidimensional separation of peptides, 279
　separation techniques, 267–268
　　peptide sequences, 280–289
Free radical, 63, 167–168
FSANZ, see Food Standards Australia and New Zealand
FSC, see Food Standards Code
FSSAI, see Food Safety and Standard Act 2005

G

γ-glutamyl cysteine peptide, 14
Gastrointestinal endopeptidases, 46
Gel-based electrophoresis, 271–272
Gel filtration chromatography (GPC-HPLC), 147
Generally Regarded as Safe (GRAS), 234, 240, 245
Germination/sprouting, 154
GFHI peptide, 471
Globulins, 155
GLSDGEWQ peptide, 64, 471
Glucagon-like peptide 1 (GLP-1), 170, 485–487
Glutamate excitotoxicity and cell death-signaling, 400–401
Glutammate-Glycine-Arginine-Proline-Argi ninepeptide, 16
Gluteomorphins, 429
Gly-Pro-Pro, 133
Golden threadfin bream *(Nemipterus virgatus)*, 114
GRAS, see Generally Regarded as Safe
Gut microbiota modulation, 401–402
GYPMYPLPR peptide, 444

H

Helicobacter pylori, 46
Heptapeptides (P-3-CG), 499
High hydrostatic pressure processing (HHP), 60
High molecular fraction (HMF), 141
High-performance liquid chromatography (HPLC), 213, 214, 273, 357
HILIC, see Hydrophilic interaction chromatography
HMF, see High molecular fraction
H-ORAC, see Hydrophilic Oxygen Radical Absorbance Capacity
HPLC, see High-performance liquid chromatography
Huntington's disease (HD), 396–397
　characterized, 396–397
　clinical symptoms, 396
Hydrolysates and peptides, algae, 41–45
　anticancer peptides, 46–47
　antidiabetic peptides, 46
　antihypertensive peptides, 40
　antioxidant peptides, 39–40
　biological activities, 47
　immunomodulatory peptides, 40, 46
Hydrolysis, 103; *see also specific hydrolysis*
Hydrolyzed vegetable proteins, 224
Hydrophilic interaction chromatography (HILIC), 304–305

Hydrophilic Oxygen Radical Absorbance
    Capacity (H-ORAC), 134
3-Hydroxy-3methyl-glutaryl-coenzyme A
    (HMGCoAR), 169
Hypertension, 14, 40, 243, 383

I
IAM, *see* Indian Ayurvedic Medicines
ICH, *see* International Council for
    Harmonisation
IEX, *see* Ion exchange chromatography
IL-6, 444
Ile-Ala-Pro, 133
Immobilization, 354
Immobilized pH gradient (IPG), 271
Immune system, 417, 442–443
Immunomodulating activity, 441–442
    action mechanism, 452–453
    immune system, 442–443
    immunomodulatory effect
        clinical trials, 451–452
        in vitro studies, 443–447
        in vivo assays, 447–451
    protein hydrolysates, 442–443
Immunomodulation activity, 40, 243–244
Immunomodulatory effect
    clinical trials, 451–452
    in vitro studies, 443–447
    in vivo assays, 447–451
Immunomodulatory peptides, 40, 46, 436
IMS, *see* Ion mobility mass spectrometers
Incretin peptide hormones, 130
Incretins, 65, 130–131
Indian Ayurvedic Medicines (IAM), 513
Inflammation, 402–403
International Council for Harmonisation
    (ICH), 515
Intracellular adhesion molecule 1
    (ICAM-1), 14
In vitro bioactivity assay-directed
    purification, 37
In vitro studies, 443–447
In vivo assays, 447–451
Ion exchange chromatography
    (IEX), 275–276
Ionization, 307
Ion mobility mass spectrometers (IMS), 308
IPEC-J2 cells, 464
IPG, *see* Immobilized pH gradient
IRLIIVLMPILMA peptide, 40
Isochrysis zhanjiangensis, 38
Isolation of, antihypertensive peptides, 378
    downstream processing, 379–380
    enzymatic hydrolysis, 378
    fermentation, 379
    genetic engineering, 379

K
κ-Casein, 79
Kefir, 86–87
KLVDASHRLATGDVAVRA peptide, 47
KRQKYDI peptide, 63

L
L1210 leukemic cells, 476
Lactic acid bacteria (LAB), 5, 103, 234, 240,
    243, 245
Lactobacillus spp., 235
    L. casei, 63
    L. helveticus, 83, 434
    L. helveticus LBK-16H fermented milk, 5
Lacto-fermentation, 234, 235
Lactoferricin, 18
Lactoferrin, 79
Lactotripeptides (LTP), 15
Large oligopeptides, 211
LC, *see* Liquid chromatography
Lectins, 155
Lentil protein hydrolysate, 168
Leu-Arg-Leu-Glu-Ser-Phe (LRLESF), 166
Linear-solvent-strength (LSS) model, 304
Lipid-lowering activity, 15–16
Liquid chromatography (LC), 38
    capillary-and nano-LC, 306
    hydrophilic interaction chromatography
        (HILIC), 304–305
    ion-exchange chromatography (IEC), 305
    linear-solvent-strength (LSS) model, 304
    reversed-phase liquid chromatography
        (RP-LC), 304
    trifluoroacetic acid (TFA), 304
Liquid chromatography-mass spectrometry
    (LC-MS)
    Bioactive Peptide Database (BioPepDB),
        301
    data analysis, 312
    liquid chromatography (LC) (*see* Liquid
        chromatography (LC))
    mass spectrometry (MS) (*see* Mass
        spectrometry (MS))
    protein precipitation, 303
    solid-phase extraction (SPE), 302–303
    ultrafiltration, 303
Liquid-state fermentation, 217
Low-density lipoprotein (LDL), 141
LSS model, *see* Linear-solvent-strength (LSS)
    model
Lunasin, 170

M

M6 peptide, 62
MAAH, *see* Microwave-assisted acid hydrolysis method
Macroalgal proteins, 32–34
MAE, *see* Microwave-assisted extraction
Marine ecosystems, 31
Marine organisms, 99
Marine protein sources, 198, 199
Mascot search engine, 38
Mass analyzers, 307–309
Mass spectrometry (MS), 86, 262–264
    data-dependent acquisition (DDA)/data-independent acquisition (DIA) mode, 311
    electrospray Ionization (ESI), 307
    ion mobility mass spectrometers (IMS), 308
    ion-traps, 309
    mass analyzers, 307
    matrix-assisted laser desorption ionization (MALDI), 307
    tandem, 309
Mass spectrophotometry, 38
Meat and meat products, 57–58
    bones, 58–59
    trimmings and cuttings, 58
Membrane fractionation, 268–270
Membrane processing, 38
Membrane separation technologies (MST), 103, 104
Membrane technology
    electrodialysis (ED), 270–271
    ultrafiltration and nanofiltration, 268–270
MFGM, *see* Milk fat globule membrane
MHW, *see* Ministry of Health and Welfare
Microalgal proteins, 34–36
Microbial fermentation, 8, 234, 416
Microbial hydrolysis, 217
Microbial proteases, 102
Microwave-assisted acid hydrolysis (MAAH) method, 264
Microwave-assisted extraction (MAE), 263–265, 380
Milk, 75–79
Milk-derived peptides, 77
Milk fat globule membrane (MFGM), 77
Mineral binding properties, 499–501
Ministry of Health and Welfare (MHW), 513
Miracle crop, 141
Miscellaneous chromatographic techniques, 277–279
Molecular docking, 331–333
Molecular dynamics tools, 333
Molecular weight cutoff (MWCO), 37, 104, 268–269
Monascus purpureus, 132
MS, *see* Mass spectrometry
MST, *see* Membrane separation technologies
Multidimensional separation of peptides, 279
Multiple enzyme digestion system, 192, 193
MWCO, *see* Molecular weight cutoff
Myosin proteins, 61

N

Nannochloropsis oculata, 37, 47
Nanofiltration (NF), 104, 268–270
National Food Authority (NFA), 513
Native chemical ligation (NCL), 357, 358, 432–433
Natural Health Product Directorate (NHPD), 514
Natural inhibitors, 40
Naturally active peptides, 98
NCDs, *see* Noncommunicable diseases
NCL, *see* Native chemical ligation
Neurodegeneration pathways, 402
Neurodegenerative diseases (NDDs), 398
    bioactive peptides of, 398
        glutamate excitotoxicity and cell death-signaling, 400–401
        gut microbiota modulation, 401–402
        inflammation, 402–403
        neurodegenerative disease, 398
        reactive oxygen species (ROS), 398–400
    classification, 392–394
        Alzheimer's disease (AD), 392, 394–395
        Amyotrophic lateral sclerosis (ALS), 397–398
        Huntington's disease (HD), 396–397
        Parkinson's disease (PD), 395–396
    definition, 391–392
New sources, 6–7
NF, *see* Nanofiltration
NFA, *see* National Food Authority
NHPD, *see* Natural Health Product Directorate
Ninhydrin assay principle, 194
NMR, *see* Nuclear magnetic resonance
Noncommunicable diseases (NCDs), 509
Nongenetically modified organisms (Non-GMOs), 141
Nuclear magnetic resonance (NMR), 99
Nutraceuticals, 9

O

Okara, 4, 6
Oligopeptide, 210

Olive oil extraction, 6
O-phthaldialdehyde (OPA) method, 194, 495
Opioid activity
    animal sources, 428
        dairy products, 429
        milk, 428–429
    bioactive peptides, 431
        chemical synthesis, 432–433
        enzymatic hydrolysis, 432
        integrated approach, 433
        microbial fermentation, 432
    food sources, 428
    functions, 434–435
        activities, 436
        antihypertensive peptides, 434
        antimicrobial peptides, 435–436
        antioxidative peptides, 435
        immunomodulatory peptides, 436
    future perspectives, 436–437
    mechanism, 433–434
    plant sources, 429
        endogenous peptides, 429–431
        exogenous peptides, 431
Opioid peptides, 220, 224
ORAC, *see* Oxygen Radical Absorbance Capacity
Oral bioavailability, 19
Oryzatensin, 133
Oxidative metabolism, 63
Oxidative stress, 63
Oxygen Radical Absorbance Capacity (ORAC), 134

P

Papain, 37, 195
Parkinson's disease (PD)
    alpha-synuclein, 395–396
    treatment strategy, 396
Parmigiano-reggiano, 86
PCA, *see* Perchloric acid
PD, *see* Parkinson's disease
PEAKS Studio 6.0 software, 38
Pea protein hydrolysate, 167, 171
Pepsin, 197
PepT1, *see* Peptide transporter 1
Peptide
    definition, 210
    identification, 38
    oral administration, 462
    and protein structure prediction tools, 329
    purification, 37–38
    sequencing, 38
    synthesis
        coupling reagents, 356
        protecting reagents, 355
PeptideMatch, 164
Peptide transporter 1 (PepT1), 218
Peptidomics, 301
Perchloric acid (PCA), 212
Permeate, 104
Pharmacokinetic data, 18
Phaseolus coccineus, 476
pH-stat technique, 194
PHWE, *see* Pressurized hot water extraction
Pigeon pea *(Cajanus cajan)*, 165
Plant-based by-products, 181–182
Plant-derived lectins, 16
Plant peptides, 222–224
Plant protein sources, 197, 199
Plant sources, 372–374
    antihypertensive activity
        amaranth and quinoa, 375
        fruits and vegetables, 376–377
        legumes, 377–378
        rice, 375
        soyabean, 375–376
        wheat, 374–375
    opioid activity, 429
        endogenous peptides, 429–431
        exogenous peptides, 431
Polypeptide, 211
Porcine skeletal muscle, 62
Porcine skin collagen, 64
Position-Specific Iterated BLAST (PSI-BLAST), 157
Potential scale and economic values, 225, 226
Precursor protein, 3
Pressurized hot water extraction (PHWE), 259
Probiotics, 79
Probiotic yogurt, 83
Production methodologies, 35
    peptide identification, 38
    peptide purification, 37–38
    protein extraction, 35
    protein hydrolysis, 35, 37
    validation activity of peptides, 38–39
Protein, 57
    casein, 85
    definition, 211
    extraction, 35
    fermentation, 217
    hydrolysates, 442–443
        based on degree of hydrolysis (DH), 192
        production, 190, 191
    structure, 211
Protein hydrolysates, animal nutrition
    bioactive peptides
        ACE-inhibitory peptides, 218

antimicrobial peptides, 219, 220, 223
antioxidative peptides, 219–221
definition, 218
opioid peptides, 220, 224
transport small peptides in small intestine, 218
future research, 226
industrial production
degree of hydrolysis, 212–214
general consideration, 212
methods
acid hydrolysis, 214
alkaline hydrolysis, 214
cell-free proteases, 214–217
microbial hydrolysis, 217
plant and animal protein hydrolysates application
animal peptides, 224–225
general consideration, 220, 222
plant peptides, 222–224
Protein hydrolysis, 35, 37, 189–190
Proteolysis, 76, 380, 432
Proteolytic
digestion of foods, 64
enzymes, 195, 234
PSI-BLAST, see Position-Specific Iterated BLAST
Pulse-based fermented products, 154
Pulses, 153–155
bioactive peptides production
anticholesterolemic, 169
antihypertensive pulse peptides, 165–167
antimicrobial, 169–170
antioxidative, 167–169
multifunctional peptides, 170–171
health benefits, 171
in silico approach, 157–165
proteins, 155–156
Purification and analysis, 357–358
Pyropia haitanensis, 38

Q

QEPV and QEPVL peptide, 244
QSAR, see Quantitative structure-activity relationship
QSAR Computer Program, 331
Quadrupole-TOF (Q-TOF), 38
Quantitative structure–activity relationship (QSAR), 330–331

R

RAAS, see Renin-angiotensin-aldosterone system
Randomized control trials (RCTs), 14

RAS, see Reninangiotensin aldosterone system
RAW264.7 macrophages, 444, 447
RCTs, see Randomized control trials
Reactive oxygen species (ROS), 39, 63, 398–400
Receptor, 433–434
Recombinant DNA technology, 60
Regulatory status, bioactive peptides, 509–510
approval by US-FDA, 509–512
approved bioactive peptides, 510–511
global market, 511
market authorization and commercialization, 515–516
regulatory perspectives, 513–515
significance, diseases treatment, 511
Reninangiotensin aldosterone system (RAS), 15, 165, 381
Retentate, 104
Reversed phase high performance liquid chromatography (RP-HPLC), 38, 99, 273–275, 357–358
Rice albumin, 130–131
RMLGQTPTK peptide, 62
ROS, see Reactive oxygen species
RP-HPLC, see Reversed phase high performance liquid chromatography

S

Sarcomeres, 62
SBP, see Systolic blood pressure
SCW, see Subcritical water
SCX, see Strong cationic exchanger
SDPI, see Spray-dried porcine intestine hydrolysate
SDS-PAGE, see Sulphate polyacrylamide gel electrophoresis
Seafood bioactive peptides
bioactive properties, 104–116
nutraceutical application, 116
development
isolating naturally bioactive peptides, 99–100
producing bioactive peptides, 100–103
purification, 103–104
sources, 97–98
inactive in parent protein, 98–99
naturally active peptides, 98
Seafood by-products, 99
Seafood protein hydrolysates (SPH), 100, 102, 103
Seafood proteins, 98, 102
Seaweed nitrogen-to-protein (SNP) conversion factor, 33
Seaweed species, crude protein values of, 33

SEC, *see* Size exclusion chromatography
Seed storage proteins (SSPs), 155
Sephadex G-15 gel filtration, 38
Sequence-based analysis
    allergenicity risk, 327
    Basic Local Alignment Search Tool (BLAST), 323
    BIOPEP, 326–327
    BIOPEP-UWM™ database, 323
    NCBI, 322
    in silico hydrolysis approach, 323, 326
    sensory properties, 327
Serine proteases, 102
Ser-Tyr, 149
SH, *see* Skin hydrolysate
Short-and medium-size peptides, 32
SHRs, *see* Spontaneously hypertensive rats
Single enzyme digestion system, 191, 192
Size exclusion chromatography (SEC), 276–277
Skin hydrolysate (SH), 261
Small oligopeptides, 210
SNP, *see* Seaweed nitrogen-to-protein conversion factor
SN-TCA, *see* Soluble nitrogen in trichloroacetic acid
Solid-phase extraction (SPE), 302–303
Solid-phase peptide synthesis (SPPS), 351–352, 432
    advantages, 355–356
    amino acids
        coupling, 354
        protection, 354
        resins linking, 352–354
    deprotection, 355
    polymer removal, 355
    schematic representation, 353
Solid-state fermentation, 217
Soluble nitrogen in trichloroacetic acid (SN-TCA), 194
Solution phase synthesis (SPS), 432
Soybean, 141–149
Soy-fermented foods, 6
Soy sauce, 217
Sparus aurata, 104
SPH, *see* Seafood protein hydrolysates
Spirulina spp., 40
    S. maxima, 46
    S. platensis, 46
Spontaneously hypertensive rats (SHRs), 62, 83, 132
SPPS, *see* Solid phase peptide synthesis
Spray-dried porcine intestine hydrolysate (SDPI), 224, 225

SPS, *see* Solution phase synthesis
SREBP2, *see* Sterol regulatory element-binding protein 2 pathway
SSPs, *see* Seed storage proteins
Starter culture, 85
STC-1 cells, 486, 487
Sterol regulatory element-binding protein 2 (SREBP2) pathway, 16
Strong cationic exchanger (SCX), 276
Structure-based analysis
    molecular docking, 331–333
    molecular dynamics (MD), 333
    peptide chemical structures, 328–329
    protein–peptide interactions, 330
    protein structure, 328
    quantitative structure–activity relationship (QSAR), 330–331
    tertiary structure, 328
Subcritical water (SCW), 259, 262
Subcritical water extraction (SWE), 259–260
Subcritical water hydrolysis (SWH), 260–262
Subcritical water processing (SWP), 261
Sulphate polyacrylamide gel electrophoresis (SDS-PAGE), 271–272
Supercritical fluid chromatography (SFC), 278–279
SWE, *see* Subcritical water extraction
SWH, *see* Subcritical water hydrolysis
SWP, *see* Subcritical water processing
Systolic blood pressure (SBP), 58

T
Target peptides, 103
TCA, *see* Trichloroacetic acid
TEAC, *see* Trolox equivalent antioxidant capacity
TFA, *see* Trifluoroacetic acid
TGs, *see* Triglycerides
Thermolysin, 133
Thrombosis, 47
TLR, *see* Toll-like receptors
TNBS, *see* Trinitro-benzene-sulfonic acid
Toll-like receptors (TLR), 452
Tomato seed protein hydrolysates (TSPH), 500, 501
Toroidal model, 419
Transcytosis, 9
Trichloroacetic acid (TCA), 212
Trididemnum solidum, 98
Trifluoroacetic acid (TFA), 275, 304
Triglycerides (TGs), 142
Trimmings, 58
Trinitro-benzene-sulfonic acid (TNBS), 194
Tripeptide (KYR), 64

Trolox equivalent antioxidant capacity (TEAC), 134
Trypsin, 196, 261–262
TSPH, see Tomato seed protein hydrolysates
Type 1 angiotensin-II receptors (AT1), 132

U
UF, see Ultrafiltration
Ultrafiltration (UF), 268–270, 303
Ultrasound, 60, 380
Umami peptides, 146–147

V
Valineproline-tyrosine (VPY), 14
Val-Val-Ser-Leu-Ser-Ile-Pro-Arg (VVSLSIPR), 165
Vascular cell adhesion molecule 1 (VCAM1), 14
VECYGPNRPQF peptide, 40
Vegetal sources, 6
Venous thromboembolism (VTE), 63
Very low-density lipoproteins (VLDLs), 146
Vigna sesquipedalis, 476
VKAGFAWTANQQLS peptide, 59
VKKVLGNP peptide, 62
VLDLs, see Very low-density lipoproteins
VLPVPQK peptide, 500
VPY, see Valineproline-tyrosine
VTE, see Venous thromboembolism

W
Water-soluble peptides, 85
Wheat germ protein (WGP), 132, 262
White-brined cheese, 85

X
Xanthine oxidase (XO), 114

Y
YGIYPR peptide, 444
Yogurt, 79–85
YQEPVLGPVRGPFPIIV peptide, 63